Fibre Optic Communication

Optical Waveguides, Devices and Applications

Sanjeev Kumar Raghuwanshi
Assistant Professor
Department of Electronics Engineering
Indian Institute of Technology (Indian School of Mines)
Dhanbad

Santosh Kumar
Assistant Professor
Department of Electronics and Communication Engineering
DIT University
Dehradun

Universities Press

FIBRE OPTIC COMMUNICATION – OPTICAL WAVEGUIDES, DEVICES AND APPLICATIONS

UNIVERSITIES PRESS (INDIA) PRIVATE LIMITED

Registered office
3-6-747/1/A & 3-6-754/1, Himayatnagar, Hyderabad 500 029, Telangana, India
info@universitiespress.com; www.universitiespress.com

Distributed by
Orient Blackswan Private limited

Registered office
3-6-752 Himayatnagar, Hyderabad 500 029, Telangana, India

Other offices
Bengaluru, Bhopal, Chennai, Guwahati, Hyderabad, Jaipur, Kolkata,
Lucknow, Mumbai, New Delhi, Noida, Patna, Vijayawada

ISBN: 978-93-86235-21-3

Typeset in Times New Roman PS 10.5/12.5 by
Cameo Corporate Services, Coimbatore, Tamil Nadu, India

Printed at Yash Printographics, Noida

Published by
Universities Press (India) Private Limited
3-6-747/1/A & 3-6-754/1, Himayatnagar, Hyderabad 500 029, Telangana, India

Dedicated to my Daughter and Wife

—Sanjeev Kumar Raghuwanshi

Dedicated to my Beloved Wife, Ragini

—Santosh Kumar

Table of Contents

Preface

Fibre Optic Communication – Optical Waveguides, Devices and Applications has been designed as a textbook for the students of electronics and communication engineering, electronics engineering, and electrical and communication engineering courses at undergraduate and postgraduate levels. Spread across nine chapters, the book deals at length with the general characteristics of optical fibre, its types and applications in various fields such as data transmission and optical communication system, while elaborating on the advantages of optical fibre over other types of communication systems.

Chapter 1 gives an overview of fibre optic technology and outlines the basic concepts related to fibre optics such as total internal reflection and its working principle. It explains briefly about single-mode and multi-mode fibres and describes the synchronous mode of fibre optic digital transmission system before providing an insight into the various elements of optical communication that make fibre optics preferable to any other communication system.

Chapter 2 describes the general characteristics and behaviour of optical waveguides. It first provides an outline of the optical waveguide and then proceeds to discuss about the graded-index fibre. The chapter spells out the basic concepts and equations of electromagnetic wave theory required for the comprehension of light wave propagation in optical waveguides. It also presents an in-depth discussion about the wave equation of the optical waveguide and graded optical waveguide, showing clearly the mathematical calculation for describing the principle of optical waveguide. In addition, the chapter also shows a detailed derivation of the basic equation of the slab waveguide and lays out the graphical method for calculation of the propagation constant for different modes.

Chapter 3 starts with a brief introduction to laser, its applications, and design-related issues. It explains population inversion, optical feedback and laser oscillation and progresses to delineate the construction of the laser. The chapter expounds on mode locking, pulse pumping, semiconductor laser and its types and characteristics and explains radiation modes with their superposition phenomena. The chapter also examines the use of optical computing hardware.

Chapter 4 describes the concept of coupled-mode theory for optical waveguide. It drives home the importance of Maxwell's equation in describing the wave theory of optics for various modes (TE, TM and Hybrid modes) for step-index fibre and shows the mathematical calculation for electric field inside the core and cladding region. The chapter also presents the derivation of coupled-mode equation based on perturbation and demonstrates the methods for calculating the coupling coefficient for several practically important devices. It describes a novel circuit for the generation of picoseconds optical pulse train and shows how the pulse train can be analysed for calculating pulse width that fits the circuit description. The chapter also shows the mathematical calculation and MATLAB plot for normalised intensity as a function of time for five-laser system.

Chapter 5 focuses on the importance of optical communication system in modern technology. It explains the optical network system and gives an insight into its historical background. The chapter also discusses the basic requirements, construction and advantages of fibre optic signal processing and proceeds to elucidate the concept of lattice structure and its advantages over other competing technology for efficient signal filtering action. It also describes the various elements required for the implementation of lattice signal processing. The chapter then shows how transform techniques are used for describing the signals of complex systems and establishes the application of z-transform for effective signal description. It presents MATLAB plot for each of the transfer functions, showing the appropriate filtering action in different cases. The last section of the chapter analyses the use of optical devices such as optical ring resonator integrated with asymmetric Mach–Zehnder interferometer. It shows the detailed derivation of its transfer function and also evaluates the function under resonance condition.

Chapter 6 considers the nonlinear effects in optical fibre. It describes various scattering effects such as stimulated Brillouin scattering, stimulated Raman scattering, self-phase modulation, cross-phase modulation and four-wave mixing, and investigates the nature of different types of optical fibres such as non-zero dispersion fibre, and positive and negative dispersion fibres. The chapter also describes the optical loop mirror and spells out its advantages and drawbacks, besides presenting a theoretical viewpoint of fibre lasers, and NOLM power transfer curve. The chapter ends with a description of the generation of soliton pulse, dark soliton and power of a dark soliton.

Chapter 7 deals with the one of the important effects in fibre optics, namely, the electro-optic effect. The chapter uses Maxwell's equations to help the student understand the electro-optic effect of propagation of waves and reveals the method of generation of ray velocity surface. It then analyses electro-optic effect in the lithium niobate crystal and explains the use of the crystal as an interferometer (the Mach–Zehnder interferometer) with its full mathematical analysis. The chapter analyses electro-optic effect in potassium dihydrogen phosphate (KDP) crystals with phase modulation phenomenon and details the concept of transit time limitation and the methods by which it is overcome in travelling time modulators. It also provides an analysis of core fields in elliptical cylindrical coordinate systems, describing both hyperbolic (H) and elliptical (E) type modes with different boundary conditions and limiting cases. The chapter concludes with a representation of modes for parabolic profile with circular refractive index contours.

Chapter 8 provides an introduction to microwave photonics. It delineates various methods for generating microwave signals such as optical heterodyning, optical injection locking, optical phase-locked loop (OPLL) and optical injection phase locking. The chapter provides an outline of photonic microwave delay-line filters, describes its various configurations and reviews frequency modulation, and explores the concept of instantaneous frequency, power of an angle-modulated wave, bandwidth of angle-modulated waves and narrow-band angle modulation. It also analyses methods of frequency modulation such as wide-band frequency modulation (WBFM) and verifies them using FM bandwidth relationship. Lastly, the chapter investigates photonically assisted microwave frequency multiplication technique based on external modulation, scrutinises electrical harmonic suppression and defines microwave frequency using two cascaded Mach–Zehnder modulators (MZMs) and four different bias combinations for two MZMs.

Chapter 9, the last chapter of the book, discusses the application of optical fibre in various fields such as remote sensing and the modification of these fibres as sensors for measuring physical quantities such as strain, pressure and temperature. It deals at length with the use of optical fibre in broadband network architecture to replace all or part of the local metal loop deploying several configurations such as Fibre to the X (FTTX) for last-mile connectivity. The chapter also elaborates on direct fibre, shared fibre, active optical network and passive optical network, and describes the application of optical fibres in medical science and medical engineering. The chapter gives an overview of the fibre optic sensor, its concept and classification before delving into the construction and applications of the Sagnac, Mach–Zehnder and Michelson interferometers, which use fibre optic sensors. It examines the working of the optical spectrum analyser (OSA) and reveals how popular OSAs such as the Fabry–Perot, diffraction grating and Fourier Transform OSA are used for monitoring the distribution and allocation of wave spectrum.

At the end of each chapter, there is a summary that effectively recapitulates the gist of the discussion for the student's benefit. In addition, there are short questions with answers, pertinent from the examination perspective. Multiple choice questions and Exercises provide practice to hone student skills. MATLAB programs for selected exercises are given in the Appendix.

While extreme care has been taken at the time of preparing the manuscript of this book, it is possible that a few errors may have crept in inadvertently. We shall be thankful if these are pointed out to us or the publisher.

Acknowledgements

It is a pleasure to acknowledge the assistance we received from several individuals at the time of writing this book. We are thankful to the book's reviewers who patiently examined its contents and provided constructive inputs for value addition. We are obliged to Prof. D C Panigrahi, Director, IIT (ISM) Dhanbad, for his encouragement and support. Special thanks go to Prof. K K Raina, Vice-chancellor, and Prof. S Swaminathan, Pro-Vice-chancellor of DIT University, Dehradun, who actively supported us while we were working on the manuscript pages. We also express our gratitude to our Ph.D. and M.Tech. students for their help in reviewing the text.

Last but not least, we thank our publisher Universities Press (India) Private Limited, for bringing out this book in its present form. In particular, we are grateful to Mr Kallol Das for his valuable support and Mr Ramesh for going through the text and suggesting significant modifications.

Sanjeev Kumar Raghuwanshi

Santosh Kumar

About the Authors

Sanjeeev Kumar Raghuwanshi is Assistant Professor, Department of Electronics Engineering, Indian Institute of Technology (Indian School of Mines)/IIT (ISM), Dhanbad. He received his bachelor's degree in electronic and instrumentation engineering from SGSITS (Autonomous Institute) Indore, and master's degree in solid-state technology from IIT Kharagpur. He then joined IIT Bombay as Research Engineer before proceeding to pursue his Ph.D. degree in the field of fibre optics at IISc Bangalore. He did his post-doctorate fellowship in instrumentation and sensor technology (School of Engineering and Mathematical Sciences), City University, London, and served as Lecturer and Assistant Professor at various institutes before joining IIT (ISM) Dhanbad.

Dr Raghuwanshi has authored three books, worked on two functional requirement specification (FRS) projects for IIT (ISM) and one project for ISRO. He has published more than 110 research papers and presented about 75 research papers in journals and conferences of international and national repute. A life member of International Academy of Physical Sciences, Institution of Electronics and Telecommunication Engineers and Institution of Engineers (India), Dr Raghuwanshi has three patents to his credit.

Santosh Kumar is Assistant Professor, Department of Electronics and Communication Engineering, DIT University, Dehradun. He was awarded a Ph.D. degree for his work in the field of fibre optics by IIT (ISM), Dhanbad. He has authored one book on the subject for the benefit of undergraduate students and published articles in more than 100 journals and conferences of international and national repute. A seasoned academician, he has attended several conventions pertaining to his area of expertise such as SPIE Photonic Europe in 2014 and 2016 at Brussels, Belgium, and Frontiers in Optics/Laser Science in 2014 and 2015 at Tucson, Arizona, and San Jose, California, respectively.

A reviewer of technical journals of SPIE, OSA, Springer, Elsevier and IEEE, Dr Kumar is a member of the IEEE, IEEE Photonics Society, OSA, and SPIE. He is a life member of Optical Society of India (OSI) and a faculty advisor of the student chapters of two Optics Societies (OSA and SPIE) at DIT University, Dehradun. A recipient of Research Excellence Award at DIT University in the years 2015 and 2016, he has completed two major projects at the university and is currently guiding six students to their Ph.D. degrees. His areas of academic interest are fibre optic communication and photonic devices.

1 Overview of Fibre Optics

Objectives

- This chapter deals with introduction to optical fibre communication.
- The various types of optical fibres are described.
- Different types of modes supported by optical fibres and their parameters are explained.
- Various types of losses that occur in optical fibres and their remedies are discussed.

1.1 INTRODUCTION

With the advent of fibre optics, data can be easily and rapidly transmitted over long distances, through glass fibres, of the size of a human hair. It is soon expected to be as common as electrical wiring in automobiles or at homes. Fibre optics technology will take its time to develop completely. However, it is increasingly replacing wire transmission lines in communication systems [1–5].

1.2 OPTICAL FIBRES

Optical fibres are the light equivalent of microwave waveguides, with an additional advantage of a very wide bandwidth [6–11]. Physically, an optical fibre is a very thin and flexible medium, having a cylindrical shape, consisting of three sections: (a) the core, (b) the cladding and (c) the jacket. Among these, the core is the innermost section and is made of glass or plastic. This is the actual fibre and has the remarkable property of propagating an optical beam. It is surrounded by its own cladding, a glass or a plastic coating that has optical properties, very different from those of the core. The outer section, called the jacket, is made of plastic or polymer and other materials. It is provided for protection against moisture, absorption, crushing, and other environmental dangers. The core acts like a continuous layer of two parallel mirrors. A signal is first encoded into a light beam, which is then passed in between the two boundaries and propagated as a result of multiple internal reflections.

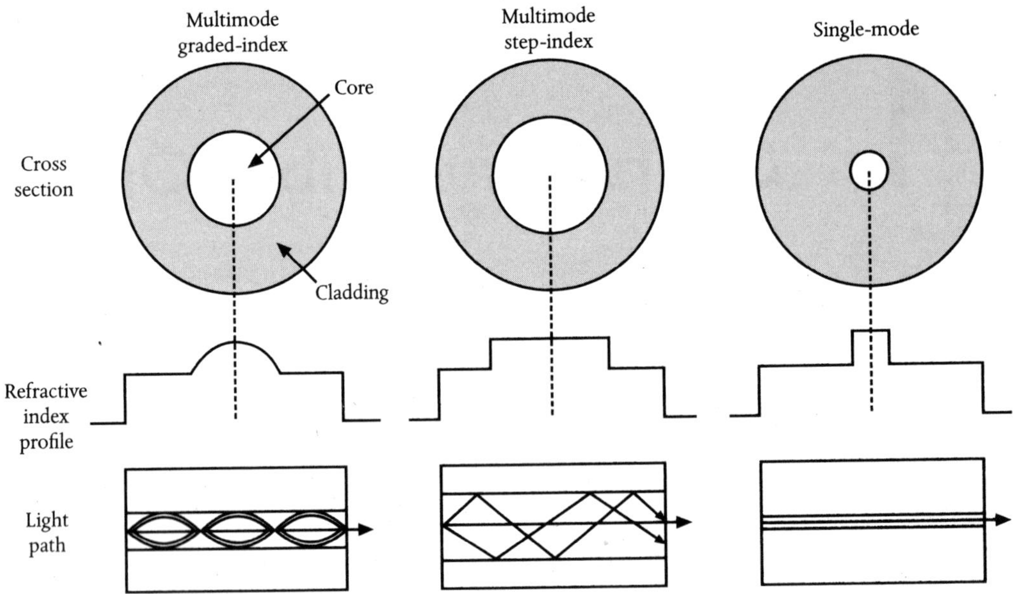

Fig. 1.1 Classification of optical fibre with their refractive index profile and light propagation

Variations in the material composition of the core give rise to two commonly used fibre types shown in Fig. 1.1. In the first case, the core refractive index is made to vary as a function of the radial distance from the centre of the fibre. This fibre is a graded-index fibre. In the second case, the refractive index of the core is uniform throughout and undergoes an abrupt change (or step) at the cladding boundary. Intermodal dispersion or modal delay can be reduced by using a graded-index profile in a fibre core. This allows graded-index fibres to have much larger bandwidths (data rate transmission capabilities) than step-index fibres. The step-and the graded-index fibres can be further divided into single-mode and multimode classes. As the name implies, a single-mode fibre sustains only one mode of propagation, whereas multimode fibres contain hundreds of modes. Multimode fibres offer several advantages over single-mode fibres. The larger core radii of multimode fibre make it easier to launch optical power into the fibre and facilitate connecting similar fibres. Another advantage is that light can be launched into a multimode fibre using a light-emitting-diode (LED) source, whereas single-mode fibres must generally be excited with laser diodes. Although LEDs have less optical output power than laser diodes, they are easier to make, are less expensive, require less complex circuitry, and have longer lifetimes than laser diodes, thus making them more desirable in certain applications.

1.2.1 Single-mode and Multimode Fibres

In fibre optic communication, a single-mode optical fibre (SMF) is an optical fibre designed to carry only a single ray of light (mode) [12–15]. Modes are the possible solutions of the Helmholtz equation for waves, which is obtained by combining Maxwell's equations and the boundary conditions. These modes define the way of wave travels through space, that is, how the wave is distributed in space. Waves can have the same mode but have different frequencies. This is the case in single-mode fibres, where we can have waves with different frequencies but of the same mode. This means that they are distributed in space in the same way and give us a single ray of light. Although the ray travels parallel to the length of the fibre, it is often called transverse mode since its electromagnetic vibrations occur perpendicular (transverse) to the length of the fibre.

Multimode optical fibre is a type of optical fibre mostly used for communication over shorter distances, such as within a building or on a campus [16–20]. Typical multimode links have data rates of 10 Mbit/s to 10 Gbit/s over link lengths of up to 600 metres. One disadvantage of multimode fibres is that they suffer from intermodal dispersion. When an optical pulse is launched into a fibre, the optical power in the pulse is distributed over all (or most) the modes of the fibre. Each mode can propagate in a multimode fibre at a slightly different velocity. This means that the modes in a given optical pulse arrive at the fibre end at a slightly different times, thus causing the pulse to spread out in time as it travels along the fibre. This effect is known as intermodal dispersion or modal delay. The single-mode fibre possesses higher bandwidth as against multimode fibre due to the absence of intermodal dispersion effect inside it. Therefore, single-mode fibres are used for long-haul communication.

1.3 IMPORTANCE OF OPTICAL FIBRES

An optical fibre has clear-cut advantages over wire or radio systems. Therefore, telecommunication industries have introduced the fibre optic systems. The main advantages of optical fibres are as follows [21–23].

1. Attenuation in a fibre is markedly lower than that in a coaxial cable or twisted pair and is constant over a very wide range. Therefore, transmission within a wide range of distance is possible without repeaters, etc.
2. As they are smaller in size and lighter, than coaxial cable or bundled twisted-pair cable, optical fibres occupy less space.
3. Electromagnetic waves generated from electrical disturbances or electrical noises do not interfere with light signals. As a result, the system is not vulnerable to interference, impulse noise or cross-talk.
4. No physical electrical connection is required between the transmitter and the receiver.
5. The fibre is much more reliable, because it can better withstand environmental conditions, such as pollution, radiation and salt and produces no corrosion. Moreover, it is nominally affected by nuclear radiation. Its life is longer in comparison to copper wire.
6. There is almost no cross-talk in optical fibres and hence transmission is more secure and private, as it is very difficult to tap into a fibre.
7. Bandwidth of optical fibre is higher than that of an equivalent wire transmission line.
8. As fibres are very good dielectrics, isolation coating is not required.
9. Data rate is much higher in a fibre and hence more information can be carried by each fibre than by equivalent copper cables. For example, at immensely high frequencies of an optical fibre, data rates of 2 Gbps over tens of kilometres have been demonstrated.
10. The cost per channel is lower than that of an equivalent wire cable system. It is expected that in the near future, the optical fibre communication system will be more economical than any other communication system.
11. Due to the non-inductive and non-conductive nature of a fibre, there is no radiation or interference on other circuits and systems.
12. Fewer repeaters indicate lower cost and fewer sources of error. It has been observed that a fibre transmission can achieve a data rate of 5 Gbps over a distance of 111 km without repeaters, whereas coaxial and twisted-pair systems generally have repeaters every few kilometres.
13. The raw material is available in plenty.

1.4 OPTICAL WINDOW IN TELECOMMUNICATION

Optical fibre communications typically operate in a wavelength region corresponding to one of the following telecom windows:

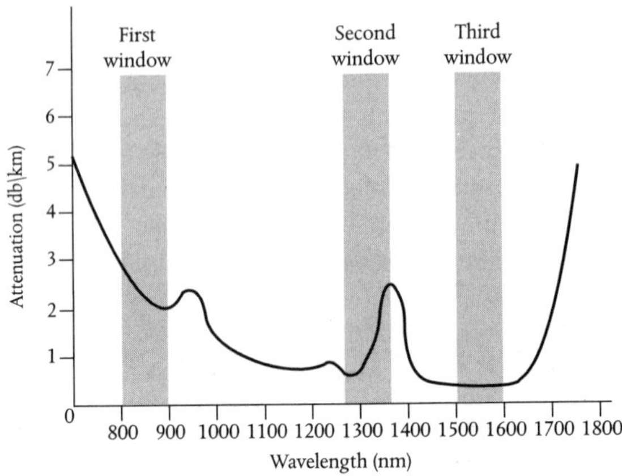

Fig. 1.2 Optical window in telecommunication

- The photodiode electrical output after amplification needs demodulation for restoring the original signal. At the time of development of fibre optic communication, the efficiencies of optical sources available peak around the wavelength of 800 nm as shown in Fig. 1.2. The transmission loss at this wavelength is around 3–4 dB/km. This working wavelength is known as the first window of telecommunication [21].
- Further purification of silica fibres leads to reduction in losses, up to 0.4 dB/km at the wavelength of 1300 nm as shown in Fig. 1.2. At this wavelength, dispersion in silica fibres is also reduced to a great extent. Due to very low dispersion losses, broadening of pulse is reduced and this wavelength is considered as second window of telecommunication. This window was originally used for long-haul transmission. However, low dispersion is not necessarily ideal for long-haul transmission as it can increase the effect of optical nonlinearities.
- On further purification in silica fibres in terms of removing OH ions, properties are improved with a shift in wavelength at 1500 nm as shown in Fig. 1.2. At this wavelength, losses are around only 0.2 dB/km. This is the wavelength region which is favoured for most telecommunication operations.

1.5 PROPAGATION OF LIGHT IN DIFFERENT MEDIA

Propagation of light is based on a single parameter which determines the optical characteristics of the material. This parameter is the refractive index of the material in which it passes. The refractive index (n) is expressed as:

$$n = \frac{\text{speed of light in vacuum}}{\text{speed of light in material concerned}} \tag{1.1}$$

Normally, $n > 1$ for all known materials. Hence, light travels more slowly in a material than in vacuum. The velocity of light in a substance is expressed as,

$$V_{sub} = \frac{c}{n} \tag{1.2}$$

where c = speed of light in vacuum. The optical wavelength in the substance is,

$$\lambda_{\text{sub}} = \frac{V_{\text{sub}}}{f} = \frac{\lambda_{\text{vacuum}}}{n} \tag{1.3}$$

Here, f is the frequency.

Optical fibre is a term used to denote any plastic or glass conduit meant to transport light. An optical fibre is a thin, flexible, transparent fibre that acts as a waveguide or light pipe, to transmit light between the two ends of the fibre. Optical fibres are widely used in fibre optic communications for transmission over longer distances at bandwidths (data rates) higher than in other forms of communication. The principles behind optical fibres are actually quite old, but in recent years, optical fibres have become an incredibly important technology because communication systems have begun to use optical fibre to transmit data at extremely high rates. Apart from fibre optic communications, optical fibre has several applications in medical instruments, consumer products and physics. Such fibres offer many advantages over traditional metal wires, the most important advantage being that there is considerably less signal degradation. Additionally, optical fibres are immune to electromagnetic interference, which can seriously impede the transmission of data along normal metal wires. This adds an added security measure as well, since optical fibres can survive an electromagnetic pulse which would destroy metal cables.

Fibre optics technology makes use of the theory of total internal reflection of light [24–27]. Total internal reflection occurs when a ray or a beam of light travels through a medium and strikes a medium with a different refractive index. If the angle of incidence (the angle at which light strikes the medium), is greater than the critical angle (the angle of incidence at which if light falls, is refracted along the boundary), the light beam does not get refracted but gets fully reflected. Fibre optics technology makes use of this fundamental principle of physics to transmit data by converting it into light signals.

1.5.1 Principle of Working

The basic principle behind optical fibre is quite simple: the fibre is coated to make it completely reflective on the inside, so that when light goes in, it gets reflected without losing any intensity, passing down the fibre to the other end. This basic idea of guiding light by refraction goes back to the 1840s. By the beginning of the 20th century, some practical applications were developed, most notably the use of optical fibre in dentistry to light up the inner sides of the mouth.

An optical fibre is a cylindrical dielectric waveguide (nonconducting waveguide) that transmits light along its axis by the process of total internal reflection. The fibre consists of a core surrounded by a cladding layer, both of which are made of dielectric materials. To confine the optical signal to the core, the refractive index of the core must be greater than that of the cladding. The boundary between the core and the cladding may either be abrupt as in a step-index fibre or gradual as in a graded-index fibre.

1.5.2 Total Internal Reflection

When light travelling in a dense medium hits a boundary at a steep angle (larger than the critical angle for the boundary), it will be completely reflected. This effect is used in optical fibres to confine light to the core. The concepts of reflection and refraction can be interpreted most easily by considering the behaviour of light rays associated with plane waves travelling in a dielectric material [28–29]. When a light ray encounters a boundary separating two different media, part of the ray is reflected back to the first medium and the remainder is bent (or refracted) as it enters the second material. This is shown in Fig. 1.3, where $n_2 < n_1$. The bending or refraction of the light ray at the interface is a result of the difference in the speed of light in two materials that have different refractive indices. The relationship at the interface is known as Snell's law and is given by

$$n_1 \sin\phi_1 = n_2 \sin\phi_2 \tag{1.4}$$

or, equivalently, as

$$n_1 \cos\theta_1 = n_2 \cos\theta_2 \tag{1.5}$$

The angle ϕ_1 between the incident ray and the normal to the surface is known as the angle of incidence.

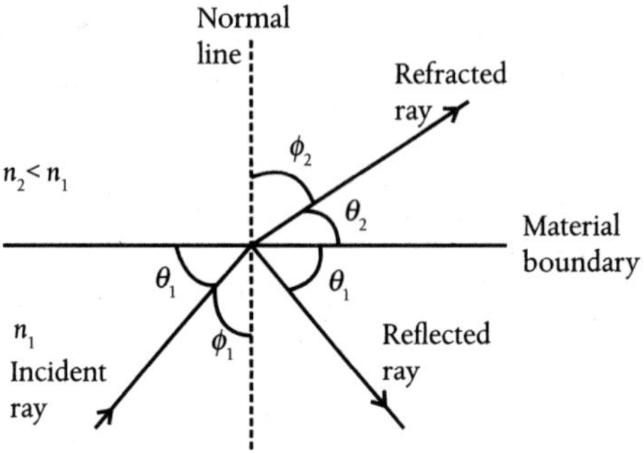

Fig. 1.3 Refraction and reflection of a light ray at a material boundary

According to the law of reflection, the angle θ_1 at which the incident ray strikes the interface is exactly equal to the angle that the reflected ray makes with the same interface.

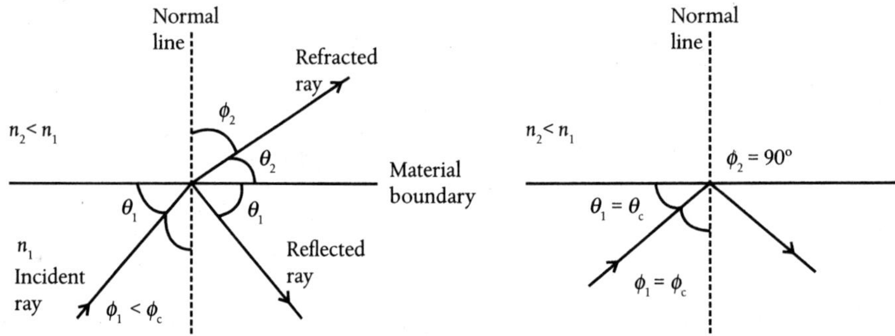

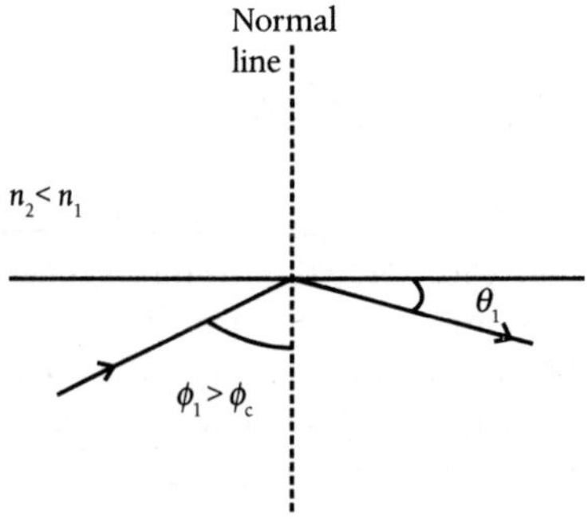

Fig. 1.4 Representation of the critical angle and total internal reflection at a glass-air interface

As the angle of incidence ϕ_1 is an optically denser material and becomes larger, the refracted angle ϕ_2 approaches $\pi/2$. Beyond this point, no refraction is possible and the light rays become totally internally reflected. The conditions required for total internal reflection can be determined by using Snell's law. Consider Fig. 1.4, which shows a glass surface in air. A light ray gets bent toward the glass surface as it leaves the glass in accordance with Snell's law. If the angle of incidence ϕ_1 is increased, a point will eventually be reached where the light ray in air is parallel to the glass surface. This point is known as the critical angle of incidence ϕ_c. When the incidence angle ϕ_1 is greater than the critical angle, the condition for total internal reflection is satisfied; that is, the light is totally reflected back into the glass with no light escaping from the glass surface. When the light ray in air is parallel to the glass surface, then $\phi_2 = 90^\circ$ so that $sin\,\phi_2 = 1$. The critical angle in the glass is thus

$$sin\,\phi_c = \frac{n_2}{n_1} \tag{1.6}$$

1.5.3 Derivation of Numerical Aperture and Multipath Dispersion in Step Index Fibres

If we want to achieve total internal reflection for each ray at the boundary of core of refractive index n_1 and cladding of refractive index n_2 where, $n_1 > n_2$, then the angle of incidence θ_1 of the incoming ray on to the end face of the fibre must be less than a certain value θ_2. In order to calculate θ_2, apply Snell's law as shown in Fig. 1.5 [21].

$$n_a \sin\theta_1 = n_1 \sin\theta = n_1 \cos\phi \tag{1.7}$$

where, $\theta_1 < \theta_2$.

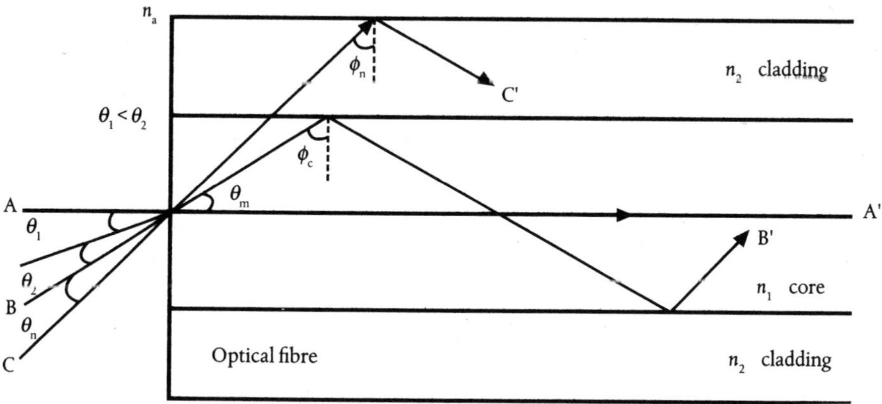

Fig. 1.5 Total internal reflection

Figure 1.5 shows three rays. AA' is the axial ray, BB' is the critical ray in core and CC' is oblique ray for any angle $\theta_1 < \theta_2$, $n_a \sin\theta_n = n_1 \sin\theta_m = n_1 \cos\phi_n$ where, $n_1 > n_2$. For critical rays (BB' w.r.t. AA'), we get,

$$n_a \sin\theta_2 = n_1 \sin\theta_m = n_1 \cos\phi_c \quad (1.8)$$

where,

$\theta_m = \pi/2 - \phi_c$ and ϕ_c is the value of critical angle.

We know that,

$$\sin\phi_c = n_2/n_1,$$

so that

$$\cos\phi_c = \left(n_1^2 - n_2^2\right)^{1/2}/n_1 \quad (1.9)$$

This implies that,

$$\sin\theta_2 = \left(n_1^2 - n_2^2\right)^{1/2}/n_a \quad (1.10)$$

So, the greater value of θ_2 means that greater proportion of the light incident on the end faces can be collected and passed by the fibre and propagated inside the fibre by total internal reflection.

Therefore, numerical aperture is $\text{NA} = n_a \sin\theta_2 = \left(n_1^2 - n_2^2\right)^{1/2} = n_1\sqrt{2\Delta}$ where, $\Delta = (n_1 - n_2)/n_1$. If the two rays are launched together and after travelling the fibre length (L), they are separated by ΔT on arrival at other end.

So far, we have considered the propagation of a single ray of light. However, a pulse of light consists of several rays, which may propagate at all values of α, varying from 0 to α_m. The paths traversed by the two extreme rays, one corresponding to $\alpha = 0$ and the other corresponding to α very nearly equal to (but less than) α_m, are shown in Fig. 1.6.

An axial ray travels a distance L inside the core of index n_1 with velocity v in time.

$$t_1 = \frac{L}{v} = \frac{Ln_1}{c} \quad (1.11)$$

since, by definition, $n_1 = c/v$.

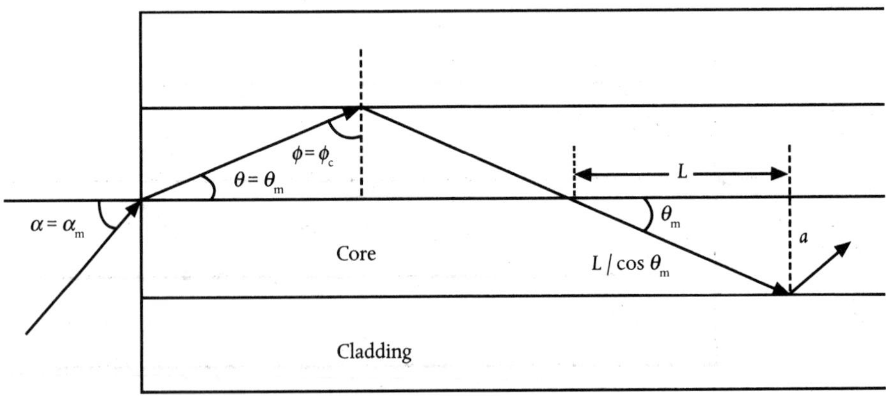

Fig. 1.6 Trajectories of two extreme rays inside the core of a step-index fibre

The most oblique ray corresponding to $\alpha = \alpha_m$ will cover the same length of fibre (axial length L, but actual distance $L/\cos\theta_m$) in time t_2 and is given by

$$t_2 = \frac{L/\cos\theta_m}{v} = \frac{Ln_1}{c\cos\theta_m} = \frac{Ln_1}{c\sin\phi_c} = \frac{Ln_1}{c(n_2/n_1)} = \frac{Ln_1^2}{cn_2} \quad (1.12)$$

The two rays are launched at the same time, but will be separated by a time interval ΔT after travelling the length L of the fibre, and is given by

$$\Delta T = t_2 - t_1 = \frac{Ln_1^2}{cn_2} - \frac{Ln_1}{c} = \frac{Ln_1}{c}\left(\frac{n_1 - n_2}{n_2}\right) \quad (1.13)$$

Thus, a light pulse consisting of rays spread over $\alpha = 0$ to $\alpha = \alpha_m$ will be broadened as it propagates through the fibre, and the pulse broadening per unit length of traversal will be given by

$$\Delta T = \frac{n_1 L}{n_2 c}(n_1 - n_2) \quad (1.14)$$

where c is the velocity of light. This is called multipath time dispersion of fibre. If $n_1 = 1.5$, $n_2 = 1(\text{air})$, $c = 3\times10^8$ m/s.

$$\frac{\Delta T}{L} = 2.5\times10^{-9} = 2.5\,\text{ns/m} = 2.5\,\mu\text{s/km}$$

when $(n_1 - n_2) \ll n$, the time dispersion of fibre becomes

$$\frac{\Delta T}{L} \simeq \frac{n_1 - n_2}{c} \quad (1.15)$$

The bandwidth × distance relation is

$$BL = \frac{L}{\Delta T} = c/\Delta n \quad (1.16)$$

The small value of Δn will give large BL product value. The multipath dispersion in step index multimode fibre reduces its use in short-range applications with modest bandwidth.

1.5.4 Derivation to Find Parameters of Monomode and Multimode Fibres

The single-mode fibre allows only transmission of monomode light which is usually known as fundamental mode (TEM_{00}). The monomode fibre has several advantages over multimode fibres which allow propagation of several modes simultaneously. The losses due to attenuation, absorption, scattering and dispersion are low in single-mode fibre; hence, monomode fibre is used for long-haul communication, whereas multimode fibres are used for short-haul application where the cost is to be kept low, especially in local area network (LAN) application. Figure 1.7 shows the typical profiles of single-mode fibre and multimode fibre.

It is necessary to know as to how a monomode fibre conducts light and how the attenuation is reduced to 0.2 dB/km or even less which is responsible for signal power loss and reduction of signal to noise (S/N) ratio. Figure 1.8 shows the index variation of a step-index fibre. The refractive indices of core and cladding in the optical fibre are related to

$$n(r) = n_2(1 - \Delta) \text{ (for step index fibre)} \quad (1.17)$$

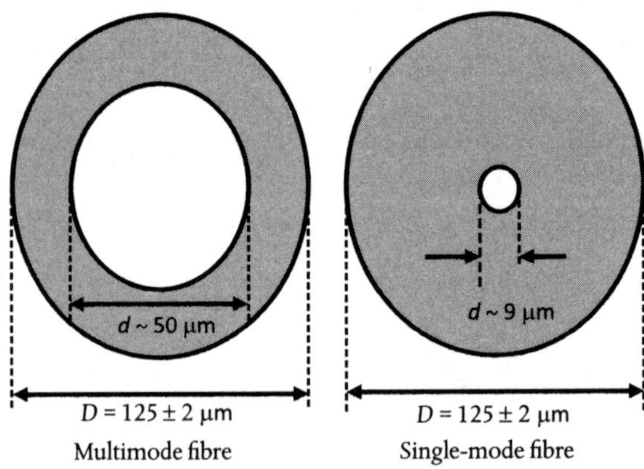

Fig. 1.7 Typical profiles of single-mode fibre and multimode fibre

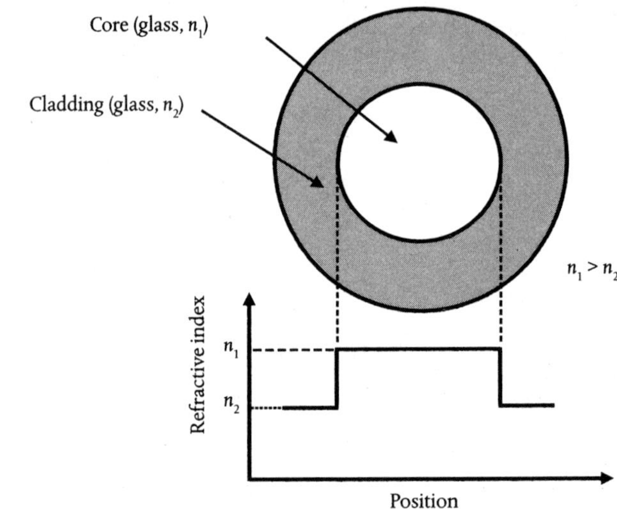

Fig. 1.8 Index variation of a step-index fibre

where Δ is the variation in refractive index. V (a dimensionless parameter) for different fibres is

$$V = \frac{\pi d}{\lambda}\sqrt{n_1^2 - n_2^2} \tag{1.18}$$

where number of modes (M) are related to V as

$$M = \frac{V^2}{2}\text{(for step index fibre)} \tag{1.19}$$

and

$$M = \frac{V^2}{4}\text{(for graded index fibre)} \tag{1.20}$$

If the V-parameter is less than or equal to 2.405, the fibre is known as single-mode fibre; in other cases, it is called a multimode fibre. Therefore, to reduce the value of V, that is, to provide monomode propagation through a fibre, some parameters have to be readjusted such as reducing the core diameter (d) and increasing the operating wavelength (λ) or making n_2 as close to n_1 as far as possible.

For a fabricated monomode (single-mode) fibre, the core diameter (d) is 10 μm and less, the range of operating wavelengths start from 1300 nm and the refractive index $\Delta = (n_1 - n_2)/n_2$ is less than 0.4%. Such a fibre rejects all higher order modes and allows transmissions of the lowest order (TEM_{00}) along the central axis of the fibre. A comparative chart of single-mode and multimode fibre characteristics is given in Table 1.1.

Table 1.1 Parametric comparison of monomode and multimode fibres.

Parameters	Monomode fibre	Multimode fibre
Core diameter (d)	<10 μm	>50 μm – 1000 μm
Wavelength (λ)	>1300 nm	850 (Typical 850 nm and below)
Refractive index (Δ)	<0.4%	>1%
Numerical aperture (NA)	0.125	0.25

1.5.5 Limitations of Single-mode Fibres

An optical fibre having a circular core that supports only the fundamental mode (LP_{01} mode or HE_{11} mode) is called a single-mode fibre. This type of fibre is designed such that all higher order modes are cut off at the operating wavelength. We have seen that the number of modes supported by a fibre is governed by the V-parameter, and the cut-off condition of various modes is also determined by V. The cut-off normalised frequency V_c for the LP_{11} mode in a step index (SI) circular fibre is 2.405. Thus, single-mode operation (of the fundamental LP_{01} mode) in an SI fibre is possible over a normalised frequency range $0 < V < 2.405$ (since there is no cut-off for the LP_{01} mode, $V_c = 0$ for this mode). We know that the normalised frequency parameter V is given by

$$V = \frac{2\pi a}{\lambda}\left(n_1^2 - n_2^2\right)^{1/2} \tag{1.21}$$

In terms of the relative refractive index difference Δ, this relation may also be written as

$$V = \frac{2\pi a n_1}{\lambda}(2\Delta)^{1/2} \tag{1.22}$$

Thus, for a particular value of λ, if the value of V is to be made lower than 2.405, either the diameter (2a) of the fibre or the relative refractive index difference Δ of the fibre has to be made smaller. Both these factors create practical problems. With small core diameter, it becomes difficult to launch the light into the fibre and also to join (splice) the fibres in the field. It is also practically difficult to get very low values of Δ.

An alternate solution for getting higher values of V_c and hence large core diameters is that instead of using an SI fibre, a graded index (GI) fibre may be used. The cut-off value of the normalised frequency V_c for single-mode operation in a GI fibre is given by

$$V_c = 2.405\left(1 + \frac{2}{\alpha}\right)^{1/2} \tag{1.23}$$

Thus, V_c will increase by a factor of $\sqrt{2}$ for a parabolic profile $(\alpha = 2)$ and by a factor of $\sqrt{3}$ for a triangular profile $(\alpha = 1)$.

Another problem with low V-values and low Δ for the single-mode operation is that the modal field of the LP_{01} mode extends well into the cladding. It has been shown that for V-values less than 1.4, more than half the modal power propagates through the cladding. Thus, the evanescent field, which decays exponentially, may extend to a considerable distance in the cladding. To ensure that this power propagating in the cladding is not lost, the thickness of the cladding should be large (cladding radius greater than 50 μm). Further, the absorption and scattering losses in the cladding should also be minimum.

Before we take up the different types of SMFs, let us discuss the parameters that characterise an SMF and those that are helpful in evaluating the dispersion, attenuation and the jointing and bending losses.

1.5.6 Limitations of Multimode Fibres

Multimode fibres support many modes. The higher order modes (corresponding to oblique rays in terms of the ray model) travel slower and hence arrive at the other end of the fibre later than the lower order modes (corresponding to axial modes). This means that different modes travel with different group velocities. Thus, a light pulse propagating through such a fibre will get broadened. This is called multipath dispersion or intermodal dispersion. The pulse dispersion per unit length for a step index fibre $(q = \infty)$, it is given by,

$$\frac{\Delta T}{L} = \frac{n_1}{n_2}\left(\frac{n_1 - n_2}{c}\right) \approx \frac{n_1 \Delta}{c} \tag{1.24}$$

Similarly, it can be shown that the pulse dispersion per unit length for a graded index fibre with a parabolic profile $(\alpha = 2)$ may be given by,

$$\frac{\Delta T}{L} = \frac{n_0}{2c}\Delta^2 \tag{1.25}$$

and that for a GI fibre with an optimum profile $(\alpha = 2 - 2\Delta)$ may be given by;

$$\frac{\Delta T}{L} \approx \frac{n_0}{8c}\Delta^2 \tag{1.26}$$

Thus, pulse broadening due to intermodal dispersion, varying from about 0.05 ns/km to 80 ns/km, depending on the value of α and the core index n_0, has been observed in multimode fibres. This severely restricts the use of such fibres in long-haul communications.

Further, a light pulse has a number of spectral components (of different frequencies), and the group velocity of a mode varies with frequency. Therefore, different spectral components in the pulse propagate with slightly different group velocities, resulting in pulse broadening. This phenomenon is called group velocity dispersion (GVD) or intramodal dispersion.

In spite of these limitations, multimode fibres are used in local area networks and short-haul communication links.

1.6 FIBRE OPTICS TRANSMISSION TECHNOLOGY

Fibre optics transmission technology has the following features:

(1) **Transmitter:** Transmitters are important fibre optics equipment used at the sender's end. The most common transmitter used for creating optical signals are LEDs and laser diodes. These are miniature transmitters capable of converting electrical signals into optical pulses.
(2) **Cladding:** The optical fibre consists of a cladding as discussed earlier. Cladding is used for total internal reflection of light rays.
(3) **Amplifiers or repeaters:** These are used after regular intervals owing to the physical limitations of optical fibres. These amplifiers receive the pulses, convert them into electrical signals, reconvert them into light pulses and transmit them again at even more intensity than before.
(4) **Photodetectors:** The concept of a photoelectric field is used at the receiving end of these signals. A photodetector is used to convert the light pulses into electric signals using photoelectric fields. A photodetector is mostly a photodiode.

1.6.1 Splicer and Connectors

When frequent connectors and disconnectors are not required, splicing is used to create long optical links. While going for fibre splicing, the geometrical differences and misalignment at joint and mechanical strength of splice factors must be considered. It includes semi-permanent or permanent joints between two fibres. While some splices are re-enterable for repairs, they are made without the purpose of even irregular connections and disconnections. The most popular type is the fusion splice, where fibres are aligned and melted until they fuse as shown in Fig. 1.9. This process is carried out either in grooved fibre holder or under a microscope, which has micromanipulators. Fusion splicing is done with the help of an electric arc or a laser. The loss is of only 0.06 dB in this technique. While performing this technique some defects may be created near the joints due to excessive heat. Splicing may also be performed using other techniques such as V-groove splicing, elastic tube splicing, and so on.

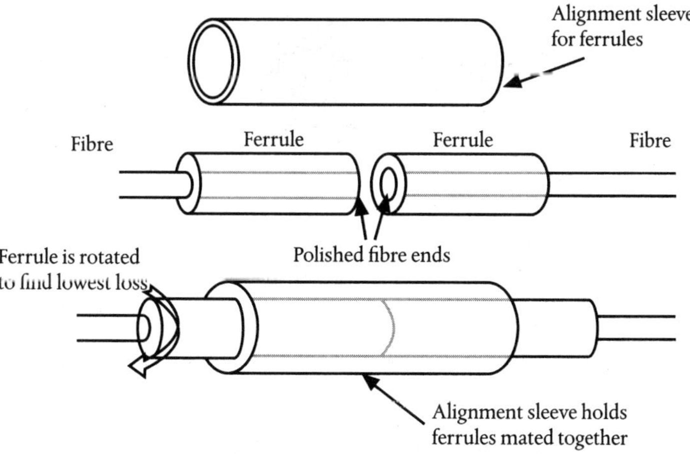

Fig. 1.9 Splicer process

Just as an electronic connector provides a pluggable connection between electronic circuits, a fibre optic connector provides an optical connection. The aim of connectors is to join the various optical components such as fibre to source, fibre to fibre itself, etc. The splicer is a permanent joint but connectors are removable joints. Joining two fibres with the help of connectors will keep a low resistance path across the junction. There are various benefits of using connectors as follows:

- Low coupling losses—the connector assembalance must have rigid alignment tolerance to ensure the low matching losses.
- Interchangeability
- Ease of assembly
- Low environment sensitivity
- Low-cost and reliable construction
- Ease of connection

Commercially, there are various types of connectors are available as shown in Fig. 1.10.

- ST connectors: It stands for straight tip. These are mostly use in LANs.
- SC connectors: It stands for subscriber connectors. These are also available for single-mode and multimode fibres. These connectors are able to replace the FC and D4 connectors in the world of telecommunication, by having the ferrule assembly.
- MT-RJ connectors: These are mostly used in telephone lines and are available with the name RJ-45. About 4–6 modular plugs and jacks are used by these connectors.
- LC connectors: These are almost the same as RJ-45 connectors but use 125 mm ceramic ferrule.

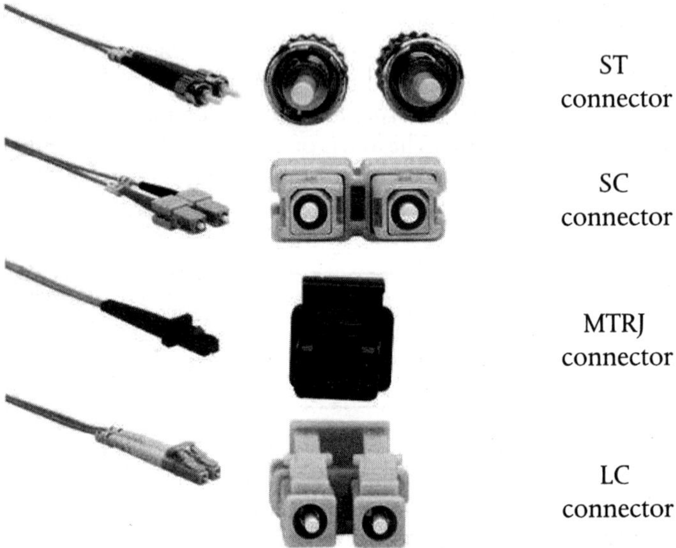

Fig. 1.10 Different types of fibre-optic cables and connectors

1.7 ONE-DIMENSIONAL RAY EQUATION

In optics, a ray is an idealised narrow beam of light. Rays are used to model the propagation of light through an optical system by dividing the real light field into discrete rays that can be computationally propagated through the system by the techniques of ray tracing [21]. This allows even very complex optical systems to be analysed mathematically or simulated by the computer. Ray tracing uses approximate solutions to

Maxwell's equations that are valid as long as the light waves propagate through and around objects whose dimensions are much greater than the light wavelength. The ray theory does not describe phenomena such as interference and diffraction, which require wave theory (involving the phase of the wave). We consider a medium in which the refractive index depends only on the *x*-coordinate as given by (1.27).

$$n = n(x) \tag{1.27}$$

A medium with continuously varying refractive index given by (1.27) can be thought of as a limiting case of a medium consisting of a set of thin layers, each characterised by a specific value of the refractive index (see Fig. 1.11a). To trace out the rays through such a stack of thin layers, we can use Snell's law according to which,

$$n_1 \sin\phi_1 = n_2 \sin\phi_2 = n_3 \sin\phi_3 = \ldots \text{constant} \tag{1.28}$$

where ϕ_1, ϕ_2, ... are the angles of incidence at various interfaces as shown in Fig. 1.11a. If ϕ_1, ϕ_2, ... are the corresponding angles that the rays make with the z-axis, then

$$n_1 \cos\phi_1 = n_2 \cos\phi_2 = n_3 \cos\phi_3 = \ldots \text{constant}, \text{ say } \tilde{\beta} \tag{1.29}$$

(We have chosen the z-axis in such a way that the ray lies in the *x*-*z* plane).

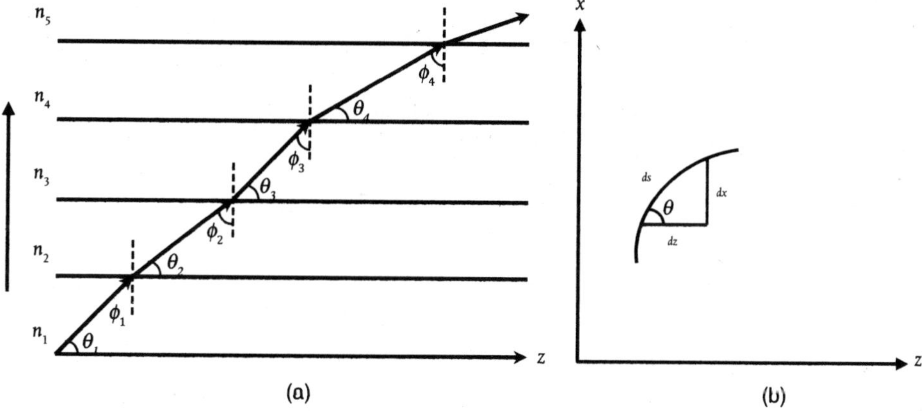

Fig. 1.11 (a) A continuously varying refractive index distribution can be approximated by a layered structure, and ray paths in such a layered structure can be obtained by using Snell's law. (b) As the number of layers increases and the thickness of each layer reduce to zero, the ray path becomes a continuous curve.

When the refractive index variation is continuous, the thickness of each layer becomes infinitesimally small and the piecewise straight lines shown in Fig. 1.11a form a continuous curve as shown in Fig. 1.11b. From (1.29), we can see that the rays bend in such a way that the product $n(x)\cos\theta(x)$ remains constant, which we denote by $\tilde{\beta}$. Thus

$$n(x)\cos\theta(x) = \tilde{\beta} \quad \text{(invariant of the ray path)} \tag{1.30}$$

Now if ds represents the infinitesimal arc-length along the ray path, then from Fig. 1.11b we have,

$$(ds)^2 = (dx)^2 + (dz)^2 \tag{1.31}$$

or

$$\left(\frac{ds}{dz}\right)^2 = \left(\frac{dx}{dz}\right)^2 + 1 \tag{1.32}$$

Since

$$\cos\theta = \frac{dz}{ds} \tag{1.33}$$

we obtain

$$\frac{ds}{dz} = \frac{1}{\cos\theta(x)} = \frac{n(x)}{\tilde{\beta}} \tag{1.34}$$

Substituting in (1.34) we get,

$$\left(\frac{dx}{dz}\right)^2 = \frac{n^2(x)}{\tilde{\beta}} - 1 \tag{1.35}$$

which represents the rigorously correct ray equation for n^2 depending only on the x-coordinate. Notice from (1.35) that ray paths are possible only in regions having $\tilde{\beta} \le n(x)$ hence, the ray will move in such a fashion as to always have $\tilde{\beta} \le n(x)$.

For a given refractive index profile, a ray will become parallel to the z-axis and turn back toward the axis when $dx/dz = 0$ that is at the value $x = x_t$ satisfying

$$n(x_t) = \tilde{\beta} \tag{1.36}$$

The point $x = x_t$ is known as the turning point of the ray. Equation (1.36) can be put in a more convenient form by differentiating it with respect to z

$$2\frac{dx}{dz}\frac{d^2x}{dz^2} = \frac{1}{\tilde{\beta}^2}\frac{dn^2}{dx}\frac{dx}{dz}$$

or,

$$\frac{d^2x}{dz^2} = \frac{1}{2\tilde{\beta}^2}\frac{dn^2}{dx} \qquad \text{ray equation} \tag{1.37}$$

this is another form of the ray equation.

1.7.1 Transit Time Calculations

An important characteristic of an optical waveguide is pulse dispersion, which is the temporal spreading of a pulse of light launched into the waveguide [27–29]. One of the mechanisms that lead to pulse dispersion is the difference in time taken by different rays. Thus, if all rays are launched simultaneously in a waveguide at the output, the rays will arrive at different times; hence, it will correspond to a temporal dispersion of the pulse. To calculate the pulse dispersion, we calculate the time taken by a ray to traverse a given length of the waveguide.

Let τ_p represent the time taken for the ray to traverse the length z_p. Now the time taken to travel an arc length ds along the ray is

$$d\tau = \frac{ds}{v(x)} = \frac{1}{c}n(x)ds \tag{1.38}$$

where $v(x) = c/n(x)$ is the velocity of ray along the ray path. Hence, the time taken for one period is

$$\tau_p = \int \frac{n(x)ds}{c}$$

where the integral is over one period. Since from Fig. 1.11b,

$$ds = \left[(dx)^2 + (dz)^2\right]^{1/2} = \left[1 + \left(\frac{dz}{dx}\right)^2\right]^{1/2} dx$$

which, using the ray equation (1.35), can be written as

$$ds = \frac{n(x)}{\left[n^2(x) - \tilde{\beta}^2\right]^{1/2}} dx$$

Hence, the time taken travel over a quarter of a period (that is, from the axis to the turning point) is $\tau_p / 4$, which is given by

$$\frac{1}{4}\tau_p = \frac{1}{c}\int_0^{x_t} \frac{n^2(x)}{\left[n^2(x) - \tilde{\beta}^2\right]^{1/2}} dx \tag{1.39}$$

where x_t is the x-coordinate at the turning point. This equation is valid for an arbitrary profile. Some specific profiles are considered in the following section.

1.7.2 Pulse Dispersion in a Parabolic Index Medium

A square law medium is characterised by the following refractive index distribution [21],

$$\left.\begin{aligned} n^2(x) &= n_1^2\left[1 - 2\Delta\left(\frac{x}{a}\right)^2\right]; \quad |x| < a \quad (\text{core}) \\ &= n_1^2(1 - 2\Delta) = n_2^2; \quad |x| > a \quad (\text{cladding}) \end{aligned}\right\} \tag{1.40}$$

We evaluate (1.39) for a parabolic index medium characterised by (1.40). Since the turning point is defined by $n(x) = \tilde{\beta}$, using (1.40) we have

$$n_1^2\left[1 - 2\Delta\left(\frac{x_t}{a}\right)^2\right] = \tilde{\beta}^2$$

or,

$$x_t = \frac{a}{n_1\sqrt{2\Delta}}\left[n_1^2 - \tilde{\beta}^2\right]^{1/2} \tag{1.41}$$

Thus,

$$\tau_p = \frac{4}{c}\int_0^{x_t} \frac{n_1^2\left[1 - 2\Delta\left(\frac{x}{a}\right)^2\right]}{\left\{n_1^2\left[1 - 2\Delta\left(\frac{x}{a}\right)^2\right] - \tilde{\beta}^2\right\}^{1/2}} dx$$

$$= \frac{4}{c}\int_0^{x_t} \frac{n_1^2 - 2\Delta n_1^2\left(\frac{x}{a}\right)^2 - \tilde{\beta}^2 + \tilde{\beta}^2}{\left\{n_1^2 - 2\Delta n_1^2\left(\frac{x}{a}\right)^2 - \tilde{\beta}^2\right\}^{1/2}} dx$$

or,

$$\tau_p = \frac{4}{c}\left[\frac{n_1\sqrt{2\Delta}}{a}\int_0^{x_t}\sqrt{x_t^2 - x^2}\,dx + \frac{\tilde{\beta}^2 a}{n_1\sqrt{2\Delta}}\int_0^{x_t}\frac{dx}{\sqrt{x_t^2 - x^2}}\right] \tag{1.42}$$

where x_t is defined through (1.41). Carrying out the elementary integration, we get

$$\tau_p = \frac{\pi}{c}\frac{a}{n_1\sqrt{2\Delta}}\left[n_1^2 - \tilde{\beta}^2\right] \tag{1.43}$$

since the ray path in square law medium is given by,

$$x(z) = \frac{a\sin\theta_1}{\sqrt{2\Delta}}\sin\left(\frac{n_1\sqrt{2\Delta}}{a\tilde{\beta}}z\right) \tag{1.44}$$

The periodic length of the sinusoidal ray path is given by,

$$z_p = \frac{2\pi}{\Gamma} = \frac{2\pi a\tilde{\beta}}{n_1\sqrt{2\Delta}} \tag{1.45}$$

where Γ is the distance the pulse has propagated and the ray makes an angle θ with the z-axis at $z = 0$, now from (1.45) it readily follows that the turning point corresponds to $z = z_p/4$ and

$$\frac{1}{4}\Gamma z_p = \frac{\pi}{2} \tag{1.46}$$

implying (see (1.45))

$$z_p = \frac{2\pi}{\Gamma} = \frac{2\pi a\tilde{\beta}}{n_1\sqrt{2\Delta}} \tag{1.47}$$

Thus, if $\tau(z)$ represents the time taken by the ray to traverse the distance z (which covers many periods) then,

$$\frac{\tau(z)}{z} = \frac{\tau_p}{z_p} = \frac{1}{2c}\left[\tilde{\beta} + \frac{n_1^2}{\tilde{\beta}}\right] \tag{1.48}$$

Since

$$n_2 < n(x) < n_1 \tag{1.49}$$

we have for guided rays,

$$n_2 < \tilde{\beta} < n_1 \tag{1.50}$$

A ray with $\tilde{\beta} = n_1$ takes the minimum time given by

$$\tau_{min}(z) = \frac{1}{2c}\left[n_1 + \frac{n_1^2}{n_1}\right]z = \frac{n_1}{c}z \tag{1.51}$$

whereas a ray with $\tilde{\beta} = n_2$ takes the maximum time given by,

$$\tau_{max}(z) = \frac{1}{2c}\left[n_2 + \frac{n_1^2}{n_2}\right]z \tag{1.52}$$

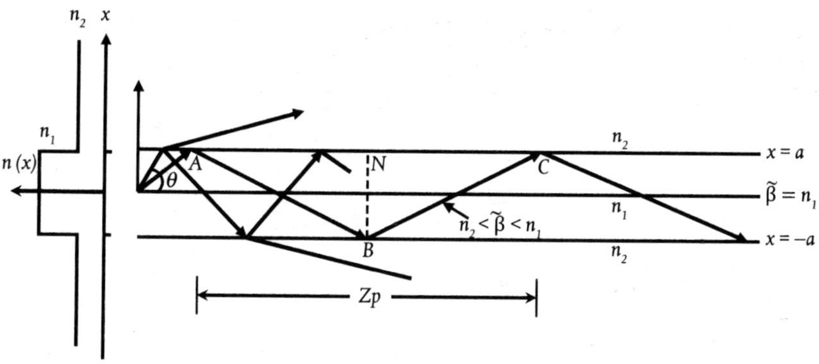

Fig. 1.12 Guided and refracting rays in a step index planar waveguide. For guided rays, $n_2 < \tilde{\beta} < n_1$ and for refracting rays, $\tilde{\beta} < n_2$

Obviously, $\tau_{\min}$ corresponds to the axial ray for which $\theta = 0$ and $\tau_{\max}$ to a ray so that it becomes parallel to the z-axis at $x = a$. Thus, the pulse dispersion is given by,

$$\Delta\tau = \tau_{\max} - \tau_{\min} = \frac{1}{2cn_2}(n_1 - n_2)^2 z \simeq \frac{n_2}{2c}\Delta^2 z \tag{1.53}$$

In the last step, we have assumed $n_1 \simeq n_2$. Thus, for $n_1 = 1.46$, $\Delta = 0.01$, we have

$$\Delta\tau \simeq 250\,\text{ps/km}$$

Note that $\Delta\tau$ is proportional to Δ^2; thus, for reducing pulse dispersion, one must have waveguides with small NA.

1.7.3 Pulse Dispersion in a Planar Step Index Waveguide

To appreciate the small pulse dispersion given by (1.53) in a square law medium, we now obtain the pulse dispersion in a step index planar waveguide characterised by (see Fig. 1.12);

$$\begin{aligned} n^2(x) &= n_1^2 && |x| < a \quad \text{(core)} \\ &= n_2^2 = n_1^2(1 - 2\Delta) && |x| > a \quad \text{(cladding)} \end{aligned} \tag{1.54}$$

As the core and cladding are homogeneous media, ray paths in both media are straight lines (see Fig. 1.12). For a ray to be guided in the core, it must suffer total internal reflection at the core-cladding interfaces at $x = \pm a$. Hence, if the angle made by the ray with the z axis is θ, then for a guided ray

$$0 < \theta < \cos^{-1}\left(\frac{n_2}{n_1}\right) \tag{1.55}$$

or,

$$n_1 > n_1 \cos\theta > n_2 \tag{1.56}$$

Since, $\tilde{\beta} = n(x)\cos\theta(x)$, for the present case $\tilde{\beta} = n_1 \cos\theta$ and again we obtain

$$n_2 < \tilde{\beta} < n_1 \qquad \text{guided rays} \tag{1.57}$$

$$\tilde{\beta} < n_2 \qquad \text{refracted rays} \tag{1.58}$$

To obtain the transit time, we must obtain the ray period z_p and the time taken τ_p to cover one period. Now, from Fig. 1.12 we have,

$$z_p = AC = 2AN = \frac{4a}{\tan\theta} \tag{1.59}$$

Also,

$$\tau_p = \frac{AB + BC}{c/n_1} = \frac{2n_1}{c} AB = \frac{4an_1}{c\sin\theta} \tag{1.60}$$

Thus, the time taken to cover one period is

$$\tau_p = \frac{n_1 z_p}{c\cos\theta} = \frac{n_1^2}{c\tilde{\beta}} z_p \tag{1.61}$$

where we have used $\tilde{\beta} = n_1 \cos\theta$. Thus, the time taken to cover distance z is

$$\tau = \frac{\tau_p}{z_p} z = \frac{n_1^2}{c\tilde{\beta}} \tag{1.62}$$

Thus,

$$\tau_{\min} = \frac{n_1 z}{c} \tag{1.63}$$

$$\tau_{\max} = \frac{n_1^2 z}{cn_2} \tag{1.64}$$

The pulse dispersion is given by

$$\Delta\tau = \tau_{\max} - \tau_{\min} = \frac{n_1}{cn_2}(n_1 - n_2) z \simeq \frac{n_1 \Delta}{c} z \tag{1.65}$$

where we have used $n_1 - n_2 \simeq n_1\Delta$ and $n_1 \simeq n_2$. For $n_1 \simeq 1.46$, $\Delta = 0.01$, we have

$$\Delta\tau \simeq 50\,\text{ns/km}$$

Comparing (1.65) with (1.53), we see that the pulse dispersion in a square law medium is reduced by a factor $\Delta/2$ compared with a step index waveguide. For $\Delta = 0.01$, this corresponds to a reduction by a factor of 200.

Although these results are derived for a slab waveguide, it so happens that (1.53) and (1.65) are rigorously valid for parabolic index and step index optical fibres, respectively. Due to this small pulse dispersion, the first and second generation optical communication systems use near parabolic index (multimode) optical fibres.

1.7.4 Transit Time Calculation in a Medium Characterised by a Power Law Profile

The pulse dispersion for rays propagating through a graded index medium is characterised by the following the refractive index profile [21–22].

$$\begin{aligned} n^2(x) &= n_1^2\left[1 - 2\Delta\left|\frac{x}{a}\right|^q\right] = n_1^2 - \gamma|x|^q \qquad &: |x| < a \\ &= n_1^2(1 - 2\Delta) = n_2^2 &: |x| > a \end{aligned} \tag{1.66}$$

where

$$\gamma = \frac{n_1^2 2\Delta}{a^q} \tag{1.67}$$

The profile described by the aforementioned equation is usually referred to as the power law profile. Obviously, $q=2$ represents the parabolic index profile and $q=\infty$ represents the step index profile. In general, it is not possible to determine the ray path analytically for an arbitrary value of q; however, we can still make some general observations:

(1) For a ray launched in the core if $\tilde{\beta}<n_2$, the ray will hit the core cladding interface and will be refracted away.
(2) Rays for which $n_2<\tilde{\beta}<n_1$ will be guided through the waveguide.
(3) Guided rays will be periodic.
(4) Guided rays will become parallel to the z-axis at $x=x_t$ where

$$\tilde{\beta}=n(x=x_t)$$

At $x=x_t$, $dx/dz=0$

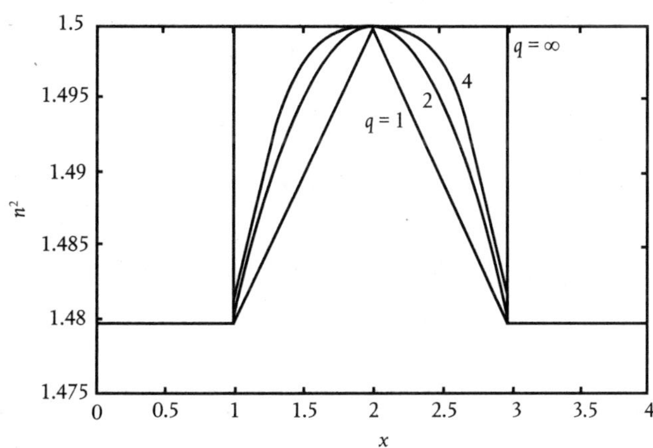

Fig. 1.13 Refractive index variation corresponding to a power law profile given by (1.66). Parabolic profile corresponds to q = 2 and step index profile corresponds to q = ∞. The figure corresponds to n_1 = 1.5, Δ = 0.01 and a = 25 μm

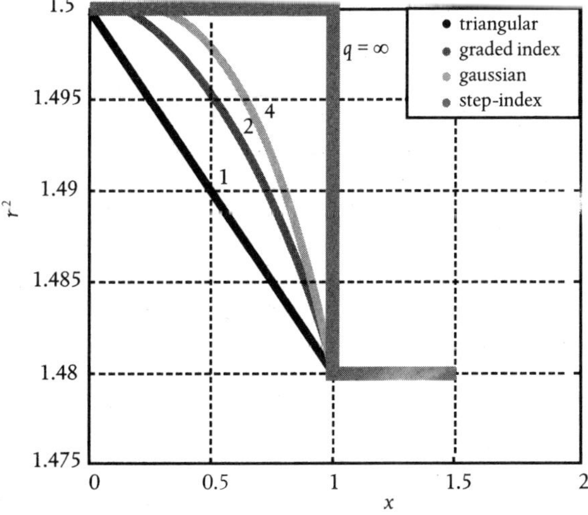

Fig.1.14 The power law profile as given (1.66)

Thus, if τ_p represents the time taken for the ray to traverse the length z_p, then

$$\frac{1}{4}\tau_p = \frac{1}{c}\int_0^{x_t} \frac{n^2(x)}{\left[n^2(x)-\tilde{\beta}^2\right]^{1/2}} dx \tag{1.68}$$

Now, the periodic length of the ray path is given by,

$$\frac{1}{4}z_p = \int_0^{x_t} \frac{dx}{dx/dz} = \tilde{\beta}\int_0^{x_t} \frac{dx}{\left[n^2(x)-\tilde{\beta}^2\right]^{1/2}} \tag{1.69}$$

Substituting (1.66) in (1.68) we get,

$$\frac{1}{4}\tau_p = \frac{1}{c}\left[n_1^2\int_0^{x_t} \frac{dx}{\left[n^2(x)-\tilde{\beta}^2\right]^{1/2}} - \gamma\int_0^{x_t} \frac{x^q}{\left[n^2(x)-\tilde{\beta}^2\right]^{1/2}} dx\right] = \frac{1}{c}\left[\frac{z_p n_1^2}{4\tilde{\beta}} - I_1\right] \tag{1.70}$$

where we have used (1.69) and

$$I_1 = \gamma\int_0^{x_t} \frac{x^q}{\sqrt{f(x)}} dx$$

with

$$f(x) = n^2(x) - \tilde{\beta}^2 = \left(n_1^2 - \tilde{\beta}^2\right) - \gamma x^q$$

We may rewrite

$$I_1 = \gamma\int_0^{x_t} \frac{x^{q+1}}{\sqrt{x^2 f(x)}} dx = \gamma\int_0^{x_t} \frac{x^{q+1}}{\sqrt{g(x)}} dx \tag{1.71}$$

where

$$g(x) = x^2 f(x) = \left(n_1^2 - \tilde{\beta}^2\right)x^2 - \gamma x^{q+2}$$

Thus,

$$x^{q+1} = \frac{1}{\gamma(q+2)}\left[\left(n_1^2 - \tilde{\beta}^2\right)2x - \frac{dg}{dx}\right] \tag{1.72}$$

Substituting in (1.71) we get,

$$I_1 = \frac{1}{q+2}\left[2\left(n_1^2 - \tilde{\beta}^2\right)\int_0^{x_t} \frac{dx}{\sqrt{f(x)}} - \int_0^{x_t} \frac{dg/dx}{\sqrt{g(x)}} dx\right]$$

$$\Rightarrow \frac{2\left(n_1^2 - \tilde{\beta}^2\right)}{q+2}\frac{z_p}{4\tilde{\beta}} - \frac{2}{q+2}\left[g^{1/2}\right]_0^{x_t}$$

But $g^{1/2} = x\sqrt{\left(n^2(x) - \tilde{\beta}^2\right)}$ vanishes at $x = 0$ and at $x = x_t$. Thus

$$I_1 = \frac{z_p}{2\tilde{\beta}}\frac{\left(n_1^2 - \tilde{\beta}^2\right)}{q+2} \tag{1.73}$$

Substituting in (1.70) and simplifying we get,

$$\frac{\tau_p}{z_p} = A\tilde{\beta} + \frac{B}{\tilde{\beta}}$$

where

$$A = \frac{2}{c(q+2)}, \quad B = \frac{qn_1^2}{c(q+2)} \tag{1.74}$$

If the ray covers many periods, then we may write for the time taken to propagate through a distance z,

$$\tau = \left[A\tilde{\beta} + \frac{B}{\tilde{\beta}} \right] z \tag{1.75}$$

1.7.5 Pulse Dispersion in Graded Index Optical Fibres

Although the afore-mentioned result has been derived for a slab waveguide, it is rigorously valid even for fibres.

If c is the velocity of light in free space, we may note the following points:

(i) The time taken is independent of the skewness parameter $\tilde{l}$; this is a characteristic of the power law profile.

(ii) The time taken depends only on the invariant $\tilde{\beta}$ and for guided rays

$$n_2 < \tilde{\beta} < n_1 \tag{1.76}$$

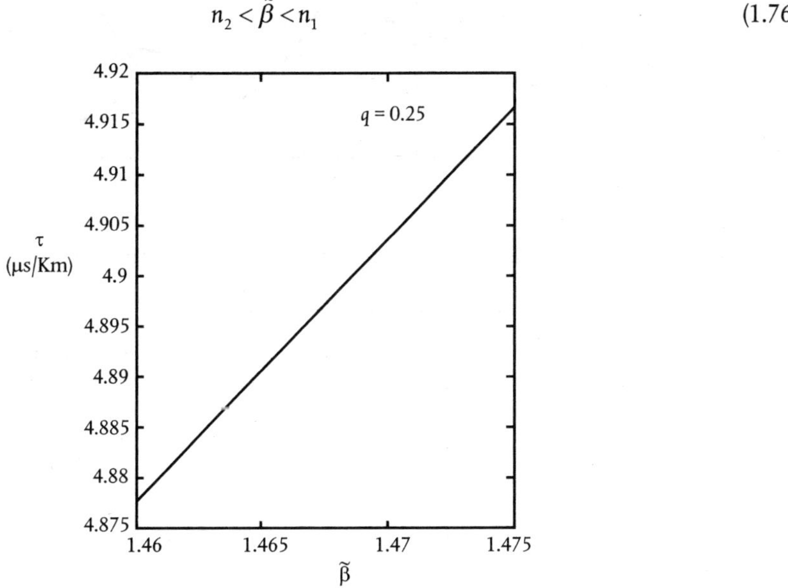

Fig. 1.15 Variation of $\tau(\tilde{\beta})$ with $\tilde{\beta}$ for $q = 0.25$

(iii) Since $\tau(z)$ is independent of $\tilde{l}$, equation(1.75) is identical to the corresponding result derived for a planar waveguide.

(iv) We have neglected material dispersion – that is, we have neglected the dependence of the refractive index on the wavelength. Thus, the temporal broadening of a pulse represents the broadening due to the fact that different rays take different amounts of time to propagate through the fibre. This is also referred to as intermodal dispersion.

In Figs 1.15–1.18, we have plotted $\tau(\tilde{\beta})$ as a function of $\tilde{\beta}$ for $n_2 = 1.46$, $\Delta = 0.01 (\Rightarrow n_1 \approx 1.4748)$ corresponding to $q = 0.25, 1.98, 2.00$, and ∞, respectively, with $n_2 < \tilde{\beta} < n_1$. The vertical axis gives the time taken by the ray to traverse a kilometre length of the fibre. We consider some specific cases:

Case 1: $q = \infty$ (step index fibres). For $q = \infty$, $A = 0$, and $B = n_1^2 / c$.
Thus,

$$\tau(\tilde{\beta}) = \frac{n_1^2}{c\tilde{\beta}} z \tag{1.77}$$

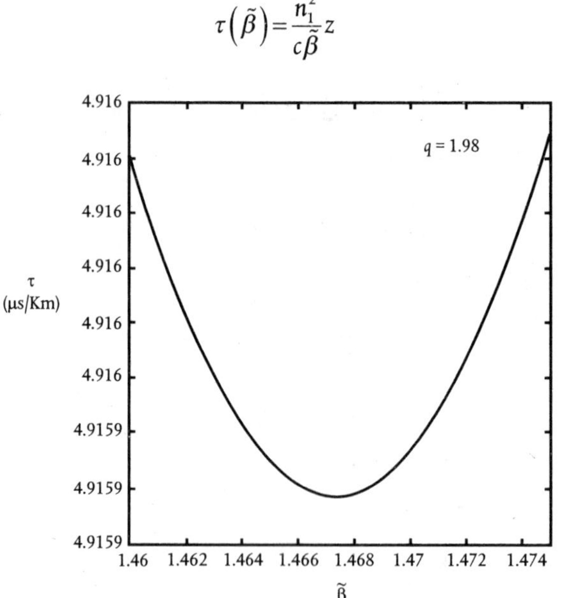

Fig. 1.16 Variation of $\tau(\tilde{\beta})$ with $\tilde{\beta}$ for $q = 1.98$

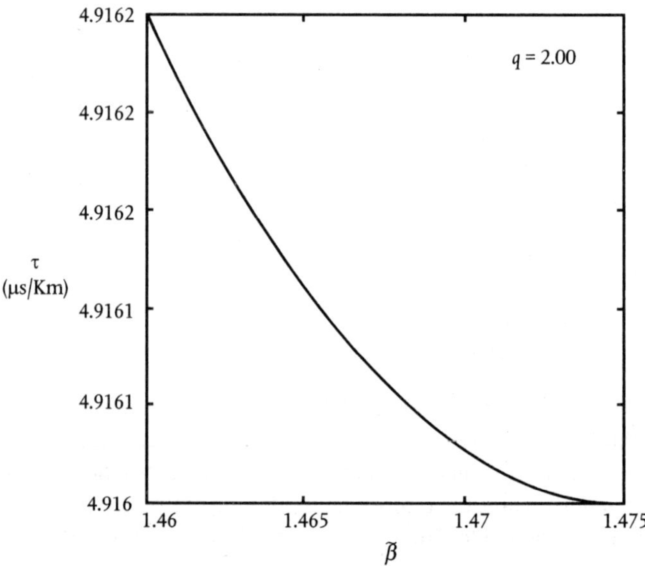

Fig. 1.17 Variation of $\tau(\tilde{\beta})$ with $\tilde{\beta}$ for $q = 2.00$

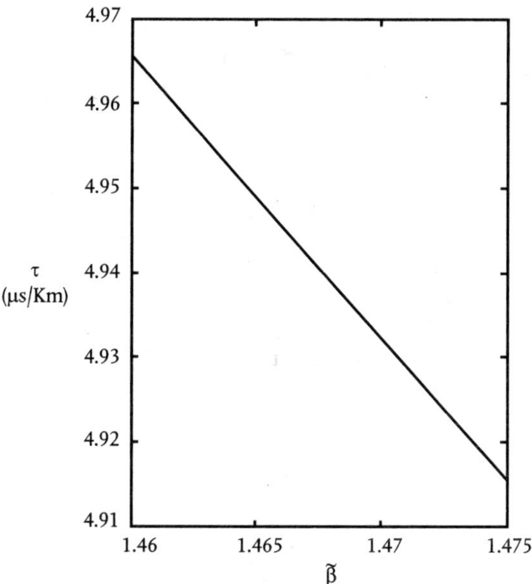

Fig. 1.18 Variation of $\tau\left(\tilde{\beta}\right)$ with $\tilde{\beta}$ for $q = \infty$

represents a monotonically decreasing function of $\tilde{\beta}$ (see Fig. 1.18). Since $n_2 < \tilde{\beta} < n_1$, we get

$$\tau_{\max} = \tau\left(\tilde{\beta} = n_2\right) = \frac{n_1^2}{cn_2} z$$

which corresponds to the ray that is incident at the core-cladding interface at the critical angle. Further,

$$\tau_{\min} = \tau\left(\tilde{\beta} = n_1\right) = \frac{z}{c/n_1}$$

which corresponds to the ray propagating parallel to the z axis. Thus, the pulse dispersion is given by

$$\Delta\tau = \tau_{\max} - \tau_{\min} = \frac{n_1}{c}\frac{\left(n_1 - n_2\right)}{n_2} L \simeq \frac{n_1\Delta}{c} L \tag{1.78}$$

Thus, for $n_1 \simeq 1.46$, $\Delta \simeq 0.01$, we obtain;

$$\Delta\tau \simeq 50 \, \text{ns/km}$$

Case 2: $q = 2$ (parabolic index fibres). For $q = 2$, $A = 1/2c$, and $B = n_1^2/2c$. Thus

$$\tau\left(\tilde{\beta}\right) = \frac{1}{2c}\left[\tilde{\beta} + \frac{n_1^2}{\tilde{\beta}}\right] z \tag{1.79}$$

which is again a monotonically decreasing function of $\tilde{\beta}$ for $n_2 < \tilde{\beta} < n_1$ (see Fig. 1.17). Indeed

$$\tau_{\max} = \tau\left(\tilde{\beta} = n_2\right) = \frac{1}{2c}\left[n_2 + \frac{n_1^2}{n_2}\right] z$$

and

$$\tau_{\min} = \tau\left(\tilde{\beta} = n_1\right) = \frac{z}{c/n_1}$$

giving the following expression for pulse dispersion

$$\Delta\tau = \tau_{\max} - \tau_{\min} \simeq \frac{n_2\Delta^2}{2c}z \tag{1.80}$$

For $n_1 \simeq 1.46,\ \Delta \simeq 0.01$, we obtain;

$$\Delta\tau \simeq \frac{1}{4}\ \text{ns/km}$$

Case 3: The optimum profile: From analysis of a parabolic index fibre, we find that there is a tremendous decrease in the pulse dispersion as we go from a step index fibre to a parabolic index fibre. To obtain the optimum value of q corresponding to minimum pulse dispersion, we first note that for $n_2 < \tilde{\beta} < n_1$, $\tau\left(\tilde{\beta}\right)$ is a monotonically increasing or decreasing function except for the values of q given by

$$2 - 4\Delta < q < 2$$

This can be easily shown by finding the value of $\tilde{\beta}$ for which $d\tau/d\tilde{\beta} = 0$. The condition immediately gives

$$\mathrm{A} - \frac{\mathrm{B}}{\tilde{\beta}_{\mathrm{m}}^2} = 0 \Rightarrow \tilde{\beta}_{\mathrm{m}} = \sqrt{\frac{\mathrm{B}}{\mathrm{A}}} = n_1\sqrt{\frac{q}{2}} \tag{1.81}$$

Since for guided rays $n_2 < \tilde{\beta} < n_1$ and $n_2 = n_1\sqrt{1-2\Delta}$, the value of $\tilde{\beta}$ given by (1.81) lies in the range $n_2 < \tilde{\beta} < n_1$ only for $2 - 4\Delta < q < 2$. For $q > 2$, $\tau\left(\tilde{\beta}\right)$ decreases monotonically as $\tilde{\beta}$ increases from n_2 to n_1 and the pulse dispersion is given by

$$\Delta\tau(q>2) = \tau\left(\tilde{\beta} = n_2\right) - \tau\left(\tilde{\beta} = n_1\right)$$

$$\Rightarrow \frac{(n_1 - n_2)}{c(q+2)}\left(\frac{qn_1}{n_2} - 2\right)z \tag{1.82}$$

$$q = \frac{2n_2}{n_1} = 2\sqrt{1-2\Delta} \simeq 2 - 2\Delta \qquad \text{optimum profile} \tag{1.83}$$

The corresponding pulse dispersion is given by

$$\Delta\tau \simeq \frac{n_1}{8c}\Delta^2 z \tag{1.84}$$

1.8 DISPERSION IN SINGLE-MODE CIRCULAR CORE FIBRES

The types of dispersion that occur in single-mode fibres are discussed in the following section.

1.8.1 Group Velocity Dispersion

In a multimode fibre, different modes at a single-spectral frequency travel with different group velocities, giving rise to pulse dispersion [27–29].

$$\frac{\Delta T}{L} = \frac{n_1}{n_2}\left(\frac{n_1 - n_2}{c}\right) \tag{1.85}$$

Equation (1.85) gives the multipath time dispersion per unit length of the optical fibre. In terms of mode theory, this is called intermodal dispersion. As expected, multimode step-index fibres exhibit large intermodal dispersion. By choosing an appropriate graded-index profile, this type of dispersion can be considerably reduced.

In single-mode fibres, intermodal dispersion is absent because the power of the launched pulse is carried by a single mode. However, pulse broadening does not vanish altogether, due to chromatic dispersion. A light pulse, although propagating as a fundamental mode, has a number of spectral components (of different frequencies), and the group velocity of the fundamental mode varies with frequency. Therefore, different spectral components in the pulse propagate with slightly different group velocities, resulting in a pulse broadening. This phenomenon is called group velocity dispersion (GVD) or intramodal dispersion. This has two components, namely material dispersion and waveguide dispersion. Let us see how the GVD limits the performance of single-mode fibres.

Suppose we have a single-mode fibre of length L and a pulse is launched at one end of this fibre. Then, a spectral component of a given frequency ω would travel with a group velocity V_g given by

$$V_g = \frac{1}{d\beta / d\omega} \tag{1.86}$$

After travelling length L, this component would arrive at the other end after time τ_g, known as the group delay time. The latter may be given by

$$\tau_g = \frac{L}{V_g} = L\frac{d\beta}{d\omega} \tag{1.87}$$

If the spectral width of the pulse is not very wide, the delay difference per unit frequency along the propagation length may be expressed as $d\tau_g / d\omega$. For the spectral components that are $\Delta\omega$ apart, the total delay difference ΔT over length L may be given by

$$\Delta T = \frac{d\tau_g}{d\omega}\Delta\omega \tag{1.88}$$

Substituting for τ_g from (1.87), we get

$$\Delta T = L\left(\frac{d^2\beta}{d\omega^2}\right)\Delta\omega - L\beta_2\Delta\omega \tag{1.89}$$

The factor $\beta_2 \equiv d^2\beta / d\omega^2$ is known as the GVD parameter. It determines the magnitude of pulse broadening as the pulse propagates inside the fibre.

If the spectral spread is measured in terms of the wavelength range $\Delta\lambda$ around the central wavelength λ, then replacing ω by $2\pi c / \lambda$ and $\Delta\omega$ by $\left(-2\pi c / \lambda^2\right)\Delta\lambda$, we can write (1.89) as follows:

$$\Delta T = \frac{d\tau_g}{d\lambda}\delta\lambda = \frac{d}{d\lambda}\left(\frac{L}{V_g}\right)\Delta\lambda = L\frac{d}{d\lambda}\left(\frac{1}{V_g}\right)\Delta\lambda \tag{1.90}$$

The factor

$$D = \frac{1}{L}\frac{d\tau_g}{d\lambda} = \frac{d}{d\lambda}\left(\frac{L}{V_g}\right) = \frac{-2\pi c}{\lambda^2}\beta_2 \tag{1.91}$$

is called the dispersion parameter. It is defined as the pulse spread per unit length per spectral width of the source and is measured in units of ps $\mathrm{nm}^{-1}\mathrm{km}^{-1}$. It is the combined effect of waveguide dispersion D_w and material dispersion D_m present together. However, we can write to a very good approximation.

$$D = D_w + D_m \tag{1.92}$$

In the following sections, we will discuss each of these separately and obtain separate expressions for D_w and D_m considering that in each case the other one is absent.

1.8.2 Waveguide Dispersion

The normalised propagation constant b of a mode is,

$$b = \left(1 - \frac{u^2}{V^2}\right) = \frac{\beta^2 - \beta_2^2}{\beta_1^2 - \beta_2^2} = \frac{\beta^2 - k^2 n_2^2}{k^2 n_1^2 - k^2 n_2^2} = \frac{(\beta / k)^2 - n_2^2}{n_1^2 - n_2^2} \tag{1.93}$$

In this expression, β / k is called the mode index n. Rewriting the above expression, we get

$$b = \frac{n^2 - n_2^2}{n_1^2 - n_2^2} = \frac{(n - n_2)(n + n_2)}{(n_1 - n_2)(n_1 + n_2)}$$

Taking $n_1 \approx n_2$, which is true for most fibres, we can write,

$$b \approx \frac{(n - n_2)}{(n_1 - n_2)} \tag{1.94}$$

or the mode index

$$n \approx n_2 + b(n_1 - n_2) \tag{1.95}$$

But β and n are related by the expression.

$$\beta = \frac{\omega}{c} n \tag{1.96}$$

Therefore, from. (1.95) and (1.96), we get

$$\beta = \frac{\omega}{c}\left[n_2 + (n_1 - n_2)b\right] \tag{1.97}$$

Let us assume that there is no material dispersion, and hence n_1 and n_2 do not vary with λ. But b is a function of V and hence of ω. Therefore, using (1.97) we can write

$$\frac{1}{V_g} = \frac{d\beta}{d\omega} = \frac{1}{c}\left[n_2 + (n_1 - n_2)b\right] + \frac{\omega}{c}(n_1 - n_2)\frac{db}{d\omega}$$

$$\Rightarrow \frac{1}{V_g} = \frac{1}{c}\left[n_2 + (n_1 - n_2)b\right] + \frac{\omega}{c}(n_1 - n_2)\frac{db}{dV}\frac{dV}{d\omega} \tag{1.98}$$

We know that for a step-index fibre of core radius a,

$$V = \frac{2\pi a}{\lambda}\left(n_1^2 - n_2^2\right)^{1/2} = \frac{\omega}{c} a\sqrt{\left(n_1^2 - n_2^2\right)} \tag{1.99}$$

Therefore, using (1.99), we get

$$\frac{dV}{d\omega} = \frac{a}{c}\sqrt{\left(n_1^2 - n_2^2\right)} = \frac{V}{\omega} \tag{1.100}$$

The same relation can also be expressed as $dV / d\lambda = V / \lambda$. Substituting the value of $dV / d\omega$ from (1.100) into (1.98), we get

$$\frac{1}{V_g}=\frac{1}{c}\left[n_2+(n_1-n_2)b\right]+\frac{1}{c}(n_1-n_2)V\frac{db}{dV}$$

$$\Rightarrow\frac{1}{V_g}=\frac{n_2}{c}+\frac{(n_1-n_2)}{c}\left[\frac{d}{dV}(bV)\right]$$

$$\Rightarrow\frac{1}{V_g}=\frac{n_2}{c}\left[1+\frac{(n_1-n_2)}{n_2}\left\{\frac{d}{dV}(bV)\right\}\right]$$

Since the relative refractive index difference Δ can be approximately given by $(n_1-n_2)/n_2$, we have

$$\Rightarrow\frac{1}{V_g}=\frac{n_2}{c}\left[1+\Delta\frac{d}{dV}(bV)\right] \tag{1.101}$$

From this expression, it is clear that the group velocity V_g of the mode varies with V and hence with ω, even in the absence of material dispersion. This is called waveguide dispersion. Thus, the group delay, that is, the time taken by a mode to travel the length L of the fibre due to waveguide dispersion will be given by

$$\tau_w=\frac{L}{V_g}=\frac{Ln_2}{c}\left[1+\Delta\frac{d}{dV}(bV)\right] \tag{1.102}$$

For source of spectral width $\Delta\lambda$, the corresponding pulse spread due to waveguide dispersion may be obtained from the derivative of the group delay with respect to wavelength, that is,

$$\Delta\tau_w=\frac{d\tau_w}{d\lambda}\Delta\lambda\approx\left[\frac{Ln_2}{c}\Delta\frac{d^2}{dV^2}(bV)\frac{dV}{d\lambda}\right]\Delta\lambda$$

Since,

$$V=\frac{2\pi a n_1\sqrt{2\Delta}}{\lambda} \tag{1.103}$$

$$\frac{dV}{d\lambda}=\frac{-2\pi a n_1\sqrt{2\Delta}}{\lambda^2}=-\frac{V}{\lambda}$$

Therefore,

$$\Delta\tau_w=-\frac{L}{c}n_2\Delta\frac{1}{\lambda}\left[V\frac{d^2}{dV^2}(bV)\right]\Delta\lambda \tag{1.104}$$

or,

$$\frac{\Delta\tau_w}{L}=-\frac{n_2\Delta}{c\lambda}\left[V\frac{d^2}{dV^2}(bV)\right]\Delta\lambda=D_w\lambda \tag{1.105}$$

where,

$$D_w=-\frac{n_2\Delta}{c\lambda}\left[V\frac{d^2}{dV^2}(bV)\right] \tag{1.106}$$

is called the waveguide dispersion parameter or simply the waveguide dispersion. Calculation of actual values of the term within square brackets is not a simple task, and hence the following easy-to-use empirical formula may be employed:

$$V\frac{d^2}{dV^2}(bV)\approx 0.080+0.549+(2.834-V)^2 \tag{1.107}$$

From (1.107), we see that D_w depends on n_2, Δ, λ, the V-parameter, and the $V\left(d^2/dV^2\right)(bV)$. The value of the last term, that is, $V\left(d^2/dV^2\right)(bV)$ is positive for V-values in the range $0.5-3$, which means that in the normal case, for single-mode fibres, D_w will remain negative for wavelengths in the range of interest, e.g., $1.0-1.7\ \mu m$.

1.8.3 Material Dispersion

The refractive index n of any transparent medium is given by $=c/v$, where c is the velocity of light in vacuum and v is its velocity in the medium [30–32]. In terms of wave theory, v is called the phase velocity v_p of the wave in the medium. Thus,

$$v = v_p = c/n \tag{1.108}$$

If $\omega(=2\pi f)$ is the angular frequency of the wave (in rad/sec), f being the frequency in hertz, and $\beta(=2\pi/\lambda_m)$ is the propagation constant, λ_m being the wavelength of light in the medium (which is equal to λ/n; λ is the free space wavelength), then the phase velocity of the wave is also given by

$$v_p = \omega/\beta \tag{1.109}$$

Thus, using (1.108) and (1.109), we get

$$v_p = c/n = \omega/\beta$$

or

$$n = \frac{c\beta}{\omega} \tag{1.110}$$

At this point, it is important to know that any signal superposed onto a wave does not propagate with the phase velocity v_p of the wave, but travels with a group velocity v_g given by the following expression:

$$v_g = \frac{d\omega}{d\beta} = \frac{1}{d\beta/d\omega} \tag{1.111}$$

now since,

$$\beta(\omega) = n(\omega)\frac{\omega}{c} = \beta_0 + \beta_1(\omega-\omega_0) + \frac{1}{2}\beta_2(\omega-\omega_0)^2 + \ldots \tag{1.112}$$

where,

$$\beta_m = \left[\frac{d^m\beta}{d\omega^m}\right]_{\omega=\omega_0} \qquad (m = 0,1,2,3\ldots) \tag{1.113}$$

$$\beta_1 = \frac{d\beta}{d\omega} = \frac{n(\omega)}{c} + \frac{\omega}{c}\frac{dn(\omega)}{d\omega} = \frac{1}{c}\left[n(\omega) + \omega\frac{dn}{d\omega}\right] \tag{1.114}$$

where,

$$\beta_1 = \frac{n_g}{c} = \frac{1}{v_g} \tag{1.115}$$

$$n_g = \frac{c}{v_g} \tag{1.116}$$

Let us substitute for v_g from (1.111) in (1.116) to get

$$n_g = c\frac{d\beta}{d\omega} = c\frac{d}{d\omega}\left(\frac{\omega n}{c}\right) = \frac{d}{d\omega}(n\omega) = n + \omega\frac{dn}{d\omega} \tag{1.117}$$

This is an important expression, relating the group index n_g with the ordinary refractive index or the phase index n.

Since,

$$\frac{dn}{d\omega}=\frac{dn}{d\lambda}\frac{d\lambda}{d\omega}$$

and

$$\omega=\frac{2\pi c}{\lambda}$$

$$\frac{d\omega}{d\lambda}=-\frac{2\pi c}{\lambda^2}$$

we have from (1.117),

$$n_g=n+\frac{2\pi c}{\lambda}\frac{dn}{d\lambda}\left(-\frac{\lambda^2}{2\pi c}\right)=n-\lambda\frac{dn}{d\lambda} \tag{1.118}$$

Thus,

$$v_g=\frac{c}{n_g}=\frac{c}{(n-\lambda dn/d\lambda)} \tag{1.119}$$

A light pulse, therefore, will travel through the core of an optical fibre of length L in time t given by

$$t=\frac{L}{v_g}=\left[n-\lambda\frac{dn}{d\lambda}\right]\frac{L}{c} \tag{1.120}$$

If the spectrum of the light source has a spread of wavelength $\Delta\lambda$ about λ and if the medium of the core is dispersive, the pulse will spread out as it propagates and will arrive at the other end of length L, over a spread of time ΔT is much smaller than the central wavelength (λ), we can write

$$\Delta t=\frac{dt}{d\lambda}\Delta\lambda=\frac{L}{c}\left[\frac{dn}{d\lambda}-\frac{dn}{d\lambda}-\lambda\frac{d^2n}{d\lambda^2}\right]\Delta\lambda=-\frac{L}{c}\lambda\frac{d^2n}{d\lambda^2}\Delta\lambda \tag{1.121}$$

If $\Delta\lambda$ is the full width at half maximum (FWHM) of the peak spectral power of the optical source at, then its relative spectral width γ is given by

$$\gamma=\left|\frac{\Delta\lambda}{\lambda}\right| \tag{1.122}$$

If an impulse of negligible width is launched into the fibre, then ΔT will be the half power width τ of the output (or the broadened pulse). Thus, the pulse broadening due to material dispersion may be given by

$$\tau=\frac{L}{c}\gamma\left|\lambda^2\frac{d^2n}{d\lambda^2}\right|$$

or,

$$\frac{\tau}{L}=\frac{\gamma}{c}\left|\lambda^2\frac{d^2n}{d\lambda^2}\right| \tag{1.123}$$

The material dispersion of optical fibres is quoted in terms of the material dispersion parameter D_m given by

$$D_m=\frac{1}{L}\frac{\tau}{\Delta\lambda}=\frac{\lambda}{c}\left|\frac{d^2n}{d\lambda^2}\right| \tag{1.124}$$

D_m has the units of ps $nm^{-1}km^{-1}$. The example of pure silica has been chosen because a majority of fibres are manufactured using silica as the host material.

It is known that D_m changes sign at $\lambda = \lambda_{ZD} = 1.32\ \mu m$ (for pure silica). This point has been frequently referred to as the wavelength of zero material dispersion. This wavelength can be changed slightly by adding other dopants to silica. Thus, the use of an optical source with a narrow spectral width (e.g., injection laser) around λ_{ZD} would substantially reduce the pulse broadening due to material dispersion.

Figure 1.19 shows the variation of D_m, D_w, and their sum $D = D_w + D_m$ for a typical conventional single-mode fibre. The effect of waveguide dispersion is to shift λ_{ZD} by about 30–40 μm so that the total dispersion D is zero near 1.32 μm. However, material dispersion is dominant above λ_{ZD}, say around 1.55 μm, giving rise to high values of D. The window around 1.55 μm is of considerable interest for fibre-optic communication systems, as silica fibres exhibit least attenuation around 1.55 μm.

We have seen in (1.106) that D_w depends on fibre parameters such as Δ, n_2, V, the core diameter, etc. Therefore, it is possible to design the fibre such that λ_{ZD} is shifted to 1.55 μm. Such fibres are called dispersion-shifted fibres (DSFs). It is also possible to modify D_w such that the total dispersion D is small over a wide range of wavelengths (typically from 1.3 to 1.6 μm). Such fibres are called dispersion flattened fibres (DFFs).

We know the conventional single-mode fibre has zero total dispersion around 1300 nm. However, the minimum fibre loss is at 1550 nm. If the zero dispersion wavelengths could be shifted to the 1550 nm region, we could have both minimum loss and very low dispersion. This would lead to very high bandwidth systems with very long (~100 km) repeater less transmission.

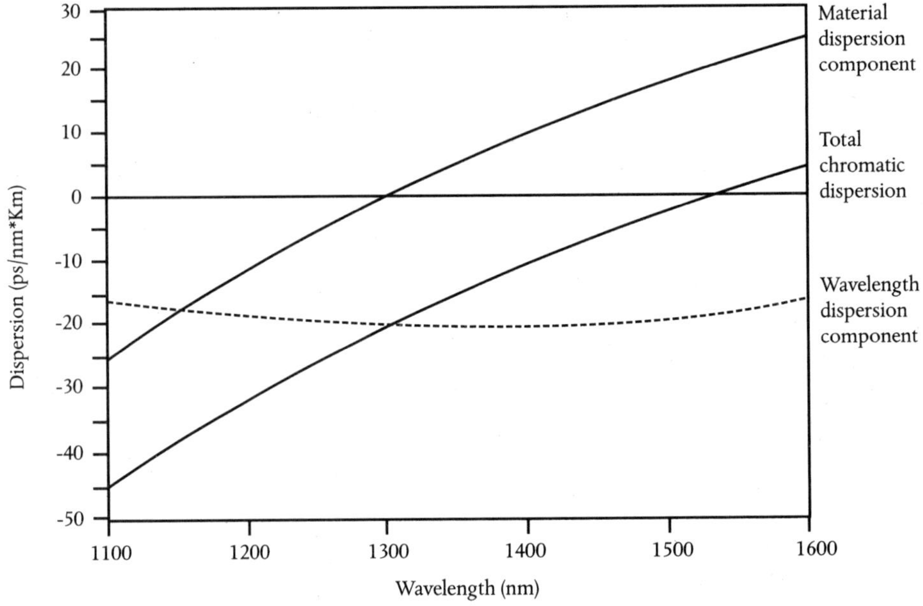

Fig. 1.19 Variation of D_w, D_m, and their sum $D = D_w + D_m$ for a typical single-mode fibre

The size of the waveguide dispersion has been found to be sensitive to the doping levels as well as the values of Δ and a (core radius). Indeed, we have achieved this by changing the fibre parameters. The resulted fibre is called dispersion-shifted fibre. The first dispersion-shifted fibres had zero dispersion shifted to 1550 nm to match their minimum absorption wavelength. The refractive index profile for zero dispersion-shifted fibre is shown in Fig. 1.20.

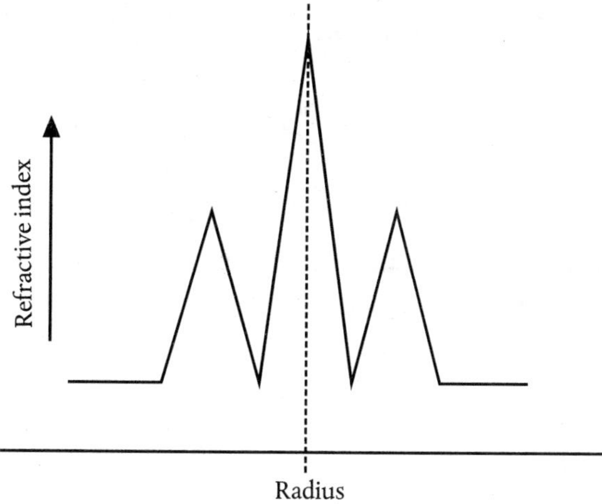

Fig. 1.20 Zero dispersion-shifted fibre

Although this type of fibre (zero dispersion-shifted fibre) worked well for single-channel systems, it proved to be unusable for wavelength division multiplexing (WDM) systems. When multiple optical channels pass through the same fibre at wavelengths where dispersion is very close to zero, they suffer from a type of crosstalk called four wave mixing. The degradation is so severe that zero dispersion-shifted fibre cannot be used for dense WDM systems.

In order to avoid four wave mixing problem in WDM systems, non-zero dispersion-shifted fibres having refractive index were invented as shown in Fig. 1.21. The idea is to move the zero dispersion wavelength outside the band used for erbium fibre doped amplifiers.

The term 'non-zero dispersion shifted' comes from the fact that the dispersion is shifted to a value that is low but not zero in the 1550 nm bed of erbium fibre amplifiers. However, you can shift the dispersion to a positive or negative value. This small dispersion is sufficient to keep signals at closely spaced wavelengths from staying in phase over long distances, thus preventing four wave mixing. Figure 1.22, shows the variation of total dispersion D for typical standard, SMF DSF and DFF.

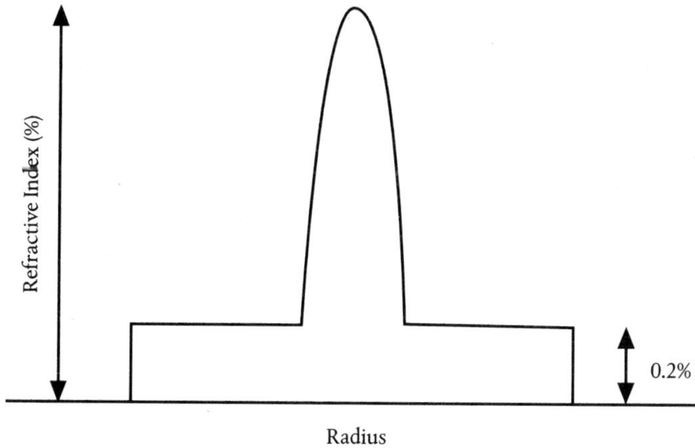

Fig. 1.21 Non-zero dispersion-shifted fibre

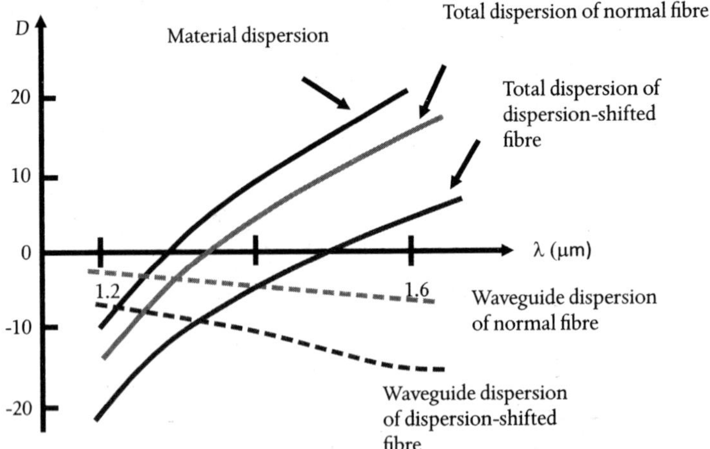

Fig. 1.22 Variation of total dispersion D for a typical standard (conventional) SMF, DSF

1.8.4 Combined Effect of Multipath and Material Dispersion

In a fibre optic communication system, the optical power generated by the optical source, for example, LED is proportional to the input current to the transmitter. The optical power received by the detector is proportional to the power launched into and propagated by the optical fibre. This, in turn, gives rise to a proportional current at the receiver end, thus giving an overall linearity to the system.

It is apparent that multipath and material dispersion lead to broadening of the pulse launched into a fibre. Thus, the received pulse represents the impulse response of the fibre. If we assume that the FWHM of the transmitted pulse is τ_0, and the impulse responses due to multipath and material dispersion lead to approximately Gaussian pulses of FWHM τ_1 and τ_2, respectively, as shown in Fig. 1.23, and that the two mechanisms are almost independent of each other, then the received pulse width at half maximum power, τ, will be given by

$$\tau = \left[\tau_0^2 + \tau_1^2 + \tau_2^2\right]^{1/2} \tag{1.125}$$

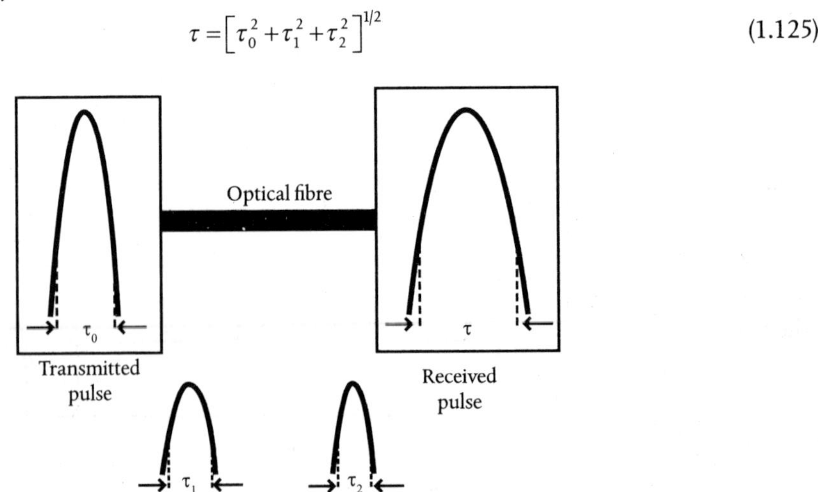

Fig. 1.23 The combined effect of multipath and material dispersion

The temporal pulse broadening per unit length due to both the effects is then given by

$$\frac{\tau}{L}=\left[\left(\frac{\tau_0}{L}\right)^2+\left(\frac{\tau_1}{L}\right)^2+\left(\frac{\tau_2}{L}\right)^2\right]^{1/2}$$

where L is the total length of the fibre. It is also possible to express the same result in terms of the root mean square (rms) pulse widths. The calculation of the rms pulse width is given in the following section. Thus, if the transmitted pulse is Gaussian in shape and has an rms width σ_0 and this pulse is broadened by both multipath and material dispersion leading to nearly Gaussian pulses of rms widths σ_1 and σ_2, respectively, then the rms width σ of the received pulse is given by,

$$\frac{\sigma}{L}=\left[\left(\frac{\sigma_0}{L}\right)^2+\left(\frac{\sigma_1}{L}\right)^2+\left(\frac{\sigma_2}{L}\right)^2\right]^{1/2}$$

The rms width σ of a pulse is defined as follows. If $p(t)$ is the power distribution in the pulse as a function of time t and the total energy in the pulse is

$$\varepsilon=\int_{-\infty}^{\infty}p(t)dt$$

then its rms width σ is given by

$$\sigma^2=\frac{1}{\varepsilon}\int_{-\infty}^{\infty}t^2p(t)dt-\left[\frac{1}{\varepsilon}\int_{-\infty}^{\infty}tp(t)dt\right]^2 \tag{1.126}$$

1.8.5 Polarisation Mode Dispersion

Birefringence is another cause for pulse broadening in a single-mode fibre. If the input pulse excites both orthogonally polarised components of the fundamental fibre mode, they will travel with different group velocities; and if the group velocities of the two components are v_{gx} and v_{gy}, respectively, the two will arrive at the other end of the fibre of length L after time L/v_{gx} and L/v_{gy}. The time delay ΔT between two orthogonally polarised components will then be given by

$$\Delta T=\left|\frac{L}{v_{gx}}-\frac{L}{v_{gy}}\right| \tag{1.127}$$

This difference in the propagation times gives rise to pulse broadening. This is called polarisation mode dispersion (PMD). This can be a limiting factor, particularly in long-haul fibre optic communication systems operating at high bit rates. In writing (1.127), we have assumed that the fibre has constant birefringence. This is true only in polarisation maintaining fibres. In conventional single-mode fibres, birefringence varies randomly and hence (1.127) is not valid in such fibres. However, an approximate estimate of PMD-induced pulse broadening can be made using the following relation:

$$\Delta T\approx D_{\text{PMD}}\sqrt{L} \tag{1.128}$$

where D_{PMD}, which is measured in units of $\text{ps}/\sqrt{\text{km}}$, is the average PMD parameter and L is the length of the fibre. Values of D_{PMD} normally vary from 0.01 to 10 $\text{ps}/\sqrt{\text{km}}$.

SUMMARY

- An optical fibre is a thin, flexible, transparent fibre that acts as a waveguide or light pipe to transmit light between the two ends of the fibre.
- Optical fibres are widely used in fibre optic communications, which permits transmission over longer distances and at higher bandwidths (data rates) than other forms of communication.
- An optical fibre is a cylindrical dielectric waveguide (nonconducting waveguide) that transmits light along its axis, by the process of total internal reflection.
- There is a maximum angle from the fibre axis at which light may enter the fibre so that it will propagate, or travel, in the core of the fibre. The size of this maximum angle is the NA of the fibre. Fibre with a larger NA requires less precision to splice. Single-mode fibre has a small NA.
- In fibre optic communication, an SMF is an optical fibre designed to carry only a single ray of light (mode). These modes define the way the wave travels through space, that is, how the wave is distributed in space.
- Multimode optical fibres are mostly used for communication over shorter distances, such as within a building or on a campus.
- A synchronous optical digital transmission system comprises a local interface unit having a phase locked loop circuit that senses and adjusts the phase of the clock signals arriving from the remote unit to match the phase of the principal or local clock so that arriving data bit streams can be directly applied to the local system for further processing.
- The most common transmitters used for creating optical signals are LEDs and laser diodes.
- Amplifiers are used at regular intervals owing to the physical limitations of optical fibres. These amplifiers receive the pulses, convert them into electrical signals, reconvert them into light pulses and transmit them again at even more intensity.
- Optical fibres exceed the data handling performance of copper wire in almost every regard. They are totally immune to electromagnetic interference and have very high bandwidths.
- The phenomenon of microbending is one weakness of the optical fibre. This is where the fibre is bent around a small radius, causing the light to escape the inner core, through the cladding.
- A light pulse emitted by the LED taking a shorter path through the fibre will arrive at the detector sooner than light pulses taking longer paths. The result is distortion of the square waves rising and falling edges, called pulse stretching.
- Fibre so designed to permit only a single mode of light is known as single-mode fibre. As a single-mode fibre escapes the problem of pulse stretching experienced in long cables, it is the best choice for long-distance (several miles or more) networks.
- Line coding is the process of representing the digital signal to be transported by an amplitude and time discrete signal that is optimally tuned for the specific properties of the physical channel (and of the receiving equipment).
- For short distances of a few hundred metres or less (e.g. within storage area networks), it is often more convenient to utilise multimode fibres, as they are cheaper to install (e.g.,due to their large core areas, they are easier to splice).
- Attenuation in fibre optics, also known as transmission loss, is the reduction in the intensity of the light beam (or signal) with respect to distance travelled through a transmission medium.

THEORY QUESTIONS AND ANSWERS

Question 1: What is the difference between micro bending and macro bending losses in fibre? Explain in brief.

Answer: Bends in the fibre can cause some modes to be lost and also result in coupling of power from one mode into another. Bends are caused by putting fibre in direct contact with a rough surface of some kind and perhaps pressure onto it. The microbends so created will cause the fibre to lose power and also to couple power between modes. Figure P1 shows the cross section of fibre.

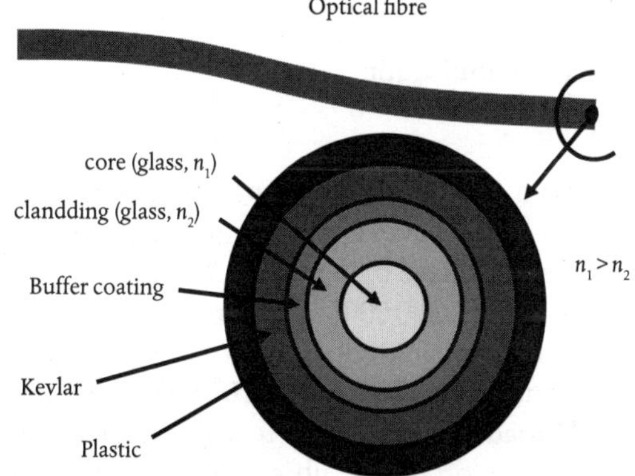

Fig. P1 Cross section of fibre

Macro bending loss: It is the loss which arises due to turning of the optical fibre in a curvature. It is negligible for large radius of curvature. As the bends are increased, the radius of curvature decreases. Macro bending becomes appreciable at some cut-off value of radius of curvature. Further bending results in reduced radius of curvature and exponential increase in macro bending losses. Figure P2 (a) shows the macro bending losses.

Micro bending loss: Micro bending loss arises due to fabrication defects at the core-cladding interface or due to insertion of fibre. It results in coupling of power between lower and higher order mode and the cladding. Figure P2 (b) shows the micro-bending losses.

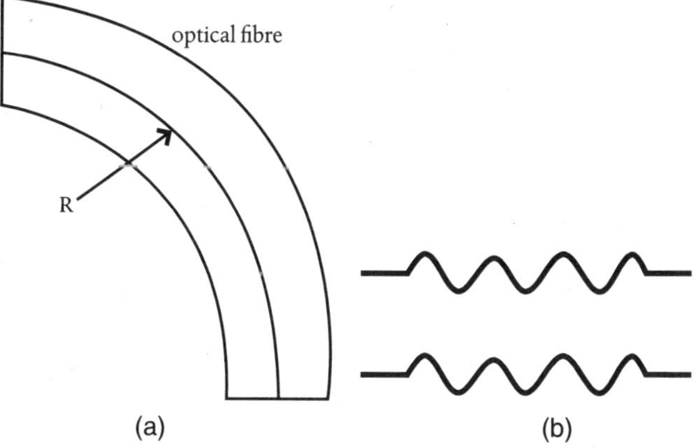

Fig. P2 (a) Macro-bending losses; (b) Micro-bending losses

Question 2: Describe at least seven advantages of optical fibre communication.

Answer: The advantages of optical fibre communication with respect to copper wire systems are as follows:

1. Broad bandwidth
2. Immunity to electromagnetic interference
3. Low attenuation loss over long distances
4. Signal security and no cross talk
5. Light weight and small diameter cables
6. Electrical insulator
7. Low cost for long-distance communication

Each advantage is described as follows:

1. **Broad bandwidth:** Broadband communication is possible over fibre optical cables which means that audio signal, video signal, microwave signal, text and data from computers can be modulated over light carrier wave and demodulated by optical receiver at the other end.

   It is possible to transmit around 300 000 two-way voice signals or 90 000 TV channels over one optical fibre. If it is a multicore cable consisting of 24 optical fibre cables, it will be possible to transmit and receive data and information up to 10^9 bits per second. By using the technique of WDM, it is possible to send broadband signals up to 10^{12} bits per second. Research is going on to increase the information capacity on optical fibre communication up to 40 tera-bits per second (Tb/s) which will meet the demand of large information capacity. Figure P3 shows the propagation of light rays in the optical fibre at different propagation directions.

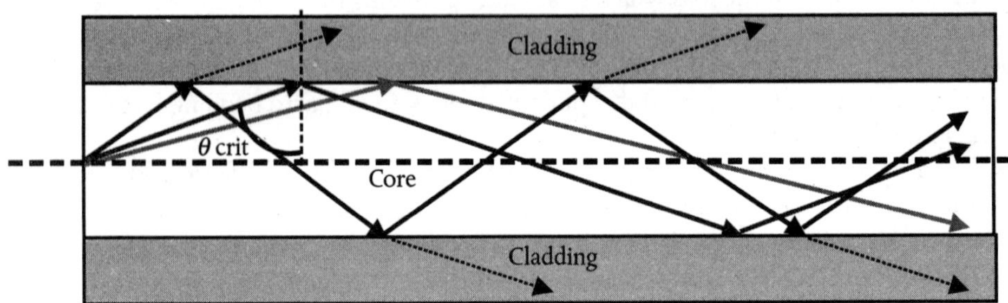

Fig. P3 Propagation of light rays in the optical fibre at different propagation directions

2. **Immunity to electromagnetic interference:** These cables carry information over light waves travelling in the fibres due to properties of the fibre materials. There are no pick up of signals from vehicles, lightning and other sources and power cables adjacent to fibre cables; there is no effect of electromagnetic pulse generated by nuclear devices. Therefore, electromagnetic interference with the light wave travelling inside the optical fibre cables does not exist.
3. **Low attenuation loss over long distances:** There are various optical windows in the optical fibre cable at which the attenuation loss is found to be comparatively low; therefore, the transmitter and receiver devices are developed and used in these low attenuation regions. Due to low attenuation of 0.2 dB/km in optical fibre cables, it is possible to achieve reliable long-distance communication efficiently even for a capacity rate of 1 Tb/s.

4. **Signal security and no cross talk:** Since optical fibres are coated with plastic layers to give them further mechanical strength along with other coating, the light signal passing through these fibres does not leak out of fibre, these fibres are safe for security reasons. The signals cannot be tapped due to the presence of an opaque jacket around the fibre cable. The cross talk from one fibre to another fibre is also not possible due to opaque coatings and total internal reflection of light travelling within the fibre cable itself.
5. **Light weight and small diameter cables:** The diameter of fibre is in the range of a few microns and the fibres are further coated with plastics and so these fibres are very light and have small size due to thin diameter. Even with six core fibre cables in which six fibres are packed together for long-distance communication, it is easy to lay fibre optical cables as the weight and size are very small, whereas copper cable laying is time consuming and expensive due to its heavy weight.
6. **Electrical insulator:** Optical fibres are made and drawn from silica glass which is nonconductor of electricity and so there are no ground loops and leakage of any type of current. Optical fibres are thus laid down along with high voltage cables on the electricity polls due to its electrical insulator behaviour.
7. **Low cost for long-distance communication:** Since optical fibres are made from silica glass which is found in abundance on the Earth, the cost of optical fibres is low as against copper cable due to lower cost of silica glass. The cost on fibres is incurred for making extremely pure materials out of earth silica and then drawing the cables, packaging and testing the cables.

Question 3: Write short notes on absorption.

Answer: A material can absorb light if the incident photon energy is equal to the bandgap of the material. This bandgap varies from one material to another.

For example, the OH^- anion of water molecule can absorb light at 950 nm, 1240 nm and 1380 nm but it is less at higher infrared wavelength of 1550 nm; therefore, in order to reduce the effect of absorption due to OH^- anion, it is necessary to shift the light wavelength more than 1380 nm which will call for designing of light transmitter and receiver at high wavelength. Thus, by changing the transmitting wavelength from 1300 nm to 1550 nm, the absorption by OH^- molecules can be avoided. The attenuation is 0.5 dB/km at 1300 nm and it is 0.20 dB/km at 1550 nm. It can be reduced to 0.001 dB/km at higher frequencies ($> 2.5\ \mu m$) in fluoride fibres.

The fibre is made from silica oxides and other oxides. During manufacturing process, some molecules of the hydroxide anion OH^-, often called hydrogen water molecules, are incorporated in silica fibre which cannot be removed and are responsible for absorption of light when it propagates through the fibre.

Question 4: Write short notes on Rayleigh scattering.

Answer: When light strikes an object, the light is reflected into different directions, that is, if the light ray strikes a drop of water in atmosphere, it diffusely reflects light in several directions. It is called light scattering. In optical fibre, due to impurity particle in the path of light in the core, the particle will scatter the light in another direction and thus affect the total internal reflection at the boundary of core-cladding. This will create loss in the amount of light. When the fibre materials are prepared, there may be some inhomogeneities or imperfections in the core layer. Suppose there is a filament type obstacle. Due to imhomogeneity, light beam propagating at an angle close to or more than critical angle will hit the obstacle and bend its direction because of scattering.

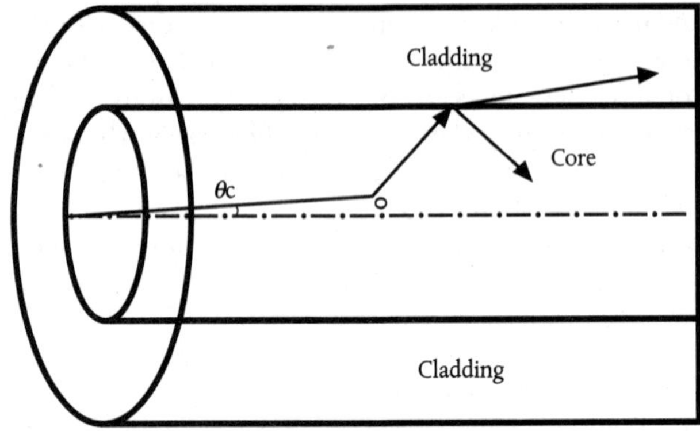

Fig. P4 Scattering loss

The beam may be refracted into the cladding layer. This means, after hitting the obstacles, light will be scattered and it will affect the total internal reflection at the core cladding boundary and the ray entering into cladding layer will create power loss due to scattering of light in the fibres. Figure P4 shows the scattering loss in optical fibre. The inhomogeneity can result in variation of refractive index and the variation of refractive index may be such that the particular location with large refractive index will act as an obstacle and cause scattering loss. The manufacturer takes care of the scattering loss by developing high quality monomode fibre of several kilometers length, where inhomogenities and imperfections have been removed to a large extent. Due to increase of wavelength from 800 nm to 1300 nm and further to 1550 nm, the scattering losses are reduced to 0.15 dB/km at 1550 nm.

Rayleigh scattering coefficient

It occurs due to inhomogenity in glass material and the attenuation loss of light is proportional to $1/\lambda^4$. The Rayleigh scattering coefficient is given by

$$y_R = \frac{8\pi^3}{3\lambda^4} n^8 p^2 B_c k T_F$$

where,

$\lambda =$ Optical wavelength

$n =$ Refractive index of medium

$p =$ Average photoelastic coefficient

$B_c =$ Isothermal compressibility

$T_F =$ Fictive temperature

$k =$ Boltzmann's constant.

The transmission loss factor (Transmissivity) of the fibre T_L is related to Rayleigh scattering coefficient by

$$T_L = \exp(-y_R L)$$

where, L is the length of the fibre in metres.

SOLVED PROBLEMS

Problem 1: For a multimode step index fibre the refractive indices of core and cladding are 1.53 and 1.48, respectively. Find out the angle of acceptance and critical angle when light is incident from air?

Solution:

Using (1.10)

$$\theta_a = \sin^{-1}(NA) = \sin^{-1}\left[\left(n_1^2 - n_2^2\right)^{1/2}\right] = 22.8^\circ$$

Using (1.6)

$$n_1 \sin\theta_c = n_2 \Rightarrow \theta_c = \sin^{-1}\left(\frac{n_2}{n_1}\right) = \sin^{-1}\left(\frac{1.48}{1.53}\right) = 75.3^\circ$$

Problem 2: For a step index single-mode fibre, the following properties are given: $a = 5\,\mu m$, $n_1 = 1.5, n_2 = 1.48$. Determine the cut off wavelength of the fibre.

Solution:

Using (1.18), we know that for V = 2.405, for single optic fibre

$$V = \frac{2\pi}{\lambda_c} a\sqrt{n_1^2 - n_2^2} \Rightarrow \lambda_c = \frac{2\pi a}{2.405}\sqrt{n_1^2 - n_2^2} = 3.18\,\mu m$$

Problem 3: Determine the number of modes for a step-index multimode fibre, if $a = 30\,\mu m, \lambda = 0.95\,\mu m,\ n_1 = 1.475$ and $n_2 = 1.472$. How many modes will propagate if λ is changed to 1.5 μm?

Solution:

Using (1.18)

$$V = \frac{2\pi}{\lambda_c} a\sqrt{n_1^2 - n_2^2} = 18.656$$

and number of modes can be calculated using (1.19)

$$M = \frac{V^2}{2} = 174$$

at $\lambda = 1.5\,\mu m$, $V = 11.81$, hence; $M = 69$

Problem 4: The mean power launched in a fibre link is 1.2 mW and the fibre has an attenuation of 0.5 dB/km. Determine the maximum possible link length when the minimum mean optical power level at the detector is 2.6 μW.

Solution:

$$P_i = 1.2\,mW,\ P_0 = 2.6\,\mu W,\ \alpha = 0.5\,dB/km$$

$$P_0 = P_i e^{-\alpha L} \Rightarrow L = \frac{10}{\alpha}\log_{10}\left(\frac{P_i}{P_0}\right) = 53.28\,km$$

Problem 5: In a multimode step index fibre, the following parameters were measured:

(a) Number of modes passing at an operating wavelength of 1500 nm are 1000.
(b) Refractive index of core: 1.50.
(c) Refractive index of cladding: 1.48.

Determine core diameter of above step index fibre.

Solution:
Given

$$M = 1000,\ n_1 = 1.50,\ n_2 = 1.48,\ \lambda = 1500\,\text{nm} = 1.5\ \mu\text{m}$$

we know for a multimode step index fibre,

$$\text{Number of modes } (M) = \frac{V^2}{2}$$

where,

$$V = \frac{\pi d}{\lambda}\sqrt{n_1^2 - n_2^2}$$

Now,

$$M = \frac{V^2}{2} = 1000$$

$$\Rightarrow V = \sqrt{2 \times 1000} = 44.72$$

The core diameter is given by

$$d = \frac{\lambda V}{\pi\left(n_1^2 - n_2^2\right)^{1/2}} = \frac{44.72 \times 1.5}{\pi\sqrt{(1.5)^2 - (1.48)^2}} = 87.46\ \mu\text{m}$$

Problem 6: If the step index fibre of Fig. 1.6 has a core of refractive index 1.52, a cladding of refractive index 1.487, and a core diameter of 110 μm, calculate, assuming that the fibre is kept in air, the (a) NA of the fibre; (b) angles α_m, θ_m, and ϕ_c; and (c) pulse broadening per unit length $(\Delta T / L)$ due to multipath dispersion.

Solution:

$$n_1 = 1.52, n_2 = 1.487, 2a = 110\ \mu\text{m and } n_a = 1$$

(a) $NA = \sqrt{n_1^2 - n_2^2} = 0.315$

(b) $NA = n_a \sin\alpha_m = 1 \times \sin\alpha_m = 0.315$

Therefore,

$$\alpha_m = 18.36^{\circ}$$

Also,

$$n_a \sin\alpha_m = n_1 \sin\theta_m$$

or,

$$0.315 = 1.52 \sin\theta_m$$

Therefore,

$$\theta_m = \sin^{-1}\left(\frac{0.315}{1.52}\right) = 11.96^\circ$$

Further,

$$\phi_c = \sin^{-1}\left(\frac{n_2}{n_1}\right) = 78.03^\circ$$

(c)

$$\frac{\Delta T}{L} = \frac{n_1}{n_2}\left(\frac{n_1 - n_2}{c}\right) = \frac{1.52}{1.487}\left(\frac{1.52 - 1.487}{3\times 10^8\ \text{ms}^{-1}}\right) = 1.124\times 10^{-10}\ \text{sm}^{-1}$$

Problem 7: F or the problem in Fig. 1.6, what are the minimum and maximum numbers of reflections per-metre for the rays guided by it?

Solution:

The ray that passes along the axis of the fibre, that is, the one for which (angle of incidence, α) $\alpha = 0$, will not be reflected, but the ray that follows the most oblique path is incident at an angle α very nearly equal to (but less than) α_m, will suffer maximum number of reflections.

$$\tan\theta_m = \frac{a}{L}$$

Hence,

$$L = \frac{a}{\tan\theta_m} = \frac{55\times 10^{-6}\ \text{m}}{\tan(11.96^\circ)} = 2.60\times 10^{-4}\ \text{m}$$

Therefore, the maximum number of reflections per metre would be

$$\frac{1}{2L} = 1923.07\ \text{m}^{-1}$$

All other rays will suffer reflections between these two extremes of 0 and 1923.07 m^{-1}.

Problem 8: A step index glass fibre has higher core index lower cladding index of 1.50 and 1.472, respectively. Find out following parameters of such a fibre (a) critical angle of the fibre, (b) the corresponding acceptance angle of fibre in air and corresponding oblique angle in fibre, and (c) numerical aperture (NA), (d) percentage of light collected by the fibre, and (e) diameter of the single-mode step index fibre operating at wavelength of 1330 nm.

Solution:

Given,

$n_1 = 1.50, n_2 = 1.472, \lambda = 1330\ \text{nm}, V = 2.405$ for single-mode step index fibre.

(a) Critical angle is given by (1.6) as,

$$\phi_c = \sin^{-1}\frac{n_2}{n_1} = \sin^{-1}\left(\frac{1.472}{1.50}\right) = \sin^{-1} 0.981 = 78.91^\circ$$

Hence, the oblique angle $\theta_m = 90^\circ - 78.91^\circ = 11.09^\circ$

(b) The equation of angles at critical angle is:

$$n_a \sin\theta_1 = n_1 \sin\theta_m = n_1 \cos\phi_c$$

Hence,

$$\frac{\sin\theta_1}{\sin\theta_m} = \frac{n_1}{n_a}$$

Since

$$n_a = 1 \text{ for air}, n_1 = 1.5 \text{ for core},$$

$$\sin\theta_1 = (1.5)\sin\theta_m = (1.5)\sin 11.09^\circ = 0.288$$

$$\Rightarrow \theta_1 = \sin^{-1}(0.288) = 16.76^\circ$$

The acceptance angle is 16.76°.

(c) Numerical aperture $NA = \sqrt{n_1^2 - n_2^2} = \sqrt{(1.50)^2 - (1.472)^2} = 0.288$

(d) Percentage of light collected in the fibre

$$= (NA)^2 \times 100$$

$$= (0.288)^2 \times 100 = 8.32\%$$

Only 8.32% of light is collected by the fibre.

(e) The diameter of the fibre is

$$d = \frac{V\lambda}{\pi\sqrt{n_1^2 - n_2^2}} = \frac{2.405 \times 1.33\ \mu\text{m}}{3.143 \times 0.288} = 3.53\ \mu\text{m}$$

The diameter of such a fibre is 3.53 μm. To increase diameter of fibre, the difference in refractive index $(n_1 - n_2)$ is to be reduced.

Problem 9: A silica optical fibre with a core diameter (2*a*) has a core refractive index of 1.50 and a cladding refractive index of 1.48. Calculate the numerical aperture of the fibre and the critical angle of core cladding interface for the silica fibre.

Solution:
Given that;

$$n_1 = 1.50,\ n_2 = 1.48$$

$$NA = \left(n_1^2 - n_2^2\right)^{1/2} = 0.244$$

$$\frac{\sin\theta_c}{\sin 90^\circ} = \frac{n_2}{n_1} \Rightarrow \frac{\sin\theta_c}{1} = \frac{n_2}{n_1} \rightarrow \theta_c = 80.63^\circ$$

Problem 10: A multimode step index fibre with a core diameter of 100 μm and a relative refractive index difference of 1.5% is operating at a wavelength of 0.75 μm. If the core refractive index is 1.48, estimate (a) normalised frequency for the fibre, and the number of guided modes.

Solution:

(a)

$$V = \frac{2\pi}{\lambda}.a.n_1(2\Delta)^{1/2} = 107.37$$

(b)

$$M_g = V^2/2 = 5764$$

Problem 11: A multimode step index fibre with a core diameter of 60 μm and a refractive index difference of 1.2 % is operating at a wavelength of 1.55 μm. If the core refractive index is 1.50, estimate (a) the normalised mode frequency for fibre, and (b) the number of guided modes due to the above.

Solution: Given:

$$\lambda = 1.55\,\mu\text{m}\ \text{, Index difference} = 1.2\,\%\text{, } n_1 = 1.50$$

$$V = \frac{2\pi}{\lambda} \times a \times n_1(2\Delta)^{1/2} = 28.25$$

$$M_g = V^2/2 = 399.30$$

Problem 12: A multimode step index fibre with a core diameter of 60 μm and a refractive index difference is 1.5 % is operating at a wavelength of 1.55 μm. If the core refractive index is 1.48, calculate (a) the normalised mode frequency for the fibre, and (b) the number of guided modes.

Solution:

$$V = \frac{2\pi}{\lambda} a n_1(2\Delta)^{1/2} = 31.17$$

$$M_g = V^2/2 = 485.90 \approx 486$$

Problem 13: A single-mode fibre of step index profile having a core diameter of 8 μm and a refractive index difference of 1.0 % is operating at a wavelength of 1.55 μm. If the core refractive index is 1.48, calculate: (a) *V* and (b) the number of modes for above value of *V*.

Solution:

$$V = \frac{2\pi}{\lambda} a n_1(2\Delta)^{1/2} = 3.393$$

$$M_g = V^2/2 = 5.758 \approx 6$$

Problem 14: Calculate radius of the core of step index whose refractive index is 1.48 and index difference is 0.015, whereas wavelength of light is 0.85 μm and value of $V = 2.405$ for single-mode fibre.

Solution:

$$a = \frac{V\lambda}{2\pi n_1(2\Delta)^{1/2}}$$

if $V = 2.405$ for single-mode fibre.

$$\Rightarrow \frac{2.405\times 0.85\times 10^{-6}}{2\pi\times 1.48\times(2\times 0.015)^{1/2}} = 1.268\ \mu\text{m}$$

Therefore, the maximum core diameter for single-mode operation = 2.53 μm. If the difference in refractive index becomes 0.0015 instead of 0.015, then $a = 4.0\ \mu\text{m}$ for single-mode propagation. The maximum core diameter for a single-mode fibre is 8 μm and the further reduction in refractive index difference will add to nonlinearity.

Problem 15: Calculate the cut-off wavelength for a step index single-mode fibre when the refractive index and radius of the core are 1.46 and 4.5 μm, respectively, whereas the relative index difference between core and cladding interface is 0.25 %.

Solution:

For a single-mode optical fibre, $V = 2.405$

The cut off wavelength is:

$$\lambda_c = \frac{2\pi a n_1 (2\Delta)^{1/2}}{2.405} = 1214\ \text{nm}$$

Problem 16: Find the core radius of a single-mode step index fibre whose core and cladding refractive indices are 1.48 and 1.478 at an operating wavelength of 1320 nm. Calculate maximum value of the angle of acceptance of the fibre.

Solution:

Given: $n_1 = 1.48,\ n_2 = 1.478,\ \lambda = 1320\ \text{nm}$

We know that for the single-mode step index fibre $= 2.405$.

The value of NA is calculated as:

$$\text{NA} = \left(n_1^2 - n_2^2\right)^{1/2} = 0.0769$$

The value of core radius a is given by

$$a = \frac{V\lambda}{2\pi\text{NA}} = 6.57\ \mu\text{m}$$

Angle of acceptance $\theta_a = \sin^{-1}(\text{NA}) = 4.41^\circ$

Problem 17: Find out the critical wavelength of a single-mode step index fibre in which the value of numerical aperture is 0.015 and core radius is 4 μm.

Solution:

Given: $\text{NA} = 0.015$, $a = 4\ \mu\text{m},\ V = 2.405$ for single-mode step index fibre.

The equation of critical wavelength is:

$$\lambda_c = \frac{2\pi\text{NA}a}{V} = \frac{2\times 3.143\times 0.015\times 4\times 10^{-6}}{2.405} = 156.75\ \text{nm}$$

Problem 18: A step index single-mode circular core fibre has a core index of 1.50, a relative refractive index difference 0.3 %, and a core diameter of 8.2 μm. Calculate the waveguide dispersion parameter for this fibre at $\lambda = 1.30\ \mu\text{m}$ and 1.55 μm.

Solution:

It is given that $n_1 = 1.50$ and $\Delta = 0.003$. Hence

$$n_2 \approx n_1(1-\Delta) = 1.50(1-0.003) = 1.4955$$

Since

$$V = \frac{2\pi a n_1 \sqrt{2\Delta}}{\lambda}$$

at $\lambda = 1.30\,\mu\text{m}$,

$$V_1 = \frac{\pi \times 8.2}{1.3} \times 1.50\sqrt{2 \times 0.003} = 2.30$$

and at $\lambda = 1.55\,\mu\text{m}$,

$$V_2 = \frac{\pi \times 8.2}{1.55} \times 1.50\sqrt{2 \times 0.003} = 1.93$$

at $V = V_1$ from (1.107)

$$V\frac{d^2}{dV^2}(bV) \approx 0.080 + 0.549 + (2.834 - V)^2$$

$$\Rightarrow V\frac{d^2}{dV^2}(bV) \approx 0.080 + 0.549 + (2.834 - 2.30)^2 = 0.914$$

and at $V = V_2$ from (1.107)

$$\Rightarrow V\frac{d^2}{dV^2}(bV) \approx 0.080 + 0.549 + (2.834 - 1.93)^2 = 1.446$$

Then, the waveguide dispersion parameter D_w at $\lambda_1 = 1.30\,\mu\text{m}$ may be given by (1.106)

$$D_w = -\frac{n_2\Delta}{c\lambda}\left[V\frac{d^2}{dV^2}(bV)\right]$$

$$\Rightarrow (D_w)_{\lambda_1 = 1.30\,\mu\text{m}} = -\frac{1.4955 \times 0.003}{3 \times 10^8\ \text{ms}^{-1} \times 1.30\,\mu\text{m}}[0.914] = -1.051\,\text{psnm}^{-1}\,\text{km}^{-1}$$

and D_w at $\lambda_2 - 1.55\,\mu\text{m}$ may be given by

$$\Rightarrow (D_w)_{\lambda_2 = 1.55\,\mu\text{m}} = -\frac{1.4955 \times 0.003}{3 \times 10^8\ \text{ms}^{-1} \times 1.55\,\mu\text{m}}[1.446] = -1.3951\,\text{psnm}^{-1}\,\text{km}^{-1}$$

Problem 19: A step index single-mode fibre exhibits material dispersion of $6\,\text{psnm}^{-1}\text{km}^{-1}$ at an operating wavelength of 1.55 μm. Assume that a core index of 1.50, a relative refractive index difference 0.5 %. Calculate the diameter of the core needed to make the total dispersion of the fibre zero at this wavelength.

Solution:

It is given that $n_1 = 1.50$ and $\Delta = 0.005$. Hence,

$$n_2 \approx n_1(1-\Delta) = 1.50(1-0.005) = 1.4925$$

The total dispersion $D = D_w + D_m = 0$. Since $D_m = 6\,\text{psnm}^{-1}\text{km}^{-1}$, $D_w = -6\,\text{psnm}^{-1}\text{km}^{-1}$. From (1.106),

$$D_w = -\frac{n_2\Delta}{c\lambda}\left[0.080+0.549+(2.834-V)^2\right]$$

$$D_w = -\frac{1.4925\times0.005}{3\times10^8\ \text{ms}^{-1}\times1.55\ \mu\text{m}}\left[0.080+0.549+(2.834-V)^2\right]$$

$$D_w = -1.60\left(\text{psnm}^{-1}\text{km}^{-1}\right)\left[0.080+0.549+(2.834-V)^2\right]$$

As the required value of $D_w = -6\ \text{psnm}^{-1}\text{km}^{-1}$, we have,

$$-6 = -1.60\left[0.080+0.549+(2.834-V)^2\right]$$

This gives,

$$V = 1.067$$

Since

$$V = \frac{2\pi a n_1\sqrt{2\Delta}}{\lambda}$$

The core diameter needed to make total dispersion zero will be given by

$$2a = \frac{V\lambda}{\pi n_1\sqrt{2\Delta}} = \frac{1.067\times1.55\ \mu\text{m}}{\pi\times1.50\sqrt{2\times0.005}} = 3.50\ \mu\text{m}$$

Problem 20: A glass fibre exhibits material dispersion by $\left|\lambda^2\left(\frac{d^2n_1}{d\lambda^2}\right)\right|$ of 0.025. Determine the material dispersion parameter at wavelength of 0.85 µm.

Solution:

Given: $\lambda = 0.85\ \mu\text{m}$

From Eq. 1.124 $\left|\lambda^2\left(\frac{d^2n_1}{d\lambda^2}\right)\right| = 0.025$

$$M = \frac{\lambda}{c}\left|\frac{d^2n_1}{d\lambda^2}\right| = \frac{1}{c\lambda}\left|\lambda^2\left(\frac{d^2n_1}{d\lambda^2}\right)\right| = \frac{0.025}{3\times10^5\left(\frac{\text{km}}{\text{s}}\right)\times850\ \text{nm}} = 98.03\ \text{ps/km}-\text{nm}$$

MULTIPLE CHOICE QUESTIONS

1. The first low-loss optical fibre was made
   (a) at Corning Glass Works in 1970.
   (b) at Standard Telecommunication Labs in 1966.
   (c) at Bell Telephone Labs in 1960.
2. Optical fibres are made of
   (a) glass coated with plastic.
   (b) ultrapure glass.
   (c) plastic.
   (d) All of these

3. Unlike wires, optical fibres are immune to
   (a) electromagnetic interference.
   (b) high frequency transmission.
   (c) signal losses.
4. An optical fibre has a core with reflective index of 1.52 and a cladding with index of 1.45. Its numerical aperture is
   (a) 0.15.
   (b) 0.20.
   (c) 0.35.
   (d) 0.46.
   (e) 0.70.
5. What is the diameter of the cores of a multimode step index fibre?
   (a) 100 μm
   (b) 200 μm
   (c) 400 μm
   (d) 1000 μm
   (e) All the above
6. Amplifiers are needed
   (a) to overcome the threshold for driving an optical fibre.
   (b) to compensate for fibre attenuation.
   (c) only with a copper-wire system.
   (d) to convert optical signal into electronic form.
7. What are the major advantage of optical fibres for long-distance communication?
   (a) Small fibre size.
   (b) Low loss when carrying high-speed signals.
   (c) Nonmetallic.
   (d) Low loss.
   (e) High-speed signal capacity.
8. Which of the following statements is true about interconnection networks?
   (a) The number of channels on a single fibre that can transmit data simultaneously refers to the degree of simultaneity.
   (b) Degree is the number of channels on a single fibre.
   (c) Point-to-point communication connections are called switches.
   (d) Diameter is the number of links in the largest path between any two nodes.
9. Light is confined within the core of a simple optical fibre by
   (a) refraction.
   (b) total internal reflection at the outer edge of the cladding.
   (c) total internal reflection at the core cladding boundary.
   (d) reflection from the fibre's plastic coating.
10. A popular light wavelength fibre-optic cable is
    (a) 0.7 micrometre
    (b) 1.3 micrometre
    (c) 1.5 micrometre
    (d) 1.8 micrometre

Answers to Multiple Choice Questions

1. (a)
2. (d)
3. (a)
4. (d)
5. (a)
6. (b)
7. (c)
8. (d)
9. (c)
10. (b)

QUESTIONS

1. Explain the principle of optical fibre communication with a block diagram.
2. What is the basic principle behind optical transmission?
3. What is the need for cladding?
4. What are external and internal reflection?
5. What is a step-index and graded index fibre? Sketch their RI profiles.
6. Define fibre attenuation.
7. What are the causes for absorption?
8. What are the types of bends?
9. What is meant by group velocity dispersion?
10. Enumerate the characteristics of an single-mode (SM) fibre.
11. What are the causes for intramodal dispersion?
12. What is meant by modes of a waveguide?
13. Sketch the structure, index profile and path taken by light ray in stepped index (SI) and graded index (GI) fibres.
14. Why are single-mode lasers used for high speed communication?
15. What is total internal reflection ?
16. Define critical angle.
17. What is meant by leaky modes?
18. What are hybrid modes?
19. Compare multimode (MM) and single-mode (SM) fibres
20. Define normalised frequency.
21. How do scattering losses arise?
22. What is the reason for material dispersion?
23. What are microbends?
24. What is waveguide dispersion?
25. Define numerial aperture and give the expression for the same in SI and GI fibres.
26. What are the attenuation mechanisms in a fibre?
27. Differentiate attenuation from dispersion.
28. Differentiate step index from graded index fibre.
29. A continuous 12 -km-long optical fibre link has a loss of 1.5 db/km. What is the minimum optical power level that must be launched into the fibre to maintain an optical power level of 0.3 μW at the receiving end?
30. The refractive index of a core is $n_1 = 1.48$ and cladding is $n_1 = 1.46$. Under what conditions will light be trapped inside the core?
31. For the plastic fibre, $n_1 = 1.495, n_2 = 1.402$. Calculate the acceptance angle and numerical aperture.

32. Distinguish between linear and nonlinear scattering.
33. Distinguish between intermodal and intramodal dispersion.
34. Explain how intermodal dispersion is reduced in GI fibre as compared to SI fibre.
35. What is mode field diameter (MFD)
36. What is polarisation mode dispersion (PMD)

REFERENCES

1. Agrawal G P , *Fiber-Optic Communication Systems,* 3rd ed., Wiley India Edition, 2008.
2. Cherlin A H, *An Introduction to Optical Fibers,* New-York: McGraw Hill Book Co, 1983.
3. Baccarelli P, Burghignoli P, Frezza F, Galli A, Lampariello P, Lovat G and Paulotto S S, Fundamental modal properties of surface waves on meta-material grounded slabs, *IEEE Trans. On Microwave Theory and Tech.* vol. 33, no. 4, pp. 1431–1442, 2005.
4. Belanger M and Yip G L, Mode conversion analysis in a single mode planar taper optical waveguide, *Appl. Sci. Res.,* vol. 41, pp. 305–314, 1984.
5. Berglund W and Gopinath A, WKB analysis of bend losses in optical waveguides, *IEEE J. Lightwave Technol,* vol. 18, no. 8, pp. 1161–1166, 2000.
6. Boucouvalas A C and Qian X, Mode dispersion and delay characteristics of optical waveguides using equivalent TL circuits, *IEEE J. Quantum Electron.,* vol. 41, no. 7, pp. 951–957, 2005.
7. Boucouvalas A C and Qian X, Optical fiber refractive index profile synthesis from nearfield, *IEEE GLOBECOM, Optical Networking and Systems,* pp. 2669–2673, Nov. 2003.
8. Boucouvalas A C and Thraskias C A, Accurate optical fiber refractive index reconstruction from near field, *IEEE Communication Systems Networks and Digital Signal Processing (CSNDSP),* vol. 3, pp. 501–505, 2008.
9. Burns W K and Milton F, Mode conversion in planar-dielectric separating waveguides, *IEEE J. of Quantum Electronics,* vol. QE-11, no. 1, pp. 32–39, Jan. 1975.
10. Cai Y, Mizumoto T and Naito Y, Analysis of the coupling characteristics of a tapered coupled waveguide system, *IEEE J. Lightwave Technol,* vol. 8, no. 1, pp. 90–98, Jan. 1990.
11. Chatak A and Thyagarajan K, *Introduction to Fiber Optics,* 1st ed., USA, 1998.
12. Cherlin A H, *An Introduction to Optical Fibers,* New-York : McGraw Hill Book Co, 1983.
13. Aschiéri P, Complex behavior of a ray in a Gaussian index profile periodically segmented waveguide, *J. Opt. A: Pure Appl. Opt.,* vol. 8, p. 386, 2006.
14. Bowers J E, Newton S A, Sorin W V and Shaw H J, Filter response of single mode fiber recirculating delay lines, *Electron. Lett.,*vol. 18, pp. 110–111, 1982.
15. Collin R E, *Field Theory of Guided Waves,* New York: McGraw Hill, 1960.
16. Marcuse D, *Light Transmission Optics,* New York: Van Nostrand Reinhold, 1982.
17. Daum W, Krauser J, Zamzou P E and Ziemann O, *Polymer Optical Fibers for Data Communications,* 2nd ed., Springer, 2005.
18. Feit M D and Fleck J A, Jr. Light propagation in graded-index optical fibers, Appl. Opt., vol. 17, pp. 3990–3998, 1978.
19. Gambling W A, Payne D N and Matsumura H, Cut-off frequency in radically inhomogeneous single mode fiber, *Electron. Letters,* vol. 13, no. 5, pp. 139–140, 1977.
20. Gambling W A, The rise and rise of optical fibers, *IEEE J.* on *Selected Topics in Quantum Electronics,* vol. 6, no. 6, pp. 1084–1093, Nov./Dec. 2000.
21. Ghatak A and Thyagarajan K, *Introduction to Fiber optics,* Cambridge University Press, 2007.
22. Ghatak A K and Thyagarjan K, *Contemporary Optics,* New York: Plenum Pub. Corp., 1978.

23. Gloge D, Dispersion in weakly guiding fibers, *Applied Opt.*, vol. 10, pp. 2442–2445, 1971.
24. Gloge D, Weakly guiding fibers, *Applied Opt.*, vol. 10, pp. 2252–2258, 1971.
25. Hagemann H J, Lade H and Warnier J, The performance of depressed-cladding single mode fibers with different b/a ratios, *J. LightwaveTechnol*, vol. 9, pp. 689–694, 1991.
26. Hosaka T, Okamoto K, Sasaki Y and Edahiro T, Single mode fibers with asymmetrical refractive index pits on both sides of core, *Electron. Letters*, vol. 17, pp. 191–193, 1981.
27. Keiser G, *Optical Fiber Communication*, 3rd ed., McGraw-Hill International Edition, Electrical Engineering Series, New York, 2000.
28. Marcuse D, Bending losses of the asymmetric slab waveguide, *Bell Syst. Tech. J.*, pp. 2551–2563, 1971.
29. Midwinter J, *Optical Fibers for Transmission*, New York: Wiley, 1979.
30. Palais J C, *Fiber Optic Communications*, 4th ed., Prentice Hall, 1998.
31. Raghuwanshi S K, Low dispersion due to square law medium profile, Annual University Magazine of Sir Padampat Singhania University, Udaipur, Rajasthan, India 2009.
32. Rahman B M A, Finite element analysis of optical waveguides, *Progress in Electromagnetic Res., PIER*, vol. 10, pp. 187–216, 1995.

2 Optical Waveguide Analysis

Objectives

- This chapter deals with various characteristics of optical fibre.
- Basic wave equations are briefly explained and numerically derived.
- Propagation of light through various structures of optical fibres is detailed.
- Effect of various optical fibre parameters on propagation of optical signal through them is spelled out.
- Parameters of optical fibre and their dependence on each other are explored.

2.1 INTRODUCTION

The rapid developments in fields such as fibre optics communication engineering and integrated optical electronics have expanded the interest and increased expectations about guided wave optics in which optical waveguides play a central role [1–5]. Optical waveguides for optical fibres and optical integrated circuits utilise a wave phenomenon that traps the light locally and guides it in any direction, although their propagation lengths differ greatly. In order to develop new optical communication systems or optical devices, we need to fully understand the principles of optical guiding while obtaining accurate quantitative propagation characteristics of waveguides and utilising them effectively in actual design [6–9]. Maxwell's equations are used to calculate the propagation characteristics of optical waveguides. However, it is rather rare to obtain a precise analytic solution, and therefore, a precise analysis of optical waveguide is generally considered to be difficult [10–11]. For this reason, various methods of optical waveguide analysis have been developed [12–19]. These methods may be broadly classified into two categories: analytical approximation solutions and numerical solution using computers. In order to analyse and design optical waveguides, it is necessary to study these analysis methods in depth and to be able to use these methods as tools. Planar optical waveguides are the key devices to construct integrated optical circuits and semiconductor lasers [20–24]. Generally, rectangular waveguides consist of a square or rectangular core surrounded by a cladding with a lower refractive index than that of the core [25–28]. Three-dimensional analysis is necessary to investigate the transmission characteristics of rectangular waveguides. However, rigorous three-dimensional analysis usually requires numerical calculations and does not always give a clear insight into the problem [29–33]. Therefore, this chapter describes two-dimensional slab waveguides to acquire a fundamental understanding of optical waveguides. Although these are approximate methods, the essential light wave transmission mechanism in rectangular waveguides can be fully investigated. The rigorous treatment of three-dimensional rectangular waveguides by the finite element method has been discussed in literature elsewhere.

2.1.1 Outline of Optical Waveguides

Optical waveguides, which trap light locally and guide it in a specific direction, can be roughly classified into optical waveguides for optical integrated circuits and optical fibres.

In the microwave band, electromagnetic fields are generally distributed within a finite area surrounded by conductors as in waveguides and coaxial lines. On the other hand, it is difficult to construct a waveguide mainly from metal in the visible waveband since the metal will behave as a substance having a complex permittivity with a large absolute value. For this reason, waveguides are usually constructed by combining appropriate dielectrics, as shown in Fig. 2.1, thus distributing an electromagnetic field over an infinite area. By loading a thin film with a higher refractive index than either the substrate or the upper cladding on the substrate surface, the light can be trapped inside this film. Furthermore, as the thickness of the film increases, the effective refractive index sensed by the light also increases.

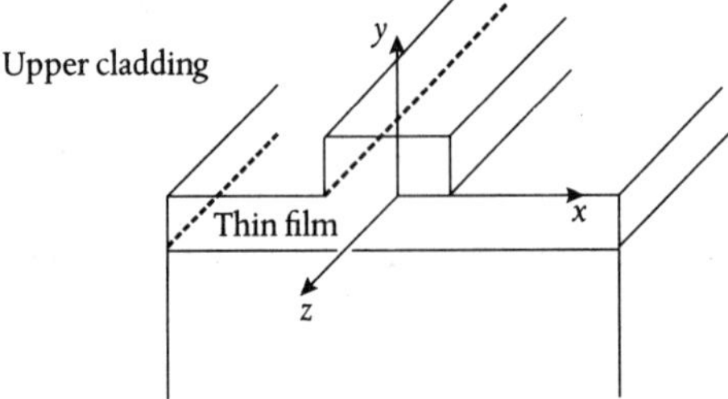

Fig. 2.1 Optical waveguide

Therefore, in the case of the waveguide shown in Fig. 2.1, the light is pulled toward the central area inside the thin film resulting in the propagation of light in the z-direction with the power of the light concentrated in a certain part of the cross-section of the waveguide.

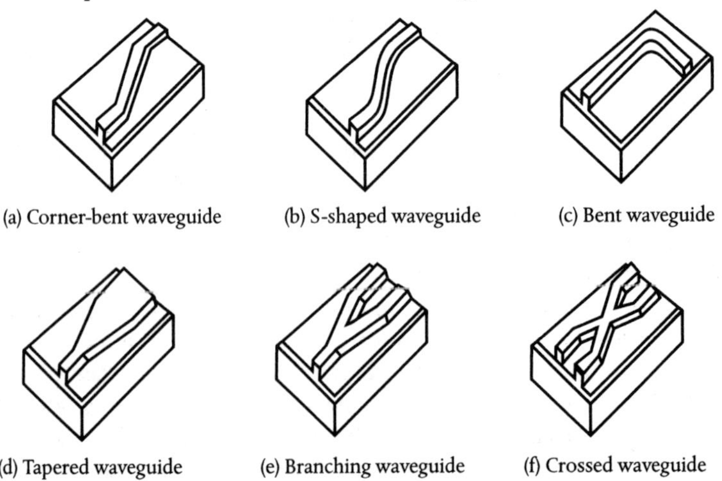

Fig. 2.2 Examples of optical waveguide usage

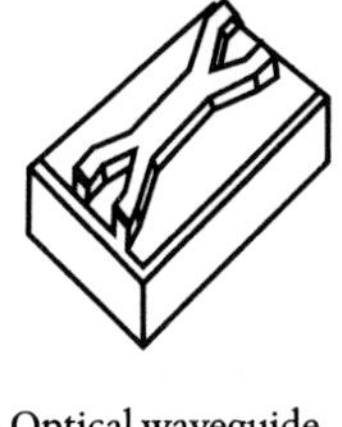
(g) Optical waveguide directional coupler

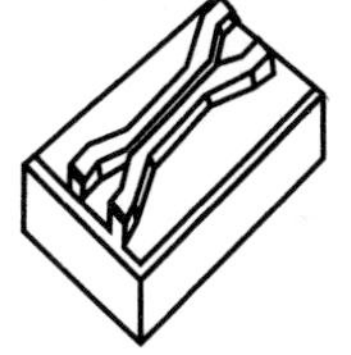
(h) Two-mode waveguide coupler

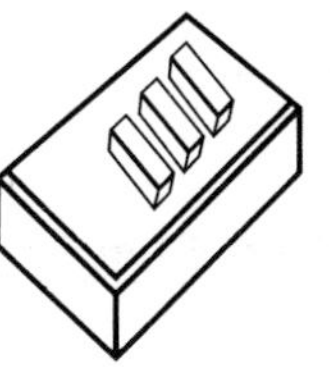
(i) Waveguide grating

Fig. 2.2 (continued)

Waveguides in which the refractive index changes in stages are called step-index (SI) optical waveguides, while those with a gradual refractive index change are called graded index (GI) optical waveguides. An optical waveguide that is uniform in the direction of propagation, as shown in Fig. 2.1, is the most basic type of waveguide; however, this alone is not sufficient for the construction of an optical integrated circuit. In reality, an appropriate combination of various forms of optical waveguides, as shown in Fig. 2.2, is placed on the substrate to construct an optical circuit with desired features. Corner-bent waveguides, S-shaped waveguides, and bent waveguides are used to change the direction of light waves. Tapered waveguides are used to change the width of waveguides; branching waveguides and crossed waveguides are used for splitting, combining, and interference; and optical waveguide directional couplers and two-mode waveguide couplers are used for coupling. Waveguide gratings, with a periodic structure in the direction of propagation, play many important roles in the optical integrated circuit such as wavelength filter, mode converter, reflector, resonator, and demultiplexer. Waveguide gratings are also used widely as a laser element, such as a distributed Bragg reflector (DBR) laser or distributed feedback (DFB) laser. Although waveguides come in various forms and with a variety of functions, the fact remains that the optical waveguide that is uniform in the direction of propagation, as shown in Fig. 2.1, is the basic form of waveguide. The information regarding the propagation characteristics of the optical waveguide shown in Fig. 2.1 is the most basic and important information required when designing the types of waveguides shown in Fig. 2.2. Optical fibre may be classified roughly into two categories: single-mode (SM) optical fibre with single-mode propagation and multimode (MM) optical fibre with multimode propagation. Normally, the core diameter of SM fibre is 5 – 10 μm, and the core diameter of MM fibre is in the 50 – 100 μm range. The difference in refractive indexes between core and cladding is 0.1 – 1%, which is minute, and the extent of the difference in the refractive indices is expressed by the specific (or relative) refractive index difference Δ, which is defined as

$$\Delta = \frac{n_1^2 - n_2^2}{2n_1^2} \simeq \frac{n_1 - n_2}{n_1} \tag{2.1}$$

where n_1 and n_2 are the refractive indices of the core and cladding, respectively.

Figure 2.3 shows the refractive index distribution of stepped optical fibre in the radial (r) direction. Generally, uniform core optical fibre, whose core has uniform refractive index is used. The dual-shape core optical fibre, whose core refractive index changes in stages is known as dispersion shifted single-mode optical fibre.

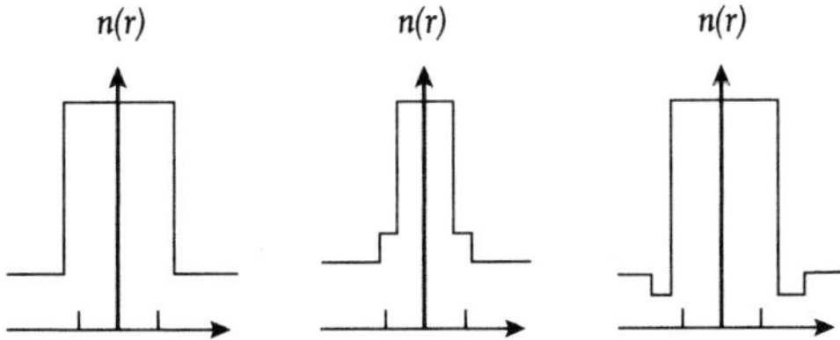

Fig. 2.3 Stepped-index optical fibre

It has excellent connecting and bending characteristics as well as dispersion control performance.

The W-type optical fibre is an optical fibre with low dispersion characteristics over a wide range of wavelengths (see Section 2.7.4 for dispersion in optical fibre).

2.1.2 Graded Optical Fibre

Typical examples of the refractive index distribution in graded optical fibre are shown in Fig. 2.4. The α- profile optical fibre (power law), stepped α-power profile optical fibre, and W-type α-power profile optical fibre are used for reducing multimode dispersion. The refractive index distribution is given as:

$$n(r) = \begin{cases} n_1\left[1 - 2\rho\Delta(r/a)^{\alpha}\right]^{1/2} & 0 \le r \le a \\ n_2 = n_1(1-2\Delta)^{1/2} & a \le r \end{cases} . \tag{2.2}$$

If $\rho < 1$, a step exists between the core and the cladding. If $1 < \rho$, there is a notch.

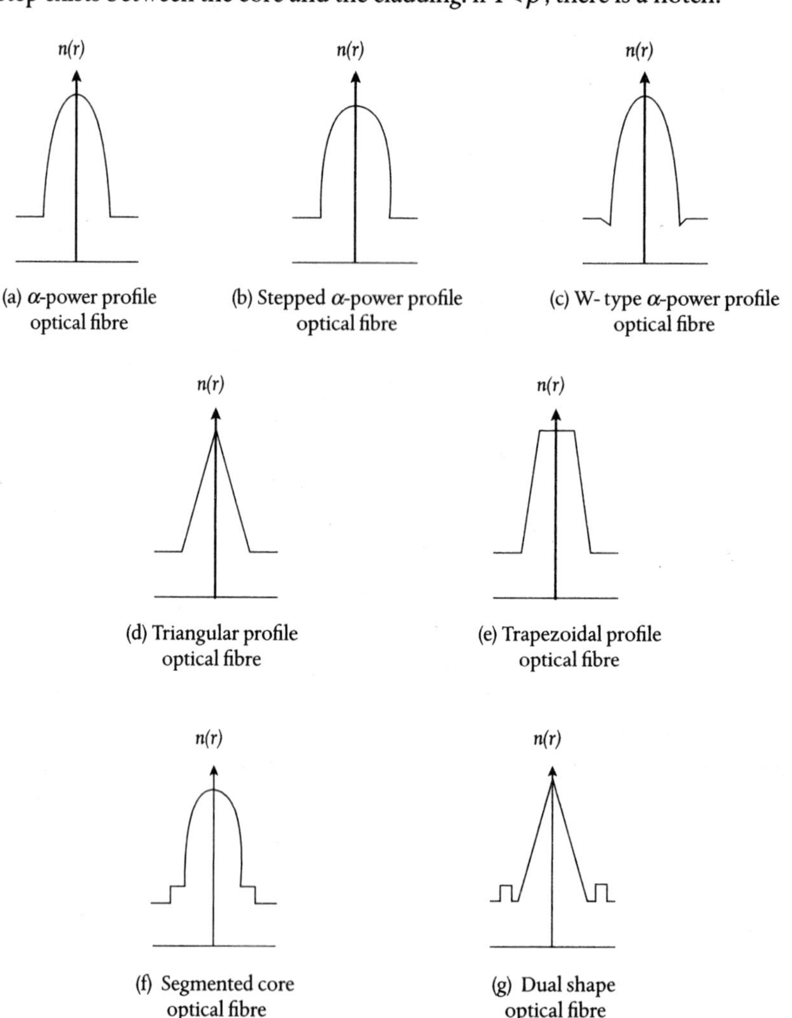

Fig. 2.4 Graded optical fibre

The trapezoidal profile optical fibre, triangular profile optical fibre, segmented core optical fibre, and dual-shape core optical fibre are all used as dispersion shifted optical fibre; however, the trapezoidal profile optical fibre and triangular profile optical fibre are inferior to the segmented core optical fibre and dual-shape core optical fibre in bending characteristics and dispersion control performance. The segmented core optical fibre as well as the dual-shape core optical fibre are known as dispersion shifted optical fibre and have excellent connecting and bending characteristics and dispersion control performance, but the connection loss is somewhat larger in the segmented core optical fibre than in the dual-shape core optical fibre.

2.2 WAVE THEORY OF OPTICAL WAVEGUIDES

The basic concepts and equations of electromagnetic wave theory required for the comprehension of light wave propagation in optical waveguides are presented. In the case of a slab waveguide, the light confinement and formation of modes in the waveguide are qualitatively explained. Optical fibres and optical waveguides consist of a core in which light is confined and a cladding or the substrate surrounding the core, as shown in Fig. 2.5. The refractive index of core n_1 is higher than that of cladding n_0. Therefore, the light beam that is coupled to the end face of the waveguide is confined in the core by total internal reflection. The condition for total internal reflection at the core–cladding interface is given as $n_1 \sin(\pi/2-\varnothing) \geq n_0$. Since angle $\varnothing$ is related with the incident angle θ by $\sin\theta = n_1 \sin\varnothing \leq \sqrt{n_1^2 - n_0^2}$, we obtain the critical condition for the total internal reflection as,

$$\theta \leq \sin^{-1}\sqrt{n_1^2 - n_0^2} \equiv \theta_{max} \tag{2.3}$$

The refractive index difference between core and cladding is of the order of $n_1 - n_0 = 0.01$. Then θ_{max} in (2.3) can be approximated by:

$$\theta_{max} \cong \sqrt{n_1^2 - n_0^2}. \tag{2.4}$$

θ_{max} denotes the maximum light acceptance angle of the waveguide and is known as the numerical aperture (NA)

The relative refractive index difference between n_1 and n_0 is defined as:

$$\Delta = \frac{n_1^2 - n_0^2}{2n_1^2} \cong \frac{n_1 - n_0}{n_1}. \tag{2.5}$$

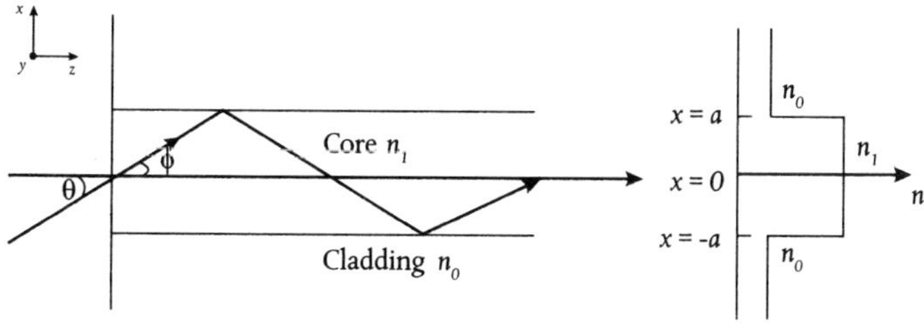

Fig. 2.5 Basic structure and refractive index profile of the optical waveguide

Δ is commonly expressed as a percentage. The numerical aperture NA is related to the relative refractive index difference Δ as:

$$\mathrm{NA} = \theta_{max} \cong n_1\sqrt{2\Delta} \tag{2.6}$$

The maximum angle for the propagating light within the core is given as $\phi_{max} \cong \theta_{max}/n_1 \cong \sqrt{2\Delta}$. For typical optical waveguides, $\mathrm{NA} = 0.21$ and $\theta_{max} = 12^0\left(\phi_{max} = 8.1^0\right)$ when $n_1 = 1.47$, $\Delta = 1\%$ (for $n_0 = 1.455$).

2.2.1 Formation of Guided Modes

We have accounted for the mechanism of mode confinement and have indicated that angle ϕ must not exceed the critical angle. Even though angle ϕ is smaller than the critical angle, light rays with arbitrary angles are not able to propagate in the waveguide. Each mode is associated with light rays at a discrete angle of propagation, as given by electromagnetic wave analysis. Here, we describe the formation of modes with the ray picture in the slab waveguide, as shown in Fig. 2.6. Let us consider a plane wave propagating along the z-direction with inclination angle ϕ. The phase fronts of the plane waves are perpendicular to the light rays. The wavelength and the wavenumber of light in the core are λ/n_1 and $kn_1\ (k = 2\pi/\lambda)$, respectively, where λ is the wave length. The propagation constants along z and x (lateral direction) are expressed as:

$$\beta = kn_1 \cos\phi \tag{2.7}$$

$$\kappa = kn_1 \sin\phi \tag{2.8}$$

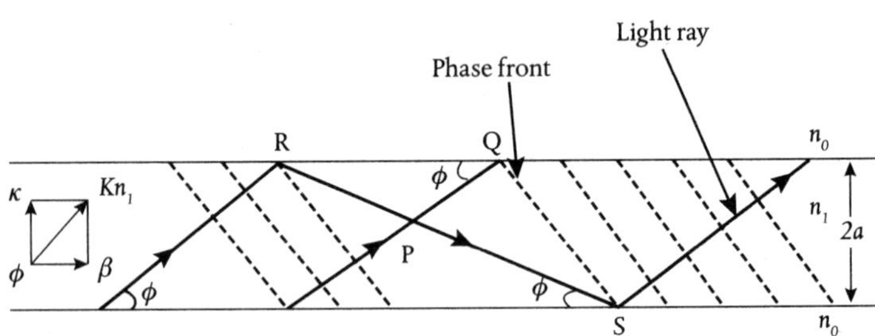

Fig. 2.6 Light rays and their phase fronts in the waveguide

Before describing the formation of modes in detail, we must explain the phase shift of a light ray that suffers total reflection. The reflection coefficient of the totally reflected light, which is polarised perpendicular to the incident plane (plane formed by the incident and reflected rays), as shown in Fig. 2.7, is given as:

$$r = \frac{A_r}{A_i} = \frac{n_1 \sin\phi + j\sqrt{n_1^2\cos^2\phi - n_0^2}}{n_1 \sin\phi - j\sqrt{n_1^2\cos^2\phi - n_0^2}}. \tag{2.9}$$

When we express the complex reflection coefficient r as $r = \exp(-j\phi)$, the amount of phase shift ϕ is obtained as:

$$\phi = -2\tan^{-1}\frac{\sqrt{n_1^2\cos^2\phi - n_0^2}}{n_1 \sin\phi} = -2\tan^{-1}\sqrt{\frac{2\Delta}{\sin^2\phi} - 1}. \tag{2.10}$$

where (2.5) has been used. The foregoing phase shift for the totally reflected light is called the Goos–Hánchen shift.

Let us consider the phase difference between the two light rays belonging to the same plane wave in Fig. 2.6. Light ray *PQ*, which propagates from point *P* to Q, does not suffer the influence of reflection. On the other hand, light ray *RS*, propagating from point *R* to *S*, is reflected two times (at the upper and lower core–cladding interfaces). Since points *P* and *R* or points Q and *S* are on the same phase front, optical paths *PQ* and *RS* (including the Goos–Hánchen shifts caused by the two total reflections) should be equal, or their difference should be an integral multiple of 2π. Since the distance between points Q and *R* is $2a/\tan\phi - 2a\tan\phi$, the distance between points *P* and Q is expressed by:

$$l_1 = \left(\frac{2a}{\tan\phi} - 2a\tan\phi\right)\cos\phi = 2a\left(\frac{1}{\sin\phi} - 2\sin\phi\right). \tag{2.11}$$

Further, the distance between points *R* and *S* is given as:

$$l_2 = \frac{2a}{\sin\phi}. \tag{2.12}$$

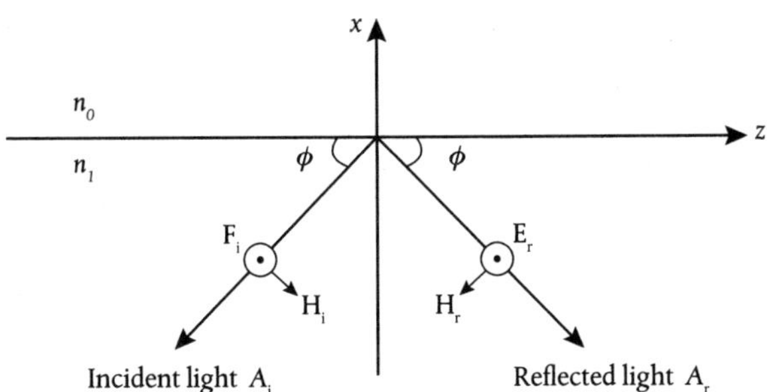

Fig. 2.7 Total reflection of a plane wave at a dielectric interface

The phase-matching condition for the optical paths *PQ* and *RS* becomes:

$$(kn_1 l_2 + 2\phi) - kn_1 l_1 = 2m\pi, \tag{2.13}$$

where m is an integer. Substituting (2.10)–(2.12) in (2.13), we obtain the condition for the propagation angle ϕ as:

$$\tan\left(kn_1 a\sin\phi - \frac{m\pi}{2}\right) = \sqrt{\frac{2\Delta}{\sin^2\phi} - 1}. \tag{2.14}$$

Equation (2.14) shows that the propagation angle of a light ray is discrete and is determined by the waveguide structure (core radius (a), refractive index (n_1), and refractive index difference (Δ)) and the wavelength (λ) of the light source (wavenumber is $k = 2\pi/\lambda$). The optical field distribution that satisfies the phase-matching condition of (2.14) is called the mode. The allowed value of propagation constant (β) is also discrete and is denoted as an eigenvalue. The mode that has the minimum angle ϕ in (2.14) $(m = 0)$ is the fundamental mode and the other modes having larger angles are higher order modes $(m \geq 1)$.

2.3 BASIC EQUATIONS FOR LIGHT WAVES: MAXWELL'S EQUATIONS

Light is an electromagnetic wave phenomenon. Its magnetic field is represented by four electromagnetic field vectors that are functions of position $r[m]$ and time $t[s]$. The four vectors electric field $E[V/m]$, magnetic field $H[A/m]$, electric flux density $D[C/m^2]$, and magnetic density $B[Wb/m^2]$ conform to Maxwell's equations:

$$\nabla \times E = -\partial B/\partial t \tag{2.15}$$

$$\nabla \times H = \partial D/\partial t + J. \tag{2.16}$$

Here, $J[A/m^2]$ is the electric current density. The following current continuity equation (or conservation of charge equation) holds for J and the charge density $\rho[C/m^3]$:

$$\nabla \cdot J = -\partial \rho/\partial t \tag{2.17}$$

From $(2.15)-(2.17)$ and the vector formulas,

$$\nabla \cdot \nabla \times E = 0$$

$$\nabla \cdot \nabla \times H = 0$$

we get

$$\nabla \cdot B = 0 \tag{2.18}$$

$$\nabla \cdot D = \rho. \tag{2.19}$$

Equations (2.15)–(2.19) are usually called Maxwell's equations.

The flow density of the energy carried by an electromagnetic field is expressed by the Poynting vector $S[W/m^2]$, which is defined as follows:

$$S = E \times H. \tag{2.20}$$

2.3.1 Constitutive Relations

If the polarisation generated in a light transmitting medium is $P[C/m^2]$ and the magnetisation is $M[A/m]$, the electric flux density D, and the magnetic flux density B are, then:

$$D = \varepsilon_0 E + P \text{ and} \tag{2.21}$$

$$B = \mu_0 (H + M). \tag{2.22}$$

Here, $\varepsilon_0[F/m]$ and $\mu_0[H/m]$ are the permittivity of vacuum and the permeability of free space, respectively, and their values are

$$\varepsilon_0 = 8.8541878 \times 10^{-12} \qquad [F/m] \tag{2.23}$$

and

$$\mu_0 = 4\pi \times 10^{-7}. \qquad [H/m]. \tag{2.24}$$

If the medium is isotropic, linear, and non-dispersive, polarisation P and magnetisation M are obtained as follows:

$$P = \varepsilon_0 \chi_e E \tag{2.25}$$

$$M = \chi_m H. \tag{2.26}$$

Here, χ_e and χ_m are the electric susceptibility and the magnetic susceptibility, respectively. If (2.25) and (2.26) are substituted in (2.21) and (2.22), respectively, then relative permittivity ε_r and the relative permeability μ_r are,

$$\varepsilon_r = 1 + \chi_e \tag{2.27}$$

$$\mu_r = 1 + \chi_m \tag{2.28}$$

Furthermore, if the permittivity ε and the permeability μ are, respectively,

$$\varepsilon = \varepsilon_0 \varepsilon_r \tag{2.29}$$

$$\mu = \mu_0 \mu_r. \tag{2.30}$$

The constitutive relations are as follows:

$$D = \varepsilon E \tag{2.31}$$

$$B = \mu H. \tag{2.32}$$

Refractive index n can be obtained from

$$n = \sqrt{\varepsilon_r \mu_r}. \tag{2.33}$$

If, as in many optical materials, $\mu = \mu_0$, that is, $\mu_r = 1$, the refractive index n is:

$$n = \sqrt{\varepsilon_r}. \tag{2.34}$$

In an isotropic medium, if the medium constant is a scalar quantity, as shown in (2.31) and (2.32), and it is dependent on the direction of the electromagnetic field vector, then the medium constant is dependent on the direction of the electromagnetic field vector and it is called an anisotropic medium; further, its constitutive relations are:

$$D = [\varepsilon] E \tag{2.35}$$

$$B = [\mu] H. \tag{2.36}$$

Here, $[\varepsilon]$ and $[\mu]$ are the rank-two permittivity and permeability tensors, respectively. In (2.35) and (2.36), tensors are indicated in a simplified form. For example, (2.35) represents the following relational expression:

$$D_i = \sum_{j=1}^{3} \varepsilon_{ij} E_j. \tag{2.37}$$

Here, $i, j = 1, 2, 3$ correspond, respectively, to rectangular coordinates, x, y, and z. If the medium is lossless,

$$\varepsilon_{ji}^* = \varepsilon_{ij} \tag{2.38}$$

and therefore, the permittivity tensor is the Hermitian matrix:

$$[\varepsilon]^{\dagger} = [\varepsilon]^{*}. \tag{2.39}$$

Here, * denotes the complex conjugate and † denotes the transpose of the complex conjugate. Until now, we have considered only linear media. In the case of a nonlinear medium, polarisation is given as:

$$P_i = \varepsilon_0 \left[\sum_{j=1}^{3} \chi_{ij}^{(1)} E_j + \sum_{j=1}^{3} \sum_{k=1}^{3} \chi_{ijk}^{(2)} E_j E_k + \sum_{j=1}^{3} \sum_{k=1}^{3} \sum_{l=1}^{3} \chi_{ijkl}^{(3)} E_j E_k E_l + \ldots \right]. \tag{2.40}$$

The first term is a linear term with $\chi_{ij}^{(1)}$ representing the rank-two linear susceptibility tensor. The second and subsequent terms are nonlinear terms: $\chi_{ijk}^{(2)} [m/V]$, representing the secondary nonlinear effect is the rank-three nonlinear susceptibility tensor, and $\chi_{ijkl}^{(3)} \left[m^2/V^2\right]$, representing the tertiary nonlinear effect is the rank-four nonlinear susceptibility tensor.

2.3.2 Boundary Conditions

Boundary conditions are conditions that must be met at the boundary surface where two different media 1 and 2 come into contact. If the unit normal vector directed from medium 2 to medium 1 is n, the boundary conditions are given as follows:

1. The tangential component of the electric field must be continuous:

$$n \times (E_1 - E_2) = 0. \tag{2.41}$$

2. The tangential component of the magnetic field must be continuous:

$$n \times (H_1 - H_2) = 0. \tag{2.42}$$

3. Only the surface current density $J_s [A/m]$, however, may be discontinuous if surface current is flowing on the boundary surface:

$$n \times (H_1 - H_2) = J_s. \tag{2.43}$$

4. The normal component of the electric flux density must be continuous:

$$n \cdot (D_1 - D_2) = 0. \tag{2.44}$$

5. Only the surface charge density $\rho_s \left[C/m^2\right]$ however, may be discontinuous if a surface charge exists on the boundary surface:

$$n \cdot (D_1 - D_2) = \rho_s. \tag{2.45}$$

6. The normal component of the magnetic flux density must be continuous:

$$n \cdot (B_1 - B_2) = 0. \tag{2.46}$$

In the above-mentioned expressions, subscripts 1 and 2 refer to media 1 and 2, respectively.

2.3.3 Wave Equations

In an isotropic lossless medium with no wave source $(J=0,\ \rho=0)$, with uniform permeability, $\mu=\mu_0$, (2.15), (2.16), (2.18), (2.19), (2.31), and (2.32) and the vector formulas are:

$$\nabla\times\nabla\times E=\nabla(\nabla\cdot E)-\nabla^2 E$$

$$\nabla\cdot E=\nabla\cdot(D/\varepsilon)=(\nabla\cdot D)/\varepsilon+D\cdot\nabla(1/\varepsilon)$$

$$\nabla\times\nabla\times H=\nabla(\nabla\cdot H)-\nabla^2 H$$

$$\nabla\times(\varepsilon E)=\varepsilon\nabla\times E+\nabla\varepsilon\times E,$$

which give us

$$\nabla^2 E-\varepsilon\mu_0\frac{\partial^2 E}{\partial t^2}+\nabla\left(\frac{\nabla\varepsilon}{\varepsilon}\cdot E\right)=0 \tag{2.47}$$

$$\nabla^2 H-\varepsilon\mu_0\frac{\partial^2 H}{\partial t^2}+\frac{\nabla\varepsilon}{\varepsilon}\times(\nabla\times H)=0. \tag{2.48}$$

If the permittivity is uniform, E and H, respectively, will satisfy the following wave equations:

$$\nabla^2 E-\varepsilon\mu_0\frac{\partial^2 E}{\partial t^2}=0 \tag{2.49}$$

$$\nabla^2 H-\varepsilon\mu_0\frac{\partial^2 H}{\partial t^2}=0. \tag{2.50}$$

If we assume an electromagnetic wave with unit angular frequency $\omega[rad/s]$ and introduce the electric field $\tilde{E}$, magnetic field $\tilde{H}$, electric flux density $\tilde{D}$, and magnetic flux density $\tilde{B}$, indicated as phasors (complex amplitude):

$$E(r,t)=Re\left[\tilde{E}(r)\exp(j\omega t)\right] \tag{2.51}$$

$$H(r,t)=Re\left[\tilde{H}(r)\exp(j\omega t)\right] \tag{2.52}$$

$$D(r,t)=Re\left[\tilde{D}(r)\exp(j\omega t)\right] \tag{2.53}$$

$$B(r,t)=Re\left[\tilde{B}(r)\exp(j\omega t)\right]. \tag{2.54}$$

Further, when we express them as $E,\ H,\ D,$ and B, Maxwell's equations become:

$$\nabla\times E=-j\omega B=-j\omega\mu_0 H \tag{2.55}$$

$$\nabla\times H=j\omega D=j\omega\varepsilon E \tag{2.56}$$

and the wave equations for expressions (2.47) and (2.48) are

$$\nabla^2 E+k^2 E+\nabla\left(\frac{\nabla\varepsilon}{\varepsilon}\cdot E\right)=0 \tag{2.57}$$

$$\nabla^2 H+k^2 H+\frac{\nabla\varepsilon}{\varepsilon}\times(\nabla\times H)=0. \tag{2.58}$$

Here, $k[\text{rad/m}]$ is

$$k = \omega\sqrt{\varepsilon\mu_0} \tag{2.59}$$

and is called the wavenumber. If $\varepsilon = \varepsilon_0$, wavenumber k_0, which is called the free-space wavenumber, is

$$k_0 = \omega\sqrt{\varepsilon_0\mu_0}. \tag{2.60}$$

From expressions (2.29) and (2.33), we get:

$$k = k_0 n. \tag{2.61}$$

If the permittivity is constant, (2.57) and (2.58), respectively, will be

$$\nabla^2 E + k^2 E = 0 \tag{2.62}$$

$$\nabla^2 H + k^2 H = 0. \tag{2.63}$$

The time average value, $\langle S \rangle$, of the energy flow density is

$$\langle S \rangle = Re\left[(1/2) E \times H^*\right]. \tag{2.64}$$

Here, $(1/2) E \times H^*$ is called the complex Poynting vector.

2.3.4 Plane Wave

As (2.62) and (2.63) show any one of the rectangular coordinate components in the electromagnetic field, say E_x satisfies the scalar wave equation,

$$\nabla^2 E_x + k^2 E_x = 0. \tag{2.65}$$

Equation (2.65) is called the Helmholtz equation. Now, consider a plane wave with wave number vector k. A plane wave is a type of wave whose wavefront is perpendicular to the direction of propagation. At any point of that plane, the amplitude and phase of the electromagnetic field are constant. If the wavenumber vector k and the position vector r are expressed in terms of the coordinate direction components as follows:

$$k = \left[k_x \; k_y \; k_z\right]^T \tag{2.66}$$

$$r = \left[x \; y \; z\right]^T. \tag{2.67}$$

The plane wave can be expressed as:

$$E_x = A\exp(-jk \cdot r) = A\exp\left[-j\left(k_x x + k_y y + k_z z\right)\right], \tag{2.68}$$

in which A represents the amplitude of the plane wave and T denotes the transpose. If (2.66) is substituted in (2.68), we get,

$$|k| = \sqrt{k_x{}^2 + k_y{}^2 + k_z{}^2}. \tag{2.69}$$

This holds that

$$k = \omega\sqrt{\varepsilon\mu_0} \tag{2.70}$$

is obtained.

The propagation velocity of the equiphase wave surface of the wave motion is called the phase velocity. For a plane wave, the phase velocity $v_p\,[m/s]$ in the direction of propagation is:

$$v_p = \omega/k = 1/\sqrt{\varepsilon\mu_{0.}} \tag{2.71}$$

If we represent the phase velocity of a plane wave in a vacuum by c, then

$$c = 1/\sqrt{\varepsilon\mu_0} = 2.99792458\times10^8 \qquad \mathrm{m/s}. \tag{2.72}$$

This is the value normally associated with the velocity of light. As shown in (2.71) and (2.72), the phase velocity of a plane wave in a medium with refractive index n is

$$v_p = c/n. \tag{2.73}$$

The propagation velocity of electromagnetic wave energy, that is, the group velocity v_g, is calculated as follows for a plane wave:

$$v_g = \left(d|k|/d\omega\right)^{-1}. \tag{2.74}$$

If (2.70) is used,

$$v_g = 1/\sqrt{\varepsilon\mu_0} = v_{p.} \tag{2.75}$$

Further, the group and phase velocities coincide. The distance in which the phase changes by 2π in the direction of propagation of wave motion is called wavelength. If the frequency is $f\,[Hz]$, the wavelength $\lambda_0\,[m]$ of a plane wave in a vacuum is:

$$\lambda_0 = 2\pi/k_0 = 2\pi/\omega\sqrt{\varepsilon_0\mu_0} = c/f. \tag{2.76}$$

The wavelength λ of a plane wave in a medium with refractive index n is:

$$\lambda = 2\pi/\omega\sqrt{\varepsilon\mu_0} = v_p/f = \lambda_0/n. \tag{2.77}$$

If the wave motion is accompanied by attenuation in a given direction, the phase constant in that direction is a complex number. If the complex number phase constant is represented by γ' and its real part and imaginary parts by $\beta[rad/m]$ and $-\alpha[Np/m]$, respectively, then

$$\gamma' = \beta - j\alpha. \tag{2.78}$$

Here, α, β, and γ' are called the attenuation constant, the phase constant, and the propagation constant, respectively. Traditionally, the propagation constant refers to the γ in

$$\gamma = j\gamma' = \alpha + j\beta. \tag{2.79}$$

However, it is also customary to call γ' the propagation constant. Therefore, when α can be ignored, β may also be called propagation constant.

2.4 POLARISATION OF LIGHT: BASIC EXPRESSIONS

The inclination in a certain direction of the vibration direction of the electric field of an electromagnetic wave is called wave polarisation. In particular, if the polarisation is in the domain of light, it is called optical polarisation. The plane defined by the direction of propagation of the electromagnetic wave and the direction of the electric field is called the plane of polarisation.

If we assume a plane wave that propagates in direction z and if $\partial/\partial x = \partial/\partial y = 0$, then

$$dE_x / dz = -j\omega\mu_0 H_y \tag{2.80a}$$

$$dH_y / dz = -j\omega\varepsilon E_x \tag{2.80b}$$

$$dE_y / dz = j\omega\mu_0 H_x \tag{2.81a}$$

$$dH_x / dz = j\omega\varepsilon E_y \tag{2.81b}$$

can be obtained from (2.55) and (2.56). As shown here, a plane wave with electric field E_x in the *x*-direction and a plane wave with electric field E_y in the *y*-direction are mutually independent. The electromagnetic fields for these plane waves are expressed as follows:

$$E_x = A_x exp(-jkz) \tag{2.82a}$$

$$H_y = E_x / Z \tag{2.82b}$$

$$E_y = A_y exp(-jkz) \tag{2.83a}$$

$$H_x = -E_y / Z \tag{2.83b}$$

where A_x and A_y are amplitudes and $Z[\Omega]$

$$Z = \sqrt{\mu_0 / \varepsilon} \tag{2.84}$$

is called the wave impedance. In particular, in a vacuum,

$$Z_0 = \sqrt{\mu_0 / \varepsilon_0} = 376.7 \ [\Omega] \tag{2.85}$$

and Z_0 is called the intrinsic (characteristic) impedance of a vacuum or of free space. The planes of polarisation of the plane waves given in (2.82) and (2.83) are *zx* and *yz* planes, respectively.

Now, if $A_x = |A_x| exp(j\phi_x)$ and $A_y = |A_y| exp(j\phi_y)$, the electric fields of (2.82) and (2.83) can be expressed by real functions representing instantaneous values as follows:

$$E_x = |A_x| \cos(\omega t - kz + \phi_x) \tag{2.86}$$

$$E_y = |A_y| \cos(\omega t - kz + \phi_y). \tag{2.87}$$

2.4.1 Elliptic Polarisation

The path of the end point of an electric field vector can be computed from expressions (2.86) and (2.87)

$$\frac{E_x^2}{|A_x|^2}+\frac{E_y^2}{|A_y|^2}-2\frac{E_x E_y}{|A_x||A_y|}\cos\phi=\sin^2\phi. \tag{2.88}$$

Here, $\phi=\phi_y-\phi_x$. Equation (2.88) represents an inclined ellipse, which is inscribed in a rectangle $2|A_x|$ wide and $2|A_y|$ high. If $\phi=\pm\pi/2$, (2.88) becomes

$$\left(E_x/|A_x|\right)^2+\left(E_y/|A_y|\right)^2=1, \tag{2.89}$$

which represents an ellipse whose principal axis coincides with the *x*- and *y*-axis. Therefore, if observed within the plane at a constant distance, the end point of the electric field vector will follow the path of an ellipse. Such polarisation is called elliptical polarisation.

2.4.2 Circular Polarisation

When $|A_x|=|A_y|=|A|$, (2.89) becomes

$$E_x^2+E_y^2=|A|^2. \tag{2.90}$$

Therefore, when observed within the plane at a constant distance z, the electric field vector rotates around the circumference at the angular frequency ω as shown in Fig. 2.8(a). This type of polarisation is called circular polarisation.

The direction of rotation for circular polarisation differs according to the phase difference, ϕ. When $\phi=+\pi/2$, the rotation is in the clockwise (right handed or positive) direction as viewed in the direction of light propagation.

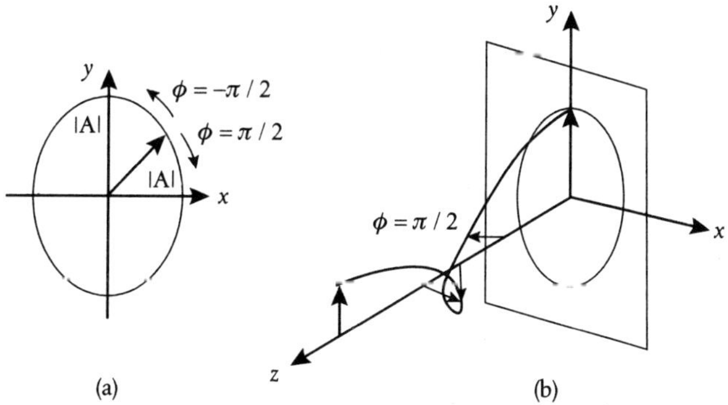

Fig. 2.8 Circular polarisation

When $\phi=+\pi/2$, it is in the counter clockwise (left handed or negative) direction. (Note that the direction of light propagation is toward you, as you face the page.) If time t is constant, the electric field vector is along a helix on a cylinder of constant diameter, as shown in Fig. 2.8(b), and its period is equal to the wavelength.

2.4.3 Linear Polarisation

If $\phi=0$, (2.88) becomes

$$E_x/|A_x|=E_y/|A_y| \tag{2.91}$$

If $\phi=\pi$,

$$E_x/|A_x|=-E_y/|A_y|. \tag{2.92}$$

Therefore, when observed within the plane at a constant distance z, the electric field vector lies on a straight line as shown in Fig. 2.9(a). This type of polarisation is called linear polarisation or plane polarisation. If time t is constant, the electric field vector lies in a plane:

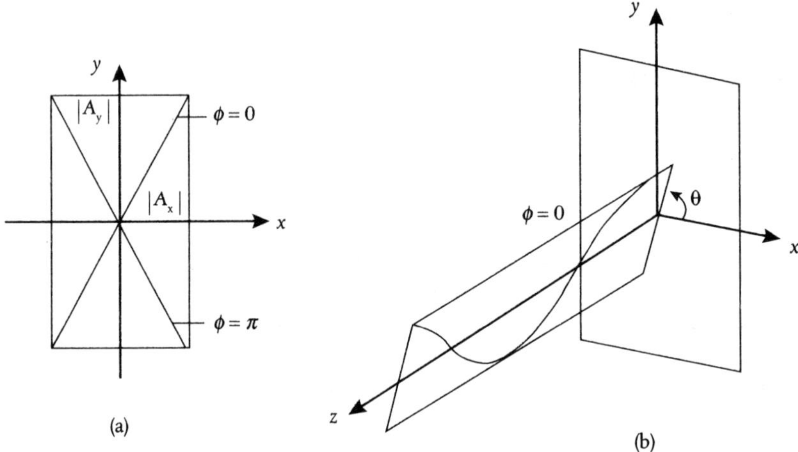

Fig. 2.9 Linear polarisation

$$\theta=\begin{cases}\tan^{-1}\left(|A_y|/|A_x|\right) & \text{when } \phi=0\\ \tan^{-1}\left(-|A_y|/|A_x|\right) & \text{when } \phi=\pi\end{cases} \tag{2.93}$$

as shown in Fig. 2.9(b).

2.5 REFLECTION AND REFRACTION

2.5.1 Basic Expressions

Consider a plane wave incident at an oblique angle on the interface between two dielectrics with different refractive indices n_1 and n_2 (as shown in Fig. 2.10) and let them contact each other at the boundary surface $y=0$.

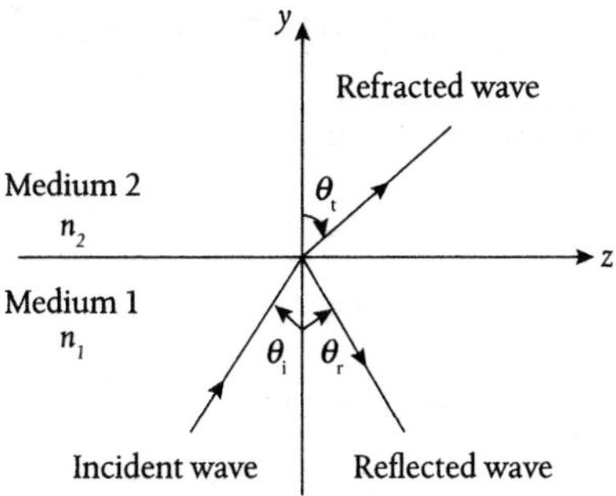

Fig. 2.10 Reflection and refraction of a plane wave at an interface between dielectrics

Now, the angles formed by the perpendicular to the boundary surface with the directions of incidence, reflection, and refraction are called the angle of incidence θ_i, the angle of reflection (θ_r), and the angle of refraction (θ_t). The wavenumber vectors of the incident wave, reflected wave, and refracted wave are, respectively,

$$k_i = \begin{bmatrix} 0 & k_0 n_1 \cos\theta_i & k_0 n_1 \sin\theta_i \end{bmatrix}^T \tag{2.94}$$

$$k_r = \begin{bmatrix} 0 & -k_0 n_1 \cos\theta_r & k_0 n_1 \sin\theta_r \end{bmatrix}^T \tag{2.95}$$

$$k_t = \begin{bmatrix} 0 & k_0 n_2 \cos\theta_t & k_0 n_1 \sin\theta_t \end{bmatrix}^T \tag{2.96}$$

where T represents the transpose. The plane that is perpendicular to the boundary surface and contains the wavenumber vector of the incident wave is called the plane of incidence.

Since, all the wavenumber vectors lie in the yz-plane, if $\partial/\partial x = 0$, then the following can be obtained from (2.55) and (2.56):

$$\partial E_x / \partial z = -j\omega\mu_0 H_y \tag{2.97a}$$

$$\partial E_x / \partial y = j\omega\mu_0 H_z \tag{2.97b}$$

$$\partial H_z / \partial y - \partial H_y / \partial z = j\omega\varepsilon_0 n^2 E_x \tag{2.97c}$$

$$\partial H_x / \partial z = j\omega\varepsilon_0 n^2 E_y \tag{2.98a}$$

$$\partial H_x / \partial y = -j\omega\varepsilon_0 n^2 E_z \tag{2.98b}$$

$$\partial E_z / \partial y - \partial E_y / \partial z = -j\omega\mu_0 H_x. \tag{2.98c}$$

Therefore, the plane wave with an electric field component (x component) perpendicular to the plane of incidence and the plane wave with electric field components (y and z components) parallel to the plane of incidence are mutually independent. They are called s-polarised light and p-polarised light, respectively.

From (2.97), the electromagnetic field of transverse electric (TE) polarised or s-polarised light in medium 1 is

$$E_{x1} = A_{s1} \exp(-jk_i \cdot r) + B_{s1} \tag{2.99a}$$

$$H_{y1} = \left[A_{s1} n_1 \sin\theta_i \exp(-jk_i \cdot r) + B_{s1} n_1 \sin\theta_r \exp(-jk_r \cdot r)\right] / Z_0 \tag{2.99b}$$

$$H_{z1} = \left[-A_{s1} n_1 \cos\theta_i \exp(-jk_i \cdot r) + B_{s1} n_1 \cos\theta_r \exp(-jk_r \cdot r)\right] / Z_0. \tag{2.99c}$$

In medium 2, it is

$$E_{x2} = A_{s2} \exp(-jk_t \cdot r) \tag{2.100a}$$

$$H_{y2} = \left[A_{s2} n_2 \sin\theta_t \exp(-jk_t \cdot r)\right] / Z_0 \tag{2.100b}$$

$$H_{z2} = \left[-A_{s2} n_2 \cos\theta_t \exp(-jk_t \cdot r)\right] / Z_0. \tag{2.100c}$$

From (2.98), the electromagnetic field of transverse magnetic (TM) polarised light in medium 1 is

$$H_{x1} = \left[A_{p1} n_1 \exp(-jk_i \cdot r) - B_{p1} n_1 \exp(-jk_r \cdot r)\right] / Z_0 \tag{2.101a}$$

$$E_{y1} = -A_{p1} n_1 \sin\theta_i \exp(-jk_i \cdot r) + B_{p1} \sin\theta_r \exp(-jk_r \cdot r) \tag{2.101b}$$

$$E_{z1} = A_{p1} \cos\theta_i \exp(-jk_i \cdot r) + B_{p1} \cos\theta_r \exp(-jk_r \cdot r). \tag{2.101c}$$

In medium 2, it is:

$$H_{x2} = \left[A_{p2} n_2 \exp(-jk_t \cdot r)\right] / Z_0 \tag{2.102a}$$

$$E_{y2} = -A_{p2} \sin\theta_t \exp(-jk_t \cdot r) \tag{2.102b}$$

$$E_{z2} = A_{p2} \cos\theta_t \exp(-jk_t \cdot r). \tag{2.102c}$$

2.5.2 Laws of Reflection and Refraction

The tangential components of the electromagnetic field $E_x, E_z, H_x, \text{and} H_z$ must be continuous at the interface $y = 0$. By $(2.94)-(2.96)$ and $(2.99)-(2.102)$, we know that the following must hold for this boundary condition to be met at any position z on the interface:

$$\theta_i = \theta_r \tag{2.103}$$

$$\sin\theta_i / \sin\theta_t = n_2 / n_1 \tag{2.104}$$

In other words, the angle of incidence and the angle of reflection are equal, and Snell's law equation (2.104) applies to the incidence and the angle of refraction.

2.5.3 Fresnel Formulas

If we take expressions (2.99) and (2.100) and consider the boundary conditions at the interface $y = 0$, along with expressions (2.103) and (2.104), the reflection coefficient R_s and the transmission coefficient T_s for TE polarised light can be calculated as follows:

$$R_s = \frac{B_{s1}}{A_{s1}} = \frac{\cos\theta_i - \sqrt{(n_2/n_1)^2 - \sin^2\theta_i}}{\cos\theta_i + \sqrt{(n_2/n_1)^2 - \sin^2\theta_i}} \tag{2.105}$$

$$T_s = \frac{A_{s2}}{A_{s1}} = \frac{2\cos\theta_i}{\cos\theta_i + \sqrt{(n_2/n_1)^2 - \sin^2\theta_i}}. \tag{2.106}$$

If we take expressions (2.101) and (2.102) and consider the boundary conditions at the interface $y=0$, along with expressions (2.103) and (2.104), the reflection coefficient R_p, and the transmission coefficient T_p for TM polarised light can be calculated as follows:

$$R_p = \frac{B_{p1}}{A_{p1}} = \frac{-(n_2/n_1)^2\cos\theta_i + \sqrt{(n_2/n_1)^2 - \sin^2\theta_i}}{(n_2/n_1)^2\cos\theta_i + \sqrt{(n_2/n_1)^2 - \sin^2\theta_i}} \tag{2.107}$$

$$T_p = \frac{A_{p2}}{A_{p1}} = \frac{2(n_2/n_1)\cos\theta_i}{(n_2/n_1)^2\cos\theta_i + \sqrt{(n_2/n_1)^2 - \sin^2\theta_i}}. \tag{2.108}$$

Expressions (2.105) – (2.106) are called the Fresnel formulas and $R_s, T_s, R_p,$ and T_p are called the Fresnel coefficients.

2.5.4 Total Reflection

If we calculate the time average values of the energy flow density using (2.64) in the $y-$direction for the incident wave, the reflected wave, and the refracted wave, defining angle θ_c as follows:

$$\theta_c = \sin^{-1}(n_2/n_1) \tag{2.109}$$

and if

$$\theta_i < \theta_c, \tag{2.110}$$

then for TE polarised light the reflection and transmission coefficients in terms of energy flow, r_s and t_s, are

$$r_s = R_s^2 \tag{2.111}$$

$$t_s = T_s^2\sqrt{\left((n_2/n_1)^2 - \sin^2\theta_i\right)}/\cos\theta_i \tag{2.112}$$

and for TM polarised light the reflection and transmission coefficients in terms of energy flow, r_p and t_p, are:

$$r_p = R_p^2 \tag{2.113}$$

$$t_p = T_p^2\sqrt{\left((n_2/n_1)^2 - \sin^2\theta_i\right)}/\cos\theta_i. \tag{2.114}$$

Further, if

$$\theta_i > \theta_c, \tag{2.115}$$

the following holds:

$$r_s = r_p = 1,\ t_s = t_p = 0. \tag{2.116}$$

The quantity θ_c in (2.109) is called the critical angle. If $n_1 < n_2$, that is, when the light is incident on the interface in the medium of lower refractive index, no critical angle exists; the reflection and transmission coefficients for energy flow are given in $(2.111)-(2.114)$. On the other hand, if $n_1 > n_2$, that is, when the light is incident on the interface in the medium of higher refractive index, and the angle of incidence is larger than the critical angle, there is no transmitted light since all light is reflected. This is called total internal reflection.

If $k_0 n_1 \sin\theta_i$, which is the phase constant in the z-direction, is represented by β, the reflection coefficient R_s for TE polarised light is

$$R_s = \exp(j2\theta_s) \tag{2.117a}$$

$$\phi_s = \tan^{-1}\left[\sqrt{\sin^2\theta_i - (n_2/n_1)^2}\,/\cos\theta_i\right] \tag{2.117b}$$

$$\phi_s = \tan^{-1}\left[\sqrt{\beta^2 - k_0^2 n_2^2}\,/\sqrt{k_0^2 n_1^2 - \beta^2}\right] \tag{2.117c}$$

and the reflection coefficient R_p for TM polarised light is

$$R_p = exp(j2\theta_p) \tag{2.118a}$$

$$\phi_p = \tan^{-1}\left[(n_1/n_2)^2\sqrt{\sin^2\theta_i - (n_2/n_1)^2}\,/\cos\theta_i\right] \tag{2.118b}$$

$$\phi_p = \tan^{-1}\left[n_1^2\sqrt{\beta^2 - k_0^2 n_2^2}\,/n_2^2\sqrt{k_0^2 n_1^2 - \beta^2}\right]. \tag{2.118c}$$

With total reflection, 100% of the light can be reflected. Optical waveguides guide light via multiple total reflections.

A detailed wave theoretic examination of the total reflection phenomenon reveals the existence of a so-called evanescent wave, which slightly penetrates medium 2 and then rapidly attenuates, as shown in Fig. 2.11. Therefore, part of the incident light penetrates a short distance y_{GH} from the interface and then it is returned to medium 1.

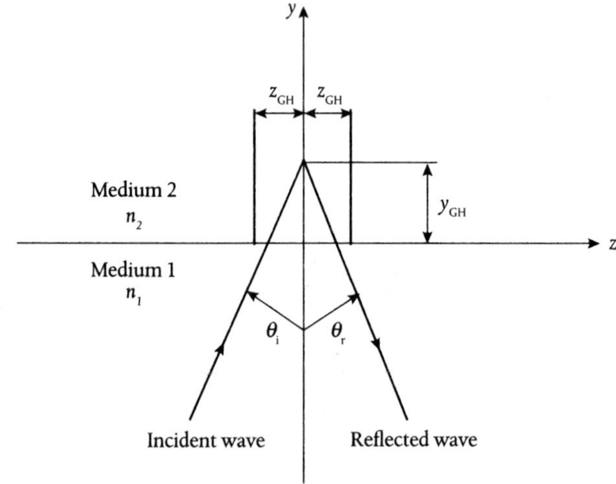

Fig. 2.11 Goos–Hánchen shift

In other words, a shift of reflection point by $2z_{GH}$ in the z-direction is observed. This shift of reflection point is called the Goos–Hánchen shift.

From (2.117c), the amount of Goos–Hánchen shift for TE polarised light is:

$$z_{GH} = \frac{d\phi_s}{d\beta} = \frac{\tan\theta_i}{k_0\sqrt{(\beta/k_0)^2 - n_2^2}}. \tag{2.119}$$

and from (2.118c), the amount of Goos–Hánchen shift for TM polarised light is

$$z_{GH} = \frac{d\phi_p}{d\beta} = \frac{\tan\theta_i}{k_0\sqrt{(\beta/k_0)^2 - n_2^2}\left[(\beta/k_0 n_1)^2 + (\beta/k_0 n_2)^2 - 1\right]}. \tag{2.120}$$

The depth of the equivalent position of reflection y_{GH} is

$$y_{GH} = z_{GH}/\tan\theta_i. \tag{2.121}$$

In multipath reflection, the above-mentioned shift may accumulate, creating a need to take it into consideration.

2.5.5 Brewster Angle

For TM polarised light, if the angle of incidence is equal to θ_B, this is defined as

$$\theta_B = \tan^{-1}(n_2/n_1). \tag{2.122}$$

The reflection coefficient R_p is 0 (2.107) and the light passes through the interface without reflection. This angle of incidence θ_B is called the Brewster angle and is used for the discrimination of polarised light and in non-reflective windows.

2.6 WAVE EQUATIONS FOR OPTICAL WAVEGUIDES

2.6.1 Basic Expressions for Optical Waveguides

Consider light with angular frequency ω and phase constant β propagated through a waveguide whose geometry and refractive index are constant in the direction of propagation. If the electric field $\tilde{E}$ and magnetic field $\tilde{H}$ are indicated by phasors as follows:

$$E(r,t) = \mathrm{Re}\{\tilde{E}(r_t)\exp[j(\omega t - \beta z)]\} \tag{2.123}$$

$$H(r,t) = \mathrm{Re}\{\tilde{H}(r_t)\exp[j(\omega t - \beta z)]\} \tag{2.124}$$

and if they are represented as E and H as functions of cross-sectional position r_t in the waveguide, then Maxwell's equations become

$$\nabla \times E = -j\omega\mu_0 H \tag{2.125}$$

$$\nabla \times H = j\omega\varepsilon_0 n^2 E \tag{2.126}$$

$$\nabla \cdot H = 0 \tag{2.127}$$

$$\nabla \cdot E + \frac{\nabla_t (n^2)}{n^2} \cdot E = 0. \tag{2.128}$$

Here, the differential operators $\nabla\times$, $\nabla\cdot$, and ∇_t are given in rectangular coordinates as follows:

$$\nabla\times = \begin{bmatrix} 0 & j\beta & \partial/\partial y \\ -j\beta & 0 & -\partial/\partial x \\ -\partial/\partial y & \partial/\partial x & 0 \end{bmatrix} \tag{2.129}$$

$$\nabla\cdot = [\partial/\partial x \quad \partial/\partial y \quad -j\beta] \tag{2.130}$$

$$\nabla_t = [\partial/\partial x \quad \partial/\partial y \quad 0]^T. \tag{2.131}$$

In cylindrical coordinates, they are

$$\nabla\times = \begin{bmatrix} 0 & j\beta & (1/r)\partial/\partial\theta \\ -j\beta & 0 & -\partial/\partial r \\ -(1/r)\partial/\partial y & \partial/\partial r + 1/r & 0 \end{bmatrix} \tag{2.132}$$

$$\nabla\cdot = [\partial/\partial r + 1/r \quad (1/r)\partial/\partial\theta \quad -j\beta] \tag{2.133}$$

$$\nabla_t = [\partial/\partial r \quad (1/r)\partial/\partial\theta \quad 0]^T. \tag{2.134}$$

For an anisotropic dielectric optical waveguide,

$$\nabla \times H = j\omega[\varepsilon]E = j\omega\varepsilon_0[\varepsilon_r]E \tag{2.135}$$

$$\nabla \cdot ([\varepsilon]E) = \varepsilon_0 \nabla \cdot ([\varepsilon_r]E) = 0 \tag{2.136}$$

are used instead of expressions (2.126) and (2.128). Here, $[\varepsilon]$ is the permittivity tensor and $[\varepsilon_r]$ is the relative permittivity tensor.

2.6.2 Wave Equations for Graded Optical Waveguides

Since, we have assumed that the refractive index is constant in the direction of propagation, if we note that $\nabla\varepsilon = \nabla_t\varepsilon$, the wave equations in (2.47) and (2.48) become

$$\nabla_t^2 E + (k_0^2 n^2 - \beta^2)E + \nabla\left[\frac{\nabla_t (n^2)}{n^2} \cdot E\right] = 0 \tag{2.137}$$

$$\nabla_t^2 H + (k_0^2 n^2 - \beta^2)H + \frac{\nabla_t (n^2)}{n^2} \times (\nabla \times H) = 0 \tag{2.138}$$

If we define Δ_t as

$$\Delta_t = \partial^2/\partial x^2 + \partial^2/\partial y^2, \tag{2.139}$$

then differential operators ∇ and ∇_t^2 become

$$\nabla = \begin{bmatrix}\partial/\partial x & \partial/\partial y & -j\beta\end{bmatrix}^T \tag{2.140}$$

$$\nabla_t^2 = \begin{bmatrix}\Delta_t & 0 & 0\\ 0 & \Delta_t & 0\\ 0 & 0 & \Delta_t\end{bmatrix} \tag{2.141}$$

in a rectangular coordinate system. In a cylindrical coordinate system, if we define Δ_t as

$$\Delta_t = \partial^2/\partial r^2 + (1/r)\partial/\partial r + (1/r^2)\partial^2/\partial\theta^2, \tag{2.142}$$

then ∇ and ∇_t^2 become

$$\nabla = \begin{bmatrix}\partial/\partial r & (1/r)\partial/\partial\theta & -j\beta\end{bmatrix}^T \tag{2.143}$$

$$\nabla_t^2 = \begin{bmatrix}\Delta_t - 1/r^2 & -(2/r^2)\partial/\partial\theta & 0\\ (2/r^2)\partial/\partial\theta & \Delta_t - 1/r^2 & 0\\ 0 & 0 & \Delta_t\end{bmatrix}. \tag{2.144}$$

2.6.3 Wave Equations for Stepped Optical Waveguides

If the refractive index is constant, (2.137) and (2.138) become

$$\nabla_t^2 E + (k_0^2 n^2 - \beta^2)E = 0 \tag{2.145}$$

$$\nabla_t^2 H + (k_0^2 n^2 - \beta^2)H = 0 \tag{2.146}$$

and (2.128) becomes

$$\nabla \cdot E = 0. \tag{2.147}$$

2.6.4 Ray Equations

If the curvilinear coordinate along the path of the ray is s and the position vector representing the position on the ray is r, the unit vector s in the direction (L) of the ray is:

$$s = \frac{dr}{ds} = \frac{\nabla L}{|\nabla L|} = \frac{\nabla L}{n}. \tag{2.148}$$

If (2.148) is multiplied by the refractive index n and differentiated with respect to s, then

$$\frac{d}{ds}\left(n\frac{dr}{ds}\right) = \frac{d}{ds}\nabla L = \left(\frac{dr}{ds}\cdot\nabla\right)\nabla L = \frac{1}{n}(\nabla L\cdot\nabla)\nabla L. \tag{2.149a}$$

Since, from literature,

$$|\nabla L|^2 = n^2. \tag{2.149b}$$

Further, from (2.149b), we get

$$\nabla|\nabla L|^2 = 2(\nabla L\cdot\nabla)\nabla L = \nabla n^2 = 2n\nabla n, \tag{2.150}$$

and therefore,

$$\frac{d}{ds}\left(n\frac{dr}{ds}\right) = \nabla n. \tag{2.151}$$

Equation (2.151) is called the ray equation and along with the eikonal equation, it is a basic equation that defines geometrical optics. In a homogeneous medium, the refractive index n is constant, and therefore, (2.151) becomes

$$d^2r / ds^2 = 0. \tag{2.152}$$

If a and b are arbitrary constant vectors, the solution of (2.151) is:

$$r = as + b. \tag{2.153}$$

In a homogeneous medium, this means that the ray is a straight line. When the ray is paraxial, that is, if the inclination of the ray to the coordinate axis z, or the optical axis is sufficiently small,

$$\frac{d}{dz}\left(n\frac{dr}{dz}\right) = \nabla n. \tag{2.154}$$

This is called the paraxial ray equation. A ray has the following characteristics:

1. It has a tendency to turn toward a larger refractive index.
2. The propagation time of light between wave fronts is constant.
3. The propagation time is shortest along the ray.

2.7 PRINCIPLE OF OPTICAL WAVEGUIDE

In the three-layer dielectric optical waveguide shown in Fig. 2.12, assume that z is the direction of light propagation and the waveguide structure are homogeneous in the z-direction. Further, assume that the relationship among refractive indexes n_f, n_s, and n_c is $n_c \le n_s < n_f$.

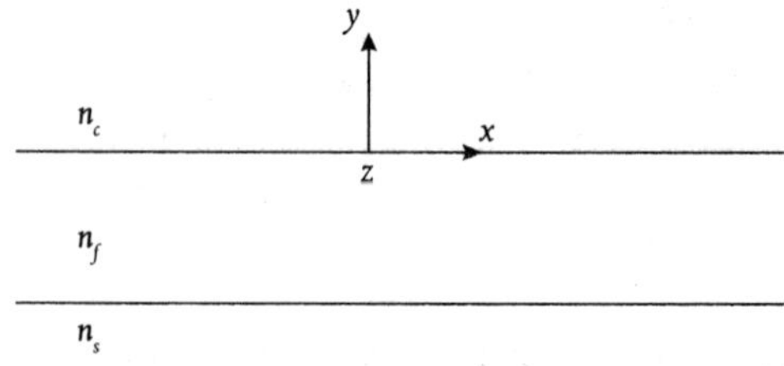

Fig. 2.12 Three-layer dielectric waveguide

Now, if the plane wave given in (2.97) or (2.98) enters the top and bottom interfaces at incident angle θ, the phase constant β in the direction of propagation is:

$$\beta = k_0 n_f \sin\theta, \tag{2.155}$$

where n_{eff}

$$n_{\text{eff}} \equiv \beta / k_0 = n_f \sin\theta \tag{2.156}$$

is called the effective refractive index (or equivalent refractive index formula). If the critical angles at the interfaces, $y = -t$ and $y = 0$ in (2.109), are, respectively,

$$\theta_s = \sin^{-1}\left(n_s / n_f\right) \tag{2.157}$$

$$\theta_c = \sin^{-1}\left(n_c / n_f\right), \tag{2.158}$$

then $\theta_c \le \theta_s$ based on the assumption that $n_c \le n_s < n_f$. Therefore, when $\theta_s < \theta$, total reflection occurs at both interfaces, resulting in the light being transmitted while trapped in the thin film, as shown in Fig 2.13(a). This mode is called the guided, bound, or trapped mode. Here, $n_s \le n_{\text{eff}} \le n_f$.

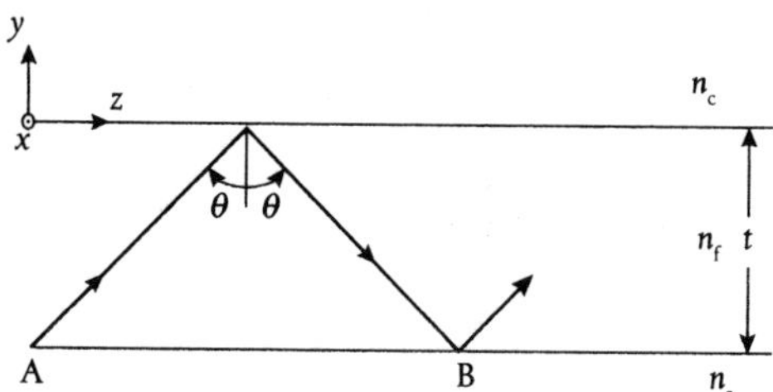

Fig. 2.13 (a) Path of ray in guided mode

On the other hand, when θ is smaller than θ_s and $\theta_c < \theta < \theta_s$, the conditions for total reflection are not met at the lower interface and part of the light is radiated into the substrate, as shown in Fig. 2.13(b). This mode is called the substrate radiation mode; here, $n_c \le n_{\text{eff}} < n_s$.

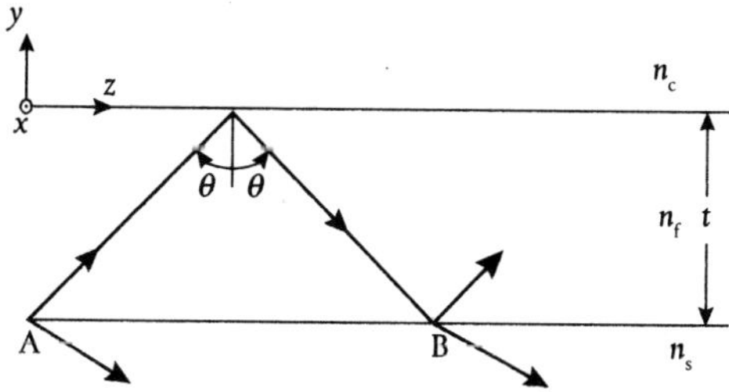

Fig. 2.13 (b) Path of ray in substrate radiation mode

When θ becomes even smaller and $\theta < \theta_c$, total reflection conditions are not met at either the upper or lower interface and the light is radiated at both sides, as shown in Fig. 2.13(c). This mode is called the substrate cladding radiation mode $n_{\text{eff}} \le n_c$.

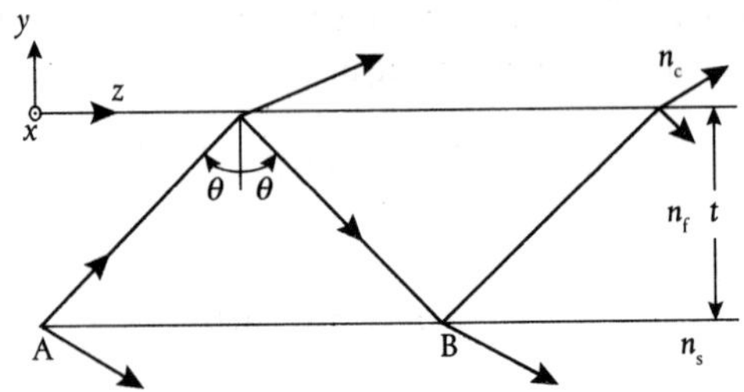

Fig. 2.13 (c) Path of ray in substrate cladding radiation mode

The following paragraphs describe how to obtain the eigenvalue equation (also called the characteristic equation or dispersion equation) for the guided mode using geometrical optics. It has already been stated that (2.97) and (2.98) are obtained from Maxwell's equations, if we assume an electromagnetic distribution, which is homogeneous in x-direction with $\partial/\partial x = 0$. The electromagnetic field in (2.97) has no electric field component E_z (longitudinal electric field), and the field in (2.98) has no magnetic field component H_z (longitudinal magnetic field). Therefore, the former is called a TE wave or the TE mode, and the latter, a TM wave or the TM mode.

If we define m_f, m_s and m_c such that

$$m_f = m_s = m_c = 1 \tag{2.159}$$

in the TE mode, and

$$m_f = 1/n_f^2, \quad m_s = 1/n_s^2, \quad m_c = 1/n_c^2 \tag{2.160}$$

in the TM mode, the phase shift due to total reflection is

$$2\phi_s = 2\tan^{-1}\left(m_s\alpha_s / m_f k_f\right) \tag{2.161}$$

at the interface between the thin film and the substrate. By (2.117) and (2.118) and

$$2\phi_c = 2\tan^{-1}\left(m_c\alpha_c / m_f k_f\right) \tag{2.162}$$

at the interface between the thin film and the upper cladding. Here, k_f, which corresponds to the phase constant in the y-direction, and α_s and α_c, which correspond to attenuation constants are

$$k_f = \sqrt{k_0^2 n_f^{\,2} - \beta^2} \tag{2.163}$$

$$\alpha_s = \sqrt{\beta^2 - k_0^2 n_s^{\,2}} \tag{2.164}$$

$$\alpha_c = \sqrt{\beta^2 - k_0^2 n_c^{\,2}}. \tag{2.165}$$

The condition for the existence of a guided mode is that the total phase shift in the y-direction during one cycle (from point A to point B in Fig. 2.13(a)) must be equal to an integral multiple of 2π. Therefore, this condition can be expressed as follows:

$$2k_f t - 2\phi_s - 2\phi_c = 2q\pi, \quad q = 0, 1, 2, \ldots \tag{2.166}$$

If we substitute (2.161) and (2.162) in (2.166), the eigenvalue equation is:

$$k_f t = \tan^{-1}\frac{m_s \alpha_s}{m_f k_f} + \tan^{-1}\frac{m_c \alpha_c}{m_f k_f} + q\pi. \tag{2.167}$$

Phase constant β of a guided mode at a given angular frequency ω can be calculated from the above-mentioned eigenvalue equation. Here, only a limited number of discrete values are allowed for β. In other words, the angle of incidence θ is discrete spectrum. In the radiation mode, on the other hand, the angle of incidence θ and phase constant β are continuous quantities. The radiation mode, therefore, is also called the continuous mode or the continuous spectrum. Since such a guided mode and radiation mode form a complete orthogonal system, the total power flux on the waveguide cross section is calculated as the sum of the powers for each mode.

The cut-off frequency f_{cutoff} for the guided mode is obtained as follows from (2.167) by considering the condition in which the angle of incidence θ is equal to the critical angle θ_s at the interface between the thin film and the substrate, that is, $\beta = k_0 n_s$:

$$f_{cutoff} = \frac{c}{2\pi t\sqrt{n_f^2 - n_s^2}}\left[\tan^{-1}\left(\frac{m_c\sqrt{n_s^2 - n_c^2}}{m_f\sqrt{n_f^2 - n_s^2}}\right) + q\pi\right]. \tag{2.168}$$

2.7.1 Uniform Core Optical Fibre

As shown in Fig. 2.14, there are two types of ray in a uniform core optical fibre with a refractive index distribution: meridional rays, which lie in a plane including the central axis of the fibre and intersect the central axis twice during one cycle, and skew rays, which do not intersect the central axis of the fibre. Here, we concentrate on meridional rays.

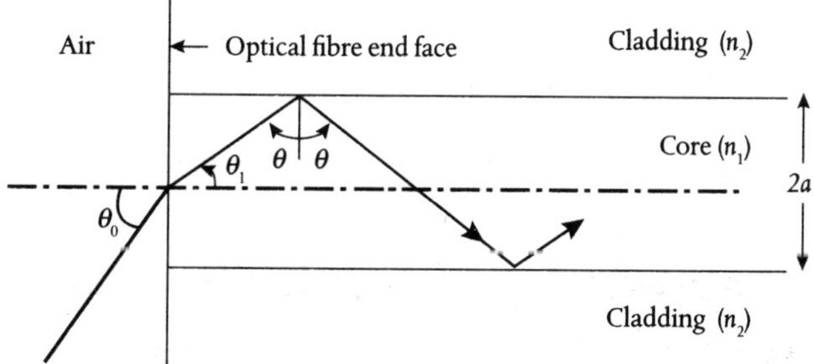

Fig. 2.14 Meridional ray and its incidence conditions

As shown in Fig. 2.14, the light that enters the fibre end face at an angle θ_0 to the central axis propagates inside the fibre in the direction of the angle θ_1 given in Snell's law:

$$\sin\theta_1 / \sin\theta_0 = 1/n_1. \tag{2.169}$$

When we take into consideration the total reflection condition at the core–cladding interface:

$$\sin^{-1}(n_2/n_1) = \theta_c < \theta. \tag{2.170}$$

The condition for the incident light to propagate as a meridional ray is:

$$\sin\theta_0 < \sqrt{n_1^2 - n_2^2}. \quad (2.171)$$

The angle of incidence $\theta_0 = \theta_{0,max}$ for which $\theta_0 = \theta_c$ is called the maximum acceptance angle. If we use the relative refractive index, difference Δ given as:

$$\Delta = \frac{n_1^2 - n_2^2}{2n_1^2} \simeq \frac{n_1 - n_2}{n_1} \quad (2.172a)$$

$$\sin\theta_{0,\ max} = n_1\sqrt{2\Delta} \equiv \text{NA}. \quad (2.172b)$$

This is called the numerical aperture (NA) of the optical fibre. For an efficient transmission of a ray into an optical fibre, a lens with numerical aperture of the greater or smaller value should be used. If $\theta_0 = \theta_{0,\max}$ and $\theta_1 = \theta_{1,\ \max}$,

$$\sin\theta_{1,\ \max} = \sqrt{2\Delta}. \quad (2.173)$$

Since different guided modes correspond to rays with different angles θ as shown in Fig. 2.14, the velocity of rays in the axial direction also differs according to the mode. In other words, since the time required for signal transmission differs according to the mode, if a light pulse enters at the transmitter end of the optical fibre and is then transmitted in many different modes, the pulse at the output end is spread out in time. This phenomenon is called multimode dispersion. If we evaluate such multimode dispersion as the spread in signal delay time generated when the pulse travels a unit length of the optical fibre, its value is the time difference between the rays with minimum required time $(\theta_1 = 0)$ and the ray with maximum required time $(\theta_1 = \theta_{1,max})$. If these required times are $t_0 = n_1/c$ and t_{max}, respectively, the difference τ is

$$\tau = t_{\max} - t_0 = t_0\left(\sec\theta_{1,\ max} - 1\right) \quad (2.174)$$

$$\simeq (t_0/2)\sin^2\theta_{1,\ \max} = t_0\Delta = (n_1/c)\Delta.$$

2.7.2 Wave Theoretic Analysis of Uniform Core Optical Fibre: Guided Mode

In a uniform core optical fibre with the refractive index distribution shown in Fig. 2.15, let us assume that the refractive indices for core and cladding are n_1 and n_2, respectively, and the core radius is a.

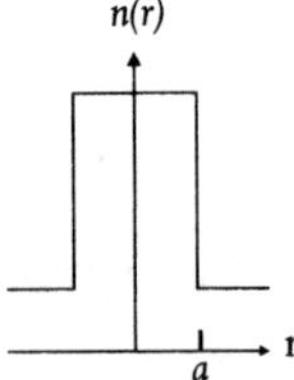

Fig. 2.15 Stepped-index optical fibre

The wave equations for the electromagnetic field components in the direction of propagation (longitudinal electromagnetic fields), E_z and H_z are, from (2.145) and (2.146), we get:

$$\frac{\partial^2 E_z}{\partial r^2}+\frac{1}{r}.\frac{\partial E_z}{\partial r}+\frac{1}{r^2}.\frac{\partial^2 E_z}{\partial \theta^2}+k_t^2 E_z=0 \tag{2.175}$$

$$\frac{\partial^2 H_z}{\partial r^2}+\frac{1}{r}.\frac{\partial H_z}{\partial r}+\frac{1}{r^2}.\frac{\partial^2 H_z}{\partial \theta^2}+k_t^2 H_z=0. \tag{2.176}$$

Here, k_t is called the lateral phase constant and it is given as:

$$k_t=\sqrt{k_0^2 n^2-\beta^2}. \tag{2.177}$$

The lateral electromagnetic fields E_r, E_θ, H_r, and H_θ can be expressed in terms of the longitudinal electric and magnetic fields alone:

$$E_r=-j\frac{1}{k_t^2}\left(\beta\frac{\partial E_z}{\partial r}+\frac{\omega\mu_0}{r}.\frac{\partial H_z}{\partial \theta}\right) \tag{2.178}$$

$$E_\theta=-j\frac{1}{k_t^2}\left(\frac{\beta}{r}.\frac{\partial E_z}{\partial \theta}-\omega\mu_0\frac{\partial H_z}{\partial r}\right) \tag{2.179}$$

$$H_r=-j\frac{1}{k_t^2}\left(\beta\frac{\partial H_z}{\partial r}-\frac{\omega\varepsilon_0 n^2}{r}.\frac{\partial E_z}{\partial \theta}\right) \tag{2.180}$$

$$H_\theta=-j\frac{1}{k_t^2}\left(\frac{\beta}{r}.\frac{\partial H_z}{\partial \theta}+\omega\varepsilon_0 n^2\frac{\partial E_z}{\partial r}\right). \tag{2.181}$$

Equations (2.175) and (2.176) can be solved by separation of variables. If we apply (2.178)–(2.181), the electromagnetic fields in the core are:

$$E_z=A_1 J_p\left(\frac{ur}{a}\right)\begin{Bmatrix}\cos p\theta\\ \sin p\theta\end{Bmatrix} \tag{2.182}$$

$$E_r=j\left[-A_1\frac{\beta a}{u}J_p'\left(\frac{ur}{a}\right)+B_1\frac{p\omega\mu_0 a^2}{u^2 r}J_p\left(\frac{ur}{a}\right)\right]\begin{Bmatrix}\cos p\theta\\ \sin p\theta\end{Bmatrix} \tag{2.183}$$

$$E_\theta=j\left[-A_1\frac{p\beta a^2}{u^2 r}J_p\left(\frac{ur}{a}\right)+B_1\frac{\omega\mu_0 a}{u}J_p'\left(\frac{ur}{a}\right)\right]\begin{Bmatrix}-\sin p\theta\\ \cos p\theta\end{Bmatrix} \tag{2.184}$$

$$H_z=B_1 J_p\left(\frac{ur}{a}\right)\begin{Bmatrix}-\sin p\theta\\ \cos p\theta\end{Bmatrix} \tag{2.185}$$

$$H_r=j\left[A_1\frac{p\omega\varepsilon_0 n_1^2 a^2}{u^2 r}J_p\left(\frac{ur}{a}\right)-B_1\frac{\beta a}{u}J_p'\left(\frac{ur}{a}\right)\right]\begin{Bmatrix}-\sin p\theta\\ \cos p\theta\end{Bmatrix} \tag{2.186}$$

$$H_\theta=j\left[-A_1\frac{\omega\varepsilon_0 n_1^2 a}{u}J_p'\left(\frac{ur}{a}\right)+B_1\frac{p\beta a^2}{u^2 r}J_p\left(\frac{ur}{a}\right)\right]\begin{Bmatrix}\cos p\theta\\ \sin p\theta\end{Bmatrix} \tag{2.187}$$

and the electromagnetic fields in the cladding are:

$$E_z=j\left[A_2 K_p\left(\frac{wr}{a}\right)\right]\begin{Bmatrix}\cos p\theta\\ \sin p\theta\end{Bmatrix} \tag{2.188}$$

$$E_r = j\left[A_2\frac{\beta a}{w}K_p'\left(\frac{wr}{a}\right) - B_2\frac{p\omega\mu_0 a^2}{w^2 r}K_p\left(\frac{wr}{a}\right)\right]\begin{Bmatrix}\cos p\theta\\ \sin p\theta\end{Bmatrix} \tag{2.189}$$

$$E_\theta = j\left[A_2\frac{p\beta a^2}{w^2 r}K_p\left(\frac{wr}{a}\right) - B_2\frac{\omega\mu_0 a}{w}K_p'\left(\frac{wr}{a}\right)\right]\begin{Bmatrix}-\sin p\theta\\ \cos p\theta\end{Bmatrix} \tag{2.190}$$

$$H_z = \left[B_2 K_p\left(\frac{wr}{a}\right)\right]\begin{Bmatrix}-\sin p\theta\\ \cos p\theta\end{Bmatrix} \tag{2.191}$$

$$H_r = j\left[-A_2\frac{p\omega\varepsilon_0 n_2^2 a^2}{w^2 r}K_p\left(\frac{wr}{a}\right) + B_2\frac{\beta a}{w}K_p'\left(\frac{wr}{a}\right)\right]\begin{Bmatrix}-\sin p\theta\\ \cos p\theta\end{Bmatrix} \tag{2.192}$$

$$H_\theta = j\left[A_2\frac{\omega\varepsilon_0 n_2^2 a^2}{w}K_p'\left(\frac{wr}{a}\right) - B_2\frac{p\beta a^2}{w^2 r}K_p\left(\frac{wr}{a}\right)\right]\begin{Bmatrix}\cos p\theta\\ \sin p\theta\end{Bmatrix}. \tag{2.193}$$

Here, J_p and K_p are Bessel functions of the first and the second kinds, respectively. $J_P'(ur/a)$ denotes the derivative of $J_p(ur/a)$ with respect to (ur/a); $K_P'(wr/a)$, the derivative of $K_p(wr/a)$ with respect to (wr/a). u and w, which are called the normalised lateral phase constant and normalised lateral attenuation constant, respectively, are obtained as follows:

$$u = k_{t1}a = \left(\sqrt{k_0^2 n_1^2 - \beta^2}\right)a \tag{2.194}$$

$$w = |k_{t2}|a = \left(\sqrt{\beta^2 - k_0^2 n_2^2}\right)a. \tag{2.195}$$

Furthermore, v is expressed in terms of u and w as follows:

$$v = \sqrt{u^2 + w^2} = k_0 n_1 a\sqrt{2\Delta} \tag{2.196}$$

is called the normalised frequency.

From the continuity conditions for the tangential electromagnetic field components at the core–cladding interface $r = a$, the following can be obtained:

$$\begin{bmatrix} J_p(u) & -K_p(w) & 0 & 0\\ 0 & 0 & J_p(u) & -K_p(w)\\ \frac{p\beta}{u^2}J_p(u) & \frac{p\beta}{w^2}J_p(u) & -\frac{\omega\mu_0}{u}J_P'(u) & -\frac{\omega\mu_0}{w}K_P'(w)\\ \frac{\omega\varepsilon_0 n_1^2}{u}J_P'(u) & \frac{\omega\varepsilon_0 n_2^2}{w}K_P'(w) & -\frac{p\beta}{u^2}J_p(u) & -\frac{p\beta}{w^2}K_p(w)\end{bmatrix}\bullet\begin{bmatrix}A_1\\ A_2\\ B_1\\ B_2\end{bmatrix} = 0. \tag{2.197}$$

Finally, from the conditions for a non-trivial solution to exist in (2.197), the eigenvalue equation for the guided mode is obtained as follows:

$$\left[\frac{J_P'(u)}{uJ_p(u)} + \frac{K_P'(w)}{wK_p(w)}\right]\left[\frac{n_1^2}{n_2^2}\cdot\frac{J_P'(u)}{uJ_p(u)} + \frac{K_P'(w)}{wK_p(w)}\right] = p^2\left(\frac{1}{u^2} + \frac{1}{w^2}\right)\left(\frac{n_1^2}{n_2^2}\cdot\frac{1}{u^2} + \frac{1}{w^2}\right). \tag{2.198}$$

As shown in (2.182)–(2.193) and (2.197), the TM mode, in which $E_\theta = H_r = H_z = 0$, and the TE mode, in which $H_\theta = E_r = E_z = 0$, are separated when $p = 0$, and the eigenvalue equations for the respective modes are as follows:

$$\frac{n_1^2}{n_2^2}\cdot\frac{J_0'(u)}{uJ_0(u)}+\frac{K_0'(w)}{wK_0(w)}=0 \tag{2.199}$$

$$\frac{J_0'(u)}{uJ_0(u)}+\frac{K_0'(w)}{wK_0(w)}=0. \tag{2.200}$$

When $H_\theta \neq 0$, it follows that $E_z \neq 0$ and $H_z \neq 0$, so that TE and TM modes do not exist, both modes being hybrid modes. With (2.197) and the Bessel functions formulas, we get

$$J_p'\left(\frac{ur}{a}\right)=\frac{1}{2}\left[J_{p-1}\left(\frac{ur}{a}\right)-J_{p+1}\left(\frac{ur}{a}\right)\right]$$

$$\frac{pa}{ur}J_p\left(\frac{ur}{a}\right)=\frac{1}{2}\left[J_{p-1}\left(\frac{ur}{a}\right)+J_{p+1}\left(\frac{ur}{a}\right)\right].$$

The electromagnetic fields in the core can be expressed as follows:

$$E_z=A_1J_p\left(\frac{ur}{a}\right)\begin{Bmatrix}\cos p\theta\\ \sin p\theta\end{Bmatrix} \tag{2.201}$$

$$E_r=-j\frac{\beta a}{u}A_1\left[\frac{1-P}{2}J_{p-1}\left(\frac{ur}{a}\right)-\frac{1+P}{2}J_{p+1}\left(\frac{ur}{a}\right)\right]\begin{Bmatrix}\cos p\theta\\ \sin p\theta\end{Bmatrix} \tag{2.202}$$

$$E_\theta=-j\frac{\beta a}{u}A_1\left[\frac{1-P}{2}J_{p-1}\left(\frac{ur}{a}\right)+\frac{1+P}{2}J_{p+1}\left(\frac{ur}{a}\right)\right]\begin{Bmatrix}-\sin p\theta\\ \cos p\theta\end{Bmatrix} \tag{2.203}$$

$$H_z=\frac{\beta P}{\omega\mu_0}A_1J_p\left(\frac{ur}{a}\right)\begin{Bmatrix}-\sin p\theta\\ \cos p\theta\end{Bmatrix} \tag{2.204}$$

$$H_r=j\frac{\beta^2 a}{\omega\mu_0 u}A_1\left[\frac{(k_0n_1/\beta)^2-P}{2}J_{p-1}\left(\frac{ur}{a}\right)+\frac{(k_0n_1/\beta)^2+P}{2}J_{p+1}\left(\frac{ur}{a}\right)\right]\begin{Bmatrix}-\sin p\theta\\ \cos p\theta\end{Bmatrix} \tag{2.205}$$

$$H_\theta=-j\frac{\beta^2 a}{\omega\mu_0 u}A_1\left[\frac{(k_0n_1/\beta)^2-P}{2}J_{p-1}\left(\frac{ur}{a}\right)-\frac{(k_0n_1/\beta)^2+P}{2}J_{p+1}\left(\frac{ur}{a}\right)\right]\begin{Bmatrix}\cos p\theta\\ \sin p\theta\end{Bmatrix}. \tag{2.206}$$

Here, P is

$$P=\frac{p\left(\frac{1}{u^2}+\frac{1}{w^2}\right)}{\frac{J_p'(u)}{uJ_p(u)}+\frac{K_p'(w)}{wK_p(w)}} \tag{2.207}$$

$$K_p'\left(\frac{wr}{a}\right)=-\frac{1}{2}\left[K_{p-1}\left(\frac{wr}{a}\right)+K_{p+1}\left(\frac{wr}{a}\right)\right]$$

$$\frac{pa}{wr}K_p\left(\frac{wr}{a}\right)=-\frac{1}{2}\left[K_{p-1}\left(\frac{wr}{a}\right)-K_{p+1}\left(\frac{wr}{a}\right)\right].$$

The electromagnetic fields in the cladding can be expressed as follows:

$$E_z = \frac{J_p(u)}{K_p(w)} A_1 K_p\left(\frac{wr}{a}\right)\begin{Bmatrix}\cos p\theta \\ \sin p\theta\end{Bmatrix} \tag{2.208}$$

$$E_r = -j\frac{\beta a}{w}\cdot\frac{J_p(u)}{K_p(w)} A_1\left[\frac{1-P}{2}K_{p-1}\left(\frac{wr}{a}\right)+\frac{1+P}{2}K_{p+1}\left(\frac{wr}{a}\right)\right]\begin{Bmatrix}\cos p\theta \\ \sin p\theta\end{Bmatrix} \tag{2.209}$$

$$E_\theta = -j\frac{\beta a}{w}\cdot\frac{J_p(u)}{K_p(w)} A_1\left[\frac{1-P}{2}K_{p-1}\left(\frac{wr}{a}\right)-\frac{1+P}{2}K_{p+1}\left(\frac{wr}{a}\right)\right]\begin{Bmatrix}-\sin p\theta \\ \cos p\theta\end{Bmatrix} \tag{2.210}$$

$$H_z = \frac{\beta P}{\omega\mu_0}\cdot\frac{J_p(u)}{K_p(w)} A_1 K_p\left(\frac{wr}{a}\right)\begin{Bmatrix}-\sin p\theta \\ \cos p\theta\end{Bmatrix} \tag{2.211}$$

$$H_r = j\frac{\beta^2 a}{\omega\mu_0 w}\cdot\frac{J_p(u)}{K_p(w)} A_1\left[\frac{(k_0 n_2/\beta)^2 - P}{2}K_{p-1}\left(\frac{wr}{a}\right)-\frac{(k_0 n_2/\beta)^2 + P}{2}K_{p+1}\left(\frac{wr}{a}\right)\right]\begin{Bmatrix}-\sin p\theta \\ \cos p\theta\end{Bmatrix} \tag{2.212}$$

$$H_\theta = -j\frac{\beta^2 a}{\omega\mu_0 w}\cdot\frac{J_p(u)}{K_p(w)} A_1\left[\frac{(k_0 n_2/\beta)^2 - P}{2}K_{p-1}\left(\frac{wr}{a}\right)+\frac{(k_0 n_2/\beta)^2 + P}{2}K_{p+1}\left(\frac{wr}{a}\right)\right]\begin{Bmatrix}\cos p\theta \\ \sin p\theta\end{Bmatrix}. \tag{2.213}$$

2.7.3 Weakly Guiding Approximation

If the relative refractive index difference Δ is small, $\Delta \ll 1$ and if $n_1 \simeq n_2$, then in the "weakly guiding approximation" (2.198) becomes

$$\frac{J_p'(u)}{uJ_p(u)}+\frac{K_p'(w)}{wK_p(w)} = \pm p\left(\frac{1}{u^2}+\frac{1}{w^2}\right). \tag{2.214}$$

In the weakly guiding approximation, the TM mode and the TE mode become degenerate, and from the Bessel function formulas:

$$J_0'(u) = -J_1(u)$$

$$K_0'(w) = -K_1(w).$$

The eigenvalue equation for the TM mode and the TE mode is:

$$\frac{uJ_0(u)}{J_1(u)} = -\frac{wK_0(w)}{K_1(w)}. \tag{2.215}$$

The hybrid modes corresponding to the eigenvalue equations with plus and minus signs in (2.214) are called, respectively, the EH mode and the HE mode. With the Bessel function formulas, we get:

$$\frac{J_p'(u)}{uJ_p(u)} = -\frac{J_{p+1}(u)}{uJ_p(u)}+\frac{p}{u^2} = \frac{J_{p-1}(u)}{uJ_p(u)}-\frac{p}{u^2}$$

$$\frac{K_p'(w)}{wK_p(w)} = -\frac{K_{p+1}(w)}{wK_p(w)}+\frac{p}{w^2} = -\frac{K_{p-1}(w)}{wK_p(w)}-\frac{p}{w^2}$$

$$J_p(u)=2\frac{p-1}{u}J_{p-1}(u)-J_{p-2}(u)$$

$$K_p(w)=2\frac{p-1}{w}K_{p-1}(w)-K_{p-2}(w).$$

The eigenvalue equation for the EH mode can be obtained as:

$$\frac{uJ_p(u)}{J_{p+1}(u)}=-\frac{wK_p(w)}{K_{p+1}(w)}, \tag{2.216}$$

and the eigenvalue equation for the HE mode as

$$\frac{uJ_{p-2}(u)}{J_{p-1}(u)}=-\frac{wK_{p-2}(w)}{K_{p-1}(w)}. \tag{2.217}$$

In (2.217), if $p=1$, that is, $p-2=-1$, from the Bessel function formulas

$$J_{-1}(u)=-J_1(u)$$

$$K_{-1}(w)=K_1(w)$$

are applied, then

$$\frac{uJ_1(u)}{J_0(u)}=\frac{wK_1(w)}{K_0(w)}. \tag{2.218}$$

By assuming that $k_2=k_0n_2$, that is, $w=0(u=v)$, the normalised cut-off frequency, v_{cutoff}, can be obtained from $(2.215)-(2.218)$ as follows:

$$v_{cutoff}=\begin{cases} j_{0,q} & TM_{0q} \text{ mode and } TE_{0q} \text{ mode} \\ j_{p,q} & EH_{pq} \text{ mode } (p=1,2,3,\ldots) \\ 0 \text{ and } j_{1,q-1} & HE_{1q} \text{ mode} \\ j_{p-2,q} & HE_{pq} \text{ mode } (p=2,3,4,\ldots). \end{cases} \tag{2.219}$$

Here, $j_{p,q}$ represents the q_{th} zero of the Bessel function of the first kind of order p, excluding 0. The fundamental mode of an optical fibre is the HE_{11} mode, whose cut-off frequency is 0. The single-mode condition for uniform core optical fibre is:

$$v=k_0n_1a\sqrt{2\Delta}<j_{0,1}=2.405\equiv v_c. \tag{2.220}$$

Here, $j_{0,1}$ is the normalised cut-off frequency for the TE_{01}, TM_{01}, and HE_{21} modes; v_c represents the single-mode condition for optical fibre and is called the cut-off v value:

$$\lambda_c=\frac{2\pi}{v_c}n_1a\sqrt{2\Delta} \tag{2.221}$$

is called the cut-off wavelength. The optical fibre operates in a single mode for wavelengths longer than λ_c. When we apply the weakly guiding approximation, P given in (2.207) is -1 for the HE mode. Therefore, the electromagnetic field components for HE_{11} mode can be expressed as follows, based on $(2.201)-(2.203)$ (only electric fields in the core are shown):

$$E_z = A_1 J_1\left(\frac{ur}{a}\right)\begin{Bmatrix}\cos\theta\\ \sin\theta\end{Bmatrix} \tag{2.222}$$

$$E_r = -j\frac{\beta a}{u}A_1 J_0\left(\frac{ur}{a}\right)\begin{Bmatrix}\cos\theta\\ \sin\theta\end{Bmatrix} \tag{2.223}$$

$$E_\theta = -j\frac{\beta a}{u}A_1 J_0\left(\frac{ur}{a}\right)\begin{Bmatrix}-\sin\theta\\ \cos\theta\end{Bmatrix}. \tag{2.224}$$

x and y components of the electric field are represented as:

$$E_x = E_r\cos\theta - E_\theta\sin\theta \tag{2.225}$$

$$E_y = E_r\sin\theta + E_\theta\cos\theta. \tag{2.226}$$

If the upper expressions in $\{\}$ in (2.223) and (2.224) are used:

$$E_x = -j\frac{\beta a}{u}A_1 J_0\left(\frac{ur}{a}\right) \tag{2.227a}$$

$$E_y = 0 \tag{2.227b}$$

and if the lower expressions are used:

$$E_x = 0 \tag{2.228a}$$

$$E_y = -j\frac{\beta a}{u}A_1 J_0\left(\frac{ur}{a}\right). \tag{2.228b}$$

Equations (2.227) and (2.228) describe linear polarisation with x and y electric field components, respectively, and show that the HE_{11} mode is doubly degenerate when the state of polarisation is in question; HE_{11} modes given in (2.227) and (2.228) are distinguished from each other by calling them HE_{11^x} and HE_{11^y} modes, respectively. However, it is not only HE_{11} mode but also EH mode and HE mode are doubly degenerate. On the other hand, there is no degeneracy due to the direction of polarisation in the TM mode and the TE mode, since they are axisymmetric.

In the weakly guiding approximation, if we define l by

$$l = \begin{cases} 1 & \text{TM mode and TE mode} \\ p+1 & \text{EH mode} \\ p-1 & \text{HE mode} \end{cases} \tag{2.229}$$

then modes with the same parameter l will satisfy the same eigenvalue equation:

$$\frac{uJ_{l-1}(u)}{J_l(u)} = -\frac{wK_{l-1}(w)}{K_l(w)} \tag{2.230}$$

and will become degenerate. Such a degenerate mode is called the linearly polarised (LP) mode. Table 2.1 shows the corresponding relationships between the LP mode and the TM, TE, EH, and HE modes, along with the degree of degeneracy of the LP mode. The intensity distribution of x or y component of the electric field for modes belonging to the same LP mode is given in the same pattern.

Table 2.1 Relationship between LP mode and TM, TE, EH, and HE modes

$\mathbf{LP_{0q}}$ **Mode** $(l=0)$	$\mathbf{LP_{lq}}$ **Mode** $(l=1)$	$\mathbf{LP_{lq}}$ **Mode** $(l=2,3,4,\ldots)$
HE_{1q} Mode	TM_{0q} mode TE_{0q} mode HE_{2q} mode	$EH_{l-1,q}$ mode $HE_{l+1,q}$ mode
Degree of Degeneracy 2	Degree of degeneracy 4	Degree of degeneracy 4

2.7.4 Dispersion in Optical Fibre

Transmission distortion in optical fibre depends greatly on the dispersion characteristics of the fibre. This imposes restrictions on the data transmission speed, or the bit rate, in optical fibre communication. There are four types of dispersion in optical fibre. They are, in addition to the multimode dispersion, material dispersion caused by the wavelength dependence of the refractive index of the optical fibre material; waveguide dispersion caused by the structure dependent curvature of the dispersion curve (also called the $\omega-\beta$ curve or $\omega-\beta$ diagram) representing the relationship between angular frequency ω and phase constant β; and polarisation mode dispersion caused by the elimination of degeneracy between HE_{11}^{x} mode and HE_{11}^{y} mode due to the non-axisymmetricity and bending that exist, even it is small in actual optical fibre. Both material dispersion and waveguide dispersion depend on wavelength and the sum of these is called wavelength dispersion or chromatic dispersion.

The extent of dispersion in optical fibre is evaluated as the spread in propagation delay time per unit length of optical fibre or the spread in group delay. If the group velocity is v_g, the group delay at a ω close to the angular frequency ω_0 is:

$$\tau=\frac{1}{v_g}=\frac{d\beta}{d\omega}=\left.\frac{d\beta}{d\omega}\right|_{\omega=\omega_0}+\delta\omega\left.\frac{d^2\beta}{d\omega^2}\right|_{\omega=\omega_0}. \tag{2.231}$$

Here, $\delta\omega=\omega=\omega_0$.

The first term of (2.231) represents the delay for a single frequency. If a number of modes are propagated in the optical fibre, multimode dispersion occurs due to the different group velocities of the modes. Using the normalised phase constant b, which is defined as

$$b=\frac{(\beta/k_0)^2-n_2^2}{n_1^2-n_2^2}\simeq\frac{(\beta/k_0)-n_2}{n_1-n_2} \tag{2.232}$$

and the group refractive index, which is defined as

$$\mathrm{N}=\frac{d(k_0 n)}{dk_0}=n+k_0\frac{dn}{dk_0}, \tag{2.233}$$

the multimode dispersion τ_{multi} can be expressed as follows:

$$\tau_{\text{multi}}=\frac{d\beta}{d\omega}\simeq\frac{1}{c}\left[\mathrm{N}_2+(\mathrm{N}_1-\mathrm{N}_2)b+(n_1-n_2)k_0\frac{db}{dk_0}\right]. \tag{2.234}$$

If we take into consideration the fact that $n_1-n_2\simeq\mathrm{N}_1-\mathrm{N}_2$ in an actual glass material and use the normalised frequency, (2.234) becomes

$$\tau_{\text{multi}}=\frac{d\beta}{d\omega}\simeq\frac{1}{c}\left[\mathrm{N}_2+(\mathrm{N}_1-\mathrm{N}_2)\frac{d(vb)}{dv}\right]. \tag{2.235}$$

That is, multimode dispersion is evaluated in terms of the group refractive index and the $v-b$ curve (which is essentially the same as the $\omega-\beta$ curve). In uniform core optical fibre, the value of τ_{multi} generally falls within the range of $(N_1/c)\Delta$, which coincides well with (2.174) obtained from geometrical optics.

The second term of (2.231) is the delay, which occurs when the signal light contains a spread of frequencies. It represents the wavelength dispersion. Here, if we apply the following relationship between $\delta\omega$ and wavelength spread $\delta\lambda_0$ corresponding to $\delta\omega$:

$$\frac{\delta\omega}{\omega}=-\frac{\delta\lambda_0}{\lambda_0}. \tag{2.236}$$

The group delay expansion τ_c due to wavelength dispersion is:

$$\tau_0=\delta\omega\frac{d^2\beta}{d\omega^2}=-\sigma_c\delta\lambda_0. \tag{2.237}$$

Here, σ_c represents the increase in group delay per unit wavelength of light spectrum and unit length of optical fibre, that is, the wavelength dispersion (or total dispersion). σ_c is given as:

$$\sigma_c=\frac{k_0^2}{2\pi c}\cdot\frac{d^2\beta}{dk_0^2} \tag{2.238}$$

and is usually expressed in units of $\left[\text{ps}/(\text{km.nm})\right]$. Further, with (2.235), σ_c is

$$\sigma_c=\frac{1}{c\lambda_0}\left[k_0\frac{dN_2}{dk_0}+k_0\frac{d(N_1-N_2)}{dk_0}\cdot\frac{d(vb)}{dv}+(N_1-N_2)v\frac{d^2(vb)}{dv^2}\right]. \tag{2.239}$$

Furthermore, if we apply the following approximation, which holds over a wide range:

$$d(vb)/dv\simeq 1. \tag{2.240}$$

Equation (2.238) becomes

$$\sigma_c=\frac{1}{c\lambda_0}\left[k_0\frac{dN_1}{dk_0}+(N_1-N_2)v\frac{d^2(vb)}{dv^2}\right]. \tag{2.241}$$

The first term of (2.239) represents the material dispersion σ_m. Since

$$k_0\frac{dN_1}{dk_0}=-\lambda_0\frac{dN_1}{d\lambda_0}=\lambda_0{}^2\frac{d^2n_1}{d\lambda_0}. \tag{2.242}$$

σ_m is

$$\sigma_m=\frac{\lambda_0}{c}\cdot\frac{d^2n_1}{d\lambda_0{}^2}. \tag{2.243}$$

The wavelength dependence of the refractive index, which is necessary for evaluation of the material dispersion, is given by the three-term Sellmeier equation:

$$n_1(\lambda)=\left[1+\sum_{i=1}^{3}\frac{a_i\lambda_0{}^2}{\lambda_0{}^2-b_i}\right]^{\frac{1}{2}} \tag{2.244}$$

or the power series

$$n_1(\lambda)=A\lambda^{-4}+B\lambda^{-2}+C+D\lambda^2+E\lambda^4. \tag{2.245}$$

The second term of (2.241) represents the waveguide dispersion σ_w. If $N_1-N_2\simeq n_1-n_2\simeq n_1\Delta$,

$$\sigma_w=\frac{n_1 v\Delta}{c\lambda_0}\cdot\frac{d^2(vb)}{dv^2}. \tag{2.246}$$

The wavelength corresponding to zero wavelength dispersion, $\sigma_c=\sigma_m+\sigma_w=0$, is the wavelength at which the group delay spreads due to material dispersion and waveguide dispersion cancel each other. This wavelength is called the zero dispersion wavelengths. A zero dispersion wavelength is normally found in the neighbourhood of $1.3\,\mu\text{m}$, but pure quartz core/fluorine-doped quartz cladding optical fibre shows a minimum loss value of $0.154\ \text{dB}/\text{km}$ at a wavelength in the neighbourhood of $1.55\ \mu\text{m}$. Therefore, dispersion shifted single-mode optical fibre with zero dispersion wavelength shifted from $1.3\ \mu\text{m}$ band to $1.55\ \mu\text{m}$ band has been developed.

So far, we have only discussed axis-symmetric optical fibre. In actual optical fibre, the core form inevitably has some degree of non-axis symmetricity. This causes the elimination of degeneracy between HE_{11}^x mode and HE_{11}^y mode, causing the coupling of these modes and a change in the state of polarisation.

If we represent HE_{11}^x mode phase constant by β_x, the electric field amplitude E_x by $A_x=|A_x|$, the HE_{11}^y mode phase constant by β_y, and the electric field amplitude E_y by $A_y=|A_y|$, the electric fields E_x and E_y can be expressed by real functions representing instantaneous values as follows:

$$E_x=|A_x|\cos(\omega t-\beta_x z) \tag{2.247}$$

$$E_y=|A_y|\cos(\omega t-\beta_y z). \tag{2.248}$$

If we assume that HE_{11}^x and HE_{11}^y modes are propagated simultaneously, the locus of the tip of the electric field vector can be calculated on the basis of (2.247) and (2.248):

$$\frac{E_x^2}{|A_x|^2}+\frac{E_y^2}{|A_y|^2}-2\frac{E_xE_y}{|A_x||A_y|}\cos(\Delta\beta\cdot z)=\sin^2(\Delta\beta\cdot z). \tag{2.249}$$

$\Delta\beta=|\beta_x-\beta_y|$ is the phase constant difference and is called polarisation birefringence or modal birefringence. When HE_{11}^x mode and HE_{11}^y mode are mutually degenerate and $\Delta\beta=0$, (2.249) takes the same form as (2.91) and the polarisation plane does not change with time becoming a constant plane. On the other hand, if the degeneracy between these polarisation modes is eliminated and $\Delta\beta\neq 0$, the polarisation plane rotates, as suggested by (2.249). In order to reduce the coupling of polarisation modes, the normalised modal birefringence

$$B=\Delta\beta/k_0 \tag{2.250}$$

may be maximised. In other words, if the beat length (or coupling length) of the interference wave created by HE_{11}^x mode and HE_{11}^y mode,

$$L_b=2\pi/\Delta\beta=\lambda_0/B \tag{2.251}$$

can be shortened sufficiently in comparison to a typical cycle of structural fluctuation in the axial direction that exists in optical fibre, and the coupling of polarisation modes can be ignored. For evaluation of polarisation mode dispersion, we use the group delay difference τ_p between HE_{11}^x mode and HE_{11}^y mode, which is given as:

$$\tau_p = d(\Delta\beta)/d\omega \tag{2.252}$$

or the normalised group delay difference σ_p

$$\sigma_p = c\tau_p = \frac{d(\Delta\beta)}{dk_0} = \frac{d(k_0 B)}{dk_0} \tag{2.253}$$

where c is the velocity of light in a vacuum.

2.8 SLAB WAVEGUIDES: DERIVATION OF BASIC EQUATIONS

In this section, the wave analysis is described for the slab waveguide (Fig. 2.16) whose propagation characteristics have been explained. By considering the fact that we treat dielectric optical waveguides, we set permittivity and permeability as $\varepsilon = \varepsilon_0 n^2$ and $\mu = \mu_0$ in Maxwell's equation (2.254a) and equation (2.254b):

$$\nabla \times \tilde{E} = -\mu \frac{\partial \tilde{H}}{\partial t} \tag{2.254a}$$

$$\nabla \times \tilde{H} = \varepsilon \frac{\partial \tilde{E}}{\partial t} \tag{2.254b}$$

and as a result, we get

$$\nabla \times \tilde{E} = -\mu_0 \frac{\partial \tilde{H}}{\partial t} \tag{2.254c}$$

$$\nabla \times \tilde{H} = \varepsilon_0 n^2 \frac{\partial \tilde{E}}{\partial t} \tag{2.254d}$$

where n is the refractive index. We are interested in planar wave propagation in the form of:

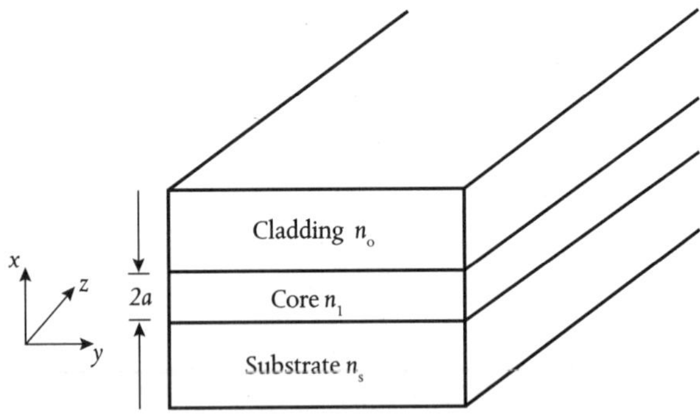

Fig. 2.16 Slab optical waveguide

$$\tilde{E} = E(x, y)e^{j(wt-\beta z)} \tag{2.255a}$$

$$\tilde{H} = H(x, y)e^{j(wt-\beta z)}. \tag{2.255b}$$

Substituting (2.255a) and (2.255b) in (2.254c) and (2.254d), we obtain the following set of equations for the electromagnetic field components:

$$\begin{cases} \frac{\partial E_z}{\partial y} + j\beta E_y = -j\omega\mu_0 H_x \\ -j\beta E_x - \frac{\partial E_z}{\partial x} = -j\omega\mu_0 H_y \\ \frac{\partial E_y}{\partial x} - \frac{\partial E_x}{\partial y} = -j\omega\mu_0 H_z \end{cases} \tag{2.256}$$

$$\begin{cases} \frac{\partial H_z}{\partial y} + j\beta H_y = j\omega\varepsilon_0 n^2 E_x \\ -j\beta H_x - \frac{\partial H_z}{\partial x} = j\omega\varepsilon_0 n^2 E_y. \\ \frac{\partial H_y}{\partial x} - \frac{\partial H_x}{\partial y} = j\omega\varepsilon_0 n^2 E_z \end{cases} \tag{2.257}$$

In the slab waveguide, as shown in Fig. 2.16, electromagnetic fields E and H do not have the y-axis dependency. Therefore, we set $\partial E/\partial y = 0$ and $\partial H/\partial y = 0$. Substituting these relations in (2.256) and (2.257), two independent electromagnetic modes are obtained, which are denoted as TE mode and TM mode, respectively. The TE mode satisfies the following wave equation:

$$\frac{d^2 E_y}{dx^2} + \left(k^2 n^2 - \beta^2\right) E_y = 0 \tag{2.258a}$$

where

$$H_x = -\frac{\beta}{\omega\mu_0} E_y \tag{2.258b}$$

$$H_z = \frac{j}{\omega\mu_0} \frac{dE_y}{dx} \tag{2.258c}$$

$$E_x = E_z = H_y = 0. \tag{2.258d}$$

Further, the tangential components E_y and H_z should be continuous at the boundaries of two different media. As shown in (2.258d), the electric field component along the z-axis is zero $(E_z = 0)$. Since the electric field lies in the plane that is perpendicular to the z-axis, this electromagnetic field distribution is called TE mode.

The TM mode satisfies the following wave equation:

$$\frac{d}{dx}\left(\frac{1}{n^2}\frac{dH_y}{dx}\right) + \left(k^2 - \frac{\beta^2}{n^2}\right) H_y = 0, \tag{2.259a}$$

where

$$E_x = \frac{\beta}{\omega\mu_0 n^2} H_y \quad (2.259b)$$

$$E_z = -\frac{j}{\omega\varepsilon_0 n^2} \frac{\partial H_y}{\partial x} \quad (2.259c)$$

$$E_y = H_x = H_z = 0. \quad (2.259d)$$

As shown in (2.259d), the magnetic field component along the z-axis is zero $(H_z = 0)$. Since the magnetic field lies in the plane that is perpendicular to the z-axis, this electromagnetic field distribution is called TM mode.

2.8.1 Dispersion Equations for TE and TM Modes

Propagation constants and electromagnetic fields for TE and TM modes can be obtained by solving (2.258) or (2.259). Here, the derivation method to calculate the dispersion equation (also called the eigenvalue equation) and the electromagnetic field distribution is given. We consider the slab waveguide with uniform refractive-index profile in the core, as shown in Fig. 2.17. By considering the fact that the guided electromagnetic fields are confined in the core and exponentially decay in the cladding, the electric field distribution is expressed as:

$$E_y = \begin{cases} A\cos(\kappa a - \phi)e^{-\sigma(x-a)} & (x > a) \\ A\cos(\kappa x - \phi) & (-a \le x \le a), \\ A\cos(\kappa a + \phi)e^{\xi(x+a)} & (x < -a) \end{cases} \quad (2.260a)$$

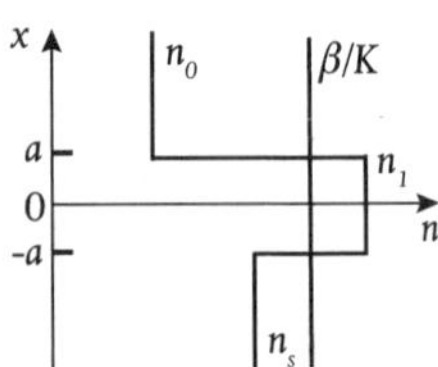

Fig. 2.17 Refractive-index profile of the slab waveguide

where k, σ, and ξ are wavenumbers along the x-axis in the core and cladding regions and are given as:

$$\begin{cases} \kappa = \sqrt{k^2 n_1^2 - \beta^2} \\ \sigma = \sqrt{\beta^2 - k^2 n_0^2}\,. \\ \xi = \sqrt{\beta^2 - k^2 n_s^2} \end{cases} \quad (2.260b)$$

The electric field component E_y in (2.260a) is continuous at the boundaries of core–cladding interfaces $(x = \pm a)$. There is another boundary condition that the magnetic field component H_z is given in (2.258c). Neglecting the terms independent of x, the boundary condition for H_z is treated by the continuity condition of dE_y / dx as:

$$\frac{dE_y}{dx}=\begin{cases}-\sigma A\cos(\kappa a-\phi)e^{-\sigma(x-a)} & (x>a)\\ -\kappa A\sin(\kappa x-\phi) & (-a\le x\le a)\\ \xi A\cos(\kappa a+\phi)e^{\xi(x+a)} & (x<-a)\end{cases} \tag{2.261}$$

From the condition that dE_y/dx are continuous at $x=\pm a$, the following equations are obtained:

$$\kappa A\sin(\kappa a+\phi)=\xi A\cos(\kappa a+\phi)$$

$$\sigma A\cos(\kappa a-\phi)=\kappa A\sin(\kappa a-\phi).$$

Eliminating constant A, we have:

$$\tan(u+\phi)=\frac{w}{u} \tag{2.262a}$$

$$\tan(u-\phi)=\frac{w'}{u}, \tag{2.262b}$$

where

$$\begin{cases}u=\kappa a\\ w=\xi a\ .\\ w'=\sigma a\end{cases} \tag{2.263}$$

From (2.262), we obtain the eigenvalue equations as:

$$u=\frac{m\pi}{2}+\frac{1}{2}\tan^{-1}\left(\frac{w}{u}\right)+\frac{1}{2}\tan^{-1}\left(\frac{w'}{u}\right) \qquad (m=0,1,2,\ldots) \tag{2.264}$$

$$\phi=\frac{m\pi}{2}+\frac{1}{2}\tan^{-1}\left(\frac{w}{u}\right)-\frac{1}{2}\tan^{-1}\left(\frac{w'}{u}\right). \tag{2.265}$$

The normalised transverse wavenumbers u, w, and w' are not independent. Using (2.260b) and (2.263), it is known that they are related by the following equations:

$$u^2+w^2=k^2a^2\left(n_1^2-n_s^2\right)\equiv v^2 \tag{2.266}$$

$$w'=\sqrt{\gamma v^2+w^2} \tag{2.267a}$$

$$\gamma=\frac{n_s^2-n_0^2}{n_1^2-n_s^2} \tag{2.267b}$$

$$v=kn_1a\sqrt{2\Delta}, \tag{2.267c}$$

where v is the normalised frequency, defined in (2.267c), and γ is a measure of the asymmetry of the cladding refractive indices. Once the wavelength of the light signal and the geometrical parameters of the waveguide are determined, the normalised frequency v and γ are determined. Therefore u, w, w', and ϕ are given by solving the eigenvalue (2.264) and (2.265) under the constraints of (2.266)–(2.267). In the asymmetrical waveguide $(n_s>n_0)$ as shown in Fig. 2.17, the higher refractive index n_s is used as the cladding refractive index, which is adopted for the definition of the normalised frequency (υ). It is preferable to use the higher refractive index n_s because the cut-off conditions are determined, when the normalised propagation constant β/k coincides with the higher cladding refractive index. Equations (2.264), (2.266), and (2.267) are the dispersion equations or eigenvalue equations for TE_m modes. When the wavelength of the light signal and the geometrical parameters of the waveguide are determined, in other words, when the normalised frequency υ and asymmetrical parameter γ are determined, the propagation constant β can be determined from these equations. As known from Fig. 2.17 or (2.260a) and (2.260b), the transverse wavenumber κ should be a real number for the main part of the optical field to be confined in the core region. Then, the following condition should be satisfied:

$$n_s \le \frac{\beta}{k} \le n_1. \tag{2.268}$$

β/k is a dimensionless value and is a refractive index itself for the plane wave. Therefore, it is called the effective index n_e and is usually expressed as:

$$n_e = \frac{\beta}{k}. \tag{2.269}$$

When $n_e < n_s$, the electromagnetic field in the cladding becomes oscillatory along the transverse direction; that is, the field is dissipated as the radiation mode. Since the condition $\beta = kn_s$ represents the critical condition under which the field is cut off and becomes the nonguided mode (radiation mode), it is called as cut-off condition. Here, we introduce a new parameter, which is defined as

$$b = \frac{n_e^2 - n_s^2}{n_1^2 - n_s^2}. \tag{2.270}$$

Then, the conditions for the guided modes are expressed, from (2.268) and (2.269), as:

$$0 \le b \le 1 \tag{2.271}$$

and the cut-off condition is expressed as:

$$b = 0. \tag{2.272}$$

b is called the normalised propagation constant. Rewriting the dispersion (2.264) by using the normalised frequency, υ and the normalised propagation constant b, we obtain:

$$2\upsilon\sqrt{1-b} = m\pi + tan^{-1}\sqrt{\frac{b}{1-b}} + tan^{-1}\sqrt{\frac{b+y}{1-b}}. \tag{2.273}$$

Further, (2.260b) is rewritten as:

$$\begin{cases} u = \upsilon\sqrt{1-b} \\ w = \upsilon\sqrt{b} \\ w' = \upsilon\sqrt{b+y} \end{cases}. \tag{2.274}$$

For the symmetrical waveguides with $n_0 = n_s$, we have $\gamma = 0$ and the dispersion (2.264) and (2.265) are reduced to

$$u = \frac{m\pi}{2} + \tan^{-1}\left(\frac{w}{u}\right) \tag{2.275a}$$

$$\phi = \frac{m\pi}{2}. \tag{2.275b}$$

Equation (2.275a) is also expressed as:

$$w = u\tan\left(u - \frac{m\pi}{2}\right) \tag{2.276}$$

or

$$\upsilon\sqrt{1-b} = \frac{m\pi}{2} + tan^{-1}\sqrt{\frac{b}{1-b}}. \tag{2.277}$$

2.8.2 Computation of Propagation Constant

First, the graphical method to obtain qualitatively the propagation constant of the symmetrical slab waveguide is shown, and then, the quantitative numerical method to calculate accurately the propagation constant is described. The relationship between u and w for the symmetrical slab waveguide, which is given in (2.276), is plotted in Fig. 2.18. Transverse wavenumbers u and w should satisfy (2.266) for a given normalised frequency υ. This relation is also plotted in Fig. 2.18 for the case of $\upsilon = 4$ as the semicircle with the radius of 4. The solutions of the dispersion equation are then given as the crossing points in Fig. 2.18.

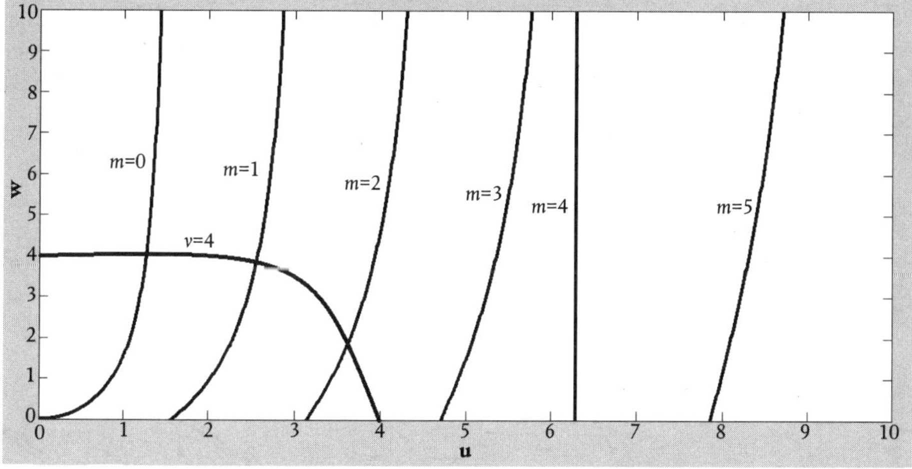

Fig. 2.18 *u–w* Relationship in the slab waveguide

For example, the transverse wavenumbers u and w for the fundamental mode are given by the crossing points of the curve tangential with $(m = 0)$ and the semicircle. The propagation constant (or eigenvalue) β is then obtained by using (2.260b) and (2.263). In Fig. 2.18, there is only one crossing point for the case of $v < \pi/2$. Then, the value of $v_c = \pi/2$ gives the critical point at which the higher order modes are cut off in the symmetrical slab waveguide. v_c is called the cut-off normalised frequency, which is obtained from the cut-off condition for $m=1$ mode

$$\begin{cases} b=w=0 & (2.278a) \\ u=\upsilon=\dfrac{\pi}{2}, & (2.278b) \end{cases}$$

where (2.272) and (2.274) have been used. Generally, cut-off υ value for the TE mode is given in (2.273) as:

$$\upsilon_{v,TE}=\frac{m\pi}{2}+\frac{1}{2}\tan^{-1}\sqrt{\gamma}. \quad (2.279)$$

Further, cut-off υ value for the TM mode is given as

$$\upsilon_{v,TM}=\frac{m\pi}{2}+\frac{1}{2}\tan^{-1}\left(\frac{n_1^2}{n_0^2}\sqrt{\gamma}\right) \quad (2.280)$$

A qualitative value can be obtained by this graphical solution for the dispersion equation. However, in order to obtain an accurate solution of the dispersion equation, we should rely on the numerical method. Here, we show the numerical treatment for the symmetrical slab waveguide so as to compare with the previous graphical method. We first rewrite the dispersion equation (2.280) in the following form:

$$f(\upsilon,m,b)=\upsilon\sqrt{1-b}-\frac{m\pi}{2}-\tan^{-1}\sqrt{\frac{b}{1-b}}=0. \quad (2.281)$$

The b-value at which $f=0$ gives the normalised propagation constant b for the given υ value. The solution of (2.281) is obtained by the Newton–Raphson method or the bisection method or a similar method. The normalised propagation constant (b) is calculated for each normalised frequency υ. Figure 2.19 shows the $(\upsilon-b)$ relationship, which is called the dispersion curve, for the TE mode. The mode number is expressed by the subscript m, such as TE_m or TM_m mode. The parameter in Fig. 2.19 is the measure of asymmetry (γ).

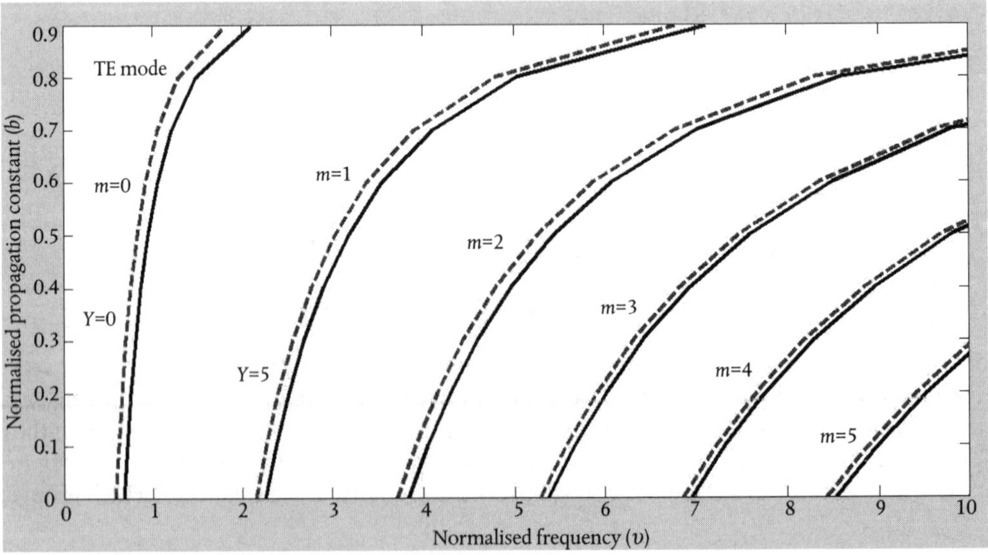

Fig. 2.19 Dispersion curves for the TE modes in the slab waveguide

It is known that there is no cut off for the lowest TE_0 mode in the symmetrical waveguide $(\gamma=0)$. On the other hand, cut off exists for the TE_0 mode in the asymmetrical waveguide $(\gamma\neq 0)$.

2.8.3 Electric Field Distribution

Once the eigenvalue of the waveguide is obtained, the electric field distribution given in (2.260a) is determined except for the arbitrary constant A. Constant A is determined when we specify the optical power P carried by the waveguide. Power P is expressed as:

$$P=\int_0^1 dy\int_{-\infty}^{\infty}\frac{1}{2}\left(E\times H^*\right)u_z dx=\int_{-\infty}^{\infty}\frac{1}{2}\left(E_xH_y^*-E_yH_x^*\right)dx. \tag{2.282}$$

For the TE mode, we can rewrite (2.282) by using (2.258) as

$$P=\frac{\beta}{2\omega\mu_0}\int_{-\infty}^{\infty}\left|E_y\right|^2 dx \tag{2.283}$$

Substituting (2.260a) in (2.283), we obtain the power fractions in the core, substrate, and cladding regions, respectively, as:

$$P_{\mathrm{core}}=\frac{\beta aA^2}{2\omega\mu_0}\left\{1+\frac{\sin^2(u+\phi)}{2w}+\frac{\sin^2(u-\phi)}{2w'}\right\}\qquad(-a\le x\le a) \tag{2.284a}$$

$$P_{\mathrm{sub}}=\frac{\beta aA^2}{2\omega\mu_0}\frac{\cos^2(u+\phi)}{2w}\qquad(x\le -a) \tag{2.284b}$$

$$P_{\mathrm{clad}}=\frac{\beta aA^2}{2\omega\mu_0}\frac{\cos^2(u-\phi)}{2w'}\qquad(x>a). \tag{2.284c}$$

For the calculation of (2.284a), we use (2.262). The total power P is then given as

$$P=P_{\mathrm{core}}+P_{\mathrm{sub}}+P_{\mathrm{clad}}=\frac{\beta aA^2}{2\omega\mu_0}\left\{1+\frac{1}{2w}+\frac{1}{2w'}\right\}. \tag{2.285}$$

Here, the constant A is determined as:

$$A=\sqrt{\frac{2\omega\mu_0 P}{\beta a(1+1/2w+1/2w')}}. \tag{2.286}$$

The power confinement factor is calculated, by using (2.284) and (2.285), as

$$\Gamma=\frac{P_{\mathrm{core}}}{P}=\frac{1+\dfrac{\sin^2(u+\phi)}{2w}+\dfrac{\sin^2(u-\phi)}{2w'}}{1+\dfrac{1}{2w}+\dfrac{1}{2w'}}. \tag{2.287}$$

2.8.4 Dispersion Equation for TM Mode

Based on (2.259a), the dispersion equation for the TM mode is obtained in a similar manner to that for the TE mode. We first express the magnetic field distribution H_y as

$$H_y = \begin{cases} A\cos(ka-\phi)e^{-\sigma(x-a)} & (x>a) \\ A\cos(ka-\phi) & (-a \le x \le a). \\ A\cos(ka+\phi)e^{-\xi(x-a)} & (x<a) \end{cases} \tag{2.288}$$

By applying the boundary condition that H_y and E_z should be continuous at $x = \pm a$, the following dispersion equation is obtained:

$$u = \frac{m\pi}{2} + \frac{1}{2}\tan^{-1}\left(\frac{n_1^2 w}{n_s^2 u}\right) + \frac{1}{2}\tan^{-1}\left(\frac{n_1^2 w'}{n_0^2 u}\right). \tag{2.289}$$

Rewriting this equation by using the normalised frequency υ and the normalised propagation constant b, we reduce it to:

$$2\upsilon\sqrt{1-b} = m\pi + \tan^{-1}\left(\frac{n_1^2}{n_s^2}\sqrt{\frac{b}{1-b}}\right) + \tan^{-1}\left(\frac{n_1^2}{n_0^2}\sqrt{\frac{b+y}{1-b}}\right). \tag{2.290}$$

The dispersion curves for the TM modes in the waveguide with $n_1 = 3.38$,

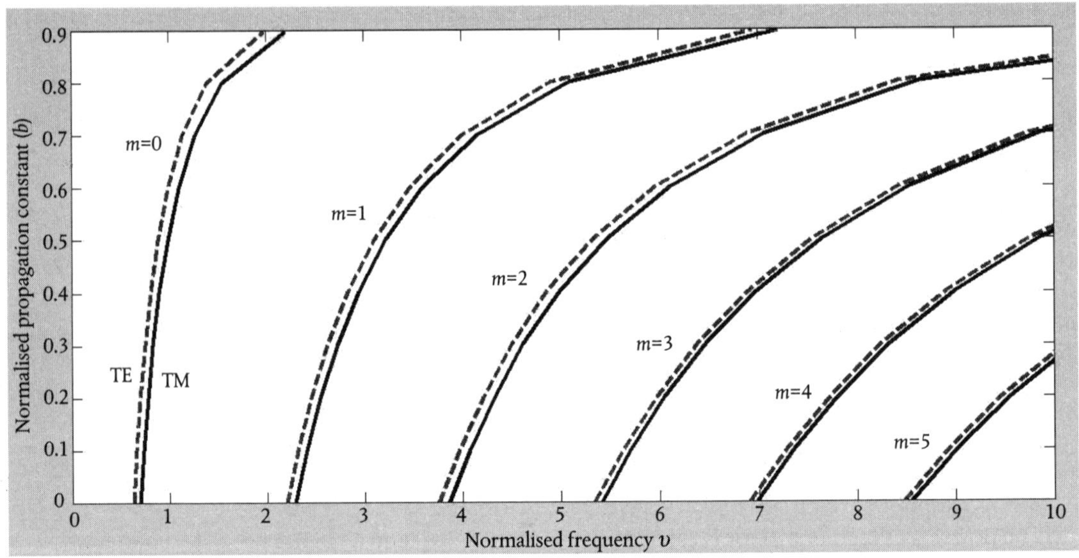

Fig. 2.20 Dispersion curves for the TE and TM modes in the slab waveguide

$n_s = n_0 = 3.17\,(\gamma = 0)$ are shown in Fig. 2.20 and compared with those for the TE modes. It is known that the normalised propagation constant b for the TM mode is smaller than that for the TE mode with respect to the same υ. This means that the TE mode is slightly better confined in the core than the TM mode. The power carried by the TM mode is obtained from (2.259a) and (2.282) by

$$P = \frac{\beta}{2\omega\varepsilon_0}\int_{-\infty}^{\infty}\frac{1}{n^2}\left|H_y\right|^2 dx. \tag{2.291}$$

2.8.5 Planar Slab Optical Waveguide

Planar optical waveguides are important components in integrated optical devices. The modal analysis of such waveguides is easier to understand. Hence, we will take up their analysis first. The simplest optical waveguide may have the geometric configuration shown in Fig. 2.21. It consists of a thin dielectric slab of refractive index n_1 and thickness $2a$ sandwiched between two symmetrical dielectric slabs of refractive index n_2 and infinite thickness $(n_2 < n_1)$. The waveguide is oriented such that the wave propagates along the z-direction. The y and z dimensions of the guide are assumed to extend to infinity. The thickness of the slabs is along the x-direction (as shown in Fig. 2.21). A ray of light launched into the guide slab or layer would progress by multiple reflections as shown in Fig. 2.22. We may assume that such a ray represents a plane TEM wave travelling at an angle θ with the z-axis. As the refractive index within the guide layer is n_1, the wavelength of light in the layer is reduced to $\lambda_m = \lambda / n_1$, where λ is the wavelength of light in vacuum and the propagation constant is increased to:

$$\beta_1 = \frac{2\pi}{\lambda_m} = \frac{2\pi n_1}{\lambda} = kn_1, \tag{2.292}$$

where $k = 2\pi / \lambda$ is the vacuum propagation constant or the propagation vector.

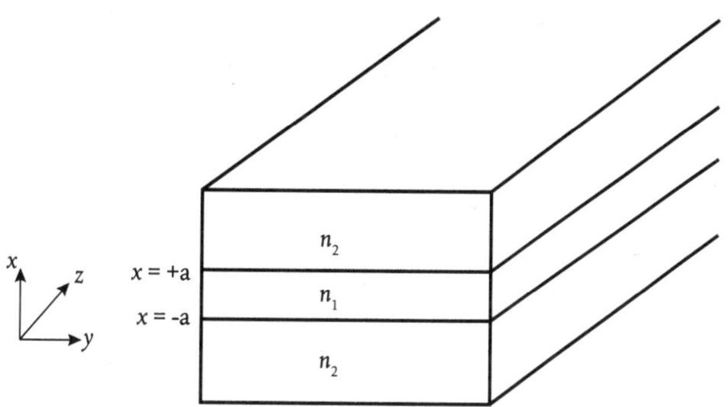

Fig. 2.21 Structure of a planar optical waveguide

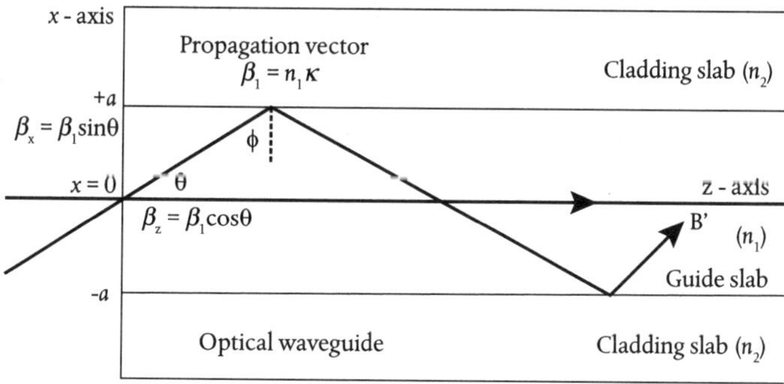

Fig. 2.22 Ray propagation in a planar waveguide

If the propagation vector β_1 makes an angle θ with the z-axis (which is same as the guide axis), the plane wave may be resolved into two-component plane waves propagating in the z - and x -directions as shown in Fig. 2.22. The component of the propagation vector in the z-direction, or in other words, the effective propagation constant of the guided wave (along the z-direction) will be:

$$\beta = \beta_z = \beta_1 \cos\theta. \tag{2.293}$$

The limiting value of θ, that is, θ_m is related to the critical angle ϕ_c at the interface of the guide layer and the cladding slabs, and it is given as:

$$\sin\phi_c = \cos\theta_m = n_2 / n_1. \tag{2.294}$$

Therefore, the minimum value of β in the z-direction, that is, β_{min}, will be determined by the maximum value of θ, that is, θ_m or

$$\beta_{min} = \beta_1 \cos\theta_m = \beta_1 \frac{n_2}{n_1} = \beta_2. \tag{2.295}$$

The maximum value that β can have β_1, which corresponds to $\theta = 0$, that is, the plane TEM waves travelling parallel to the guide axis: $\beta_{max} \approx \beta_1$. We, therefore, expect β to lie between β_1 and β_2, or $\beta_2 < \beta < \beta_1$. The component of the propagation vector β_1 in the x-direction is

$$\beta_x = \beta_1 \sin\theta = n_1 k \sin\theta, \tag{2.296}$$

where

$$k = \frac{2\pi}{\lambda}$$

or

$$\beta_x = \frac{2\pi}{\lambda_m}\sin\theta. \tag{2.297}$$

This component of the plane wave is reflected at the interface between the guide layer and the cladding slabs. When the total phase change after two successive reflections at the upper and lower interfaces is equal to $2i\pi$ radians, where i is an integer (0, 1, 2, 3 ...), constructive interference will occur and a standing wave pattern will be formed in the x-direction. This stable field pattern in the x-direction with only a periodic z-dependence is known as a mode. Therefore, only a finite number of discrete modes, which satisfy the above-mentioned condition, will propagate through the guide, that is, $4a\beta_x = 2i\pi$

or

$$4a\sin\theta_i = i\lambda_m. \tag{2.298}$$

Each value of θ_i corresponds to a particular mode with its own characteristic field pattern and its own propagation constant β_i in the z-direction. Obviously, β_i lies between β_1 and β_2. Since the maximum value that θ_i can take is θ_m, the number of guided modes is limited to

$$M = i_{max} = \frac{4a\sin\theta_m}{\lambda_m} = \frac{4an_1\sin\theta_m}{\lambda} = \frac{4a}{\lambda}\left(n_1^2 - n_2^2\right)^{1/2} = \frac{2V}{\pi}. \tag{2.299}$$

It should be noted that the requirement for the ith mode to be propagated is that $i \le (4a/\lambda)\left(n_1^2 - n_2^2\right)^{1/2}$. The mode corresponding to the highest value of i, that is, i_{max} does not meet the condition for total internal reflection, as the value of θ_m corresponds exactly to the critical angle ϕ_c, and is refracted at the interface. However, it can propagate freely in the cladding slabs and is said to be a radiation mode.

2.9 RECTANGULAR WAVEGUIDES

2.9.1 Basic Equations

In this section, the analytical method, which was proposed by Marcatili, to deal with the three-dimensional optical waveguide, as shown in Fig. 2.23, is described. The important assumption of this method is that the electromagnetic field in the shaded area in Fig. 2.23 can be neglected, since the electromagnetic field of the well-guided mode decays quite in the cladding region. Then, we do not impose the boundary condition for the electromagnetic field in the shaded area.

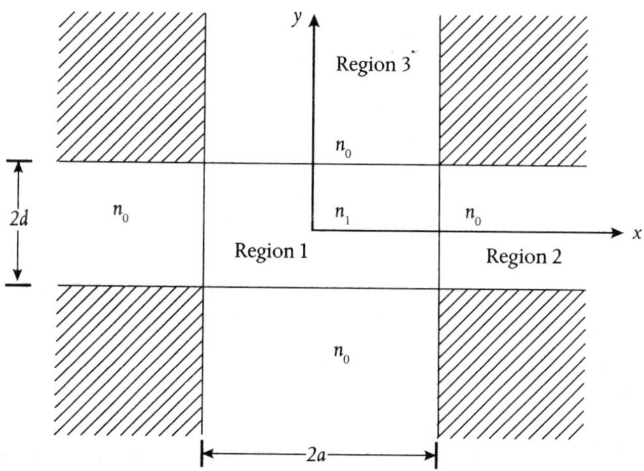

Fig. 2.23 Three-dimensional rectangular waveguide

We first consider the electromagnetic mode in which E_x and H_y are predominant. According to Marcatili's treatment, we set $H_x = 0$ in (2.256) and (2.257). Then, the wave equation and electromagnetic field representation are obtained as:

$$\frac{\partial^2 H_y}{\partial x^2} + \frac{\partial^2 H_y}{\partial y^2} + \left(k^2 n^2 - \beta^2\right) H_y = 0, \tag{2.300}$$

where

$$\begin{cases} H_x = 0 \\ E_x = \dfrac{\omega\mu_0}{\beta} H_y + \dfrac{1}{\omega\varepsilon_0 n^2 \beta} \dfrac{\partial^2 H_y}{\partial x^2} \\ E_y = \dfrac{1}{\omega\varepsilon_0 n^2 \beta} \dfrac{\partial^2 H_y}{\partial x \partial y} \\ E_z = -\dfrac{j}{\omega\varepsilon_0 n^2} \dfrac{\partial H_y}{\partial x} \\ H_z = -\dfrac{j}{\beta} \dfrac{\partial H_y}{\partial y} \end{cases} . \tag{2.301}$$

On the other hand, we set $H_y = 0$ in (2.256) and (2.257) to consider the electromagnetic field in which E_y and H_x are predominant. The wave equation and electromagnetic fields representation are given as:

$$\frac{\partial^2 H_x}{\partial x^2}+\frac{\partial^2 H_x}{\partial y^2}+\left(k^2n^2-\beta^2\right)H_x=0, \quad (2.302)$$

where

$$\begin{cases} H_y=0 \\ E_x=-\dfrac{1}{\omega\varepsilon_0 n^2\beta}\dfrac{\partial^2 H_x}{\partial x\partial y} \\ E_y=-\dfrac{\omega\mu_0}{\beta}H_x-\dfrac{1}{\omega\varepsilon_0 n^2\beta}\dfrac{\partial^2 H_x}{\partial y^2}. \\ E_z=\dfrac{j}{\omega\varepsilon_0 n^2}\dfrac{\partial H_x}{\partial y} \\ H_z=-\dfrac{j}{\beta}\dfrac{\partial H_x}{\partial x} \end{cases} \quad (2.303)$$

The modes in (2.300) and (2.301) are described as E^x_{pq} (p and q are integers) since E_x and H_y are the dominant electromagnetic fields. On the other hand, the modes in (2.302) and (2.303) are called E^y_{pq} , since E_y and H_x are the dominant electromagnetic field. In the following section, the solution method of the dispersion equation for the E^x_{pq} mode is described in detail and the results are shown only for the E^y_{pq} mode.

2.9.2 Dispersion Equations for E^x_{pq} and E^Y_{Pq} Modes

Since the rectangular waveguide shown in Fig. 2.23 is symmetrical with respect to the x- and y-axis, we analyse only regions 1–3. We first express the solution fields, which satisfy the wave (2.300), as:

$$H_y=\begin{cases} A\cos\left(k_x x-\phi\right)\cos\left(k_y y-\psi\right) & \textit{region } 1 \\ A\cos\left(k_x a-\phi\right)e^{-\gamma_x(x-a)}\cos\left(k_y y-\psi\right) & \textit{region } 2, \\ A\cos\left(k_x x-\phi\right)e^{-\gamma_y(y-d)}\cos\left(k_y d-\psi\right) & \textit{region } 3 \end{cases} \quad (2.304)$$

where the transverse wavenumber $k_x, k_y, \gamma_x,$ and γ_y and optical phase ϕ and ψ are given as

$$\begin{cases} -k_x^2-k_y^2+k^2n_1^2-\beta^2=0 & \text{region } 1 \\ \gamma_x^2-k_y^2+k^2n_0^2-\beta^2=0 & \text{region } 2, \\ -k_x^2+\gamma_y^2+k^2n_0^2-\beta^2=0 & \text{region } 3 \end{cases} \quad (2.305)$$

and

$$\begin{cases} \phi=(p-1)\dfrac{\pi}{2} & (p=1,2,\ldots) \\ \psi=(q-1)\dfrac{\pi}{2} & (q=1,2,\ldots) \end{cases}. \quad (2.306)$$

We should note here that the integers p and q start from 1 because we follow the mode definition by Marcatili. To the contrary, the mode number m in (2.264) for the slab waveguide starts from zero. By the conventional mode definition, the lowest mode in the slab waveguide is the $TE_{m=0}$ mode, which has one electric field peak. On the other hand, the lowest mode in rectangular waveguide is $E^x_{p=1,q=1}$ or $E^y_{p=1,q=1}$ mode, which has only one electric field peak along both the *x*-axis and *y*-axis direction as shown in Fig. 2.24. Therefore, in the mode definition by Marcatili, integers p and q represent the number of local electric field peaks along the *x*- and *y*-axis directions. When we apply the boundary conditions that the electric field $E_z \propto (1/n^2)\dfrac{\partial H_y}{\partial x}$ should be continuous at $x=a$ and the magnetic field $H_z \propto \dfrac{\partial H_y}{\partial y}$ should be continuous at $y=d$, we obtain the following dispersion equation:

$$k_x a = (p-1)\frac{\pi}{2} + \tan^{-1}\left(\frac{n_1^2 \gamma_x}{n_0^2 k_x}\right), \tag{2.307a}$$

$$k_y d = (q-1)\frac{\pi}{2} + \tan^{-1}\left(\frac{\gamma_y}{k_y}\right). \tag{2.307b}$$

Transverse wavenumbers k_x, k_y, γ_x, and γ_y, are related by (2.305) as:

$$\gamma_x^2 = k^2\left(n_1^2 - n_0^2\right) - k_x^2, \tag{2.308}$$

$$\gamma_y^2 = k^2\left(n_1^2 - n_0^2\right) - k_y^2, \tag{2.309}$$

where k_x is obtained from (2.307a) and (2.308) and k_y is determined from (2.307b) and (2.309), respectively. The propagation constant β is then obtained from:

$$\beta^2 = k^2 n_1^2 - \left(k_x^2 + k_y^2\right). \tag{2.310}$$

In order to calculate the dispersion equation for the E^y_{pq} mode, we express the magnetic field H_x as

$$H_x = \begin{cases} A\cos(k_x x - \phi)\cos(k_y y - \psi) & \text{region 1} \\ A\cos(k_x a - \phi)e^{-\gamma_x(x-a)}\cos(k_y y - \psi) & \text{region 2.} \\ A\cos(k_x a - \phi)e^{-\gamma_y(y-d)}\cos(k_y d - \psi) & \text{region 3} \end{cases} \tag{2.311}$$

By applying the boundary conditions that the magnetic field $H_z \propto \dfrac{\partial H_x}{\partial x}$ should be continuous at $x=a$ and electric field $E_z = (1/n^2)\dfrac{\partial H_x}{\partial y}$ should be continuous at $y=d$, we obtain the following dispersion equations:

$$k_x a = (p-1)\frac{\pi}{2} + \tan^{-1}\left(\frac{\gamma_x}{k_x}\right) \tag{2.312}$$

$$k_y d = (q-1)\frac{\pi}{2} + \tan^{-1}\left(\frac{n_1^2 \gamma_y}{n_0^2 k_y}\right). \tag{2.313}$$

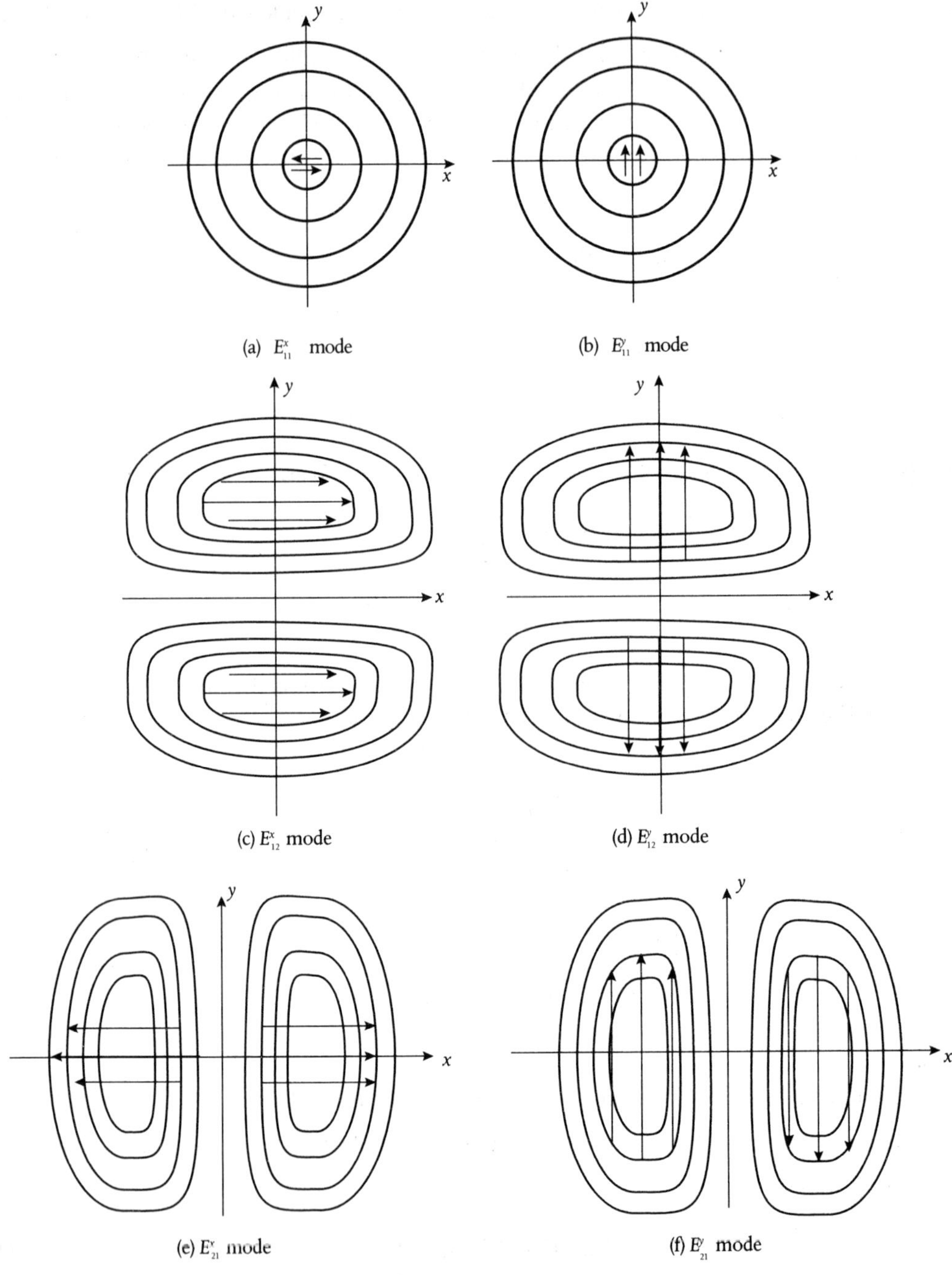

Fig. 2.24 Mode definition and electric field distribution in Marcatili's method

2.9.3 Power Distribution and Confinement Factor for Symmetric Waveguide

We know that the electric field distribution of the guided modes extends beyond the boundary of the guide layer and the extent of penetration into the cladding layer depends on the thickness and the mode number. It is important to know, under many situations, that the fraction of guided optical power confined within the guide layer. This fraction is called the confinement factor. The power flow is given by Poynting vector that is defined as:

$$S = E \times H, \tag{2.314}$$

where E and H are normally expressed in the complex form. However, the actual fields are the real part of the complex form. Therefore, taking the time average of the Poynting vector, we get:

$$\langle S \rangle = \langle Re\, E \times Re\, H \rangle \Rightarrow \langle S \rangle = \frac{1}{2} Re \langle E \times H^* \rangle, \tag{2.315}$$

where H^* is the complex conjugate of H. Therefore, the time average of S along the z-direction will be given as:

$$\langle S_z \rangle = \frac{1}{2} \langle (E_x H_y - H_x E_y) \rangle. \tag{2.316}$$

We have seen that, for TE modes, the field components E_x and E_z vanish and

$$H_x = -\frac{\beta}{\mu_0 \omega} E_y.$$

Therefore,

$$S_z = \frac{1}{2} \frac{\beta}{\mu_0 \omega} \left| E_y^2 \right|. \tag{2.317}$$

For a particular mode, the power associated per unit area per unit length in the y-direction will, thus, be given as:

$$P = \frac{1}{2} \frac{\beta}{\mu_0 \omega} \int_{x=-\infty}^{\infty} \left| E_y^2 \right| dx. \tag{2.318}$$

The power inside the guide layer will be given as:

$$P_{in} = \frac{1}{2} \frac{\beta}{\mu_0 \omega} \int_{-a}^{a} E_y^2 dx \tag{2.319}$$

and the power outside the guide layer will be given as:

$$P_{out} = \frac{1}{2} \frac{\beta}{\mu_0 \omega} \left[\int_{-\infty}^{-a} E_y^2 dx + \int_{a}^{\infty} E_y^2 dx \right]. \tag{2.320}$$

For symmetric modes, the electric field distribution takes the form

$$E_y(x) = \left. \begin{matrix} A \cos ux, & |x| < a \\ C \exp(-w|x|), & |x| > a \end{matrix} \right\}. \tag{2.321}$$

For symmetric TE modes, as shown in Fig. 2.24, substituting for E_y from (2.321) in (2.319) and (2.320), we get:

$$P_{in} = \frac{1}{2}\frac{\beta}{\mu_0\omega}2\int_0^a (A\cos ux)^2\,dx = \frac{\beta}{\mu_0\omega}A^2\int_0^a \frac{1}{2}(1+\cos 2ux)dx$$

or

$$P_{in} = \frac{1}{2}\frac{\beta}{\mu_0\omega}A^2\left[x+\frac{1}{2u}\sin 2ux\right]_0^a = \frac{\beta}{2\mu_0\omega}A^2\left[a+\frac{1}{2u}\sin 2ua\right] \quad (2.322)$$

and from (2.321), we get

$$P_{out} = \frac{1}{2}\frac{\beta}{\mu_0\omega}2\int_a^\infty \left(Ce^{-wx}\right)^2 dx = \frac{\beta}{\mu_0\omega}C^2\left[-\frac{1}{2w}e^{-2wx}\right]_a^\infty$$

or

$$P_{out} = \frac{\beta}{\mu_0\omega}C^2\left[\frac{1}{2w}e^{-2wa}\right]. \quad (2.323)$$

From (2.322) and (2.323), we can derive the final expression for P as follows:

$$P = P_{in} + P_{out}$$

$$\Rightarrow P = \frac{\beta}{2\mu_0\omega}A^2\left[a+\frac{1}{2u}\sin 2ua\right] + \frac{\beta}{\mu_0\omega}C^2\left[\frac{1}{2w}e^{-2wa}\right]$$

$$\Rightarrow P = \frac{\beta A^2}{2\mu_0\omega}\left[a+\frac{1}{2u}\sin 2ua + \left(\frac{C}{A}\right)^2\frac{1}{w}e^{-2wa}\right].$$

Substituting the value of $C/A = e^{wa}\cos ua$ from (2.323), we get

$$P = \frac{\beta A^2}{2\mu_0\omega}\left[a+\frac{1}{2u}\sin 2ua + \frac{1}{w}\cos^2 ua\right]$$

$$\Rightarrow P = \frac{\beta A^2}{4\mu_0\omega}\left[2a+\frac{2}{w}+\frac{2\sin ua\cos ua}{u} - \frac{2\sin^2 ua}{w}\right]$$

$$\Rightarrow P = \frac{\beta A^2}{4\mu_0\omega}\left[\left(2a+\frac{2}{w}\right)+\frac{2\sin ua\cos ua}{uwa}\{wa - ua\tan ua\}\right]. \quad (2.324)$$

From (2.324), the term within the braces becomes zero. This gives the following expression for P:

$$P = \frac{\beta A^2}{4\mu_0\omega}\left(a+\frac{1}{w}\right). \quad (2.325)$$

A similar expression may be derived for power carried by antisymmetric modes.

The fraction of power confined within the guide layer, called confinement factor G, may be calculated as follows:

$$G = \frac{P_{in}}{P_{in}+P_{out}} \quad (2.326)$$

$$\Rightarrow G=\frac{\left[a+\frac{1}{2u}\sin 2ua\right]}{\left[a+\frac{1}{2u}\sin 2ua\right]+\left[\frac{1}{w}e^{-2wa}\right]\left(\frac{C}{A}\right)^2}. \tag{2.327}$$

Substituting the value of $C/A=e^{wa}cosua$ from (2.323), in (2.327), we get:

$$\Rightarrow G=\frac{\left[a+\frac{1}{2u}\sin 2ua\right]}{\left[a+\frac{1}{2u}\sin 2ua\right]+\left[\frac{1}{w}e^{-2wa}\left(e^{wa}\cos ua\right)^2\right]}$$

$$\Rightarrow G=\left[\frac{a+\frac{1}{2u}\sin 2ua+\frac{1}{w}\cos^2 ua}{a+\frac{1}{2u}\sin 2ua}\right]^{-1}$$

$$G=\left[1+\frac{\cos^2 ua}{wa\{1+[\sin(ua)\cos(ua)]/ua\}}\right]^{-1}. \tag{2.328}$$

A similar expression may be derived for antisymmetric modes. Some important points may be noted from (2.328).

(i) As $a\to 0$, $G\to 0$, that is, for a very thin guide layer, there is no guidance of light. Rather, almost all the power is lost in the cladding.

(ii) As the G-factor depends on u and w and both these parameters in turn depend on m, G will vary with the mode number.

(iii) It can be shown that G increases, first rapidly and then slowly, with an increase in the thickness of the guide layer for each mode.

2.10 MODAL ANALYSIS OPTICAL FIBRE

2.10.1 Modal Analysis of an Ideal Step-index Circular Core Optical Fibre

Consider an ideal step-index fibre consisting of a uniform cylindrical dielectric core of radius a and refractive index n_1, surrounded by an infinitely thick uniform dielectric cladding of refractive index n_2. We attempt a solution of the wave equation for electric and magnetic fields for the modes propagated by such a fibre. The equation for an electric field in an isotropic, linear, nonconducting, and non-magnetic, but inhomogeneous medium is given as

$$\nabla^2 E+\nabla\left\{\frac{1}{\varepsilon_r}\nabla(\varepsilon_r)\cdot E\right\}-\mu_0\varepsilon_0\varepsilon_r\frac{\partial^2 E}{\partial t^2}=0. \tag{2.329}$$

Similarly, the equation for a magnetic field is given as:

$$\nabla^2 H+\frac{1}{\varepsilon_r}\{\nabla(\varepsilon_r)\times(\nabla\times H)\}-\mu_0\varepsilon_0\varepsilon_r\frac{\partial^2 H}{\partial t^2}=0. \tag{2.330}$$

From (2.329) and (2.330), it is evident that the different components of E as well as H are coupled. In the present case, the refractive index $n=\sqrt{\varepsilon_r}$ is constant $(=n_1)$ up to $r \le a$ and equal to n_2 for points $r>a$, but there is a discontinuity at $r=a$. Let us assume that this discontinuity is small (i.e., $n_1 \approx n_2$). This is called a weakly guiding approximation. With this approximation, the second term on the left-hand side (LHS) of (2.329) and (2.330) may be neglected and each Cartesian component of E and H satisfies the scalar wave equation. Substituting $\varepsilon_r = n^2$, we get

$$\nabla^2 \psi = \varepsilon_0 \mu_0 n^2 \frac{\partial^2 \psi}{\partial t^2}, \tag{2.331}$$

where ψ represents the scalar E or H field. The fibre boundary conditions have cylindrical symmetry and we assume that the direction of propagation of the electromagnetic waves is along the axis of the fibre, which we take to be the z-axis. In the scalar wave approximation, the modes may be assumed to be nearly transverse and they may possess an arbitrary state of polarisation. These linearly polarised modes are referred to as LP modes. The propagation constants of the TE and TM modes are nearly equal.

In cylindrical coordinates (r,ϕ,z), (2.331) may be written as

$$\nabla^2 \psi - \varepsilon_0 \mu_0 n^2 \frac{\partial^2 \psi}{\partial t^2} = \frac{\partial^2 \psi}{\partial r^2} + \frac{1}{r}\frac{\partial \psi}{\partial r} + \frac{1}{r^2}\frac{\partial^2 \psi}{\partial \phi^2} + \frac{\partial^2 \psi}{\partial z^2} - \varepsilon_0 \mu_0 n^2 \frac{\partial^2 \psi}{\partial t^2}. \tag{2.332}$$

Since n may depend on the transverse coordinates (r,ϕ), it usually depends only on r, and the wave is propagating along the z-direction, we may write the solution of (2.332) as:

$$\psi(r,\phi,z,t) = \psi(r,\phi) e^{i(\omega t - \beta z)}. \tag{2.333}$$

Substituting the value of ψ from (2.333) in (2.332), we get:

$$e^{i(\omega t-\beta z)} \frac{\partial^2 \psi}{\partial r^2} + \frac{1}{r} e^{i(\omega t-\beta z)} \frac{\partial \psi}{\partial r} + \frac{1}{r^2} e^{i(\omega t-\beta z)} \frac{\partial^2 \psi}{\partial \phi^2} + \psi(-\beta^2) e^{i(\omega t-\beta z)} - \varepsilon_0 \mu_0 n^2 (-\omega^2) \psi e^{i(\omega t-\beta z)} = 0$$

or

$$\frac{\partial^2 \psi}{\partial r^2} + \frac{1}{r}\frac{\partial \psi}{\partial r} + \frac{1}{r^2}\frac{\partial^2 \psi}{\partial \phi^2} + \left[\varepsilon_0 \mu_0 \omega^2 n^2 - \beta^2\right]\psi = 0.$$

Substituting $\varepsilon_0 \mu_0 = 1/c^2$ and $\omega/c = k$, the free-space wavenumber, in the equation, we get:

$$\frac{\partial^2 \psi}{\partial r^2} + \frac{1}{r}\frac{\partial \psi}{\partial r} + \frac{1}{r^2}\frac{\partial^2 \psi}{\partial \phi^2} + \left[n^2 k^2 - \beta^2\right]\psi = 0. \tag{2.334}$$

Since the fibre under construction has cylindrical symmetry, the variables can be separated:

$$\psi(r,\phi) = R(r)\Phi(\phi), \tag{2.335}$$

where R is a function of only r and Φ is a function of only ϕ. Substituting ψ from (2.335) in (2.334), we get:

$$\Phi \frac{\partial^2 R}{\partial r^2} + \frac{1}{r}\Phi \frac{\partial R}{\partial r} + \frac{R}{r^2}\frac{\partial^2 \Phi}{\partial \phi^2} + \left[n^2 k^2 - \beta^2\right] R\Phi = 0.$$

Since the derivatives involved are dependent either on r or ϕ only, the partial derivatives may be replaced by full derivatives. Further, dividing the entire LHS by Rf, we get:

$$\frac{1}{R}\frac{\partial^2 R}{\partial r^2} + \frac{1}{Rr}\frac{\partial R}{\partial r} + \frac{1}{r^2}\frac{1}{\Phi}\frac{\partial^2 \Phi}{\partial \phi^2} + \left[n^2 k^2 - \beta^2\right] = 0$$

or

$$\frac{r^2}{R}\left(\frac{\partial^2 R}{\partial r^2}+\frac{1}{r}\frac{\partial R}{\partial r}\right)+r^2\left[n^2k^2-\beta^2\right]=-\frac{1}{\Phi}\frac{\partial^2\Phi}{\partial\phi^2}=l^2, \quad (\text{say}) \tag{2.336}$$

where l is a constant, which is known as an azimuthal eigenvalue.

The dependence of Φ on ϕ will be of the form $e^{il\varnothing}$. For the function to be single valued, that is, $\Phi(\phi+2\pi)=\Phi(\phi)$, the constant l is required to be an integer, that is,

$$l=0, 1, 2, 3, ... \tag{2.337}$$

Therefore, the complete transverse field will be given as:

$$\psi(r,\phi,z,t)=R(r)e^{il\varnothing}e^{i(\omega t-\beta z)}. \tag{2.338}$$

The radial part of (2.336) may be written as:

$$\frac{r^2}{R}\left(\frac{\partial^2 R}{\partial r^2}+\frac{1}{r}\frac{\partial R}{\partial r}\right)+r^2\left[n^2k^2-\beta^2\right]=l^2,$$

which can be rearranged to give

$$r^2\frac{\partial^2 R}{\partial r^2}+r\frac{\partial R}{\partial r}+\left[r^2\left(n^2k^2-\beta^2\right)-l^2\right]R=0. \tag{2.339}$$

We know that $n=n_1$ for $r\le a$ and $n=n_2$ for $r>a$. Therefore, substituting the value of n in (2.339), we obtain for the case of a step-index fibre:

$$r^2\frac{\partial^2 R}{\partial r^2}+r\frac{\partial R}{\partial r}+\left[r^2\left(k^2n_1^2-\beta^2\right)-l^2\right]R=0, \qquad r\le a \tag{2.340}$$

and

$$r^2\frac{\partial^2 R}{\partial r^2}+r\frac{\partial R}{\partial r}+\left[r^2\left(k^2n_2^2-\beta^2\right)-l^2\right]R=0 \qquad r>a. \tag{2.341}$$

In order to simplify the above-mentioned equations, we substitute

$$u^2\equiv\left(k^2n_1^2-\beta^2\right)a^2 \tag{2.342}$$

and

$$w^2\equiv\left(\beta^2-k^2n_2^2\right)a^2. \tag{2.343}$$

The normalised waveguide parameter V for the fibre is defined as:

$$V=\left(u^2+w^2\right)^{1/2}=ka\left(n_1^2-n_2^2\right)^{1/2}=\frac{2\pi a}{\lambda}\left(n_1^2-n_2^2\right)^{1/2}. \tag{2.344}$$

Substituting the values of u and w in (2.339) and (2.341), we get:

$$r^2\frac{\partial^2 R}{\partial r^2}+r\frac{\partial R}{\partial r}+\left[\frac{u^2r^2}{a^2}-l^2\right]R=0, \qquad r\le a \tag{2.345}$$

and

$$r^2\frac{\partial^2 R}{\partial r^2}+r\frac{\partial R}{\partial r}-\left[\frac{w^2r^2}{a^2}+l^2\right]R=0, \qquad r>a. \tag{2.346}$$

Equations (2.345) and (2.346) are second-order equations, and hence, they should possess two independent solutions. The solutions corresponding to (2.345) are the Bessel function of the first kind and the modified Bessel function of the first kind. The solutions corresponding to (2.346) are the Bessel function of the second kind and the modified Bessel function of the second kind. The modified Bessel function of the first kind has a discontinuity at the origin and the Bessel function of the second kind has an asymptotic form. Hence, these are discarded in the solutions for fibre modes. For the solutions to be well-behaved, that is, be finite at $r=0$ and tend to zero as $r \to \infty$, it is essential that both u and w are real. Therefore, a valid solution of (2.345) would be given by the first kind of Bessel function of order l and that of (2.346) would be given by the second kind of modified Bessel function of order l. Therefore,

$$R(r) = AJ_l\left(\frac{ur}{a}\right), \quad r<a \tag{2.347}$$

and

$$R(r) = BK_l\left(\frac{wr}{a}\right), \quad r>a. \tag{2.348}$$

The Bessel function of the first kind of order l and argument x, denoted by $J_l(x)$, is defined in terms of an infinite series as follows:

$$J_l(x) \sum_{n=0}^{\infty} \frac{(-1)^n}{n!\Gamma(n+l+1)}\left(\frac{x}{2}\right)^{2n+l} \tag{2.349}$$

where $x = ur/a$ and the gamma function $\Gamma(n+l+1) = (n+l)!$. For $l=0$,

$$J_0(x) = 1 - \frac{\frac{1}{4}x^2}{(1!)^2} + \frac{\left(\frac{1}{4}x^2\right)^2}{(2!)^2} + \frac{\left(\frac{1}{4}x^2\right)^3}{(3!)^2} + \dots$$

and for $l=1$,

$$J_1(x) = \frac{1}{2}x - \frac{\left(\frac{1}{2}x\right)^3}{2!} - \frac{\left(\frac{1}{2}x\right)^5}{3!} - \dots$$

and so on for higher values of l.

The second kind of modified Bessel function of order l is given as:

$$K_l(\tilde{x}) = (\pi/2) i^{-(l+1)} H_l(-i\tilde{x}), \quad \tilde{x} = \frac{wr}{a}, \tag{2.350}$$

where $H_l(-i\tilde{x})$ is a Hankel function, which is a linear combination of Bessel functions of the first (J_l) and second (Y_l) kind. These functions have been chosen because for $x \ll 1$,

$$J_l(x) = \frac{1}{l!}\left(\frac{1}{2}x\right)^l, \quad l=0,1,\dots \tag{2.351}$$

$$K_l(\tilde{x}) = (l-1)!\,2^{l-1}\tilde{x}^{-1}, \quad l \geq 1 \tag{2.352}$$

and for $x \gg 1$,

$$J_l(x) = \sqrt{\frac{2}{\pi x}} \cos\left[x - \frac{\pi(2l+1)}{4}\right] \tag{2.353}$$

and

$$K_l(\tilde{x}) = \sqrt{\frac{2}{\pi x}} e^{-\tilde{x}} \left[1 + \frac{4l^2 - 1}{8\tilde{x}} \right]. \tag{2.354}$$

Therefore, $J_l(x)$ is a well-behaved function for $r < a$ and $K_l(\tilde{x})$ is well behaved for $r > a$. For further analysis, some recurrence relations for these functions (with argument x) and some asymptotic forms are given as follows:

$$J_{-l} = (-1)^l J_l \tag{2.355}$$

$$J_l' = \frac{1}{2}(J_{l-1} + J_{l+1}) = \pm J_{l\mp1} \mp \frac{lJ_l}{x} \tag{2.356}$$

$$J_{l\mp1} = \frac{2lJ_l}{x} - J_{l\pm1} \tag{2.357}$$

$$J_{l\mp2} = \frac{2(l\mp1)J_{l\mp1}}{x} - J_l \tag{2.358}$$

$$K_l = K_{-l} \tag{2.359}$$

$$K_l' = -\frac{1}{2}(K_{l-1} + K_{l+1}) = \mp\frac{lK_l}{x} - K_{l\mp1} \tag{2.360}$$

$$K_{l\mp1} = \mp\frac{2lK_l}{x} + K_{l\pm1} \tag{2.361}$$

$$K_{l\mp2} = \mp\frac{2(l\mp1)K_{l\mp1}}{x} + K_l. \tag{2.362}$$

Here, prime denotes the first derivative. For $\tilde{x} \ll 1$,

$$\frac{K_0}{K_1} = \tilde{x} \ln \frac{2}{1.782\tilde{x}} \tag{2.363}$$

$$\frac{K_{l-1}}{K_l} = \frac{\tilde{x}}{2(l-1)}, \quad l \geq 2 \tag{2.364}$$

$$\frac{K_{l+1}}{K_l} = \frac{2l}{\tilde{x}}, \quad l \geq 1. \tag{2.365}$$

For $\tilde{x} \gg 1$,

$$\frac{K_{l+1}}{K_l} = 1 + \frac{1 \mp 2l}{2\tilde{x}}. \tag{2.366}$$

Since ψ is continuous at $r = a$, $R(r)$ must be continuous at $r = a$. By imposing this condition, we can get the values of constants A and B. Therefore, from (2.347) and (2.348), we have:

$$A = \frac{R(a)}{J_l(u)} \tag{2.367}$$

$$B = \frac{R(a)}{K_l(w)}. \tag{2.368}$$

Substituting the values of R and Φ in (2.335), we get the transverse dependence of the modal fields as follows:

$$\psi(r,\phi) = AJ_l\left(\frac{ur}{a}\right)\begin{bmatrix}\cos(l\phi) \\ \sin(l\phi)\end{bmatrix} \qquad r<a \tag{2.369}$$

and

$$\psi(r,\phi) = BK_l\left(\frac{wr}{a}\right)\begin{bmatrix}\cos(l\phi) \\ \sin(l\phi)\end{bmatrix} \qquad r>a. \tag{2.370}$$

Now, $\partial\psi/\partial r$ is also continuous at $r=a$:

$$\left.\frac{\partial\psi}{\partial r}\right|_{r=a} = A\frac{u}{a}J_l'\left(\frac{ur}{a}\right)\cos l\phi\Big|_{r=a} = A\frac{u}{a}J_l'(u)\cos l\phi$$

$$\left.\frac{\partial\psi}{\partial r}\right|_{r=a} = B\frac{w}{a}K_l'\left(\frac{wr}{a}\right)\cos l\phi\Big|_{r=a} = B\frac{w}{a}K_l'(w)\cos l\phi.$$

Therefore, substituting the value of A and B from (2.367) and (2.368)

$$\frac{R(a)}{J_l(u)}\frac{u}{a}J_l'(u)\cos l\phi = \frac{R(a)}{K_l(w)}\frac{w}{a}K_l'(w)\cos l\phi$$

or

$$\frac{uJ_l'(u)}{J_l(u)} = \frac{wK_l'(w)}{K_l(w)}.$$

Therefore, the continuity of ψ and $\partial\psi/\partial r$ at the core–cladding interface $(r=a)$ leads to an eigenvalue equation of the form:

$$\frac{uJ_l'(u)}{J_l(u)} = \frac{wK_l'(w)}{K_l(w)}. \tag{2.371}$$

Substituting the values of J_l' and K_l' from (2.356) and (2.360), respectively, in (2.371), we get:

$$\frac{u}{J_l(u)}\left[\pm J_{l\mp1}(u) \mp \frac{lJ_l(u)}{u}\right] = \frac{w}{K_l(w)}\left[\mp\frac{lK_l(w)}{w} - K_{l\mp1}(w)\right]$$

or

$$\pm u\frac{J_{l\mp1}(u)}{J_l(u)} \mp l = \mp l - \frac{wK_{l\mp1}(w)}{K_l(w)}.$$

This equation may be written in either of the following two forms:

$$u\frac{J_{l+1}(u)}{J_l(u)} = \frac{wK_{l+1}(w)}{K_l(w)} \tag{2.372}$$

or

$$u\frac{J_{l-1}(u)}{J_l(u)} = -w\frac{K_{l-1}(w)}{K_l(w)} \tag{2.373}$$

One can obtain, from (2.372) and (2.373), the values of u and w for various values of l and the corresponding values of the propagation constant β. For β-values lying within the range:

$$\beta_2^2\left(=n_2^2k^2\right) < \beta^2 < \beta_1^2\left(=n_1^2k^2\right) \tag{2.374}$$

The radial part of the field, $R(r)$, in the core is given by the Bessel function $J_l(x)$, $x = ur/a$, which is oscillatory in nature. Hence, there exist m allowed solutions of β for each value of l. Therefore, each allowed value of β is characterised by two integers l and m. The first integer l is associated with two circular functions $\cos(l\phi)$ and $\sin(l\phi)$ corresponding to the azimuthal part of the solution, and the second integer m is associated with the mth root of the eigenvalue equation corresponding to the radial part of the solution. These are known as guided modes. From (2.344), we know that $V^2 = u^2 + w^2$. Therefore, the solution of the transcendental equations (for given values of l and V) will give universal curves describing the dependence of u or w on V. The value of β can be calculated by substituting the values of u (or w) and V in the defining equations. Alternatively, we can define the normalised propagation constant b as follows:

$$b = \frac{\beta^2 - \beta_2^2}{\beta_1^2 - \beta_2^2} = \frac{\beta^2 - n_2^2k^2}{n_1^2k^2 - n_2^2k^2} = \frac{w^2}{V^2} = 1 - \frac{u^2}{V^2}. \tag{2.375}$$

Since β lies between $\beta_1(=n_1k)$ and $\beta_2(=n_2k)$ for the guided modes, the value of b will lie between 0 (for $\beta=\beta_2$) and 1 (for $\beta=\beta_1$). The plots of b as a function of V, shown in Fig. 2.25, from universal curves.

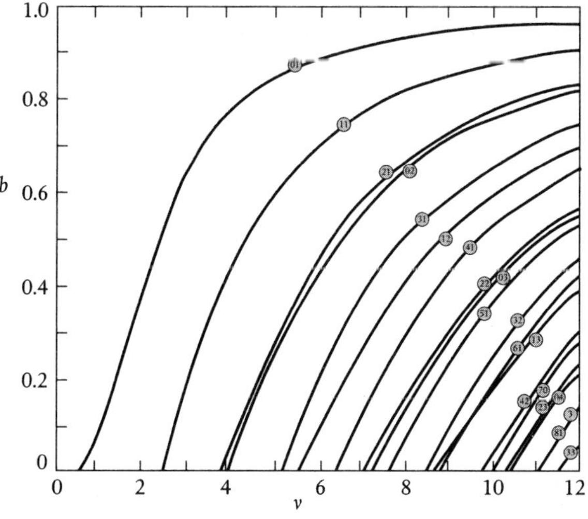

AQ3

Fig. 2.25 Plots of the normalised propagation constant b as a function of the normalised frequency parameter V for a SI fibre

A mode ceases to be guided when $\beta^2 < \beta_2^2$. Such modes are called radiation modes. In terms of the ray model, these modes correspond to rays that undergo refractions rather than total internal reflection at the core–cladding interface. When such modes are excited, they quickly leak away from the core. The condition $\beta = \beta_2$ corresponds to what is known as the cut-off mode. Therefore, $\beta = \beta_2$; $b = 0$, $w = 0$, and $u = V = V_c$, where V_c is the cut-off frequency. It is important to mention here that

$$\lim_{w \to 0} w \frac{K_{l-1}(w)}{K_l(w)} \to 0, \quad l = 0, 1, 2, 3, \ldots.$$

Therefore, the RHS of (2.373) may be used for finding the cut-off frequencies of different modes. For $l = 0$, we get the following form from (2.373):

$$u \frac{J_{-1}(u)}{J_0(u)} = -w \frac{K_{-1}(w)}{K_0(w)}.$$

Using (2.355) and (2.359), we get:

$$J_{-1}(u) = -J_1(u) \text{ and } K_{-1}(w) = K_1(w).$$

Substituting the value of $J_{-1}(u)$ and $K_{-1}(w)$ in the above-mentioned equation, we obtain:

$$\frac{uJ_1(u)}{J_0(u)} = w \frac{K_1(w)}{K_0(w)}. \tag{2.376}$$

At the cut-off frequency, $w = 0$, $u = V = V_c$, the RHS = 0, and (2.376) is transformed into

$$\frac{V_c J_1(V_c)}{J_0(V_c)} = 0$$

or

$$J_1(V_c) = 0. \tag{2.377}$$

The roots of (2.377) give the values of the cut-off frequency for $l = 0$ and $m = 1, 2, 3, \ldots$.
Similarly, for $l = 1$, we get the following form from (2.373):

$$\frac{uJ_0(u)}{J_1(u)} = -w \frac{K_0(w)}{K_1(w)}. \tag{2.378}$$

At the cut-off frequency, $w = 0$, $u = V = V_c$, the RHS = 0, and we obtain:

$$J_0(V_c) = 0. \tag{2.379}$$

The roots of (2.379) give V_c for $l = 1$ and $m = 1, 2, 3, \ldots$.
For $l \geq 2$ modes, the following equation gives the values of V_c:

$$J_{l-1}(V_c) = 0, \quad V_c \neq 0.$$

It should be mentioned here that, for $l \geq 2$, the root $V_c = 0$ must not be included because

$$\lim_{V \to 0} V \frac{J_{l-1}(V)}{J_l(V)} \neq 0, \quad \text{for } l \geq 2.$$

The values of the cut-off frequencies for the first few LP modes are given in Table 2.2. Figure 2.26 shows the oscillatory nature of Bessel functions J_o and J_1 and their roots.

Table 2.2 Cut-off frequencies of the first few lower order LP_{lm} modes in a SI fibre

l	m			
	1	**2**	**3**	**4**
0	0	3.832	7.106	10.173
1	2.405	5.520	8.654	11.790
2	3.832	7.016	10.173	13.324
3	5.136	8.417	11.620	14.796

Before proceeding further, a few important points must be mentioned about the modes:

(i) The $l=0$ modes have two-fold degeneracy corresponding to two orthogonal linearly polarised states.

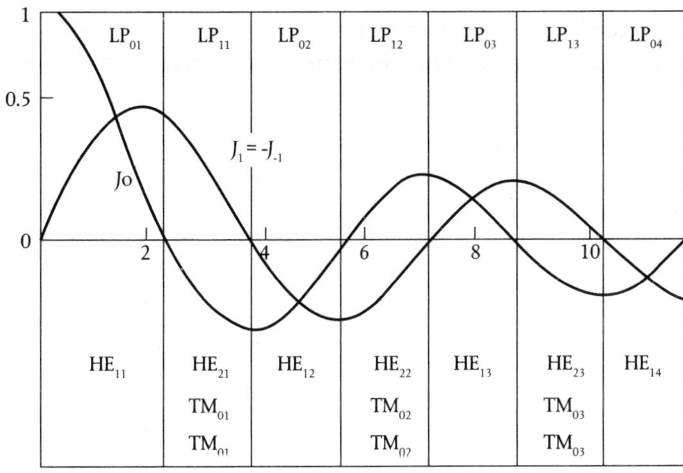

Fig. 2.26 Plot of Bessel functions J_0 and J_1, indicating the range of allowed values of u for lower order modes

(ii) The $l \geq 1$ modes have four-fold degeneracy as each polarisation state may have ϕ-dependence of the $\cos l\phi$ type or the $\sin l\phi$ type.

(iii) We know that the total number of modes (when $V \geq 1$) guided along a step index fibre is given approximately by $M = V^2/2$. As we know from (2.363) that the V-value of the fibre is dependent on the dimensions, the numerical aperture of the fibre and the wavelength of the light signal; the total number of guided modes in a particular fibre at a specific wavelength is fixed.

2.10.2 Fractional Modal Power Distribution

Once again, using the scalar approximation, let us calculate the fractional modal power distribution in the core and the cladding of an SI fibre. The power in the core is given by the integral:

$$P_{core} = (\text{constant}) \int_{r=0}^{a} \int_{\phi=0}^{2\pi} |\psi(r,\phi)|^2 \, r dr d\phi$$

$$= \text{constant}\{R(a)\}^2 \int_{r=0}^{a} \frac{J_l^2(ur/a) r dr}{J_l^2(u)} \int_{\phi=0}^{2\pi} \cos^2(l\phi) d\phi \quad (2.380)$$

where the transverse dependence of the modal fields as follows:

$$\psi(r,\phi) = AJ_l\left(\frac{ur}{a}\right)\begin{bmatrix}\cos(l\phi); & r<a \\ \sin(l\phi) & \end{bmatrix} \quad (2.381)$$

and

$$\psi(r,\phi) = BK_l\left(\frac{wr}{a}\right)\begin{bmatrix}\cos(l\phi); & r>a \\ \sin(l\phi) & \end{bmatrix}, \quad (2.382)$$

where we have substituted the value of $\psi(r,\phi)$ from (2.380) in (2.382) and taken the ϕ -dependence to be of the form $\cos(l\phi)$. The solution of the above-mentioned integral can be shown to be:

$$P_{core} = C\pi a^2 \left[1 - \frac{J_{l-1}(u) J_{l+1}(u)}{J_l^2(u)}\right], \quad (2.383)$$

where C is a constant.

Similarly, the power distribution in the cladding may be obtained by solving the integral:

$$P_{cladding} = (\text{constant}) \int_{r=a}^{\infty} \int_{\phi=0}^{2\pi} |\psi(r,\phi)|^2 \, r dr d\phi, \quad (2.384)$$

where we have substituted the value of $\psi(r,\phi)$ from (2.381) in (2.382) and taken the ϕ -dependence to be of the form $\cos(l\phi)$. The solution of the above-mentioned integral can be shown to be:

$$P_{cladding} = \text{constant}\{R(a)\}^2 \int_{r=a}^{\infty} \frac{K_l^2(wr/a) r dr}{K_l^2(w)} \int_{\phi=0}^{2\pi} \cos^2 l\phi \, d\phi \quad (2.385)$$

$$P_{cladding} = C\pi a^2 \left[\frac{K_{l-1}(w) K_{l+1}(w)}{K_l^2(w)} - 1\right]. \quad (2.386)$$

Adding (2.383) and (2.386), we get an expression for the total power P_T as follows:

$$P_T = P_{core} + P_{cladding}$$

$$\Rightarrow P_T = C\pi a^2 \left[1 - \frac{J_{l-1}(u) J_{l+1}(u)}{J_l^2(u)}\right] + C\pi a^2 \left[\frac{K_{l-1}(w) K_{l+1}(w)}{K_l^2(w)} - 1\right]$$

$$\Rightarrow P_T = C\pi a^2 \left[\frac{K_{l-1}(w) K_{l+1}(w)}{K_l^2(w)} - \frac{J_{l-1}(u) J_{l+1}(u)}{J_l^2(u)}\right]. \quad (2.387)$$

The eigenvalue equation of an ideal step-index optical fibre may be written in either of the following two forms:

$$u\frac{J_{l+1}(u)}{J_l(u)}=\frac{wK_{l+1}(w)}{K_l(w)} \tag{2.388}$$

or

$$u\frac{J_{l-1}(u)}{J_l(u)}=-w\frac{K_{l-1}(w)}{K_l(w)}. \tag{2.389}$$

Multiplying the eigenvalue in (2.388) and (2.389), we get:

$$\frac{u^2 J_{l-1}(u)J_{l+1}(u)}{J_l^2(u)}=-\frac{w^2 K_{l-1}(w)K_{l+1}(w)}{K_l^2(w)}. \tag{2.390}$$

Using (2.390), (2.387) may be written as:

$$P_T=C\pi a^2\left(\frac{K_{l-1}(w)K_{l+1}(w)}{K_l^2(w)}\right)\left(1+\frac{w^2}{u^2}\right) \tag{2.391}$$

$$\Rightarrow P_T=C\pi a^2\frac{V^2}{u^2}\left[\frac{K_{l-1}(w)K_{l+1}(w)}{K_l^2(w)}\right]. \tag{2.392}$$

Again using (2.390), (2.383) may be written as:

$$P_{core}=C\pi a^2\left[1+\frac{w^2}{u^2}\frac{K_{l-1}(w)K_{l+1}(w)}{K_l^2(w)}\right]. \tag{2.393}$$

Dividing (2.393) by (2.392), we obtain the fractional power propagating in the core as:

$$\frac{P_{core}}{P_T}=\frac{C\pi a^2\left(1+\frac{w^2}{u^2}\frac{K_{l-1}(w)K_{l+1}(w)}{K_l^2(w)}\right)}{C\pi a^2\frac{V^2}{u^2}\left[\frac{K_{l-1}(w)K_{l+1}(w)}{K_l^2(w)}\right]}$$

$$\Rightarrow\frac{P_{core}}{P_T}=\left[\frac{u^2}{V^2}\frac{K_l^2(w)}{K_{l-1}(w)K_{l+1}(w)}+\frac{w^2}{V^2}\right]. \tag{2.394}$$

The fractional power propagating in the cladding is given as:

$$\frac{P_{cladding}}{P_T}=1-\frac{P_{core}}{P_T}=\frac{u^2}{V^2}\left[1-\frac{K_l^2(w)}{K_{l-1}(w)K_{l+1}(w)}\right]. \tag{2.395}$$

It is interesting to note that for the first two lower order modes, the power flow is mostly in the cladding near cut off. Prime denotes the first derivative. For $\tilde{x}\ll 1$,

$$\frac{K_0}{K_l}=\tilde{x}\ln\frac{2}{1.782\tilde{x}} \tag{2.396}$$

$$\frac{K_{l-1}}{K_l}=\frac{\tilde{x}}{2(l-1)},\quad l\geq 2 \tag{2.397}$$

$$\frac{K_{l+1}}{K_l} = \frac{2l}{\tilde{x}}, \quad l \geq 1 \tag{2.398}$$

Using (2.396)–(2.398), it can be shown that as $w \to 0$, and $u \to V_c$,

$$\frac{P_{core}}{P_T} \to \left. \begin{matrix} 0 & \text{for } l = 0 \text{ and } 1 \\ \frac{(l-1)}{l} & \text{for } l \geq 2 \end{matrix} \right\}. \tag{2.399}$$

However, for larger values of l, the power remains in the core even at or just beyond cut off. Another point to be mentioned is that the power associated with a particular mode is mostly confined in the core for large values of V.

2.10.3 Modal Analysis of an Ideal Graded Index Fibre

We know that the refractive index profile for a multimode graded index (GI) fibre is given as:

$$n(r) = \left. \begin{matrix} n_0 \left\{1 - 2\Delta (r/a)^{\alpha}\right\}^{1/2}, & r \leq a \\ n_0 \left\{1 - 2\Delta\right\}^{1/2} = n_c, & r \geq a \end{matrix} \right\}, \tag{2.400}$$

where

$$\Delta = \left(n_0^2 - n_c^2\right) / 2n_0^2 \approx \frac{(n_0 - n_c)}{n_0}$$

when $\Delta \ll 1$. n_0 is the refractive index at $r = 0$, n_c is the refractive index of the cladding, and α is called the profile parameter. The equation for an electric field in an isotropic, linear, non-conducting, and non-magnetic; however, inhomogeneous medium is given as:

$$\nabla^2 E + \nabla \left\{ \frac{1}{\varepsilon_r} \nabla(\varepsilon_r) \cdot E \right\} - \mu_0 \varepsilon_0 \varepsilon_r \frac{d^2 E}{dt^2} = 0. \tag{2.401}$$

Similarly, the equation for a magnetic field is given as:

$$\nabla^2 H + \frac{1}{\varepsilon_r} \left\{ \nabla(\varepsilon_r) \times (\nabla \times H) \right\} - \mu_0 \varepsilon_0 \varepsilon_r \frac{d^2 H}{dt^2} = 0. \tag{2.402}$$

In general, for a cylindrical dielectric waveguide, for example, an optical fibre, the electric and magnetic fields are governed by (2.401) and (2.402), respectively. The exact solutions of these equations for graded-index fibres are difficult to obtain. However, using the Wentzel–Kramers–Brillouin (WKB) approximation (Gloge and Marcatili 1973), it is possible to study the propagation characteristics of these fibres. The propagation constant β_p of the pth mode in a GI fibre with an α-profile may be given as:

$$\beta_p \approx \beta_0 \left[1 - 2\Delta \left(\frac{P}{M_g} \right) \right]^{1/2}, \tag{2.403}$$

where $p = 1, 2, 3, \ldots,$ M_g and $\beta_0 = kn_0$. Here, M_g represents the total number of guided modes and it is given as:

$$M_g = k^2 n_0^2 a^2 \Delta \left(\frac{\alpha}{\alpha + 2} \right). \tag{2.404}$$

Substituting the values of Δ and k, we get:

$$M_g = \frac{1}{2}\left(\frac{\alpha}{\alpha+2}\right)\left[\frac{2\pi a}{\lambda}\left(n_0^2 - n_c^2\right)^{1/2}\right]^2. \tag{2.405}$$

Equation (2.405) gives the approximate modal volume or the number of modes guided by the α -profile graded index fibre. For a step-index fibre $(\alpha = \infty)$, $n = n_1$ in the core and $n = n_2$ in the cladding. Therefore, the modal volume M_s for such a fibre will be given as:

$$M_s \approx \frac{1}{2}\left[\frac{2\pi a}{\lambda}\left(n_1^2 - n_2^2\right)^{1/2}\right]^2. \tag{2.406}$$

We know that the normalised frequency parameter is given as:

$$V = \frac{2\pi a}{\lambda}\left(n_1^2 - n_2^2\right)^{1/2}.$$

Therefore,

$$M_s = \frac{V^2}{2}. \tag{2.407}$$

For a graded-index fibre, the numerical aperture (NA) for the guided rays varies with r as follows:

$$NA(r) = \left.\begin{matrix} \left\{n^2(r) - n_c^2\right\}^{1/2} \\ 0 \quad \text{for} \quad r \geq a \end{matrix}\right\}. \tag{2.408}$$

However, for small variation of $n(r)$ with r:

$$NA \approx \left(n_0^2 - n_c^2\right)^{1/2}$$

and

$$V \approx \frac{2\pi a}{\lambda}\left(n_0^2 - n_c^2\right)^{1/2}.$$

With this approximation, we get:

$$M_g \cong \left(\frac{\alpha}{\alpha+2}\right)\frac{V^2}{2}. \tag{2.409}$$

Care must be taken when using (2.403)–(2.409) because the WKB analysis is valid for highly multimode fibres with V much greater than 1. For a parabolic profile, $\alpha = 2$ and

$$M_g \approx \frac{V^2}{2}.$$

It can be shown that the cut-off value of the normalised frequency V_c to support a single mode in a graded-index fibre is given as:

$$V_c = 2.405\left(1 + \frac{2}{\alpha}\right)^{1/2}. \tag{2.410}$$

Therefore, it is possible to determine the structural and/or operational parameters of the fibre, which give single-mode operation. We have obtained a similar result for an SI fibre $(\alpha = \infty)$ in our earlier discussion.

SUMMARY

- Optical waveguides for optical fibres and optical integrated circuits utilise a wave phenomenon that traps the light locally and guides it in any direction, although their propagation lengths differ greatly.
- In the microwave band, electromagnetic fields are generally distributed within a finite area surrounded by conductors, as in waveguides and coaxial lines. On the other hand, it is difficult to construct a waveguide mainly from metal in the visible wave band, since the metal will behave as a substance having a complex permittivity with a large absolute value. For this reason, waveguides are usually constructed by combining appropriate dielectrics.
- Tapered waveguides are used to change the width of waveguides; branching waveguides and crossed waveguides are used for splitting, combining, and interference; and optical waveguide directional couplers and two-mode waveguide couplers are used for coupling.
- Light is an electromagnetic wave phenomenon. Its magnetic field is represented by four electromagnetic field vectors that are functions of position $r[\mathrm{m}]$ and time $t[\mathrm{s}]$. The four vectors electric field $E[\mathrm{V/m}]$, magnetic field $H[\mathrm{A/m}]$, electric flux density $D[\mathrm{C/m^2}]$, and magnetic density $B[\mathrm{wb/m^2}]$ conform to Maxwell's equations. Here, $J[\mathrm{A/m^2}]$ is the electric current density. These equations are usually called Maxwell's equations.

$$\nabla \cdot B = 0$$
$$\nabla \cdot D = \rho$$
$$\nabla \times E = -\partial B/\partial t$$
$$\nabla \times H = \partial D/\partial t + J$$

- The flow density of the energy carried by an electromagnetic field is expressed by the Poynting vector $S[\mathrm{W/m^2}]$, which is defined as follows:
$S = E \times H$.
- If the polarisation generated in a light transmitting medium is $P[\mathrm{C/m^2}]$ and the magnetisation is $M[\mathrm{A/m}]$, $\varepsilon_0[\mathrm{F/m}]$ and $\mu_0[\mathrm{H/m}]$ are the permittivity of vacuum and the permeability of free space, respectively. The electric flux density D and the magnetic flux density B are, respectively,

$$D = \varepsilon_0 E + P$$
$$B = \mu_0 (H + M)$$

If the medium constant is dependent on the direction of the electromagnetic field vector, then the medium constant is dependent on the direction of the electromagnetic field vector, which is called an anisotropic medium; here, $[\varepsilon]$ and $[\mu]$ are the rank-two permittivity and permeability tensors, respectively and its constitutive relations are

$$D = [\varepsilon]E$$
$$B = [\mu]H$$

- In the case of a nonlinear medium, polarisation is given as:

$$P_i = \varepsilon_0 \left[\sum_{j=1}^{3} \chi_{ij}^{(1)} E_j + \sum_{j=1}^{3}\sum_{k=1}^{3} \chi_{ijk}^{(2)} E_j E_k + \sum_{j=1}^{3}\sum_{k=1}^{3}\sum_{l=1}^{3} \chi_{ijkl}^{(3)} E_j E_k E_l + \ldots \right]$$

The first term is a linear term with $\chi_{ij}^{(1)}$ representing the rank-two linear susceptibility tensor. The second and subsequent terms are nonlinear terms; $\chi_{ijk}^{(2)}[m/V]$, representing the secondary nonlinear effect, is the rank-three nonlinear susceptibility tensor, and $\chi_{ijkl}^{(3)}[m^2/V^2]$, representing the tertiary nonlinear effect, is the rank-four nonlinear susceptibility tensor.

- If the permittivity is uniform, E and H, respectively, will satisfy the following wave equations:

$$\nabla^2 E - \varepsilon\mu_0 \frac{\partial^2 E}{\partial t^2} = 0$$

$$\nabla^2 H - \varepsilon\mu_0 \frac{\partial^2 H}{\partial t^2} = 0$$

- In particular, in a vacuum,

$$Z_0 = \sqrt{\mu_0 / \varepsilon_0} = 376.73[\Omega]$$

and Z_0 is called the intrinsic (characteristic) impedance of a vacuum or of free space.

- The path of the end point of an electric field vector can be computed as:

$$\frac{E_x^2}{|A_x|^2} + \frac{E_y^2}{|A_y|^2} - 2\frac{E_x E_y}{|A_x||A_y|}\cos\phi = \sin^2\phi$$

Here, $\phi = \phi_y - \phi_x$. The afore-mentioned equation represents an inclined ellipse, which is inscribed in a rectangle $2|A_x|$ wide and $2|A_y|$ high.

- If $\phi = 0$,

$$E_x / |A_x| = E_y / |A_y|$$

If $\phi = \pi$,

$$E_x / |A_x| = -E_y / |A_y|$$

Therefore, when observed within the plane at a constant distance z, the electric field vector lies on a straight line.

- The angle of incidence and the angle of reflection are equal and Snell's law applies to the incidence and the angle of refraction.

$$\sin\theta_i / \sin\theta_t = n_2 / n_1$$

- Critical angle θ_c is:

$$\theta_c = \sin^{-1}(n_2 / n_1)$$

- The amount of Goos–Hánchen shift for TE polarised light is:

$$z_{GH} = \frac{d\phi_s}{d\beta} = \frac{\tan\theta_i}{k_0\sqrt{(\beta / k_0)^2 - n_2^2}}$$

- The amount of Goos–Hánchen shift for TM polarised light is:

$$z_{GH} = \frac{d\phi_s}{d\beta} = \frac{\tan\theta_i}{k_0\sqrt{(\beta / k_0)^2 - n_2^2\left[(\beta / k_0 n_1)^2 + (\beta / k_0 n_2)^2 - 1\right]}}$$

- For TM polarised light, if the angle of incidence is equal to θ_B, this is defined as:

$$\theta_B = \tan^{-1}(n_2 / n_1)$$

is called the Brewster angle and is used for the discrimination of polarised light and in non-reflective windows.

- Wave equations for graded optical waveguides is:

$$\nabla_t^2 E + \left(k_0^2 n^2 - \beta^2\right)E + \nabla\left[\frac{\nabla_t\left(n^2\right)}{n^2}.E\right] = 0$$

$$\nabla_t^2 H + \left(k_0^2 n^2 - \beta^2\right)H + \frac{\nabla_t\left(n^2\right)}{n^2} \times (\nabla \times H) = 0.$$

- Wave equations for stepped optical waveguides is:

$$\nabla_t^2 E + \left(k_0^2 n^2 - \beta^2\right)E = 0$$

$$\nabla_t^2 H + \left(k_0^2 n^2 - \beta^2\right)H = 0.$$

- The cut-off frequency f_{cutoff} for the guided mode is obtained as follows:

$$f_{cutoff} = \frac{c}{2\pi t\sqrt{n_f^2 - n_s^2}}\left[\tan^{-1}\left(\frac{m_c\sqrt{n_s^2 - n_c^2}}{m_f\sqrt{n_f^2 - n_s^2}}\right) + q\pi\right].$$

- The modal dispersion σ_m is:

$$\sigma_m = \frac{\lambda_0}{c}.\frac{d^2 n_1}{d\lambda_0^2}.$$

- The waveguide dispersion σ_w. If $N_1 - N_2 \simeq n_1 - n_2 \simeq n_1\Delta$,

$$\sigma_w = \frac{n_1 v\Delta}{c\lambda_0}.\frac{d^2(vb)}{dv^2}.$$

- $\Delta\beta = \left|\beta_x - \beta_y\right|$ is the phase constant difference and is called polarisation birefringence or modal birefringence.
- The normalised group delay difference σ_p is:

$$\sigma_p = c\tau_p = \frac{d(\Delta\beta)}{dk_0} = \frac{d(k_0\beta)}{dk_0}$$

SOLVED PROBLEMS

Problem 1: A multimode graded index fibre exhibits a total pulse broadening of 1.2 ns over a distance of 30 km. Determine the pulse dispersion per unit length.

Solution:

Pulse broadening $= 1.2\,\text{ns}$, Distance = 30 km

Dispersion per unit length

$$= \frac{1.2}{30}\left(\frac{\text{ns}}{\text{km}}\right) = 40\left(\frac{\text{ps}}{\text{km}}\right).$$

Problem 2: Consider a planar slab waveguide of infinite extent in the *y*- and *z*-directions. A planar waveguide is made with $n_f = 1.48, n_s = 1.46$, and $n_c = 1.46$. The thickness of the guiding film is $10\,\mu\text{m}$. The waveguide excited with a 1.3-μm source. For TE modes:

(a) What is the range of allowed β values for this waveguide?
(b) What is the numerical aperture for this waveguide?
(c) What is the longest wavelength that can be carried in a TE mode in this waveguide?
(d) How many modes will this waveguide carry if the excitation wavelength is 0.6 μm?

Solution:

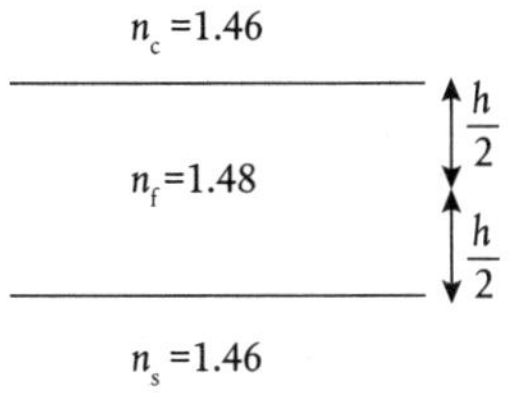

$n_c = 1.46$

$n_f = 1.48$ $\frac{h}{2}$ $\frac{h}{2}$

$n_s = 1.46$

Fig. P1 Planar slab waveguide

Consider Fig. P1; this is a symmetric planar slab waveguide since $n_s = n_c$. Wavelength of excitation λ. = 1.3 μm.
For TE mode:
We know that

$$k_0 n_s \le \beta \le k_0 n_f,$$

$$k_0 = \left|\frac{2\pi}{\lambda}\right| = \frac{2\pi}{1.3\times 10^{-6}} = 4.8\frac{1}{\mu\text{m}}$$

Hence

(a) $7.05\frac{1}{\mu\text{m}} \le \beta \le 7.104\frac{1}{\mu\text{m}}$
(b) using 1.10, $\text{NA} \Rightarrow \sin i = \sqrt{n_f^2 - n_s^2} = 0.242$
(c) The longest wavelength would be at the smallest value of β. When $\beta = k_0 n_s$,

$$\frac{2\pi}{\lambda} = \frac{2\pi}{\lambda_0}\times n_s \Rightarrow \lambda = \frac{\lambda_0}{n_s} = \frac{1.3\,\mu\text{m}}{1.46} = 0.8904\,\mu\text{m}$$

(d) The approximate number of modes can be found from

$$m = \text{Int}\left[\frac{\kappa_{max} h}{\pi}\right] = \text{Int}\left[\frac{hk_0\left(n_f^2 - n_s^2\right)^{1/2}}{\pi}\right]$$

where

$$k = \sqrt{k_0^2 n_f^2 - \beta^2}$$

$$\gamma = \sqrt{\beta^2 - k_0^2 n_s^2}\,.$$

Hence,

$$k_{max} = k_o\sqrt{n_f^2 - n_s^2}\,.$$

Every time $\kappa_{max}h$ increases by π, another mode is allowed. Therefore, approximation for "m" is most accurate when "m" is a large number. Note that the mode count increases with the thickness h of the guide with the difference in index $(n_f^2 - n_s^2)$ between the core and the cladding and as the wavelength of the guided light gets shorter.

We usually characterise a waveguide by its normalised frequency defined as $V = hk_o(n_f^2 - n_s^2)^{1/2}$. In terms of the normalised frequency, the approximate number of modes "m" in a waveguide is $m \approx \dfrac{V}{\pi}$.

For this case,

$$V = 10 \times 10^{-6} \times \left(\frac{2\pi}{0.6 \times 10^{-6}}\right)\sqrt{(1.48)^2 - (1.46)^2} = 25.3.$$

Hence,

$$m \approx \frac{V}{\pi} = \frac{25.3}{3.14} \approx 8 \text{ modes}.$$

Problem 3: For a symmetric waveguide of thickness h as shown in Fig. P2, show that the coefficient A is given as:

$$A^2 = \frac{4\omega\mu_0 P}{\left(h + \frac{2}{\gamma}\right)\beta},$$

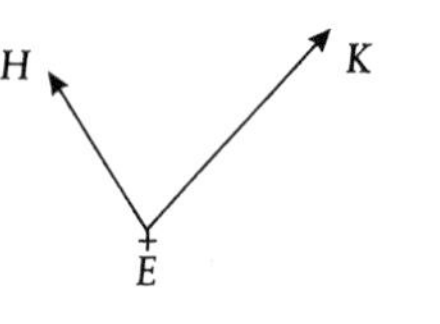

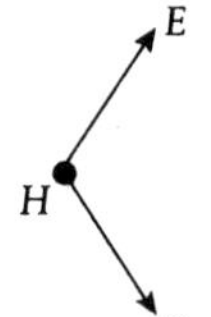

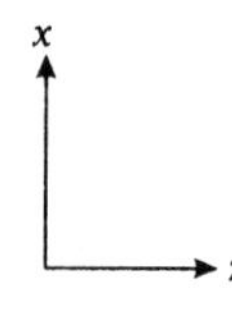

Fig. P2 Symmetric waveguide

where P is the power per unit length carried by the waveguide and the meaning of the various terms as usable. Further, show that $H_x = \dfrac{\beta}{\omega\mu} E_y$ for TE wave. In Fig. P2, differentiate between electric (E_x, E_y, E_z) and magnetic (H_x, H_y, H_z) field components into TE and TM modes.

Solution:

Note that the power associated with TE mode, the power flow is given as:

$$S = \frac{1}{2}Re(E \times H^*) \Rightarrow S_z = -\frac{1}{2}Re(E_y \times H_x^*) = \frac{\beta}{2w\mu_0}|E_y|^2.$$

Hence, the total power associated with the mode per unit length is given in (2.283):

$$P = \frac{1}{2}\frac{\beta}{w\mu_0}\int_{-\infty}^{\infty}|E_y|^2\,dx$$

For the symmetric waveguide case,

$$P=\frac{1}{2}\frac{\beta}{w\mu_0}\left(\frac{A^2}{\cos^2\frac{\kappa h}{2}}\int_0^{\frac{h}{2}}\cos^2\kappa x dx+A^2e^{\gamma h}\int_{\frac{h}{2}}^{\infty}e^{-2\gamma x}dx\right).$$

In this expression,

$$k=\sqrt{k_0^2n_f^2-\beta^2}$$

$$\gamma=\sqrt{\beta^2-k_0^2n_s^2}$$

$$\Rightarrow\frac{1}{2}\frac{\beta A^2}{w\mu_0}\left(\frac{h}{2}+\frac{\sin\kappa h}{2\kappa}+\frac{1}{\gamma}\cos^2\frac{\kappa h}{2}\right)$$

$$\Rightarrow\frac{1}{4}\frac{\beta A^2}{w\mu_0}\left[h+\frac{2\sin\frac{\kappa h}{2}\cos\frac{\kappa h}{2}}{\kappa}+\frac{2}{\gamma}\left(1-\sin^2\frac{\kappa h}{2}\right)\right]$$

$$\Rightarrow\frac{1}{4}\frac{\beta A^2}{w\mu_0}\left[h+\frac{2}{\gamma}+\frac{2\sin\frac{\kappa h}{2}\cos\frac{\kappa h}{2}}{\gamma\kappa}\left(\gamma-\kappa\tan\frac{\kappa h}{2}\right)\right].$$

Since

$$\tan\frac{\kappa h}{2}=\frac{\gamma}{\kappa},$$

we get

$$P_{\text{total}}=\frac{1}{4}\frac{\beta A^2}{w\mu_0}\left(h+\frac{2}{\gamma}\right).$$

or

$$A^2=\frac{4w\mu_0P_{\text{total}}}{\left(h+\frac{2}{\gamma}\right)\beta}$$

For TE mode:

E_y, H_x, and H_z components exist and $H_y=0=E_x=E_z$.

Since

$$\nabla\times E=-\frac{\partial\beta}{\partial t}=-\mu\frac{\partial H}{\partial t}=-\mu j\omega\left(H_xa_x+H_ya_y+H_za_z\right),$$

$$\begin{vmatrix}a_x & a_y & a_z\\ \frac{\partial}{\partial x} & \frac{\partial}{\partial y} & \frac{\partial}{\partial z}\\ 0 & E_y & 0\end{vmatrix}=-j\omega\mu\left(H_xa_x+H_ya_y+H_za_z\right).$$

By comparing a_x, a_y, and a_z on both sides, we get

$$-\frac{\partial}{\partial z}E_y = -j\omega\mu H_x.$$

Since the field dependent on z is $e^{j(\omega t-\beta z)}$, we get:

$$\frac{\partial}{\partial z} \to -j\beta.$$

Then,

$$|H_x| = \left|\frac{\beta}{\omega\mu}\right|.$$

For TM mode, $H_y E_x E_z$ components exist; hence, $H_x = H_z = E_y = 0$, for TM mode.

Problem 4: Given a symmetric waveguide with $n_f = 1.5, n_s = 1.47$, and $\lambda = 1.0\,\mu\text{m}$. Determine the fractional power carried in the cladding if the guiding layer is $3-\mu\text{m}$ thick. Further, show that the eigenvalue equation for β for the asymmetric waveguide reduces to that of symmetric waveguide when $n_s = n_c$.

Solution:
Since we know that the power carried by the guiding layer is

$$\frac{P_{\text{core}}}{P_{\text{total}}} = \frac{\int_{-h/2}^{h/2} E_y(x)H_x^*(x)dx}{\int_{-\infty}^{\infty} E_y(x)H_x^*(x)dx}.$$

Hence, using (2.284a) and (2.284c), we get:

$$P_{\text{core}} = \frac{\frac{1}{2}\beta}{\omega\mu_0}\left(\frac{A^2}{\cos^2\left(\frac{\kappa h}{2}\right)}\int_0^{h/2}\cos^2\kappa x dx\right) = \frac{\beta A^2}{2\omega\mu_0}\frac{(h\kappa + \sin^2\kappa h)}{(1+\cos\kappa h)}.$$

Therefore, we have

$$\left|\frac{P_{\text{core}}}{P_{\text{total}}}\right| = \frac{\gamma(h\kappa + \sin\kappa h)}{(h\gamma + 2)\kappa}.$$

Since we know from previous example

$$P_{\text{total}} = \frac{1}{4}\frac{\beta A^2}{\omega\mu_0}\left(h + \frac{2}{\gamma}\right).$$

Now,

$$\left|\frac{P_{\text{cladding}}}{P_{\text{total}}}\right| = 1 - \left|\frac{P_{\text{core}}}{P_{\text{total}}}\right| = 1 - \frac{\gamma(h\kappa + \sin\kappa h)}{(h\gamma\kappa + 2\kappa)} = \frac{h\gamma\kappa + 2\kappa - h\gamma\kappa - \gamma\sin\kappa h}{(h\gamma\kappa + 2\kappa)}.$$

Fractional power carried by cladding is

$$= \frac{2\kappa - \gamma\sin\kappa h}{(h\gamma\kappa + 2\kappa)}.$$

Hence, we can find out the fractional power carried by allowed values of k and γ. For asymmetric waveguide with $n_s \neq n_c$ have the eigenvalue equation:

$$\tan(\kappa h) = \frac{\gamma_s + \gamma_c}{\kappa_f \left[1 - \frac{\gamma_c \gamma_s}{\kappa_f^2}\right]}.$$

Problem 5: Consider the situation shown in Fig. P3. A TE wave (polarised along $\hat{y}$) with wavelength of $1\,\mu m$ is incident from air onto the GaAs air interface at an angle of incidence of $45°$. The index of refraction of GaAs equals 3.4 at $1\,\mu m$. Describe the electric fields in all regions surrounding the interface.

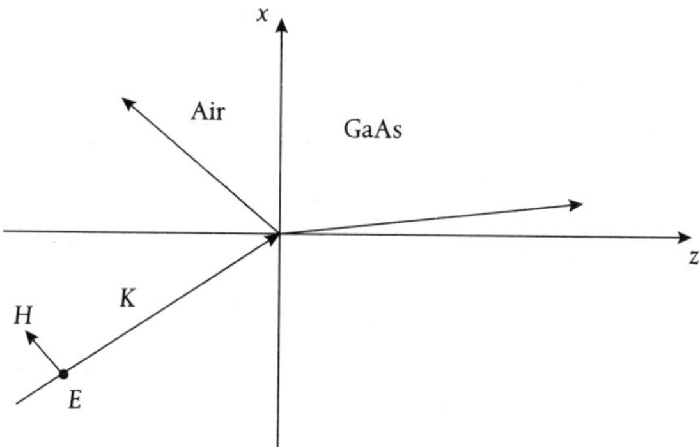

Fig. P3 A TE wave polarised along axis

Solution:

This is the case of TE wave. Hence,
reflection coefficient

$$r = \frac{E_{ro}}{E_i} = \frac{n_1 \cos\theta_i - n_2 \cos\theta_t}{n_1 \cos\theta_i + n_2 \cos\theta_t} \tag{P1}$$

and transmission coefficient

$$T = \frac{E_{to}}{E_i} = \frac{2n_1 \cos\theta_i}{n_1 \cos\theta_i + n_2 \cos\theta_t}. \tag{P2}$$

Now, since

$$n_1 \sin\theta_1 = n_2 \sin\theta_2,$$

$\sin 45° = 3.4 \times \sin\theta_2$, $\theta_t = \theta_2 = 12°$ and $E_r = rE_i$, and $E_t = TE_i$,
the incident electric field:

$$\begin{cases} E_i = E_{i0} e^{j(\omega t - \kappa r)} \hat{a}_y \\ E_t = E_{t0} e^{j(\omega t - \kappa r)} \hat{a}_y \\ E_r = E_{r0} e^{j(\omega t - \kappa r)} \hat{a}_y \end{cases}. \tag{P3}$$

From (P1), $\theta_i = 45°$, $\theta_t = 12°$ & $n_1 = 1$, $n_2 = 3.4$, we can find reflection coefficient (*r*) and transmission coefficient (*T*) from (P2), and finally, (P3) can be evaluated.

Problem 6: Write down the eigenvalue equation for the TM mode in a symmetric waveguide as shown in Fig. P4. Which is the lowest order mode in an asymmetric/symmetric waveguide: the TE_0 mode or the TM_0 mode?

Solution:

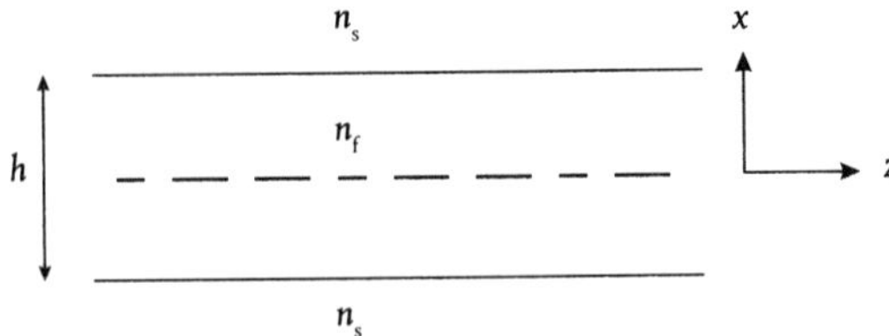

Fig.P4 Refractive index profile of symmetric waveguide

The characteristic equation for TM modes is

$$\tan\frac{\kappa h}{2} = \frac{n_f^2}{n_s^2}\cdot\frac{\gamma}{\kappa},$$

where

$$\kappa = \sqrt{\kappa_0^2 n_f^2 - \beta^2}$$

$$\gamma = \sqrt{\beta^2 - \kappa_0^2 n_f^2}\ .$$

The lowest order mode in a symmetric/asymmetric waveguide is TE_{01} mode.

Problem 7: Consider a multimode fibre that has core refractive index $n_1 = 1.48$ and cladding index $n_2 = 1.46$. Find

(a) Critical angle
(b) Numerical aperture
(c) Acceptance angle in air

Solution:

(a) Using (1.6), $n_1 \sin\theta_1 = n_2 \Rightarrow \theta_c = \sin^{-1}\left(\frac{n_2}{n_1}\right) = \sin^{-1}\left(\frac{1.46}{1.48}\right) = 80.56°$.

(b) Using (1.10), $\text{NA} = \left(n_1^2 - n_2^2\right)^{1/2} \Rightarrow \text{NA} = 0.24$.

(c) $\theta_a = \sin^{-1}(\text{NA}) = \sin^{-1}\left[\left(n_1^2 - n_2^2\right)^{1/2}\right] = 13.88$.

Problem 8: A multimode optical fibre has a V-number 26.6 at 1300 nm wavelength. Find the number of modes supported by multimode fibre?

Solution: Number of modes [using (1.19)],

$$M = \frac{V^2}{2} = 353.78$$

Problem 9: An optical fibre has a core refractive index of 1.5 and a cladding refractive index of 1.47. Find the following:

(a) Critical angle at core–cladding interface
(b) Numerical aperture NA of the fibre
(c) The acceptance angle in air for the fibre

Solution:
Refractive index of the core $n_1 = 1.5$
Cladding $n_2 = 1.47$

(a) $\sin\theta_c = \left(\frac{n_2}{n_1}\right) = \left(\frac{1.47}{1.5}\right) = 0.98 \Rightarrow \theta_c = 78.52^{\circ}$

(b) $\text{NA} = \sqrt{n_1^2 - n_2^2} = 0.298$

(c) Acceptance angle, $\theta_A = \sin^{-1}(\text{NA}) = 17.36°$

Problem 10: Determine the cut-off wavelength for a step-index fibre to exhibit single-mode operation when the core refractive index and radius were 1.46 and 4.5 μm, respectively, with the relative index difference being 0.25%.

Solution:
Core refractive index, $n_1 = 1.46$
Core radius, $a = 4.5\ \mu\text{m}$
Relative index, $\Delta = 0.25\%$
So,

$$\lambda_c = \frac{2\pi a}{V_c}(\text{NA}) = \frac{2\pi a n_1}{V_c}(2\Delta)^{1/2}$$

$$\text{NA} = n_1(2\Delta)^{1/2}$$

$$= \frac{2\pi \times 4.5 \times 10^{-6} \times 1.46}{2.405}\left(2 \times \frac{0.25}{100}\right)^{1/2} = 1.214\ \mu\text{m}.$$

Problem 11: A multimode graded index fibre has an acceptance angle in air of 9°. Calculate the relative refractive index difference between the core axis and the cladding when the refractive index at the core axis is 1.45.

Solution:
Numerical aperture is NA and it is equal to

$$\text{NA} = n_0 \sin\theta_a$$

where $n_0 \rightarrow$ refractive index of air $(\cong 1)$

$$\theta_a \rightarrow \text{Acceptance angle}$$

$$\Rightarrow \text{NA} = \sin 9^0 = 0.1564$$

Further, $\text{NA} = n_1(2\Delta)^{1/2}$

$$n_1 \rightarrow \text{Refractive index of the core}$$

$\Delta \rightarrow$ Relative refractive index difference

$$\Rightarrow 0.1564 = 1.45(2\Delta)^{1/2}$$

$$\Rightarrow 2\Delta = (0.10786)^2$$

$$\Rightarrow \Delta = 5.817 \times 10^{-3}$$

$$\Rightarrow \Delta \cong 0.582\%$$

Problem 12: An optical fibre has a core refractive index of 1.45 and a cladding refractive index of 1.4. Assuming ray theory of analysis, determine the following:

(a) Numerical aperture in air for fibre.
(b) Critical angle at core–cladding interface.
(c) Acceptance angle in air for fibre.

Solution:
Given that
Core Refractive index, $n_1 = 1.45$
Cladding Refractive index, $n_2 = 1.40$
Assuming ray theory of analysis, we get

(a) Numerical aperture (N.A) of the fibre, $\text{NA} = \sqrt{n_1^2 - n_2^2} = 0.3774$
(b) Using (1.6), $\sin\theta_c = \frac{n_2}{n_1} = 0.965$
$\Rightarrow \theta_c$ = critical angle at core–cladding interface.

$$\Rightarrow \theta_c = \sin^{-1} 0.965 = 74.9^\circ$$

Further, maximum angle of obliqueness $(\phi_m) = 90^\circ - \theta_c = 15.09^\circ$
(c) Using (1.8), $n_a \sin\alpha_m = n_1 \sin\varnothing_m$
where n_a = refractive index of air = 1

$$\sin\alpha_m = 1.45\sin(15.09^\circ)$$

$$\Rightarrow \alpha_m = \sin^{-1}(0.3774)$$

$$\Rightarrow \text{Acceptance angle in air} = 22.178^\circ.$$

Problem 13: An optical fibre cable has a core of 1.45 refractive index and $10-\mu\text{m}$ diameter, with a fractional index difference of 0.3%. Determine its numerical aperture, permissible wavelengths for single-mode and multimode operations, acceptance, and critical angles.

Solution:
Refractive index of the core, $n_1 = 1.45$
Diameter of core, $D = 10\,\mu\text{m}$
Fractional index difference, $\Delta = 0.3\%$

Using (1.10), $\text{NA} = n_1 (2\Delta)^{1/2} = 0.1123$
Further, $\sin\theta_a = \text{NA}$

$$\theta_a = \text{acceptance angle}$$

$$\Rightarrow \theta_a = \sin^{-1}(\text{NA}) = 6.448^\circ$$

$$\text{NA} = \sqrt{n_1^2 - n_2^2} \Rightarrow n_2 = 1.4456$$

$$\sin\theta_c = \frac{n_2}{n_1} \Rightarrow \theta_c = 85.535^\circ$$

$$\theta_c = \text{critical angle}$$

Problem 14: Assume a step-index multimode fibre having a core refractive index $n_1 = 1.46$ and the refractive index of the cladding $n_2 = 1.43$. Calculate the following:

(a) the critical angle for total reflections;
(b) the maximum acceptance angle for the fibre, assuming the outer medium as air;
(c) numerical aperture;
(d) relative core–cladding difference.

Solution:

(a) The critical angle is given as $\theta = \theta_c = \sin^{-1}(n_2 / n_1) = 78.36^\circ$
(b) The maximum acceptance angle

$$\theta_M = \sin^{-1}\left(\sqrt{n_1^2 - n_2^2}\right) = \sin^{-1}\left(\sqrt{1.46^2 - 1.43^2}\right) = 17.12^\circ$$

(c) The numerical aperture of the fibre NA is $\sin(\theta_M) = \sin(17.12) = 0.29$.
(d) The percent of relative core–cladding difference is [using (1.16)]

$$\Delta = \frac{n_1 - n_2}{n_1} \times 100 = \frac{1.46 - 1.43}{1.43} \times 100 = 2\%.$$

Problem 15: Consider a single-mode fibre having $n_1 = 1.46$ and $n_2 = 1.43$.

(a) Find the cut-off wavelength of the fibre if the core radius is $4\,\text{m}$.

(b) Find the minimum core diameter of the fibre for single-mode operation at $\lambda_c = 1350\,\text{nm}$.

Solution:

The maximum acceptance angle from equation is

$$\theta_M = \sin^{-1}\left(\sqrt{n_1^2 - n_2^2}\right) = \sin^{-1}\left(\sqrt{1.46^2 - 1.43^2}\right) = 17.12^\circ.$$

The numerical aperture of the fibre NA is $\sin(\theta_M) = \sin(17.12) = 0.29$
NA for the given fibre is 0.29. The cut-off wavelength of the fibre is

$$\lambda_c = \frac{d\pi\text{NA}}{2.4048} = \frac{4 \times \pi \times 0.2433}{2.4048}\,\mu\text{m} = 1.27\,\mu\text{m}.$$

The minimum core diameter for single-mode operation is

$$d \le \frac{\lambda_c \times 2.4048}{\pi \times \text{NA}} \le \frac{1.350 \times 2.4048}{\pi \times 0.29} = 0.29\,\mu\text{m}.$$

Problem 16: Consider a step-index multimode fibre having a core diameter of $50\,\text{m}$ and $\text{NA} = 0.2$. Determine the number of modes propagating in this fibre at an operating wavelength of $900\,\text{nm}$.

Solution:

$$\text{V} = \left(\frac{d \times \pi \times \text{NA}}{\lambda}\right)$$

$$\text{V} = \left(\frac{50\,\text{m} \times \pi \times 0.29}{0.90\,\mu\text{m}}\right) = 50.61$$

The numbers of modes propagating in a fibre are $\text{N} = \frac{\text{V}^2}{2} = 1280.68$.

Problem 17: Find the derivation of propagation and multipath dispersion in graded index fibre.

Solution:

The fibre refractive index variation

$$n(r) = \left(\frac{r}{a}\right)^{a}.$$

This is called α-profile. Since $\alpha \simeq 2(1-\Delta)$ approximately, multipath dispersion in this case becomes (discussed in previous chapter):

$$\frac{\Delta T_{\min}}{L} = \frac{n_0 \Delta^2}{8c}.$$

If $\Delta = 0.0135, \alpha = 1.973,$ parabolic profile, where $n_0 = 1.47,$ and $n_c = 1.45$, $\frac{\Delta T_{\min}}{L} = 110\,\text{ps/km}$ is minimum multipath dispersion.

The bandwidth × distance is 9 GHzkm. This is definitely an improvement by three orders of magnitude when compared to equivalent step-index fibre.

Well-produced graded index fibre of overall multipath dispersion less than 1 ns/km are widely available and they are having their bandwidth in the region of around 1 GHzkm. Therefore, multimode graded index fibres are widely used in LAN and WAN applications due to better performance than step-index fibres as around 1000 modes may propagate together in a multimode fibre and graded index fibre having less modal dispersion.

Single-mode fibre

The mode number is proportional to core area and square of numerical aperture. If we decrease the product of these two parameters considerably by reducing the core diameter to around $5\,\mu\text{m}$ to 8 μm only, it allows the propagation of only one mode, and therefore, multimode dispersion of modal dispersion becomes zero. The condition of single-mode operation in that situation is

$$2a < \frac{2.405\lambda}{\pi\sqrt{n_1^2 - n_2^2}}.$$

For a single-mode (or monomode) fibre, there is a cut-off wavelength λ_c, which is given as

$$\lambda_C = \frac{2\pi a\left(n_1^2 - n_2^2\right)^{1/2}}{2.405}.$$

Therefore, at wavelength $\lambda < \lambda_C$, higher order modes may propagate. As an example, if we are using wavelength of $\lambda = 1300\,\text{nm}$ for fibre optical communication, λ_C should be around $1200\,\text{nm}$. This means $2a \simeq 8.5\ \mu\text{m}$ and $\Delta n = 0.004$, and therefore, core diameter shall be around $8\ \mu\text{m}$ and the refractive index variation in core region should be less than Δn. Single-mode step-index fibres are, thus, used in long haul communication due to zero modal dispersion and low loss.

Problem 18: Write down short notes on: attenuation loss in optical fibre.

Solution:
In 1970, fibre of attenuation below 20 dB/km was manufactured by corning glass at a wavelength of 820 nm. In 1987, fibre of attenuation below 0.2 dB/km was reported working at a wavelength of 1550 nm. Once the attenuation is reduced in optical fibres, the increase in bandwidth is considered by reducing the spatial dispersion within the fibre. Dispersion less fibres and dispersion compensated fibres were successfully developed and the bandwidth was increased to 10 Gb/s for a long distance fibre communication. As optical fibre cables of very low attenuation and dispersion were manufactured, it became possible to do long distance communication underneath the sea and terrestrial communication on earth; for example, in USA, Europe, and Japan. Now, optical fibre has become the backbone of long haul communication. Therefore, attenuation is basically a loss in signal power over the length of the fibre.

If P_{in} is the input power in fibre and P_{out} is the output power in fibre received at the other end, then attenuation loss dB/km is given as

$$A = \frac{10}{L}\log_{10}\left(\frac{P_{in}}{P_{out}}\right)$$

where A is the signal attenuation in dB/km and L is the fibre length in km. The power loss is given as

$$\text{Power loss (dB)} = -10\log\left(\frac{P_{out}}{P_{in}}\right)$$

Attenuation = power loss/fibre length.
By substituting value of power loss,

$$\log\left(\frac{P_{out}}{P_{in}}\right) = \frac{[-A \times L]}{10}$$

$$P_{out} = P_{in} \times 10^{-A\times L/10}$$

where P_{out} and P_{in} are in W and L is in km.
The length in km can be derived from

$$L = \frac{10}{A}\log_{10}\frac{P_{in}}{P_{out}}.$$

Problem 19: Calculate the output power with the help of the following data:
Input power $P_{in} = 1\,\text{mW}$
Attenuation $A = 0.5\ \text{dB/km}$
Fibre link length = 15 km.

Solution:

$$P_{out} = P_{in} \times 10^{-A\times L/10}$$

Substituting the given values in the above-mentioned relation, we have

$$P_{out} = 1\times 10^{-0.5\times 15/10} = 0.177 \text{ mW}$$

Problem 20: If $A = 0.5\,\text{dB/km}$, $P_{in} = 1\text{mW}$, and $P_{out} = 50\,\mu\text{W}$. Calculate the maximum transmission distance for a fibre link.

Solution:
Given:
$P_{in} = 1\text{mW},\ A = 0.5\,\text{dB/km},\ P_{out} = 50\,\mu\text{W},\ \text{and}\ L = \frac{10}{A}\log_{10}\frac{P_{in}}{P_{out}} = 26.02\text{ km}$

Problem 21: Calculate the output power (P_{out}) for various values of attenuation and length of the fibre link. Assume $P_{in} = 1\text{mw}$, $AL = 10\text{dB}$, and 20 dB.

Solution: Given

(a) Input power $P_{in} = 1\text{mW}$

$$AL = 10\,\text{dB}$$

$$P_{out} = P_{in} \times 10^{-A\times L/10} = P_{in}/10 = 0.1\,\text{mW}$$

(b) Input power, $P_{in} = 1\text{mw}$

$$AL = 20\,\text{dB}$$

$$P_{out} = P_{in} \times 10^{-A\times L/10} = 10\ \mu\text{w}$$

Problem 22: Calculate the rms pulse width σ of the rectangular pulse shown in Fig. P5 in terms of its FWHM τ.

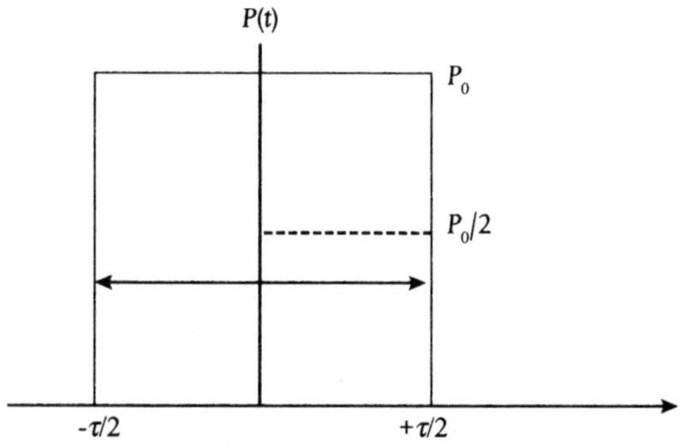

Fig. P5 Rectangular pulse

Solution:

The total energy in the pulse

$$\varepsilon = \int_{-\infty}^{\infty} p(t)dt = \int_{-\tau/2}^{\tau/2} p_0 dt = p_0 \tau$$

$$\sigma^2 = \frac{1}{\varepsilon}\int_{-\tau/2}^{\tau/2} t^2 p_0 dt - \left[\frac{1}{\varepsilon}\int_{-\tau/2}^{\tau/2} t p_0 dt\right]^2$$

$$\Rightarrow \frac{1}{p_0\tau}\left[p_0 \int_{-\tau/2}^{\tau/2} t^2 dt\right] - \left[\frac{1}{p_0\tau} p_0 \int_{-\tau/2}^{\tau/2} t dt\right]^2$$

$$\Rightarrow \frac{1}{\tau}\left[\frac{\tau^3}{12}\right] - 0$$

$$\Rightarrow \sigma^2 = \frac{\tau^2}{12}$$

Therefore,

$$\sigma = \frac{\tau}{\sqrt{12}} = \frac{\tau}{2\sqrt{3}}.$$

Problem 23: In a typical fibre-optic communication system, the received power distribution as a function of time t is given by a symmetrical triangular pulse, as shown in Fig. P6. Calculate (a) the total energy in the pulse, (b) the mean pulse arrival time, and (c) the rms pulse width.

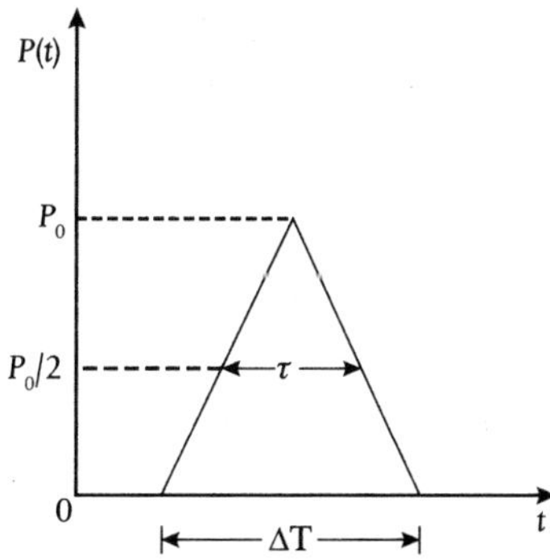

Fig. P6 Triangular pulse

Solution:

The power $p(t)$ in the pulse varies as follows:

$$p(t) = \begin{cases} \frac{p_0}{\tau} t, & 0 < t < \tau \\ -\frac{p_0}{\tau} t + 2p_0, & \tau < t < 2\tau \end{cases}.$$

(a) The total energy in the pulse

$$\varepsilon=\int_{-\infty}^{\infty} p(t)dt=\int_{0}^{\tau}\frac{p_0}{\tau}tdt+\int_{\tau}^{2\tau}p_0\left(2-\frac{t}{\tau}\right)dt$$

$$\Rightarrow\left[\frac{p_0}{\tau}\frac{t^2}{2}\right]_0^{\tau}+\left[2p_0t-\frac{p_0}{\tau}\frac{t^2}{2}\right]_{\tau}^{2\tau}=p_0\tau.$$

(b) The mean pulse arrival time $\bar{t}$ is given as

$$\bar{t}=\frac{1}{\varepsilon}\int_{-\infty}^{\infty}p(t)tdt=\frac{1}{p_0\tau}\int_0^{\tau}\frac{p_0}{\tau}t^2dt+\frac{1}{p_0\tau}\int_{\tau}^{2\tau}\left[2p_0-\frac{p_0}{\tau}t\right]tdt$$

$$\Rightarrow\frac{1}{p_0\tau}\left(\frac{p_0}{\tau}\right)\left[\frac{t^3}{3}\right]_0^{\tau}+\frac{1}{p_0\tau}(2p_0)\left[\frac{t^2}{2}\right]_{\tau}^{2\tau}-\frac{1}{p_0\tau}\left(\frac{p_0}{\tau}\right)\left[\frac{t^3}{3}\right]_{\tau}^{2\tau}.$$

(c) The rms pulse width σ is determined as follows:

$$\sigma^2=\frac{1}{\varepsilon}\int_{-\infty}^{\infty}t^2p(t)dt-[\bar{t}]^2$$

$$\Rightarrow\frac{1}{p_0\tau}\int_0^{\tau}\frac{p_0}{\tau}t^3dt+\frac{1}{p_0\tau}\int_{\tau}^{2\tau}\left[2p_0-\frac{p_0}{\tau}t\right]t^2dt-\tau^2$$

$$\Rightarrow\frac{1}{4}\tau^2+\frac{2}{\tau}\left[\frac{t^3}{3}\right]_{\tau}^{2\tau}-\frac{1}{\tau^2}\left[\frac{t^4}{4}\right]_{\tau}^{2\tau}-\tau^2$$

$$\Rightarrow\frac{\tau^2}{6}.$$

Therefore,

$$\sigma=\frac{\tau}{\sqrt{6}}.$$

Problem 24: A symmetric step-index (SI) planar waveguide is made of glass with $n_1=1.5$ and $n_2=1.49$. The thickness of the guide layer is $9.83\,\mu\text{m}$ and the guide is excited by a source of wavelength $\lambda=0.85\,\mu\text{m}$. What is the range of the propagation constants? What is the maximum number of modes supported by the guide?

Solution:

The phase propagation constant β lies between β_1 and β_2. Here,

$$\beta_1=kn_1=\frac{2\pi n_1}{\lambda}=\frac{2\pi}{(0.85\times10^{-6})}\times1.50=11.087\times10^6\text{ m}^{-1}$$

and

$$\beta_2 = kn_2 = \frac{2\pi n_2}{\lambda} = \frac{2\pi}{\left(0.85\times10^{-6}\right)}\times1.49 = 11.014\times10^6\ \text{m}^{-1}.$$

The maximum number of modes that the guide can support is given as

$$\text{M} = \frac{2\times2a}{\lambda}\left(n_1^2 - n_2^2\right)^{1/2} = \frac{2\times(9.83)}{0.85}\left[(1.5)^2 - (1.49)^2\right]^{1/2} \approx 4\,.$$

Problem 25: What should be the maximum thickness of the guide slab of a symmetrical SI planar waveguide so that it supports only the fundamental TE mode? Consider $n_1 = 3.6$, $n_2 = 3.56$, and $\lambda = 0.85\,\mu\text{m}$.

Solution:

For the waveguide to support only the fundamental mode, V should be less than $\pi/2$ [using (1.21)], that is,

$$\frac{2\pi a}{\lambda}\left(n_1^2 - n_2^2\right)^{1/2} < \pi/2$$

or

$$2a < \frac{\lambda}{2\left(n_1^2 - n_2^2\right)^{1/2}}$$

$$\Rightarrow 2a < 0.793\,\mu\text{m}.$$

Problem 26: A planar waveguide is formed from a $10.11-\mu\text{m}$ thin film of dielectric material of refractive index 1.50 sandwiched between two infinite dielectric slabs of refractive index 1.48.

(a) Find the number of TE modes of propagation that the guide can support at a wavelength of $1.55\,\mu\text{m}$.
(b) Estimate approximately the propagation parameters u_m, w_m, b_m, and β_m for different values of m.

Solution:

(a) Using (1.21), we get

$$\text{V} = \frac{2\pi a}{\lambda}\left(n_1^2 - n_2^2\right)^{1/2} \approx 5\,.$$

We can conclude that for this value of V, the number of TE modes of propagation is four, corresponding to $m = 0, 1, 2, \text{and } 3$. Therefore, there are two symmetric modes corresponding to $m = 0$ and $m = 2$ and two antisymmetric modes corresponding to $m = 1$ and $m = 3$.

(b) The abscissae of the points of intersection of the ua versus wa curves for $m = 0,1,2,\text{and } 3$ with the quadrant of a circle of radius $V = 5$ give the values of $u_m a$, and the corresponding ordinates give the values of $w_m a$. Dividing these values of $u_m a$ and $w_m a$ by a gives the values u_m and w_m. In the present case, $a = 5.055\times10^{-6}\,\text{m}$. Here, the subscript m simply denotes the value of u, w, or b for the mth mode. The results are given in Table P1.

Table P1 Values of β_m

m	$u_m a$ rad	$u_m\left(m^{-1}\right)$	$w_m a$ rad	$w_m\left(m^{-1}\right)$	$b_m = \left(\frac{w_m a}{V}\right)^2$
0	1.30644	2.58445×10^5	4.82630	9.54757×10^5	0.93172
1	2.59574	5.13499×10^5	4.27342	8.45384×10^5	0.73048
2	3.83747	7.59143×10^5	3.20529	6.34083×10^5	0.41095
3	4.90630	9.70583×10^5	0.963466	1.90596×10^5	0.03713

The values of β_m may be obtained using the relation

$$\beta_m = \left[b_m\left(\beta_1^2 - \beta_2^2\right) + \beta_2^2\right]^{1/2}$$

where $\beta_1 = 2\pi n_1/\lambda = 6.08\times10^6\,\text{m}^{-1}$ and $\beta_2 = 2\pi n_2/\lambda = 5.99\times10^6\,\text{m}^{-1}$. Therefore, the phase propagation constant for the mth mode will be given as

$$\beta_m = \left[b_m\left(0.9764\times10^{12}\right) + 35.99\times10^{12}\right]^{1/2}\ \text{m}^{-1}$$

$$\Rightarrow \beta_m = \left[0.9764 b_m + 35.99\right]^{1/2}\times10^6\ \text{m}^{-1}.$$

Problem 27: Calculate the G-factor for the fundamental TE mode supported by the waveguide of a symmetrical SI planar waveguide. Consider $n_1 = 3.6$, $n_2 = 3.56$, $2a = 0.793\ \mu\text{m}$, and $\lambda = 0.85\ \mu\text{m}$.

Solution:
For thickness $2a = 0.793\,\mu\text{m}$, and $V = \pi/2$. The point of intersection of an arc of $V = \pi/2$ with the curve $ua\tan(ua) = wa$ for $m = 0$ gives

$$ua = 0.934$$

and

$$wa = 1.262.$$

The confinement factor G may be calculated as follows:

$$G = \left[1 + \frac{\cos^2(0.934)}{1.262\left\{1 + \left[\sin(0.934)\cos(0.934)\right]/0.934\right\}}\right]^{-1} = 0.8436.$$

This means that only 84.36% of the total power carried by the fundamental mode is confined within the guide layer; the remaining 15.64 % extends into the cladding slabs.

Problem 28: A step-index fibre has a core diameter of $7.2\ \mu\text{m}$, a core index of 1.46, and a relative refractive index difference of 1%. A light of wavelength $1.55\ \mu\text{m}$ is used to excite the modes in the fibre. Find (a) the normalised frequency parameter V, (b) the propagation constants β for these modes, and (c) the phase velocity of each mode.

Solution:

(a)

$$V=\frac{2\pi a}{\lambda}n_1\sqrt{2\Delta}=\frac{\pi\times7.2\times10^{-6}}{1.55\times10^{-6}}\times1.46\sqrt{2\times0.01}=3.01.$$

(b) Draw a vertical line on the graph of b versus V (the plots of Fig. P7) at $V=3$. The line intersects the curves for only two modes, namely the LP_{01} and LP_{11} modes. This implies that the fibre is supporting only these two modes.

(c) The β-values can be obtained:

$$b_{\text{lm}}=\frac{\beta_{\text{lm}}^2-\beta_2^2}{\beta_1^2-\beta_2^2}$$

or

$$\beta_{\text{lm}}=\left[\beta_2^2+b_{\text{lm}}\left(\beta_1^2-\beta_2^2\right)\right]^{1/2}$$

$$\beta_1=kn_1=\frac{2\pi}{\lambda}n_1=\frac{2\pi}{1.55\times10^{-6}}\times1.46=5.915\times10^6\ \text{m}^{-1}$$

$$\beta_2=kn_2=\frac{2\pi}{\lambda}n_2=\frac{2\pi}{1.55\times10^{-6}}\times1.4454=5.855\times10^6\ \text{m}^{-1}$$

[since $n_2=n_1(1-\Delta)=1.46(1-0.01)=1.4454$].

$$\beta_{\text{lm}}=\left[34.281\times10^{12}+b_{\text{lm}}\left(0.7062\times10^{12}\right)\right]^{1/2}\ \text{m}^{-1}$$

We can get approximately $b_{01}=0.62$ and $b_{11}=0.18$. Therefore,

$$\beta_{01}=[34.281+0.62\times0.7062]^{1/2}\times10^6\ \text{m}^{-1}=5.8658\times10^6\ \text{m}^{-1}$$

and

$$\beta_{11}=[34.281+0.18\times0.7062]^{1/2}\times10^6\ \text{m}^{-1}=5.8658\times10^6\ \text{m}^{-1}$$

(d) Phase velocity

$$V_p=\frac{\omega}{\beta}=\frac{2\pi c}{\lambda\beta}.$$

Therefore,

$$\left(V_p\right)_{01}=\frac{2\pi\times3\times10^8}{1.55\times10^{-6}\times5.8922\times10^6}=2.06\times10^8\ \text{ms}^{-1}$$

and

$$\left(V_p\right)_{02}=\frac{2\pi\times3\times10^8}{1.55\times10^{-6}\times5.8658\times10^6}=2.07\times10^8\ \text{ms}^{-1}.$$

Problem 29: A step-index fibre has a core diameter of $7.2\,\mu m$, a core index of 1.46, and a relative refractive index difference of 1%. A light of wavelength $1.55\,\mu m$ is used to excite the modes in the fibre. The normalised frequency parameter $V = 3$, the propagation constants is $\beta_{01} = 5.8922 \times 10^6\ m^{-1}$, and $\beta_{11} = 5.8658 \times 10^6\ m^{-1}$ for these modes, and the phase velocity of each mode is $\left(V_P\right)_{01} = 2.06 \times 10^8\ ms^{-1}$ and $\left(V_P\right)_{02} = 2.07 \times 10^8\ ms^{-1}$. What is the fractional power propagating in the cladding for the two modes?

Solution:

The step-index fibre has a V-parameter approximately equal to 3. For this value of V, from Fig. P7, we get for modes LP_{01} and LP_{11}.

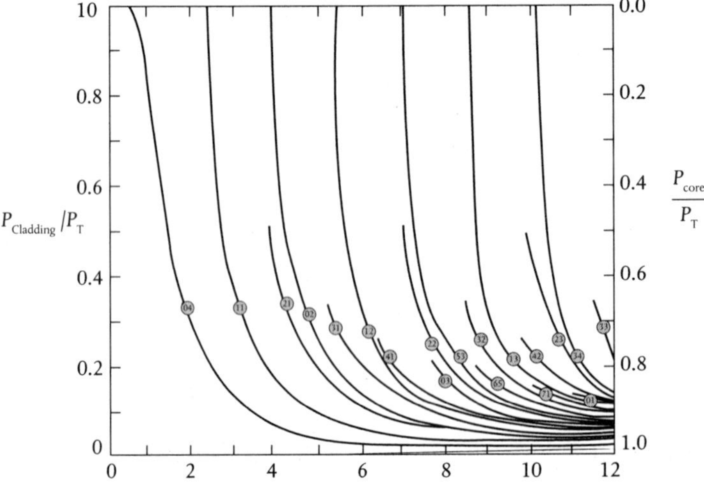

Fig. P7 Plots of fractional power contained in the core and cladding of an SI fibre as a function of *V*

$$\left(\frac{P_{cladding}}{P_T}\right)_{01} = 0.11$$

$$\left(\frac{P_{cladding}}{P_T}\right)_{11} = 0.347$$

That is, for the LP_{01} and LP_{11} modes, respectively, 11 % and ≈ 35 % power flows in the cladding.

Problem 30: A step-index fibre has a numerical aperture of 0.17 and a core diameter of $100\,\mu m$. Determine the normalised frequency parameter of the fibre when light of wavelength $0.85\,\mu m$ is transmitted through it. Further, estimate the number of guided modes propagating in the fibre.

Solution:

$$V = \frac{2\pi a}{\lambda}\left(n_1^2 - n_2^2\right)^{1/2} = \frac{2\pi a(NA)}{\lambda} = \frac{\pi(100\,\mu m)(0.17)}{(0.85\,\mu m)} = 62.83$$

Therefore,

$$M_s = \frac{V^2}{2} = 1974\,.$$

Problem 31: A multimode step-index fibre has a relative refractive index difference of 2% and a core refractive index of 1.5. The number of modes propagating at a wavelength of 1.3 μm is 1000. Calculate the diameter of the fibre core.

Solution:

$$M_s = \frac{V^2}{2} = \frac{1}{2}\left[\frac{2\pi a}{\lambda} n_1 \sqrt{2\Delta}\right]^2 = \frac{1}{2}\frac{4\pi^2 a^2 n_1^2 2\Delta}{\lambda^2}$$

Therefore,

$$2a = \frac{\lambda}{\pi n_1}\left(\frac{M_s}{\Delta}\right)^{1/2} = \frac{1.3}{\pi \times 1.5}\left(\frac{1000}{0.02}\right)^{1/2} \approx 62\ \mu\text{m}\,.$$

Problem 32: A graded-index fibre with a parabolic profile supports the propagation of 700 guided modes. The fibre has a relative refractive index difference of 2%, a core refractive index of 1.45, and a core diameter of 75 μm. Calculate the wavelength of light propagating in the fibre. Further, estimate the maximum diameter of the fibre core that can give single-mode operation at the same wavelength.

Solution:
Since

$$M_g \cong \left(\frac{\alpha}{\alpha+2}\right)\frac{V^2}{2}\,.$$

With $\alpha = 2$ for the parabolic profile.

$$V = \sqrt{4M_g} = \sqrt{4 \times 700} = 52.91\,.$$

and

$$\lambda = \frac{2\pi a}{V} n_1 \sqrt{2\Delta} = \frac{\pi \times (75\,\mu\text{m})}{52.91} \times 1.45 \times (2 \times 0.02)^{1/2} = 1.29\,\mu\text{m}$$

The cut-off value of the normalised frequency, V_c, for single-mode operation in a graded index fibre is given as

$$V_c = 2.405\left(1 + \frac{2}{\alpha}\right)^{1/2}.$$ Therefore, with $\alpha = 2$.

The maximum core diameter may be obtained as follows:

$$2a = \frac{V_c \lambda}{\pi n_1 \sqrt{2\Delta}} = \frac{2.405 \times \sqrt{2} \times (1.3\,\mu\text{m})}{\pi \times 1.45 \times \sqrt{2 \times 0.02}} = 4.85\,\mu\text{m}\,.$$

MULTIPLE CHOICE QUESTIONS

1. A dielectric waveguide for the propagation of electromagnetic energy at light frequencies:
   (a) Stripline
   (b) Microstrip
   (c) Laser beam
   (d) Fibre optics
2. Who proposed the use of clad glass fibre as a dielectric waveguide?
   (a) Karpon and Keck
   (b) Karpon and Bockham
   (c) Bockham and Kao
   (d) Kao and Keck
3. The core of an optical fibre has a
   (a) lower refracted index than air.
   (b) lower refractive index than the cladding.
   (c) higher refractive index than the cladding.
   (d) similar refractive index with the cladding.
4. Between silicon and gallium arsenide, which has the greatest index of refraction?
   (a) Gallium arsenide
   (b) Neither silicon nor gallium arsenide
   (c) Silicon
   (d) They are equal
5. Which is the minimum optical power a light detector can receive and still produce a usable electrical output signal?
   (a) Light responsivity
   (b) Light sensitivity
   (c) Light collectivity
   (d) Illumination
6. The time taken by a light-induced carrier to travel across the depletion region of the semiconductor is called
   (a) dispersion.
   (b) response time.
   (c) irradiance.
   (d) transit time.
7. The range of wavelength values that a given photodiode will respond is
   (a) spectral response.
   (b) presence.
   (c) dark current.
   (d) reluctance.
8. The dispersion of light in fibre-optic cable caused by a portion of the light energy travelling in the cladding is called
   (a) modal dispersion.
   (b) material dispersion.
   (c) waveguide dispersion.
   (d) cable dispersion.
9. What are the advantages of using parallel optical fibre interconnections within board-to-board or frame-to-frame connections in computer systems?

(a) No bandwidth limit
(b) Space saving
(c) Avoids electrical crosstalk
(d) All the above

10. What method allows large number of independent, selectable channels to exist on a single fibre?
(a) Frequency division multiplexing
(b) Time division multiplexing
(c) Frequency modulation
(d) Phase modulation

Answers to Multiple Choice Questions

1. (d)
2. (c)
3. (c)
4. (a)
5. (b)
6. (d)
7. (a)
8. (c)
9. (d)
10. (a)

Questions

1. A silica optical fibre with a core diameter large enough to be considered by ray theory analysis has a core refractive index of 1.50 and a cladding refractive index of 1.47. Determine (a) the critical angle at the core–cladding interface, (b) the NA for the fibre, and (c) the acceptance angle in air for the fibre.
2. A typical relative refractive index difference for an optical fibre designed for long distance transmission is 1%. Estimate the NA and the solid acceptance angle in air for the fibre when the core index is 1.46. Further, calculate the critical angle at the core–cladding interface within the fibre. It may be assumed that the concepts of geometric optics hold for the fibre.
3. An optical fibre in air has an NA of 0.4. Compare the acceptance angle for meridional rays with that for skew rays that change direction by 100° at each reflection.
4. A multimode step-index fibre with a core diameter of 80 μm and a relative index difference of 1.5% is operating at a wavelength of 0.85 μm. If the core refractive index is 1.48, estimate (a) the normalised frequency for the fibre and (b) the number of guided modes.
5. A graded index fibre has a core with a parabolic refractive index profile that has a diameter of 50 μm. The fibre has a numerical aperture of 0.2. Estimate the total number of guided modes propagating in the fibre when it is operating at a wavelength of 1 μm.
6. Estimate the maximum core diameter for an optical fibre having relative refractive index difference of 1.5% and core refractive index of 1.48. It may be assumed that the fibre is operating at wavelength of 0.85 μm. Further, estimate the new maximum core diameter for single-mode operation when the relative refractive index difference is reduced by a factor of 10.
7. A graded index fibre with a parabolic refractive index profile core has a refractive index at the core axis of 1.5 and a relative index difference of 1%. Estimate the maximum possible core diameter that allows single-mode operation at a wavelength of 1.3 μm.
8. Determine the cut-off wavelength for a step-index fibre to exhibit single-mode operation when the core refractive index and radius are 1.46 and 4.5 μm, respectively, with the relative index difference being 0.25%.

9. Given that a useful approximation for the eigenvalue of the single-mode step-index fibre cladding W is $W(V) \simeq 1.1428\,V - 0.9960$. Deduce an approximation for the normalised propagation constant $b(V)$.
10. Estimate the fibre core diameter for a single-mode step-index fibre that has an MFD of 11.6 μm when the normalised frequency is 2.2.
11. A parabolic profile graded index single-mode fibre designed for operation at a wavelength of 1.30 μm has a cut-off wavelength of 1.08 μm. From experimental measurement, it is established that the first minimum in the diffraction pattern occurs at an angle of 12°. Using an ESI technique, determine the spot size at the operating wavelength.
12. Obtain the ESI relative refractive index difference for a graded index fibre, which has a cut-off wavelength and spot size of 1.190 μm and 5.2 μm, respectively. The maximum refractive index of the fibre core is 1.485.
13. Using simple ray theory, describe the mechanism for the transmission of light within an optical fibre. Briefly discuss with the aid of a suitable diagram what is meant by the acceptance angle for an optical fibre. Show how this is related to the fibre numerical aperture and the refractive indices for the fibre core and cladding. An optical fibre has a numerical aperture of 0.20 and a cladding refractive index of 1.59. Determine
    (a) The acceptance angle for the fibre in water, which has a refractive index of 1.33;
    (b) The critical angle at the core–cladding interface.
    Comment on any assumptions made on the fibre.
14. The velocity of light in the core of a step-index fibre is $2.01 \times 10^{8}\,\text{ms}^{-1}$, and the critical angle at the core–cladding interface is 80°. Determine the numerical aperture and the acceptance angle for the fibre in air, assuming it has a core diameter suitable for consideration by ray analysis. The velocity of light in a vacuum is $2.998 \times 10^{3}\,\text{ms}^{-1}$.
15. Define the relative refractive index difference for an optical fibre and show how it may be related to the numerical aperture.
16. A step-index fibre with a large core diameter when compared with the wavelength of the transmitted light has an acceptance angle in air of 22° and a relative refractive index difference of 3%. Estimate the numerical aperture and the critical angle at the core–cladding interface for the fibre.
17. A step-index fibre has a solid acceptance angle in air of 0.115 radians and a relative refractive index difference of 0.9%. Estimate the speed of light in the fibre core.
18. Briefly indicate with the aid of suitable diagrams the difference between meridional and skew ray paths in step-index fibres.
19. Derive an expression for the acceptance angle for a skew ray that changes direction by an angle 2γ at each reflection in a step-index fibre in terms of the fibre NA and γ. It may be assumed that the ray theory holds for the fibre.
20. A step-index fibre with a suitably large core diameter for ray theory considerations has core and cladding refractive indices of 1.44 and 1.42, respectively. Calculate the acceptance angle in air for skew rays that change direction by 150° at each reflection.
21. Skew rays are accepted into a large core diameter (when compared to the wavelength of the transmitted light) step-index fibre in air at a maximum axial angle of 42°. Within the fibre, they change direction by 90° at each reflection. Determine the acceptance angle for meridional rays for the fibre in air.
22. Explain the concept of electromagnetic modes in relation to a planar optical waveguide. Discuss the modifications that may be made to electromagnetic mode theory in a planar waveguide in order to describe optical propagation in a cylindrical fibre.

23. Briefly discuss, with the aid of suitable diagrams, the following concepts in optical fibre transmission:
    (a) The evanescent field;
    (b) Goos–Hánchen shift;
    (c) Mode coupling.
    Describe the effects of these phenomena on the propagation of the light in optical fibres.
24. Define the normalised frequency for an optical fibre and explain its use in the determination of the number of guided modes propagating within a step-index fibre.
25. A step-index fibre in air has a numerical aperture of 0.16, a core refractive index of 1.45, and a core diameter of 60 μm. Determine the normalised frequency for the fibre when light at a wavelength of 0.9 μm is transmitted. Further, estimate the number of guided modes propagating in the fibre.
26. Describe with the aid of simple ray diagrams:
    (a) The multimode step-index fibre;
    (b) The single-mode step-index fibre.
    Compare the advantages and disadvantages of these two types of fibre for using an optical channel.
27. A multimode step-index fibre has a relative refractive index difference of 1% and a core refractive index of 1.5. The number of modes propagating at a wavelength of 1.3 μm is 1100. Estimate the diameter of the fibre core.
28. Explain what is meant by a graded index optical fibre giving an expression for the possible refractive index profile. Using simple ray theory concepts, discuss the transmission of light through the fibre. Indicate the major advantage of this type of fibre with regard to multimode propagation.
29. The relative refractive index difference between the core axis and the cladding of a graded index fibre 0.7% when the refractive index at the core axis is 1.45. Estimate values for the numerical aperture of the fibre when:
    (a) the index profile is not taken into account and
    (b) the index profile is assumed to be triangular
    Comment on the results.
30. A multimode graded index fibre has an acceptance angle in air of 8°. Estimate the relative refractive index difference between the core axis and the cladding when the refractive index at the core axis is 1.52.
31. A graded index fibre with a parabolic index profile supports the propagation of 742 guided modes. The fibre has a numerical aperture in air of 0.3 and a core diameter of 70 μm. Determine the wavelength of the light propagating in the fibre. Further, estimate the maximum diameter of the fibre that gives single-mode operation at the same wavelength.
32. A graded index fibre with a core axis refractive index of 1.5 has a characteristic index of 40 μm Estimate the total number of guided modes propagating in the fibre when the transmitted light has a wavelength of 1.55 μm, and determine the cut-off value of the normalised frequency for single-mode transmission in the fibre.
33. A single-mode step-index fibre has a core diameter of 7 μm and a core refractive index of 1.49. Estimate the shortest wavelength of light that allows single-mode operation when the relative refractive index difference for the fibre is 1%.
34. In Problem 33, it is required to increase the fibre core diameter to 10 μm whilst maintaining single-mode operation at the same wavelength. Estimate the maximum possible relative refractive index difference for the fibre.
35. Show that the maximum value of a/λ is approximately 1.4 times larger for a parabolic refractive index profile single-mode fibre than for a single-mode step-index fibre. Hence, sketch the

relationship between the maximum core diameter and the propagating optical wavelength that will facilitate single-mode transmission in the parabolic profile fibre.

36. A single-mode step-index fibre, which is designed for operation at a wavelength of 1.3 μm, has core and cladding refractive indices of 1.447 and 1.442, respectively. When the core diameter is 7.2 μm, confirm that the fibre will permit single-mode transmission and estimate the range of wavelengths over which this will occur.
37. A single-mode step-index fibre has a core and cladding refractive indices of 1.498 and 1.495, respectively. Determine the core diameter required for the fibre to permit its operation over the wavelength range 1.48–1.60 μm. Calculate the new fibre core diameter to enable single-mode transmission at a wavelength of 1.30 μm.
38. A single-mode fibre has a core refractive index of 1.47. Sketch a design characteristic of relative refractive index difference Δ against core radius for the fibre to operate at a wavelength of 1.30 μm. Determine whether the fibre remains single mode at a transmission wavelengths of 0.85 μm when its core radius is 4.5 μm.
39. The cut-off method employed to obtain the ESI parameters for a graded index single-mode fibre. If the ESI relative index difference was found to be 0.30% when the spot size and cut-off wavelength were 4.6 μm and 1.29 μm, respectively, calculate the maximum refractive index of the fibre core.

REFERENCES

1. Abramowitz M and Stegun I A, *Handbook of Mathematical Functions,* 2nd ed., New York: Dover, 1971.
2. Adams M J, *An Introduction to Optical Waveguides,* Chichester: John Wiley, 1981.
3. Snyder A W and Love J D, *Optical Waveguide Theory,* New York: Chapman and Hall, 1983.
4. Baccarelli P, Burghignoli P, Frezza F, Galli A, Lampariello P, Lovat G and Simone S, Fundamental modal properties of surface waves on meta-material grounded slabs, *IEEE Trans. Microw. Theory Tech.*, vol. 33, no. 4, pp. 1431–1442, 2005.
5. Baccarelli P, Burghignoli P, Lovat G and Paulotto S, Surface-wave suppression in a double-negative metamaterial grounded slab, *IEEE Antenn. Wireless Propag. Lett.*, vol. 2, pp. 269 – 272, 2003.
6. Baets R and Lagasse P E, Calculation of radiation loss in integrated optic tapers and Y-junctions, *App. Opt.*, vol. 21, pp. 1972–1978, 1982.
7. Banai A and Hashemi A, A hybrid multimode contour integral method of analysis of the H-plane waveguide discontinuities, *Prog. Electromagn. Res.*, PIER 81, pp. 167–182, 2008.
8. Bienstman P, Six E, Roelens M, Vanwolleghem M and Baets R, Calculation of bending loss in dielectric waveguides using eigenmode expansion and perfectly matched layers, *IEEE Photonics Tech. Lett.*, vol. 14, no. 2, pp. 164–166, 2002.
9. Chakraborty A and Sanyal G S, Transmission matrix of a linear double taper in rectangular waveguides, *IEEE Trans. Microw. Theory Tech.*, vol. MTT-28, no. 6, pp. 577–579, Jun. 1980.
10. Chiang K S, Review of numerical and approximate methods for modal analysis of general dielectric waveguides, *Opt. Quant. Electron.*, vol. 26, pp. 113–134, 1994.
11. Dwari S, Chakraborty A and Sanyal S, Analysis of linear tapered waveguide by two approaches, *Prog. Electromagn. Res. (PIER)*, vol. 64, pp. 219–238, 2006.
12. Engguist B and Kriegsmann G A, *Computational Wave Propagation,* New York: Springer, 1997.
13. Findakly T and Chen C L, Diffused optical waveguides with exponential profile: Effects of metal-clad and dielectric overlay, *Appl. Opt.*, vol. 17, pp. 469–474, 1978.
14. Ghatak A K, Thyagarajan K and Shenoy M R, Numerical analysis of planar optical waveguide using matrix approach, *J. Lightwave Technol.*, vol. LT-5, pp. 660–666, 1987.
15. Haus H A, Huang W P and Whitaker N M, Optical waveguide dispersion characteristics from the

scalar wave equation, *IEEE J. Lightwave Technol.*, vol. 5, no. 12, pp. 1748–1754, 1987.

16. He Y, Cao Z and Shen Q, Guided optical modes in asymmetric left-hand waveguides, *Opt. Commun.*, vol. 245, pp. 125–135, 2005.
17. Hirama K and Koshiba M, A new low loss structure of abrupt bends in dielectric waveguides, *IEEE J. Lightwave Technol.*, vol. 10, no. 5, pp. 563–569, May 1992.
18. Janma R, Jatan R, Singh S P, Dwivedi P, Dwivedi V and Singh V, Modal analysis of a double clad optical waveguide having a cross section of the shape of a lemniscates, *J. Infrared Milli. Terahz. Waves*, vol. 30, pp. 1012–1019, 2009.
19. Jeff H, *Understanding Fibre Optics*, 4th ed., Upper Saddle River, NJ: Prentice-Hall, 2002.
20. Kassim N M, Mohammad A B and Ibrahim M H, Optical waveguide modelling based on scalar finite difference scheme, *J. Teknologi*, vol. 42 (D), pp. 41–54, 2005.
21. Jose J, Controlled coupling of planar waveguides using a negative refractive-index medium, *J. Phys. B At. Mol. Opt.Phys.*, vol. 40, pp. 497–505, 2007.
22. Kogelnik H, 'Theory of Optical Waveguides' in *Guided-Wave Optoelectronics*, T. Tamir, Ed., New York: Springer-Verlag, pp. 7–88, 1990.
23. Marcatili E A, Bends in optical dielectric guides, *Bell Syst. Tech. J.*, pp. 2103–2132, 1969.
24. Marcuse D, *Theory of Dielectric Waveguides, 2nd ed.*, New York: McGraw Hill Inc, 2003.
25. Matsumaru K, Reflection coefficient of E-plane tapered waveguides, *IEEE Trans. Microw. Theory Tech.*, vol. MTT-6, pp. 143–149, April 1958.
26. Melloni A, Carniel F, Costa R and Martinelli M, Determination of bend mode characteristics in dielectric waveguides, *IEEE J. Lightwave Technol.*, vol. 19, no. 4, pp. 571–577, 2001.
27. Moayedi S K and Rostami A, Exact treatment of light in inhomogeneous slab waveguide using super-symmetric quantum mechanics, *IL NuovoCimentoB*, vol. 118, 729, 2003.
28. Nyquist D P, Johnson D R and Hsu S V, Orthogonality and amplitude spectrum of radiation modes along open boundary waveguides, *J. Opt. Soc. Am.*, vol. 71, pp. 49–54, 1981.
29. Pregla R, The method of lines for the analysis of dielectric waveguide bends, *IEEE J. Lightwave Technol.*, vol. 14, no. 4, pp. 634–639, April 1996.
30. Raghuwanshi S K and Kumar V, Analysis of double clad single-mode step-index fibres having depressed versus raised inner Cladding, *Int. Conf. Comm., Computation Control and Nano-tech. (ICN 2010)*, 29–30 Nov. 2010, Bhalki India.
31. Checcacci P F, Applicability of an asymptotic numerical method to the determination of whispering and bouncing modes in elliptical fibres, *J. Opt. Soc. Am.*, vol. 70, no. 12, 1980.
32. Haus H A, Huang W P and Whitaker N M, Optical waveguide dispersion characteristics from the scalar wave equation, IEEE J. Lightwave Technol., LT-5, No. 12, December 1987.
33. Gloge D and Marcatift E A I, Multimode theory of graded-core fibers, *Bell Syst. Tech.*, vol. 1, no. 52, pp. 1563–78, 1973.

3 Laser Optics and Related Devices

Objectives

- This chapter gives an insight into the different optical sources that are used with optical fibres.
- Working principle of the optical sources is elucidated.
- Applications of optical sources with different optical fibres are revealed.
- The application of optical sources for WDM is described.

3.1 LASER

The expanded form of laser is light amplification by stimulated emission of radiation. A device that creates and amplifies electromagnetic radiation of a specific frequency through the process of stimulated emission. The radiation emitted by a laser consists of a coherent beam of photons, all in phase and having the same polarisation. Lasers have many uses, such as cutting hard or delicate substances, reading data from compact disks and other storage devices, and establishing straight lines in geographical surveying [1–5].

A laser is a device that emits light (electromagnetic radiation) through a process of optical amplification based on the stimulated emission of photons. The emitted laser light is notable for its high degree of spatial and temporal coherence; however, this is unattainable using other technologies [6–11]. Spatial coherence, typically, is expressed through the output being a narrow beam, which is diffraction limited, often a so-called pencil beam. Laser beams can be focused to very tiny spots achieving a very high irradiance or they can be launched into a beam of very low divergence in order to concentrate their power at a large distance. Temporal (or longitudinal) coherence implies a polarised wave at a single frequency whose phase is correlated over a relatively large distance (the coherence length) along the beam. A beam produced by a thermal or other incoherent light source has an instantaneous amplitude and phase that vary randomly with respect to time and position, and thus, a very short coherence length. The most so-called single-wavelength lasers actually produce radiation in several modes having slightly different frequencies (wavelengths), often not in a single polarisation; although temporal coherence implies monochromaticity, there are even lasers that emit a broad spectrum of light or emit different wavelengths of light simultaneously. There are some lasers that are not single spatial mode, and consequently, their light beams diverge more than required by the diffraction limit. However, all such devices are classified as lasers based on their method of producing that light: stimulated emission. Lasers are employed in applications where light of the required spatial or temporal coherence could not be produced using simple technologies.

A laser emits a thin, intense beam of nearly monochromatic visible or infrared light that can travel long distances without diffusing. Most light beams consist of many waves travelling in roughly the same direction, but the phases and polarisations of each individual wave (or photon) are randomly distributed. In laser light, the waves are all precisely in step, or in phase, with each other and have the same polarisation. Such light is called coherent. All of the photons that make up a laser beam are in the same quantum state. Lasers produce coherent light through a process called stimulated emission. The laser contains a chamber in which atoms of a medium such as a synthetic ruby rod or a gas are excited, bringing their electrons into higher orbits with higher energy states. When one of these electrons jumps down to a lower energy state (which can happen spontaneously), it gives off its extra energy as a photon with a specific frequency. While this photon encountering another atom with an excited electron will stimulate that electron to jump; furthermore, it will emit another photon with the same frequency as that of the first photon and also in phase with it. This effect cascades through the chamber constantly stimulating other atoms to emit yet more coherent photons. Mirrors at both ends of the chamber cause the light to bounce back and forth in the chamber, sweeping across the entire medium. If a sufficient number of atoms in the medium are maintained by some external energy source in the higher energy state, that is, a condition called population inversion, then emission is continuously stimulated and a stream of coherent photons will develop. One of the mirrors is partially transparent allowing the laser beam to exit from that end of the chamber. Lasers have many industrial, military, and scientific uses, including welding, target detection, microscopic photography, fibre optics, surgery, and optical instrumentation for surveying [12–16].

3.2 FIBRE LASER

A fibre laser is a laser in which the active gain medium is an optical fibre doped with rare earth elements such as erbium, ytterbium, neodymium, dysprosium, praseodymium, and thulium. They are related to doped fibre amplifiers that provide light amplification without lasing. Fibre nonlinearities such as stimulated Raman scattering or four-wave mixing can also provide gain, and thus, serve as gain media for a fibre laser.

3.2.1 Applications

Applications of fibre lasers include material processing, telecommunications, spectroscopy, and medicine. The advantages of fibre lasers over other types include the following:

1. **Light is already coupled into a flexible fibre**: The fact that the light is already in a fibre allows it to be easily delivered to a movable focusing element. This is important for laser cutting, welding, and folding of metals and polymers.
2. **High output power**: Fibre lasers can have active regions several kilometres long and so can provide very high optical gain. They can support kilowatt levels of continuous output power because of the fibre's high surface area to volume ratio, which allows efficient cooling.
3. **High optical quality**: The fibre wave guiding properties reduce or eliminate thermal distortion of the optical path, typically producing diffraction limited and high-quality optical beam.
4. **Compact size**: Fibre lasers are compact when compared to rod or gas lasers of comparable power because the fibre can be bent and coiled to save space. Fibre lasers exhibit high vibrational stability, extended lifetime, and maintenance-free turnkey operation.

3.3 DESIGN AND MANUFACTURE

Unlike most other types of lasers, the laser cavity in fibre lasers is constructed monolithically by fusion splicing different types of fibre; fibre Bragg gratings replace conventional dielectric mirrors to provide optical feedback. Another type is the single longitudinal mode operation of ultra narrow distributed feedback (DFB) lasers where a phase-shifted Bragg grating overlaps the gain medium. Fibre lasers are pumped by semiconductor laser diodes or by other fibre lasers.

3.3.1 Double Clad Fibres

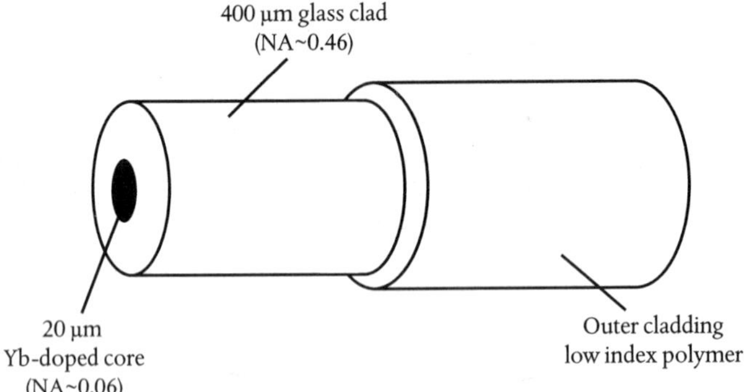

Fig. 3.1 Double clad fibres

Many high-power fibre lasers are based on double clad fibre. The gain medium forms the core of the fibre, which is surrounded by two layers of cladding. The lasing mode propagates in the core, while a multimode pump beam propagates in the inner cladding layer. The outer cladding keeps this pump light confined. This arrangement allows the core to be pumped with a much higher power beam than could otherwise be made to propagate in it and allows the conversion of pump light with relatively low brightness into a much higher brightness signal. As a result, fibre lasers and amplifiers are occasionally referred to as brightness converters. Double clad fibre is shown in Fig. 3.1.

There is an important question about the shape of the double clad fibre; a fibre with circular symmetry seems to be the worst possible design. The design should allow the core to be small enough to support only a few (or even one) mode. It should provide sufficient cladding to confine the core and optical pump section over a relatively short piece of the fibre.

3.3.2 Dark Soliton Fibre Lasers

In the non-mode locking regime, the first dark soliton fibre laser has been successfully achieved in an all-normal dispersion region. However, experimentally finding that apart from the bright pulse emission under appropriate conditions, the fibre laser could also emit single or multiple dark pulses. Based on numerical simulations, we interpret the dark pulse formation in the laser as a result of dark soliton shaping.

3.3.3 Multi-wavelength Fibre Lasers

Recently, multi-wavelength dissipative soliton in an all-normal dispersion fibre laser passively mode locked with a semiconductor saturable absorber mirror (SESAM) has been generated. It is found that depending on the cavity birefringence, stable single, dual and triple wavelength dissipative solitons can be formed in the laser. Its generation mechanism can be traced to the nature of dissipative soliton.

3.3.4 Fibre Disc Lasers

Another type of fibre laser is the fibre disc laser. In such, the pump is not confined within the cladding of the fibre (as in the double clad fibre), but pump light is delivered across the core multiple times because the core is coiled on itself like a rope. This configuration is suitable for power scaling in which many pump sources are used around the periphery of the coil.

3.4 BASIC CONCEPTS: OPTICAL SOURCES

To gain an understanding of the light generating mechanisms within the major optical sources used in optical fibre communications, it is necessary to consider both the fundamental atomic concepts and the device structure. In this context, the requirement for the laser source are far more stringent than those for the LED. Unlike the LED, strictly speaking, the laser is a device that amplifies the light. Hence, the derivation of the term LASER as an acronym for Light Amplification by Stimulated Emission of Radiation. Lasers, however, are seldom used as amplifiers since there are practical difficulties in relation to the achievement of high gain whilst avoiding oscillation from the required energy feedback. Therefore, the practical realisation of the laser is as an optical oscillator. The operation of the device may be described by the formation of an electromagnetic standing wave within a cavity (or optical resonator) that provides an output of monochromatic highly coherent radiation. By contrast, the LED provides optical emission without an inherent gain mechanism. This results in incoherent light output.

In this section, we elaborate on the basic principles that govern the operation of both these optical sources. It is clear, however, that the operation of the laser must be discussed in some detail in order to provide an appreciation of the way it functions as an optical source. Hence, we concentrate first on the general principles of laser action. Figure 3.2 shows the mechanism of light generation in optical devices.

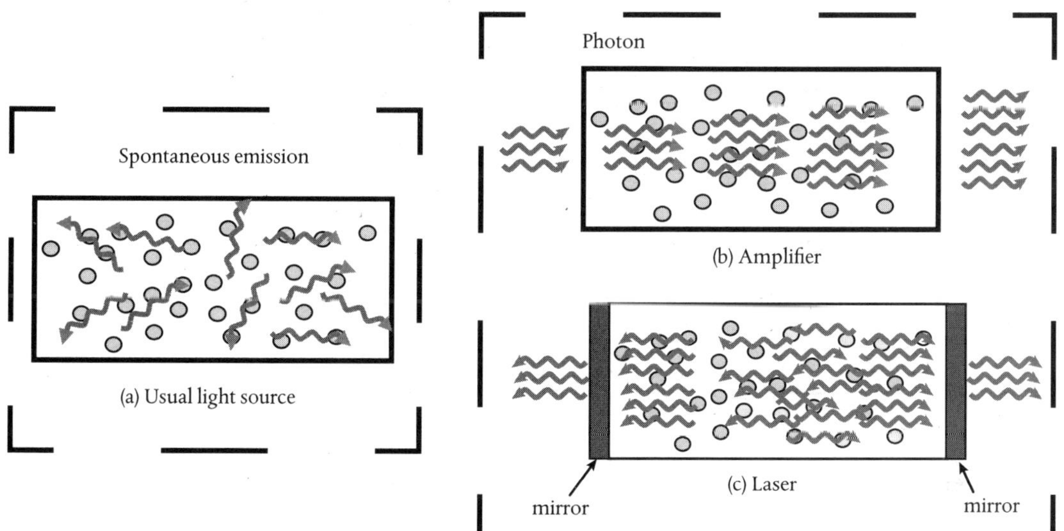

Fig. 3.2 (a) In a usual light source, there is only spontaneous emission; (b) In an optical amplifier, stimulated emission occurs, but there is no cavity. Light is amplified through a single pass; (c) in a laser, stimulated emission occurs and the light bounces forward and backward in the cavity, and therefore, light amplification occurs for many trips in the cavity

3.4.1 Absorption and Emission of Radiation

The interaction of light with matter takes place in discrete packets of energy or quanta called photons. Furthermore, the quantum theory suggests that atoms exist only in certain discrete energy states such that absorption and emission of light cause them to make a transition from one discrete energy state to another. The frequency of the absorbed or emitted radiation f is related to the difference in energy E between the higher energy state E_2 and the lower energy state E_1 by the expression:

$$E = E_2 - E_1 = hf, \tag{3.1}$$

where $h = 6.626 \times 10^{-34}$ Js is Planck's constant. These discrete energy states for the atom may be considered to correspond to electrons occurring in particular energy levels relative to the nucleus. Hence, different energy states for the atom correspond to different electron configurations, and a single electron transition between two energy levels within the atom will provide a change in energy suitable for the absorption or emission of a photon. It must be noted, however, that modern quantum theory gives a probabilistic description that specifies the energy levels in which electrons are most likely to be found. Nevertheless, the concept of stable atomic energy states and electron transitions between energy levels is still valid.

Figure 3.3(a) illustrates a two-energy state or level atomic system where an atom is initially in the lower energy state E_1. When a photon with energy $(E_2 - E_1)$ is incident on the atom, it may be excited into the higher energy state E_2 through absorption of the photon. This process is sometimes referred to as stimulated absorption. Alternatively, when the atom is initially in the higher energy state E_2, it can make a transition to the lower energy state E_1 providing the emission of a photon at a frequency corresponding to (3.1) and as shown in Fig. 3.3 (b).

Fig. 3.3 (a) A photon is absorbed by the atomic system, causing the electron to transit from a low-energy level to a high-energy level; (b) The electron in the atomic system transits from a high-energy level to a low-energy level, emitting a photon

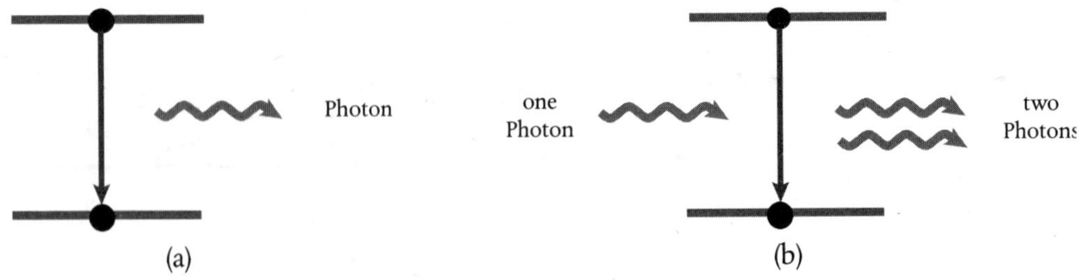

Fig. 3.4 (a) Spontaneous emission; (b) Stimulated emission

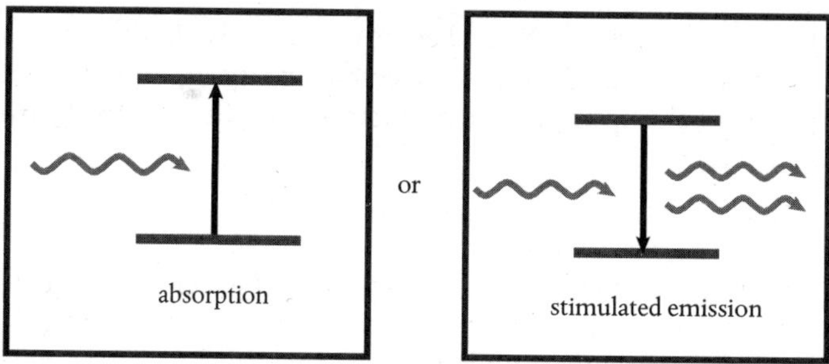

Fig. 3.5 When a photon is incident on the atomic system, it will be absorbed or cause stimulated emission

This emission process can occur in two ways:

(a) by spontaneous emission in which the atom returns to the lower energy state in an entirely random manner;
(b) by stimulated emission when a photon having an energy equal to the energy difference between the two states $(E_2 - E_1)$ interacts with the atom in the upper energy state causing it to return to the lower state with the creation of a second photon.

These two emission processes are illustrated in Fig. 3.4 (a) and 3.4 (b), respectively. The random nature of the spontaneous emission process where light is emitted by electronic transitions from a large number of atoms gives incoherent radiation. A similar emission process in semiconductors provides the basic mechanism for light generation within the LED.

It is the stimulated emission process, which gives the laser its special properties as an optical source. Firstly, the photon produced by stimulated emission is generally of an identical energy to the one which caused it, and hence, the light associated with them is of the same frequency. Secondly, the light associated with the stimulating and stimulated photon is in phase and has the same polarisation. Therefore, in contrast to spontaneous emission, coherent radiation is obtained. Furthermore, this means that when an atom is stimulated to emit light energy by an incident wave, the liberated energy can add to the wave in a constructive manner providing amplification. Absorption and stimulated emission process are shown in Fig. 3.5.

3.4.2 Einstein Relations

Prior to discussion of laser action in semiconductors, it is useful to consider optical amplification in the two-level atomic systems shown in Fig. 3.5. In 1917, Einstein demonstrated that the rates of the three transition processes of absorption, spontaneous emission, and stimulated emission were related mathematically. He achieved this by considering the atomic system to be in thermal equilibrium such that the rate of the upward transitions must be equal to the rate of the downward transitions. The populations of the two energy levels of such a system are described by Boltzmann statistics that give:

$$\frac{N_1}{N_2} = \frac{g_1 exp(-E_1/kT)}{g_2 exp(-E_2/kT)} = \frac{g_1}{g_2} exp\left[(E_2 - E_1)/kT\right] = \frac{g_1}{g_2} exp(hf/kT), \quad (3.2)$$

where N_1 and N_2 represent the density of atoms in energy levels E_1 and E_2, respectively, with g_1 and g_2 being the corresponding degeneracies of the levels, k is Boltzmann's constant, and T is the absolute temperature.

As the density of atoms in the lower or ground energy state E_1 is N_1, the rate of upward transition or absorption is proportional to both N_1 and the spectral density ρ_f of the radiation energy at the transition frequency f. Hence, the upward transition rate R_{12} (indicating an electron transition from level 1 to level 2) may be written as:

$$R_{12} = N_1 \rho_f B_{12}, \tag{3.3}$$

where the constant of proportionality B_{12} is known as the Einstein coefficient of absorption. By contrast, the atoms in the higher or excited energy state can undergo electron transitions from level 2 to level 1 either spontaneously or through stimulation by the radiation field. For spontaneous emission, the average time an electron exists in the excited state before a transition occurs is known as the spontaneous life-time τ_{21}. If the density of atoms within the system with energy E_2, is N_2, then the spontaneous emission rate is given by the product of N_2 and $1/\tau_2$. This may be written as $N_2 A_{21}$, where A_{21}, is the Einstein coefficient of spontaneous emission and is equal to the reciprocal of the spontaneous lifetime.

The rate of stimulated downward transition of an electron from level 2 to level 1 may be obtained in a similar manner to the rate of stimulated upward transition. Hence, the rate of stimulated emission is given as $N_2 \rho_f B_{21}$, where B_{21} is the Einstein coefficient of stimulated emission. The total transition rates from level 2 to level 1, R_{21}, is the sum of the spontaneous and stimulated contributions. Hence,

$$R_{21} = N_2 A_{21} + N_2 \rho_f B_{21}. \tag{3.4}$$

For a system in thermal equilibrium, the upward and downward transition rates must be equal, and therefore, $R_{12} = R_{21}$, or

$$N_1 \rho_f B_{12} = N_2 A_{21} + N_2 \rho_f B_{21}. \tag{3.5}$$

It follows that:

$$\rho_f = \frac{N_2 A_{21}}{N_1 B_{12} - N_2 B_{21}}$$

and

$$\rho_f = \frac{A_{21} / B_{21}}{(B_{12} N_1 / B_{21} N_2) - 1} \tag{3.6}$$

Substituting (3.2) in (3.6) gives:

$$\rho_f = \frac{A_{21} / B_{21}}{\left[(B_{12} g_1 / B_{21} g_2) \exp(hf / KT)\right] - 1}. \tag{3.7}$$

However, since the atomic system under consideration is in thermal equilibrium, it produces a radiation density that is identical to black-body radiation. Planck showed that the radiation spectral density for a black body radiating within a frequency range f to $f + df$ is given as:

$$\rho_f = \frac{8\pi h f^3}{c^3} \left(\frac{1}{exp(hf / KT) - 1} \right). \tag{3.8}$$

Comparing (3.8) with (3.7), we obtain the Einstein relations:

$$B_{12} = \left(\frac{g_2}{g_1} \right) B_{21} \tag{3.9}$$

and

$$\frac{A_{21}}{B_{21}} = \frac{8\pi hf^3}{c^3}. \tag{3.10}$$

It may be observed from (3.9) that when the degeneracies of the two levels are equal $(g_1 = g_2)$, then the probabilities of absorption and stimulated emission are equal. Furthermore, ratios of the stimulated emission rate to the spontaneous emission rate is given as:

$$\frac{\text{Stimulated emission rate}}{\text{Spontaneous emission rate}} = \frac{B_{21}\rho_f}{A_{21}} = \frac{1}{\exp(hf / KT) - 1}. \tag{3.11}$$

From (3.11), it may be noted that for stimulated emission to dominate over absorption and spontaneous emission in a two-level system; both the radiation density and the population density of the upper energy level N_2 must be increased in relation to the population density of the lower energy level N_1.

3.4.3 Population Inversion

Under the conditions of thermal equilibrium given by the Boltzmann distribution, the lower energy level E_1 of the two-level atomic system contains more atoms than the upper energy level E_2. This situation, is normal for structures at room temperature. However, to achieve optical amplification, it is necessary to create a non-equilibrium distribution of atoms such that the population of the upper energy level is greater than that of the lower energy level (i.e., $N_2 > N_1$). This condition, which is known as population inversion. Figure 3.6 shows the distribution number of electrons versus energy and number of electrons (N_2) in high-energy level (E_2) versus lower energy level (E_1), respectively.

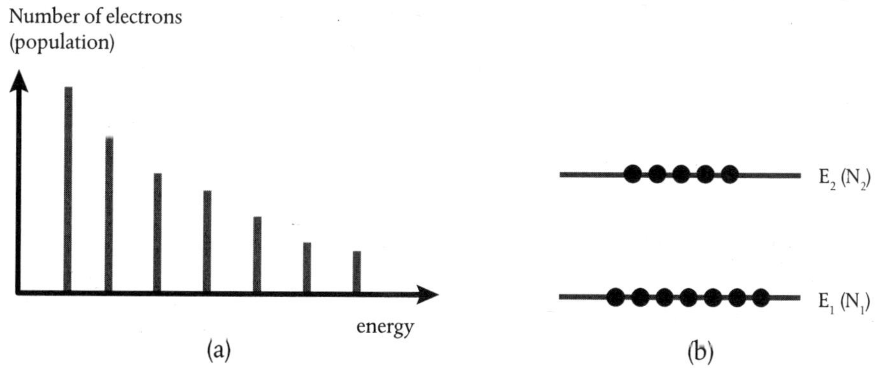

Fig. 3.6 (a) Illustration of the distribution of number of electrons versus energy; (b) Number of electrons (N_2) in high-energy level (E_2) is smaller than that (N_1) in low-energy level (E_1)

In order to achieve population inversion, it is necessary to excite atoms into the upper energy level E_2, and hence, obtain a nonequilibrium distribution. This process is achieved using an external energy source and it is referred to as 'pumping'. A common method used for pumping involves the application of intense radiation (e.g., from an optical flash tube or high-frequency radio field). In the former case, atoms are excited into the higher energy state through stimulated absorption. However, the two-level system discussed earlier does not lend itself to suitable population inversion. When the two levels are equally degenerated (or not degenerated), then $B_{12} = B_{21}$. Therefore, the probabilities of absorption and stimulated emission are equal, providing at best equal populations in the two levels.

Population inversion, however, may be obtained in systems with three or four energy levels. The energy level diagrams for two such systems, which correspond to two non-semiconductor lasers. To aid attainment of population inversion both systems display a central metastable state in which the atoms spend an unusually long time. It is from this metastable level that the stimulated emission or lasing takes place. Figure 3.7 shows the amplification of light through a material with population inversion.

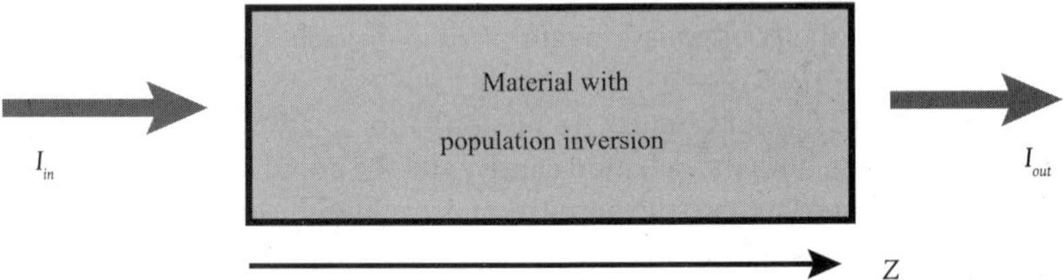

Fig. 3.7 Amplification of light through a material with population inversion

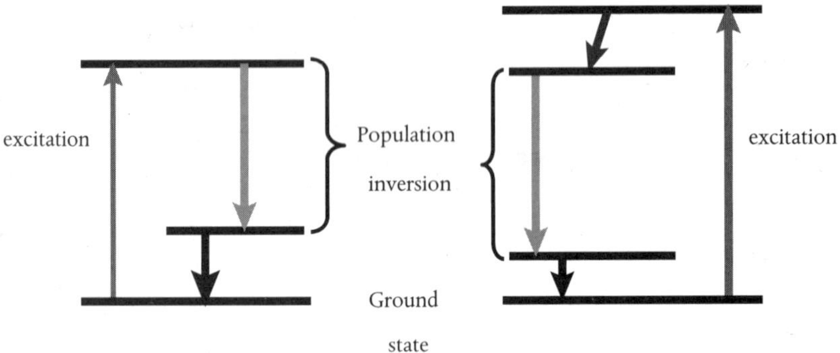

Fig. 3.8 A schematic showing that population inversion can be achieved at three-level and four-level systems

Figure 3.8 shows a schematic that population inversion can be achieved at three-level and four-level systems. The three-level system (Fig. 3.9) consists of a ground level E_0, a metastable level E_1, and a third level above the metastable level E_2. Initially, the atomic distribution will follow Boltzmann's law. However, with suitable pumping, the electrons in some of the atoms may be excited from the ground state into the higher level E_2. Since E_2 is a normal level, the electrons will rapidly decay by non-radiative process to either E_1 or directly to E_0. Hence, empty states will always be provided in E_2. The metastable level E_1 exhibits a much longer lifetime than E_2, which allows a large number of atoms to accumulate at E_1. Over a period, the density of atoms in the metastable state N_1 increases above those in the ground state N_0 and a population inversion is obtained between these two levels. Stimulated emission and lasing can then occur creating radiative electron transitions between levels E_1 and E_0. A drawback with the three-level systems like the ruby laser is that it generally requires very high pump powers because the terminal state of the laser transition is the ground state. Hence, more than half the ground state atoms must be pumped into the metastable state to achieve population inversion. Figure 3.9 shows a three-level system that is used to explain laser rate equation.

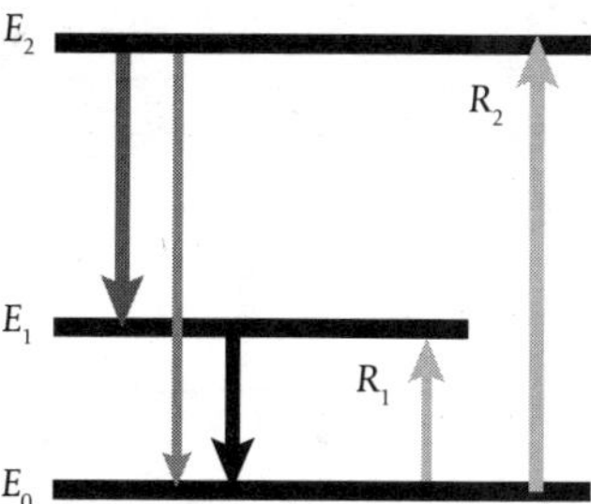

Fig. 3.9 A three-level system used to explain laser rate equation

By contrast, a four-level system like the He–Ne laser is characterised by much lower pumping requirements. In this case, the pumping excites the atoms from the ground state into energy level E_3 and they decay rapidly to the metastable level E_2. However, since the populations of E_3 and E_1 will remain essentially unchanged, a small increase in the number of atoms in energy level E_2 creates population inversion and lasing takes place between this level and level E_1. Figure 3.10 shows the four-level system that is used to describe population inversion.

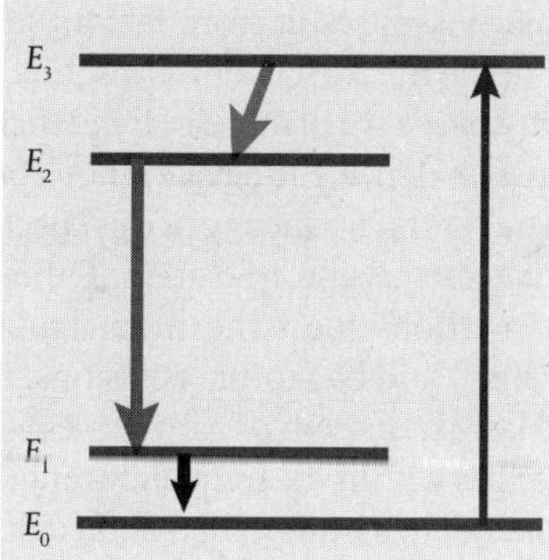

Fig. 3.10 A four-level system used to describe population inversion

3.4.4 Optical Feedback and Laser Oscillation

Light amplification in the laser occurs when a photon colliding with an atom in the excited energy state causes the stimulated emission of a second photon and both these photons release two more photons. Continuation of this process effectively creates avalanche multiplication, and when the electromagnetic waves associated with these photons are in phase, amplified coherent emission is obtained. To achieve this laser action, it is necessary to contain photons within the laser medium and maintain the conditions for coherence. This is accomplished by placing or forming mirrors (plane or curved) at either end of the amplifying medium. Figure 3.11 shows the schematic of a laser system. It consists of three major parts: (i) gain material for light amplification, (ii) cavity formed by mirrors and (iii) excitation system to maintain the gain material in population inversion.

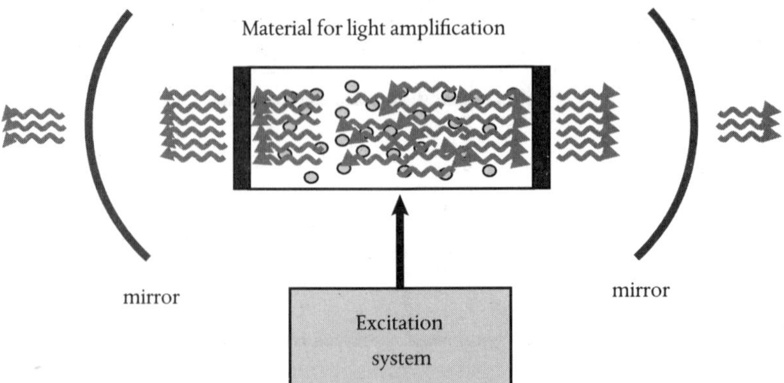

Fig. 3.11 Schematic of a laser system

The optical cavity formed is more analogous to an oscillator than an amplifier as it provides positive feedback of the photons by reflection at the mirrors at either end of the cavity. Hence, the optical signal feedback happens many times whilst receiving amplification as it passes through the medium. Therefore, the structure acts as a Fabry–Perot resonator. Although the amplification of the signal from a single pass through the medium is quite small, after multiple passes the net gain can be large. Furthermore, if one mirror is made partially transmitting, useful radiation may escape from the cavity. A stable output is obtained at saturation when the optical gain is exactly matched by the losses incurred in the amplifying medium. The major losses result from factors such as absorption and scattering in the amplifying medium, diffraction at the mirrors, and non-useful transmission through the mirrors.

Oscillations occur in the laser cavity over a small range of frequencies where the cavity gain is sufficient to overcome the above-mentioned losses. Hence, the device is not a perfectly monochromatic source but emits over a narrow spectral band. The central frequency of this spectral band is determined by the mean energy-level difference of the stimulated emission transition. Other oscillation frequencies within the spectral band result from frequency variations due to the thermal motion of atoms within the amplifying medium (known as Doppler broadening) and by atomic collisions. Hence, the amplification within the laser medium results in a broadened laser transition or gain curve over a finite spectral width. Therefore, the spectral emission from the device lies within the frequency range dictated by this gain curve.

Since the structure forms a resonant cavity and when sufficient population inversion exists in the amplifying medium, the radiation builds up and becomes established as standing waves between the mirrors. These standing waves exist only at frequencies for which the distance between the mirrors is an integral number of half wavelengths. Therefore, when the optical spacing between the mirrors is L, the resonance condition along the axis of the cavity is given as:

$$L = \frac{\lambda q}{2n}, \tag{3.12}$$

where λ is the emission wavelength, n is the refractive index of the amplying medium, and q is an integer. Alternatively, discrete emission frequencies f are defined as:

$$f = \frac{qc}{2nL} \tag{3.13}$$

where c is the velocity of light. The different frequencies of oscillation within the laser cavity are determined by the various integer values of q and each constitutes a resonance or mode. Since (3.12) and (3.13) apply for the case when L is along the longitudinal axis of the structure. The frequencies given by (3.13) are known as the longitudinal or axial modes. Furthermore, from (3.13), it may be observed that these modes are separated by a frequency interval δf where

$$\delta f = \frac{c}{2nL}. \tag{3.14}$$

The mode separation in terms of the free-space wavelength, assuming $\delta f \ll f$ and as $f = c/\lambda$, is given as:

$$\delta\lambda = \frac{\lambda \delta f}{f} = \frac{\lambda^2}{c}\delta f. \tag{3.15}$$

Hence, substituting for δf from (3.14) gives:

$$\delta\lambda = \frac{\lambda^2}{2nL}. \tag{3.16}$$

In addition, it should be noted that (3.15) can be used to determine the device spectral line width as a function of wavelength when it is quoted in Hertz or vice-versa. A large number of modes may be generated within the laser cavity; the spectral output from the device is defined by the gain curve. Hence, the laser emission will only include the longitudinal modes contained within the spectral width of the gain curve. This situation is illustrated in Fig 3.12, where several modes are shown to be present in the laser output. Such a device is said to be multimode.

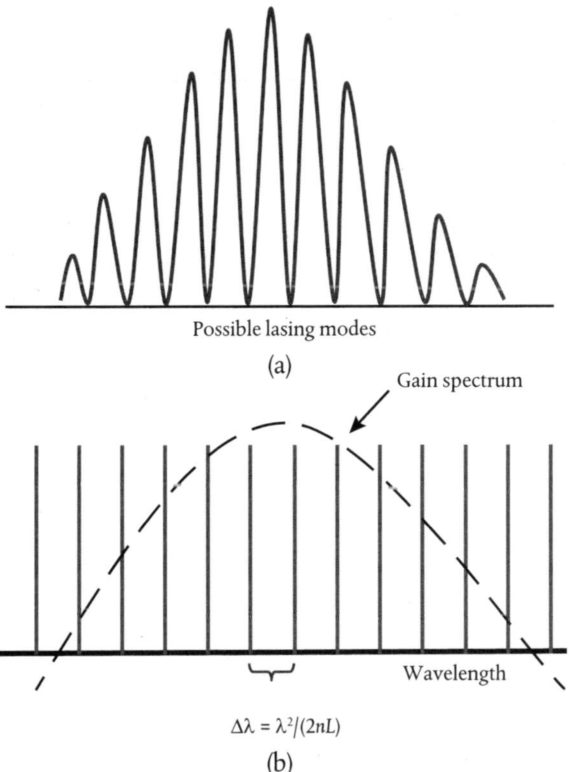

Fig. 3.12 (a) The possible lasing spectrum that contains those lasing modes; (b) Longitudinal modes that are within the gain line shape

Laser oscillation may also occur in a direction that is transverse to the axis of the cavity. This gives rise to resonant modes that are transverse to the direction of propagation. These transverse electromagnetic modes are designed in a similar manner to transverse modes in waveguides by TEM_{lm} where the integers l and m indicate the number of transverse modes. Unlike the longitudinal modes that contribute only a single spot of light to the laser output, transverse modes that contribute only a single spot of light to the laser output and they may give rise to a pattern of spots at the output. This may be observed from the low-order transverse mode patterns in which the direction of the electric field is also indicated. In the case of the TEM_{00} mode, all parts of the propagating wavefront are in phase. However, this is not so with higher order modes (TEM_{10}, TEM_{11}, etc.), where phase reversals produce the various mode patterns. Therefore, the greatest degree of coherence, together with the highest level of spectral purity, may be obtained from a laser that operates in only TEM_{00} mode. Higher order transverse modes only occur when the width of the cavity is sufficient for them to oscillate. Consequently, they may be eliminated by suitable narrowing of the laser cavity.

3.4.5 Threshold Condition for Laser Oscillation

It has been indicated that steady-state conditions for laser oscillation are achieved when the gain in the amplifying medium exactly balances the total losses. Hence, although population inversion between the energy levels providing the laser transition is necessary for oscillation to be established, it is not alone sufficient for lasing to occur. In addition, a minimum or threshold gain within the amplifying medium must be attained such that laser oscillations are initiated and sustained. This threshold gain may be determined by considering the change in energy of a light beam as it passes through the amplifying medium. For simplicity, all the losses except those due to transition through the mirrors may be included in a single loss coefficient per unit length $\bar{\alpha}\,\text{cm}^{-1}$. Again, we assume the amplifying medium occupies a length L completely filling the region between two mirrors that have reflectivities r_1 and r_2. On each round trip, the beam passes through the medium twice. Hence, the fractional loss incurred by the light beam is:

$$\text{Fractional loss} = r_1 r_2 exp(-2\bar{\alpha}L) \tag{3.17}$$

Furthermore, it is found that the increase in beam intensity resulting from stimulated emission is exponential. Therefore, if the gain coefficient per unit length produced by stimulated emission is $\bar{g}\,\text{cm}^{-1}$, the fractional round trip gain is given as:

$$\text{Fractional gain} = \exp(2\bar{g}L) \tag{3.18}$$

Hence,

$$\exp(2\bar{g}L) \times r_1 r_2 \exp(-2\bar{\alpha}L) = 1$$

and

$$r_1 r_2 \exp 2(\bar{g} - \bar{\alpha})L = 1 \tag{3.19}$$

The threshold gain per unit length may be obtained by rearranging the aforementioned expression to give:

$$\bar{g}_{\text{th}} = \bar{\alpha} + \frac{1}{2L}\ln\frac{1}{r_1 r_2} \tag{3.20}$$

The second term on the right-hand side of (3.20) represents the transmission loss through the mirrors. For laser action to be easily achieved, it is clear that a high threshold gain per unit length is required in order to balance the losses from the cavity. However, it must be noted that the parameters displayed in (3.20) are totally dependent on the laser type.

3.5 LASER CONSTRUCTION

A laser is constructed from three principal parts, which is shown in Fig. 3.13.

1. An energy source (usually referred to as the pump or pump source),
2. A gain medium or laser medium, and
3. Two or more mirrors that form an optical resonator.

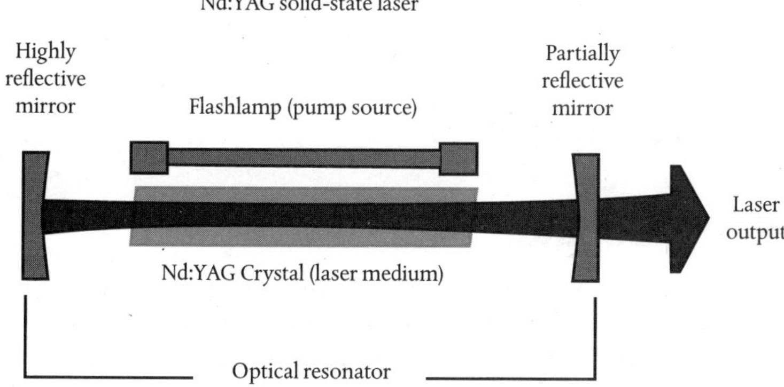

Fig. 3.13 Schematic diagram of a typical laser, showing the three major parts

3.5.1 Pump Source

The pump source is the part that provides energy to the laser system. Examples of pump sources include electrical discharges, flash lamps, arc lamps, light from another laser, chemical reactions, and even explosive devices. The type of pump source used principally depends on the gain medium and this also determines how the energy is transmitted to the medium. A helium–neon (He–Ne) laser uses an electrical discharge in the helium–neon gas mixture, an Nd:YAG laser uses either light focused from a xenon flash lamp or diode lasers, and excimer lasers use a chemical reaction.

3.5.2 Gain Medium/Laser Medium

The gain medium is the major determining factor of the wavelength of operation and other properties of the laser. Gain media in different materials have linear spectra or wide spectra. Gain media with wide spectra allow the tuning of laser frequency. There are hundreds of different gain media in which laser operation has been achieved. The gain medium is excited by the pump source to produce a population inversion and it is in the gain medium that spontaneous and stimulated emission of photons takes place leading to the phenomenon of optical gain or amplification.

3.5.3 Examples of Different Gain Media

1. Liquids like dye lasers. These are usually organic chemical solvents, such as methanol, ethanol, or ethylene glycol to which chemical dyes are added, such as coumarin, rhodamine, and fluorescein. The exact chemical configuration of the dye molecules determines the operation wavelength of the dye laser.
2. Gases such as carbon dioxide, argon, and krypton and mixtures like helium–neon. These lasers are often pumped by electrical discharge.
3. Solids such as crystals and glasses. The solid-host materials are usually doped with an impurity such as chromium, neodymium, erbium, or titanium ions. Typical hosts include yttrium aluminium garnet (YAG), yttrium lithium fluoride (YLF), sapphire (aluminium oxide), and various glasses. Examples of solid-state laser media include Nd:YAG, Ti:sapphire, Cr:sapphire (usually known as ruby), chromium-doped lithium strontium aluminium fluoride (Cr:LiSAF), Er:YLF, Nd:glass, and Er:glass. Solid-state lasers are usually pumped by flash lamps or light from another laser.
4. Semiconductors, a type of solid, crystal with uniform dopant distribution or material with differing dopant levels in which the movement of electrons can cause laser action. Semiconductor lasers are typically very small and can be pumped with a simple electric current enabling them to be used in consumer devices like compact disc players.

3.5.4 Optical Resonator

The optical resonator or optical cavity in its simplest form is two parallel mirrors placed around the gain medium that provide feedback of the light. The mirrors are given optical coatings that determine their reflective properties. Typically, one will be a high reflector and the other will be a partial reflector. The latter is called the output coupler because it allows some of the light to leave the cavity to produce the laser's output beam. Light from the medium produced by spontaneous emission is reflected by the mirrors into the medium where it may be amplified by stimulated emission. The light may reflect from the mirrors, and thus, it passes through the gain medium many hundreds of times before exiting the cavity. In more complex lasers, configurations with four or more mirrors forming the cavity are used. The design and alignment of the mirrors with respect to the medium is crucial for determining the exact operating wavelength and other attributes of the laser system.

Other optical devices, such as spinning mirrors, modulators, filters, and absorbers may be placed within the optical resonator to produce a variety of effects on the laser output, such as altering the wavelength of operation or the production of pulses of laser light. Some lasers do not use an optical cavity, but instead rely on very high optical gain to produce significant amplified spontaneous emission (ASE) without needing feedback of the light into the gain medium. Such lasers are said to be super luminescent, and they emit light with low coherence but with high bandwidth. Since they do not use optical feedback, these devices are often not categorised as lasers. Resonator modes have been described in Section 3.6.

The gain medium of a laser is a material of controlled purity, size, concentration, and shape, which amplifies the beam by the process of stimulated emission. It can be of any state: gas, liquid, solid, or plasma. The gain medium absorbs pump energy, which raises some electrons into higher energy (excited) quantum states. Particles can interact with light by either absorbing or emitting photons. Emission can be spontaneous or stimulated. In the latter case, the photon is emitted in the same direction as the light that is passing by. When the number of particles in one excited state exceeds the number of particles in some lower energy state, population inversion is achieved and the amount of stimulated emission due to light that passes through is larger than the amount of absorption. Hence, the light is amplified.

By itself, this makes an optical amplifier. When an optical amplifier is placed inside a resonant optical cavity, one obtains a laser.

The light generated by stimulated emission is very similar to the input signal in terms of wavelength, phase, and polarisation. This gives laser light its characteristic coherence, and allows it to maintain the uniform polarisation and often mono-chromaticity established by the optical cavity design. The optical resonator is sometimes referred to as an optical cavity but this is a misnomer: lasers use open resonators as opposed to the literal cavity that would be employed at microwave frequencies in a maser. The resonator typically consists of two mirrors between which a coherent beam of light travels in both directions, reflecting on it so that an average photon will pass through the gain medium repeatedly before it is emitted from the output aperture or lost to diffraction or absorption. If the gain (amplification) in the medium is larger than the resonator losses, then the power of the re-circulating light can rise exponentially. However, each stimulated emission event returns an atom from its excited state to the ground state, reducing the gain of the medium. With increasing beam power, the net gain (gain times loss) reduces to unity and the gain medium is said to be saturated. In a continuous-wave (CW) laser, the balance of pump power against gain saturation and cavity losses produces an equilibrium value of the laser power inside the cavity; this equilibrium determines the operating point of the laser. If the applied pump power is too small, the gain will never be sufficient to overcome the resonator losses, and laser light will not be produced. The minimum pump power needed to begin laser action is called the lasing threshold. The gain medium will amplify any photons passing through it, regardless of direction; but only the photons in a spatial mode supported by the resonator will pass more than once through the medium and receive substantial amplification.

The beam in the cavity and the output beam of the laser, when travelling in free space (or a homogenous medium) rather than waveguides (as in an optical fibre laser), can be approximated as a Gaussian beam in most lasers; such beams exhibit the minimum divergence for a given diameter. However, some high-power lasers may be multimode with the transverse modes often approximated using Hermite Gaussian or Laguerre Gaussian functions. It has been shown that unstable laser resonators (not used in most lasers) produce fractal-shaped beams. Near the beam waist (or focal region), it is highly collimated: the wavefronts are planar, normal to the direction of propagation with no beam divergence at that point. However, due to diffraction, it can only remain true well within the Rayleigh range. The beam of a single transverse mode (Gaussian beam) laser eventually diverges at an angle that varies inversely with the beam diameter, as required by diffraction theory.

Therefore, the pencil beam directly generated by a common helium–neon laser would extend to a size of perhaps 500 km when shone on the Moon (from the distance of the earth). On the other hand, the light from a semiconductor laser typically exits the tiny crystal with a large divergence: up to 50°. However, even such a divergent beam can be transformed into a similarly collimated beam by means of a lens system, as is always included, for instance, in a laser pointer whose light originates from a laser diode. That is possible due to the light being of a single spatial mode. This unique property of laser light, spatial coherence, cannot be replicated using standard light sources (except by discarding most of the light) as can be appreciated by comparing the beam from a flashlight (torch) or spotlight to that of almost any laser.

The mechanism of producing radiation in a laser relies on stimulated emission, where energy is extracted from a transition in an atom or a molecule. This is a quantum phenomenon discovered by Einstein who derived the relationship between the (a) coefficient describing spontaneous emission and the (b) coefficient that applies to absorption and stimulated emission. However, in the case of the free-electron laser, atomic energy levels are not involved; it appears that the operation of this rather exotic device can be explained without reference to quantum mechanics.

3.5.5 Stimulated Emission

If an atom is already in the excited state, it may be perturbed by the passage of a photon that has a frequency v_{21} corresponding to the energy gap ΔE of the excited state to ground state transition. In this case, the excited atom relaxes to the ground state and is induced to produce a second photon of frequency v_{21}. The original photon is not absorbed by the atom and so the result is two photons of the same frequency. This process is known as stimulated emission.

Specifically, an excited atom will act like a small electric dipole that will oscillate with the external field provided. One of the consequences of this oscillation is that it encourages electrons to decay to the lowest energy state. When this happens due to the presence of the electromagnetic field from a photon, a photon is released in the same phase and direction as the stimulating photon, and this process is called stimulated emission. Figure 3.14 shows the stimulated emission process.

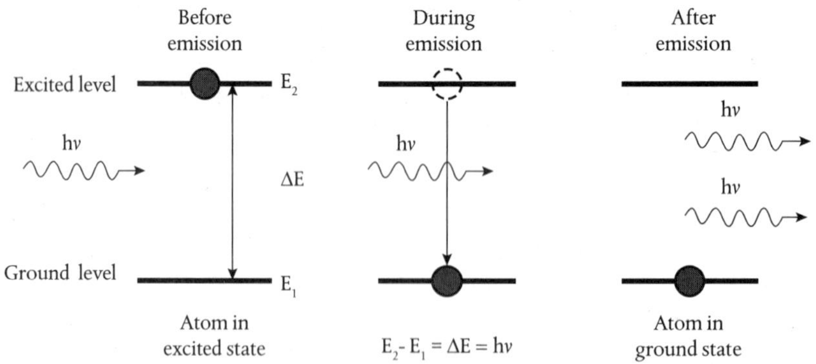

Fig. 3.14 Stimulated emission diagram

The rate at which stimulated emission occurs is proportional to the number of atoms N_2 in the excited state and the radiation density of the light. The base probability of a photon causing stimulated emission in a single excited atom was shown by Albert Einstein to be exactly equal to the probability of a photon being absorbed by an atom in the ground state shown in (3.9). Therefore, when the numbers of atoms in the ground and excited states are equal, the rate of stimulated emission is equal to the rate of absorption for a given radiation density.

The critical detail of stimulated emission is that the induced photon has the same frequency and phase as the incident photon. In other words, the two photons are coherent. It is the property that allows optical amplification and the production of a laser system. During the operation of a laser, all three light matter interactions described earlier are taking place. Initially, atoms are energised from the ground state to the excited state by a process called pumping, which is described in Section 3.4.3. Some of these atoms decay via spontaneous emission, releasing incoherent light as photons of frequency.

These photons are fed back into the laser medium usually by an optical resonator. Some of these photons are absorbed by the atoms in the ground state and the photons are lost to the laser process. However, some photons cause stimulated emission in excited state atoms releasing another coherent photon, and this results in optical amplification.

If the number of photons being amplified per unit time is greater than the number of photons being absorbed, then the net result is a continuously increasing number of photons being produced; the laser medium is said to have a gain of greater than unity.

From the earlier descriptions of absorption and stimulated emission, we discussed that the rates of these two processes are proportional to the number of atoms in the ground and excited states, N_1 and

N_2, respectively. If the ground state has a higher population than the excited state $N_1 > N_2$, the process of absorption dominates and there is a net attenuation of photons. If the populations of the two states are the same $N_1 = N_2$, the rate of absorption of light exactly balances the rate of emission; the medium is then said to be optically transparent. If the higher energy state has a greater population than the lower energy state $N_1 < N_2$, then the emission process dominates and light in the system undergoes a net increase in intensity. Therefore, it is clear that to produce a faster rate of stimulated emissions than absorptions, and it is required that the ratio of the populations of the two states is such that $N_2 / N_1 > 1$. In other words, a population inversion is required for laser operation.

3.6 MODES OF OPERATION

A laser can be classified as operating in either continuous or pulsed mode, depending on whether the power output is essentially continuous over time or whether its output takes the form of pulses of light on one or another time scale. Of course, even a laser whose output is normally continuous can be intentionally turned on and off at some rate in order to create pulses of light. When the modulation rate is on time scales much slower than the cavity lifetime and the time period over which energy can be stored in the lasing medium or pumping mechanism, then it is still classified as a modulated or pulsed CW laser. Most laser diodes used in communication systems fall in that category.

3.6.1 Continuous Wave Operation

Some applications of lasers depend on a beam whose output power is constant over time. Such a laser is known as continuous wave or CW. Many types of lasers can be made to operate in CW mode to satisfy such an application. Many of these lasers operate in several longitudinal modes at the same time, and beats between the slightly different optical frequencies of those oscillations will, in fact, produce amplitude variations on time scales shorter than the round-trip time (the reciprocal of the frequency spacing between modes), typically a few nanoseconds or less. In most cases, these lasers are still termed CW as their output power is steady when averaged over any longer time periods with the very high-frequency power variations having little or no impact in the intended application. However, the term is not applied to mode locked lasers, where the intention is to create very short pulses at the rate of the round-trip time.

For CW operation, it is required for the population inversion of the gain medium to be continually replenished by a steady pump source. In some lasing media, this is impossible. In some other lasers, it would require pumping the laser at a very high continuous power level that would be impractical or would destroy the laser by producing excessive heat. Such lasers cannot be run in CW mode.

3.6.2 Pulsed Operation

Pulsed operation of lasers refers to any laser not classified as CW, so that the optical power appears in pulses of some duration at some repetition rate. This encompasses a wide range of technologies addressing a number of different motivations. Some lasers are pulsed simply because they cannot be run in continuous mode.

In other cases, the application requires the production of pulses having energy as large as possible. Since the pulse energy is equal to the average power divided by the repetition rate, this goal can sometimes be satisfied by lowering the rate of pulses so that more energy can be built up in between pulses. In laser ablation, for example, a small volume of material at the surface of a work piece can be evaporated if it is heated in a very short time, whereas supplying the energy gradually would allow for the heat to be absorbed into the bulk of the piece, never attaining a sufficiently high temperature at a particular point.

Other applications rely on the peak pulse power (rather than the energy in the pulse), especially in order to obtain nonlinear optical effects. For a given pulse energy, this requires creating pulses of the shortest possible duration utilising techniques like Q-switching.

The optical bandwidth of a pulse cannot be narrower than the reciprocal of the pulse width. In the case of extremely short pulses, that implies lasing over a considerable bandwidth, quite contrary to the very narrow bandwidths typical of CW lasers. The lasing medium in some dye lasers and vibronic solid-state lasers produces optical gain over a wide bandwidth, making a laser possible that can generate pulses of light as short as a few femto-seconds $\left(10^{-15}\,\mathrm{s}\right)$.

3.6.3 Q-Switching

In a Q-switched laser, the population inversion is allowed to build-up by introducing loss inside the resonator that exceeds the gain of the medium; this can also be described as a reduction of the quality factor or 'Q' of the cavity. Then, after the pump energy stored in the laser medium has approached the maximum possible level, the introduced loss mechanism (often an electro- or acousto-optical element) is rapidly removed (or that occurs by itself in a passive device) allowing lasing to begin, which rapidly obtains the stored energy in the gain medium. This results in a short pulse incorporating that energy, and thus, a high peak power.

3.6.4 Mode Locking

A mode-locked laser is capable of emitting extremely short pulses on the order of 10 ps down to less than 10 fs. These pulses will repeat at the round-trip time, that is, the time that it takes light to complete one round trip between the mirrors comprising the resonator. Due to the Fourier limit (also known as energy time uncertainty), a pulse of such short temporal length has a spectrum spread over a considerable bandwidth. Therefore, such a gain medium must have a gain bandwidth sufficiently broad to amplify those frequencies. An example of a suitable material is titanium doped, artificially grown sapphire (Ti:sapphire) that has a very wide gain bandwidth and can, thus, produce pulses of only a few femtoseconds duration.

Such mode-locked lasers are a most versatile tool for researching processes occurring on extremely short time scales (known as femtosecond physics, femtosecond chemistry, and ultrafast science) for maximising the effect of nonlinearity in optical materials (e.g., in second harmonic generation, parametric down conversion, and optical parametric oscillators) due to the large peak power and in ablation applications. Again, due to the extremely short pulse duration, such a laser will produce pulses that achieve an extremely high peak power.

3.6.5 Pulsed Pumping

Another method of achieving pulsed laser operation is to pump the laser material with a source that is itself pulsed either through electronic charging in the case of flash lamps or another laser that is already pulsed. Pulsed pumping was historically used with dye lasers where the inverted population lifetime of a dye molecule was so short that a high energy, fast pump was needed. One way to overcome this problem was to charge up large capacitors that are then switched to discharge through flash lamps, producing an intense flash. Pulsed pumping is also required for three-level lasers in which the lower energy level rapidly becomes highly populated preventing further lasing until those atoms relax to the ground state. These lasers such as the excimer laser and the copper vapour laser can never be operated in CW mode.

3.7 SEMICONDUCTOR LASERS

Semiconductor lasers are lasers based on semiconductor gain media, where optical gain is usually achieved by stimulated emission at an inter-band transition under conditions of a high carrier density in the conduction band.

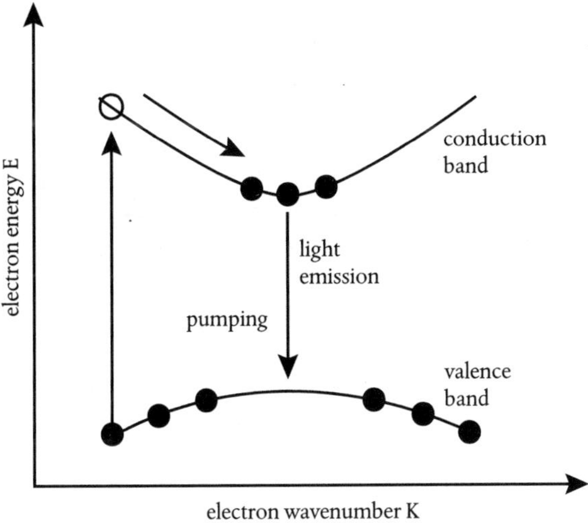

Fig. 3.15 Physical origin of gain in a semiconductor

The physical origin of gain in a semiconductor (for the usual case of an inter-band transition) is illustrated in Fig. 3.15. Without pumping, most of the electrons are in the valence band. A pump beam with photon energy slightly above the band gap energy can excite electrons into a higher state in the conduction band from where they quickly decay to states near the bottom of the conduction band. At the same time, the holes generated in the valence band move to the top of the valence band. Electrons in the conduction band can recombine with these holes, emitting photons with energy near the band gap energy. This process can also be stimulated by incoming photons with suitable energy. A quantitative description can be based on the Fermi–Dirac distributions for electrons in both bands.

Most semiconductor lasers are laser diodes, which are pumped with an electrical current in a region where an *n*-doped and a *p*-doped semiconductor materials meet. However, there are also optically pumped semiconductor lasers, where carriers are generated by absorbed pump light and quantum cascade lasers and intra-band transitions are utilised. Common materials for semiconductor lasers (and for other optoelectronic devices) are as follows:

- GaAs (gallium arsenide)
- AlGaAs (aluminium gallium arsenide)
- GaP (gallium phosphide)

- InGaP (indium gallium phosphide)
- GaN (gallium nitride)
- InGaAs (gallium indium arsenide)
- GaInNAs (indium gallium arsenide nitride)
- InP (indium phosphide)
- GaInP (gallium indium phosphide)

These are all direct band gap semiconductors; indirect band gap semiconductors such as silicon and germanium do not exhibit strong and efficient light emission. As the photon energy of a laser diode is close to the band gap energy, compositions with different band gap energies allow for different emission of wavelengths. For the ternary and quaternary semiconductor compounds, the band gap energy can be continuously varied in some substantial range. In $AlGaAs = Al_xGa_{1-x}As$, for example, an increased aluminium content (increased x) causes an increase in the band gap energy.

While the most common semiconductor lasers are operating in the near infrared spectral region, some others generate red light (e.g., in GaInP-based laser pointers) or blue or violet light (with gallium nitrides). For mid-infrared emission, there are lasers such as lead selenide (PbSe) lasers (lead salt lasers) and quantum cascade lasers.

Apart from the above-mentioned inorganic semiconductors, organic semiconductor compounds might also be used for semiconductor lasers. The corresponding technology is by far not mature, but its development is pursued because of the attractive prospect of finding a way for cheap mass production of such lasers. So far, only optically pumped organic semiconductor lasers have been demonstrated; however, for various reasons, it is difficult to achieve a high efficiency with electrical pumping.

3.7.1 Types of Semiconductor Lasers

There is a great variety of different semiconductor lasers, spanning wide parameter regions, and different application areas:

1. Small edge-emitting laser diodes generate a few milliwatts (or up to 0.5 W) of output power in a beam with high beam quality. For example, they are used in laser pointers, in CD players, and for optical fibre communications.
2. External-cavity diode lasers (ECDLs) contain a laser diode as the gain medium of a longer laser cavity. They are often wavelength tunable and exhibit a small emission line width.
3. Both monolithic and external cavity low-power levels can also be mode locked for ultrashort pulse generation.
4. Broad-area laser diodes generate up to a few watts of output power, but with significantly poorer beam quality.
5. High-power diode bars contain an array of broad area emitters, generating tens of watts with poor beam quality.
6. High-power stacked diode bars contain stacks of diode bars for the generation of extremely high powers of hundreds or thousands of watts.
7. Vertical-cavity surface-emitting lasers (VCSELs) emit the laser radiation in a direction perpendicular to the wafer, delivering a few milliwatts with high beam quality.
8. Optically pumped surface-emitting external cavity semiconductor lasers (VECSELs) are capable of generating multi-watt output powers with excellent beam quality, even in mode locked operation.
9. Quantum cascade lasers operate on intra-band transitions (rather than inter-band transitions) and usually emit in the mid-infrared region, and sometimes in the terahertz region. For example, they are used for trace gas analysis.

3.7.2 Typical Characteristics and Applications

Some typical aspects of semiconductor lasers are as follows:

1. Electrical pumping with moderate voltages and high efficiency is possible particularly for high-power diode lasers and allows their use as pump sources for highly efficient solid-state lasers.
2. A wide range of wavelengths are accessible with different devices, covering much of the visible, near infrared and mid-infrared spectral region. Some devices also allow for wavelength tuning.
3. Small laser diodes allow fast switching and modulation of the optical power, allowing their use in transmitters of fibre optic links.

Such characteristics have made semiconductor lasers as the technologically most important type of lasers. Their applications are extremely widespread, including areas as diverse as optical data transmission, optical data storage, metrology, spectroscopy, material processing, pumping solid-state lasers, and various kinds of medical treatments.

3.8 SUPERPOSITION OF MODES

In laser physics, it is usually convenient to describe the radiation within a resonator as a superposition of light in one or several resonator modes. Some examples are as follows:

1. Single-frequency operation of a laser means that only a single resonator mode (nearly always a Gaussian one) is excited; this leads to a much lower emission bandwidth than in cases where multiple resonator modes are excited.
2. Pulse generation with mode locking can also be explained in terms of resonator modes; in that case, several or even many fundamental modes are simultaneously excited with a fixed (locked) phase relationship of the corresponding fields. Under such conditions, a periodic pulse train is emitted by the laser.
3. When a laser has a poor beam quality, this is usually (although not always) the result of the excitation of higher order transverse cavity modes.

Note that it is important to distinguish coherent and incoherent mode superposition, which can have very different properties. If different modes of a laser resonator are simultaneously excited, there is usually the phenomenon of mode competition.

3.8.1 Properties of the Modes of Laser Resonators

Modes of laser resonator can differ significantly from those of an empty resonator because they are subjected to transversely varying gain and loss. This not only results in some deformation of the spatial shape but also in the resonator modes that are no longer mutually orthogonal. Instead, there is a set of adjoint modes that are related to the actual resonator modes by some bi-orthogonality relations. This bi-orthogonality (non-normal) nature has a number of peculiar implications. For example, the total power circulating in the laser is no longer simply the sum of the powers propagating in the different modes. There are also effects on the laser noise.

3.8.2 Single-mode/Frequency Laser

A single-frequency laser (sometimes called a single wavelength laser) is a laser that operates on a single resonator mode, so that it emits quasi-monochromatic radiation with a very small line width and low-phase noise. Because any mode distribution noise is eliminated, single-frequency lasers also have the potential to have very low-intensity noise. In nearly all cases, the excited mode is a Gaussian mode so that the output is diffraction limited.

Particularly, in low-power single-frequency lasers like laser diodes, there is some small amount of optical power in various resonator modes, even though one mode is clearly dominating. This is because such modes may be only slightly below the laser threshold so that spontaneous emission can already generate some substantial power. The mode suppression ratio (MSR) is then defined as the power of the lasing mode divided by that in the next strongest mode. It can be optimised by making the laser resonator more frequency selective.

Single-frequency lasers can be very sensitive to optical feedback. Even if less than a millionth of the output power is sent to the laser, this may, in some cases, cause strongly increased phase noise and intensity noise or even chaotic multimode operation. Therefore, single-frequency lasers have to be carefully protected against any back reflections, often using one or two Faraday isolators.

3.8.3 Operation

The line width of the laser radiation can be very small, that is, often only a few kilohertz, while the coherence length can accordingly be long. In contrast, multimode operation causes the line width to be a multiple of the mode spacing (free spectral range) of the resonator. Single-frequency operation is required to drive resonant cavities (e.g., for efficient frequency doubling in an external enhancement cavity), for high-resolution spectroscopy, for coherent beam combining of laser outputs, and for applications where the intensity noise must be very low.

Single-frequency operation is usually obtained when the bandwidth of the net gain is smaller than the frequency spacing of the resonator modes. The laser should operate on the fundamental transverse mode only, that is, it will emit a diffraction limited beam in order to eliminate the different mode frequencies of higher order transverse modes. For gain media with large bandwidth, single-frequency operation can be achieved by using some optical filters (e.g., an etalon) in the resonator and/or by making the resonator short (thus increasing the mode spacing to obtain a large-free spectral range).

Note that in reduced mode competition and inhomogeneous gain saturation caused by spatial hole burning in a standing wave resonator, it will be difficult to achieve single-frequency operation. Therefore, single-frequency operation is often realised with unidirectional ring lasers or sometimes with the twisted-mode technique. In fibre lasers and amplifiers, additional problems arise from stimulated Brillouin scattering. It might be expected that intra-cavity frequency doubling is not readily compatible with single-frequency operation, but the opposite is true; through sum frequency generation, the intra-cavity nonlinearity even protects the lasing modes by introducing higher losses for other modes. The concept of DFB lasers is often used for achieving single-frequency operation, particularly for semiconductor lasers and fibre lasers. This usually leads to fairly short laser resonators.

3.9 TYPES OF SINGLE-FREQUENCY LASERS

The following are the most important types of single-frequency lasers, which differ very much in terms of output power, line width, wavelength, complexity, and cost:

1. Some low-power laser diodes, in particular, index-guided lasers usually emit on a single mode. Stable single-mode operation is often achieved with DFB lasers or distributed Bragg reflector lasers (DBR lasers). Typical line widths are in the megahertz region. Significantly, smaller line widths are

possible by extending the resonator with a single-mode fibre containing a narrow bandwidth fibre Bragg grating or with external cavity diode lasers.

2. Special kinds of fibre lasers allow for single-frequency operation. Some of these are based on unidirectional ring laser designs; others have linear resonators and very short (highly doped) fibres. In any case, at least one fibre Bragg grating is usually used. Linear fibre lasers are sometimes realised as DFB lasers. Particularly, using devices having somewhat longer resonator can achieve very narrow line widths of a few kilohertz; on the other hand, in a single-frequency fibre amplifier, output powers varying between a few milliwatts and several watts in combination can be achieved.
3. Diode-pumped solid-state bulk lasers can be forced to operate on a single mode; this is often achieved with unidirectional ring laser designs, often with an intra-cavity filter and sometimes with the twisted-mode technique. Output powers can reach the multi-watt level and the line width can be as low as a few kilohertz.
4. Vertical-cavity surface-emitting lasers (VCSELs) have very short monolithic laser resonators, and hence, they result in huge cavity mode spacing and easily emit a few milliwatts on a single mode. The line width is at least a few megahertz, but the tuning range (without mode hops) can be very large.
5. A helium–neon laser can easily emit a single frequency, if its laser resonator is made short enough (of the order of 20 cm) because the gain bandwidth is small.

Most single-frequency lasers operate continuously, but there are also Q-switched single-frequency lasers, which do not exhibit mode beating, and thus, they exhibit very clean pulse shapes and low noises.

3.9.1 Distributed Feedback Lasers

A DFB laser is a laser where the whole resonator consists of a periodic structure, which acts as a distributed reflector in the wavelength range of laser action and contains a gain medium. Typically, the periodic structure is made with a phase shift in its middle. This structure is essentially the direct concatenation of two Bragg gratings with internal optical gain. It has multiple axial resonator modes, but there is typically one mode, which is favoured in terms of losses. This property is related to the above-mentioned phase shift. Therefore, single-frequency operation is often easily achieved, despite spatial hole burning due to the standing wave pattern in the gain medium. Due to the large free spectral range, wavelength tuning without mode hops may be possible over a range of several nanometres.

However, the tuning range may not be as large as for a DBR laser. Most distributed feedback lasers are either fibre lasers or semiconductor lasers operating on a single resonator mode (single-frequency operation). In the case of a fibre laser, the distributed reflection occurs in a fibre Bragg grating, typically with a length of a few millimetres or centimetres. Efficient pump absorption can be achieved only with a high doping concentration of the fibre, and unfortunately, it is often not easy to write Bragg gratings into fibres with a composition (e.g., phosphate glass) that allows for a high doping concentration. Therefore, the output power is usually fairly limited (e.g., to a few tens of milliwatts). However, this kind of single-frequency fibre laser is very simple and compact.

Fig. 3.16 DFB fibre laser, containing a fibre Bragg grating with a phase change in the middle, directly written into a rare earth-doped fibre

Semiconductor DFB lasers can be built with an integrated grating structure, for example, a corrugated waveguide. Such devices are available in a wide spectral range at least between 0.8 μm and 2.8 μm. Typical output powers are some tens of milliwatts. The linewidth is typically a few hundred megahertz, and wavelength tuning is often possible over several nanometres. Temperature-stabilised devices, as used in DWDM systems, can exhibit a high wavelength stability. Figure 3.16 shows the schematic of DFB Fibre laser.

3.9.2 Distributed Bragg Reflector Lasers

A DBR laser is a laser, where the laser resonator is made with at least one DBR outside the gain medium (the active region). A DBR is a Bragg mirror, that is, a light reflecting device (a mirror) based on Bragg reflection at a periodic structure. In most cases, the Bragg mirror is more specifically a quarter wave mirrors, providing the maximum amount of reflection for the given number of layers.

Lasers of DBR type are usually laser diodes, but the term is also sometimes used for fibre lasers containing fibre Bragg gratings. The following describes both laser types. Most solid-state bulk lasers actually also use laser mirrors that are Bragg mirrors; nevertheless, such lasers are not called DBR lasers. A DBR laser is different from a distributed feedback laser, where the whole active medium is embedded in a single-distributed reflector structure.

3.9.3 DBR Laser Diodes

A DBR laser diode contains some corrugated waveguide structure (a grating section) providing wavelength-dependent feedback to define the emission wavelength. Another section of the laser waveguide acts as the amplifying medium (active region) and the other end of the resonator may have another DBR.

The DBR laser diodes are usually single-frequency lasers with diffraction limited output, and often they are wavelength tunable (tunable lasers). Tuning within the free spectral range of the laser resonator may be accomplished with a separate phase section, which can be electrically heated or simply by varying the temperature of the gain region via the drive current. If the temperature of the whole device is varied, the wavelength response is significantly smaller than for an ordinary single-mode laser diode, since the reflection band of the grating is shifted less than the gain maximum. Electro-optic tuning can also be accomplished. Mode hop-free tuning over a larger wavelength region is possible by coordinated tuning of the Bragg grating and the gain structure.

There are more sophisticated device designs, exploiting a kind of Vernier effect with sampled gratings (SG DBR laser), that offer a tuning range as wide as 40 nm, although not without mode hops. The line width of a DBR diode is typically a few megahertz. Due to the relatively short laser resonator, it is larger than that of an external-cavity diode laser.

The actual DBR laser is than the seed laser. Output powers above 100 mW can be achieved with such devices. It is also possible to directly generate high powers (>10 W) with a broad-area laser diode having a surface Bragg grating within its resonator. The emission may then no longer be in a single mode, but still with a relatively small bandwidth. Vertical-cavity surface-emitting lasers (VCSELs) are actually also DBR lasers, even though the term "DBR laser diodes" is normally used for edge-emitting semiconductor lasers.

Applications of DBR laser diodes include optical fibre communications, free-space optical communications, laser cooling, optical metrology and sensors, and high-resolution spectroscopy. The DBR lasers actually compete with ECDLs, which also offer wavelength tunable single-frequency output with potentially better performance in terms of noise; however, it also requires a significantly more complex setup. Chips containing DBR laser arrays can serve as very compact sources for use in wavelength-division multiplexing systems.

3.9.4 DBR Fibre Lasers

Fig. 3.17 Short DBR fibre laser for narrow line-width emission

A fibre laser of DBR type usually has a linear laser resonator formed by an active (rare earth doped) fibre between two fibres Bragg gratings, as shown in Fig. 3.17. When compared to a fibre DFB laser, which has the gain within a single grating structure, a DBR fibre laser has a longer laser resonator, and thus, the potential for higher output power, higher power efficiency, and narrower line width. On the other hand, this can also lead to less robust single-frequency operation or to multimode operation with a correspondingly much larger emission bandwidth. Single-frequency DBR fibre lasers offer similar output powers as DBR laser diodes: tens of milliwatts or sometimes >100 mW.

3.9.5 Quantum Well Laser

A quantum well laser is a laser diode in which the active region of the device is so narrow that quantum confinement occurs. The wavelength of the light emitted by a quantum well laser is determined by the width of the active region rather than just the band gap of the material from which it is constructed. This means that much shorter wavelengths can be obtained from quantum well lasers than from conventional laser diodes using a particular semiconductor material. The efficiency of a quantum well laser is also greater than a conventional laser diode due to the stepwise form of its density of states function.

3.9.6 Methods for Higher Output Powers

For higher output powers, master oscillator power amplifier configurations are often used. An alternative with potentially lower laser noise is to use injection locking of a high-power laser with a single-frequency low-power seed laser.

3.9.7 Applications

Typical applications of single-frequency lasers occur in the areas of optical metrology (e.g., with fibre optic sensors) and inter-ferometry, optical data storage, high-resolution spectroscopy (e.g., LIDAR), and optical fibre communications. In some cases like spectroscopy, the narrow spectral width of the output is directly important. In other cases like optical data storage, a low-intensity noise is required, and thus, the absence of any mode beating noise. Single-frequency sources are also attractive because they can be used for driving resonant enhancement cavities; for example, nonlinear frequency conversion and coherent beam combining. The latter technique is currently used to develop laser systems with very high output powers and good beam quality.

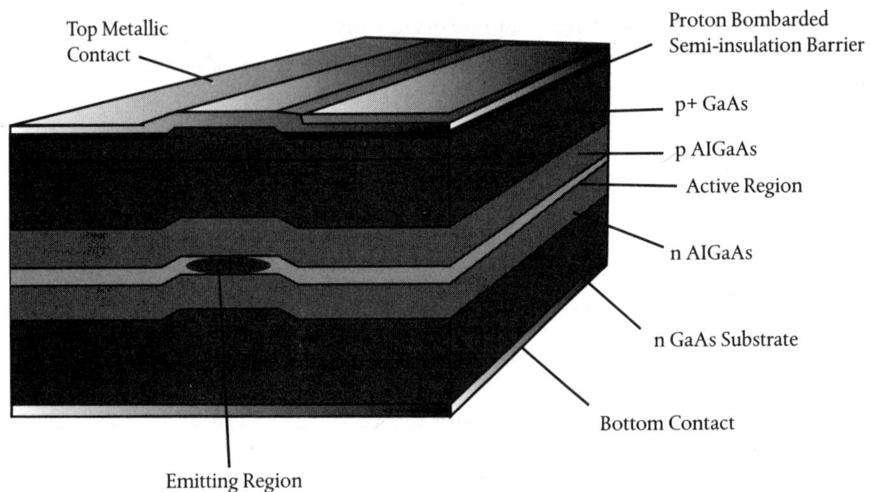

Fig. 3.18 Diagram of a Fabry–Perot laser

Figure 3.18 shows the structure of a typical edge-emitting laser. The dimensions of the active region are 200 mm in length, 210 mm in lateral width and 0.1 mm in transverse dimension. In reality, there are many different designs of edge-emitting lasers. Current flows from the *p*- to *n*-type semiconductor with electrons and holes being injected in the active region. A further advantage of the double heterostructure is that the large band gap semiconductor has a lower refractive index than that in the active region giving index guiding in the transverse direction. In the plain of the active region, the emission is confined by gain guiding, where the refractive index is modified by the carrier density. The formation of a stripe contact separated by a semi-insulating region of proton-bombarded semiconductor allows the current to flow through a restricted portion of the active region. This also aids the optical confinement in the plane of the active region. We have seen in the last section that *p–n* junctions can be used to create a laser. The current required to achieve lasing is enormous. This is because there is in defined region for radiative recombination to occur. Electrons and holes can just drift through the junction without recombining. A more efficient solution is to use heterostructures and double heterostructures.

3.10 LASER DIODE

A laser diode is a laser where the active medium is a semiconductor that is similar to that found in a light-emitting diode. The most common type of laser diode is formed from a *p–n* junction and powered by injected electric current. The former devices are sometimes referred to as injection laser diodes to distinguish them from optically pumped laser diodes.

A laser diode, like many other semiconductor devices, is formed by a very thin layer on the surface of a crystal wafer. As in other diodes, when *p–n* junction structure is forward biased, holes from the *p*-region are injected into the *n*-region, where electrons are the majority carriers. Similarly, electrons from the *n*-region are injected into the *p*-region, where holes are the majority carriers. When an electron and a hole are present in the same region, they may recombine spontaneous emission, that is, the electron may re-occupy the energy state of the hole, emitting a photon with energy equal to the difference between the electron and hole states involved. These injected electrons and holes represent the injection current of the diode and spontaneous emission gives the laser diode below the lasing threshold. Spontaneous emission is necessary to initiate laser oscillation but it is a source of inefficiency once the laser is oscillating.

Under suitable conditions, the electron and the hole may coexist in the same area for quite sometime (on the order of microseconds) before they recombine. Then, a near-by photon with energy equal to the recombination energy can cause recombination by stimulated emission. This generates another photon of the same frequency travelling in the same direction with the same polarisation and phase as the first photon. This means that stimulated emission causes gain in an optical wave (of the correct wavelength) in the injection region and the gain increases as the number of electrons and holes injected across the junction increases. The spontaneous and stimulated emission processes are vastly more efficient in direct band gap semiconductors than in indirect band gap semiconductors, and thus, silicon is not a common material for laser diodes.

3.10.1 Homojunction

The simplest *p–n* junction structure is called a homojunction, which is a basic *p–n* junction. The same semiconductor material like GaAs is on both sides of the junction. Starting with a heavily doped larger area *n*-type sample, a *p*-region is formed on it; this region is stopped by diffusing zinc atoms into *n*-type GaAs. A heavily zinc-doped layer constitutes the heavily doped *p*-layer. Grooves are cut along the length of the sample and small *p–n* junction chips are broken apart. For forming optical resonator cavity, external mirrors are not needed in a diode laser. A pair of parallel plane cleaved at the two ends of the *p–n* junction provides the required reflection to form the cavity. The two remaining sides of the diode are roughened to eliminate lasing in the directions other than the main ones that may be polished to increase reflectivity. This structure is the Fabry–Perot cavity.

3.10.2 Heterojunction

For a homostructure GaAs *p–n* junction, the threshold current density $\left(J_{\text{th}}\right)$ increases with temperature. A typical value of J_{th} is about $500\ \text{A}/\text{mm}^2$ at room temperature. Such large current density imposes serious difficulties in continuous operation of these lasers, and therefore, only pulsed operation is possible. The threshold current density can be reduced by lowering the junction temperature. The threshold current density can be reduced by using heterojunction instead of homojunction. At the heterojunction, there is carrier confinement as well as optical confinement.

When two different semiconductor crystals are joined together, the resulting structure is called a heterojunction in the semiconductor literature. In discussing the properties of heterojunction, it is generally assumed that the crystal potential and all other electronic properties of each of the constituent materials remain uncharged right up to the interface, and then change abruptly to take up the bulk properties of the other semiconductor. In most of the discussions on heterojunctions, attention is paid to the pairs, which are almost perfectly lattice matched.

Heterojunction is commonly grown with semiconductor that may be un-doped or doped *n*-type or *p*-type. Let us consider two semiconductors *A* and *B* with conduction and valence band edges E_c and E_v having different magnitudes of forbidden gap. An electron at conduction band edge in *A* will have lower energy than in material *B* when junction is made between (A) (A) and (B) (B), a potential barrier is formed at the interface. This restricts the motion of electrons incident from (A) (A) side. This type of situations are found in the case of GaAs and AlGaAs junction. The electron in GaAs experiences as a barrier to enter in AlGaAs portion of the structure as it is reflected by the potential barrier.

3.10.2.1 Single heterojunction laser

A single heterojunction (SH) laser is made by growing a layer of GaAlAs on the GaAs. In a GaAs homojunction laser, the laser action occurs primarily on the *p*-side of the junction due to a higher efficiency for electron injection than for hole. Electrons diffuse into the *p*-material and population inversion occurs for only part of the electrons near the junction. If the *p*-material is made narrow and terminated in a barrier, the injected electrons can be confined near the junction. The electrons injected from the *n*-GaAs layer to the *p*-GaAs layer are blocked from diffusing over a large distance by the potential barrier provided by the higher band gap *p*-GaAs layer. Now, the active region thickness is the thickness of *p*-GaAs layer. If its thickness is made small, a smaller drive current can lead to population inversion and lasing. Furthermore, the change of the refractive index at the heterojunction provides a waveguide effect that confines light. However, SH lasers do not confine light well enough to emit light continuously at room temperature.

3.10.2.2 Double heterojunction laser

It is possible to have more than one heterojunction in a structure. When there are two heterojunctions, it is called double heterojunction (DH) structure. This DH structure further confines injected electrons to the active region and reduces the threshold current required for lasing. In a DH laser, two heterojunctions are used to sandwich a layer of material with narrow band gap between the materials of wide band gap. The slight energy differences between the band gap of two semiconductors set up a potential barrier that serves to confine the charge carriers in the very narrow region near the junction. In double heterostructure, the heterojunction potential barriers on both sides confine the carriers in the active low band gap region. This confinement enhances the stimulated emission and substantially reduces the threshold current density, J_{th} for double heterostructure lasers can be as low as $12\ \text{A}/\text{mm}^2$ or less at $300°\text{K}$ and so continuous room temperature operation has been achieved.

Whenever III–V semiconductors are used for either the sources or the detectors in optical communication system, they invariably take the form of the five-layer double heterostructure. The two outermost layers are essentially contact layers, and one is normally the substrate on which the other layers are grown epitaxial. The middle three layers are from the DH and consists of an active layer sandwiched between two confining layers made from wider band gap material.

3.11 TYPES OF LASERS

The simple laser diode structure, described earlier, is extremely inefficient. Such devices require so much power that they can only achieve pulsed operation without damage. Although historically important and easy to explain, such devices are not practical.

- Double heterostructure lasers
- Quantum well lasers
- Quantum cascade lasers
- Separate confinement hetero structure lasers
- Distributed feedback lasers
- VCSELs
- VECSELs
- ECDLs

3.11.1 Double Heterostructure Lasers

In these devices, a layer of low band gap material is sandwiched between two high band gap layers. One commonly used pair of materials is gallium arsenide (GaAs) with aluminium gallium arsenide

$\left[Al_xGa_{(1-x)}\ As\right]$. Each of the junctions between different band gap materials is called a heterostructure, and hence, the name double heterostructure laser or DH laser. The advantage of a DH laser is that the region where free electrons and holes exist simultaneously and the active region is confined to the thin middle layer. This means that many more of the electron–hole pairs can contribute to amplification; however, not so many are left in the poorly amplifying periphery. In addition, light is reflected from the heterojunction; hence, the light is confined to the region where the amplification takes place. Figure 3.19 shows the schematic of double heterostructure laser diode.

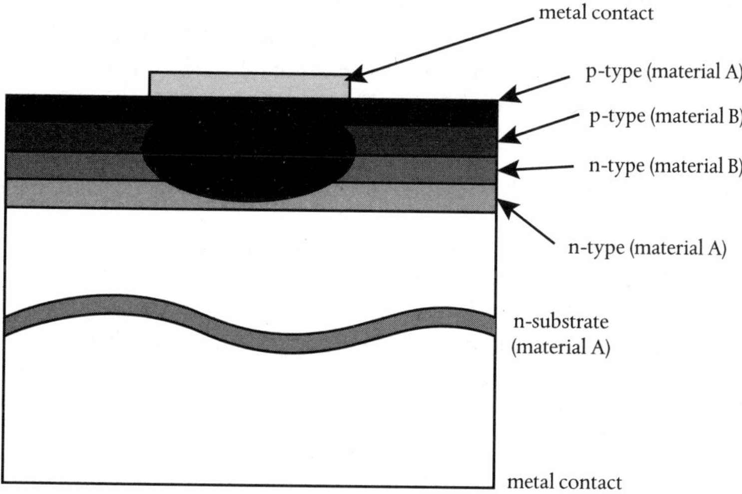

Fig. 3.19 Double heterostructure laser diode

3.11.2 Quantum Well Lasers

If the middle layer is made thin enough, it acts as a quantum well. This means that the vertical variation of the electrons wave function, and thus, a component of its energy is quantised. The efficiency of a quantum well laser is greater than that of a bulk laser because the density of states function of electrons in the quantum well system has an abrupt edge that concentrates electrons in energy states that contribute to laser action.

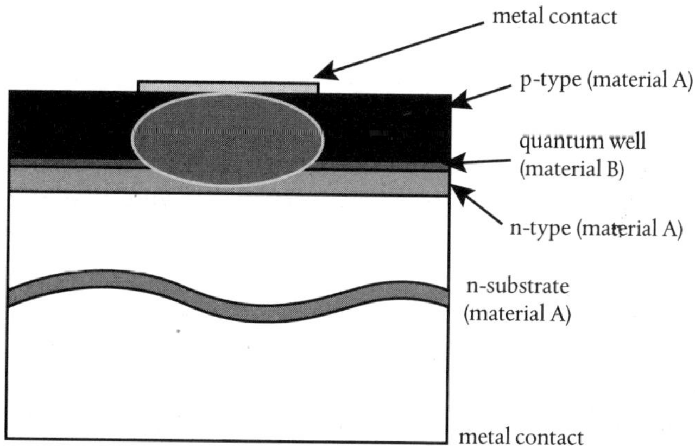

Fig. 3.20 A simple quantum well laser diode

Lasers containing more than one quantum well layer are known as multiple quantum well lasers. Multiple quantum wells improve the overlap of the gain region with the optical waveguide mode. Further improvements in the laser efficiency have also been demonstrated by reducing the quantum well layer to a quantum wire or to a sea of quantum dots. Figure 3.20 shows the schematic of quantum well laser diode.

3.11.3 Quantum Cascade Lasers

In a quantum cascade laser, the difference between quantum well energy levels is used for the laser transition instead of the band gap. This enables laser action at relatively long wavelengths, which can be tuned simply by altering the thickness of the layer. They are heterojunction lasers.

3.11.4 Separate Confinement Heterostructure Lasers

The problem with the simple quantum well diode described earlier is that the thin layer is simply too small to effectively confine the light. To compensate, another two layers are added on the outside of the first three layers. These layers have a lower refractive index than the centre layers, and hence, they confine the light effectively. Such a design is called a separate confinement heterostructure (SCH) laser diode.

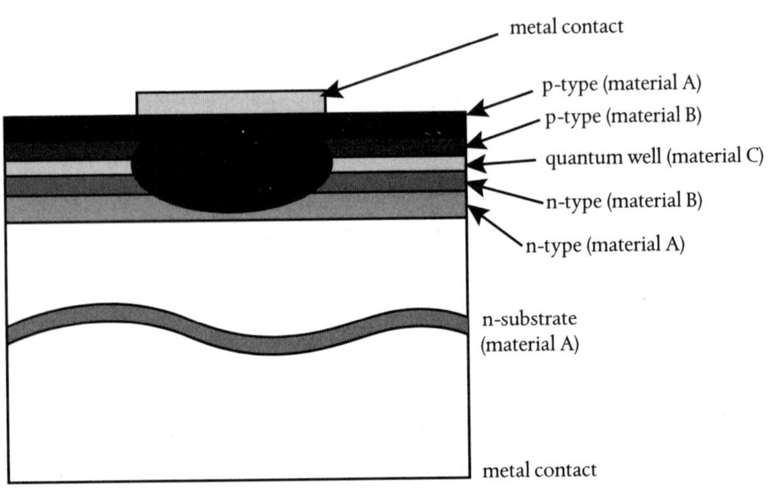

Fig. 3.21 Separate confinement heterostructure quantum well laser diode

Almost all commercial laser diodes since the 1990s have been SCH quantum well diodes. Figure 3.21 shows the schematic of separately confinement heterostructure quantum well laser diode.

3.11.5 Distributed Feedback Lasers

Distributed feedback lasers are the most common transmitter type in DWDM systems. To stabilise the lasing wavelength, a diffraction grating is etched close to the *p–n* junction of the diode. This grating acts like an optical filter, causing a single wavelength to be fed back into the gain region and laser. Since the grating provides the feedback that is required for lasing, reflection from the facets is not required.

Therefore, at least one facet of a DFB is anti-reflection coated. The DFB laser has a stable wavelength that is set during manufacturing by the pitch of the grating and can only be tuned slightly with temperature. The DFB lasers are widely used in optical communication applications, where a precise and stable wavelength is critical.

3.11.6 Vertical-cavity Surface-emitting Lasers

Vertical-cavity surface-emitting lasers (VCSELs) have the optical cavity axis along the direction of current flow rather than perpendicular to the current flow as in conventional laser diodes. The active region length is very short when compared with the lateral dimensions so that the radiation emerges from the surface of the cavity rather than from its edge as shown in Fig. 3.22. The reflectors at the ends of the cavity are dielectric mirrors made from alternating high and low refractive-index quarter-wave thick multilayer.

Such dielectric mirrors provide a high degree of wavelength selective reflectance at the required free surface wavelength λ if the thicknesses of alternating layers d_1 and d_2 with refractive indices n_1 and n_2 are such that $n_1 d_1 + n_2 d_2 = \lambda / 2$, which then leads to the constructive interference of all partially reflected waves at the interfaces. However, there is a disadvantage; because of the high mirror reflectivity, VCSELs have lower output powers when compared to edge-emitting lasers.

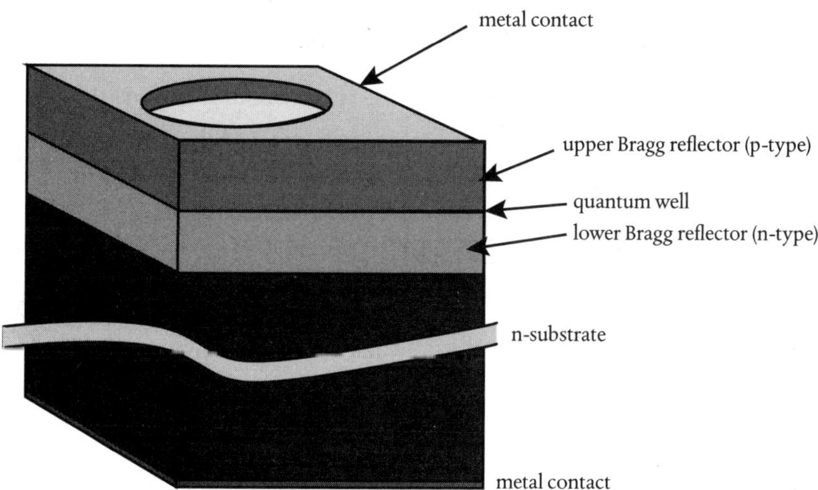

Fig. 3.22 Simple VCSEL structure

There are several advantages for the production of VCSELs when compared with the production process of edge-emitting lasers. Edge emitters cannot be tested until the end of the production process. If the edge emitter does not work due to bad contacts or poor material growth quality, the production time and the processing materials have been wasted. Additionally, because VCSELs emit the beam perpendicular to the active region of the laser as opposed to parallel as with an edge emitter, tens of thousands of VCSELs can be processed simultaneously on a three-inch Gallium Arsenide wafer. Furthermore, even though the VCSEL production process is of more labour and material intensive, the yield can be controlled to a more predictable outcome. However, they normally show a low-power output level.

3.11.7 Vertical External Cavity Surface-emitting Lasers

Vertical external cavity surface-emitting lasers, or VECSELs, are similar to VCSELs. In VCSELs, the mirrors are typically grown epitaxially as part of the diode structure or grown separately and bonded directly to the semiconductor containing the active region. The VECSELs are distinguished by a construction in which one of the two mirrors is external to the diode structure. As a result, the cavity includes a free-space region. A typical distance from the diode to the external mirror would be 1 cm.

One of the most interesting features of any VECSEL is the small thickness of the semiconductor gain region in the direction of propagation, that is, less than 100 nm. In contrast, a conventional in plane semiconductor laser entails light propagation over distances of from 250 μm upward to 2 mm or longer. The significance of the short propagation distance is that it causes the effect of "anti-guiding" nonlinearities in the diode laser gain region to be minimised. The result is a large cross-section single-mode optical beam, which is not attainable from in plane (edge-emitting) diode lasers.

Several workers demonstrated optically pumped VECSELs, and they continue to be developed for many applications including high-power sources for use in industrial machining (cutting, punching, etc.) because of their unusually high power and efficiency when pumped by multimode diode laser bars. Electrically pumped VECSELs have also been demonstrated. Applications for electrically pumped VECSELs include projection displays, served by frequency doubling of near IR VECSEL emitters to produce blue and green light.

3.11.8 External-cavity Diode Lasers

External-cavity diode lasers are tunable lasers that use mainly double heterostructures diodes of the $Al_xGa_{(1-x)}$ As type. The first ECDLs used intracavity etalons and simple tuning Littrow gratings. Other designs include gratings in grazing incidence configuration and multiple prism grating configurations.

3.12 APPLICATIONS OF LASER

Lasers are sources of light with very special properties. For that reason, there is a great variety of laser applications. The following sections give an overview.

3.12.1 Manufacturing

Lasers are widely used in manufacturing, for example, cutting, drilling, welding, cladding, soldering, hardening, ablating, surface treatment, marking, engraving, micromachining, pulsed laser deposition, lithography, alignment, and so on. In most cases, relatively high optical intensities are applied to a small spot, leading to intense heating possibly evaporation and plasma generation. Essential aspects are the high spatial coherence of laser light, allowing for strong focusing, and often also the potential for generating intense pulses.

Laser processing methods have many advantages when compared with mechanical approaches. They allow the fabrication of very fine structures with high quality, thus avoiding mechanical stress caused by mechanical drills and blades. A laser beam with high beam quality can be used to drill very fine and deep holes like injection nozzles. A high processing speed is often achieved in the fabrication of filter sieves. Furthermore, the lifetime limitation of mechanical tools is removed. It can also be advantageous to process materials without touching them.

3.12.2 Medical Applications

There is a wide range of medical applications. Often these relate to the outer parts of the human body, which are easily reached with light; examples are eye surgery and vision correction (LASIK), dentistry, dermatology (e.g., photodynamic therapy of cancer), and various kinds of cosmetic treatment such as tattoo removal and hair removal. Lasers are also used for surgery (e.g., of the prostate), exploiting the possibility to cut tissues while causing minimal bleeding.

Very different types of lasers are required for medical applications, depending on the optical wavelength, output power, pulse format and so on. In many cases, the laser wavelength is chosen such that certain substances (e.g., pigments in tattoos or caries in teeth) absorb light more strongly than surrounding tissue, so that they can be more precisely targeted. Medical lasers are not always used for therapy. Some of them rather assist the diagnosis; for example, methods of ocular imaging, laser microscopy, or spectroscopy (refer the following section).

3.12.3 Metrology

Lasers are widely used in optical metrology for extremely precise position measurements, optical surface profiling with interferometers, long distance range finding, and navigation.

Laser scanners are based on collimated laser beams, which can read bar codes or other graphics over some distance. It is also possible to scan three dimensional objects in the context of crime scene investigation (CSI).

Optical sampling is a technique applied for the characterisation of fast electronic microcircuits, microwave photonics, terahertz science, and so on. Lasers also allow for extremely precise time measurements, and therefore, they are essential components of optical clocks that are beginning to outperform the currently used caesium atomic clocks. Fibre optic sensors, often probed with laser light, allow for the distributed measurement of temperature, stress, and other quantities; for example, in oil pipelines and wings of airplanes.

3.12.4 Data Storage

Optical data storage in compact discs (CDs), DVDs, Blue-ray discs, and magneto-optical disks nearly always relies on a laser source, which has a high spatial coherence and can be used to address very tiny spots in the recording medium, allowing a very high-density data storage. Another case is holography, where the temporal coherence can also be important.

3.12.5 Communications

Optical fibre communication, extensively used particularly for long distance optical data transmission, mostly relies on laser light in optical glass fibres. Free-space optical communications like inter-satellite communications is based on higher power lasers generating collimated laser beams that propagate over large distances with small beam divergence.

3.12.6 Displays

Laser projection displays containing RGB sources can be used for cinemas, home videos, flight simulators, and so on; they are often superior to other displays with respect to possible screen dimensions, resolution, and colour saturation. However, further reductions in manufacturing costs will be essential for deep market penetration.

3.12.7 Spectroscopy

Laser spectroscopy is used in many different forms and in a wide range of applications. For example, atmospheric physics and pollution monitoring profits from trace gas sensing with differential absorption LIDAR technology. Solid materials can be analysed with laser-induced breakdown spectroscopy. Laser spectroscopy also plays a role in medicine (e.g., cancer detection), biology, and various types of fundamental research, partly related to metrology.

3.12.8 Microscopy

Laser microscopes and setups for optical coherence tomography (OCT) provide images of biological samples with very high resolution often in three dimensions. It is also possible to realise functional imaging.

3.12.9 Various Scientific Applications

Laser cooling makes it possible to bring clouds of atoms or ions to extremely low temperatures. This has applications in fundamental research and also for industrial purposes. Particularly in biological and medical research, optical tweezers can be used for trapping and manipulating small particles, such as bacteria or parts of living cells.

Laser guide stars are used in astronomical observatories in combination with adaptive optics for atmospheric correction. They allow substantially increased image resolution even in cases where a sufficiently close by natural guide star is not available.

3.12.10 Energy Technology

In the future, high-power laser systems might play a role in electricity generation. Laser-induced nuclear fusion is investigated as an alternative to other types of fusion reactors. High-power lasers can also be used for isotope separation.

3.12.11 Military Applications

There are a variety of military laser applications. In relatively few cases, lasers are used as weapons; the laser sword has become popular in movies, but not in practice. Some high-power lasers are currently developed for potential use as directed energy weapons on the battle field or for destroying missiles, projectiles, and mines.

In other cases, lasers function as target designators or laser sights (essentially laser pointers emitting visible or invisible laser beams) or as irritating or blinding (normally not directly destroying) counter-measures against heat-seeking anti-aircraft missiles. It is also possible to blind soldiers temporarily or permanently with laser beams, although the latter is forbidden by rules of war. There are also many laser applications that are not specific for military use; for example, in areas such as range finding, LIDAR, and optical communications. Table 3.1 shows the emission wavelengths of various types of laser diodes.

Table 3.1 Emission wavelengths of various types of laser diodes

Laser diode material (Active region/substrate)	Typical emission wavelengths	Typical application
InGaN	380 405 450 470 nm	Data storage
AlGaInP	635 650 670 nm	Laser pointers, DVD players
AlGaAs/GaAs	720–850 nm	CD players, laser printers
InGaAs/GaAs	900–1100 nm	Pumping EDFA; high-power VECSEL
InGaAsP/InP	1000–1650 nm	Optical fibre communications

3.13 SOME SPECIFIC APPLICATIONS OF INTEGRATED DFB LASER

Integrated DFB lasers' main application is in the area of optical fibre communications, particularly in fibre optic networks, but they can also be used for optical sensors and in metrology.

3.13.1 DWDM

Dense wavelength division multiplexing, or DWDM for short, refers originally to optical signals multiplexed within the 1550 nm band so as to leverage the capabilities (and cost) of erbium-doped fibre amplifiers (EDFAs), which are effective for wavelengths between approximately 1525–1565 nm (C-band), or 1570–1610 nm (L-band). The EDFAs were originally developed to replace SONET/SDH optical-electrical-optical (OEO) regenerators, which they have made practically obsolete. The EDFAs can amplify any optical signal in their operating range, regardless of the modulated bit rate. In terms of multi-wavelength signals, so long as the EDFA has enough pump energy available to it, it can amplify as many optical signals as can be multiplexed into its amplification band (although signal densities are limited by choice of modulation format). Therefore, EDFAs allow a single-channel optical link to be upgraded in bit rate by replacing only equipment at the ends of the link, while retaining the existing EDFA or series of EDFAs through a long-haul route. Furthermore, single wavelength links using EDFAs can similarly be upgraded to WDM links at reasonable cost. The EDFAs cost is, thus, leveraged across as many channels as can be multiplexed into the 1550-nm band.

Fibre optic sensors (also called optical fibre sensors) are fibre and laser-based devices for sensing some quantity, typically temperature or mechanical strain, but sometimes also displacements, vibrations, pressure, acceleration, rotations (measured with optical gyroscopes based on the Sagnac effect), or concentrations of chemical species. The general principle of such devices is that light from a laser (often a single-frequency fibre laser) or from a super luminescent source is sent through an optical fibre experiences changes of its parameters either in the fibre or in one or several fibre Bragg gratings, and then, reaches a detector arrangement that measures these changes.

On comparison to other types of sensors, fibre optic sensors exhibit the following advantages:

1. They consist of electrically insulating materials (no electric cables are required), which makes possible their use in high-voltage environments.
2. They can be safely used in explosive environments because there is no risk of electrical sparks even in the case of defects.
3. They are immune to electromagnetic interference (EMI), even to nearby lightning strikes, and do not themselves electrically disturb other devices.
4. Their materials can be chemically passive, that is, they do not contaminate their surroundings and are not subject to corrosion.
5. They have a very wide operating temperature range (much wider than is possible for many electronic devices).
6. They have multiplexing capabilities: multiple sensors in a single fibre line can be interrogated with a single optical source.

3.13.2 Optical Metrology

Optical metrology is the science and technology concerning measurements with light. Such measurements can either target properties of light itself or other properties like some distance. Some examples of optical metrology are as follows:

1. Optical distance measurements with lasers can be based on interferometers or measurements of the time of flight of light pulses.
2. Optical frequency metrology deals with high precision measurements of optical frequencies (→ optical clocks).
3. There are devices for optical wavelength and linewidth measurements based on self-heterodyne interferometers.
4. Optical powers can be measured with photodiodes, thermal power meters, or other equipment.
5. Optical temperatures measurements may be based on the analysis of the thermal emission of hot bodies, or rely on the measurement of occupation probabilities for energy levels of atoms or molecules.
6. Optical irradiation and illumination measurements can address either some pure physical quantity such as an optical intensity (power per unit area) (radiometry) or something like a perceived brightness (photometry).

3.14 OPTICAL BI-STABILITY

In optics, optical bi-stability is an attribute of certain optical devices where two resonant transmissions states are possible and stable depending on the input. Optical devices with a feedback mechanism, for example, a laser provides two methods of achieving bi-stability.

1. Absorptive bi-stability utilises an absorber to block light inversely dependent on the intensity of the source light. The first bi-stable state resides at a given intensity where no absorber is used. The second state resides at the point where the light intensity overcomes the absorber ability to block light.
2. Refractive bi-stability utilises an optical mechanism that changes its refractive index inversely dependent on the intensity of the source light. The first bi-stable state resides at a given intensity where no optical mechanism is used. The second state resides at the point where a certain light

intensity causes the light to resonate to the corresponding refractive index.

This effect is caused by two factors:

1. Nonlinear atom field interaction
2. Feedback effect of mirror

Important cases that might be regarded are as follows:

- Atomic detuning
- Cooperating factor
- Cavity mistuning

Applications of this phenomenon include its use in optical transmitters, memory elements, and pulse shapers. A phenomenon exhibited by certain resonant optical structures whereby it is possible to have two stable-steady transmission states for the device, depending upon the history of the input. Such a bi-stable device may be useful for optical computing elements because of its memory characteristics. The bi-stability can result from the intrinsic properties of the optical device or from some external feedback like an electrical voltage supplied by another device. This second type, extrinsic or hybrid optical bi-stability, is not true optical bi-stability.

Optical bi-stability is an inherently steady-state phenomenon, and typically, any cycling of the device through its hysteresis cycle must be done adiabatically; that is, changes in the propagating light amplitude, envelope phase, and profile must occur sufficiently slowly that their impact on the evolution of the system may be neglected. This requirement imposes some rather severe frequency response limitations on the use of intrinsically bi-stable devices in optical circuits. The two primary types of intrinsic optical bi-stability, each arising from a distinct physical mechanism, are absorptive bi-stability and refractive bi-stability.

3.14.1 Types of Bi-stability

3.14.1.1 Absorptive optical bi-stability

Absorptive optical bi-stability is based upon coupling the feedback mechanism inherent in an optical cavity with an absorbing nonlinear optical medium in which the absorption coefficient decreases with increasing light intensity (a saturable absorber). The basic theory of operation is: the saturable absorber is placed in the cavity, and the cavity is resonantly pumped. For low-light intensities, the transmission coefficient for the cavity is small because of the presence of the highly absorbing medium inside the cavity. As the pump intensity is increased, the absorption of the nonlinear medium decreases. Finally, for some threshold pump intensity, the cavity switches into a high transmission state because the absorption coefficient is reduced sufficiently that the intrinsic cavity feedback mechanism dominates. The threshold is very sharp because when the cavity is in a highly transitive state, the built-up intensity inside the cavity becomes very large when compared to the pump intensity (due to the feedback) and effectively bleaches virtually all of the absorption in the nonlinear medium. The intense pump is then largely transmitted, although some energy is stored in the cavity to bleach the absorber as in laser, optical pumping.

This device exhibits two characteristics that constrain its usefulness in particular applications. The device is based on an absorption mechanism, so the energy absorbed from the pump light must be dissipated in the bi-stable element or heat sinked elsewhere. It is highly frequency sensitive because its operation is based on the switching characteristics of a resonant cavity.

3.14.1.2 Refractive optical bi-stability

Refractive optical bi-stability is based on coupling the feedback mechanism inherent in an optical cavity with a nonlinear optical medium that exhibits a change in the refractive index as a function of light intensity. The nonlinear refractive medium is placed inside the optical cavity and the cavity is pumped slightly off resonance so that the transmission coefficient is small when compared to unity. However, a small amount of light intensity does exist inside the cavity and changes the effective optical path length inside the cavity by inducing change in the refractive index of the nonlinear medium. As the pump intensity is increased, this change in the effective path length becomes larger, until at some point the cavity switches into, and possibly past, resonance. The transmission coefficient switches abruptly to a value close to unity, and the built-up intensity inside the cavity increases abruptly. If the pump intensity is increased further, it is possible to switch the cavity through a second resonance with an additional threshold in the transmission coefficient.

3.14.2 Fabry–Perot Etalon: Implementation Scheme for a Bi-stable Optical Device

The most common implementation scheme for a bi-stable optical device is the nonlinear Fabry–Perot etalon. The device is typically fabricated from a semiconductor and consists of a slab of material of approximately 1-µm thickness. On each surface of the semiconductor, a highly reflective coating may be deposited to increase the bandwidth of the Fabry–Perot cavity. The choice of a proper nonlinear material is based upon the operating wavelength and the temporal response time desired and possibly other considerations. Typically, for applications in the far infrared, near infrared, and visible wavelengths, the proper materials are indium antimonide (InSb), gallium arsenide (GaAs), and zinc selenide (ZnSe), respectively.

OmniGuide fibres are a new type of cylindrical multilayer dielectric fibres that have only very recently been implemented experimentally. Their cladding is an Omnidirectional multilayer mirror that reflects light of any polarisation and any direction. These photonic band gap fibres can have a hollow core and a guiding mechanism that depends only on the cladding. We propose to exploit these facts to obtain much stronger axial optical modulation when compared to that in conventional fibres through insertion of material (e.g., spheres) into the core. Moreover, due to strong transverse confinement, much smaller transverse modal areas are possible when compared with usual low-index contrast fibres. In this way, we show how optimal ultrafast bi-stable devices can be achieved with operating powers less than 40 mW , whose highly nonlinear input/output power relation is key to many applications (e.g., all optical pulse reshaping, optical limiting, logic gates, etc.). Our device retains all the advantages of similar photonic crystal or high-index contrast devices in terms of power, size, and speed. On the other hand, the fact that in fibre device, it is easier to produce and to couple with another fibre. In this section, we solve the full nonlinear Maxwell's equations numerically to demonstrate optical bi-stability in this class of systems. Moreover, we present an analytical model that excellently describes their behaviour and is very useful in predicting optimal designs. For concreteness, our analysis focuses on OmniGuide fibres; nevertheless, one should be able to implement all the principles described in this section in any hollow photonic crystal fibre that has a large enough lateral band gap.

SUMMARY

- Laser is light amplification by stimulated emission of radiation. A device that creates and amplifies electromagnetic radiation of a specific frequency through the process of stimulated emission. The radiation emitted by a laser consists of a coherent beam of photons, all in phase and having the same polarisation.
- A laser emits a thin, intense beam of nearly monochromatic visible or infrared light that can travel long distances without diffusing. In laser light, the waves are all precisely in step, or in phase, with each other, and have the same polarisation. Such light is called coherent. All of the photons that make up a laser beam are in the same quantum state. Lasers produce coherent light through a process called stimulated emission.
- Lasers have many industrial, military, and scientific uses, including welding, target detection, microscopic photography, fibre optics, surgery, and optical instrumentation for surveying.
- A fibre laser is a laser in which the active gain medium is an optical fibre doped with rare earth elements such as erbium, ytterbium, neodymium, dysprosium, praseodymium, and thulium.
- Many high-power fibre lasers are based on double-clad fibre. The gain medium forms the core of the fibre, which is surrounded by two layers of cladding. The lasing mode propagates in the core, while a multimode pump beam propagates in the inner cladding layer.
- Generally, multimode fibres have a larger core diameter and are used for short-distance communication links and for applications where high power must be transmitted. Single-mode fibres are used for most communication links longer than 1050 m (3440 ft).
- For short-distance applications like creating a network within an office building fibre optic cabling can be used to save space in cable ducts. This is because a single fibre can often carry much more data than many electrical cables, for example, 4-pair Cat-5 Ethernet cabling.
- Optical fibres can be used as sensors to measure strain, temperature, pressure, and other quantities by modifying a fibre so that the quantity to be measured modulates the intensity, phase, polarisation, and wavelength or transit time of light in the fibre.
- Fibres are widely used in illumination applications. They are used as light guides in medical and other applications where bright light needs to be shown on a target without a clear line of sight path. In some buildings, optical fibres are used to route sunlight from the roof to other parts of the building.
- Optical fibre is also used in imaging optics. A coherent bundle of fibres is used, sometimes along with lenses, for a long, thin imaging device called an endoscope, which is used to view objects through a small hole.
- In spectroscopy, optical fibre bundles are used to transmit light from a spectrometer to a substance that cannot be placed inside the spectrometer itself in order to analyse its composition. A spectrometer analyses substances by bouncing light off and through them. By using fibres, a spectrometer can be used to study objects that are too large to fit inside, or gases, or reactions that occur in pressure vessels.
- An optical fibre doped with certain rare earth elements like erbium can be used as the gain medium of a laser or optical amplifier. Rare earth-doped optical fibres can be used to provide signal amplification by splicing a short section of doped fibre into a regular (un-doped) optical fibre line. Optical fibres doped with a wavelength shifter are used to collect scintillation light in physics experiments.
- A helium–neon (He–Ne) laser uses an electrical discharge in the helium–neon gas mixture, an Nd:YAG laser uses either light focused from a xenon flash lamp or diode lasers, and excimer lasers use a chemical reaction.

- The gain medium is the major determining factor of the wavelength of operation, and other properties of the laser. Gain media in different materials have linear spectra or wide spectra. Gain media with wide spectra allow the tuning of laser frequency.
- The optical resonator, or optical cavity, in its simplest form is two parallel mirrors placed around the gain medium that provides feedback of the light. The mirrors are given optical coatings that determine their reflective properties. Typically, one will be a high reflector, and the other will be a partial reflector. The latter is called the output coupler because it allows some of the light to leave the cavity to produce the laser's output beam.
- The beam in the cavity and the output beam of the laser, when travelling in free space (or a homogenous medium) rather than waveguides (as in an optical fibre laser), can be approximated as a Gaussian beam in most lasers; such beams exhibit the minimum divergence for a given diameter. However, some high-power lasers may be multimode with the transverse modes often approximated using Hermite Gaussian or Laguerre Gaussian functions.
- The pencil beam directly generated by a common helium–neon laser would spread out to a size of perhaps 500 km when shone on the Moon (from the distance of the earth). On the other hand, the light from a semiconductor laser typically exits the tiny crystal with a large divergence of upto 50°.
- An excited atom will act like a small electric dipole that will oscillate with the external field provided. One of the consequences of this oscillation is that it encourages electrons to decay to the lowest energy state. When this happens due to the presence of the electromagnetic field from a photon, a photon is released in the same phase and direction as the stimulating photon and this process is called stimulated emission.
- The rate at which stimulated emission occurs is proportional to the number of atoms N_2 in the excited state and the radiation density of the light. The base probability of a photon causing stimulated emission in a single excited atom was shown by Albert Einstein to be exactly equal to the probability of a photon being absorbed by an atom in the ground state.
- If the number of photons being amplified per unit time is greater than the number of photons being absorbed, then the net result is a continuously increasing number of photons being produced; the laser medium is said to have a gain of greater than unity.
- Some applications of lasers depend on a beam whose output power is constant over time. Such a laser is known as continuous wave or CW. Many types of lasers can be made to operate in CW mode to satisfy such an application. Many of these lasers actually operate in several longitudinal modes at the same time, and the beats between the slightly different optical frequencies of those oscillations will, in fact, produce amplitude variations on time scales shorter than the round-trip time (the reciprocal of the frequency spacing between modes), typically a few nanoseconds or less.
- In a Q-switched laser, the population inversion is allowed to build-up by introducing loss inside the resonator that exceeds the gain of the medium; this can also be described as a reduction of the quality factor or 'Q' of the cavity.
- A mode-locked laser is capable of emitting extremely short pulses on the order of tens of picoseconds down to less than 10 fs. These pulses will repeat at the round-trip time, that is, the time that it takes light to complete one round trip between the mirrors comprising the resonator.
- Pulsed laser operation is to pump the laser material with a source that is itself pulsed, either through electronic charging in the case of flash lamps or another laser that is already pulsed. Pulsed pumping was historically used with dye lasers where the inverted population lifetime of a dye molecule was so short that a high energy, fast pump was needed.
- Semiconductor lasers are lasers based on semiconductor gain media, where optical gain is usually achieved by stimulated emission at an inter-band transition under conditions of a high-carrier density in the conduction band.

- Most semiconductor lasers are laser diodes, which are pumped with an electrical current in a region where an *n*-doped and a *p*-doped semiconductor material meet. However, there are also optically pumped semiconductor lasers, where carriers are generated by absorbed pump light, and quantum cascade lasers, where intra band transitions are utilised.
- GaAs (gallium arsenide) is direct band gap semiconductors; indirect band gap semiconductors like silicon do not exhibit strong and efficient light emission. As the photon energy of a laser diode is close to the band gap energy, compositions with different band gap energies allow for different emission wavelengths. For the ternary and quaternary semiconductor compounds, the band gap energy can be continuously varied in some substantial range.
- The most common semiconductor lasers are operating in the near infrared spectral region, some others generate red light (e.g., in GaInP-based laser pointers) or blue or violet light (with gallium nitrides). For mid-infrared emission, there are lead selenide (PbSe) lasers (lead salt lasers) and quantum cascade lasers.
- Small edge-emitting laser diodes generate a few milliwatts (or up to 0.5 W) of output power in a beam with high-beam quality. They are used in laser pointers, in CD players, and for optical fibre communications.
- Quantum cascade lasers operate on intra-band transitions (rather than inter-band transitions); usually, they emit in the mid-infrared region, and sometimes, they emit in the terahertz region. They are used for trace gas analysis.
- Resonator modes are the modes of an optical resonator (cavity), that is, field distributions that reproduce themselves (apart from a possible loss of power) after one round trip. They can exist whether or not the resonator is geometrically stable, but the mode properties of unstable resonators are fairly sophisticated.
- Single-frequency operation of a laser means that only a single resonator mode (nearly always a Gaussian one) is excited; this leads to a much lower emission bandwidth than in cases where multiple resonator modes are excited.
- Modes of laser resonator can differ significantly from those of an empty resonator because they are subject to transversely varying gain and loss. This not only results in some deformation of the spatial shape; it is also because the resonator modes are no longer mutually orthogonal.
- A single-frequency laser (sometimes called a single-wavelength laser) is a laser that operates on a single resonator mode, so that it emits quasi-monochromatic radiation with a very small line width and low-phase noise.
- A distributed feedback laser is a laser where the whole resonator consists of a periodic structure, which acts as a distributed reflector in the wavelength range of laser action, and it contains a gain medium.
- A DBR laser is a laser, where the laser resonator is made with at least one DBR outside the gain medium (the active region).
- A DBR laser diode contains some corrugated waveguide structure (a grating section) providing wavelength-dependent feedback to define the emission wavelength.
- A fibre laser of DBR type usually has a linear laser resonator formed by an active (rare earth doped) fibre between two-fibre Bragg gratings.
- Quantum well laser is a laser diode in which the active region of the device is so narrow that quantum confinement occurs. The wavelength of the light emitted by a quantum well laser is determined by the width of the active region rather than just the band gap of the material from which it is constructed.

- Typical applications of single-frequency lasers occur in the areas of optical metrology (e.g., with fibre optic sensors) and interferometer, optical data storage, high-resolution spectroscopy, and optical fibre communications.
- A laser diode is a laser where the active medium is a semiconductor similar to that found in a light-emitting diode. The most common type of laser diode is formed from a *p–n* junction and powered by injected electric current. The former devices are sometimes referred to as injection laser diodes to distinguish them from optically pumped laser diodes.
- The simplest *p–n* junction structure is called a homojunction, which is a basic *p–n* junction. The same semiconductor material like GaAs is on both sides of the junction.
- When two different semiconductor crystals are joined together, the resulting structure is called a heterojunction in the semiconductor literature.
- A single heterojunction laser is made by growing a layer of GaAlAs on the GaAs. In a GaAs homojunction laser, the laser action occurs primarily on the *p*-side of the junction due to a higher efficiency for electron injection than for hole.
- It is possible to have more than one heterojunction in a structure. When there are two heterojunctions, it is called double heterostructure (DH). This DH structure further confines injected electrons to the active region and reduces the threshold current required for lasing.
- Whenever III–V semiconductors are used for either the sources or the detectors in optical communication system, they invariably take the form of the five-layer double heterostructure.
- In double heterostructure lasers, a layer of low band gap material is sandwiched between two high band gap layers. One commonly used pair of materials is gallium arsenide (GaAs) with aluminium gallium arsenide.
- A quantum well is a potential well with only discrete energy values. One technology to create quantisation is to confine particles that were originally free to move in two dimensions or three dimensions, thus forcing them to occupy a planar region.
- A potential well is the region surrounding a local minimum of potential energy. Energy captured in a potential well is unable to convert to another type of energy (kinetic energy in the case of a gravitational potential well) because it is captured in the local minimum of a potential well.
- The following equation shows the relationship between the energy level and the dimension spacing;
- $$\psi_{n_x, n_y, n_z} = \sqrt{\frac{8}{L_x L_y L_z}} \sin\left(\frac{n_x \pi x}{L_x}\right) \sin\left(\frac{n_y \pi y}{L_y}\right) \sin\left(\frac{n_z \pi z}{L_z}\right).$$
- $$E_{n_x, n_y, n_z} = \frac{h^2 \pi^2}{2m}\left[\left(\frac{n_x}{L_x}\right)^2 + \left(\frac{n_y}{L_y}\right)^2 + \left(\frac{n_z}{L_z}\right)^2\right].$$
- Laser diodes can be arrayed to produce very high-power (CW or pulsed) outputs. Such arrays may be used to efficiently pump solid-state lasers for inertial confinement fusion or high average power drilling or burning applications. Another common use is in barcode readers. Visible lasers, typically, red but later also green are common as laser pointers.
- Photonic integrated circuits are usually fabricated with a wafer scale technology (involving lithography) on substrates (often called chips) of silicon, silica, or a nonlinear crystal material like lithium niobate.
- Lithium niobate as a nonlinear crystal material is suitable for devices performing nonlinear functions, for example, electro-optic modulators or acousto-optic transducers.

- Optical metrology is the science- and technology-concerning measurements with light. Such measurements can either target properties of light itself or other properties like distance.
- In optics, optical bi-stability is an attribute of certain optical devices where two resonant transmission states are possible and stable depending on the input. Optical devices with a feedback mechanism, for example, a laser, provide two methods of achieving bi-stability.

THEORY QUESTIONS AND ANSWERS

Question 1: Explain the concept of LED. Give a strong comparison between LED and LASER.

Answer: A light-emitting diode (LED) is a semiconductor light source. The LEDs are used as indicator lamps in many devices and are increasingly used for other lighting. It is introduced as a practical electronic component in 1962; early LEDs emitted low-intensity red light, but modern versions are available across the visible, ultraviolet, and infrared wavelengths with very high brightness.

When an LED is forward biased (switched on), electrons are able to recombine with electron holes within the device, releasing energy in the form of photons. This effect is called electroluminescence and the colour of the light (corresponding to the energy of the photon) is determined by the energy gap of the semiconductor. The LEDs are often small in area (less than $1\,mm^2$), and integrated optical components may be used to shape its radiation pattern. The LEDs present many advantages over incandescent light sources including lower energy consumption, longer lifetime, improved robustness, smaller size, faster switching, greater durability, and reliability. Light-emitting diodes are used in applications as diverse as replacements for aviation lighting, automotive lighting (particularly brake lamps, turn signals, and indicators) as well as in traffic signals.

Comparison between LED and LASER

Both produce light from the changes in energy states of the electron shells in atoms. The main difference is that in a laser, the energy transitions are "synchronised" so that the emitted light is coherent and can produce interference patterns. It can also be focused to a very small spot. The LED's light is fairly pure, but the energy transitions producing it are independent and each emission is out of phase with the others. The LED light will not produce interference patterns.

Question 2: Discuss the application of VCSELs in optics.

Answer: Most optical interconnect schemes envision using vertical-cavity surface-emitting lasers (VCSELs) as their light source along with one of two types of semiconductor detectors. All three technologies use III–V materials like gallium arsenide rather than silicon, and so they cannot be directly integrated with the silicon CMOS devices with which they communicate. Nonetheless, developing silicon CMOS-based lasers and a detector is an active research area that could smooth the introduction of optical interconnects.

Many detector designs use *p–i–n* photodiodes, a sandwich of *p*-type, intrinsic, and *n*-type semiconductor material that generate current when struck by a photon. A sandwich of metal–semiconductor–metal creates MSM photodetectors that work much the same way at a higher speed than *p–i–n* photodiodes but with less sensitivity. Researchers use both in their designs depending on which attributes are most important to them. GaAs VCSELs have been commercially available for the last few years. The devices are diode lasers that, unlike most used in long haul telecommunications, emit from the surface of the laser rather than from the edge; hence, they can be placed on top of a chip. It is relatively easy to build an array of hundreds or thousands together in a plane and to make them close to the size of transistors, a few hundred nanometres across. However, a small percentage of the individual VCSELs in an array may not work properly. Bond a bum array to an expensive microprocessor and you have wasted a lot of money, so if optical interconnects are to take off, VCSEL manufacturers must reduce the failure rate. As chips continue to shrink, transistors are squeezed closer together, as would be the VCSELs communicating with them. Such spacing raises the spectre of a problem shared with electrical wires: as they get closer and operate at higher rates, the threat of electrical interference between adjacent channels increases. Although the light from the lasers itself is not the subject to crosstalk, the electric current that drives it is. The power driving the VCSELs also creates heat, which is more debilitating for lasers than for electronic circuitry. As the GaAs heats up, its material properties change and the lasers can detune wandering to a different wavelength. Beyond about 80°C, they stop working altogether. Therefore, the challenge is to increase their efficiency so they can be driven at even lower powers. Perhaps, the biggest difficulty with VCSELs is that the GaAs, and in fact, any lasing semiconductor has a lattice that does not match with silicon lasing. This means they have to be manufactured in a separate process adding enormously to production expense. They can be bonded to CMOS chips, but they will undergo thermal expansion at a different rate from the silicon, stretching out of alignment or even breaking the device. The photodetectors are also based on GaAs, and therefore, they suffer from the same drawbacks, although some researchers are looking into silicon-based detectors. Unlike semiconductors like GaAs, silicon has an indirect band gap, making a free electron more likely to give up its energy as heat than to combine with a hole and emit a photon.

Question 3: Write short notes on optical amplifier, laser amplifiers, and doped-fibre amplifiers.

Answer:

Optical Amplifier

An optical amplifier is a device that amplifies an optical signal directly without the need to first convert it to an electrical signal. An optical amplifier may be thought of as a laser without an optical cavity or one in which the feedback from the cavity is suppressed. Optical amplifiers are important in optical communication and laser physics.

There are several different physical mechanisms that can be used to amplify a light signal, which correspond to the major types of optical amplifiers. In doped-fibre amplifiers and bulk lasers, stimulated emission in the amplifier's gain medium causes amplification of incoming light. In semiconductor optical amplifiers (SOAs), electron–hole recombination occurs. In Raman amplifiers, Raman scattering of incoming light with phonons in the lattice of the gain medium produces photons coherent with the incoming photons. Parametric amplifiers use parametric amplification.

Laser Amplifiers

Almost any laser active gain medium can be pumped to produce gain for light at the wavelength of a laser made with the same material as its gain medium. Such amplifiers are commonly used to produce high-power laser systems. Special types such as regenerative amplifiers and chirped-pulse amplifiers are used to amplify ultra-short pulses.

Doped Fibre Amplifiers

Doped fibre amplifiers (DFAs) are optical amplifiers that use a doped-optical fibre as a gain medium to amplify an optical signal. They are related to fibre lasers. The signal to be amplified and a pump laser are multiplexed into the doped fibre, and the signal is amplified through interaction with the doping ions. The most common example is the Erbium-Doped Fibre Amplifier (EDFA), where the core of a silica fibre is doped with trivalent Erbium ions and can be efficiently pumped with a laser at a wavelength of 980 nm or 1480 nm and exhibits gain in the 1550 nm regions.

Amplification is achieved by stimulated emission of photons from dopant ions in the doped fibre. The pump laser excites ions into a higher energy from where they can decay via stimulated emission of a photon at the signal wavelength back to a lower energy level. The excited ions can also decay spontaneously (spontaneous emission) or even through non-radiative processes involving interactions with phonons of the glass matrix. These last two decay mechanisms compete with stimulated emission reducing the efficiency of light amplification.

The amplification window of an optical amplifier is the range of optical wavelengths for which the amplifier yields a usable gain. The amplification window is determined by the spectroscopic properties of the dopant ions, the glass structure of the optical fibre, and the wavelength and power of the pump laser.

Although the electronic transitions of an isolated ion are very well defined, broadening of the energy levels occurs when the ions are incorporated into the glass of the optical fibre, and thus, the amplification window is also broadened. This broadening is both homogeneous (all ions exhibit the same broadened spectrum) and inhomogeneous (different ions in different glass locations exhibit different spectra). Homogeneous broadening arises from the interactions with phonons of the glass, while inhomogeneous broadening is caused by differences in the glass sites where different ions are hosted. Different sites expose ions to different local electric fields, which shifts the energy levels via the Stark effect. In addition, the Stark effect also removes the degeneracy of energy states having the same total angular momentum (specified by the quantum number *J*). Therefore, for example, the trivalent Erbium ion Er^{+3} has a ground state with $J=15/2$, and in the presence of an electric field, it splits into $J+1/2=8$ sublevels with slightly different energies. The first excited state has $J=13/2$, and therefore, a Stark manifold with seven-sublevels. Transitions from the $J=13/2$ excited state to the $J=15/2$ ground state are responsible for the gain at 1.5-μm wavelength. The gain spectrum of the EDFA has several peaks that are smeared by the above-mentioned broadening mechanisms. The net result is a very broad spectrum (typically, 30 nm in silica). The broad gain-bandwidth of fibre amplifiers make them particularly useful in wavelength-division multiplexed communications systems as a single amplifier can be utilised to amplify all signals being carried on a fibre and whose wavelengths fall within the gain window.

SOLVED PROBLEMS

Problem 1: Calculate the ratio of the stimulated emission rate to the spontaneous emission rate for an incandescent lamp operating at a temperature of 1000 K. It may be assumed that the average operating wavelength is $0.5\,\mu\text{m}$.

Solution:

The average operating frequency is given as:

$$f=\frac{c}{\lambda}=\frac{2.998\times10^{8}}{0.5\times10^{-6}}\simeq6.0\times10^{14}\,\text{Hz}.$$

Using (3.11) the ratio is:

$$\frac{\text{Stimulated emission rate}}{\text{Spontaneous emission rate}}=\frac{1}{\exp\left(\dfrac{6.626\times10^{-34}\times6\times10^{14}}{1.381\times10^{-23}\times1000}\right)}=\exp(-28.8)=3.1\times10^{-13}.$$

The result obtained in Problem P1 indicates that for systems in thermal equilibrium spontaneous emission is by far the dominant mechanism. Furthermore, it illustrates that the radiation emitted from ordinary optical sources in the visible spectrum occurs in a random manner, proving these sources are incoherent.

Problem 2: A ruby laser contains a crystal length 4 cm with a refractive index of 1.78. The peak emission wavelength from the device is 0.55 μm. Determine the number of longitudinal modes and their frequency separation.

Solution:

The number of longitudinal modes supported within the structure may be obtained from (3.12) where:

$$q = \frac{2nL}{\lambda} = \frac{2 \times 1.78 \times 0.04}{0.55 \times 10^{-6}} = 2.6 \times 10^{5}.$$

Using (3.14), the frequency separation of the modes is:

$$\delta f = \frac{2.998 \times 10^{8}}{2 \times 1.78 \times 0.04} = 2.1 \text{ GHz}.$$

Problem 3: It is given that for a GaAs LED, the relative spectral width γ at $\lambda = 0.85$ μm is 0.035. This source is coupled to a pure silica fibre (with $\left|\lambda^2 \left(d^2n/d\lambda^2\right)\right| = 0.021$ for $\lambda = 0.85$ μm). Calculate the pulse broadening per kilometer due to material dispersion.

Solution:

The pulse broadening per unit length due to material dispersion is given as (as discussed in Chapter 2):

$$\frac{\tau}{L} = \frac{\gamma}{c}\left|\lambda^2 \frac{d^2n}{d\lambda^2}\right| = \frac{0.035}{3 \times 10^{8}} \times 0.021 = 2.45 \times 10^{-12} \text{ sm}^{-1}$$

$$\Rightarrow \frac{\tau}{L} = 2.45 \text{ ns km}^{-1}$$

Problem 4: Calculate the total pulse broadening due to material dispersion for a graded index fibre of total length 80 km when an LED emitting at (a) $\lambda = 850\,\text{nm}$ and (b) $\lambda = 1300\,\text{nm}$ is coupled to the fibre. In both the cases, assume that $\Delta\lambda \approx 30\,\text{nm}$. The material dispersion parameters of the fibre for the two wavelengths are $-105.5\,\text{ps nm}^{-1}\text{km}^{-1}$ and $-2.8\,\text{ps nm}^{-1}\text{km}^{-1}$, respectively.

Solution:

(a) Using

$$\tau = D_m L \Delta\lambda = 105.5 \times 80 \times 30 = 253{,}200 \; ps = 253.2 \text{ ns}$$

(b)

$$\tau = 2.8 \times 80 \times 30 = 6720 \; ps = 6.72 \text{ ns}$$

This example shows that proper selection of wavelength can reduce pulse broadening considerably.

Problem 5: Calculate the potential difference V_0 in a given germanium diode that has the following properties. Acceptor impurity level $(N_a)=10^{24}\ m^{-3}$, donor impurity level $(N_d)=10^{22}\ m^{-3}$, intrinsic electron concentration $(n_i)=2.4\times10^{19}\ m^{-3}$, and room temperature $(T)=290K$.

Solution:

Given: $N_a=10^{24}\ m^{-3}$, $N_d=10^{22}\ m^{-3}$, $n_i=2.4\times10^{19}\ m^{-3}$, $T=290K$

We know:

$$V_0=\frac{KT}{e}\ln\left(\frac{N_aN_d}{n_i^2}\right)$$

$$\Rightarrow 0.025\ln\left(\frac{10^{22}\times10^{24}}{\left(2.4\times10^{19}\right)^2}\right)=0.42$$

The potential difference varies with the variation of N_a ,N_d, and n_i.

Problem 6: Two fibres are joined together after polishing whose core refractive index is 1.5 and there is little air gap between the fibre-end faces. Calculate the loss of light signal in dB at the joint area.

Solution:

The reflection at the fibre air interface is:

$$R=\left(\frac{n_1-n}{n_1+n}\right)^2=0.04.$$

The light loss due to fibre air interface is

$$\text{loss}=-10\log_{10}(1-R)=-10\log_{10}(0.96)=0.18\ \text{dB}.$$

This loss is from face due to one fibre. The other fibre surface will also give a loss of 0.18 dB. Hence, the total loss due to joining of two fibre ends is 0.36 dB when they are polished perfectly. If the polish on fibre is not like mirror finish, it will have more loss, that is, 0.5 dB per joint or even more.

Problem 7: An optical fibre is made from core and cladding whose refractive indices are 1.48 and 1.46, respectively. Calculate the numerical aperture (NA). Furthermore, calculate the maximum angle of entrance of light into air where refractive index is 1.

Solution:

Given: $n_1=1.48$, $n_2=1.46$

The numerical aperture is given as $\text{NA}=\left(n_1^2-n_2^2\right)^{1/2}=0.242$

The maximum entrance angle is

$$\left(\theta_e\right)_{max}=\sin^{-1}\left(n_1^2-n_2^2\right)^{1/2}=14^\circ.$$

Problem 8: A GaAs injection DH laser has the length of cavity equal to $200\,\mu m$ emitting light output at $0.85\,\mu m$. If the refractive index of GaAs is 3.6, then calculate the wavelength separation between longitudinal modes?

Solution:

Given: $n_1=3.6$, $L=200\,\mu m$, $\lambda=0.85\,nm$

The wavelength separation between two modes is given by (3.16)

$$\Delta\lambda=\frac{\lambda^2}{2n_1L}=0.5\ \text{nm}$$

Problem 9: In Problem 4, if the laser diodes emitting at $\lambda = 850\,\text{nm}$ and $\lambda = 1300\,\text{nm}$ are replaced with $\Delta\lambda \approx 2\,\text{nm}$, what will be the magnitude of pulse broadening?

Solution:

(a) $\tau = D_m L\Delta\lambda = 105.5 \times 80 \times 2 = 16{,}880\ ps = 16.88\ ns$

(b) $\tau = D_m L\Delta\lambda = 2.8 \times 80 \times 2 = 448\ ps = 0.448\ ns$

This example demonstrates the superiority of laser diodes over LEDs. With laser diodes, pulse broadening can be reduced considerably.

MULTIPLE CHOICE QUESTIONS

1. Who developed the first laser?
   (a) Charles Townes
   (b) Theodore Maiman
   (c) Gordon McKenzie
   (d) Albert Einstein
2. In the dynamic response of injection laser diode (ILD), the delay that is followed by ____ frequency damped oscillations give rise to the generation of relaxation oscillations.
   (a) Low
   (b) Medium
   (c) High
   (d) All of these
3. Which type of injection laser involves the use of geometry for fabrication of the multimode injection laser with a single or small number of lateral modes?
   (a) Gain-guided laser
   (b) Index-guided laser
   (c) Quantum well laser
   (d) Quantum dot laser
4. Which among the following characteristics of laser light specifies the precise movement of all individual light waves together through time and space?
   (a) Monochromatic
   (b) Directional
   (c) Coherent
   (d) Brightness
5. Which type of laser is the simplest to modulate directly by changing its excitation?
   (a) Semiconductor
   (b) Ruby
   (c) Helium–neon
   (d) Neodymium–YAG
6. The laser frequency when the light has the wavelength 800 nm is
   (a) 375×10^{12} Hz
   (b) 475×10^{15} Hz
   (c) 375×10^{9} Hz
   (d) 375×10^{18} Hz

7. Laser light is ______ emission.
   (a) coherent
   (b) stimulated
   (c) spontaneous
   (d) coherent and stimulated
8. Which is the proper measurement of average power emitted by a pulsed laser?
   (a) Energy × time
   (b) Pulse energy × repetition rate
   (c) Pulse energy/repetition rate
   (d) Peak power × pulse length
9. Which is a non-coherent light source for optical communications system?
   (a) ILD
   (b) LED
   (c) APD
   (d) PIN diode
10. Which of the following considerations is important when deciding between using a diode laser or an LED?
    (a) Response time
    (b) Power levels
    (c) Temperature sensitivity
    (d) Failure characteristics
11. Recent laser developments for fibre-optic communication include
    (a) Distributed feedback (DFB)
    (b) Vertical-cavity surface-emitting laser (VCSEL)
    (c) Distributed feedback (DFB) and hetero-junction
    (d) Distributed feedback (DFB) and vertical-cavity surface-emitting laser (VCSEL)

Answers to Multiple Choice Questions

1. (b)
2. (c)
3. (a)
4. (c)
5. (a)
6. (a)
7. (d)
8. (b)
9. (b)
10. (a)
11. (d)

QUESTIONS

1. Explain briefly the three processes involved in the laser action. Furthermore, for a Fabry–Perot resonator laser diode, describe the modes and threshold conditions. Obtain its rate equations for steady-state output.
2. Briefly outline the general requirements for a source in optical fibre communications. Discuss the areas in which the injection laser fulfils these requirements and comment on any drawbacks of using this device as an optical fibre communication source.
3. An injection laser has a GaAs active region with band gap energy of 1.43 eV. Estimate the wavelength of optical emission from the device and determine its linewidth in hertz when the measured spectral width is 0.1 nm.

4. List the properties of lasers and how it fulfils the major requirements for an optical fibre emitter.
5. Calculate the potential difference V_0 in a given germanium diode that has the following properties. Acceptor impurity level $(N_a) = 10^{24}\ m^{-3}$, donor impurity level $(N_d) = 10^{22}\ m^{-3}$, intrinsic electron concentration $(n_i) = 2.4 \times 10^{19}\ m^{-3}$, room temperature $(T) = 290K$.
6. Derive Einstein's relation for the rates of the three transition processes of absorption, spontaneous emission, and simulated emission for two-level atomic systems.
7. Explain (a) population inversion, (b) optical feedback, and (c) laser oscillation.
8. Explain the construction and operation of ruby laser with a neat diagram.
9. The coated mirror reflectivity at either ends of the 350-μm μm long optical cavity of an injection laser is 0.5 and 0.65. At normal operating temperature, the threshold current density for the device is $2 \times 10^3\ A cm^{-2}$ and the gain factor $\beta$is $22 \times 10^{-3}\ cm\ A^{-1}$. Estimate the loss coefficient in the optical cavity.
10. Explain the construction and working of semiconductor laser with a neat diagram.
11. (a) Draw a typical double heterostructure light emitter and explain.
    (b) Draw the schematic of surface-emitting and edge-emitting LED and explain.
12. (a) Derive the threshold condition for LASER.
    (b) Explain the laser action with neat diagrams.
13. A GaAs laser operating at 850 nm has a 500-μm length and a refractive index $n = 3.7$. What are the frequency and wavelength spacing?
14. (a) What type of materials is used for optical source? What are the advantages of double heterostructure? Compare surface-emitting and edge-emitting LED structures.
    (b) Derive an expression for the internal optical power level generated in LEDs.
15. A GaAs with a cavity of length 500 μm has a loss coefficient of 20 cm^{-1}. The measured differential external quantum efficiency of the device is 45%. Calculate the internal quantum efficiency of the laser. The refractive index of GaAs is 3.6.
16. Calculate the ratio of the stimulated emission rate to the spontaneous emission rate for an incandescent lamp operating at a temperature of 1000 K. It may be assumed that the average operating wavelength is 0.5 μm.
17. A ruby laser contains a crystal of length 4 cm with a refractive index of 1.78. The peak-emission wavelength from the device is 0.55 μm. Determine the number of longitudinal modes and their frequency separation.
18. An injection laser has an active cavity with losses of $30\ cm^{-1}$ and the reflectivity of each cleaved laser facet is 30%. Determine the laser gain coefficient for the cavity when it has a length of 600 μm.
19. Briefly describe what is meant by the following terms when they are used in relation to injection lasers: (a) relaxation oscillations; (b) frequency chirp; (c) partition noise; (d) mode hopping.
20. Compare the approximate radiative minority carrier lifetimes in gallium arsenide and silicon when the minority carriers are electrons injected into the *p*-type region that has a hole concentration of $10^{18}\ cm^{-1}$. The injected electron density is small when compared with the majority carrier density.
21. A GaAs injection laser has an optical cavity of length 250 μm and width 100. μm. At normal operating temperature, the gain factor *B* is $21 \times 10^{-3}\ A\ cm^{-3}$ and the loss coefficient A per cm is 10. Determine the threshold current density, and hence, the threshold current for the device. It may be assumed that the cleaved mirrors are uncoated and that the current is restricted to the optical cavity. The refractive index of GaAs may be taken as 3.6.
22. The total efficiency of an injection laser with a GaAs active region is 18%. The voltage applied to the device is 2.5 V and the band gap energy for GaAs is 1.43 eV. Calculate the external power efficiency of the device.

23. Discuss linewidth narrowing and wavelength tunability associated with single-frequency injection lasers. Outline the major techniques that are being adopted to facilitate these characteristics.

REFERENCES

1. Weber H P and Ulrich R, A thin-film ring laser, *Appl. Phys. Lett.*, vol. 19, no. 2, pp. 38–40, July. 1971.
2. Archambault J L, Reekie L, and Russell P S J, 100-Percent reflectivity bragg reflectors produced in optical fibres by single excimer-laser pulses, *Electron. Lett.*, vol. 29, no. 5, pp. 453–455, 1993.
3. Brinkman W F, Koch T L, Lang D V and Wilt D P, The lasers behind the communications revolution, *Bell Labs Techn. J.*, pp. 150–167, Jan/Mar 2000.
4. Nagel S R, MacChesney J B and Walker K L, An overview of the Modified Chemical Vapor Deposition (MCVD) process and performance, *IEEE J. Quant. Electron.*, vol. QE-18, no. 4, p. 459–476, April 1982.
5. Raghuwanshi S K and Devendra C, "Dispersion Study of an Inhomogeneous Dielectric Planar Slab Optical Waveguide", in *National Conference on Frontiers in Electronics Communication and Instrumentation Technology (FECIT-2011)*, Dhanbad (Nov. 3–4, 2011), India.
6. Raghuwanshi S K and Talabattula S, "Multilayer thin-film filters," in *Proc. 2nd International Conference on Resource Utilisation & Intelligent Systems (INCRUIS-2008)*, pp. 699–703, Perundurai, Erode, Tamil Nadu, India, 3–5 Jan. 2008.
7. Raghuwanshi S K and Talabattula S, *Numerical Study of Propagation in Optical Waveguides and Devices*, Germany: VDM Verlag, 2009.
8. Raghuwanshi S K, Comparative study of asymmetric versus symmetric planar slab dielectric optical waveguides, *Indian J. Phys.*, vol. 84, no. 7, pp. 831–846, 2010.
9. Canning J, Fibre Gratings and Devices for Sensors and Lasers, *Laser Photon. Rev.*, vol. 2, no. 4, pp. 275–289, Wiley, USA, 2008.
10. Bayrak O and Boztosum I, Analytical solutions to the Hulthen and Morse potential by using the asymptotic iteration method, *J. Phys. A: Math. Gen.*, vol. 802, p. 17–30, 2007.
11. Raghuwanshi S K, Kumar V and Pandey R R, "Dispersion study of cylindrical dielectric waveguide without computing $(d^{\wedge}2\beta)/(dk^{\wedge}2)$ numerically", in *13th International Conference of International Academy of Physical Sciences (CONIAPS-XIII)*, Dehradun, India, June 14–16, 2011.
12. Raghuwanshi S K, Kumar V, and Pandey R R, Guided and leaky modes of a multilayer planar slab waveguide, *I-manager's J. Electron. Eng.*, vol. 1, no. 3, March–May 2011.
13. Samuel E P and Patil D S, Analysis of wavefunction distribution in quantum well biased laser diode using transfer matrix method, *Prog. Electromagn. Res. Lett.*, vol. 1, pp. 119–128, 2006.
14. Saitoh T and Mukai T, Chapter 7, in *Coherence, Amplification and Quantum Effects in Semiconductor Lasers*, Ed. Y. Yamamoto, New York: Wiley, 1991.
15. Thyagarjan K and Ghatak A K, *Lasers: Theory and Application*, New York: Plenum Press, 1981.
16. Tsutsumi K, Imada Y, Hirai H, and Yuba Y, Analysis of single mode optical Y- junction by the bounded step and bend approximation, *J. Lightwave Technol.*, vol. LT-6, pp. 590–600, 1988.

4 Coupled-mode Theory for Optical Fibre

Objectives

- This chapter deals with the coupled-mode theory for optical fibre communication.
- Modes supported by optical fibres are examined using Maxwell's equations.
- Characteristics of picosecond pulse are delineated.
- Various optical devices using directional coupler are described.
- Propagation of beam through one-dimensional waveguides is analysed numerically.

4.1 INTRODUCTION

Silica-based optical fibres are the most important transmission medium for long-distance and large-capacity optical communication systems. The most distinguished feature of optical fibre is its low-loss characteristics [1–3]. The lowest transmission loss ever achieved is 0.154 dB/km at $\lambda = 1.55\,\mu\text{m}$ wavelength. This means that signal intensity of light becomes half of the original strength after propagating 20 km along the optical fibre. Together with such low-loss properties, low dispersion is also required for signal transmission. Signal distortion due to dispersion of the fibre is closely related to the guiding structure of optical fibres [4–7]. In order to realise low-dispersion fibres, it is necessary to understand the transmission characteristics of fibres and to design and analyse the arbitrarily shaped guiding structures. In this chapter, first a rigorous analysis of step index fibre is presented to understand the basic properties of optical fibres [9–13]. The linearly polarised (LP) modes, which are quite important mode designation for practical weakly guiding fibres, will be described. Before coming to the coupled-mode theory, we would review the fundamental of optical fibre technology.

4.2 MAXWELL'S EQUATIONS

The electromagnetic fields in optical fibres are expressed in cylindrical coordinates as [4]:

$$\tilde{E} = E(r,\theta)e^{j(\omega t - \beta z)} \tag{4.1}$$

$$\tilde{H} = H(r,\theta)e^{j(\omega t - \beta z)}. \tag{4.2}$$

Maxwell's equations are written in terms of cylindrical coordinates as:

$$\begin{cases} \dfrac{1}{r}\dfrac{\partial E_z}{\partial \theta}+j\beta E_\theta=-j\omega\mu_0 H_r \\ -j\beta E_r-\dfrac{\partial E_z}{\partial r}=-j\omega\mu_0 H_\theta \\ \dfrac{1}{r}\dfrac{\partial (rE_\theta)}{\partial r}-\dfrac{1}{r}\dfrac{\partial E_r}{\partial \theta}=-j\omega\mu_0 H_z \\ \dfrac{1}{r}\dfrac{\partial H_z}{\partial \theta}+j\beta H_\theta=j\omega\varepsilon_0 n^2 E_r \\ -j\beta H_r-\dfrac{\partial H_z}{\partial r}=j\omega\varepsilon_0 n^2 E_\theta \\ \dfrac{1}{r}\dfrac{\partial (rH_\theta)}{\partial r}-\dfrac{1}{r}\dfrac{\partial H_r}{\partial \theta}=j\omega\varepsilon_0 n^2 E_z \end{cases} \tag{4.3}$$

Substituting (4.1) and (4.2) in Maxwell's equations (4.3), we obtain two sets of wave equations:

$$\frac{\partial^2 E_z}{\partial r^2}+\frac{1}{r}\frac{\partial E_z}{\partial r}+\frac{1}{r^2}\frac{\partial^2 E_z}{\partial \theta^2}+\left[k^2 n(r,\theta)^2-\beta^2\right]E_z=0 \tag{4.4}$$

$$\frac{\partial^2 H_z}{\partial r^2}+\frac{1}{r}\frac{\partial H_z}{\partial r}+\frac{1}{r^2}\frac{\partial^2 H_z}{\partial \theta^2}+\left[k^2 n(r,\theta)^2-\beta^2\right]H_z=0 \tag{4.5}$$

In axially symmetric optical fibres, the refractive index distribution is not dependent on θ and is expressed by $n(r)$. Then, the transverse electromagnetic fields are related to E_z and H_z as follows:

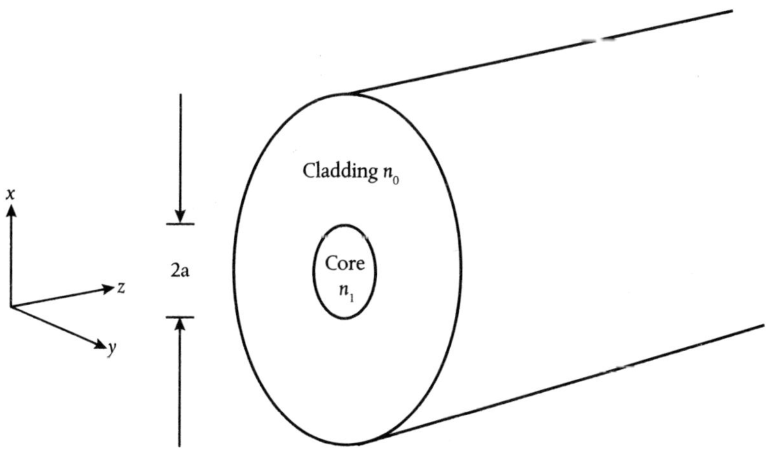

Fig. 4.1 Waveguide structure of optical fibre

$$\begin{cases} E_r = -\dfrac{j}{\left[k^2 n(r)^2 - \beta^2\right]}\left(\beta \dfrac{\partial E_z}{\partial r} + \dfrac{\omega\mu_0}{r}\dfrac{\partial H_z}{\partial \theta}\right) & (4.6) \\ E_\theta = -\dfrac{j}{\left[k^2 n(r)^2 - \beta^2\right]}\left(\dfrac{\beta}{r}\dfrac{\partial E_z}{\partial \theta} - \omega\mu_0 \dfrac{\partial H_z}{\partial r}\right) & (4.7) \\ H_r = -\dfrac{j}{\left[k^2 n(r)^2 - \beta^2\right]}\left(\beta \dfrac{\partial H_z}{\partial r} - \dfrac{\omega\varepsilon_0 n(r)^2}{r}\dfrac{\partial E_z}{\partial \theta}\right) & (4.8) \\ H_\theta = -\dfrac{j}{\left[k^2 n(r)^2 - \beta^2\right]}\left(\dfrac{\beta}{r}\dfrac{\partial H_z}{\partial \theta} + \omega\varepsilon_0 n(r)^2 \dfrac{\partial E_z}{\partial r}\right) & (4.9) \end{cases}$$

The azimuthal dependency of the electromagnetic fields in axially symmetric fibres is expressed as $\cos(n\theta+\psi)$ or $\sin(n\theta+\psi)$, where n is an integer and ψ denotes the phase. The modes in an optical fibre consist of TE modes $(E_z = 0)$, TM modes $(H_z = 0)$, and hybrid modes $(E_z \neq 0, H_z \neq 0)$, respectively. In the following section, electromagnetic field, dispersion equation, and propagation characteristics of optical fibres are described in details for step index fibres as shown in Fig. (4.1), which has a uniform refractive index in their core.

4.2.1 Wave Theory of Step Index Fibres

4.2.1.1 TE modes

When we have $E_z = 0$ in (4.4)–(4.9), we have the following set of equations for the TE mode:

$$\frac{\partial^2 H_z}{\partial r^2} + \frac{1}{r}\frac{\partial H_z}{\partial r} + \left[k^2 n(r)^2 - \beta^2 - \frac{n^2}{r^2}\right] H_z = 0 \qquad (4.10)$$

$$\begin{cases} E_r = -\dfrac{j\omega\mu_0}{\left[k^2 n(r)^2 - \beta^2\right]}\dfrac{1}{r}\dfrac{\partial H_z}{\partial r} & (4.11) \\ E_\theta = \dfrac{j\omega\mu_0}{\left[k^2 n(r)^2 - \beta^2\right]}\dfrac{\partial H_z}{\partial \theta} & (4.12) \\ H_r = -\dfrac{j\beta}{\left[k^2 n(r)^2 - \beta^2\right]}\dfrac{\partial H_z}{\partial r} & (4.13) \\ H_\theta = -\dfrac{j\beta}{\left[k^2 n(r)^2 - \beta^2\right]}\dfrac{1}{r}\dfrac{\partial H_z}{\partial \theta} & (4.14) \end{cases}$$

The magnetic field in the core and cladding are expressed as:

$$H_z = \begin{Bmatrix} g(r) \\ h(r) \end{Bmatrix} \cos(n\theta+\psi) \begin{matrix} (0 \le r \le a) \\ (r > a) \end{matrix}. \qquad (4.15)$$

From the boundary condition that the tangential field components H_z and H_θ should be continuous at the core–cladding interface $r=a$, we obtain the following two conditions:

$$g(a)=h(a). \tag{4.16}$$

$$\frac{j\beta}{\left[k^2n(a)^2-\beta^2\right]}\frac{n}{a}g(a)\sin(n\theta+\psi)=\frac{j\beta}{\left[k^2n_0^2-\beta^2\right]}\frac{n}{a}h(a)\sin(n\theta+\psi). \tag{4.17}$$

Since in the step index fibre, the refractive index at the boundary of the core $n(a)=n_1$ (n_1 is the refractive index of the core), where $n(a)$ is not equal to the refractive index of the cladding n_0. Then, it is known as that integer n should be zero for (4.17) to be satisfied to arbitrary values of θ. Consequently, the azimuthal dependency in the TE modes $\frac{\partial}{\partial\theta}=0$ and E_r and H_θ are found from (4.11) and (4.14) to be:

$$E_r=H_\theta=0. \tag{4.18}$$

In graded index fibres, there may be the case that $n(a)=n_0$. Therefore, we cannot immediately derive $n=0$ for the TE modes even in graded index fibres. Taking into account the azimuthal condition of $n=0$, the wave equation and related electromagnetic fields for the TE modes are obtained as:

$$\frac{d^2H_z}{dr^2}+\frac{1}{r}\frac{dH_z}{dr}+\left[k^2n(r)^2-\beta^2\right]H_z=0. \tag{4.19}$$

$$E_\theta=\frac{j\omega\mu_0}{\left[k^2n(r)^2-\beta^2\right]}\frac{dH_z}{dr} \tag{4.20}$$

$$H_r=-\frac{j\beta}{\left[k^2n(r)^2-\beta^2\right]}\frac{dH_z}{dr}. \tag{4.21}$$

$$E_r=H_\theta=0 \tag{4.22}$$

Here, we define the wave number in the core and cladding along the transversal direction as:

$$\kappa=\sqrt{k^2n_1^2-\beta^2} \tag{4.23}$$

$$\sigma=\sqrt{\beta^2-k^2n_0^2}. \tag{4.24}$$

The wave equation (4.19) for the field in the core $H_z=g(r)$ is obtained as:

$$\frac{d^2g}{dr^2}+\frac{1}{r}\frac{dg}{dr}+\kappa^2g=0 \quad (0\le r\le a), \tag{4.25}$$

and the wave equation for the field in the cladding $H_z=h(r)$ is given as:

$$\frac{d^2h}{dr^2}+\frac{1}{r}\frac{dh}{dr}-\sigma^2 \quad h=0(r>a). \tag{4.26}$$

The solution for (4.25) are the 0th-order Bessel function $J_0(\kappa r)$ and the 0th-order Neumann function $N_0(\kappa r)$, respectively. However, $N_0(\kappa r)$ diverges infinitely at $r=0$. Therefore, $J_0(\kappa r)$ is the proper solution in the core. The solution for (4.26) are the modified Bessel function of the first kind $I_0(\sigma r)$ and modified Bessel function of the second kind $K_0(\sigma r)$, respectively. However, $I_0(\sigma r)$ diverges infinitely at $r=\infty$. Therefore, $K_0(\sigma r)$ is the proper solution in the cladding. Then, the magnetic field for the TE modes is given as:

$$H_z = \begin{cases} AJ_0(\kappa r) & (0 \le r \le a) \\ BK_0(\sigma r) & (r > a) \end{cases}, \tag{4.27}$$

where A and B are constants. The boundary conditions are given by the conditions that H_z and E_θ should be continuous at $r = a$.

$$AJ_0(\kappa a) = BK_0(\sigma a) \tag{4.28}$$

$$\frac{A}{\kappa} J_0'(\kappa a) = -\frac{B}{\sigma} K_0'(\sigma a) \tag{4.29}$$

By using the normalised transverse wave numbers, we get:

$$u = \kappa a = a\sqrt{k^2 n_1^2 - \beta^2} \tag{4.30}$$

$$w = \sigma a = a\sqrt{\beta^2 - k^2 n_0^2}\,. \tag{4.31}$$

Equations (4.28) and (4.29) can be reduced to the following dispersion equation:

$$\frac{J_0'(u)}{uJ_0(u)} = -\frac{K_0'(w)}{wK_0(w)}\,. \tag{4.32}$$

Based on the following Bessel function formulas:

$$J_0'(u) = -J_1(u) \tag{4.33}$$

$$K_0'(w) = -K_1(w). \tag{4.34}$$

Equation (4.32) is rewritten as:

$$\frac{J_1(u)}{uJ_0(u)} = -\frac{K_1(w)}{wK_0(w)}. \tag{4.35}$$

Transverse wave numbers u and w are related, which are determined from (4.30) to (4.31), are related as:

$$u^2 + w^2 = k^2\left(n_1^2 - n_0^2\right)a^2 = v^2. \tag{4.36}$$

Therefore, when the normalised frequency v is given, the transverse wave number u and w are determined from (4.35) to (4.36). Substituting (4.27) and (4.28) in (4.20) and (4.22), the electromagnetic fields for TE modes are obtained:

$$E_r = E_z = H_\theta = 0\,. \tag{4.37}$$

1. Fields in the core $(0 \le r \le a)$

$$E_\theta = -j\omega\mu_0 \frac{a}{u} AJ_1\left(\frac{u}{a} r\right) \tag{4.38}$$

$$H_r = j\beta \frac{a}{u} AJ_1\left(\frac{u}{a}r\right) \tag{4.39}$$

$$H_z = AJ_0\left(\frac{u}{a}r\right) \tag{4.40}$$

2. Fields in the cladding $(r > a)$

$$E_\theta = j\omega\mu_0 \frac{a}{w}\frac{J_0(u)}{K_0(w)} AK_1\left(\frac{w}{a}r\right) \tag{4.41}$$

$$H_r = -j\beta \frac{a}{w}\frac{J_0(u)}{K_0(w)} AK_1\left(\frac{w}{a}r\right) \tag{4.42}$$

$$H_z = \frac{J_0(u)}{K_0(w)} AK_0\left(\frac{w}{a}r\right) \tag{4.43}$$

Constant A is determined in the relationship form the power P carried by the mode.

4.2.1.2 TM modes

When we set $H_z = 0$ in (4.4)–(4.5) and (4.6)–(4.9) for TM modes and we follow the same procedure as in the TE modes, we know that $n = 0$ holds also true for the TM mode. Then, the wave equation and other electromagnetic field components are expressed as:

$$\frac{d^2 E_z}{dr^2} + \frac{1}{r}\frac{dE_z}{dr} + \left[k^2 n(r)^2 - \beta^2\right] E_z = 0. \tag{4.44}$$

$$\begin{cases} E_r = \dfrac{-j\beta}{\left[k^2 n(r)^2 - \beta^2\right]} \dfrac{dE_z}{dr}, & (4.45) \\ H_\theta = \dfrac{-j\omega\varepsilon_0 n^2}{[k^2 n(r)^2 - \beta^2]} \dfrac{dE_z}{dr}, & (4.46) \\ E_\theta = H_r = 0 & (4.47) \end{cases}$$

The solution for (4.44) is given by the 0th-order Bessel function as:

$$E_z = \begin{cases} AJ_0(\kappa r) & (0 \le r \le a) \quad (4.48) \\ BK_0(\sigma r) & (r > a) \quad , \quad (4.49) \end{cases}$$

where A and B are constants. Applying the boundary condition that E_z and H_θ should be continuous at the core–cladding interface $r = a$, we obtain:

$$\frac{J_0'(u)}{uJ_0(u)} = -\left(\frac{n_0}{n_1}\right)^2 \frac{K_0'(w)}{wK_0(w)}. \tag{4.50}$$

Using the recurrence relation of Bessel functions (4.33)–(4.34), (4.50) can be rewritten into the dispersion equation for the TM modes:

$$\frac{J_1(u)}{uJ_0(u)}=-\left(\frac{n_0}{n_1}\right)^2\frac{K_1(w)}{wK_0(w)}. \tag{4.51}$$

The electromagnetic fields for TM modes are summarised as:

$$E_\theta=H_r=H_z=0. \tag{4.52}$$

1. **Field in the core $(0\le r\le a)$**

$$E_r=j\beta\frac{a}{u}AJ_1\left(\frac{u}{a}r\right) \tag{4.53}$$

$$E_z=AJ_0\left(\frac{u}{a}r\right) \tag{4.54}$$

$$H_\theta=j\omega\varepsilon_0 n_1^2 AJ_1\left(\frac{u}{a}u\right) \tag{4.55}$$

2. **Fields in the cladding $(r>a)$**

$$E_r=-j\beta\frac{a}{w}\frac{J_0(u)}{K_0(w)}AK_1\left(\frac{w}{a}r\right) \tag{4.56}$$

$$E_z=\frac{J_0(u)}{K_0(w)}AK_0\left(\frac{w}{a}r\right) \tag{4.57}$$

$$H_\theta=-j\omega\varepsilon_0 n_0^2\frac{a}{w}\frac{J_0(u)}{K_0(w)}AK_1\left(\frac{w}{a}r\right) \tag{4.58}$$

4.2.1.3 Hybrid modes

In hybrid modes, the axial electromagnetic field components E_z and H_z are not zero. Therefore, solutions for (4.4)–(4.5) are given by the product of the nth- order Bessel functions and $\cos(n\theta+\psi)$ or $\sin(n\theta+\psi)$. E_z and H_z should be continuous at $r=a$. Further, it is known from (4.6) to (4.9) that $\frac{\partial E_z}{\partial r}$ and $\frac{\partial H_z}{\partial \theta}$ (or $\frac{\partial E_z}{\partial \theta}$ and $\frac{\partial H_z}{\partial r}$) have the same θ dependences. By considering these, the z-components of the electromagnetic fields are expressed as:

$$E_z=\begin{cases} AJ_n\left(\frac{u}{a}r\right)\cos(n\theta+\psi) & (0\le r\le a)\\ A\frac{J_n(u)}{K_n(w)}K_n\left(\frac{w}{a}r\right)\cos(n\theta+\psi) & (r>a)\end{cases} \tag{4.59}$$

(4.60)

$$H_z = \begin{cases} CJ_n\left(\frac{u}{a}r\right)\sin(n\theta+\psi) & (0 \le r \le a) \quad (4.61) \\ C\frac{J_n(u)}{K_n(w)}K_n\left(\frac{w}{a}r\right)\sin(n\theta+\psi) & (r > a) \quad (4.62) \end{cases}$$

The transverse components are obtained by substituting (4.59)–(4.60) and (4.61)–(4.62) in (4.6)–(4.9) as follows:

1. Core region $(0 \le r \le a)$:

$$E_r = -\frac{ja^2}{u^2}\left[A\beta\frac{u}{a}J_n'\left(\frac{u}{a}r\right) + C\omega\mu_0\frac{n}{r}J_n\left(\frac{u}{a}r\right)\right]\cos(n\theta+\psi) \tag{4.63}$$

$$E_\theta = -\frac{ja^2}{u^2}\left[-A\beta\frac{n}{r}J_n\left(\frac{u}{a}r\right) - C\omega\mu_0\frac{u}{a}J_n'\left(\frac{u}{a}r\right)\right]\sin(n\theta+\psi) \tag{4.64}$$

$$H_r = -\frac{ja^2}{u^2}\left[A\omega\varepsilon_0 n_1^2\frac{n}{r}J_n\left(\frac{u}{a}r\right) + C\beta\frac{u}{a}J_n'\left(\frac{u}{a}r\right)\right]\sin(n\theta+\psi) \tag{4.65}$$

$$H_\theta = -\frac{ja^2}{u^2}\left[A\omega\varepsilon_0 n_1^2\frac{u}{a}J_n'\left(\frac{u}{a}r\right) + C\beta\frac{n}{r}J_n\left(\frac{u}{a}r\right)\right]\cos(n\theta+\psi) \tag{4.66}$$

2. Cladding region $(r > a)$:

$$E_r = \frac{ja^2}{w^2}\left[A\beta\frac{w}{a}K_n'\left(\frac{w}{a}r\right) + C\omega\mu_0\frac{n}{r}K_n\left(\frac{w}{a}r\right)\right]\frac{J_n(u)}{K_n(w)}\cos(n\theta+\psi) \tag{4.67}$$

$$E_\theta = \frac{ja^2}{w^2}\left[-A\beta\frac{n}{r}K_n\left(\frac{w}{a}r\right) - C\omega\mu_0\frac{w}{a}K_n'\left(\frac{w}{a}r\right)\right]\frac{J_n(u)}{K_n(w)}\sin(n\theta+\psi) \tag{4.68}$$

$$H_r = \frac{ja^2}{w^2}\left[A\omega\varepsilon_0 n_0^2\frac{n}{r}K_n\left(\frac{w}{a}r\right) + C\beta\frac{w}{a}K_n'\left(\frac{w}{a}r\right)\right]\frac{J_n(u)}{K_n(w)}\sin(n\theta+\psi) \tag{4.69}$$

$$H_\theta = \frac{ja^2}{w^2}\left[A\omega\varepsilon_0 n_0^2\frac{w}{a}K_n'\left(\frac{w}{a}r\right) + C\beta\frac{n}{r}K_n\left(\frac{w}{a}r\right)\right]\frac{J_n(u)}{K_n(w)}\cos(n\theta+\psi) \tag{4.70}$$

The boundary condition at $r = a$ shows that E_θ and H_θ should be continuous and we obtain two relations; one is from (4.64) and (4.68):

$$A\beta\left(\frac{1}{u^2}+\frac{1}{w^2}\right)n = -C\omega\mu_0\left[\frac{J_n'(u)}{uJ_n(u)}+\frac{K_n'(w)}{wK_n(w)}\right] \tag{4.71}$$

and the other is from the (4.66) and (4.70):

$$A\omega\varepsilon_0\left[n_1^2\frac{J_n'(u)}{uJ_n(u)}+n_0^2\frac{K_n'(w)}{wK_n(w)}\right]=-C\beta\left(\frac{1}{u^2}+\frac{1}{w^2}\right)n. \quad (4.72)$$

We obtain the dispersion equation from (4.71) and (4.72) in the form:

$$\left[\frac{J_n'(u)}{uJ_n(u)}+\frac{K_n'(w)}{wK_n(w)}\right]\left[n_1^2\frac{J_n'(u)}{uJ_n(u)}+n_0^2\frac{K_n'(w)}{wK_n(w)}\right]=\frac{\beta^2}{k^2}\left(\frac{1}{u^2}+\frac{1}{w^2}\right)^2 n^2. \quad (4.73)$$

Substituting the following relation, this is derived from (4.30) to (4.31):

$$\frac{\beta^2}{k^2}\left(\frac{1}{u^2}+\frac{1}{w^2}\right)=\frac{n_1^2}{u^2}+\frac{n_0^2}{w^2}. \quad (4.74)$$

Equation (4.73) may be rewritten as:

$$\left[\frac{J_n'(u)}{uJ_n(u)}+\frac{K_n'(w)}{wK_n(w)}\right]\left[\frac{J_n'(u)}{uJ_n(u)}+\left(\frac{n_0}{n_1}\right)^2\frac{K_n'(w)}{wK_n(w)}\right]=n^2\left(\frac{1}{u^2}+\frac{1}{w^2}\right)\left[\frac{1}{u^2}+\left(\frac{n_0}{n_1}\right)^2\frac{1}{w^2}\right]. \quad (4.75)$$

The propagation constant of the hybrid modes is calculated by solving (4.75) using the $u-w$ relation of (4.36). Although (4.75) is the strict solution of the hybrid modes in the step index fibres, it is rather difficult to investigate the propagation properties of optical fibres with (4.75). In practical fibres, the refractive index difference Δ is of the order of 1%, and therefore, we can approximate $n_1 \cong n_0$ in some cases. In these cases, (4.75) can be simplified and much practical and useful information can be obtained.

The constants C in the electromagnetic field expressions (4.63)–(4.66) and (4.67)–(4.70) can be written from (4.71).

$$C=-A\frac{\beta}{\omega\mu_0}s, \quad (4.76)$$

where

$$s=\frac{n\left(\frac{1}{u^2}+\frac{1}{w^2}\right)}{\left[\frac{J_n'(u)}{uJ_n(u)}+\frac{K_n'(w)}{wK_n(w)}\right]}. \quad (4.77)$$

When we apply the following recurrence for the Bessel functions:

$$J_n'(z)=\frac{1}{2}\left[J_{n-1}(z)-J_{n+1}(z)\right], \quad (4.78)$$

$$\frac{n}{z}J_n(z)=\frac{1}{2}\left[J_{n-1}(z)+J_{n+1}(z)\right], \quad (4.79)$$

$$K_n'(z)=-\frac{1}{2}\left[K_{n-1}(z)+K_{n+1}(z)\right] \quad (4.80)$$

$$\frac{n}{z}K_n(z) = -\frac{1}{2}\left[K_{n-1}(z) - K_{n+1}(z)\right] \tag{4.81}$$

Equations (4.63)–(4.66) and (4.67)–(4.70) become the following.

1. **Core region $(0 \le r \le a)$**

$$E_r = -jA\beta\frac{a}{u}\left[\frac{(1-s)}{2}J_{n-1}\left(\frac{u}{a}r\right) - \frac{(1+s)}{2}J_{n+1}\left(\frac{u}{a}r\right)\right]\cos(n\theta+\psi) \tag{4.82}$$

$$E_\theta = jA\beta\frac{a}{u}\left[\frac{(1-s)}{2}J_{n-1}\left(\frac{u}{a}r\right) + \frac{(1+s)}{2}J_{n+1}\left(\frac{u}{a}r\right)\right]\sin(n\theta+\psi) \tag{4.83}$$

$$E_z = AJ_n\left(\frac{u}{a}r\right)\cos(n\theta+\psi) \tag{4.84}$$

$$H_r = -jA\omega\varepsilon_0 n_1^2\frac{a}{u}\left[\frac{(1-s_1)}{2}J_{n-1}\left(\frac{u}{a}r\right) + \frac{(1+s_1)}{2}J_{n+1}\left(\frac{u}{a}r\right)\right]\sin(n\theta+\psi) \tag{4.85}$$

$$H_\theta = -jA\omega\varepsilon_0 n_1^2\frac{a}{u}\left[\frac{(1-s_1)}{2}J_{n-1}\left(\frac{u}{a}r\right) - \frac{(1+s_1)}{2}J_{n+1}\left(\frac{u}{a}r\right)\right]\cos(n\theta+\psi) \tag{4.86}$$

$$H_z = -A\frac{\beta}{\omega\mu_0}sJ_n\left(\frac{u}{a}r\right)\sin(n\theta+\psi) \tag{4.87}$$

2. **Cladding region $(r > a)$:**

$$E_r = -jA\beta\frac{aJ_n(u)}{wK_n(w)}\left[\frac{(1-s)}{2}K_{n-1}\left(\frac{w}{a}r\right) + \frac{(1+s)}{2}K_{n+1}\left(\frac{w}{a}r\right)\right]\cos(n\theta+\psi) \tag{4.88}$$

$$E_\theta = jA\beta\frac{aJ_n(u)}{wK_n(w)}\left[\frac{(1-s)}{2}K_{n-1}\left(\frac{w}{a}r\right) - \frac{(1+s)}{2}K_{n+1}\left(\frac{w}{a}r\right)\right]\sin(n\theta+\psi) \tag{4.89}$$

$$E_z = A\frac{J_n(u)}{K_n(w)}K_n\left(\frac{w}{a}r\right)\cos(n\theta+\psi) \tag{4.90}$$

$$H_r = -j\omega\varepsilon_0 n_0^2\frac{aJ_n(u)}{wK_n(w)}\left[\frac{(1-s_0)}{2}K_{n-1}\left(\frac{w}{a}r\right) - \frac{(1+s_0)}{2}K_{n+1}\left(\frac{w}{a}r\right)\right]\sin(n\theta+\psi) \tag{4.91}$$

$$H_\theta = -jA\omega\varepsilon_0 n_0^2\frac{aJ_n(u)}{wK_n(w)}\left[\frac{(1-s_0)}{2}K_{n-1}\left(\frac{w}{a}r\right) + \frac{(1+s_0)}{2}K_{n+1}\left(\frac{w}{a}r\right)\right]\cos(n\theta+\psi); \tag{4.92}$$

$$H_z = -A\frac{\beta}{\omega\mu_0}s\frac{J_n(u)}{K_n(w)}K_n\left(\frac{w}{a}r\right)\sin(n\theta+\psi), \tag{4.93}$$

where

$$s_1 = \frac{\beta^2}{k^2 n_1^2}s \tag{4.94}$$

$$s_0 = \frac{\beta^2}{k^2 n_0^2}s. \tag{4.95}$$

4.2.1.4 Fundamental HE$_{11}$ mode

It has been clarified in the literature that the fundamental mode of the optical fibre is the HE_{11} mode. Therefore, it is indispensable to understand the propagation characteristics of the HE_{11} mode in order to grasp the signal transmission properties of single-mode fibres. Here, the dispersion characteristics and electromagnetic field distribution of the HE_{11} mode are investigated in detail. In order to rewrite (4.75) and (4.77) to easily understandable expression, we use the recurrence formulas of $J' = J_0 - J_1/u$ and $K' = -K_0 - K_1/w$. Noticing that $\Delta \ll 1$ in practical fibres and also that the dispersion equation in the form can be adopted for the HE_{11} mode, (4.75) and (4.77) to reduce to:

$$\frac{J_1'(u)}{uJ_1(u)} + \frac{K_1'(w)}{wK_1(w)} \cong -\left(\frac{1}{u^2}+\frac{1}{w^2}\right)$$

$$\frac{J_0(u)}{uJ_1(u)} = (1-\Delta)\frac{K_0(w)}{wK_1(w)} \tag{4.96}$$

$$s = -1 + \frac{u^2w^2}{v^2}\frac{J_0(u)}{uJ_1(u)}\Delta + O(\Delta^2). \tag{4.97}$$

Since the maximum value of $(uw/v)^2 J_0/uJ_1$ is about 0.7, s can be well approximated by $s \cong -1$. By assigning $n=1$ in (4.82)–(4.87) and (4.88)–(4.93), and using the transformation relation in the forms:

$$\begin{cases} E_x = E_r\cos\theta - E_\theta\sin\theta \\ E_y = E_r\sin\theta + E_\theta\cos\theta' \end{cases} \tag{4.98}$$

$$\begin{cases} H_x = H_r\cos\theta - H_\theta\sin\theta \\ H_y = H_r\sin\theta + H_\theta\cos\theta \end{cases}. \tag{4.99}$$

The electromagnetic field distribution of HE_{11} mode in the Cartesian coordinates becomes the following.

1. Core region $(0 \le r \le a)$:

$$E_x = -jA\beta\frac{a}{u}\left[\frac{(1-s)}{2}J_0\left(\frac{u}{a}r\right)\cos\psi - \frac{(1+s)}{2}J_2\left(\frac{u}{a}r\right)\cos(2\theta+\psi)\right], \tag{4.100}$$

$$E_y = jA\beta\frac{a}{u}\left[\frac{(1-s)}{2}J_0\left(\frac{u}{a}r\right)\sin\psi + \frac{(1+s)}{2}J_2\left(\frac{u}{a}r\right)\sin(2\theta+\psi)\right] \quad (4.101)$$

$$E_z = AJ_1\left(\frac{u}{a}r\right)\cos(\theta+\psi) \quad (4.102)$$

$$H_x = -jA\omega\varepsilon_0 n_1^2\frac{a}{u}\left[\frac{(1-s_1)}{2}J_0\left(\frac{u}{a}r\right)\sin\psi + \frac{(1+s_1)}{2}J_2\left(\frac{u}{a}r\right)\sin(2\theta+\psi)\right] \quad (4.103)$$

$$H_y = -jA\omega\varepsilon_0 n_1^2\frac{a}{u}\left[\frac{(1-s_1)}{2}J_0\left(\frac{u}{a}r\right)\cos\psi - \frac{(1+s_1)}{2}J_2\left(\frac{u}{a}r\right)\cos(2\theta+\psi)\right] \quad (4.104)$$

$$H_z = -A\frac{\beta}{\omega\mu_0}sJ_1\left(\frac{u}{a}r\right)\sin(\theta+\psi) \quad (4.105)$$

2. Cladding region $(r > a)$:

$$E_x = -jA\beta\frac{aJ_1(u)}{wK_1(w)}\left[\frac{(1-s)}{2}K_0\left(\frac{w}{a}r\right)\cos\psi + \frac{(1+s)}{2}K_2\left(\frac{w}{a}r\right)\cos(2\theta+\psi)\right] \quad (4.106)$$

$$E_y = jA\beta\frac{aJ_1(u)}{wK_1(w)}\left[\frac{(1-s)}{2}K_0\left(\frac{w}{a}r\right)\sin\psi - \frac{(1+s)}{2}K_2\left(\frac{w}{a}r\right)\sin(2\theta+\psi)\right] \quad (4.107)$$

$$E_z = A\frac{J_1(u)}{K_1(w)}K_1\left(\frac{w}{a}r\right)\cos(\theta+\psi) \quad (4.108)$$

$$H_x = -jA\omega\varepsilon_0 n_0^2\frac{aJ_1(u)}{wK_1(w)}\left[\frac{(1-s_0)}{2}K_0\left(\frac{w}{a}r\right)\sin\psi - \frac{(1+s_0)}{2}K_2\left(\frac{w}{a}r\right)\sin(2\theta+\psi)\right]; \quad (4.109)$$

$$H_y = -jA\omega\varepsilon_0 n_0^2\frac{aJ_1(u)}{wK_1(w)}\left[\frac{(1-s_0)}{2}K_0\left(\frac{w}{a}r\right)\cos\psi + \frac{(1+s_0)}{2}K_2\left(\frac{w}{a}r\right)\cos(2\theta+\psi)\right]; \quad (4.110)$$

$$H_z = -A\frac{\beta}{\omega\mu_0}s\frac{J_1(u)}{K_1(w)}K_1\left(\frac{w}{a}r\right)\sin(\theta+\psi) \quad (4.111)$$

There are two independent polarisation modes in (4.100)–(4.105) and (4.106)–(4.111), which correspond to $\psi = 0$ and $\psi = \pi/2$. The electric field component in the core with $\psi = 0$ are expressed as:

$$E_x = -jA\beta\frac{a}{u}\left[\frac{(1-s)}{2}J_0\left(\frac{u}{a}r\right) - \frac{(1+s)}{2}J_2\left(\frac{u}{a}r\right)\cos(2\theta)\right] \quad (4.112)$$

$$E_y = jA\beta\frac{a}{u}\frac{(1+s)}{2}J_2\left(\frac{u}{a}r\right).\ \sin(2\theta) \tag{4.113}$$

$$E_z = AJ_1\left(\frac{u}{a}r\right)\cos(\theta) \tag{4.114}$$

and those corresponding to $\psi = \pi/2$ are:

$$E_x = -jA\beta\frac{a}{u}\left[\frac{(1+s)}{2}J_2\left(\frac{u}{a}r\right)\sin(2\theta)\right] \tag{4.115}$$

$$E_y = jA\beta\frac{a}{u}\left[\frac{(1-s)}{2}J_0\left(\frac{u}{a}r\right)+\frac{(1+s)}{2}J_2\left(\frac{u}{a}r\right)\cos(2\theta)\right] \tag{4.116}$$

$$E_z = -AJ_1\left(\frac{u}{a}r\right)\sin(\theta). \tag{4.117}$$

Equations (4.100)–(4.105) and (4.115)–(4.117) are rigorous electromagnetic field distribution for the HE_{11} mode. It is known that E_x is the dominant mode in (4.112)–(4.114), and E_y is the dominant mode in (4.115)–(4.117), since we know that the relation $(1-s)/2 \cong 1$ and $(1+s)/2 \cong \Delta \ll 1$ from (4.97). Therefore, the polarisation mode in (4.112)–(4.114) with $\psi = 0$ is called HE_{11}^x mode and that in (4.115)–(4.117) with $\psi = \pi/2$ is called HE_{11}^y mode, respectively. It is shown that the two perpendicularly polarised HE_{11} modes are degenerated in the so-called single-mode fibre. Although $E_x\,(E_y)$ mode is the dominant mode in the $HE_{11}^x\,(HE_{11}^y)$ mode, a quite small $E_y\,(E_x)$ component having the perpendicular polarisation exists in the strict sense. The amplitude ratio of such a minor component to the major one is of the order of $\Delta(\ll 1)$, and thus, it can be neglected in most cases. Actually, in the LP mode (weakly guiding approximation), those minor components are neglected. Under this LP mode approximation, the electric field of the $HE_{11}^x\,(HE_{11}^y)$ mode becomes only the $E_x\,(E_y)$ component in the cross-sectional plane, and thus, the electric field vector becomes linear. This is the reason that the mode with a weakly guiding approximation is called a linearly polarised mode. Substituting (4.100)–(4.105) and (4.106)–(4.111), having the approximation of $s \cong s_1 \cong s_0 \cong -1$, in the transmission power:

$$P = \frac{1}{2}\int_0^{2\pi}\int_0^{\infty}\left(E_x H_y^* - E_y H_x^*\right) r\,dr\,d\theta, \tag{4.118}$$

and considering the transformation relation of (4.98)–(4.99), the amplitude coefficient A of the field is given as:

$$A = \frac{uw}{\beta a^2 v J_1(u)}\sqrt{\frac{2P}{\pi\varepsilon_0 n_1 c}}, \tag{4.119}$$

where c is the velocity of light in a vacuum. Although the eigenvalue of the HE_{11} mode is given by the numerical solution of (4.96), a useful approximation is derived as:

$$u = v\sqrt{1-b} = 2.4048e^{-0.8985/v}. \tag{4.120}$$

When we compare the eigenvalue u, which is obtained by numerically solving (4.96) by assigning $\Delta = 0$, with that of (4.120), the eigenvalue u, by (4.120), gives a slightly smaller value than the rigorous one. The error of the eigenvalue u given by (4.120) is less than 0.75% in the range of $v = 0.9 - 10$ and less than 0.5% in the range of $v = 0.9 - 2.4048$, respectively.

4.3 DERIVATION OF COUPLED-MODE EQUATION BASED ON PERTURBATION THEORY

The transmission characteristics of independent planar optical waveguide and optical fibres have been investigated widely in literature. For the construction of practical optical devices, it is very important to utilise the mutual light wave interaction between the two co-propagating light beams in the adjacent waveguide or interaction between the co-propagating two beams in the corrugated optical waveguide. Coupled-mode theory deals with the mutual light wave interaction between the two propagation modes. In this section, derivation of coupled-mode equation based on perturbation theory is first presented and then concrete methods calculating the coupling coefficient for several practically important devices are explained in detail. Finally, several waveguide devices using directional couplers such as Mach–Zehnder interferometers, ring resonators, and bi-stable devices are described.

In axially uniform optical waveguides, a number of propagation modes exist. These propagation modes are specific to each waveguide and satisfy the orthogonality condition between the modes. If two waveguides are brought close together as shown in Fig. 4.2, optical modes of each waveguide either couple or interface with each other. When the electromagnetic field distribution after mode coupling do not differ substantially from those before coupling, the propagation characteristics of the coupled waveguide can be analysed by the perturbation method. When we denote the eigenmodes in each optical waveguide before mode coupling as $\tilde{E}_P$, $\tilde{H}_p$ (P = 1, 2), they satisfy the following Maxwell's equations.

$$\begin{cases} \nabla \times \tilde{E}_P = -j\omega_0 \tilde{H}_p \\ \nabla \times \tilde{H}_P = j\omega_0 N_p{}^2 \tilde{E}_p \end{cases} \{P=1,2\}, \tag{4.121}$$

where $N_p{}^2$ represents the refractive index distribution of each waveguide.

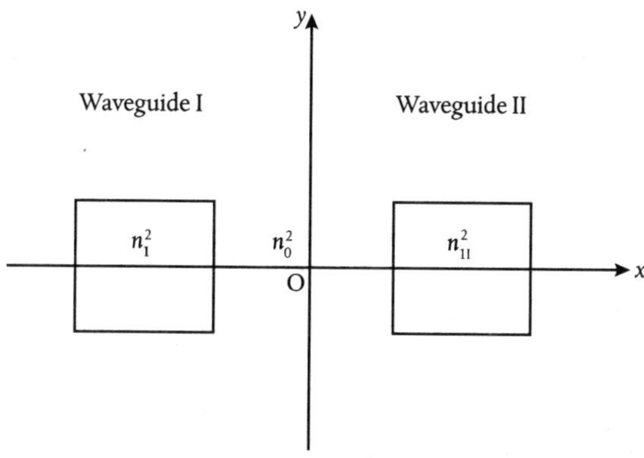

Fig. 4.2 Directionally coupled optical waveguide

We assume that the electromagnetic fields of the coupled waveguide can be expressed as the sum of the eigenmodes in each waveguide:

$$\begin{cases} \tilde{E} = A(z)\tilde{E}_1 + B(z)\tilde{E}_2 \\ \tilde{H} = A(z)\tilde{H}_1 + B(z)\tilde{H}_2 \end{cases}. \tag{4.122}$$

The electromagnetic fields in the coupled waveguide $\tilde{E}$ and $\tilde{H}$ also should satisfy Maxwell's equation. Then, substituting (4.122) in (4.121):

$$\begin{cases} \nabla\times\tilde{E} = -j\omega\mu_0\tilde{H} \\ \nabla\times\tilde{H} = j\omega N^2\varepsilon_0\tilde{E} \end{cases} \tag{4.123}$$

Further, using (4.121) and the vector formula:

$$\nabla\times(AE) = A\nabla\times E + \nabla A\times E = A\nabla\times E + \frac{dA}{dz}u_z\times E$$

We obtain the following relations:

$$\left(u_z\times\tilde{E}_1\right)\frac{dA}{dz} + \left(u_z\times\tilde{E}_2\right)\frac{dB}{dz} = 0 \tag{4.124}$$

$$(u_z\times\tilde{H}_1)\frac{dA}{dz} - j\omega\varepsilon_0\left(N^2 - N_1^{\,2}\right)A\tilde{E}_1 + \left(u_z\times\tilde{H}_2\right)\frac{dB}{dz} - j\omega\varepsilon_0\left(N^2 - N_2^{\,2}\right)B\tilde{E}_2 = 0\,. \tag{4.125}$$

Here, $N^2(x,y)$ denotes the refractive index distribution in the entire coupled waveguide. Substituting (4.125) and (4.124) in the following integral equation:

$$\int_{-\infty}^{\infty}\int_{-\infty}^{\infty}\left[\tilde{E}_1^*.(\text{eq.}(4.61)) - \tilde{H}_1^*.(\text{eq.}(4.60))\right]dxdy = 0. \tag{4.126}$$

The term in (4.126) to be integrated is expressed as:

$$I_1 = \tilde{E}_1^*.(eq.(4.125)) - \tilde{H}_1^*(eq.(4.124)) = \frac{dA}{dz}\left[\tilde{E}_1^*.\left(u_z\times\tilde{H}_1\right) - \dot{H}^*_1.\left(u_z\times E_1\right)\right] +$$
$$\frac{dB}{dz}\left[\tilde{E}_1^*.\left(u_z\times\tilde{H}_2\right) - \dot{H}^*_1.\left(u_z\times\tilde{E}_2\right)\right] - j\omega\varepsilon_0 A\left(N^2 - N^2_{\ 1}\right)\tilde{E}_1^*\tilde{E}_1 - j\omega\varepsilon_0 B\left(N^2 - N^2_{\ 2}\right)\tilde{E}_1^*\tilde{E}_2,$$

where we use the following vectorial formulas:

$$\tilde{E}_1^*.\left(u_z\times\tilde{H}_1\right) = -u_z.\left(\tilde{E}_1^*\times\tilde{H}_1\right)$$
$$H^*_{\ 1}.\left(u_z\times\tilde{E}_1\right) = u_z.\left(\tilde{E}_1\times H_1^*\right),$$

Then, I_1 can be written as:

$$I_1 = -\frac{dA}{dz}.u_z\left[\tilde{E}_1^*\times\tilde{H}_1 + \tilde{E}_1\times H^*_{\ 1}\right] - \frac{dB}{dz}u_z\left[\tilde{E}_1^*\times\tilde{H}_2 + \tilde{E}_2\times\tilde{H}_1^*\right] - j\omega\varepsilon_0 A\left(N^2 - N^2_{\ 1}\right)\tilde{E}_1^*\tilde{E}_1$$
$$-j\omega\varepsilon_0 B\left(N^2 - N^2_{\ 2}\right)\tilde{E}_1^*\tilde{E}_2.$$

Similarly,

$$\int_{-\infty}^{\infty}\int_{-\infty}^{\infty}\left[\tilde{E}_2^{*}.(eq.(4.125))-\tilde{H}_2^{*}.(eq.(4.124))\right]dxdy=0. \tag{4.127}$$

The term in (4.127) to be integrated as:

$$I_2=E_2^{*}.(eq4.125)-H_2^{*}.(eq4.124)=-\frac{dA}{dz}u_z.\left[\tilde{E}_2^{*}\times\tilde{H}_1+\tilde{E}_1\times H_2^{*}\right]-\frac{dB}{dz}u_z\left[\tilde{E}_2^{*}\times\tilde{H}_2+\tilde{E}_2\times H_2^{*}\right]$$
$$-j\omega\varepsilon_0A\left(N^2-N^2{}_1\right)\tilde{E}_2^{*}\tilde{E}_1-j\omega\varepsilon_0B\left(N^2-N_2{}^2\right)\tilde{E}_2^{*}\tilde{E}_2.$$

We obtain from (4.126):

$$\frac{dA}{dz}+\frac{dB}{dz}\frac{\int_{-\infty}^{\infty}\int_{-\infty}^{\infty}u_z.\left(\tilde{E}_1^{*}\times\tilde{H}_2+\tilde{E}_2\times\tilde{H}_1^{*}\right)dxdy}{\int_{-\infty}^{\infty}\int_{-\infty}^{\infty}u_z.\left(\tilde{E}_1^{*}\times\tilde{H}_1+\tilde{E}_1\times\tilde{H}_1^{*}\right)dxdy}+jA\frac{\omega\varepsilon_0\int_{-\infty}^{\infty}\int_{-\infty}^{\infty}\left(N^2-N_1{}^2\right)\tilde{E}_1^{*}.\tilde{E}_1dxdy}{\int_{-\infty}^{\infty}\int_{-\infty}^{\infty}u_z.\left(\tilde{E}_1^{*}\times\tilde{H}_1+\tilde{E}_1\times\tilde{H}_1^{*}\right)dxdy}$$
$$+jB\frac{\omega\varepsilon_0\int_{-\infty}^{\infty}\int_{-\infty}^{\infty}\left(N^2-N_2{}^2\right)\tilde{E}_1^{*}.\tilde{E}_2dxdy}{\int_{-\infty}^{\infty}\int_{-\infty}^{\infty}u_z.\left(\tilde{E}_1^{*}\times\tilde{H}_1+\tilde{E}_1\times\tilde{H}_1^{*}\right)dxdy}=0 \tag{4.128}$$

and similarly, we obtain from (4.127),

$$\frac{dB}{dz}+\frac{dA}{dz}\frac{\int_{-\infty}^{\infty}\int_{-\infty}^{\infty}u_z.\left(\tilde{E}_2^{*}\times\tilde{H}_1+\tilde{E}_1\times\tilde{H}_2^{*}\right)dxdy}{\int_{-\infty}^{\infty}\int_{-\infty}^{\infty}u_z.\left(\tilde{E}_2^{*}\times\tilde{H}_2+\tilde{E}_2\times\tilde{H}_2^{*}\right)dxdy}+jA\frac{\omega\varepsilon_0\int_{-\infty}^{\infty}\int_{-\infty}^{\infty}\left(N^2-N_1{}^2\right)\tilde{E}_2^{*}.\tilde{E}_1dxdy}{\int_{-\infty}^{\infty}\int_{-\infty}^{\infty}u_z.\left(\tilde{E}_2^{*}\times\tilde{H}_2+\tilde{E}_2\times\tilde{H}_2^{*}\right)dxdy}+$$
$$jB\frac{\omega\varepsilon_0\int_{-\infty}^{\infty}\int_{-\infty}^{\infty}\left(N^2-N_2{}^2\right)\tilde{E}_2^{*}.\tilde{E}_2dxdy}{\int_{-\infty}^{\infty}\int_{-\infty}^{\infty}u_z.\left(\tilde{E}_2^{*}\times\tilde{H}_2+\tilde{E}_2\times\tilde{H}_2^{*}\right)dxdy}=0. \tag{4.129}$$

Here, we separate the transverse and axial dependencies of the electromagnetic fields:

$$\begin{cases}\tilde{E}_p=E_pexp\left(-j\beta_pz\right)\\ \tilde{H}_P=H_pexp\left(-j\beta_pz\right)\end{cases}(P=1,2). \tag{4.130}$$

Substituting (4.130) in (4.128) and (4.129), we obtain:

$$\frac{dA}{dz}+c_{12}\frac{dB}{dz}\exp\left[-j(\beta_2-\beta_1)z\right]+j\chi_1A+j\kappa_{12}Bexp\left[-j(\beta_2-\beta_1)z\right]=0 \tag{4.131}$$

$$\frac{dB}{dz}+c_{21}\frac{dA}{dz}\exp\left[+j(\beta_2-\beta_1)z\right]+j\chi_2B+j\kappa_{21}A\exp\left[+j(\beta_2-\beta_1)z\right]=0, \tag{4.132}$$

where

$$\kappa_{pq}=\frac{\omega\varepsilon_0\int_{-\infty}^{\infty}\int_{-\infty}^{\infty}\left(N^2-N_q^2\right)E_p^*.E_q dxdy}{\int_{-\infty}^{\infty}\int_{-\infty}^{\infty}u_z.\left(E_p^*\times H_p+E_p\times H_p^*\right)dxdy} \tag{4.133}$$

$$c_{pq}=\frac{\int_{-\infty}^{\infty}\int_{-\infty}^{\infty}u_z.\left(E_p^*\times H_q+E_q\times H_p^*\right)dxdy}{\int_{-\infty}^{\infty}\int_{-\infty}^{\infty}u_z.\left(E_p^*\times H_p+E_p\times H_p^*\right)dxdy} \tag{4.134}$$

$$\chi_p=\frac{\omega\varepsilon_0\int_{-\infty}^{\infty}\int_{-\infty}^{\infty}\left(N^2-N_p^2\right)E_p^*.E_p dxdy}{\int_{-\infty}^{\infty}\int_{-\infty}^{\infty}u_z.\left(E_p^*\times H_p+E_p\times H_p^*\right)dxdy} \tag{4.135}$$

and the pair of p and q are either $(p, q) = (1, 2)$ or $(2, 1)$, respectively. Here, κ_{pq} is a mode-coupling coefficient of the directional coupler. The meaning of c_{pq} is described as follows.

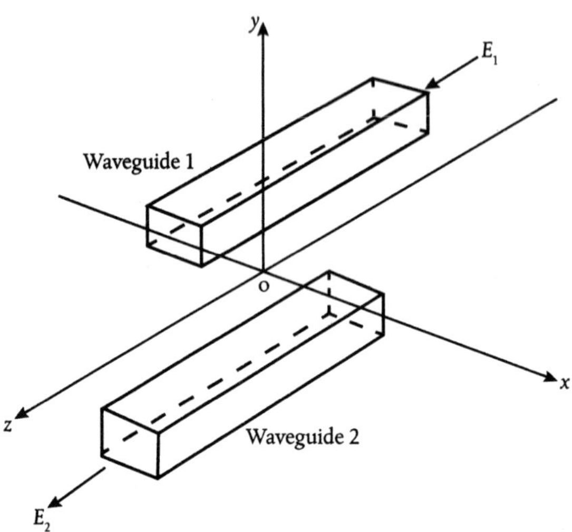

Fig. 4.3 Explanation of butt-coupling coefficients c_{12}.

Let us consider the waveguide configuration shown in Fig. 4.3 where waveguide 1 exists only in the region $z<0$ and waveguide 2 in $z>0$. When the eigenmode (E_1,H_1) of waveguide 1 propagates from the negative z-direction to $z=0$, the electromagnetic field in the cladding excites the eigenmode (E_2,H_2) at point $z=0$. This excitation efficiency is considered to be c_{12}. Therefore, c_{pq} represents the butt-coupling coefficient between two waveguides.

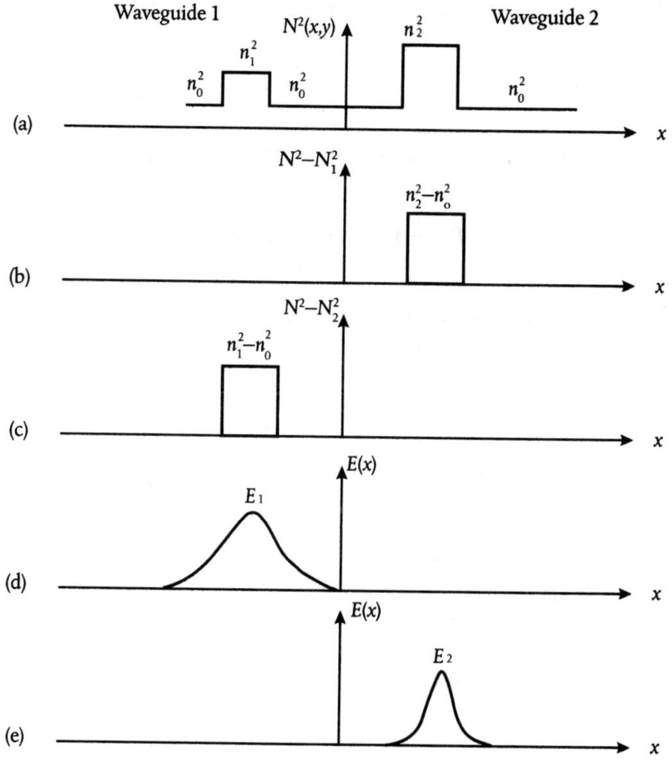

Fig. 4.4 Refractive index distribution $N^2(x,y)$ of the coupled waveguide difference of the refractive index distribution $\left(N^2 - N_p^2\right)$ and electric field E_p appearing in (4.69) and (4.71)

In most of the conventional analyses of the directional couplers, c_{pq} and χ_p were neglected and they are assumed to be $c_{pq} = \chi_p = 0$. However, both c_{pq} and χ_p should be taken into account in order to analyse the mode-coupling effect strictly.

The optical power carried by the eigenmode in the waveguide p, $(p = 1,2)$ is expressed as:

$$P_p = \frac{1}{2}\int_{-\infty}^{\infty}\int_{-\infty}^{\infty}\left(E_p \times H^*_p\right).u_z dxdy \quad (p = 1,2) \tag{4.136}$$

It is known from this equation that the denominators of (4.133)–(4.135) are equal to $4P_p$. Henceforth, we assume that the eigenmodes in both waveguides are normalised to satisfy the condition.

$$\int_{-\infty}^{\infty}\int_{-\infty}^{\infty}\left(E^*_p \times H_p + E_p \times H^*_p\right).u_z dxdy = 4P_p = 1 \qquad (p = 1, 2). \tag{4.137}$$

Then, we obtain from (4.134):

$$c_{21} = c^*_{12}. \tag{4.138}$$

Straightforwardly, we obtained from (4.135):

$$\chi_p = \chi_p^*. \tag{4.139}$$

Express the difference of the propagation constants between waveguides as:

$$\delta = \frac{(\beta_2 - \beta_1)}{2}. \tag{4.140}$$

The optical power in the entire coupled waveguide is expressed as:

$$P = \frac{1}{2}\int_{-\infty}^{\infty}\int_{-\infty}^{\infty}\left(\tilde{E} \times \tilde{H}^*\right).u_z dxdy. \tag{4.141}$$

Substituting the value from (4.130) in (4.122), we can obtain:

$$\tilde{E} = A.E_1 e^{(-j\beta_1 z)} + B.E_2 e^{(-j\beta_2 z)}$$
$$\tilde{H} = A.H_1 e^{(-j\beta_1 z)} + B.H_2 e^{(-j\beta_2 z)}.$$

and also

$$\tilde{H}^* = A^*.H^*_1 e^{(+j\beta_1 z)} + B^*.H^*_2 e^{(+j\beta_2 z)}.$$

Hence, from (4.141), we get:

$$P = \frac{1}{2}\int_{-\infty}^{\infty}\int_{-\infty}^{\infty}\left(\left(A.E_1 e^{(-j\beta_1 z)} + B.E_2 e^{(-j\beta_2 z)}\right) \times \left(A^*.H^*_1 e^{(+j\beta_1 z)} + B^*.H^*_2 e^{(+j\beta_2 z)}\right)\right)u_z.dx.dy,$$

or

$$P = \frac{1}{2}\int_{-\infty}^{\infty}\int_{-\infty}^{\infty}\left[A.A^*\left(E_1 \times H^*_1\right) + A.B^*\left(E_1 \times H^*_2\right)e^{-j(\beta_1-\beta_2)z} + A^*.B\left(E_2 \times H^*_1\right)e^{j(\beta_1-\beta_2)z} + B.B^*\left(E_2 \times H^*_2\right)u_z.dx.dy\right]$$

$$\Rightarrow P = \frac{1}{2}\int_{-\infty}^{\infty}\int_{-\infty}^{\infty}\left[|A|^2\left(E_1 \times H^*_1\right) + A.B^*\left(E_1 \times H^*_2\right)e^{+j2\delta z} + A^*.B\left(E_2 \times H^*_1\right)e^{-j2\delta z} + |B|^2\left(E_2 \times H^*_2\right)u_z.dx.dy\right] \tag{4.142}$$

Now, as we know that:

$$\int_{-\infty}^{\infty}\int_{-\infty}^{\infty}\left(E^*_p \times H_p + E_p \times H^*_p\right)u_z.dx.dy = 4P_p = 1.$$

Hence, we can write:

$$\int_{-\infty}^{\infty}\int_{-\infty}^{\infty}\left(E^*_p \times H_p\right)u_z.dx.dy = \int_{-\infty}^{\infty}\int_{-\infty}^{\infty}\left(E_p \times H^*_p\right)u_z.dx.dy = \frac{1}{2}.$$

We know that:

$$c_{pq} = \frac{\int_{-\infty}^{\infty}\int_{-\infty}^{\infty}\left(E^*_p \times H_q + E_q \times H^*_p\right)u_z.dx.dy}{\int_{-\infty}^{\infty}\int_{-\infty}^{\infty}\left(E^*_p \times H_p + E_p \times H^*_p\right)u_z.dx.dy}.$$

Thus,

$$c_{12} = \frac{\int_{-\infty}^{\infty}\int_{-\infty}^{\infty}\left(E^{*}_{1} \times H_{2} + E_{2} \times H^{*}_{1}\right)u_{z}.dx.dy}{\int_{-\infty}^{\infty}\int_{-\infty}^{\infty}\left(E^{*}_{1} \times H_{1} + E_{1} \times H^{*}_{1}\right)u_{z}.dx.dy} = 2\int_{-\infty}^{\infty}\int_{-\infty}^{\infty}\left(E^{*}_{1} \times H_{2}\right)u_{z}.dx.dy,$$

or

$$\int_{-\infty}^{\infty}\int_{-\infty}^{\infty}\left(E^{*}_{1} \times H_{2}\right)u_{z}.dx.dy = \frac{c_{12}}{2}.$$

Similarly,

$$\int_{-\infty}^{\infty}\int_{-\infty}^{\infty}\left(E^{*}_{2} \times H^{*}_{1}\right)u_{z}.dx.dy = \frac{c^{*}_{12}}{2}.$$

Substituting all these values in (4.142) we get:

$$P = \frac{1}{4}\left[|A|^{2} + |B|^{2} + A^{*}Bc_{12}e^{(-j2\delta z)} + AB^{*}c^{*}_{12}e^{(j2\delta z)}\right]. \tag{4.143}$$

In the derivation of this last equation, we used (4.134), (4.137), and (4.138). In the loss-less waveguides, the optical power remains constant because we have:

$$\frac{dP}{dz} = 0. \tag{4.144}$$

Equation (4.131) can be written as:

$$A^{*}\frac{dA}{dz} + c_{12}A^{*}\frac{dB}{dz}e^{(-2j\delta z)} + jk_{12}BA^{*}e^{(-2j\delta z)} = 0$$

and

$$A\frac{dA^{*}}{dz} + Ac^{*}_{12}\frac{dB^{*}}{dz}e^{2j\delta z} \quad jk^{*}_{12}B^{*}Ae^{(2j\delta z)} = 0.$$

Thus, from (4.143) and (4.144):

$$\frac{d|A|^{2}}{dz} = -c_{12}A^{*}\frac{dB}{dz}e^{(-2j\delta z)} - jk_{12}BA^{*}e^{(-2j\delta z)} - Ac^{*}_{12}\frac{dB^{*}}{dz}e^{(2j\delta z)} + jk^{*}_{12}B^{*}Ae^{(2j\delta z)}. \tag{4.145}$$

Similarly, from (4.132), we get

$$\left.\begin{aligned} B^{*}\frac{dB}{dz} + c_{21}B^{*}\frac{dA}{dz}e^{(2j\delta z)} + jk_{21}B^{*}Ae^{(2j\delta z)} &= 0 \\ B\frac{dB^{*}}{dz} + Bc^{*}_{21}\frac{dA^{*}}{dz}e^{(-2j\delta z)} - jk^{*}_{21}A^{*}Be^{(-2j\delta z)} &= 0 \end{aligned}\right\}$$

Thus, from (4.143) and (4.145),

$$\frac{d|B|^{2}}{dz} = -B^{*}c_{21}\frac{dA}{dz}e^{(2j\delta z)} - jk_{21}B^{*}Ae^{(2j\delta z)} - Bc^{*}_{21}\frac{dA^{*}}{dz}e^{(-2j\delta z)} + jk^{*}_{21}A^{*}Be^{(-2j\delta z)}. \tag{4.146}$$

Adding (4.145) and (4.146), we get from (4.144):

$$-2j\delta c_{12}A^{*}Be^{(-2j\delta z)}+2j\delta AB^{*}c^{*}_{12}e^{(2j\delta z)}+c_{12}B\frac{dA^{*}}{dz}e^{(-2j\delta z)}+c_{12}A^{*}\frac{dB}{dz}e^{(-2j\delta z)}+$$
$$c^{*}_{12}B^{*}\frac{dA}{dz}e^{(2j\delta z)}+c^{*}_{12}A\frac{dB^{*}}{dz}e^{(j2\delta z)}-jk_{12}BA^{*}e^{(-2j\delta z)}+jk^{*}_{12}B^{*}Ae^{(2j\delta z)}-jk_{21}B^{*}Ae^{(2j\delta z)}+$$
$$jk^{*}_{21}A^{*}Be^{(-2j\delta z)}-c_{12}A^{*}\frac{dB}{dz}e^{(-2j\delta z)}-Ac_{12}\frac{dB^{*}}{dz}e^{(2j\delta z)}-Bc_{21}\frac{dA}{dz}e^{(2j\delta z)}-Bc^{*}_{21}\frac{dA^{*}}{dz}e^{(-2j\delta z)}.$$

Since, $\left.\begin{matrix} c_{21}=c^{*}_{12} \\ c^{*}_{21}=c_{12}\end{matrix}\right\}.$

Hence,

$$jA^{*}B\left(k^{*}_{21}-k_{12}-2\delta c_{12}\right)e^{(-2j\delta z)}-jAB^{*}\left(k_{21}-k^{*}_{12}-2\delta c^{*}_{12}\right)e^{(j2\delta z)}=0$$

or

$$A^{*}B\left(\kappa_{21}^{*}-\kappa_{12}-2\delta c_{12}\right)\exp(-2j\delta z)-jAB^{*}\left(\kappa_{21}-\kappa_{12}^{*}-2\delta c_{12}^{*}\right)\exp(j2\delta z)=0. \tag{4.147}$$

In order that (4.147) is satisfied independent of z, we should have:

$$\kappa_{21}=\kappa^{*}_{12}+2\delta c^{*}_{12}\,. \tag{4.148}$$

In the most of the conventional analyses, reciprocity of the coupling co-efficient was expressed by $\kappa_{21}=\kappa^{*}_{12}$, since c_{12} was assumed to be zero. However, $\kappa_{21}=\kappa^{*}_{12}$ holds true only when (1) the propagation constants of the two waveguide are equal ($\delta=0$) or (2) waveguides are sufficiently separated ($c_{12}\triangleq 0$). Especially, when two waveguides with different core geometric shapes are placed close together, care should be taken, since the second term of the right-hand side of (4.148) cannot be neglected. Next, we derive the coupling mode equation using (4.131) and (4.132). From the equality of (4.131)–(4.132) $\times c_{12}\exp(-j2\delta z)=0$, we obtain:

$$\frac{dA}{dz}=-j\kappa_{a}B\exp(-2j\delta z)+j\alpha_{a}A. \tag{4.149}$$

Similarly, from the equality of (4.132)–(4.131) $\times c_{12}\exp(-j2\delta z)=0$, we obtain:

$$\frac{dB}{dz}=-j\kappa_{b}A\exp(2j\delta z)+j\alpha_{b}B. \tag{4.150}$$

We assume the solution of the coupled-mode (4.149) and (4.150) in the forms:

$$A(z)=\left[a_{1}e^{jQz}+a_{2}e^{-JQz}\right]\exp\left[-j\delta z+j\frac{\alpha_{a+}\alpha_{b}}{2}z\right]. \tag{4.151}$$

$$B(z)=\left[b_{1}e^{jQz}+b_{2}e^{-JQz}\right]\exp\left[j\delta z+j\frac{\alpha_{a+}\alpha_{b}}{2}z\right]. \tag{4.152}$$

Substituting the equation in (4.149) and (4.150), the following relations are obtained:

$$\chi_a b_1 = -\left(Q - \hat{\delta}\right) a_1, \tag{4.153}$$

$$\chi_a b_2 = \left(Q + \hat{\delta}\right) a_2, \tag{4.154}$$

$$\chi_b a_1 = -\left(Q + \hat{\delta}\right) b_1, \tag{4.155}$$

$$\chi_b a_2 = \left(Q - \hat{\delta}\right) b_2, \tag{4.156}$$

where

$$\hat{\delta} = \delta + \frac{\alpha_a - \alpha_b}{2}. \tag{4.157}$$

In order that (4.153) to (4.156) has nontrivial solution except for $a_1 = a_2 = b_1 = b_2 = 0$, Q (>0) should be given as:

$$Q = \sqrt{\chi_a \chi_b + \hat{\delta}^2}. \tag{4.158}$$

Considering the boundary condition at z = 0 as:

$$a_1 + a_2 = A(0) \tag{4.159}$$

$$b_1 + b_2 = B(0). \tag{4.160}$$

Coefficients $a_1, a_2, b_1,$ and b_2 are given as:

$$a_1 = \frac{1}{2}\left\{\left(1 + \frac{\hat{\delta}}{Q}\right) A(0) - \frac{\chi_a}{Q} B(0)\right\}, \tag{4.161}$$

$$a_2 = \frac{1}{2}\left\{\left(1 - \frac{\hat{\delta}}{Q}\right) A(0) + \frac{\chi_a}{Q} B(0)\right\}, \tag{4.162}$$

$$b_1 = \frac{1}{2}\left\{-\frac{\chi_b}{Q} A(0) + \left(1 - \frac{\hat{\delta}}{Q}\right) B(0)\right\}, \tag{4.163}$$

$$b_2 = \frac{1}{2}\left\{\frac{\chi_b}{Q} A(0) + \left(1 + \frac{\hat{\delta}}{Q}\right) B(0)\right\}. \tag{4.164}$$

Substituting (4.161)–(4.164) in (4.151) and (4.152), we obtain the solution for the co-directional coupled-mode equation:

$$A(z)=\left\{\left[\cos(Qz)+j\frac{\hat{\delta}}{Q}\sin(Qz)\right]A(0)-j\frac{\chi_a}{Q}\sin(Qz)B(0)\right\}\exp\left[-j\left(\hat{\delta}-\alpha_a\right)z\right] \quad (4.165)$$

$$B(z)=\left\{-j\frac{\chi_b}{Q}\sin(Qz)A(0)+\left[\cos(Qz)-j\frac{\hat{\delta}}{Q}\sin(Qz)\right]B(0)\right\}\exp\left[j\left(\hat{\delta}+\alpha_b\right)z\right]. \quad (4.166)$$

In (4.149) and (4.150), $\kappa_a,\kappa_b,\alpha_a,\alpha_b$ parameters are defined as:

$$\kappa_a=\frac{\kappa_{12}-c_{12}\chi_2}{1-|c_{12}|^2}, \quad (4.167)$$

$$\kappa_b=\frac{\kappa_{21}-c_{12}^{*}\chi_1}{1-|c_{12}|^2}, \quad (4.168)$$

$$\alpha_a=\frac{\kappa_{21}c_{12}-\chi_1}{1-|c_{12}|^2}, \quad (4.169)$$

$$\alpha_b=\frac{\kappa_{12}c_{12}^{*}-\chi_2}{1-|c_{12}|^2}. \quad (4.170)$$

In the following, the solution method of the coupled mode (4.149) and (4.150) for the co-directional coupler ($\beta_1>0,\beta_2>0$) will be described. However, we assume $c_{pq}=\chi_p=0(p,q=1,2)$ and rewrite (4.149) and (4.150) into more simplified form, since the rigorous analysis becomes very complicated. Of course, the strict analysis based on (4.149) and (4.150) is possible. With the assumption of $c_{pq}=\chi_p=0;\ (p,q=1,2)$, we can rewrite (4.149) and (4.150) as:

$$\frac{dA}{dz}=-j\kappa_{12}\ \exp\left[-j(\beta_2-\beta_1)z\right], \quad (4.171)$$

$$\frac{dB}{dz}=-j\kappa_{21}A\exp\left[+j(\beta_2-\beta_1)z\right], \quad (4.172)$$

where κ_{pq} is given as (4.133). The reciprocity relation of the coupling coefficients

$$\kappa_{21}=\kappa_{12}. \quad (4.173)$$

In most of the directional couplers, κ_{pq} is real. Therefore, we express $\kappa=\kappa_{21}=\kappa_{12}$. The reciprocity relation for the contra-directional coupler is expressed as:

$$\kappa_{21}=-\kappa_{12}^{*}. \quad (4.174)$$

Equation of the co-directional couplers ($\beta_1>0,\beta_2>0$) are assumed in the form:

$$A(z)=\left[a_1e^{jqz}+a_2e^{-jqz}\right]\exp(-j\delta z), \quad (4.175)$$

$$B(z)=\left[b_1e^{jqz}+b_2e^{-jqz}\right]\exp(j\delta z), \quad (4.176)$$

where q is an unknown parameter to be determined. Constants a_1, a_2, b_1, and b_2 should satisfy the initial conditions:

$$a_1 + a_2 = A(0), \tag{4.177}$$

$$b_1 + b_2 = B(0). \tag{4.178}$$

Substituting (4.175)–(4.176) in coupled mode (4.171) and (4.172) and applying the initial conditions of (4.177)–(4.178), we obtain:

$$A(z) = \left\{\left[\cos(qz) + j\frac{\delta}{z}\sin(qz)\right]A(0) - j\frac{\kappa}{q}\sin(qz)B(0)\right\}\exp(-j\delta z), \tag{4.179}$$

$$B(z) = \left\{-j\frac{\kappa}{q}\sin(qz)A(0) + \left[\cos(qz) - j\frac{\delta}{q}\sin(qz)\right]B(0)\right\}\exp(j\delta z), \tag{4.180}$$

where q is given as

$$q = \sqrt{\kappa^2 + \delta^2}\ . \tag{4.181}$$

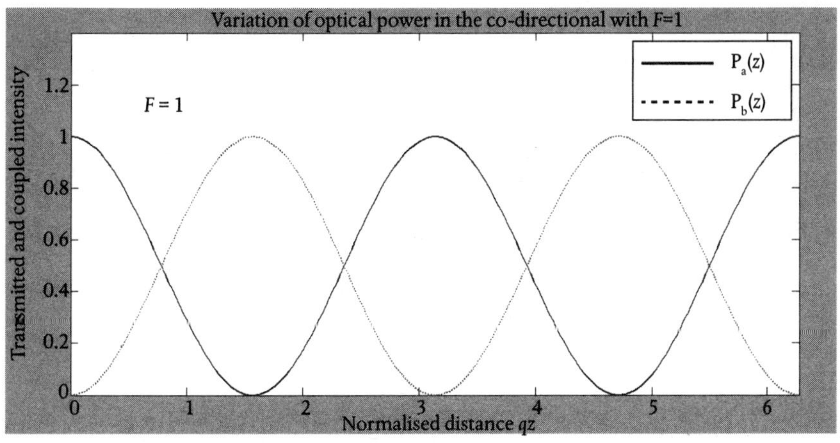

Fig. 4.5 Variation of optical power in the co-directional coupler with $F = 1.0$

For the most practical case, in which light is coupled into waveguide 1 only at $z = 0$, we have the condition of $A(0) = A_0$ and $B(0) = 0$. Then, the optical power flow along the z-direction is given as

$$P_a(z) = \frac{|A(z)|^2}{|A_0|^2} = 1 - F\sin^2(qz), \tag{4.182}$$

$$P_b(z) = \frac{|B(z)|^2}{|A_0|^2} = F\sin^2(qz), \tag{4.183}$$

where F denotes the maximum power-coupling efficiency, and it is defined as:

$$F = \left(\frac{\kappa}{q}\right)^2 = \frac{1}{1+\left(\frac{\delta}{\kappa}\right)^2}. \qquad (4.184)$$

Figure (4.5) and Fig. (4.6) shows the dependence of P_a and P_b on the normalised length qz for the two directional configurations $F=1$ and $F=0.2$, respectively. The power-coupling efficiency from excited waveguide 1 to waveguide 2 reaches maximum at:

$$z = \frac{\pi}{2q}(2m+1)(m=0,1,2,....). \qquad (4.185)$$

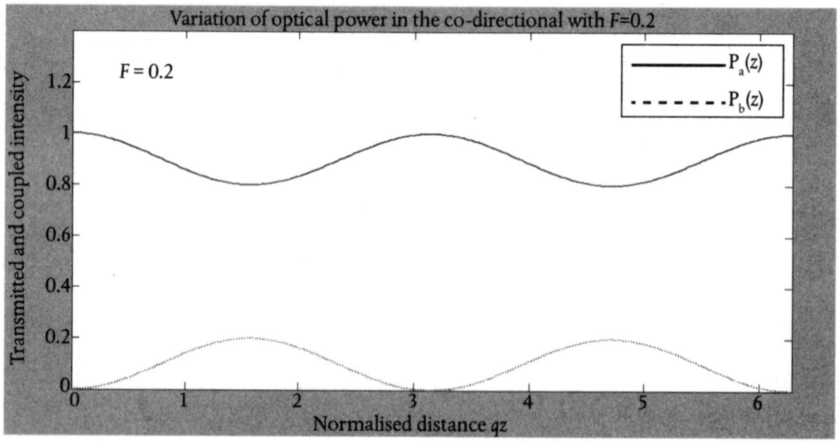

Fig. 4.6 Variation of optical power in the co-directional coupler with $F = 0.2$

The length z at $m = 0$ is called the coupling length and is given as:

$$L_C = \frac{\pi}{2q} = \frac{\pi}{2\sqrt{\kappa^2+\delta^2}}. \qquad (4.186)$$

When the propagation constants of the two waveguides are equal ($\beta_1 = \beta_2$ and then $\delta = 0$), 100% power coupling occurs at the coupling length of:

$$L_c = \frac{\pi}{2\kappa}. \qquad (4.187)$$

4.4 PICOSECONDS PULSE GENERATION SCHEME

In this section, we describe a novel circuit for the generation of picoseconds optical pulse train and analyse the same for the calculation of the pulse width [14–17]. The pulse generation circuit is shown in Fig. 4.7. The DFB laser diode (LD) at the input is the master laser that emits the light of wavelength λ. This light is then phase modulated in an optical phase modulator (OPM). The PM light wave is then coupled to the arrayed waveguide grating (AWG) having a channel spacing equal to the modulation frequency f_m. The AWG de-multiplexes the sidebands of the optical PM light wave. We select three light waves from the output of the AWG that have the frequencies $f_c - nf_m$, f_c, and $f_c + nf_m$. Here, f_c is the master laser frequency and n has a typical value of four. These three light waves are then amplified by injection-locked DFB LD_s $-LD_2$, LD_0, and LD_1, respectively. The free-running oscillation frequencies of these LDs are made equal to $f_c - nf_m$, f_c, and $f_c + nf_m$, respectively.

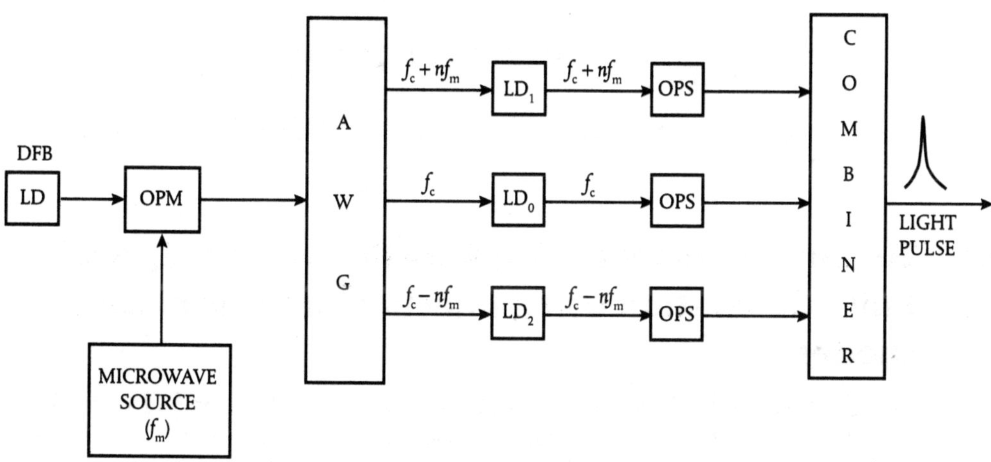

Fig. 4.7 Schematic circuit diagram of the high repetition-rate optical pulse generator LD: Laser diode; OPM: Optical phase modulator; AWG: Arrayed waveguide grating; OPS: Optical phase shifter

The phases of these amplified light waves can be adjusted by placing a voltage-tunable electro-optic phase shifter (OPS) after each injection-locked LD. The three light waves are then combined in an optical power combiner to generate an optical pulse. The light wave output from the master DFB laser is expressed as:

$$a(t) = a_0 e^{j2\pi f_c t} . \tag{4.188}$$

The actual electric field of the light wave is the real part of this complex representation. The output light wave from the kth LD is expressed as:

$$a_k(t) = a_{k0} e^{j[(\omega_c + kn\omega_m)t + \varphi_k]}, \tag{4.189}$$

where $k=1$ indicates LD_1, $k=0$ indicates LD_0, and $k=-1$ indicates LD_2. a_{k0} is the amplitude of the output light from the injection-locked k^{th} LD, ω_c is the carrier frequency, ω_m is the OPM drive frequency, all in radians, and φ_k is the phase angle of the output of the kth LD. The output light wave from the combiner can be written as:

$$Y(t) = \sum_{k=-1}^{k=1} a_k(t). \tag{4.190}$$

Substituting $a_{-10} = a_{20}$, we get:

$$|Y(t)|^2 = Y(t)Y^*(t)$$

$$=a^2_{00}+a^2_{10}+a^2_{20}+2a_{00}a_{10}\cos\left(n\omega_m+\varphi_1-\varphi_0\right)+2a_{00}a_{20}\cos\left(n\omega_m+\varphi_0-\varphi_2\right)$$
$$+2a_{10}a_{20}\cos\left(2n\omega_m t-\varphi_2+\varphi_1\right).$$

Adjusting the phases ($\varphi_0, \varphi_1, \varphi_2$) of the combining light waves, we can generate different kinds of optical pulses. For obtaining a good pulse shape, we substitute $a_{10}=a_{00}$ and $a_{20}=a_{00}/2$.

Further, to obtain the shape, we substitute $\varphi_0=\varphi_1=\varphi_2$. Then,

$$\frac{|Y(t)|^2}{|Y(0)|^2}=\frac{1}{25}\left(9+12\cos\left(n\omega_m t\right)+4\cos\left(2n\omega_m t\right)\right). \quad (4.191)$$

Let us consider n = 4 and $f_m=\dfrac{\omega_m}{2\pi}=25\text{GHz}$, we plot the normalised intensity of the output light wave from this circuit.

4.4.1 Tunable High-repetition Rate Picosecond Optical Pulse Generation Using Injection Locking of Laser Diodes

Generation of high repetition rate and short optical pulse is an important topic of research at the present time. This is because in order to realise a high-capacity optical communication system, we need a combination of wavelength-division multiplexing (WDM) and optical time-division multiplexing (OTDM) techniques [18–22]. The number of channels that can be transmitted over the same fibre is limited by fibre dispersion. The OTDM systems are limited by the limited modulator speed and by the high-repetition rate requirement of the pulsed optical source. A combination of WDM and OTDM can result in a high-capacity optical communication system within existing limitations. The electro-optic modulators can be operated at a rate of 40 Gb/s at present. To utilise this modulation rate, we must generate optical pulses having a repetition rate much greater than 40 GHz. Thus, to realise a high-speed OTDM system, we need a high repetition rate and short optical pulses. Various methods of optical pulse generation exist in the literature. Some of these are mode locking of semiconductor lasers, harmonically mode-locked fibre laser, soliton-assisted time lens compression, short-pulse generation by interferometers, and repetition-rate multiplication of a low-rate pulse source. Attempts are going on to generate optical pulses having repetition rates in the range of several hundred gigahertz. In this section, we propose a scheme of high-repetition rate, picoseconds optical pulse generation where it is possible to achieve a repetition rate of a few hundred gigahertz. The repetition rate can also be conveniently tuned over a wide range by varying the millimetre-wave drive frequency applied to the modulator. A schematic circuit diagram of the picoseconds optical pulse generator is shown in Fig. 4.8. Here, the output of a DFB laser is phase modulated in an overdriven optical phase modulator (ODPM). The phase modulator can be fabricated on an $LiNbO_3$ substrate. There are five laser diodes LD_0, LD_1, LD_2, LD_3, and LD_4. The free-running oscillation frequencies of these LDs are made to coincide with the carrier and sideband angular frequencies $\omega_c, \omega_c+n\omega_m, \omega_c+2n\omega_m, \omega_c-n\omega_m$, and $\omega_c-2n\omega_m$, respectively, where 'n' is the integer corresponding to the selected optical sideband. This can be done by proper tuning of the LDs. The injection of the light waves at the free-running frequency makes the LD injection locked to the injected light wave. All other sidebands will be suppressed by the injection-locked LD. Thus, at the output of the five LD_s — LD_0, LD_1, LD_2, LD_3, and LD_4 — we get light waves at angular frequencies $\omega_c, \omega_c+n\omega_m, \omega_c+2n\omega_m, \omega_c-n\omega_m$, and $\omega_c-2n\omega_m$, respectively. These five light waves are combined in an (5×1) optical power combiner, and at the output, we get an optical pulse.

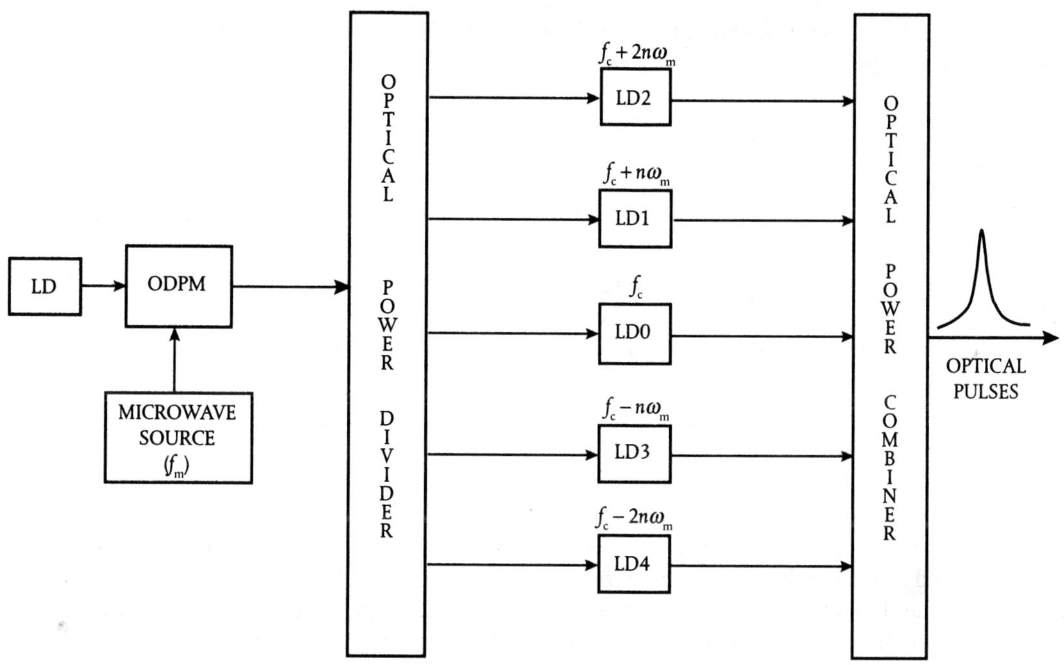

Fig. 4.8 Schematic optical circuit diagram of the pulse generation LD, OPDM: overdriven optical phase modulator

Let the output of the five injection-locked LDs be expressed in complex representation as:

$$a_k(t) = a_k e^{j[(\omega_c + kn\omega_m)t + \theta_k]}, \tag{4.192}$$

where $k = -2, -1, 0, 1, 2. k = -1$ corresponds to LD_3 and $k = -2$ corresponds to LD_4. ω_c is the carrier angular frequency and ω_m is the microwave drive angular frequency. θ_k is the constant phase angle. 'n' is the fixed integer corresponding to the selected sidebands that determine the pulse repetition rate. a_k is the electric field amplitude of the output light wave of the kth LD. Using complex representation, the sum of the electric fields of the five coherent light waves is represented as:

$$Y(t) = \sum_{k=-2}^{k=2} a_k(t), \tag{4.193}$$

where $a_k(t)$ is given in (4.192). If T is the period of the resultant pulse, we can write:

$$Y(t+T) = Y(t). \tag{4.194}$$

Equation (4.194) yields:

$$\sum_{k=-2}^{2} a_k e^{j[(\omega_c + kn\omega_m)t + \theta_k]} \left[e^{j(\omega_c + kn\omega_m)T} - 1 \right] = 0. \tag{4.195}$$

Since (4.195) holds good for any time 't', we have:

$$e^{j(\omega_c + kn\omega_m)T} = 1$$

$$(\omega_c + kn\omega_m)T = 2\pi \times \text{some integer}$$

for $k=-2,-1,0,1,2$. This integer has different values for different values of 'k'. After some mathematical manipulations, we get:

$$n\omega_m t = 2\pi. \quad (4.196)$$

Hence,

$$T = \frac{1}{nf_m}. \quad (4.197)$$

Thus, the repetition frequency of the pulse is nf_m. If $n=5$ and $f_m = 40$ GHz, then the repetition rate is 200 GHz. To find the pulse width and its dependence on the drive frequency, we have to calculate the intensity of the output pulse waveform. Now, the output intensity $I(t)$, is proportional to $|Y(t)|^2$. On carrying out the mathematical calculations, we get:

$$|Y(t)|^2 = \sum_{k=-2}^{2}\sum_{p=-2}^{2} a_k a_p \cos\left[(k-p)n\omega_m t + \theta_k - \theta_p\right]. \quad (4.198)$$

The phase of the particular optical sideband appearing at the output of the injection-locked LD can be adjusted by tuning the free-running frequency of the slave LD. In particular, if the free-running frequency of the slave DFB LD is tuned to the locking signal frequency, the phase error becomes zero, assuming symmetric locking characteristics of the DFB LD. In this case, θ_k (for $k=-2,-1,0,1,2$) can be set to be zero. In addition to this method of phase adjusting, we can place a voltage-tunable electro-optic phase shifter after each injection-locked LD, and the phases of the optical sideband signals can be accurately adjusted by varying the dc voltage applied to the electro-optic phase shifter. Thus, we can set θ_k and θ_p equal to zero for $k, p=-2,-1,0,1,2$ without any loss of generality.

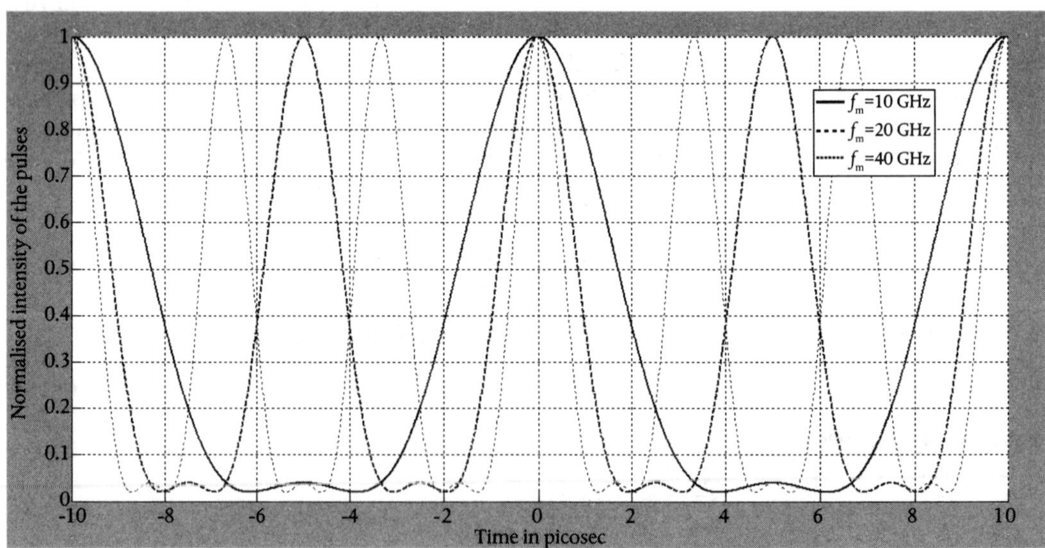

Fig. 4.9 Plot of the normalised intensity $\left(\frac{I(t)}{I(0)}\right)$ of the output light pulses as a function of time (t) using the modulator drive frequency (f_m) as a parameter for $n(5)$.

The normalised intensity of the generated light pulse is given as:

$$\frac{I(t)}{I(0)}=\frac{|Y(t)|^2}{|Y(0)|^2}. \tag{4.199}$$

The output powers of the injection-locked LDs are selected so that $a_1=a_{-1}=a_0$. This condition will lead to a good pulse shape. Thus, the free-running output powers of LD_0, LD_1, and LD_3 are equal. Further, the free-running output powers of LD_2 and LD_4 are the same and equal to one-fourth of the free-running output powers of LD_0. Now, the output intensity $I(t)$ is proportional to $|Y(t)|^2$. On carrying out mathematical calculation by using (4.192) and (4.193), we get:

$$|Y(t)|^2=\sum_{k=-2}^{2}\sum_{p=-2}^{2}a_k a_p \cos\left[(k-p)n\omega_m t+\theta_k-\theta_p\right]. \tag{4.200}$$

By assuming $\theta_k=\theta_p$,

$$\begin{aligned}|Y(t)|^2=\sum_{k=-2}^{2}a_k\left[a_{-2}\cos\{(k+2)n\omega_m t\}+a_{-1}\cos\{(k+1)n\omega_m t\}\right]+a_0\cos\{n\omega_m t\}\\ +a_1\cos\{(k-1)n\omega_m t\}+a_2\cos\{(k-2)n\omega_m t\}.\end{aligned} \tag{4.201}$$

Since,

1. For $k=-2$,

$$\begin{aligned}|Y(t)|^2=a_{-2}^2+a_{-2}a_{-1}\cos(n\omega_m t)+a_{-2}a_0\cos(2n\omega_m t)\\ +a_{-2}a_1\cos(3n\omega_m t)+a_{-2}a_2\cos(4n\omega_m t).\end{aligned} \tag{4.202}$$

2. For $k=-1$,

$$\begin{aligned}|Y(t)|^2=a_{-1}a_{-2}\cos(n\omega_m t)+a_{-1}^2+a_{-1}a_0\cos(n\omega_m t)\\ +a_{-1}a_1\cos(2n\omega_m t)+a_{-1}a_2\cos(3n\omega_m t).\end{aligned} \tag{4.203}$$

3. For $k=0$,

$$|Y(t)|^2=a_0a_{-2}\cos(2n\omega_m t)+a_0a_{-1}\cos(n\omega_m t)+a_0^2+a_0a_1\cos(n\omega_m t)+a_0a_2\cos(2n\omega_m t). \tag{4.204}$$

4. For $k=1$

$$|Y(t)|^2=a_1a_{-2}\cos(3n\omega_m t)+a_1a_{-1}\cos(2n\omega_m t)+a_1a_0\cos(n\omega_m t)+a_1^2+\cos(n\omega_m t) \tag{4.205}$$

5. For $k=2$

$$|Y(t)|^2=a_2a_{-2}\cos(4n\omega_m t)+a_2a_{-1}\cos(3n\omega_m t)+a_2a_0\cos(2n\omega_m t)+a_2a_1\cos(n\omega_m t)+a_2^2 \tag{4.206}$$

Now, by using (4.200) and (4.201), we get:

$$\begin{aligned}|Y(t)|^2 &= \left(a_{-2}a_{-2} + a_{-1}a_{-1} + a_0^2 + a_1^2 + a_2^2\right) \\ &+ \left(a_{-2}a_{-1} + a_{-1}a_{-2} + a_{-1}a_0 + a_0a_{-1} + a_0a_1 + a_1a_2 + a_1a_0 + a_2a_1\right)\cos(n\omega_m t) \\ &+ \left(a_{-2}a_0 + a_{-1}^2 + a_0a_{-2} + a_0a_2 + a_{-1}^2 + a_2a_0\right)\cos(2n\omega_m t) \\ &+ \left(a_{-2}a_1 + a_{-1}a_2 + a_{-1}a_2 + a_2a_{-1}\right)\cos(3n\omega_m t) \\ &+ \left(a_{-2}a_2 + a_2a_{-2}\right)\cos(4n\omega_m t).\end{aligned} \tag{4.207}$$

From proper selection of LDs, we can get the following equations:

$$a_{-1} = a_1 = a_0$$

$$a_2 = a_{-2} = \frac{a_0}{2}$$

Hence, from (4.207),

$$\begin{aligned}|Y(t)|^2 &= 3a^2{}_0 + 2a^2{}_2 + 4\left(a^2{}_0 + a_0a_2\right)\cos(n\omega_m t) + \left(2a^2{}_0 + 4a_0a_2\right)\cos(2n\omega_m t) \\ &+ 4a_0a_2\cos(3n\omega_m t) + 2a^2{}_2\cos(4n\omega_m t)\end{aligned} \tag{4.208}$$

$$|Y(0)|^2 = \left(3a_0 + 2a_2\right)^2. \tag{4.209}$$

On substituting $a_2 = \frac{a_0}{2}$ in (4.208) and (4.209), we get

$$\frac{I(t)}{I(0)} = \left|\frac{Y(t)}{Y(0)}\right|^2.$$

$$= \frac{3a_0^2 + 2\frac{a_0^2}{4} + 4\left\{a_0^2 + \frac{a_0^2}{2}\right\}\cos(n\omega_m t) + \left\{2a_0^2 + 4\frac{a_0^2}{2}\right\}\cos(2n\omega_m t) + 4\frac{a_0^2}{2}\cos(3n\omega_m t) + 2\frac{a_0^2}{4}\cos(4n\omega_m t)}{\left(3a_0^2 + 2\frac{a_0}{2}\right)^2}$$

$$\frac{I(t)}{I(0)} = \frac{1}{16}\left[3.5 + 6\cos(n\omega_m t) + 4\cos(2n\omega_m t) + 2\cos(3n\omega_m t) + 0.5\cos(4n\omega_m t)\right]; \tag{4.210}$$

The normalised intensity of the output light pulse is plotted in Fig. 4.9 using the drive frequency $f_m\left(=\frac{\omega_m}{2\pi}\right)$ of ODPM as a parameter. The higher the modulator drive frequency, the sharper is the pulse generated. By taking the value of the fixed integer $n = 5$, we get a value of repetition frequency of the pulse equal to $n\ f_m = 200$ GHz where $f_m = 40\,\text{GHz}$. The frequency bandwidth of the pulse is equal to $4n\ f_m$. The pulse width and frequency bandwidth product have been calculated to have a value equal to 0.8.

4.5 OPTICAL WAVEGUIDE DEVICE USING DIRECTIONAL COUPLER: MACH–ZEHNDER INTERFEROMETERS

Directional couplers are very important component in the fabrication of various optical devices. In this subsection, several optical utilising directional couplers are described. We consider the case in which light is coupled into the upper of a Mach–Zehnder interferometer. Hence, we assume that both arms have the same waveguide structure, that is, we have $\delta = 0$ and $q = k$ in (4.179)–(4.181). Substituting $A(0) = A_0$ and $B(0) = 0$ in (4.179) and (4.180), the output of the first directional coupler is given as:

$$\begin{cases} A_1 = A_0 \cos(\kappa l) & (4.211) \\ B_1 = -jA_0 \sin(\kappa l), & (4.212) \end{cases}$$

where l denotes the coupling length. It should be noted here that mode coupling takes place not only in the straight coupling region but also in the curved regions; therefore, the coupling length l in this subsection is an effective straight coupling length, which includes the entire mode-coupling effect in the straight and curved coupling regions. When the first coupler is a 3-dB coupler with $\kappa l = \frac{\pi}{4}$ splitting ratios are $A_1 = \frac{A_0}{\sqrt{2}}$ and $B_1 = \frac{-jA_0}{\sqrt{2}}$. After passing through the interferometer straight arms, A_2 and B_2 become:

$$\begin{cases} A_2 = A_1 \exp(-j\beta L) = \frac{A_0}{\sqrt{2}} \exp(-j\beta L) & (4.213) \\ B_2 = B_1 \exp(-j\beta L + j\phi) = -j\frac{A_0}{\sqrt{2}} \exp(-j\beta L + j\phi), & (4.214) \end{cases}$$

where ϕ denotes excess phase shift in the lower arms, as shown in Fig. 4.10. Then, the outputs of the interferometers are given by substituting (4.213)–(4.214) in (4.179) and (4.180):

$$\begin{cases} A_3 = -jA_0 \sin\left(\frac{\phi}{2}\right) \exp\left(-j\beta L + \frac{j\phi}{2}\right) & (4.215) \\ B_3 = -jA_0 \cos\left(\frac{\phi}{2}\right) \exp\left(-j\beta L + \frac{j\phi}{2}\right). & (4.216) \end{cases}$$

Here, we assumed that the second coupler is also a 3-dB coupler having $\kappa l = \frac{\pi}{4}$. The optical intensity in each output port is given as:

$$\begin{cases} |A_3|^2 = |A_0|^2 \sin^2\left(\frac{\phi}{2}\right) & (4.217) \\ |B_3|^2 - |A_0|^2 \cos^2\left(\frac{\phi}{2}\right). & (4.218) \end{cases}$$

It is seen from these equations that light can be switched from port A to port B or vice versa by changing phase ϕ from π to 0. When the Mach–Zehnder interferometer is used as an optical modulator, optical phase is modulated in proportion to the input signal. When phase ϕ is slightly modulated with

$\delta\phi$, output intensity in (4.217)–(4.218) becomes $|A_3|^2 = |A_0|^2\left(\frac{\delta\phi}{2}\right)^2$ and $|B_3|^2 = |A_0|^2$. It is shown that linear modulation is not obtained with the present condition. If we add the phase bias of $\frac{\pi}{2}$ and apply the phase modulation of $\delta\phi$, (4.217) becomes:

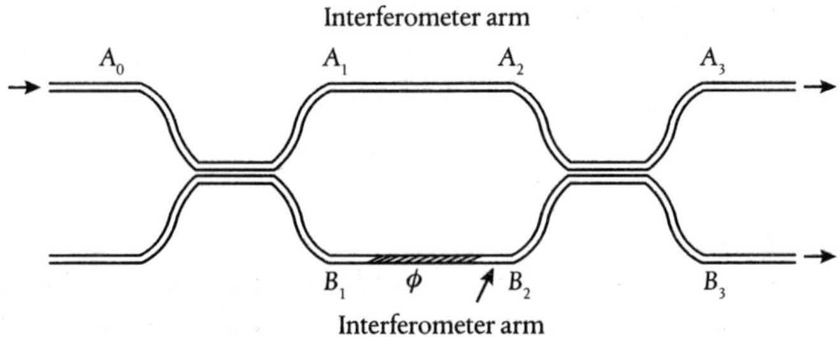

Fig. 4.10 Mach–Zehnder optical interferometer

$$|A_3|^2 = |A_0|^2 \sin^2\left(\frac{\pi}{4}+\frac{\delta\phi}{2}\right) = \frac{1}{2}|A_0|^2\left[1+\sin(\delta\phi)\right]\frac{1}{2}|A_0|^2(1+\delta\phi). \tag{4.219}$$

Then, optical intensity becomes linearly dependent on the input signal $\delta\phi$.

4.5.1 Ring Resonator

The steady-state input–output relation of the ring resonator in Fig. 4.11 are expressed as:

$$\begin{cases} A = (1-\gamma)^{\frac{1}{2}}\left[A_0\cos(\kappa l) - jB_0\sin(\kappa l)\right] & (4.220) \\ B = (1-\gamma)^{\frac{1}{2}}\left[-jA_0\sin(\kappa l) + B_0\cos(\kappa l)\right], & (4.221) \end{cases}$$

where $\kappa, l,$ and γ denote the mode-coupling coefficient of the directional coupler, the coupling length, and the intensity insertion loss coefficient, respectively. In (4.220)–(4.221), we assumed that input–output and resonator waveguide have the same propagation constant β. When we denote the intensity attenuation coefficient of the waveguide as ρ, B_0 is expressed as:

$$B_0 = B\exp\left(\rho\frac{L}{2} + j\beta L\right). \tag{4.222}$$

The amplitude transmittance of the optical ring resonator is then given, from (4.220) and (4.221), as:

$$\frac{A}{A_0} = (1-\gamma)^{\frac{1}{2}}\left[\frac{\cos(\kappa l) - (1-\gamma)^{\frac{1}{2}}\exp\left(-\frac{\rho}{2}L - j\beta L\right)}{1-(1-\gamma)^{\frac{1}{2}}\cos(\kappa l)\exp\left(-\frac{\rho}{2}L - j\beta L\right)}\right]. \tag{4.223}$$

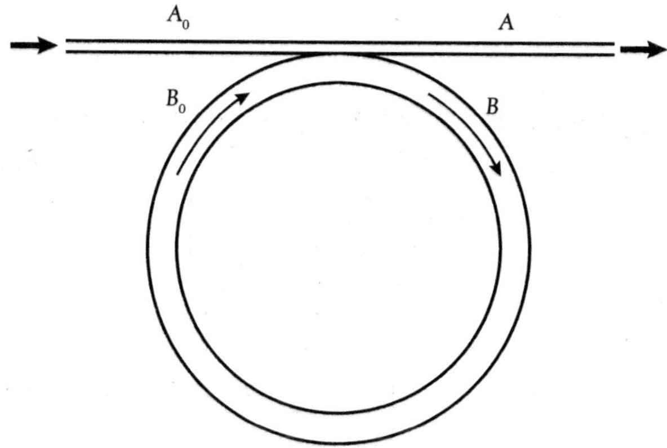

Fig. 4.11 Optical ring resonator

Since, we can write from (4.222):

$$B = B_0 exp\left(\rho\frac{L}{2} + j\beta L\right). \tag{4.224}$$

Now, substituting this value of B in (4.221), we get:

$$B_0 e^{\left(\rho\frac{L}{2}+j\beta L\right)} = (1-\gamma)^{1/2}\left[-jA_0 \sin(kl) + B_0 \cos(kl)\right]$$

or

$$B_0 = \frac{(1-\gamma)^{1/2}(-jA_0)\sin(kl)}{e^{\left(\rho\frac{L}{2}+j\beta L\right)} - (1-\gamma)^{1/2}\cos(kl)} \tag{4.225}$$

Substituting the value of B_0 from (4.225) in (4.220), we get

$$A = (1-\gamma)^{1/2}\left[A_0 \cos(kl) - \frac{j(1-\gamma)^{1/2}(-jA_0)\sin^2(kl)}{e^{\left(\rho\frac{L}{2}+j\beta L\right)} - (1-\gamma)^{1/2}\cos(kl)}\right]$$

or

$$\frac{A}{A_0} = (1-\gamma)^{1/2}\left[\frac{e^{\left(\rho\frac{L}{2}+j\beta L\right)}\cos(kl) - (1-\gamma)^{1/2}}{e^{\left(\rho\frac{L}{2}+j\beta L\right)} - (1-\gamma)^{1/2}\cos(kl)}\right]$$

or

$$\frac{A}{A_0} = (1-\gamma)^{1/2}\left[\frac{\cos(kl) - (1-\gamma)^{1/2} e^{\left(-\rho\frac{L}{2}-j\beta L\right)}}{1-(1-\gamma)^{1/2}\cos(kl)e^{\left(-\rho\frac{L}{2}-j\beta L\right)}}\right].$$

Then, we introduce new parameters x, y, and ϕ, and they are defined as $\begin{cases} x=(1-\gamma)^{\frac{1}{2}}\exp\left(-\frac{\rho}{2}L\right) \\ y=\cos(\kappa l) \\ \phi=\beta L \end{cases}$.
We get

$$\frac{A}{A_0}=(1-\gamma)^{1/2}\left[\frac{y-xe^{-j\phi}}{1-xye^{-j\phi}}\right]$$

or

$$\frac{A}{A_0}=(1-\gamma)^{1/2}\left[\frac{(y-x\cos\phi)+jx\sin\phi}{(1-xy\cos\phi)+jxy\sin\phi}\right]$$

or

$$\left|\frac{A}{A_0}\right|^2=(1-\gamma)\left[\frac{(y-x\cos\phi)^2+(x\sin\phi)^2}{(1-xy\cos\phi)^2+x^2y^2\sin^2\phi}\right]$$

or

$$\left|\frac{A}{A_0}\right|^2=(1-\gamma)\left[\frac{x^2+y^2-2xy\cos\phi}{1+x^2y^2-2xy\cos\phi}\right]$$

or

$$\left|\frac{A}{A_0}\right|^2=(1-\gamma)\left[1+\frac{x^2+y^2-1-x^2y^2}{1+x^2y^2-2xy\cos\phi}\right]$$

or

$$\left|\frac{A}{A_0}\right|^2=(1-\gamma)\left[1-\frac{1+x^2y^2-x^2-y^2}{1+x^2y^2-2xy\cos\phi}\right]$$

or

$$\left|\frac{A}{A_0}\right|^2=(1-\gamma)\left[1-\frac{(1-x^2)(1-y^2)}{1+x^2y^2-2xy\cos\phi}\right]$$

or

$$\left|\frac{A}{A_0}\right|^2=(1-\gamma)\left[1-\frac{(1-x^2)(1-y^2)}{1+x^2y^2-2xy-2xy\cos\phi+2xy}\right]$$

or

$$\left|\frac{A}{A_0}\right|^2=(1-\gamma)\left[1-\frac{(1-x^2)(1-y^2)}{(1-xy)^2+4xy\sin^2\left(\phi/2\right)}\right]. \tag{4.226}$$

Then, the intensity transmittance of the optical ring resonator is obtained as follows:

$$T(\phi)=\left|\frac{A}{A_0}\right|^2=(1-\gamma)\left[1-\frac{(1-x^2)(1-y^2)}{(1-xy)^2+4xysin^2\left(\frac{\phi}{2}\right)}\right]. \quad (4.227)$$

For minimum value of $T(\phi)$,

$$\phi=0$$

$$\left|\frac{A}{A_0}\right|^2=(1-\gamma)\left[1-\frac{(1-x^2)(1-y^2)}{(1-xy)^2}\right]$$

$$\left|\frac{A}{A_0}\right|^2=(1-\gamma)\frac{(x-y)^2}{(1-xy)^2}. \quad (4.228)$$

For maximum value of $T(\phi)$, we get:

$$\phi=\frac{\pi}{2}$$

$$\left|\frac{A}{A_0}\right|^2=(1-\gamma)\left[1-\frac{(1-x^2)(1-y^2)}{(1-xy)^2+4xy}\right]$$

$$\left|\frac{A}{A_0}\right|^2=(1-\gamma)\frac{(x+y)^2}{(1+xy)^2}. \quad (4.229)$$

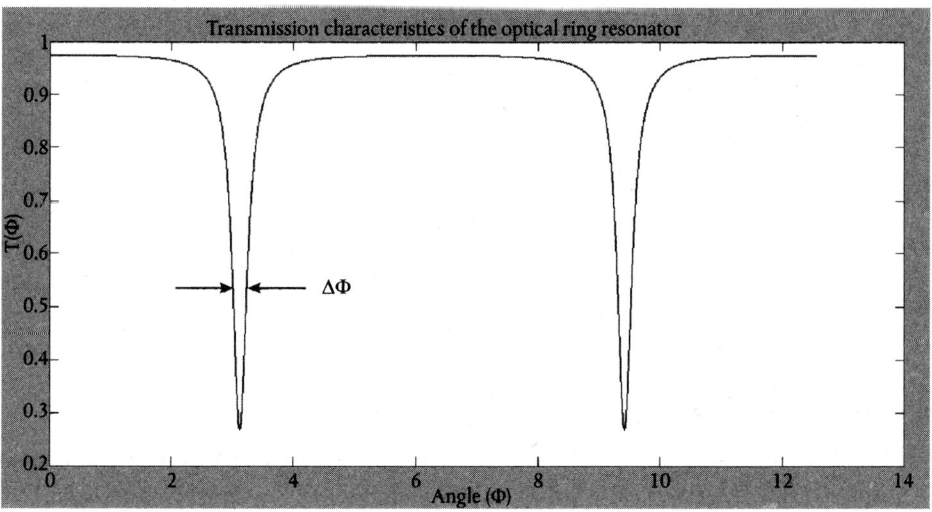

Fig. 4.12 Transmission characteristics of the optical ring resonator

Figure 4.12 shows the transmission characteristics of the optical ring resonator as a function of ϕ. The maximum and minimum transmittances are given as:

$$T_{max} = (1-\gamma)\frac{(x+y)^2}{(1+xy)^2} \tag{4.230}$$

$$T_{min} = (1-\gamma)\frac{(x-y)^2}{(1-xy)^2}. \tag{4.231}$$

It is seen from these equations that $x = y \cong 1$ should be satisfied in order to maximise T_{max} while minimising T_{min} as much as possible.

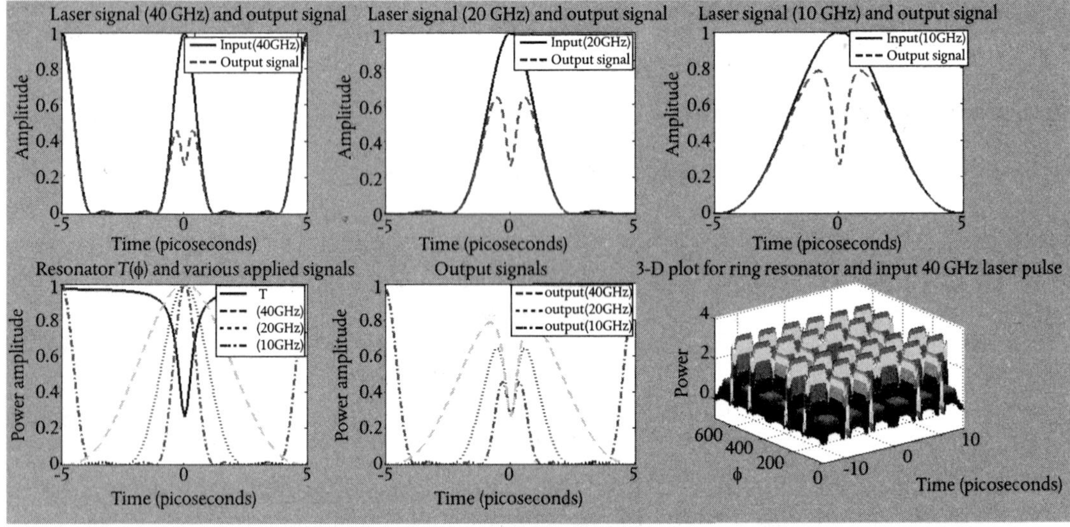

Fig. 4.13 The output waveform for general ring resonator when laser signals of different frequencies are allowed to pass through the resonator

Figure 4.13 shows the output waveform computed from (4.227), when the input signal A_0 is assumed to be a picoseconds pulse generated by circuit shown in Fig. (4.8) at a drive frequency of 10, 20, and 40 GHz, respectively. It is apparent from Fig. 4.13 that there is pulse distortion at the peak of the input signal. In this case, it seems that there is no further pulse compression effect whatsoever. However, the pulse distortion is symmetric in this case.

4.5.2 Bi-stable Ring Resonator

Here, we introduce the open type of ring resonator as shown in Fig. 4.14 consisting of the optical waveguide whose refractive index changes in accordance with optical intensity that is the Kerr medium.

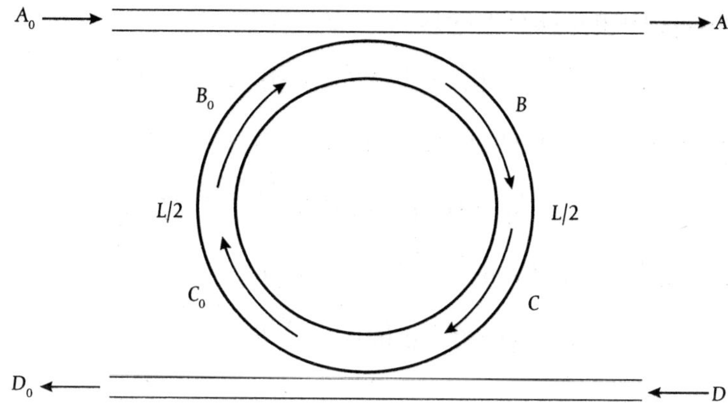

Fig. 4.14 Bi-stable ring resonator

From Fig. (4.14), it can be shown that:

$$A=(-\gamma)^{1/2}\left[A_0\cos(kl)-jB_0\sin(kl)\right] \tag{4.232}$$

$$B=(1-\gamma)^{1/2}\left[-jA_0\sin(kl)+B_0\cos(kl)\right]. \tag{4.233}$$

$$C=Be^{\left(-\frac{\rho}{4}L-j\beta\frac{L}{2}\right)} \tag{4.234}$$

$$D_0=(-\gamma)^{1/2}\left[D\cos(kl)-jC\sin(kl)\right] \tag{4.235}$$

$$B_0=C_0e^{\left(-\frac{\rho}{4}L-j\beta\frac{L}{2}\right)} \tag{4.236}$$

$$C_0=(1-\gamma)^{1/2}\left[-jD\sin(kl)+C\cos(kl)\right]. \tag{4.237}$$

By assuming $D=0$ in Fig. (4.14), we can calculate the intensity transmittance $T=\left|\frac{D_0}{A_0}\right|^2$ of the open type of ring resonator; and from (4.235), we get:

$$D_0=-jC(1-\gamma)^{1/2}\sin(kl). \tag{4.238}$$

By substituting the value of C from (4.234) in (4.238),

$$D_0=-j(1-\gamma)^{1/2}\sin(kl)e^{\left(-\frac{\rho}{4}L-j\frac{\beta}{2}L\right)}B. \tag{4.239}$$

By substituting the value of B in (4.239) from (4.233),

$$D_0=-j(1-\gamma)^{1/2}\sin(kl)e^{\left(-\frac{\rho}{4}L-j\frac{\beta}{2}L\right)}(1-\gamma)^{1/2}\left[-jA_0\sin(kl)+B_0\cos(kl)\right]$$

$$D_0=-j(1-\gamma)e^{\left(-\frac{\rho}{4}L-j\frac{\beta}{2}L\right)}\sin(kl)\left[-jA_0\sin(kl)+B_0\cos(kl)\right]. \tag{4.240}$$

Now, by substituting $D=0$ in (4.237), we get

$$C_0 = C(1-\gamma)^{1/2}\cos(kl). \qquad (4.241)$$

By substituting the value of C_0 from (4.241) in (4.236),

$$B_0 = C(1-\gamma)^{1/2}\cos(kl)e^{\left(-\frac{\rho}{4}L - j\frac{\beta}{2}L\right)}. \qquad (4.242)$$

Now, by substituting the value of C from (4.234) in (4.242),

$$B_0 = (1-\gamma)^{1/2}\cos(kl)e^{\left(-\frac{\rho}{4}L - j\frac{\beta}{2}L\right)}Be^{\left(-\frac{\rho}{4}L - j\beta\frac{L}{2}\right)}$$

or

$$B_0 = (1-\gamma)^{1/2}\cos(kl)e^{\left(-\frac{\rho}{2}L - j\beta L\right)}B. \qquad (4.243)$$

By substituting the value of B from (4.233) in (4.243),

$$B_0 = (1-\gamma)^{1/2}\cos(kl)e^{\left(-\frac{\rho}{2}L - j\beta L\right)}(1-\gamma)^{1/2}\left[-jA_0\sin(kl) + B_0\cos(kl)\right]$$

or

$$B_0 = (1-\gamma)e^{\left(-\frac{\rho}{2}L - j\beta L\right)}\cos(kl)\left[-jA_0\sin(kl) + B_0\cos(kl)\right]$$

or

$$B_0 = \frac{-jA_0(1-\gamma)e^{\left(-\frac{\rho}{2}L - j\beta L\right)}\cos(kl)\sin(kl)}{1-(1-\gamma)e^{\left(-\frac{\rho}{2}L - j\beta L\right)}\cos^2(kl)}. \qquad (4.244)$$

By substituting the value of B_0 from (4.244) in (4.239)

$$D_0 = -j(1-\gamma)e^{\left(-\frac{\rho}{4}L - j\frac{\beta}{2}L\right)}\sin(kl)\left[-jA_0\sin(kl) + \frac{-jA_0(1-\gamma)e^{\left(-\frac{\rho}{2}L - j\beta L\right)}\cos(kl)\sin(kl)}{1-(1-\gamma)e^{\left(-\frac{\rho}{2}L - j\beta L\right)}\cos^2(kl)}\cos(kl)\right]$$

or

$$\frac{D_0}{A_0} = -j(1-\gamma)e^{\left(-\frac{\rho}{4}L - j\frac{\beta}{2}L\right)}\sin(kl)\left[\frac{-j\sin(kl) + j(1-\gamma)e^{\left(-\frac{\rho}{2}L - j\beta L\right)}\cos^2(kl)\sin(kl) - j(1-\gamma)e^{\left(-\frac{\rho}{2}L - j\beta L\right)}\cos^2(kl)\sin(kl)}{1-(1-\gamma)e^{\left(-\frac{\rho}{2}L - j\beta L\right)}\cos^2(kl)}\right]$$

or

$$\frac{D_0}{A_0} = \frac{-(1-\gamma)e^{\left(-\frac{\rho}{4}L - j\frac{\beta}{2}L\right)}\sin^2(kl)}{1-(1-\gamma)e^{\left(-\frac{\rho}{2}L - j\beta L\right)}\cos^2(kl)}$$

or

$$\frac{D_0}{A_0}=\frac{-(1-\gamma)e^{\left(-\frac{\rho_L}{4}\right)}\sin^2(kl)\left[\cos\left(\beta L/2\right)-j\sin\left(\beta L/2\right)\right]}{1-(1-\gamma)e^{\left(-\frac{\rho_L}{2}\right)}\cos^2(kl)\cos(\beta L)+j\sin(\beta L)(1-\gamma)e^{\left(-\frac{\rho_L}{2}\right)}\cos^2(kl)} \tag{4.245}$$

We can assume $\begin{cases} x=\left(\sqrt{1-\gamma}\right)exp\left(-\rho\frac{L}{4}\right) \\ y=\cos(kl) \\ \phi=\beta L \end{cases}$

By substituting these values in (4.245), we get:

$$\frac{D_0}{A_0}=\frac{-\sqrt{(1-\gamma)}x\left(1-y^2\right)\left[\cos\left(\phi/2\right)-j\sin\left(\phi/2\right)\right]}{\left[1-x^2y^2\cos(\phi)\right]+j\left[x^2y^2\sin(\phi)\right]}$$

or intensity:

$$\left|\frac{D_0}{A_0}\right|^2=\frac{(1-\gamma)x^2\left(1-y^2\right)^2}{\left[1-x^2y^2\cos(\phi)\right]^2+\left[x^2y^2\sin(\phi)\right]^2}$$

or

$$\left|\frac{D_0}{A_0}\right|^2=(1-\gamma)\frac{x^2\left(1-y^2\right)^2}{1+x^4y^4\cos^2(\phi)-2x^2y^2\cos(\phi)+x^4y^4 sin^2(\phi)}$$

or

$$\left|\frac{D_0}{A_0}\right|^2=(1-\gamma)\frac{x^2\left(1-y^2\right)^2}{1+x^4y^4-2x^2y^2\cos(\phi)}$$

or

$$\left|\frac{D_0}{A_0}\right|^2=(1-\gamma)\frac{x^2\left(1-y^2\right)^2}{1+x^4y^4-2x^2y^2-2x^2y^2\cos(\phi)+2x^2y^2}$$

or

$$\left|\frac{D_0}{A_0}\right|^2=(1-\gamma)\frac{x^2\left(1-y^2\right)^2}{\left(1-x^2y^2\right)^2+2x^2y^2\left(1-\cos(\phi)\right)}$$

or

$$T = \left|\frac{D_0}{A_0}\right|^2 = (1-\gamma)\frac{x^2\left(1-y^2\right)^2}{\left(1-x^2y^2\right)^2 + 4x^2y^2 \sin^2\left(\frac{\phi}{2}\right)}. \tag{4.246}$$

Equation (4.246) shows the intensity transmittance of the open type of ring resonator. The dependence of intensity transmittance on optical phase ϕ is shown in Fig. (4.15).

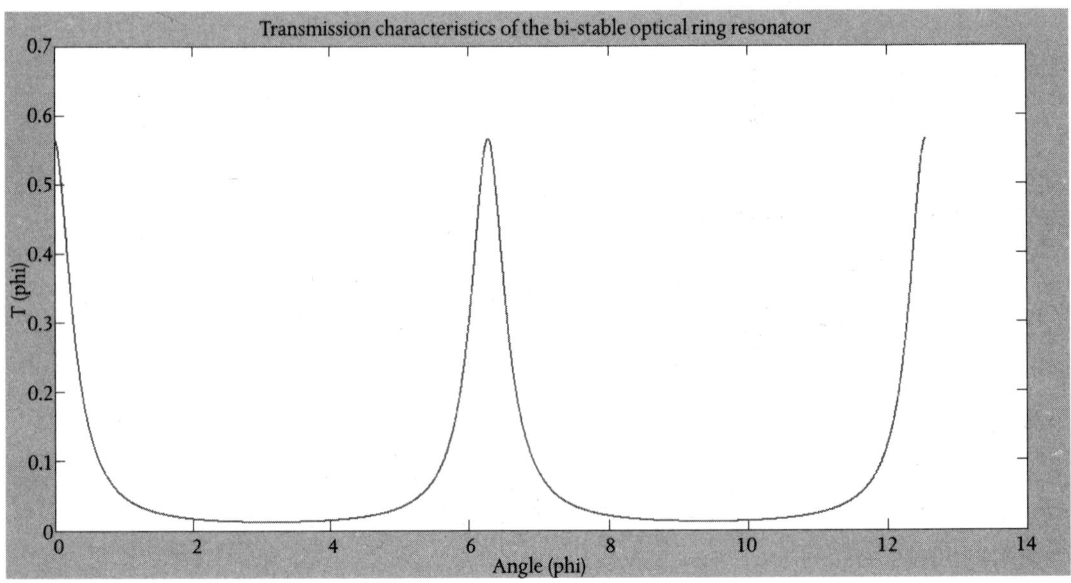

Fig. 4.15 The transmission characteristics curve for bi-stable ring resonator

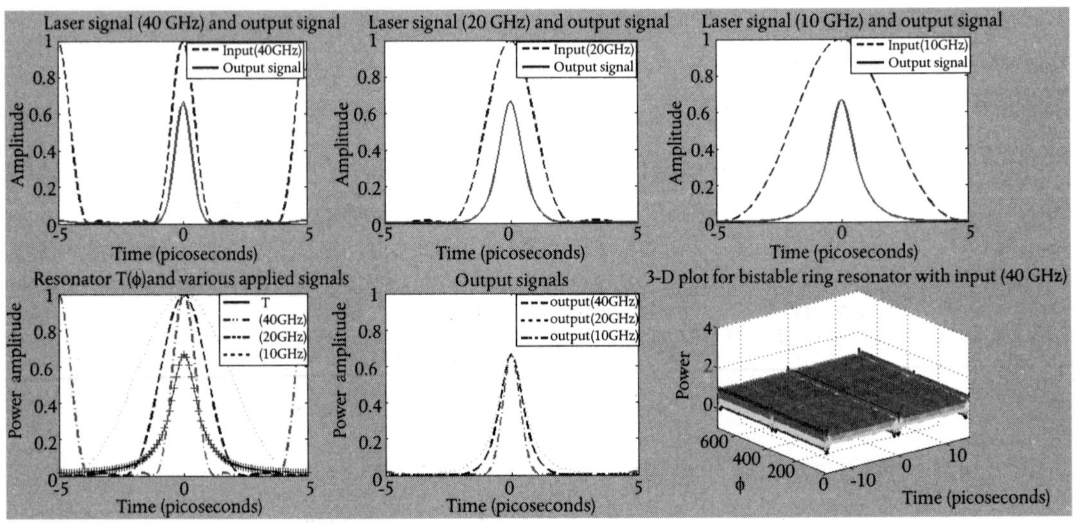

Fig. 4.16 The output waveform for bi-stable ring resonator when laser signals of different frequencies are allowed to pass through the resonator

Figure 4.16 shows the output waveform calculated from (4.246), when the input signal A_0 is assumed to be a picoseconds pulse generated by the circuit shown in Fig. (4.13) at a drive frequency of 10, 20, and 40 GHz. Apparently, in this case, the pulse shape remains preserved in entire range of input signal. From Fig. 4.16, it is also apparent that the input signal (A_0) pulse width, which is 1 ps at 40 GHz, can be further compressed up to femtoseconds pulse width while it is passing through bi-stable micro-ring resonator. The results may be further improved while replacing bi-stable micro-ring resonator discussed in this section with highly asymmetric Mach–Zehnder interferometer coupled micro-ring resonator.

4.6 DERIVATION OF COUPLING COEFFICIENTS FOR SLAB WAVEGUIDE

The mode-coupling coefficient is very important for the calculation of the coupling length in the co-directional coupler, the reflectivity in the contra-directional coupler, and so on. In this section, the derivation of the coupling coefficient for several typical optical waveguide couplers will be described.

4.6.1 Coupling Coefficient for Slab Waveguides

Here, we derive the coupling coefficient of the TE mode in the directional coupler consisting of symmetrical slab waveguides. The coupling coefficient $\kappa(=\kappa_{12}=\kappa_{21})$ is expressed from (4.133) as:

$$\kappa=\frac{\omega\varepsilon_0\int_{-\infty}^{+\infty}(N^2-N_2^2)E_1^*.E_2dx}{\int_{-\infty}^{+\infty}u_z.\left(E_1^*\times H_1+E_1\times H_1^*\right)dx}, \tag{4.247}$$

where N and N_2 denote the refractive index distribution in the entire coupled waveguide and refractive index distribution of waveguide 2, respectively. The definition of the other terms are all explained earlier. The electromagnetic field components of the TE mode is supposed to be expressed as $E_x=H_y=0$ and $H_x=-\left(\frac{\beta}{\omega\mu_0}\right)E_y$. Therefore, we have the following equalities:

$$u_z.\left(E_1^*\times H_1+E_1\times H_1^*\right)=\frac{\beta}{\omega\mu_0}\left|E_{1y}\right|^2, \tag{4.248}$$

$$E_1^*.E_2=E_{1y}^*.E_{2y}. \tag{4.249}$$

Since $(N^2-N_2^2)$ is zero outside of waveguide 1, the integration of (4.247) needs to be done only inside of waveguide 1, as shown in Fig. 4.17. The origin of the x-axis is taken at the centre of waveguide 1 and the core centre separation is obtained by D. The core width is an "a". Substituting (4.248) and (4.249) in (4.247), we obtain:

$$\kappa=\frac{\omega\varepsilon_0\left(n_1^2-n_0^2\right)\int_{-a}^{+a}E_{1y}^*.E_{2y}dx}{\frac{2\beta}{\omega\mu_0}\int_{-\infty}^{+\infty}\left|E_{1y}\right|^2dx}. \tag{4.250}$$

The electric field component in the slab waveguides are given as:

$$E_{1y} = \begin{cases} A\cos\left(\frac{u}{a}x\right) \ (|x| \le a) \\ A\cos(u)\exp\left[-\frac{w}{a}(|x|-a)\right], \ (|x| > a) \end{cases}. \tag{4.251}$$

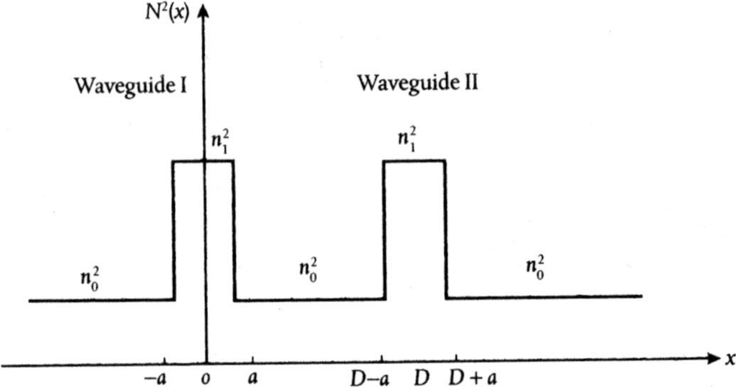

Fig. 4.17 Directional coupler consisting of slab optical waveguides

and

$$E_{2y} = A\cos(u)\exp\left[\frac{w}{a}(x-D+a)\right], (|x| \le a), \tag{4.252}$$

where $u = \kappa a = a\sqrt{k^2 n_1^2 - \beta^2}$ and $w = \sigma a = a\sqrt{\beta^2 - k^2 n_0^2}$. Here, E_{2y} is inside of core of waveguide 1.

Substituting (4.251) and (4.252) in (4.250) and utilising the eigenvalue equation for the TE mode of the form:

$$w = u\tan(u). \tag{4.253}$$

The denominator of (4.250) can be calculated as:

$$\int_{-\infty}^{+\infty} |E_{1y}|^2 dx = \int_{-a}^{+a} A^2 cos^2\left(\frac{u}{a}x\right)dx + \int_{a}^{+\infty} A^2 cos^2(u) exp\left[-\frac{2w}{a}(x-a)\right]dx + \int_{-\infty}^{-a} A^2 cos^2(u) exp\left[\frac{2w}{a}(x+a)\right]dx \tag{4.254}$$

or

$$\int_{-\infty}^{+\infty} |E_{1y}|^2 dx = A^2 a\left[1 + \frac{\sin u \cos u}{u} + cos^2(u)\left(\frac{a}{2w}\right) + cos^2(u)\left(\frac{a}{2w}\right)\right] = A^2 a\left(1 + \frac{sin^2 u}{w} + \frac{cos^2 u}{w}\right). \tag{4.255}$$

By using (4.253), it can be shown that:

$$\int_{-\infty}^{+\infty} |E_{1y}|^2 dx = A^2 a\left(\frac{1+w}{w}\right). \tag{4.256}$$

The numerator of (4.250) can be calculated as follows:

$$\int_{-a}^{+a} E_{1y}^* . E_{2y} dx = \int_{-a}^{+a} A^2 \cos\left(\frac{u}{a}x\right) \cos(u) \exp\left[\frac{w}{a}x\right] \exp\left[-\frac{w}{a}(D-a)\right]dx \tag{4.257}$$

or

$$=A^2\cos(u)\exp\left[-\frac{w}{a}(D-a)\right]\left[\int_0^{+a}\cos\left(\frac{u}{a}x\right)\exp\left[\frac{w}{a}x\right]dx+\int_{-a}^{0}\cos\left(\frac{u}{a}x\right)\exp\left[\frac{w}{a}x\right]dx\right] \tag{4.258}$$

or

$$=A^2\cos(u)\exp\left[-\frac{w}{a}(D-a)\right]\left[\frac{1}{2}\int_0^{+a}\left[\exp\left\{\frac{x}{a}(w+ju)\right\}+\exp\left\{\frac{x}{a}(w-ju)\right\}\right]dx+\right.$$
$$\left.\int_{-a}^{0}\frac{1}{2}\left[\exp\left\{\frac{x}{a}(w+ju)\right\}+\exp\left\{\frac{x}{a}(w-ju)\right\}\right]dx\right] \tag{4.259}$$

or

$$=\left(\frac{A^2}{2}\right)\cos(u)\exp\left[-\frac{w}{a}(D-a)\right]$$
$$\left[\frac{a}{(w+ju)}\left[\exp\{(w+ju)\}-1\right]+\frac{a}{(w-ju)}\left[\exp\{(w-ju)\}-1\right]+\right.$$
$$\left.\frac{a}{(w+ju)}\left[1-\exp\{-(w+ju)\}\right]+\frac{a}{(w-ju)}\left[1-\exp\{-(w-ju)\}\right]\right] \tag{4.260}$$

or

$$=\left(\frac{A^2}{2}\right)\cos(u)exp\left[-\frac{w}{a}(D-a)\right]\left[\frac{a}{(w+ju)}\left[\exp\{(w+ju)\}-\exp\{-(w+ju)\}\right]+\right.$$
$$\left.\frac{a}{(w-ju)}\left[\exp\{(w-ju)\}-\exp\{-(w-ju)\}\right]\right] \tag{4.261}$$

or

$$=\left(\frac{A^2}{2}\right)a\cos(u)exp\left[-\frac{w}{a}(D-a)\right]\left[\exp(w)\left[\frac{\exp(ju)}{(w+ju)}+\frac{\exp(-ju)}{(w-ju)}\right]-\exp(-w)\left[\frac{\exp(-ju)}{(w+ju)}+\frac{\exp(ju)}{(w-ju)}\right]\right] \tag{4.262}$$

$$=\left(\frac{A^2}{2}\right)aA^2\cos(u)exp\left[-\frac{w}{a}(D-2a)\right]\left(\frac{1}{v^2}\right)\left[\left[(w-ju)\exp(ju)+(w+ju)\exp(-ju)\right]-\right.$$
$$\left.\exp(-2w)\left[\left[(w-ju)\exp(-ju)+(w+ju)\exp(ju)\right]\right]\right] \tag{4.263}$$

$$=2awA^2\cos^2(u)exp\left[-\frac{w}{a}(D-2a)\right]\left(\frac{1}{v^2}\right)\Rightarrow 2awA^2\exp\left[-\frac{w}{a}(D-2a)\right]\left(\frac{u^2}{v^4}\right). \tag{4.264}$$

From (4.250), (4.256), and (4.264), it can be shown that:

$$\kappa=\frac{\omega\varepsilon_0\left(n_1^2-n_0^2\right)2awA^2\exp\left[-\frac{w}{a}(D-2a)\right]\left(\frac{u^2}{v^4}\right)}{\frac{2\beta}{\omega\mu_0}A^2a\left(\frac{1+w}{w}\right)} \tag{4.265}$$

$$\Rightarrow \kappa = \frac{k^2\left(n_1^2 - n_0^2\right)u^2 w^2 \exp\left[-\frac{w}{a}(D-2a)\right]}{\beta(1+w)v^4}. \quad (4.266)$$

Here, (4.266) can be rewritten further by using $\beta \approx kn_1$ and $\left(n_1^2 - n_0^2\right) = 2n_1^2\Delta$:

$$\kappa = \frac{\sqrt{2\Delta}\,u^2 w^2 exp\left[-\frac{w}{a}(D-2a)\right]}{a(1+w)v^3}. \quad (4.267)$$

For example, the mode-coupling coefficient of the directional coupler consisting of slab waveguides with $2a = 6\mu\text{m}$, $\Delta = 0.3\%, v = 1.5$, and the core centre separation of $D = 4a$ is calculated to be $\kappa = 0.39\text{mm}^{-1}$. The coupling length of this coupler is obtained as $L_c = 4mm$

4.6.2 Coupling Coefficient for Rectangular Waveguides

Here, we consider the directional coupler consisting of three-directional rectangular waveguides as shown in Fig. 4.18. When we deal with E_{11}^x mode, we have $H_x = 0$ from (2.259). Since, we can assume $|E_x| \gg |E_y|$ for the E_{11}^x mode in (4.133), the mode-coupling coefficient is expressed as:

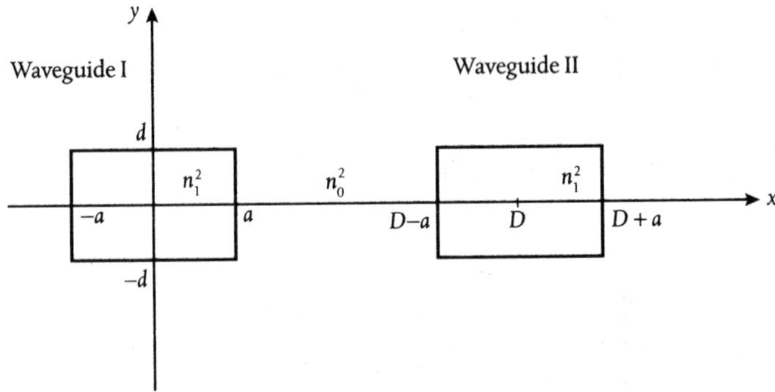

Fig. 4.18 Directional coupler consisting of three-dimensional rectangular waveguides

$$\kappa = \frac{\omega\varepsilon_0 \int_{-a}^{a}\int_{-d}^{+d}\left(n_1^2 - n_0^2\right)E_{1x}^{*}.E_{2x}dxdy}{2\int_{-\infty}^{\infty}\int_{-\infty}^{+\infty}E_{1x}^{*}.H_{1y}dx}. \quad (4.268)$$

Rewriting (2.296) in the form:

$$H_y = \begin{cases} A\cos(\kappa_x x)\cos(\kappa_y y) & -a \le x \le a, -d \le y \le d \\ A\cos(\kappa_x a)e^{-\gamma_x(|x|-a)}\cos(\kappa_y y) & |x| \ge a, -d \le y \le d \\ A\cos(\kappa_x x)e^{-\gamma_y(|y|-d)}\cos(\kappa_y d) & -a \le x \le a, |y| \ge d \end{cases}. \quad (4.269)$$

Hence, substituting (4.269) in (4.268) and using the relation:

$$E_x \approx \frac{\omega\mu_0}{\beta} H_y. \tag{4.270}$$

Then, denominator and numerator of (4.268) are calculated as follows:

$$\begin{aligned}
2\int_{-\infty}^{\infty}\int_{-\infty}^{\infty} E_{1x}^{*}.H_{1y}dxdy &\approx \frac{2\omega\mu_0}{\beta}\int_{-\infty}^{\infty}\int_{-\infty}^{\infty}\left|H_{1y}\right|^2 dxdy \\
&= \frac{2\omega\mu_0}{\beta}\left[\int_{-a}^{a} A^2\cos^2(\kappa_x x)dx + \int_{a}^{\infty} A^2\cos^2(\kappa_x a)e^{-2\gamma_x(x-a)}dx + \int_{-\infty}^{-a} A^2\cos^2(\kappa_x a)e^{2\gamma_x(x+a)}dx\right] \\
&\times\left[\int_{-d}^{d} A^2\cos^2(\kappa_y y)dy + \int_{d}^{\infty} A^2\cos^2(\kappa_y d)e^{-2\gamma_y(y-d)}dy + \int_{-\infty}^{-d} A^2\cos^2(\kappa_y d)e^{2\gamma_y(y+d)}dy\right] \\
&= \frac{2\omega\mu_0}{\beta}A^2\left(a+\frac{1}{\gamma_x}\right)\left(d+\frac{1}{\gamma_y}\right)
\end{aligned} \tag{4.271}$$

and

$$\omega\varepsilon_0\int_{-a}^{a}\int_{-d}^{+d}\left(n_1^2-n_0^2\right)E_{1x}^{*}.E_{2x}dxdy$$

$$\approx \frac{2\omega\mu_0}{\beta^2}A^2\left(d+\frac{1}{\gamma_y}\right)\left(\frac{\kappa_x^2\gamma_x a^2}{v^2}\right)exp\left[-\gamma_x(D-2a)\right]. \tag{4.272}$$

Here, normalised frequency is defined as $v = kn_1a\sqrt{2\Delta}$. The mode-coupling coefficient is then given as:

$$\kappa = \frac{\sqrt{2\Delta}(\kappa_x a)^2(\gamma_x a)^2 exp\left[-\gamma_x(D-2a)\right]}{a(1+\gamma_x a)v^3}. \tag{4.273}$$

Parameters κ_x and γ_x are obtained by solving eigenvalue equation (2.299). For example, we obtain $k_x a = 0.995$and $\gamma_x a = 1.523$ for the rectangular waveguide with $2a = 8\,\mu m$, $2d = 8\,\mu m$, $\Delta = 0.3\%$, and $v = 1.819$ at $\lambda = 1.55\,\mu m$. The mode-coupling coefficient of the directional coupler with the core centre separation of $D = 3a$ is calculated to be $\kappa = 0.638\,mm^{-1}$. The coupling length of this coupler is obtained as $L_c = \pi/2\kappa = 2.46\,mm$.

4.6.3 Derivation of Coupling Coefficient Based on Mode Interference

The mode-coupling effect in the directional coupler can be analysed by the interference phenomena between the even and the odd modes in the five-layer slab waveguide, as shown in Fig. 4.19. When we neglect the higher order modes, the electric field in the directional coupler can be approximated by the summation of the even mode (first order mode in the five-layer waveguide) and the odd mode (second-order mode in the five-layer waveguide):

$$E(x,z)=E_e(x)e^{-j\beta_e z}+E_o(x)e^{-j\beta_o z}, \tag{4.274}$$

where $E_e(x)$ and β_e denote electric field and propagation constant of the even mode and $E_o(x)$ and β_o denote those of the odd mode, respectively. The incident electric field, which is coupled to waveguide 1 at $z=0$, is expressed as:

$$|E(x,0)|=|E_e(x)+E_o(x)|=E_1(x). \tag{4.275}$$

Here, $E_1(x)$ denotes the eigenmode of waveguide 1. When the cores of waveguides 1 and 2 are not so close, this expression holds with good accuracy. The electric field amplitude at z is given, from (4.274), as:

$$|E(x,z)|=\left|E_e(x)+E_o(x)e^{[j(\beta_e-\beta_o)z]}\right|. \tag{4.276}$$

Electric field distribution at $z=\pi/(\beta_e-\beta_o)$ is then given as:

$$|E(x,z)|=|E_e(x)-E_o(x)|=E_2(x), \tag{4.277}$$

where $E_2(x)$ denotes the eigenmode of waveguide 2. Equation (4.277) means that the incident field coupled to waveguide 1 shifted to a waveguide 2 at the distance:

$$L_c=\frac{\pi}{\beta_e-\beta_o}. \tag{4.278}$$

L_c is a coupling length. Then, mode-coupling co-efficient is obtained as:

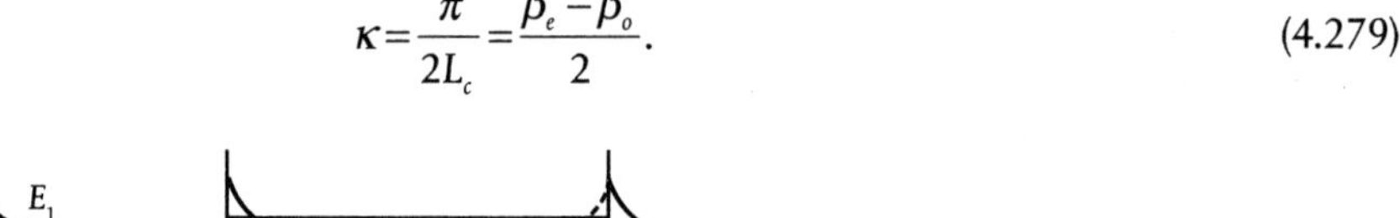

$$\kappa=\frac{\pi}{2L_c}=\frac{\beta_e-\beta_o}{2}. \tag{4.279}$$

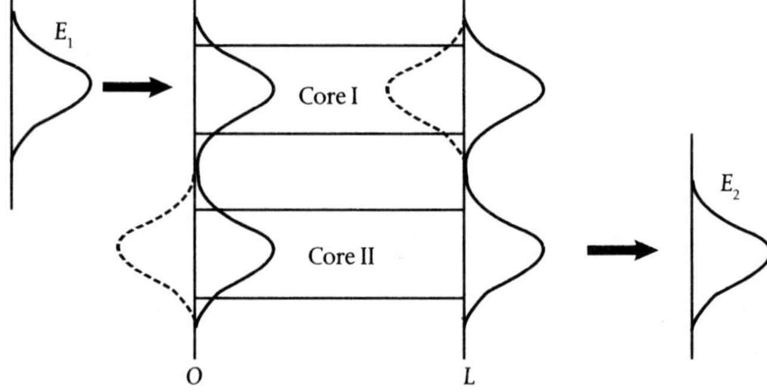

Fig. 4.19 Even (solid line) and odd (dotted line) modes in the five-layer slab waveguide

Let us calculate the mode-coupling co-efficient κ for the TE mode in the five-layer slab waveguide. The eigenvalue equation of the five-layer slab waveguide. The eigenvalue of the five-layer slab waveguide in the coordinate system shown in Fig. 4.20 are given as:

$$2u=\tan^{-1}\left(\frac{w}{u}\right)+\tan^{-1}\left\{\frac{w}{u}\tanh\left[\left(\frac{D}{2a}-1\right)w\right]\right\} \text{ (even mode)} \tag{4.280}$$

$$2u = \tan^{-1}\left(\frac{w}{u}\right) + \tan^{-1}\left\{\frac{w}{u}\coth\left[\left(\frac{D}{2a}-1\right)w\right]\right\} \text{ (odd mode)}, \tag{4.281}$$

where

$$u = a\sqrt{k^2 n_1^2 - \beta^2}, \tag{4.282}$$

$$w = a\sqrt{\beta^2 - k^2 n_0^2}. \tag{4.283}$$

Equation (4.280) has been derived in Section 4.6 of this chapter. When the core centre separation D approaches infinity $(D \to \infty)$, (4.280) and (4.281) both reduce to the eigenmode equation of the single-slab waveguide:

$$u_0 = \tan^{-1}\left(\frac{w_0}{u_0}\right). \tag{4.284}$$

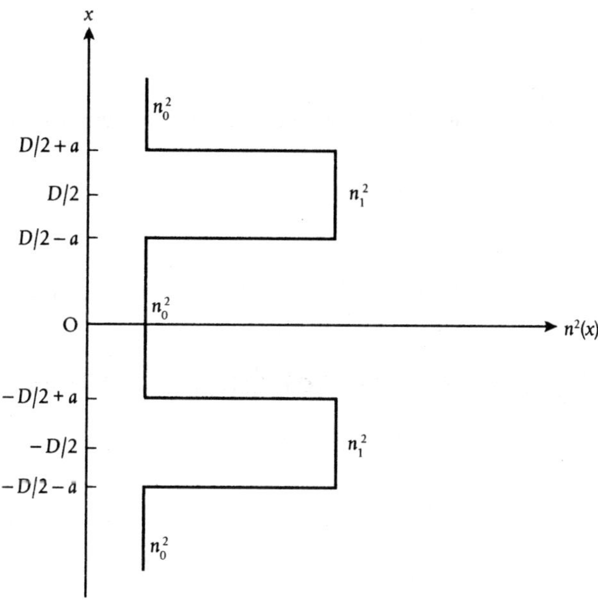

Fig. 4.20 Refractive index distribution of the five-layer slab waveguide

We denote the propagation constant derived by (4.284) as $\beta^{(0)}$. When the mode coupling between the two waveguides is weak, eigenmode (4.280) and (4.281) are considered to be a perturbation from (4.284). Then, the propagation constant for the even mode is expressed as:

$$\beta_e = \beta^{(0)} + \delta\beta_e. \tag{4.285}$$

Substituting (4.285) in (4.282), we obtain:

$$2u = 2a\sqrt{k^2 n_1^2 - \left(\beta^{(0)} + \delta\beta_e\right)^2} \cong 2u_0 - \frac{2\beta^{(0)} a^2}{u_0}\delta\beta_e. \tag{4.286}$$

In a similar manner, each term of the right-hand side of (4.280) can be approximated as:

$$\tan^{-1}\left(\frac{w}{u}\right) \cong u_o + \frac{\beta^{(0)}a^2}{u_0 w_0}\delta\beta_e \tag{4.287}$$

$$\tan^{-1}\left\{\frac{w}{u}\tanh\left[\left(\frac{D}{2a}-1\right)w\right]\right\} \cong -u_0 + \frac{\beta^{(0)}a^2}{u_0 w_0}\delta\beta_e - \frac{2u_0 w_0}{v_0^2}e^{\left[-2w_0\left(\frac{D}{2a}-1\right)\right]}, \tag{4.288}$$

where the approximation of $\tanh(z) \cong 1 - 2e^{-2z}$ has been used. Substituting (4.286) and (4.287)–(4.288) in (4.280), we obtain:

$$\frac{(1+w_0)}{u_0 w_0}\beta^{(0)}a^2\delta\beta_e = u_0 + \frac{u_0 w_0}{v_o^2}e^{\left[-\frac{w_0}{a}(D-2a)\right]}. \tag{4.289}$$

In a similar manner, the expression for $\delta\beta_0 = \beta_0 - \beta^{(0)}$ is obtained as:

$$\frac{(1+w_0)}{v_0 w_0}\beta^{(0)}a^2\delta\beta_0 = u_0 - \frac{u_0 w_0}{v_0^2}e^{\left[-\frac{w_0}{a}(D-2a)\right]}. \tag{4.290}$$

Substituting (4.289) and (4.290) into (4.279), the mode-coupling coefficient κ is obtained as:

$$\kappa = \frac{(\delta\beta_e - \delta\beta_0)}{2} = \frac{u_0^2 w_0^2}{\beta^{(0)}a^2(1+w_0)v_0^2}e^{\left[-\frac{w_0}{a}(D-2a)\right]}. \tag{4.291}$$

This equation reduces to (4.267) when we substitute $\beta^{(0)} \cong kn_1$ and $v_0^2 = k^2a^2\left(n_1^2 - n_0^2\right)$ in it.

4.6.4 Coupling Coefficients for Optical Fibres

We rewrite the mode-coupling coefficient in (4.247) as:

$$\kappa = \frac{\omega\varepsilon_0 \int_{-\infty}^{\infty}\int_{-\infty}^{\infty}\left(N^2 - N_2^2\right)E_1^* E_2\,dxdy}{\int_{-\infty}^{\infty}\int_{-\infty}^{\infty} u_z\left(E_1^* \times H_1 + E_1 \times H_1^*\right)dxdy}. \tag{4.292}$$

Substituting the relation of $s = s_1 = s_2 = -1$ and replacing A by $C = -j(\beta a/u)A$ in (4.100)–(4.105) and (4.106)–(4.111), electromagnetic fields of the fundamental HE_{11} mode are expressed as:

1. **Core region ($0 \le r \le a$):**

$$E_x = CJ_0\left(\frac{u}{a}r\right)\cos\psi, \tag{4.293}$$

$$E_y = -CJ_0\left(\frac{u}{a}r\right)\sin\psi, \tag{4.294}$$

$$E_z = j\frac{u}{\beta a}CJ_1\left(\frac{u}{a}r\right)\cos(\theta + \psi), \tag{4.295}$$

$$H_x = \frac{\omega\varepsilon_0 n_1^2}{\beta} CJ_0\left(\frac{u}{a}r\right)\sin\psi \tag{4.296}$$

$$H_y = \frac{\omega\varepsilon_0 n_1^2}{\beta} CJ_0\left(\frac{u}{a}r\right)\cos\psi \tag{4.297}$$

$$H_z = j\frac{u}{\omega\varepsilon_0 a} CJ_1\left(\frac{u}{a}r\right)\sin(\theta+\phi) \tag{4.298}$$

2. Cladding region ($r > a$)

$$E_x = C\frac{J_0(u)}{K_0(w)}K_0\left(\frac{w}{a}r\right)\cos\psi \tag{4.299}$$

$$E_y = -C\frac{J_0(u)}{K_0(w)}K_0\left(\frac{w}{a}r\right)\sin\psi, \tag{4.300}$$

$$E_z = j\frac{u}{\beta a}C\frac{J_1(u)}{K_1(w)}K_1\left(\frac{w}{a}r\right)\cos(\theta+\psi) \tag{4.301}$$

$$H_x = \frac{\omega\varepsilon_0 n_0^2}{\beta}C\frac{J_0(u)}{K_0(w)}K_0\left(\frac{w}{a}r\right)\sin\psi \tag{4.302}$$

$$H_y = \frac{\omega\varepsilon_0 n_0^2}{\beta}C\frac{J_0(u)}{K_0(w)}K_0\left(\frac{w}{a}r\right)\cos\psi \tag{4.303}$$

$$H_z = j\frac{u}{\omega\varepsilon_0 a}C\frac{J_1(u)}{K_1(w)}K_1\left(\frac{w}{a}r\right)\sin(\theta+\phi) \tag{4.304}$$

Constant C is related to optical power P by (4.119):

$$|C| = \frac{w}{avJ_1(u)}\sqrt{\frac{2P\sqrt{\mu_0/\varepsilon_0}}{\pi n_1}}. \tag{4.305}$$

where light velocity c was replaced by $1/\sqrt{\mu_0\varepsilon_0}$. The denominator of (4.292) is normalised as:

$$\int_{-\infty}^{\infty}\int_{-\infty}^{\infty} u_z.\left(E_1^* \times H_1 + E_1 \times H_1^*\right)dxdy = 4P. \tag{4.306}$$

Next, we consider the integral of the numerator of (4.292), as shown in Fig. 4.21, $N^2(r,\theta) - N_2^2(r,\theta)$ is zero outside of core 1. Noting that the electric field of the core is used for E_1 and the field of the cladding is used for E_2 for the integral inside core 1, and $E_1^*.E_2$ expressed as:

$$E_1^*.E_2 = E_{x1}^*E_{x2} + E_{y1}^*E_{y2} + E_{z1}^*E_{z2}$$

$$=|C|^2 \frac{J_0(u)}{K_0(w)} J_0\left(\frac{u}{a}r\right) K_0\left(\frac{w}{a}r\right) + \left(\frac{u}{\beta a}\right)^2 |C|^2 \frac{J_1(u)}{K_1(w)} J_1\left(\frac{u}{a}r\right) K_1\left(\frac{w}{a}r\right) \cos(\theta+\psi)\cos(\Theta+\psi), \quad (4.307)$$

where (R,Θ) is a coordinate system having the origin at the centre of core 2 as shown in Fig. 4.22. The separation of the two-core centre is denoted by D. When $D \gg r$ holds, radius R can be approximated as:

$$R = \left(D^2 + r^2 - 2Dr\cos\theta\right)^{1/2} \cong D - r\cos\theta. \quad (4.308)$$

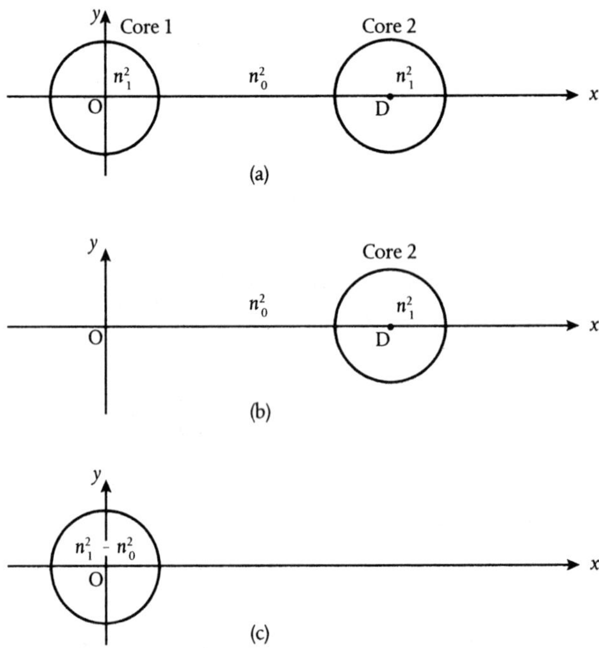

Fig. 4.21 Geometries for the calculation of the mode-coupling coefficient. Here, (a), (b), and (c) represent $N^2(r,\theta)$, $N_2^2(r,\theta)$, and $N^2(r,\theta) - N_2^2(r,\theta)$, respectively.

Since the second term of the right-hand side of (4.307) sufficiently smaller than the first term, the integration of the numerator in (4.292) becomes:

$$S = \int_{-\infty}^{\infty}\int_{-\infty}^{\infty} \left(N^2 - N_2^2\right) E_1^* E_2 \, dxdy$$

$$= \int_0^{2\pi}\int_0^{a} \left(n_1^2 - n_0^2\right) |C|^2 \frac{J_0(u)}{K_0(w)} J_0\left(\frac{u}{a}r\right) K_0\left(\frac{w}{a}R\right) r\,dr\,d\theta. \quad (4.309)$$

When the argument of the modified Bessel function $K_0(z)$ in (4.309) is large, it can be approximated as:

$$K_0(z) \cong \sqrt{\frac{\pi}{2z}} e^{(-z)}. \quad (4.310)$$

Substitution of (4.310) in (4.309) gives:

$$S=\left(n_1^2-n_0^2\right)|C|^2\frac{J_0(u)}{K_0(w)}\sqrt{\frac{\pi a}{2wD}}e^{\left(-\frac{w}{a}D\right)}\int_0^{2\pi}\int_0^{a}J_0\left(\frac{u}{a}r\right)e^{\left(\frac{w}{a}r\cos\theta\right)}rdrd\theta. \tag{4.311}$$

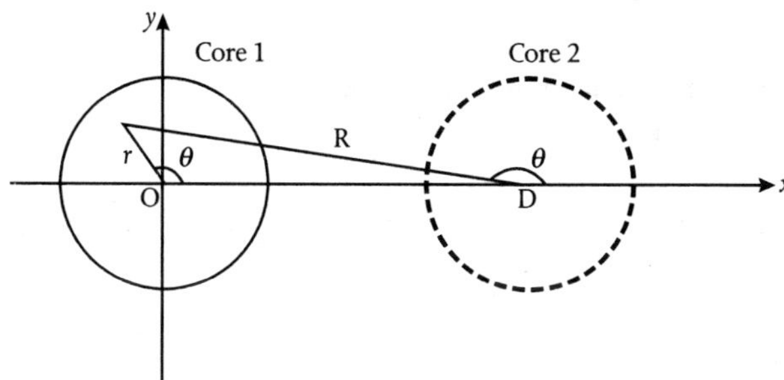

Fig. 4.22 Coordinate system for the integration of $E_1^*.E_2$.

Equation (4.311) is further rewritten by using the integral formulas of the Bessel functions.

$$I_0(z)=\frac{1}{\pi}\int_0^{\pi}e^{(z\cos\theta)}d\theta, \tag{4.312}$$

$$\int_0^1 J_0(uz)I_0(wz)zdz=\frac{J_0(u)wI_1(w)+I_0(w)uJ_1(u)}{u^2+w^2} \tag{4.313}$$

$$\begin{aligned}S=2\pi a^2\left(n_1^2-n_0^2\right)|C|^2\frac{J_0(u)}{K_0(w)}\sqrt{\frac{\pi a}{2wD}}e^{\left(-\frac{w}{a}D\right)}\times\frac{uJ_1(u)}{v^2K_1(w)}\\ \times\left[\frac{J_0(u)}{uJ_1(u)}wK_1(w)I_1(w)+K_1(w)I_0(w)\right].\end{aligned} \tag{4.314}$$

If we use the eigenvalue equation of the HE_{11} mode and the formula of the modified Bessel function

$$K_0(w)I_1(w)+K_1(w)I_0(w)=\frac{1}{w}, \tag{4.315}$$

then the mode-coupling coefficient of the optical fibre directional coupler is given as:

$$\kappa=\frac{\sqrt{\Delta}}{a}\frac{u^2}{v^3K_1^2(w)}\sqrt{\frac{\pi a}{wD}}e^{\left(-\frac{w}{a}D\right)}. \tag{4.316}$$

SUMMARY

- Silica-based optical fibres are the most important transmission medium for long-distance and large-capacity optical communication systems.
- The lowest transmission loss ever achieved is 0.154 dB/km at $\lambda=1.55$ μm wavelength. This means that the signal intensity of light becomes half of the original strength after propagating 20 km along the optical fibre.

- Maxwell's equations represent one of the most elegant and concise ways to state the fundamentals of electricity and magnetism. Maxwell's equations are a set of partial differential equations that, together with the Lorentz force law, form the following equations.
Gauss' law of electricity: $\nabla . D = \rho$
$D = \varepsilon_0 E + P$ General case
$D = \varepsilon_0 E$ Free space
Gauss' law of magnetism: $\nabla . B = 0$
Faraday's law of electromagnetic induction: $\nabla \times E = -\frac{\partial B}{\partial t}$
Ampere's law: $\nabla \times H = J + \frac{\partial D}{\partial t}$
- The electromagnetic fields in optical fibres are expressed in cylindrical coordinates as:

$$\tilde{E} = E(r, \theta) e^{j(\omega t - \beta z)}$$

$$\tilde{H} = H(r, \theta) e^{j(\omega t - \beta z)}$$

- Substituting the above-mentioned two equations in Maxwell's equations, we can get two sets of wave equations:

$$\frac{\partial^2 E_z}{\partial r^2} + \frac{1}{r}\frac{\partial E_z}{\partial r} + \frac{1}{r^2}\frac{\partial^2 E_z}{\partial \theta^2} + \left[k^2 n(r,\theta)^2 - \beta^2\right] E_z = 0$$

$$\frac{\partial^2 H_z}{\partial r^2} + \frac{1}{r}\frac{\partial H_z}{\partial r} + \frac{1}{r^2}\frac{\partial^2 H_z}{\partial \theta^2} + \left[k^2 n(r,\theta)^2 - \beta^2\right] H_z = 0$$

- In axially symmetric optical fibres, the refractive index distribution is not dependent on θ and is expressed by $n(r)$. Then, the transverse electromagnetic fields are related E_z and H_z as follows:

$$E_r = -\frac{j}{\left[k^2 n(r)^2 - \beta^2\right]}\left(\frac{\beta}{r}\frac{\partial E_z}{\partial r} + \omega\mu_0 \frac{\partial H_z}{\partial \theta}\right)$$

$$E_\theta = -\frac{j}{\left[k^2 n(r)^2 - \beta^2\right]}\left(\beta\frac{\partial E_z}{\partial \theta} - \frac{\omega\mu_0}{r}\frac{\partial H_z}{\partial r}\right)$$

$$H_r = -\frac{j}{\left[k^2 n(r)^2 - \beta^2\right]}\left(\beta\frac{\partial H_z}{\partial r} - \frac{\omega\varepsilon_0 n(r)^2}{r}\frac{\partial E_z}{\partial \theta}\right)$$

$$H_\theta = -\frac{j}{\left[k^2 n(r)^2 - \beta^2\right]}\left(\frac{\beta}{r}\frac{\partial H_z}{\partial \theta} + \omega\varepsilon_0 n(r)^2 \frac{\partial E_z}{\partial r}\right)$$

- Modes are different energy components (or rays) of light traveling down an optical fibre waveguide.
- *Transverse Electric (TE) modes*—no electric field in the direction of propagation.
- *Transverse Magnetic (TM) mode*—no magnetic field in the direction of propagation.
- *Transverse Electromagnetic (TEM) modes*—neither electric nor magnetic field in the direction of propagation.
- *Hybrid modes*—nonzero electric and magnetic fields in the direction of propagation.

➢ **For TE modes, substitute $E_z = 0$:**

We get $\dfrac{\partial^2 H_z}{\partial r^2} + \dfrac{1}{r}\dfrac{\partial H_z}{\partial r} + \left[k^2 n(r)^2 - \beta^2 - \dfrac{n^2}{r^2}\right] H_z = 0$

$$E_r = -\frac{j\omega\mu_0}{\left[k^2 n(r)^2 - \beta^2\right]}\frac{1}{r}\frac{\partial H_z}{\partial r}$$

$$E_\theta = \frac{j\omega\mu_0}{\left[k^2 n(r)^2 - \beta^2\right]}\frac{\partial H_z}{\partial \theta}$$

$$H_r = -\frac{j\beta}{\left[k^2 n(r)^2 - \beta^2\right]}\frac{\partial H_z}{\partial r}$$

$$H_\theta = -\frac{j\beta}{\left[k^2 n(r)^2 - \beta^2\right]}\frac{1}{r}\frac{\partial H_z}{\partial \theta}$$

➢ The magnetic field in the core and cladding are expressed as:

$$H_z = \begin{Bmatrix} g(r) \\ h(r) \end{Bmatrix} \cos(n\theta + \psi) \begin{matrix} (0 \le r \le a) \\ (r > a) \end{matrix}.$$

Fields in the core $(0 \le r \le a)$

$$E_\theta = -j\omega\mu_0 \frac{a}{u} A J_1\left(\frac{u}{a} r\right)$$

$$H_r = j\beta \frac{a}{u} A J_1\left(\frac{u}{a} r\right)$$

$$H_z = A J_0\left(\frac{u}{a} r\right)$$

Fields in the cladding $(r > a)$:

$$E_\theta = j\omega\mu_0 \frac{a}{w}\frac{J_0(u)}{K_0(w)} A K_1\left(\frac{w}{a} r\right)$$

$$H_r = -j\beta \frac{a}{w}\frac{J_0(u)}{K_0(w)} A K_1\left(\frac{w}{a} r\right)$$

$$H_z = \frac{J_0(u)}{K_0(w)} A K_0\left(\frac{w}{a} r\right)$$

➢ Substituting $H_z = 0$ in the wave equation, we can get the following equation:

$$\frac{d^2 E_z}{dr^2} + \frac{1}{r}\frac{dE_z}{dr} + \left[k^2 n(r)^2 - \beta^2\right] E_z = 0$$

$$E_r = \frac{-j\beta}{\left[k^2 n(r)^2 - \beta^2\right]}\frac{dE_z}{dr},$$

$$H_\theta = \frac{-j\omega\varepsilon_0 n^2}{[k^2 n(r)^2 - \beta^2]}\frac{dE_z}{dr},$$

$$E_\theta = H_r = 0$$

Field in the core ($0 \le r \le a$)

$$E_r = j\beta\frac{a}{u}AJ_1\left(\frac{u}{a}r\right)$$

$$E_z = AJ_0\left(\frac{u}{a}r\right)$$

$$H_\theta = j\omega\varepsilon_0\frac{a}{u}nr_1^2 AJ_1\left(\frac{u}{a}r\right)$$

- In hybrid modes, the axial electromagnetic field components E_z and H_z are not zero. Therefore, solutions for wave equation are given by the product of the nth– order Bessel functions and $\cos(n\theta+\psi)$ or $\sin(n\theta+\psi)$. E_z and H_z should be continuous at $r=a$.
- The fundamental mode of the optical fibre is the HE_{11} mode. Therefore, it is indispensible to understand the propagation characteristics of the HE_{11} mode in order to grasp the signal transmission properties of single-mode fibres.
- Coupled-mode theory deals with the mutual light wave interaction between the two propagation modes.

  The optical power carried by the eigenmode in the waveguide p, $(p=1,2)$, is expressed as:

$$P_p = \frac{1}{2}\int_{-\infty}^{\infty}\int_{-\infty}^{\infty}\left(E_p \times H^*_p\right).u_z dxdy \quad (p=1,2)$$

- The solution for the co-directional coupled mode equation:

$$A(z) = \left\{\left[\cos(Qz) + j\frac{\hat{\delta}}{Q}\sin(Qz)\right]A(0) - j\frac{\chi_a}{Q}\sin(Qz)B(0)\right\}\exp\left[-j\left(\hat{\delta} - \alpha_a\right)z\right]$$

- The pulse generation circuit consists of the numbers of DFB lasers.
- The light wave output from the master DFB laser is expressed as:

$$a(t) = a_0 e^{j2\pi f_c t}.$$

- Output light wave from the kth LD is expressed as:

$$a_k(t) = ak_0 e^{j[(\omega_c + kn\omega_m)t + \varphi_k]}$$

  Normalised intensity can be given as:

$$\frac{|Y(t)|^2}{|Y(0)|^2} = \frac{1}{25}\left(9 + 12\cos(n\omega_m t) + 4\cos(2n\omega_m t)\right)$$

- Generation of high repetition rate and short optical pulse is an important topic of research at the present time. This is because in order to realise a high-capacity optical communication system, we need a combination of wavelength-division multiplexing (WDM) and optical time-division multiplexing (OTDM) techniques.
- The number of channels that can be transmitted over the same fibre is limited by fibre dispersion.
- The output of the five-injection locked LDs be expressed in complex representation as:

$$a_k(t)=a_k e^{j[(\omega_c+kn\omega_m)t+\theta_k]}$$

where $k=-1,-2,0,1,2$ for five-laser system.

- The output intensity $I(t)$ is proportional to $|Y(t)|^2$. On carrying out mathematical calculations, we get:

$$|Y(t)|^2=\sum_{k=-2}^{2}\sum_{p=-2}^{2}a_k a_p \cos\left[(k-p)n\omega_m t+\theta_k-\theta_p\right].$$

- Intensity of the generated light pulse is given as:

$$\frac{I(t)}{I(0)}=\frac{|Y(t)|^2}{|Y(0)|^2}.$$

- For the five-laser system, the normalised intensity ratio can be represented as follows:

$$\frac{I(t)}{I(0)}=\frac{1}{16}\left[3.5+6\cos(n\omega_m t)+4\cos(2n\omega_m t)+2\cos(3n\omega_m t)+0.5\cos(4n\omega_m t)\right].$$

- Directional couplers are very important component in the fabrication of various optical devices. We consider the case in which light is coupled into the upper of a Mach–Zehnder interferometer.
- The output equation can be represented as follows:

$$\begin{aligned}A_1&=A_0\cos(\kappa l)\\ B_1&=-jA_0\sin(\kappa l)\end{aligned}$$

- The steady-state input–output relation of the ring resonator:

$$\begin{aligned}A&=(1-\gamma)^{\frac{1}{2}}\left[A_0\cos(\kappa l)-jB_0\sin(\kappa l)\right]\\ B&=(1-\gamma)^{\frac{1}{2}}\left[-jA_0\sin(\kappa l)+B_0\cos(\kappa l)\right]\end{aligned}.$$

- The amplitude transmittance of the optical ring resonator is given as:

$$\frac{A}{A_0}=(1-\gamma)^{\frac{1}{2}}\left[\frac{\cos(\kappa l)-(1-\gamma)^{\frac{1}{2}}\exp\left(-\frac{\rho}{2}L-j\beta L\right)}{1-(1-\gamma)^{\frac{1}{2}}\cos(\kappa l)\exp\left(-\frac{\rho}{2}L-j\beta L\right)}\right].$$

- The maximum and minimum transmittances are given as:

$$T_{max}=(1-\gamma)\frac{(x+y)^2}{(1+xy)^2}$$

$$T_{min}=(1-\gamma)\frac{(x-y)^2}{(1-xy)^2},$$

where

$$x=(1-\gamma)^{\frac{1}{2}}\exp\left(-\frac{\rho}{2}L\right)$$

$$y=\cos(\kappa l).$$

$$\phi=\beta L$$

- Bi-stable ring resonator is just the extension of the general ring resonator. It mainly consists of the two inputs and two outputs.
- The intensity ratio can be represented as follows:

$$\left|\frac{D_0}{A_0}\right|^2=(1-\gamma)\frac{x^2(1-y^2)^2}{(1-x^2y^2)^2+4x^2y^2\sin^2\left(\phi/2\right)}.$$

- The mode-coupling coefficient is very important for the calculation of the coupling length in the co-directional coupler, the reflectivity in the contra-directional coupler.
- The coupling coefficient $\kappa(=\kappa_{12}=\kappa_{21})$ is expressed as:

$$\kappa=\frac{\omega\varepsilon_0\int_{-\infty}^{+\infty}(N^2-N_2^2)E_1^*.E_2dx}{\int_{-\infty}^{+\infty}u_z.(E_1^*\times H_1+E_1\times H_1^*)dx},$$

where N and N_2 denote the refractive index distribution in the entire coupled waveguide and refractive index distribution of waveguide 2, respectively.
- The electric field component in the slab waveguides are given as:

$$E_{1y}=\begin{cases}A\cos\left(\frac{u}{a}x\right) & (|x|\le a)\\ A\cos(u)\exp\left[-\frac{w}{a}(|x|-a)\right], & (|x|>a)\end{cases}$$

$$E_{2y}=A\cos(u)\exp\left[\frac{w}{a}(x-D+a)\right],\quad (|x|\le a)$$

- $$\kappa=\frac{\sqrt{2\Delta}u^2w^2\exp\left[-\frac{w}{a}(D-2a)\right]}{a(1+w)v^3}.$$

For example, the mode-coupling coefficient of the directional coupler consisting of slab waveguides with $2a=6\mu m, \Delta=0.3\%, v=1.5$, and the core centre separation of $D=4a$ is calculated to be $\kappa=0.39mm^{-1}$. The coupling length of this coupler is obtained as $L_c=4mm$.
- Dispersion is the phenomenon in which the phase velocity of a wave depends on its frequency. Any source of light would have a certain spectral width and each spectral component would

travel with different group velocity and would always have dispersion, this is referred as material dispersion.

- The material dispersion can be specified in terms of the picoseconds per kilometre (length) per nanometre (spectral width).

$$D_m = \frac{\Delta\tau}{L\Delta\lambda_0} = -\frac{1}{\lambda_0 c}\left(\lambda_0^{\,2}\frac{d^2n}{d\lambda_0^{\,2}}\right)\times 10^3\left(\frac{ps}{km.nm}\right).$$

THEORY QUESTIONS AND ANSWERS

Question 1: Explain the characteristic parameters of single-mode fibre.

Answer: There are following types of parameter in single-mode fibre;

1. Mode field diameter:

In multimode fibres, the numerical aperture and core diameters are considered important from the point of view of predicting the performance of these fibres. However, in single-mode fibres, the radial distribution of the optical power in propagating fundamental mode plays an important role. Therefore, the mode field diameter (MFD) of the propagating mode constitutes a fundamental parameter characteristic of a single-mode fibre. This is also known as mode spot size. It may be defined as follows. For step index and graded index (parabolic profile) single-mode fibres operating near the cut-off wavelength λ_c, the field distribution of the fundamental mode is approximately Gaussian and may be expressed by the following equation:

$$\psi(r) = \psi_0 exp\left(-r^2/w^2\right), \tag{P1}$$

where $\psi(r)$ is the electric or magnetic field at the radius r, ψ_0 is the axial field (at $r=0$), and w is the mode field radius, which is the radial distance from the axis at which ψ_0 drops to ψ_0/e. Thus, the MFD is $2w$. As the power is proportional to ψ^2, the MFD may also be defined as the radial distance between the $1/e^2$ power points (in the power versus radius graph).

If the power per unit area at a radial distance r in the LP_{01} mode is $P(r)$, the power in the annular ring between r and $(r+dr)$ will be $2\pi rP(r)dr$. Therefore, the MFD may be defined as:

$$\text{MFD} = 2w = 2\left[\frac{\int_0^\infty r^2 P(r)2\pi dr}{\int_0^\infty P(r)2\pi r dr}\right]^{1/2}$$

$$\Rightarrow \text{MFD} = 2w = 2\left|\frac{\int_0^\infty r^2\psi^2(r)dr}{\int_0^\infty r\psi^2 dr}\right|^{1/2}. \tag{P2}$$

Here, $\psi(r)$ is the field distribution in the LP_{01} mode. For a step index fibre, the mode field radius w can be approximately given by the following expression:

$$w \approx a\left[0.65 + \frac{1.619}{V^{3/2}} + \frac{2.879}{V^6}\right], \tag{P3}$$

where a is the core radius. This formula gives the value of w to within about 1% for V-values nearly in the range 0.8–2.5.

Figure P1 shows the plot of normalised spot size (w/a) as a function of the V-parameter calculated using (P3). It is quite clear that for a fibre of given radius, the normalised spot size increases as V becomes smaller (or as λ becomes longer). However, as the wavelength increases, the modal field is less well confined within the core. Therefore, single-mode fibres are so designed that the cut-off wavelength λ_c is not too far from the wavelength for which the fibre is designed.

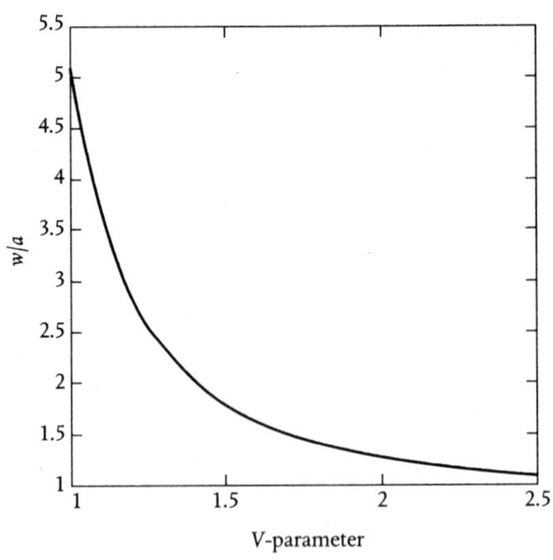

Fig. P1 Normalised spot size (*w*/*a*) as a function of V

There is another definition of MFD that is more complex and is called the Peterman-2 spot size. This is defined as:

$$(\text{MFD})_P = 2w_P = 2\left[\frac{2\int_0^{\infty} r\psi^2(r)dr}{\int_0^{\infty} r(d\psi/dr)^2\, dr}\right]^{1/2}, \qquad \text{(P4)}$$

where w_P is called the Peterman-2 mode radius. There is an empirical relation for w_P, which is accurate to within about 1% for a step index fibre with V-values between 1.5 and 2.5, and it is given as:

$$w_P \approx w - a\left[0.016 + \frac{1.567}{V^7}\right], \qquad \text{(P5)}$$

where w is mode field radius given in (P3). This definition has the advantage that the fibre dispersion and loss caused by the offset at a joint can be related to w_P. It can also be related to the far field pattern of the single-mode fibre.

2. Fibre birefringence

Single-mode fibre actually supports two orthogonal linearly polarised modes that are degenerated. In a single-mode fibre of a perfectly cylindrical core of uniform diameter, these modes have the same mode index n, and hence, the same propagation constant β. Therefore, they travel with the same velocity. However, real fibres are not perfectly cylindrical. They exhibit variation in the diameter due to the presence of non-uniform stress, bends, twists, and so on along the length of the fibre. Hence, the propagation constants of the two polarisation components become different and the fibre becomes birefringent. Modal birefringence is defined as:

$$\delta n = \left| n_x - n_y \right|, \tag{P6}$$

Let us assume arbitrarily that the two modes are polarised in the x and y direction and that n_x and n_y are the mode indices for the two polarisation components. The corresponding change in the propagation constants may be given as:

$$\delta\beta = \frac{2\pi}{\lambda}\delta n, \tag{P7}$$

where λ is the wavelength of light in vacuum. This leads to a periodic exchange of modal power between the two components, which can be understood in a simple manner as follows.

Let us suppose linearly polarised light is launched into a single-mode fibre such that both the modes are excited. We assume that the two polarisation components have the same amplitude and there is no phase difference at the launching end. However, as the light propagates down the fibre length, one mode goes out of phase with respect to the other due to different phase propagation constants. Thus, at any point along the fibre length (for a random phase difference), the two components will produce elliptically polarised light. At a phase difference of $\pi/2$, circularly polarised light will be produced. In this way, the polarisation progresses from linear to elliptical to circular to elliptical and back to linear. This sequence of alternating states of polarisation continues along the fibre length.

$$L_P = \frac{2\pi}{\delta\beta}, \tag{P8}$$

In conventional single-mode fibres, birefringence varies along the length of the fibre randomly in both magnitude as well as direction. Therefore, linearly polarised light launched in such a fibre quickly attains an arbitrary state of polarisation. The dispersion arises because of fibre birefringence.

Question 2: Discuss the coupling of modes between waveguides.

Answer: There can be significant loss in optical connections due to misalignment or mismatch of the modes between the two devices. Misalignment between a source and a single-mode waveguide by dimensions of less than 1 μm can cause coupling loss exceeding 1 dB. Coupling problems are exaggerated by the small dimensions of typical waveguides and sources, which make alignment a critical and challenging task. In this question, we will establish the fundamental rules for coupling optical power between two waveguides and between a source and a waveguide. The coupling techniques are based on the concepts on coupled-mode theory. Calculating the coupling between two optical waveguides is based on a modal description of the waveguides and it also depends on alignment, dimensional differences, and geometric shape. Following this section, one should be able to calculate the coupling efficiency between any two waveguides. Large core waveguides that have many modes are not well served by this theory and require a different approach.

Consider the problem of coupling two single-mode slab waveguides. Figure P2 shows the electric field distributions of the source and input waveguides. To efficiently excite the TE_0 mode of the input waveguide, the incident field should spatially overlap the mode profile in the waveguide as closely as possible. Any deviation between the input field and the guided mode shape will simply excite other waveguides and radiation modes.

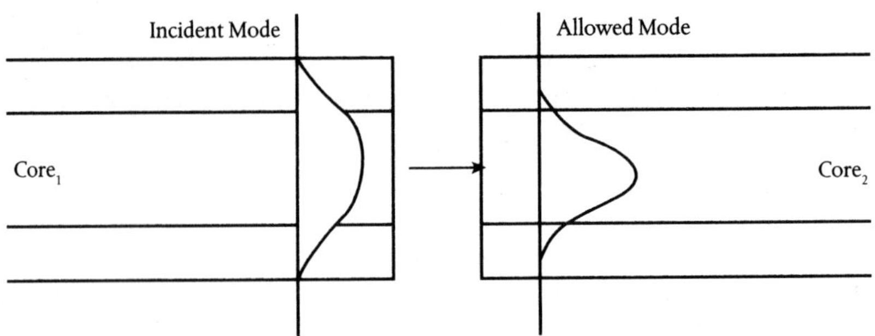

Fig. P2 An incident mode can have a different spatial profile than the field in the input waveguide. The overlap of the two modes determines the degree of coupling between the input and the guided mode

To formally calculate the coupling efficiency between waveguides, we apply the requirement for continuity of the transverse electric and magnetic fields at a dielectric interface. There are several fields involved in the coupling: the forward waves of the waveguide, the reflected wave of the incident field, and radiation modes for both the reflected and the transmitted field due to unguided propagation. If we define the normalised input spatial mode as $\varepsilon_i(x)$, the normalised transmitted mode as $\varepsilon_t(x)$, and radiation modes as $\varepsilon_\beta(x)$, the fields must satisfy the equation:

$$\varepsilon_i(x)+r\varepsilon_i(x)+\int_0^\infty r(\beta)\varepsilon_\beta(x)d\beta=t\varepsilon_t(x)+\int_0^\infty t(\beta)\varepsilon_\beta(x)d\beta \quad \text{(P9)}$$

and

$$H_i(x)-rH_i(x)-\int_0^\infty r(\beta)H_\beta(x)d\beta=tH_t(x)+\int_0^\infty t(\beta)H_\beta(x)d\beta,$$

where r and t are the amplitude reflection and transmission coefficients for the guided modes, and $r(\beta)$ and $t(\beta)$ are the reflection and transmission coefficients for the radiation modes, respectively. If the phase of the mode is described by $exp\{-j(\beta z-\omega t)\}$, then the guided power per unit length flowing in the z-direction is:

$$\frac{1}{2}\int_{-\infty}^{\infty}\varepsilon_n(x)\varepsilon_m(x)dx=\frac{\omega\mu}{\beta_n}\delta_{nm}. \quad \text{(P10)}$$

This follows from the relation:

$$H_x=\frac{-j}{\omega\mu}\frac{\partial\varepsilon_y}{\partial z}=-\frac{\beta}{\omega\mu}\varepsilon_y. \quad \text{(P11)}$$

It is impossible to obtain exact solutions to the expressions in (P9) because there are an infinite number of radiation modes. We can make an approximate solution by assuming that the amount of energy scattering into radiation modes is negligible, which will be true when the incident field and waveguide mode are not too dissimilar. Thus, we drop the integral terms from (P9). We can isolate the transmission terms, t, by taking advantage of the orthonormality of the modes. We multiply both sides of the electric filed equation by ε_t^* and integrate overall space. We get:

$$t = \frac{\beta_t}{\omega\mu}(1+r)\int_{-\infty}^{\infty}\varepsilon_i(x)\varepsilon_t^*(x)dx \tag{P12}$$

We can repeat the process for the magnetic field equation, using (P11), to get

$$t = \frac{\beta_i}{2\omega\mu}(1-r)\int_{-\infty}^{\infty}\varepsilon_i(x)\varepsilon_t^*(x)dx. \tag{P13}$$

Now, we have two equations and two unknowns. Solving t and r, respectively, yields:

$$t = \frac{2\beta_r\beta_t}{2\beta_i+\beta_t}\frac{1}{2\omega\mu}\int_{-\infty}^{\infty}\varepsilon_i(x)\varepsilon_t^*(x)dx$$

$$r = \frac{\beta_i-\beta_t}{\beta_i+\beta_t}. \tag{P14}$$

It is important to realise that $|t|^2+|r|^2 \neq 1$ for most situations because the incident power can couple into modes other than ε_i and ε_t. From (P14), we see an overlap integral is used to determine the coupling between the incident mode and the exiting mode. The overlap integral simply calculates the fraction of the incident field that looks like the desired mode. Further, the reflection coefficient, r, is very similar to the Fresnel reflection formula, when the wave is incident normal to the surface. By replacing β by k_0n_{eff} in "r" in fact produces the Fresnel expression. The power coupling efficiency is equal to the square of the amplitude coupling:

$$\eta = |t|^2. \tag{P15}$$

Often, mode amplitudes will not be formally normalised. In such a case, a working formula for coupling efficiency can be derived from (P16) to be:

$$\eta = \left[\frac{4\beta_i\beta_t}{(\beta_i+\beta_t)^2}\right]\frac{\left[\int E_t(r,\varphi)E_i^*(r,\varphi)drd\varphi\right]^2}{\int E_t(r,\varphi)E_i^*(r,\varphi)rdrd\varphi\int E_i(r,\varphi)E_i^*(r,\varphi)rdrd\varphi}, \tag{P16}$$

where $E_i(r,\varphi)$ is the input field amplitude and $E_t(r,\varphi)$ is the transmitted field amplitude.

Question 3: Discuss the power coupling between two fibres with different MFDs.

Answer: Let us calculate the coupling between two single-mode fibres as a function of their relative mode field diameters (MFD). To simplify the calculation, assume that spatial HE_{11} modes can be approximated by Gaussian profiles. Define one fibre as having a beam radius ω_1 and amplitude A_2. We will assume that these fibres have nearly identical propagation coefficients, and therefore, we will neglect the effect of β mismatch. The power coupling efficiency is given by (P16).

$$\eta = \frac{\left[\int_0^{\infty}\int_0^{2\pi}A_1e^{-r^2/\omega_1^2}A_2e^{-r^2/\omega_2^2}rdrd\varphi\right]^2}{\int_0^{\infty}\int_0^{2\pi}A_1^2e^{-2r^2/\omega_1^2}rdrd\varphi\int_0^{\infty}\int_0^{2\pi}A_2^2e^{-2r^2/\omega_2^2}rdrd\varphi}$$

$$= \frac{\left[2\pi A_1 A_2 \int_0^{\infty} re^{-r^2\left(1/\omega_1^2 + 1/\omega_2^2\right)} dr \right]^2}{(2\pi)^2 A_1^2 A_2^2 \int_0^{\infty} re^{-2r^2/\omega_1^2} dr \int_0^{\infty} re^{-2r^2/\omega_2^2} dr}$$

This integral can be evaluated in closed form:

$$\eta = 4 \frac{\omega_1^2 \omega_2^2}{\left(\omega_1^2 + \omega_2^2\right)^2}.$$

Note that the amplitudes of the modes drop out of the equation.

The coupling efficiency is plotted in Fig. P3 as a ratio of ω_1 / ω_2. Coupling is maximised when the fibres have identical mode field diameters and decrease for all other ratios. It is especially interesting to note that the coupling is reciprocal, which means that it does not matter which fibre has the larger mode. Geometrically, one might expect that a smaller mode would couple more efficiently into a larger mode. Physically, the coupling symmetry arises because we are discussing coupling between two modes not waveguides. A small mode may couple very efficiently into a large core fibre, but it does so by exciting a superposition of modes in the larger core fibre. The coupling between the two fundamental HE_{11} modes is reciprocal in this and every case. It makes no difference in which way the energy flows. If the mode field diameters are not identical, there will be a reduced overlap integral.

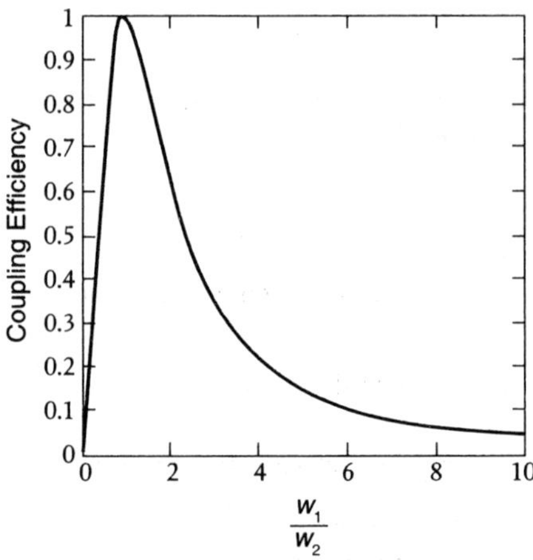

Fig. P3 Coupling efficiency between two fibres. Notice that the horizontal axis is the log of the ratio of w_1 / w_2

Question 4: What are the Maxwell's equations and their different forms?

Answer: Maxwell's equations represent one of the most elegant and concise ways to state the fundamentals of electricity and magnetism. Maxwell's equations are a set of partial differential equations that, together with the Lorentz force law, form the foundation of classical electrodynamics, classical optics, and electric circuits. These fields in turn underlie modern electrical and communications technologies.

Maxwell's equations can be represented in the differential and integral forms:

Differential forms:

1. Gauss' law of electricity: $\nabla.D = \rho$

   $D = \varepsilon_0 E + P$ (General case)

   $D = \varepsilon_0 E$ (Free space)

2. Gauss' law of magnetism: $\nabla.B = 0$
3. Faraday's law of electromagnetic induction: $\nabla \times E = -\frac{\partial B}{\partial t}$
4. Ampere's law: $\nabla \times H = J + \frac{\partial D}{\partial t}$

Integral forms:

1. Gauss' law for electricity: $\oint \vec{E}.d\vec{A} = \frac{q}{\varepsilon_0}$
2. Gauss' law for magnetism: $\oint \vec{B}.d\vec{A} = 0$
3. Faraday's law of induction: $\oint \vec{E}.\, d\vec{S} = -\frac{d\phi_B}{dt}$
4. Ampere's law: $\oint \vec{B}.d\vec{s} = \mu_0 i + \frac{1}{c^2}\frac{\partial \int \vec{E}.d\vec{A}}{\partial t}$,

   where
   E = electric field, ρ = charge density, i = electric current
   B = magnetic field, ε_0 = permitivity, J = current density
   D = electric displacement, μ_0 = permeability, c = speed of light
   H – Magnetic field strength, P = polarisation

Question 5: What is mode?

Answer: Modes are different energy components (or rays) of light traveling on an optical fibre waveguide. The modes are basically defined by the properties of coherence and orthogonality: modes are orthogonal solutions of the wave equation, and they do not interfere (the energy or optical power of a linear superposition of modes is equal to the sum the energy or the optical power of the individual modes). Only the light (photons) within one and the same mode is coherent and does interfere. Two types of modes are distinguished: spatial modes transverse to the direction of propagation (cross-section and divergence) and temporal modes in the direction of propagation (time and frequency).

Basic properties of modes:

1. Light within one mode is coherent (for identical polarisation).
2. Light from different modes is incoherent, and it does not interfere (orthogonality).
3. Photons within one mode cannot be distinguished (uncertainty relation).

Question 6: What are the different types of modes present in the optical fibre?

Answer: Transverse modes are classified into different types:

1. *Transverse electric (TE) modes:* They do not have electric field in the direction of propagation.
2. *Transverse magnetic (TM) modes:* They do not have magnetic field in the direction of propagation.
3. *Transverse electromagnetic (TEM) modes:* They have neither electric nor magnetic field in the direction of propagation.
4. *Hybrid modes:* They have nonzero electric and magnetic fields in the direction of propagation.
5. *H-modes:* They have a magnetic field component in the direction of propagation. H-modes are equivalent to TE modes.
6. *E-modes:* They have an electric field component in the direction of propagations. E-modes are equivalent to TM modes.

Question 7: What is the method of calculating the number of modes in a step index multimode optical fibre?

Answer: The number of modes that can propagate in a fibre depends on the optical fibre's numerical aperture as well as on its core diameter and the wavelength of the light. When V is larger than 2.405, the estimated total number of modes M can be calculated as:

$$M \approx \frac{V^2}{2} \; for\,(V > 2.45)$$

$$\approx (kan_1)^2 \Delta = \left(\frac{2\pi a n_1}{\lambda}\right)^2 \Delta,$$

where

$$\Delta = \frac{n_1^2 - n_2^2}{2n_1^2} \approx \frac{n_1 - n_2}{n_1}$$

V number (or called normalised frequency) is defined as:

$$V = \left(\frac{2\pi a}{\lambda}\right)\sqrt{n_1^2 - n_2^2} = \frac{2\pi a}{\lambda} n_1 \sqrt{2\Delta}$$

a = Radius of fibre core
λ = Vacuum wavelength of the light
n_1 = Core refractive index
n_2 = Cladding refractive index

Question 8: Describe the Mach–Zehnder interferometer.

Answer: The Mach–Zehnder interferometer is a device used to determine the relative phase shift between two collimated beams from a coherent light source. The design and use of optical interferometers include precise measurement of wavelength, measurement of very small distances, and thickness.

Question 9: Explain the ring resonator in brief.

Answer: An optical ring resonator is a set of waveguides in which at least one is a closed loop coupled to some sort of light input and output. When light of the resonant wavelength is passed through the loop from input waveguide, it builds up in intensity over multiple round trips due to constructive interference and is output to the output bus waveguide that serves as a detector waveguide. Because only a select few wavelengths will be at resonance within the loop, the optical ring resonator functions as a filter. Additionally, as implied earlier, two or more ring waveguides can be coupled to each other to form an add/drop optical filter. One more feature about optical coupling is the critical coupling. The critical coupling shows that no light is passing through the waveguide after the light beam is coupled into the optical ring resonator.

Question 10: What are dispersion phenomena?

Answer: In optics, dispersion is the phenomena in which the phase velocity of a wave depends on its frequency. Media having such a property are termed as the dispersive media. Dispersion is sometimes called chromatic dispersion to emphasise its wavelength-dependent nature or group velocity dispersion. Dispersion is most often described for light wave, but it may occur for any kind of wave that interacts with a medium or passes through an inhomogeneous geometry, for example, sound wave.

Dispersion of the transmitted optical signal causes distortion for both digital and analogue transmission along the optical fibre. The most familiar example of dispersion is probably a rainbow in which dispersion causes the spatial separation of a white light into components of different wavelengths (different colours). However, dispersion also has an effect in many other circumstances: for example, GVD causes pulses to spread in optical fibres, degrading signals over long distances; further, a cancellation between group velocity dispersion and nonlinear effects leads to soliton waves.

Question 11: What are the different types of dispersion? Explain.

Answer: Intramodal or chromatic dispersion may occur in all types of optical fibre and results from the finite spectral line width of the optical sources. Since optical sources do not emit just a single frequency but band of frequencies, then there may be propagation delay differences between the different spectral components of the transmitted signal. This causes the broadening of each transmitted mode and hence intramodal dispersion.

Pulse broadening due to intermodal dispersion (sometimes referred to simply as modal or mode dispersion) results from the propagation delay differences between modes within a multimode fibre. As the different modes, which constitute a pulse in a multimode fibre, travel along the channel at different group velocities, the pulse width at the output is dependent upon the transmission times of the slowest and fastest modes.

Pulse broadening due to material dispersion results from the different group velocities of the various spectral components launched into the fibre from the optical sources. It occurs when the phase velocity of a plane wave propagating in the dielectric medium varies nonlinearly with wavelength and a material is said to exhibit material dispersion when the second order of the refractive index with respect to wavelength $\left(d^2n/d\lambda^2\right)$ is not zero.

SOLVED PROBLEMS

Problem 1: Derive the following wave equation in cylindrical coordinate system by using basic Maxwell's equations.

$$\frac{\partial^2 E_z}{\partial r^2}+\frac{1}{r}\frac{\partial E_z}{\partial r}+\frac{1}{r^2}\frac{\partial^2 E_z}{\partial \theta^2}+q^2 E_z=0,$$

where $q^2=\omega^2\varepsilon\mu_0-\beta^2$.

Solution:

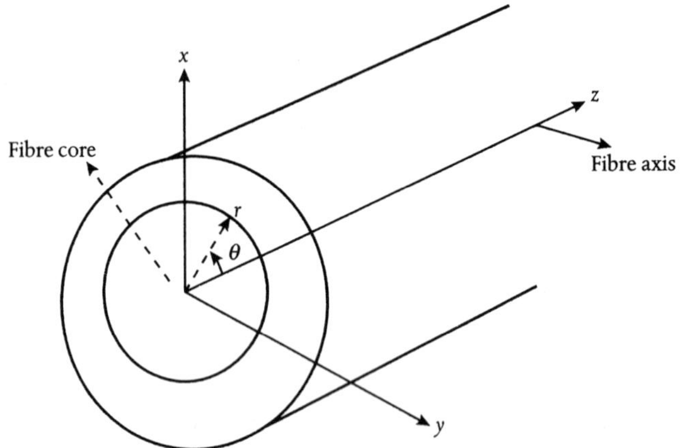

Fig. P4 Electromagnetic waves propagation along the cylindrical fibre

Consider electromagnetic waves propagation along the cylindrical fibre, as shown in Fig. P4. For this fibre, a cylindrical coordinate system (r,θ,z) is defined with the z-axis lying along the axis of the wavelength. If the electromagnetic waves are to propagate along the z-axis, they will have a functional dependence of the form:

$$E=E_0(r,\theta)e^{j(\omega t-\beta z)}$$

$$H=H_0(r,\theta)e^{j(\omega t-\beta z)}.$$

The parameter β is the z-component of the propagation vector and will be determined by the boundary conditions on the electromagnetic fields at the core–cladding interface. In cylindrical coordinate system, we know that

$$\nabla\times E=-\mu_0\frac{\partial H}{\partial t}$$

$$\begin{vmatrix} \frac{1}{r}a_r & a_\theta & \frac{1}{r}a_z \\ \frac{\partial}{\partial r} & \frac{\partial}{\partial \theta} & \frac{\partial}{\partial z} \\ E_r & rE_\theta & E_z \end{vmatrix}=-\mu_0 j\omega\left(H_r a_r+H_\theta a_\theta+H_z a_z\right).$$

Comparing both sides of the above-mentioned equation, we have:

$$\frac{1}{r}\left(\frac{\partial E_z}{\partial\theta}+jr\beta E_\theta\right)=-j\omega\mu_0 H_r \tag{P17}$$

$$j\beta E_r+\frac{\partial E_z}{\partial r}=j\omega\mu_0 H_\theta \tag{P18}$$

$$\frac{1}{r}\left[\frac{\partial}{\partial r}(rE_\theta)-\frac{\partial E_r}{\partial\theta}\right]=-j\omega\mu_0 H_\theta \tag{P19}$$

and also from

$$\nabla\times H=\varepsilon\frac{\partial E}{\partial t}$$

$$\frac{1}{r}\left(\frac{\partial H_z}{\partial\theta}+jr\beta H_\theta\right)=j\omega\varepsilon E_r \tag{P20}$$

$$j\beta H_r+\frac{\partial H_z}{\partial r}=-j\omega\varepsilon E_\theta \tag{P21}$$

$$\frac{1}{r}\left[\frac{\partial}{\partial r}(rH_\theta)-\frac{\partial H_r}{\partial\theta}\right]=j\omega\varepsilon E_z, \tag{P22}$$

where $\varepsilon=\varepsilon_0 n^2$. By eliminating variables, these equations can be rewritten such that, when E_z and H_z are known, the remaining transverse components E_r, E_θ, H_r, and H_θ can be determined.

From (P17) to (P22):

$$E_\theta=\frac{j}{q^2}\left(\frac{\beta}{r}\frac{\partial E_z}{\partial\theta}-\mu_0\omega\frac{\partial H_z}{\partial r}\right) \tag{P23}$$

$$H_r=-\frac{j}{q^2}\left(\frac{\beta}{r}\frac{\partial H_z}{\partial r}-\frac{\omega\varepsilon}{r}\frac{\partial E_z}{\partial\theta}\right). \tag{P24}$$

Similarly, component H_θ or E_r can be formed in terms of E_z or H_z. This results in the following equations:

$$E_r=-\frac{j}{q^2}\left(\beta\frac{\partial E_z}{\partial r}+\frac{\mu_0\omega}{r}\frac{\partial H_z}{\partial\theta}\right) \tag{P25}$$

$$H_\theta=-\frac{j}{q^2}\left(\frac{\beta}{r}\frac{\partial H_z}{\partial\theta}+\omega\varepsilon\frac{\partial E_z}{\partial r}\right), \tag{P26}$$

where $q^2=\omega^2\mu_0\varepsilon-\beta^2$.

Substituting (P24) and (P26) in (P22) results in the wave equation in cylindrical coordinates.

$$\frac{\partial^2 E_z}{\partial r^2}+\frac{1}{r}\frac{\partial E_z}{\partial r}+\frac{1}{r^2}\frac{\partial^2 E_z}{\partial\theta^2}+q^2E_z=0.$$

Problem 2: A single-mode step index fibre with a core refractive index of 1.49 has a critical bending radius of 10.4 mm when illuminated with light at a wavelength of 1.3 μm. If the cut-off wavelength of the fibre is 1.15 μm, calculate its refractive index difference.

Solution:

Core refractive index, $n_1 = 1.49$.
Critical bending radius, $R_{cs} = 10.4$ mm.
Wavelength, $\lambda = 1.3$ μm.
Cut-off wavelength, $\lambda_c = 1.15$ μm.
From the literature, for a single-mode fibre:

$$R_{cs} \approx \frac{20\lambda}{(n_1 - n_2)^{3/2}} \left(2.748 - 0.996\frac{\lambda}{\lambda_c}\right)^{-3},$$

where n_2 = refractive index of cladding.

Substituting the values:

$$10.4 \times 10^3 \approx \frac{20 \times 1.3}{(1.49 - n_2)^{3/2}} \left(2.748 - 0.996 \times \frac{1.3}{1.15}\right)^{-3}$$

$$1.49 - n_2 = 7 \times 10^{-3}$$

$$\Rightarrow n_2 = 1.483.$$

Relative refractive index difference (Δ),

$$\Delta = \frac{n_1^2 - n_2^2}{2n_1^2}.$$

$$\Rightarrow \Delta = \frac{(1.49)^2 - (1.483)^2}{2(1.49)^2} = 4.687 \times 10^{-3} = 0.4687\%$$

Problem 3: An Si fibre has NA = 0.173 and supports 400 modes at a wavelength of 1.3 μm. Calculate its core area. To what value of its core area should be reduced so that it is single-mode fibre at $\lambda \geq 1.3$ μm?

Solution:

(a) NA = 0.173
Modes M = 400
$\lambda = 1.3$ μm
We know that number of modes:

$$M = \frac{V^2}{2}$$

$$\Rightarrow 400 = \frac{V^2}{2}$$

$$\Rightarrow V = 28.28$$

$$V = \frac{2\pi}{\lambda} a(\text{NA})$$

$$\Rightarrow 28.28 = \frac{2\pi}{1.3\ \mu\text{m}} \times a \times 0.173.$$

$$\text{Radius of the core} = 33.82\ \mu\text{m}$$

$$\text{Core area } A = 3.59\ \text{nm}^2$$

(b) Cut-off V for single-mode $V_c = 2.405$

$$\Rightarrow 2.405 = \frac{2\pi a}{1.3 \times 10^{-6}} \times 0.173$$

$$\Rightarrow a = 2.876\ \mu\text{m}.$$

Problem 4: Describe the derivation of modes in monomode and multimode fibres.

Solution:

The Maxwell's equations for linear, isotropic dielectric material having no currents and free charge are written as:

$$\nabla \times E = -\frac{\partial B}{\partial t}$$

$$\nabla \times H = \frac{\partial D}{\partial t}$$

$$\nabla . D = 0$$

$$\nabla . B = 0$$

where $D = \varepsilon E$ and $B = \mu H$ where ε is the permittivity of medium and μ is the permeability of the medium. Therefore,

$$\nabla \times (\nabla \times E) = -\mu\varepsilon \times \frac{\partial^2 E}{\partial t^2}$$

Using the vector identity, we get:

$$\nabla^2 E = \varepsilon\mu \frac{\partial^2 E}{\partial t^2}. \quad \text{(P27)}$$

Similarly,

$$\nabla^2 H = \varepsilon\mu \frac{\partial^2 H}{\partial t^2}. \quad \text{(P28)}$$

Equations (P27) and (P28) are wave equations of electric field and magnetic field. Since optical fibre is cylindrical in shape, wave equations using cylindrical coordinates are:

$$\frac{\partial^2 E_z}{\partial r^2}+\frac{1}{r}\frac{\partial E_z}{\partial r}+\frac{1}{r^2}\frac{\partial^2 E_z}{\partial \varphi^2}+q^2 E_z=0$$

and

$$\frac{\partial^2 H_z}{\partial r^2}+\frac{1}{r}\frac{\partial H_z}{\partial r}+\frac{1}{r^2}\frac{\partial^2 H_z}{\partial \varphi^2}+q^2 H_z=0,$$

where z is the fibre axis, and r and φ are cylindrical coordinates.

In a step index fibre, there is a homogeneous core of index n_1 and radius a, which is surrounded by an infinite cladding of index n_2. The infinite cladding is assumed so that guided modes are confined within cladding layer. In practice, the cladding thickness is sufficiently large and is coated by such a material that all the modes are well within the cladding layer. The V-parameter at cut-off condition is defined as:

$$V^2=\left(\frac{2\pi a}{\lambda}\right)^2\left(n_1^2-n_2^2\right),$$

which is a dimensionless parameter that determines the number of modes a fibre can support. For a single-mode fibre, we can choose a, n_1, and n_2 values and the value of V is:

$$V=\frac{2\pi a}{\lambda}\left(n_1^2-n_2^2\right)^{1/2}\leq 2.405, \text{ where } n_1>n_2.$$

At this value of V, all modes except the lowest mode HE_{11} are cut-off.

Further,

$$V=\frac{2\pi a}{\lambda}(NA).$$

If $V=26.6$ at $\lambda=1300\,\text{nm}$, $NA=0.22$.

Similarly, total number of modes in multimode optical fibre can be calculated for large number of modes (M). The numerical aperture NA is given as:

$$NA=\sin\theta=\left(n_1^2-n_2^2\right)^{1/2}.$$

For practical case, $NA=\sin\theta$ is small so that $\sin\theta=\theta$.

The solid acceptance angle for fibre is:

$$\Omega=\pi\theta^2=\pi\left(n_1^2-n_2^2\right).$$

The modes per unit solid angle for laser source or a waveguide is given as $2A/\lambda^2$, where A is the area in which the mode is passing. Since $A=\pi a^2$, the total number of modes M entering the fibre is defined as:

$$M=\frac{2A}{\lambda^2}\Omega=\frac{2\pi^2 a^2}{\lambda^2}\left(n_1^2-n_2^2\right)=\frac{V^2}{2}.$$

Problem 5: A silica optical fibre with a core diameter large enough to be considered by ray theory analysis has a core refractive index of 1.50 and a cladding refractive index of 1.47. Determine (a) the critical angle at the core–cladding interface; (b) the NA for the fibre; and (c) the acceptance angle in air for the fibre.

Solution:
Given,

$n_1 = 1.50$, $n_2 = 1.47$

(a) Critical angle,

$$\theta_c = \sin^{-1}\left(\frac{n_2}{n_1}\right) = 78.5°$$

(b)

$$\text{NA} = \left(n_1^2 - n_2^2\right)^{1/2} = 0.30$$

(c)

$$\theta_a = \sin^{-1}\text{NA} = 17.4°$$

Problem 6: A typical step index fibre has a core refractive index of 1.46, a relative refractive index difference of 0.003, and a core radius of $4\,\mu\text{m}$. Calculate w and w_P at $\lambda = 1.30\,\mu\text{m}$ and $1.55\,\mu\text{m}$.

Solution:
The normalised frequency parameter of the fibre is:

$$V = \frac{2\pi a n_1}{\lambda}(2\Delta)^{1/2} = \frac{2\pi \times 4\times 10^{-6}\times 1.46\sqrt{2\times 0.003}}{\lambda(m)} = \frac{2.842\times 10^{-6}}{\lambda(m)}.$$

Therefore, for $\lambda = 1.30\ \mu\text{m}$, $V = 2.186$, and for $\lambda = 1.55\ \mu\text{m}$, $V = 1.834$. Using approximate expressions, we get at $\lambda = 1.30\,\mu\text{m}$. From (P3)

$$w \approx a\left[0.65 + \frac{1.619}{V^{3/2}} + \frac{2.879}{V^6}\right]$$

$$w \approx 4\times 10^{-6}\left[0.65 + \frac{1.619}{(2.186)^{3/2}} + \frac{2.879}{(2.186)^6}\right] = 4.7058\ \mu\text{m}$$

and from (P5),

$$w_P \approx 4.7508\,\mu\text{m} - \left(4\times 10^{-6}\right)\left[0.016 + \frac{1.567}{(2.186)^7}\right] = 4.6155\ \mu\text{m}.$$

Similarly, at $\lambda = 1.55\ \mu\text{m}$,

$$w \approx 4\times 10^{-6}\left[0.65 + \frac{1.619}{(1.834)^{3/2}} + \frac{2.879}{(1.834)^6}\right] = 5.5098\ \mu\text{m}$$

and

$$w_P \approx 5.5098\ \mu\text{m} - \left(4\times10^{-6}\right)\left[0.016 + \frac{1.567}{(1.834)^7}\right] = 5.3559\ \mu\text{m}.$$

Problem 7: A beat length of 10 cm is observed in a typical single-mode fibre when light of wavelength 1.0 μm is launched into it. Calculate (a) the difference between the propagation constants for the two orthogonally polarised modes and (b) the modal birefringence.

Solution:

(a) Using (P8):

$$\delta\beta = \frac{2\pi}{L_P} = \frac{2\pi}{10\times10^{-2}\ m} = 62.83\,\text{m}^{-1}$$

(b) Modal birefringence from (P7):

$$\delta n = \delta\beta\frac{\lambda}{2\pi} = 62.83\,\text{m}^{-1} \times \frac{1\times10^{-6}\ \text{m}}{2\pi} = 9.999\times10^{-6}.$$

and respectively yields

Problem 8: Describe the coupling a single-mode fibre to a Gaussian beam. The output (1.52 μm) from an HeNe laser is to be coupled into a commercial fused silica single-mode optical fibre. The laser beam has a Gaussian amplitude profile with a characteristic beam radius $\omega_0 = 0.70\,\text{mm}$. The fibre has a mode field diameter of 10 μm. What focal length of lens should be used to focus the beam into the fibre to maximise coupling efficiency? What is the maximum possible coupling efficiency?

Solution:

Thus, the problem describes the coupling a single-mode fibre to a Gaussian beam. To answer this, we must know some features of Gaussian optics. When a Gaussian beam is focused using a high-quality lens with focal length f, the focal spot size is approximately described by the formula:

$$\omega_1 = \frac{2\pi}{\lambda}\frac{f}{2\omega_0}.$$

Plugging in numbers, we find a relation between the spot size at the focal point of the lens:

$$\omega_1 = 2.952\times10^{-4} f.$$

To maximise the coupling between the laser and the fibre, the spot size of the focused beam should exactly match the mode field radius of the fibre.

$$\left(2.952\times10^{-4} f\right) = 5\times10^{-6}\ \text{cm}$$

Therefore,

$$f = 1.693\times10^{-15}\,\text{cm}.$$

This is a moderately short focal length lens and is difficult to achieve with a simple lens. The examination of several optics catalogues reveals that a 10× microscope objective has a 0.43-cm focal length. The 20× objective would be the best choice for this fibre. Using the results of the last section, the power coupling efficiency, η, would equal 0.98. The Fresnel reflection at the input surface will further reduce the coupling efficiency. In this case, the reflection coefficient will be given as:

$$R=\left(\frac{\beta_1-\beta_2}{\beta_1-\beta_2}\right)^2=\left(\frac{k_0 n_{\text{eff}}-k_0}{k_0 n_{\text{eff}}+k_0}\right)^2=\left(\frac{n_{\text{eff}}-1}{n_{\text{eff}}+1}\right)^2.$$

Since n_{eff} is not exactly defined, we cannot proceed. However, note that this is, in fact, the Fresnel reflection that would be expected when crossing a dielectric interface and $n_{eff} \approx n_{sio_2} \approx 1.5$. For fused silica with index $n=1.5$, the power reflection is on the order of 4 %, , so and therefore, the maximum coupling will be approximately $0.96\times 0.98 = 0.94$. Practically speaking, this limit is rarely reached. By achieving 80% coupling efficiency between a laser and a single-mode fibre is considered excellent by most practitioners n_{eff}.

Problem 9: A glass fibre exhibits material dispersion given by $|\lambda^2 d^2 n_1 / d\lambda^2|$ of 0.025. Determine the material dispersion parameter at a wavelength of $0.85\ \mu\text{m}$, and estimate the rms pulse broadening per kilometre for a good LED source with an rms spectral width of 20 nm at this wavelength.

Solution:

The material dispersion parameter may be obtained as:

$$M=\frac{\lambda}{C}\left|\frac{d^2 n_1}{d\lambda^2}\right|=\frac{1}{c\lambda}\left|\lambda^2\frac{d^2 n_1}{d\lambda^2}\right|=\frac{0.025}{2.998\times 10^5\times 850}\text{snm}^{-1}\text{km}^{-1}=98.1\ \text{psnm}^{-1}\text{km}^{-1}.$$

The rms pulse broadening is given as:

$$\sigma_m=\frac{\sigma_\lambda L}{c}\left|\lambda^2\frac{d^2 n_1}{d\lambda^2}\right|.$$

Therefore, in terms of the material dispersion parameter, M is defined as:

$$\sigma_m = \sigma_\lambda LM.$$

Hence, the rms pulse broadening per kilometres due to material dispersion:

$$\sigma_m(1\ \text{km}) = 20\times 1\times 98.1\times 10^{-12} = 1.96\ \text{nskm}^{-1}$$

Problem 10: A 6-km optical link consists of multimode step index fibre with a core refractive index of 1.5 and a relative refractive index difference of 1%. Estimate:

1. The delay difference between the slowest and the fastest modes at the fibre outputs;
2. The rms pulse broadening due to intermodal dispersion on the link;
3. The maximum bit rate that may be obtained without substantial errors on the link assuming only intermodal dispersion;
4. The bandwidth–length product corresponding to (3).

Solution:

1. The delay difference is given as:

$$\delta T_s \cong \frac{L n_1 \Delta}{c} = \frac{6\times 10^3\times 1.5\times 0.01}{2.998\times 10^8} = 300\text{ns}.$$

2. The rms pulse broadening due to intermodal dispersion may be obtained as:

$$\sigma_s = \frac{Ln_1\Delta}{2\sqrt{3}c} = \frac{1}{2\sqrt{3}}\frac{6\times10^3\times1.5\times0.01}{2.998\times10^8} = 86.7\text{ns}.$$

3. The maximum bit rate can be estimated in two ways. Firstly, to get an idea of the maximum bit rate when assuming no pulse overlap can be used:

$$B_T(\max) = \frac{1}{2\tau} = \frac{1}{2\delta T_s} = \frac{1}{600\times10^{-9}} = 2.3\text{ Mbitss}^{-1}.$$

4. Using the most accurate estimate of the maximum bit rate from (3), and assuming return to zero pulse, the bandwidth–length product is

$$B_{opt}\times L = 2.3\text{ MHz}\times6\text{km} = 13.8\text{ MHzkm}.$$

Problem 11: Estimate the rms pulse broadening per kilometre for the fibre in (P22), when the optical source used is an injection laser with a relative spectral width σ_λ/λ of 0.0012 at a wavelength of $0.85\ \mu\text{m}$.

Solution:

The rms spectral width may be obtained from the relative spectral width as:

$$\sigma_\lambda = 0.0012\lambda = 0.0012\times0.85\times10^{-6} = 1.02\text{nm}$$

The rms pulse broadening in terms of the material dispersion parameter is given as:

$$\sigma_m = \sigma_{»}LM.$$

where M is the material dispersion, and therefore, the rms pulse broadening per kilometre due to material dispersion is:

$$\sigma_m = 1.02\times1\times98.1\times10^{-12} = 0.10\text{nskm}^{-1}.$$

Hence, in this example, rms pulse broadening is reduced by a factor of around 20 (i.e., equivalent to the reduced rms spectral width of the injection laser sources) when compared with that obtained with the LED source.

MULTIPLE CHOICE QUESTIONS

1. A bare silica fibre having a refractive index of 1.45 has air as the cladding. The maximum angle of acceptance of the fibre would be
   (a) less than 90°
   (b) equal to 90°
   (c) greater than 90°
2. The acceptance angle of an optical fibre placed in air is 12°. When the fibre tip is immersed in water, the acceptance angle of the fibre will
   (a) increase
   (b) decrease
   (c) remain same

3. An optical fibre has core and cladding refractive indices of 1.45 and 1.44, respectively. State which of the following effective indices correspond to guided modes and radiation modes. Also indicate the ones that are not allowed.
   (a) 1.446
   (b) 1.449
   (c) 1.43
   (d) 1.46
4. An optical fibre supports only two guided modes namely LP_{10} and LP_{11} modes. Light is launched so that both modes have equal power at the input. By assuming no attenuation, if fibre is ideal, then which of the following statements are correct?
   (a) The power will flow from one mode to the other periodically along the length.
   (b) The powers in individual modes will remain constant with distance.
   (c) The power from LP_{11} mode will flow into LP_{10} mode.
   (d) The transverse intensity pattern will vary with distance.
5. A depressed clad optical fibre consists of three regions: region between r = 0 and a has a refractive index n1, region between $r = a$ and $b(>a)$ has a refractive index n3 and the region beyond r = b has a refractive index n2 with $n_1 > n_2 > n_3$. Which of the following conditions define loss-less guided modes?
   (a) $n_1 > n_{\text{eff}} > n_3$
   (b) $n_2 > n_{\text{eff}} > n_3$
   (c) $n_1 > n_{\text{eff}} > n_2$
6. An optical fibre with an attenuation coefficient of 0.2 dB/km is 100-km long. If the input power is 1 mW, then the power coming from the fibre would be 0.
   (a) 100 μW
   (b) 10 μW
   (c) 1 μW
7. An optical fibre communication system has two splices each with a loss of 0.1 dB, two connectors each with a loss of 0.9 dB. If the fibre used in the system has an attenuation coefficient of 0.28 dB/km, then the maximum permissible length whose output power is at least 10 μW and the input power is 10 mW is
   (a) 10 km
   (b) 100 km
   (c) 104 km
   (d) 200 km
8. Silica optical fibres have the lowest loss at a wavelength of
   (a) 850 nm
   (b) 1000 nm
   (c) 1310 nm
   (d) 1550 nm
9. A single-mode fibre has a cut off of 1250 nm. It will be single mode
   (a) only at 1250 nm
   (b) at all wavelengths below 1250 nm
   (c) at all wavelengths above 1250 nm
10. Two fibres have the same cut-off wavelength of 1200 nm. The fibre with a higher refractive index difference would have
   (a) larger core radius
   (b) smaller core radius
   (c) same core radius as the other

11. Two single-mode optical fibres have identical core radii but different NA. Which of the following statements are true?
    (a) Both fibres will have the same spot size.
    (b) Fibre having a larger NA would have larger spot size.
    (c) Fibre having a larger NA would have smaller spot size.
12. The Gaussian spot size of a given fibre
    (a) increases with an increase in wavelength
    (b) decreases with an increase in wavelength
    (c) is independent of wavelength
13. By increasing the Gaussian spot size of a single-mode fibre,
    (a) it is possible to increase the tolerance to both transverse and angular misalignment
    (b) the tolerance to transverse misalignment would reduce while that for angular misalignment would increase
    (c) the tolerance to transverse misalignment would increase while that for angular misalignment would reduce
14. Material dispersion in silica optical fibres becomes zero at around
    (a) 850 nm
    (b) 1270 nm
    (c) 1310 nm
    (d) 1550 nm

Answers to Multiple Choice Questions

1. (b)
2. (b)
3. (a) guided modes
4. (b) and (d) are correct
5. (c)
6. (b)
7. (b)
8. (d)
9. (c)
10. (b)
11. (c)
12. (a)
13. (c)
14. (b)

QUESTIONS

1. Discuss the approximations inherent in the derivation of the coupled mode equations for a directional coupler. Are they valid? If not, can you suggest the likely consequences for device performance?
2. A fibre directional coupler is to be made by polishing two glass blocks, which have fibres embedded in curved grooves. (a) Suggest how the polishing process can be monitored, and therefore, it can be terminated when the core is reached. (b) When the coupler is assembled, it is found that the amount of cross-coupled light is very small, despite attempts to design it for 100% coupling. Suggest two possible explanations and the remedy in each case.
3. Explain what is meant by the critical bending radius for an optical fibre.
4. A Bragg reflector is to be used with an integrated optic channel waveguide. Past measurements with this system suggest that the coupling coefficient is $\kappa = 200/\text{m}$ at the design wavelength. How long should the grating be for $\geq 99\%$ reflectivity?

5. Compare the wavelength selectivity of waveguide filters based on (a) co-directional coupling between dissimilar waveguides and (b) contra-directional coupling via a grating. What are the factors limiting selectivity in each case?
6. Derive coupled-mode theory and explain.
7. (a) Show that for collinear and co-directional coupling, the free-space wavelength λ_0 for maximum diffraction efficiency is given by $\lambda_0 = \lambda_{0c} = (n_2 - n_1)\wedge$.
   When n_1 and n_2 are the refractive indices seen by the incident and diffracted waves, respectively.
   (b) Show that when $\sigma L = \frac{1}{2}$, for wavelengths $\lambda_0 = \lambda_{0c} \pm \Delta\lambda = \lambda_{0c} \pm \frac{\sqrt{3}}{2}\frac{\wedge}{L}\lambda_0$.
   The diffraction efficiency becomes zero. The quantity $\Delta\lambda$ is known as the bandwidth of the filter.
   For lithium niobate at $\lambda_0 = 0.6\ \mu\text{m}$, $n_1 = n_e = 2.2082$, $n_2 = n_0 = 2.2967$, where the subscripts *o* and *e* refer to the *o*- and *e*-waves, respectively. Calculate $\wedge$ required for a λ_{0c} of 0.6 μm and obtain the corresponding bandwidth for L = 3.6 cm.
8. State the principle of operation of Mach–Zehnder Interferometer.
9. Explain the principle of ring resonator.
10. Consider an N-node star network in which 0 dBm of optical power is coupled from any given transmitter into the star. Let the fibre loss be 0.3 dB/km . If we assume that the stations are located 2 km from the star, the receiver sensitivity is -38 dBm , each connector has a $1-$dB loss, the excess loss in the star coupler is 3 dB, and the link margin is 3 dB.
    (a) Determine the maximum number of stations N that can be incorporated on this network.
    (b) How many stations can be attached if the receiver sensitivity is -32 dBm ?
11. Derive the mode theory of circular and rectangular waveguide.
12. Derive the calculation for material/waveguide dispersion.
13. Illustrate the finite element analysis of inhomogeneous planar slab waveguides.
14. Draw schematic of dispersion characteristics of graded/linear chirp-type refractive index profile of planar slab waveguides.

REFERENCES

1. Raghuwanshi S K, Kumar V and Pandey R R, Derivation of eigenvalue equation by using equivalent transmission line method for the case of symmetric/asymmetric planar slab waveguide structure, *J. Int. Acad. Phys. Sci.*, vol. 15, No.1, pp. 1–14, 2011.
2. Raghuwanshi S K, Kumar V and Pandey R R, "Performance study of Exponential varying refractive index planar slab optical waveguide", *Proceeding of IEEE International Conference on Computer, Communication & Electrical Technology (ICCCET)*, Tamilnadu, India, pp. 16–20, 18 and 19 March 2011.
3. Raghuwanshi S K, Comparative study of asymmetric versus symmetric planar slab dielectric optical waveguides, *Indian J. Phys.*, vol. 84, No. 7, pp. 831–846, 2010.
4. Raghuwanshi S K, Kumar V, Chack D and Pandey R R, "Dispersion study of even mode thin planar slab waveguide without computing (*d* ^ 2β)/(*dk* ^ 2) numerically", submitted to *World Conference on Information Technology*, Turkey, 23–27 Nov. 2011.
5. Raghuwanshi S K, Tiwari A and Pandey R, "Study of Mode Cut-off Condition of Single Mode Planar Slab Optical Waveguide", *12th International Conf. of International Academy of Physical Sciences (CONIAPS XII)*, 22–24 Dec. 2010, Jaipur.
6. Rausch E O, Efurd R B and Corbin M A, A fibre-optic pulse compression device for high resolution radars, *IEEE Conf. Radar*, London, England, 1982.

7. Bender C M and Wang Q, A class of exactly-solvable eigenvalue problems, *J. Phys. A*, vol. 34, 9835–9851, 2001.
8. Ho M, Lai F S, Tan S W and Chem P W, Numerical simulation of propagation of propagation of EM pulse through lossless non-uniform dielectric slab using characteristic-based method, *Prog. Electromagn. Res.*, PIER, vol. 81, pp.197–212, 2008.
9. Renner H, Polarisation characteristics of optical waveguides with separable symmetric refractive index profiles, *J. Op. Soc. Am. A*, vol. 15, No. 5, pp. 1401–1410, 1998.
10. Rostami A and Moayedi S K, Exact solutions for the TM mode in inhomogeneous slab waveguide, *Laser Phys.*, vol. 14, No. 12, pp. 1492–1498, 2004.
11. Rostami A and Moayedi S K, TM mode in inhomogeneous slab waveguide as an exactly solvable oscillator-like Hamiltonian, *J. Opt. A*: *Pure Appl. Op.*, vol. 5, 380, 2003.
12. Rostami A and Motavali H, Asymptotic iteration method: A power approach for analysis of inhomogeneous dielectric slab waveguides, *Prog. Electromagn. Res. B*, vol. 4, pp. 171–182, 2008.
13. Schwelb O and Frigyes I, All optical tunable filters built with discontinuity-assisted ring resonator, *J. Lightwave Technol.*, vol. 19, no. 3, pp. 380–386, Mar. 2001.
14. Shu W and Song J M, Complete mode spectrum of a grounded dielectric slab with double negative metamaterials, *Prog. Electromagn. Res.*, PIER, vol. 65, pp. 103–123, 2006.
15. Sharma E K, Goyal I C and Ghatak A K, Calculation of cut-off frequencies in optical fibres for arbitrary profiles using matrix method, *IEEE J. Quant. Electron.*, vol. 17, no.12, pp. 2317–2320, 1981.
16. Raghuwanshi S K, "Analysis of integrated optical micro-ring resonator," in *National Conference on Information and Communication Technology, NCICT-2009*, Mumbai.
17. Rostami A and Moayedi S K, Exactly modal analysis of inhomogeneous slab waveguide using similarity transformed harmonic-oscillator algebra, *IL NuovoCimento* B, vol. 120, 285, 2005.
18. Sammut R A, Orthogonality and normalisation of radiation modes in dielectric waveguides, *J. Opt. Soc. Am.*, vol. 72, pp. 1335–1337, 1982.
19. Sepulveda B and Lechuga L M, Matrix analysis of discontinuities in nonreciprocal waveguides: Analytical description for magneto optical slab waveguides, *IEEE J. Lightwave Technol.*, vol. 22, no. 7, pp. 1772–1781, July 2007.
20. Wang Z J and Dong J F, Analysis of guided modes in asymmetric left-handed slab waveguides, *Prog. Electromagn. Res.*, PIER, vol. 62, pp. 203–215, 2006.
21. Weisshaar A and Tripathi V K, Modal analysis of step discontinuities in graded index dielectric slab waveguides, *IEEE J. Lightwave Technol.*, vol. 10, no. 5, pp. 593–602, May 1992.
22. Attiya A M and Kishk A A, Modal analysis of a two-dimensional dielectric grating slab excited by an obliquely incident plane wave, *Prog. Electromagn. Res.*, PIER, vol. 60, pp. 221–243, 2006.

5 Overview of Optical Signal Processing and Optical Filter Design

Objectives

- This chapter focuses on generation of optical communication.
- Various optical networks are examined.
- The need for optical signal processing is explained.
- The operations of different coupler systems are numerically analysed.

5.1 INTRODUCTION

Optical fibre transmission has played a key role in increasing the bandwidth of telecommunication [1–3]. The access mechanism is based on the processing of signals with a so-called optical signal processor, abbreviated as OSP [4]. The OSP has the ability to modify the information content of data signals while, at the same time, preserving certain properties of the physical carrier. Applications are generally found in optical communication networks. Before we focus on the OSP, its functionality, performance, and issues, we first give an overview of the area of optical communication in which this finds its basis [5–7]. High-speed signal processing is needed to cope with the continuing rapid growth in internet traffic. However, signal processing at data rates greater than 100 Gbits/s is difficult to provide using conventional electronic processing. Therefore, all-optical signals processing at a wavelength of $1.55\,\mu m$ will be necessary [8–11].

5.1.1 Optical Communication in Signal Processing

Optical communication is any form of telecommunication that uses light as the transmission medium. An optical communication system consists of a transmitter, which encodes a message into an optical signal, a channel, which carries the signal to its destination, and a receiver, which reproduces the message from the received optical signal [12–16]. Optical fibre communication is the most common type of channel for optical communications; however, other types of optical waveguides are used within computers or communications gear, and have even formed the channel of very short distance (e.g., chip to chip, intrachip) links in laboratory trials. The transmitters in optical fibre links are generally light-emitting diodes (LEDs) or laser diodes. Infrared light, rather than visible light, is used more commonly because optical fibres transmit infrared wavelengths with less attenuation and dispersion. The signal encoding is typically simple intensity modulation, although historically optical phase and frequency modulation have been demonstrated in the lab. The need for periodic signal regeneration was largely superseded by the introduction of the erbium-doped fibre amplifier (EDFA), which extended link distances at significantly lower cost.

In 1970s, a major breakthrough was made on both the fibres and the sources so that practical applications became attractive [17–18]. The performance of a communication system is often measured in the bit rate–distance product. In Fig. 5.1, the progress of the bit rate-distance product over the last 150 years has been given, together with the new technologies, techniques, and engineering that emerged during that time. In 1978, the first-generation commercial lightwave systems operating at 0.8 μm were deployed in an operational environment. In 1980s, it was followed by the second-generation fibre optic communication systems operating at 1.3 μm.

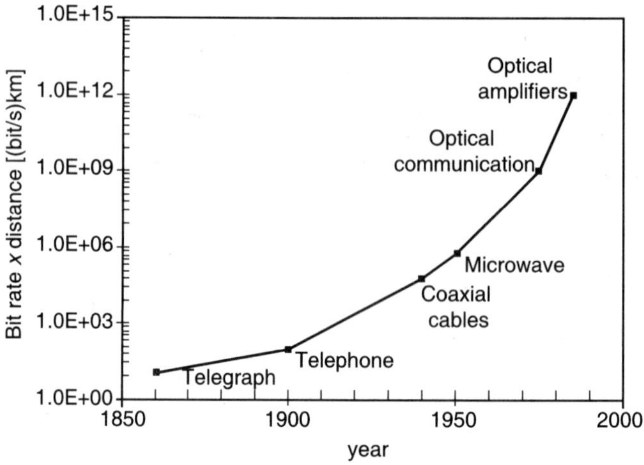

Fig. 5.1 Increase in the bit rate-distance product during 1850–2000

These systems provided larger repeater spacing. In 1990, third-generation fibre optic systems were introduced operating at the minimum loss window of silica fibres, which is around 1.55 μm. Fibre losses of approximately 0.2 dB/km and the development of single-mode lasers again resulted in a longer distance that could be bridged. The fourth generation of lightwave systems is concerned with an increase of the bit rate through frequency division multiplexing and optical amplification to extend the repeater spacing even further. The commercial introduction of these systems began in 1996 and its success is accelerating at a great pace. Currently, the fifth generation of fibre optic communication systems is in the stage of research and development. It is based on fibre solitons, optical pulses that preserve their shape during propagation in a fibre.

5.2 OPTICAL NETWORK

Optical networks are high-capacity telecommunications networks based on optical technologies and components that provide routing, grooming, and restoration at the wavelength-level as well as wavelength-based services. Telecommunication networks have evolved during a century long history of technological advances and social changes. The networks that once basic telephone services through a friendly local operator are now transmitting the equivalent of thousands of encyclopaedias per second.

Throughout this history, the digital network has evolved in three fundamental stages: asynchronous, synchronous, and optical.

1. **Asynchronous:** The first digital networks were asynchronous networks. In asynchronous networks, each network element's internal clock timed its transmitted signal. Because each clock has certain amount of variations.
2. **Synchronous:** The need for optical standards led to the creation of the synchronous optical network (SONET). The SONET standardises the line rate, coding schemes, bit-rate hierarchies, and operation maintenance functionality. The SONET also defined the types of network elements required and required network architecture that vendors could implement.
3. **Optical:** The one aspect of SONET that has allowed it to survive during a time of tremendous change in network capacity needs is its scalability. Based on its open growth plan for higher bit rates, theoretically no upper limit exists for SONET bit rate. However, as higher bit rates are used, physical limitation in the laser sources and optical fibre begin to make the practice of endlessly increasing the bit rate on each impractical solution.

As a general rule, optical networks can be visualised as user network interfaces that only perform electro-optical conversions at the borders of the network. Within the network, the data signal is routed to the destination without undergoing opto-electronic conversions or vice-versa. Hence, these signals are transported through 'transparent optical ether' with a virtually unlimited bandwidth. This interpretation of an optical network should not be considered as a strict definition, its purpose is merely to indicate how such a network can be visualised (see also Fig. 5.2)

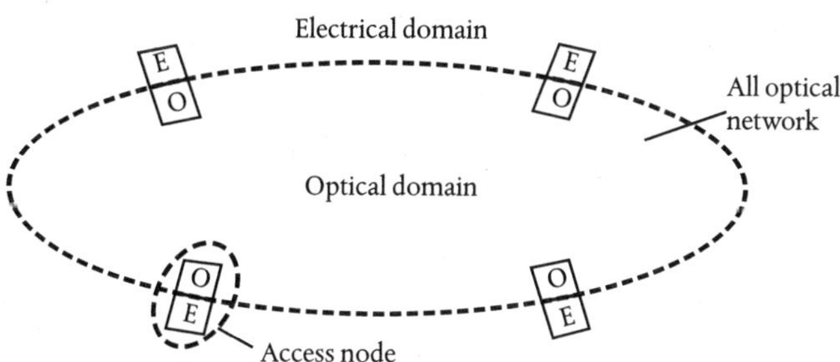

Fig. 5.2 General representation of all-optical networks

5.2.1 Principle of Operation

The principle of operation can easily be explained if we assume that the optical signal is modulated with a Manchester channel code. Recall that for Manchester coding the signal transition denotes the data symbol that is encoded. Hence, a data ZERO is transmitted as '10' and a data ONE as '01'. The main task of the OSP is to provide temporal reversion of the biphase-encoded input signal, in case this is required. In other words, independent of the input sequence, the OSP must be able to output a ONE ('10') or a ZERO ('01').

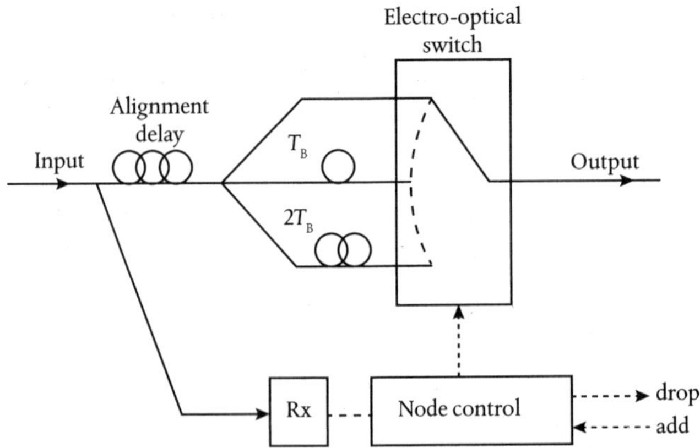

Fig. 5.3 Illustration of the optical signal processor

The configuration of Fig. 5.3 can be used to perform this task. Assume that T_B equals the duration of an encoded signal (i.e., the duration of a data ONE or ZERO equals $2T_B$). First, the encoded signal is split into three branches and is delayed by $0T_B$, $1T_B$, and $2T_B$, respectively. This procedure effectively represents an optical version of serial to parallel conversion. The outcome of this conversion is that a '1' signal and a '0' signal are both simultaneously accessible at one of the electro-optical switch inputs at every timeslot. By correct control of this switch, the required signal bit can be directed to the OSP output.

If the 'drop' data equals the 'add' data, the input data sequence is simply delayed by one-bit period T_B and forwarded to the OSP output using the electro-optical switch. If the 'drop' data and the 'add' data are not equal, the Manchester representation at the input should arrive at the output in the reversed temporal order. This can be achieved by the adjustment of the gate such that the second bit is selected from the undelayed branch (thereby appearing as the first bit at the output) and the first bit is selected from the branch having a delay of $2T_B$ (and thus emerging as the second bit at the output).

The aim of the processing operation is to preserve the optical transparency to the physical signals that are processed. In most of the communication links, the characteristics of the physical signal that transports the information message (such as signal wavelength, signal power, modulation format, and bit rate) are established at the transmitter and absorbed at the receiver. For example, visualise an optical transmitter and receiver in a point-to-point transmission link. The properties of the information-carrying signal are defined at the transmitter. At the receiver, upon reception of the information, the information carrying signal is absorbed and its properties are lost when not constantly being monitored or adequately measured. Figure 5.4 shows the block diagram of the communication channel.

In many circumstances, the physical signal characteristics at the receiver are still of sufficiently high quality to be reused for other communication sessions. The OSP provides the means to reuse the physical signal while, at the same time, it allows one to substitute a new digital information content on it. Hence, the OSP can be regarded as optically transparent to (or independent of) certain physical signal characteristics.

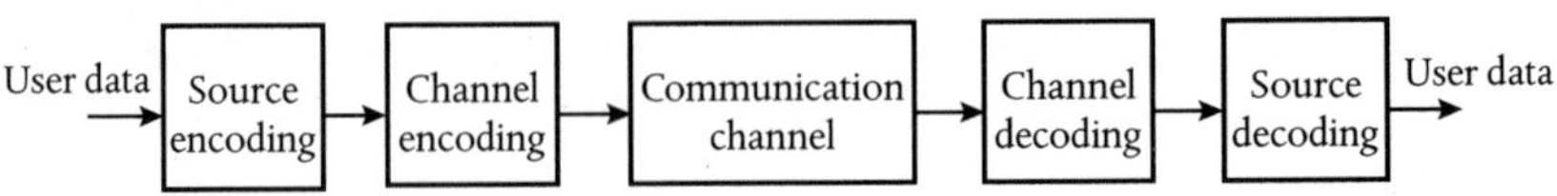

Fig. 5.4 Block diagram of the communication channel, including source and channel coding

5.3 FIBRE OPTIC LATTICE SIGNAL PROCESSING

The low loss (fractional decibels/kilometre) and large bandwidth-length product of optical fibres together with an advance in the manufacturing of fibre optic components provide an attractive technology for processing broadband signal.

Fibre-optic signal processing devices can be constructed to perform various functions, such as convolution, correlation, matrix operation, and frequency filtering, pulse train generation, pulse compression, and matched filtering are also achievable.

Previous work has concentrated on classical tapped delay-line forms (transverse filters). Various mechanisms such as radiation due to bending and evanescent coupling by polishing the cladding down very close to fibre core have been used to realise the taps. In this chapter, attention is on different fibre optic structure, namely lattice (or ladder) forms, which can be used as alternatives for performing optical signal processing.

Lattice structure has found increasing applications in signal processing during the past decades. This interest has been for the following reasons:

1. They allow a systematic approach to filter synthesis.
2. They have a very straightforward implementation.
3. They are both regular and modular. That is, the implementation is accomplished by interconnecting section having the same basic topology, making the assembly easily expandable.
4. They have convenient numerical properties that make them suitable for application to digital signal processing.
5. They provide a natural model for many important physical and mathematical phenomena such as speech signal generation and are useful in estimation problems.

In optic fibre form, we use single mode fibres and directional couplers in conjunction with semiconductor laser sources and optical detectors to realise a fibre-optic lattice processor. Higher order lattices are simply linear extensions of the basic sections.

The signals to be processed are modulated as intensity variation on optical carriers whose coherence time is less than the shortest relevant time delay in the system. This restriction is to avoid environmentally sensitive optical interference (phase) effects. In this chapter, we consider only fibre optic systems, which are linear in the intensities of the propagating waves. It should be mentioned that we can use a short coherence length of the multimode semiconductor laser as the optical source. Furthermore, the output light intensities were detected by photodiodes whose outputs can be connected to measurements systems such as an oscilloscope, a network analyser, or a spectrum analyser and for which the output voltage is proportional to input intensities.

In this section, we introduced the elementary concepts that we needed to construct fibre optic signal processing devices. Using the concepts, we demonstrate two structures that are basic to fibre signal processing: two-coupler non-recirculating and two-coupler recirculating delay lines. These basic fibre structures are employed further to implement two different types (feed-forward and feed-backward configuration) of so-called lattice structure in their fibre optic forms.

We also discuss mathematical tools for the analysis of the lattice structure. These tools include transformation techniques and a modern control theory formulation. The transfer matrix and chain matrix characterisation are useful for the analysis of cascaded two-port system that include our fibre lattice structure. We utilise the Z-transform method, together with the transfer matrix and chain matrix formulation, to describe these structures for filtering applications.

The modern control theory formulation for these lattice structures uses the notion of state vector to represent the system with two-matrix equation: a state update equation and an output equation. This formulation is well-suited for the theory of positive systems that will be used further to draw some conclusion about the behaviour of optic fibre system incorporating positive-valued quantities. We shall see that this knowledge alone, namely, the fact that all the system quantities are non-negative in value, enables us to make some very strong statements about the behaviour of the system, such as pole–zero position, system stability, and sensitivity.

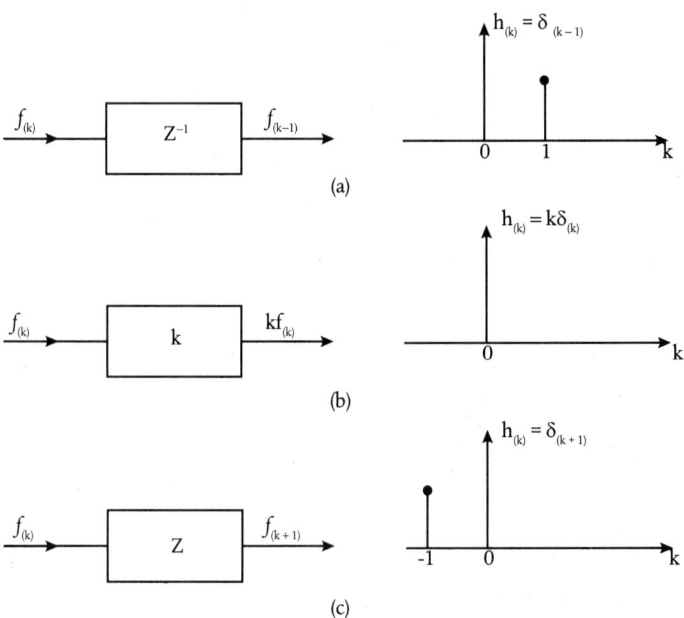

Fig. 5.5 Elementary concepts for signal processing: (a) time delay element; (b) tapping element; (c) time advance element.

Frequency domain signal processing application of fibre lattice structure is also presented. In this section, we use Z-transform methods to study the properties of fibre optic lattice structure when used as frequency filters with possible pole–zero patterns of various lattice orders are shown with a discussion of their implication for the analysis and design of various filter types. We also examine a few fibre optics filtering device, such as all poles, all zeros, and all pass (phase equaliser) filters, together with the presentation of some experimental and theoretical results and optical power efficiency calculations. To perform the various signal processing operations described, we need to employ the following elements.

5.3.1 Time Delay

The output of the delay elements is the input delayed signal by a unit of the time. That is, for a discrete-time input signal $f_{(k)}$, the output is $f_{(k-1)}$, where k represents the time. A block diagram representation and the impulse response (response to an intensity impulse input) of this element are shown in Fig. 5.5(a). Note that z^{-1} inside the block diagram represents a unit time delay.

Here, we have used a notation from the Z-transform employed to perform the time delay function. Assuming a refractive index of about 1.5 for the fibre, the propagation delay is $= 5\,\text{ns}/\text{m}$

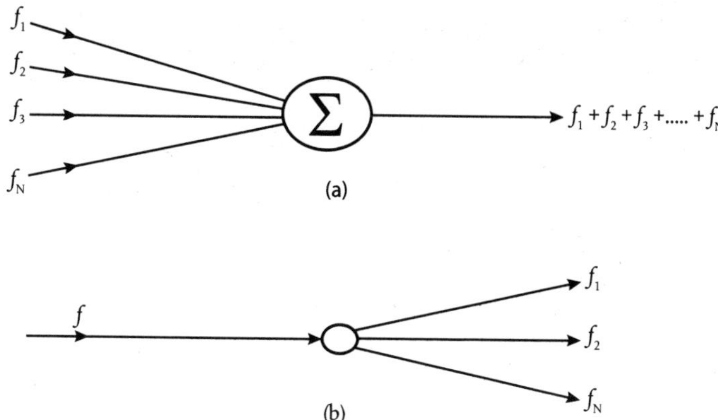

Fig. 5.6 Summing and branching elements: (a) summing elements; (b) branching element

5.3.2 Tapping

Various mechanisms, such as evanescent directional coupler or ("kinks"), can be used to accomplish tapping. Mathematically, the output of tapping elements is a constant multiple, *k*, of the input as shown in Fig. 5.5 (b).

5.3.3 Time Advance

The signal is advanced one unit of time rather than delayed as shown in Fig. 5.5 (c). Note that this element has only mathematical meaning and those systems requiring units advanced in their construction are not physically realisable.

5.3.4 Summing Element

This subsystem sums incoming signals. In fibre optic system, a directional coupler can be used to do the summing operation, which is shown in Fig. 5.6(a).

5.3.5 Branching Element

This subsystem divides the incoming signal into several parts. A directional coupler can also be used to divide the incoming light signal into two parts. This is shown in Fig. 5.6(b).

Recirculating (feed backward) and non-recirculating (feed forward) delay lines are two structures that are basic to fibre optic signal processing. Using single-mode fibres and directional coupler in conjunction with semiconductor for laser sources, one can construct a two-coupler recirculating delay line. In the two-coupler recirculating delay line, a single-mode fibre loop (with loop delay *T*) is closed upon itself by two-directional coupler with the second coupler feeding some portion of the light to the first coupler through the fibre loop as shown in Fig. 5.7 (a).

Therefore, optical signal sent into the input port, say at x_1, circulates repeatedly around the loop, sending a portion of the recirculating intensity to output port y_1 and y_2. The impulse response of the system consists of a series of decaying impulses equally spaced in time by the loop delay T, as shown in Fig. 5.7(c). A two-coupler recirculating delay line has already been used for frequency filtering.

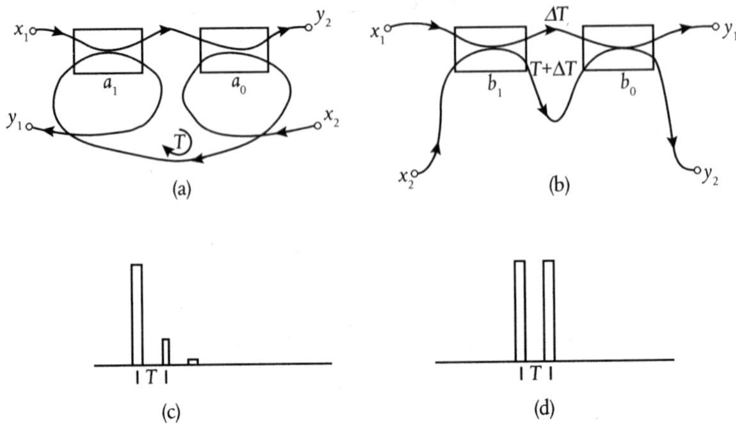

Fig. 5.7 Basic fibre structure: (a) two-coupler recirculating (feed backward) delay line; (b) two-coupler non-recirculating (feed forward) delay line; (c) impulse response of the two-coupler recirculating delay line; (d) impulse response of the two-coupler non-recirculating delay line.

Figure 5.7 (b) shows the two-coupler non-recirculating delay line also consists of two directional couplers, but in this case, the output of the first coupler are both fed forward to the second coupler where they are recombined after a time delay T, which is the time delay difference between the two feed-forward fibre lines. The impulse response of the system is composed of just two impulses spaced by the amount T in time as shown in Fig. 5.7(d). The input–output relationship of the directional coupler, as shown in Fig. 5.8, can describe by a 2×2 complex transfer matrix.

$$\begin{pmatrix} E_3 \\ E_4 \end{pmatrix} = (\gamma)^{\frac{1}{2}} \begin{pmatrix} A & B \\ C & D \end{pmatrix} \cdot \begin{pmatrix} E_1 \\ E_2 \end{pmatrix} \tag{5.1}$$

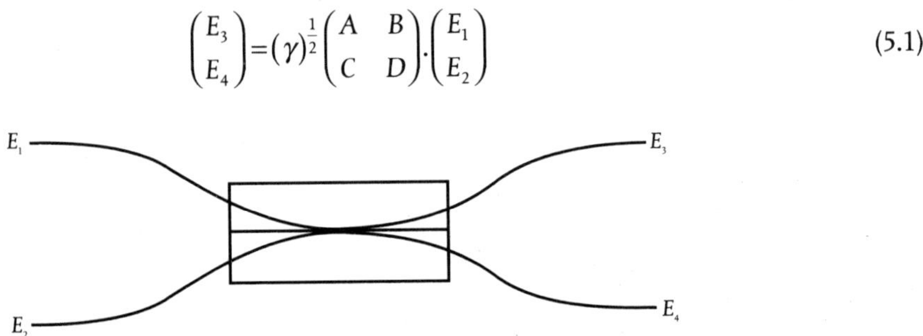

Fig. 5.8 Schematic diagram of a fibre optic directional coupler

where (E_1, E_2) and (E_3, E_4) are, respectively, the input and output complex amplitude of the vector fields. Hence, E_1, E_2, E_3, and E_4 are assumed, for simplicity, to have the same state of polarisation. It should be mentioned that the coupler itself exhibits very little dependence on the state of the polarisation of the input fields. Since backward reflection is negligible, the coupler matrix can be approximated by the product of an overall amplitude transmission factor $(\gamma)^{\frac{1}{2}}$ $(0.9<\gamma<1)$ and a unitary matrix. This property of the matrix implies that $|A|^2+|B|^2=|C|^2+|D|^2=1$ and $CA^*+DB^*=0$ (the star denotes the complex conjugation). For a symmetric coupler, ($A=D$ and $B=C$) can be rewritten as:

$$\begin{pmatrix} E_3 \\ E_4 \end{pmatrix}=(\gamma)^{\frac{1}{2}}\begin{pmatrix} \sqrt{1-\kappa} & j\sqrt{\kappa} \\ j\sqrt{\kappa} & \sqrt{1-\kappa} \end{pmatrix}\begin{pmatrix} E_1 \\ E_2 \end{pmatrix}. \tag{5.2}$$

Here, (5.2) is same as (5.1) but in matrix forms, same for the weakly coupled directional coupler, one can assume $\kappa l \approx \sqrt{\kappa}$, where square root of κ is the amplitude coupling coefficient and $j=\sqrt{-1}$ indicates a $90°$ phase shift for the coupled signal. Furthermore, in the case, when a temporarily incoherent light source is used, the input and output light intensities can be related to each other by the following expression:

$$\begin{pmatrix} I_3 \\ I_4 \end{pmatrix}=(\gamma)\begin{pmatrix} 1-\kappa & \kappa \\ \kappa & 1-\kappa \end{pmatrix}\begin{pmatrix} I_1 \\ I_2 \end{pmatrix}, \tag{5.3}$$

where (I_1,I_2) and (I_3,I_4) are the input and output intensities, respectively, γ is the overall intensity transmission factor, and κ is the intensity coupling coefficient. The above-mentioned two equations will be used in later section for our mathematical analysis.

The basic recirculating and non-recirculating delay lines described previously are used as elementary section of feed-backward and feed-forward lattice structures, respectively. These structures are the subject of our discussion in the next section.

5.3.6 Lattice Structure

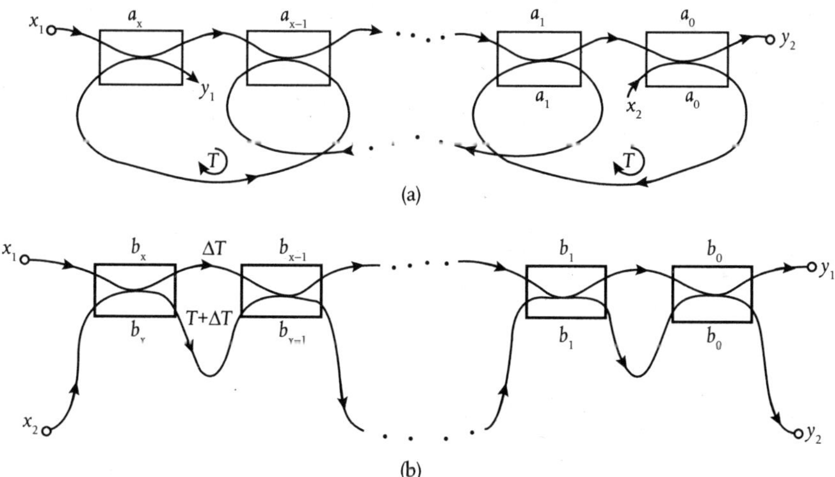

Fig. 5.9 Schematic diagrams of the N_{th} order fibre optic lattice structure with $(N+1)$ coupler and N delay elements (a) feed-backward configuration and (b) feed-forward configuration.

Lattice structures are suitable forms for performing the various signal processing operations described earlier. In comparison to other forms, lattice structure has some advantageous characteristics, such as modularity, regularity, ease of implementation, and good sensitivity. In this section, we introduce two fibre optic lattice configurations; recirculating (feed backward) and non-recirculating forms shown an N_{th} order feed-forward forms. Figure 5.9(a) shows an N_{th} order feed-backward (recirculating) fibre optic lattice structure. In this chapter, we are interested in this case where *T* is the same for all sections, but this restriction would not be necessary in general. This structure is obtained by interconnecting N two-coupler recirculating delay lines in tandem. As seen in Fig. 5.9(a), this structure has one feed-forward and one feed-backward fibre line. In contrast, Fig. 5.9(b) illustrates an N_{th} order feed-forward (non-recirculating) fibre lattice structure, which is implemented by cascading N two-coupler non-recirculating delay lines. In this configuration, there are two feed-forward lines with no feed-backward line.

Both of these lattice configurations have two ports with two terminals at each port (two inputs, two outputs). Figure 5.10 shows the equivalent diagram of the feed-backward and feed-forward lattice structure. These two lattice forms can also be combined to construct more sophisticated fibre optic structures.

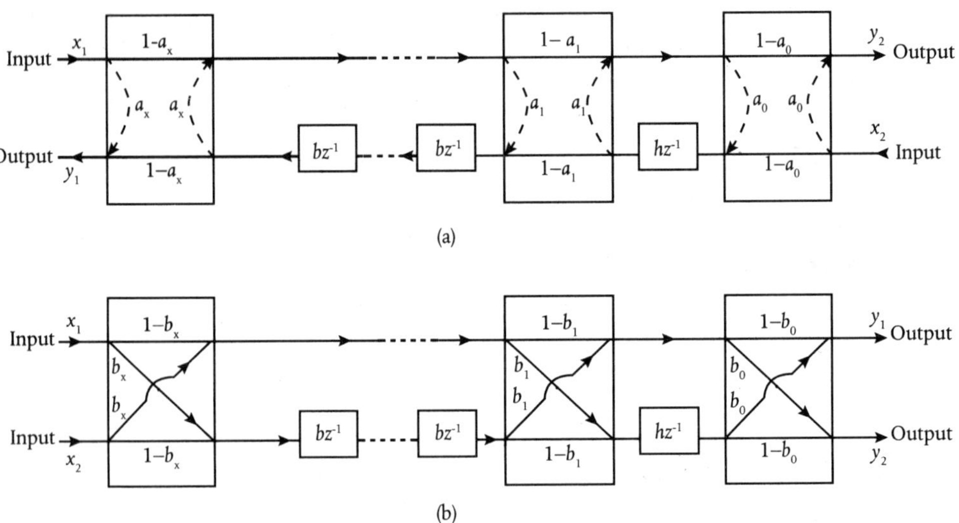

Fig. 5.10 Equivalent diagrams of the N_{th} order fibre optic lattice structure with (N+1) couplers and N delay elements (a) feed-backward configuration; (b) feed-forward configuration (I_1 is the intensity of the i_{th} section).

In the later section, we discuss the application of the various structures introduced here. Before that, however, we present different methods for mathematical analysis of the lattice structure.

5.4 MATHEMATICAL ANALYSIS: Z-TRANSFORMS TECHNIQUES

In order to increase our ability to handle more complex system problems, we utilise transform techniques for describing the signal. We particularly use the Z-transform, and later, we convert Z-transform into Laplace domain. This usage is justified since all the systems of interest here are, first, linear and time invariant and second discrete in time. The latter is due to the fact that we can define a basic time delay T such that any other relevant system delays are integer multiples of the basic time delay. That is, the impulse response of the system is the series of the impulses that are equally spaced in time. As a result, we can simplify the mathematical analysis of such system by considering the values of the system signals only at discrete instants in time. Since the input and output are described by in terms of the discrete samples, these systems are called sample data systems.

The mathematical analysis of discrete systems is essentially similar to that of continuous data system, but it is simpler and more intuitive in terms of physical interpretation. The Z-transform, $F(z)$ of a signal, $f(k)$, is defined by the following expression:

$$F(z) \triangleq \sum_{k=-\infty}^{\infty} f(k).z^{-k}, \tag{5.4}$$

where k is the time index, which is an integer multiple of the basic time delay ($k = nT$), and z is the transform variable that represents a unit time advance (z^{-1} represents a unit time delay).

We also describe the input–output relationship by system's transfer function (the ratio of the Z-transform of the input) whose poles and zeros are very important in the design and analysis frequency selective filters. The frequency response is obtained when we evaluate the transfer function at $z = e^{j\omega T}$. Further, since we know that $s = j\omega$, $z = e^{sT}$. Hence, we can easily connect Z-transform domain into s-domain. Here, s-domain is Laplace transform and we can easily apply control theory concepts on these transfer formulas.

This $e^{j\omega T}$ describes a unit circle centred at the origin of the Z-plane. The unit circle also plays an important role in problems regarding stability. Figure 5.11 shows z-plane with the unit circle.

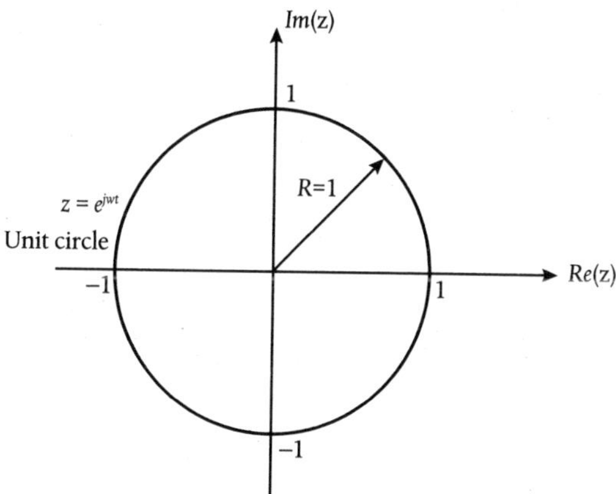

Fig. 5.11 Z-plane with the unit circle.

To ensure stability, all the poles of the system's transfer function must be inside the unit circle or in s-domain, all the poles–zeros should be in entire left half of s-plane. Fibre lattice filters incorporating passive elements (no gain) are sure to be stable. Furthermore, if all the zeros are also inside the unit circle, then the system is called a minimum phase system; on the other hand, if any zero lies outside the unit circle, the system is called the non-minimum phase system. It should be mentioned that for a minimum phase filter, the phase filter, the phase response, and the logarithm of the magnitude response are a Hilbert transform pair. This is, in addition, to the property of all real, stable, causal systems that the real and imaginary parts of the frequency response are Hilbert transformation of each other.

Using the definition of the Z-transform, we find the following expression for the system's transfer functions, $H(z)$, of the two-coupler delay lines introduced earlier.

5.4.1 Case (A)

Two-coupler recirculating delay line are X_1 as the input and Y_1 as the output. The relation between input and output for two-coupler recirculating system, as shown in Fig. 5.12, given as:

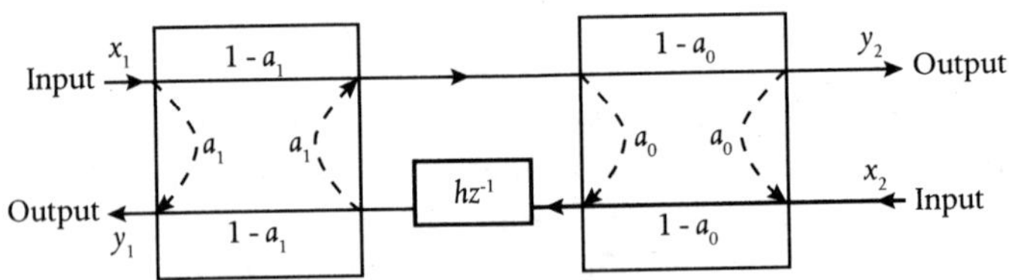

Fig. 5.12 Two-coupler recirculating delay line

$$Y_1 = a_1 X_1 + (1-a_1)^2 a_0 l_1 X_1 z^{-1} + (1-a_1)(1-a_0) l_1 z^{-1} X_2$$

When $X_2 = 0$,

$$Y_1 = a_1 X_1 + (1 - 2a_1 + a_1^2) a_0 l_1 X_1 z^{-1}$$

$$Y_1 = a_1 X_1 + (1 - 2a_1) a_0 l_1 X_1 z^{-1} + a_1^2 a_0 l_1 X_1 z^{-1}$$

Now, we can write:

$$Y_1 = a_1 X_1 + (1 - 2a_1) a_0 l_1 X_1 z^{-1} + a_1 a_0 l_1 Y_1 z^{-1}$$

$$\frac{Y_1}{X_1} = \frac{a_1 + (1-2a_1) a_0 l_1 z^{-1}}{1 - a_1 a_0 l_1 z^{-1}}$$

Hence,

$$H_{11}(z) = \frac{a_1 + (1-2a_1) a_0 l_1 z^{-1}}{1 - a_1 a_0 l_1 z^{-1}}, \quad (5.5)$$

where a_1 and a_0 are the intensity coupling coefficient of the couplers, and l_1 is the loop intensity transmittance of the system. As shown, this system has one zero at:

$$z = \frac{(2a_1 - 1)}{a_1} a_0 l_1. \quad (5.6)$$

Figure 5.13 shows the plot of normalised magnitude and phase response for this case.

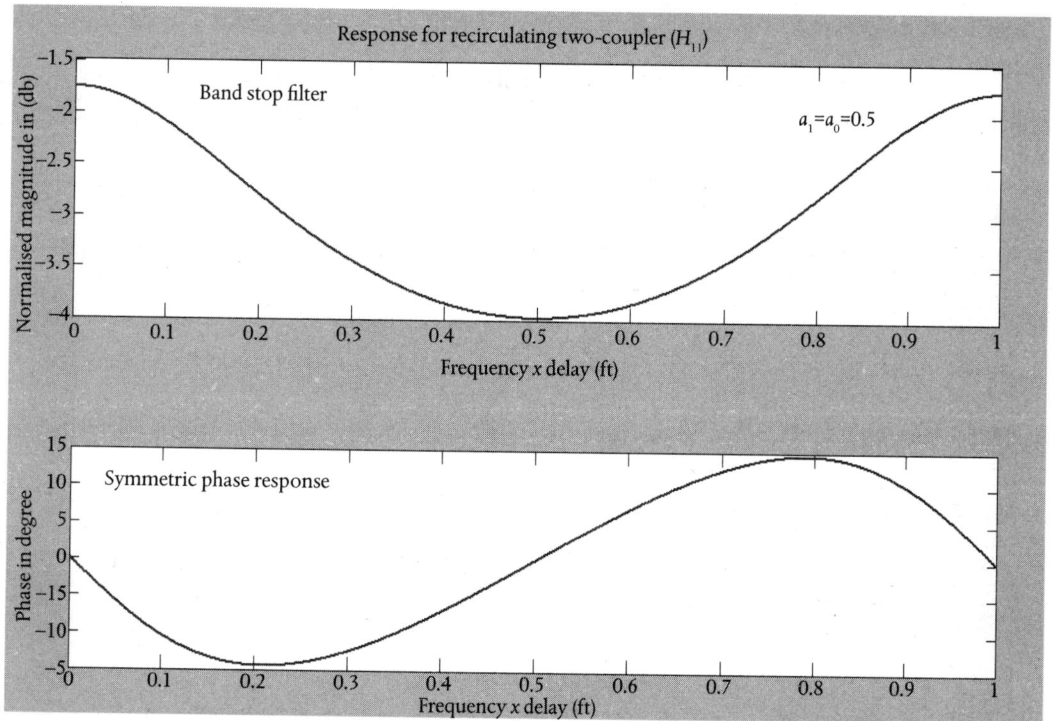

Fig. 5.13 Plot of normalised magnitude and phase response for recirculating two-coupler systems (H_{11}).

5.4.2 Case (B)

Two-coupler recirculating delay line with X_1 as the input and Y_2 as the output as shown in Fig. 5.12.

We can write the relation between the input X_1, X_2 and output Y_2 as follows:

$$Y_2 = (1-a_1)(1-a_0)l_{11}X_1 + a_1(1-a_0)^2 l_1 X_2 z^{-1} + a_0 X_2$$

$$Y_2 = (1-a_1)(1-a_0)l_{11}X_1 + (1-2a_0 + a_0^{\ 2})a_1 l_1 X_2 z^{-1} + a_0 X_2$$

$$Y_2 = (1-a_1)(1-a_0)l_{11}X_1 + (1-2a_0)a_1 l_1 X_2 z^{-1} + a_0^{\ 2} a_1 l_1 X_2 z^{-1} \tag{5.7}$$

Substituting $a_0 X_2 = Y_2$ in third term and $X_2 = 0$ in second term of above-mentioned equation:

$$Y_2 = (1-a_1)(1-a_0)l_{11}X_1 + a_0 a_1 l_1 Y_2 z^{-1}. \tag{5.8}$$

Therefore,

$$\frac{Y_2}{X_1} = \frac{(1-a_1)(1-a_0)\,l_{11}}{1-a_1 a_0 l_1 z^{-1}}.$$

Hence,

$$H_{21}(z)=\frac{(1-a_1)(1-a_0)\iota_{11}}{1-a_1a_0\iota_1z^{-1}}. \tag{5.9}$$

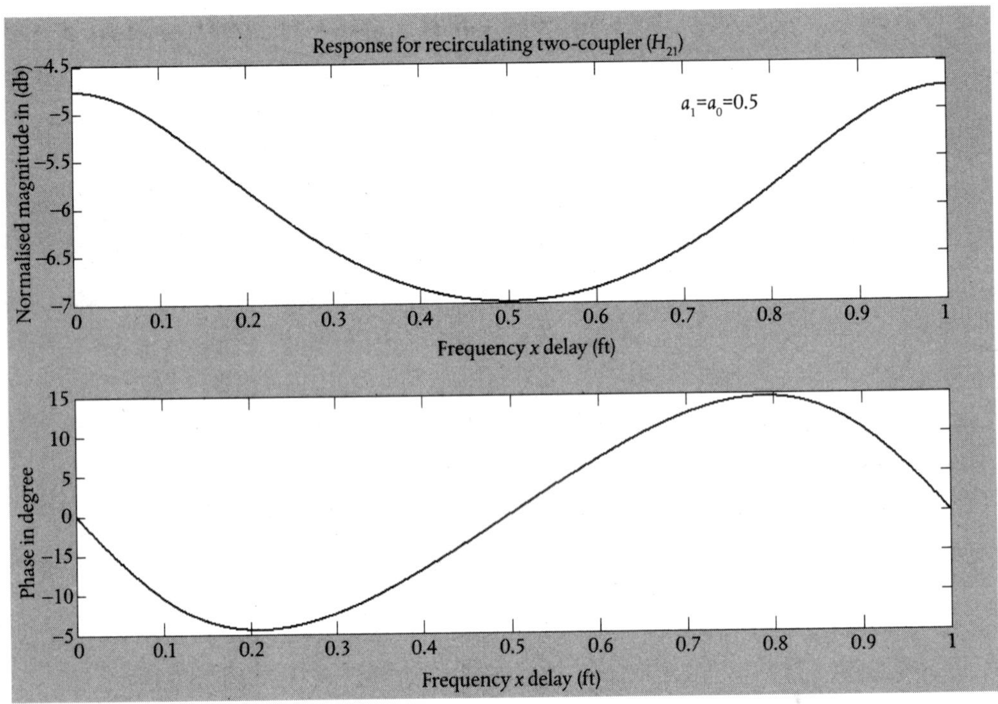

Fig. 5.14 Plot of normalised magnitude and phase response for recirculating two-coupler systems (H_{21}).

where l_{11} is the intensity transmission factor of the forward fibre line. In this case, the system has one zero at the origin and one pole at $z=a_1a_0\iota_1$. Figure 5.14 shows the plot of normalised magnitude and phase response for this case.

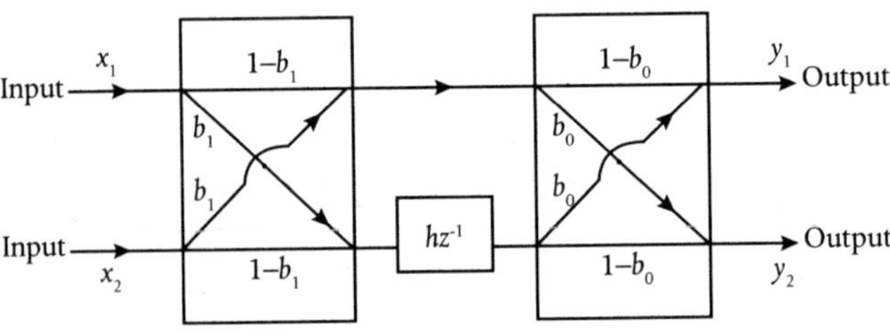

Fig. 5.15 Two-coupler non-recirculating delay line

5.4.3 Case (C)

Two-coupler non-recirculating delay line, is shown in Fig. 5.15, with X_1 as the input and Y_1 as the output. As we know that relation between the input and the output in the case of two-coupler systems.

$$Y_1 = (1-b_1)(1-b_0)X_1 l_{11} + b_1 b_0 X_1 z^{-1} + (1-b_1)b_0 l_{12} z^{-1} X_2 + (1-b_0)b_1 l_{11} X_2$$

As X_2 is zero, thus

$$Y_1 = (1-b_1)(1-b_0)X_1 l_{11} + b_1 b_0 X_1 z^{-1}$$

$$\frac{Y_1}{X_1} = (1-b_1)(1-b_0)\,l_{11} + b_1 b_0\, l_{12} z^{-1}$$

Hence,

$$H_{11}(z) = (1-b_1)(1-b_0)\,l_{11} + b_1 b_0\, l_{12} z^{-1}, \tag{5.10}$$

where b_1 and b_0 are the intensity coupling coefficient of the couplers, and l_{11} and l_{12} are the intensity transmittances in the two feed-forward fibres lines. This system has one pole at the origin and one zero at:

$$z = -\frac{b_1 b_0\, l_{12}}{(1-b_1)(1-b_0)\,l_{11}}. \tag{5.11}$$

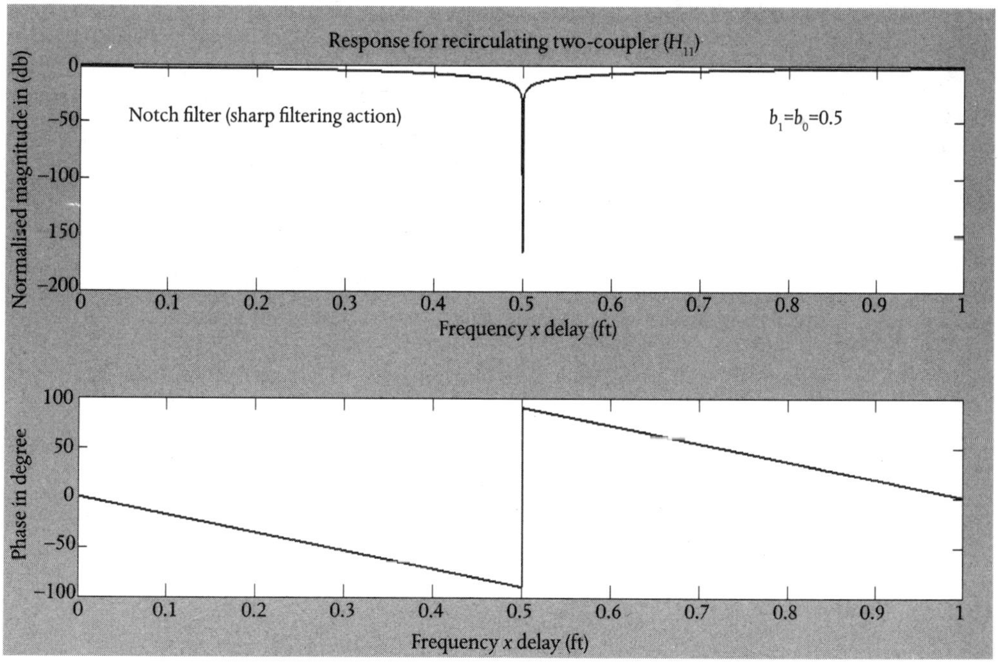

Fig. 5.16 Plot of normalised magnitude and phase response for non-recirculating two-coupler system (H_{11})

Note that those zeros or poles that are located at the origin do not affect the frequency response of the system (except for a linear phase factor). For this reason, we refer to the two-coupler recirculating delay line, with Y_2 as the output, as a first-order (one pole) all-pole system. Figure 5.16 shows the plot of normalised magnitude and phase response for this case. The two-coupler non-recirculating delay lines, with Y_2 as the output, are referred to as first-order (one zero) all zeros and first-order (one pole) pole–zero system, respectively. The pole–zero diagram of the system discussed in this section is illustrated in Fig. 5.17.

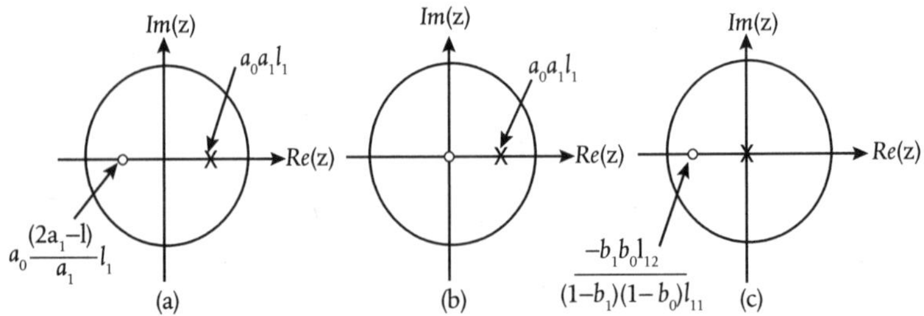

Fig. 5.17 Pole–zero diagrams of the two-coupler delay lines with X_1 as the input. (a) Two-coupler recirculating delay line, with Y_1 as the output. (b) Two-coupler recirculating delay line with Y_2 as the output. (c) Two-coupler non-recirculating delay line with Y_1 as the output.

This diagram contains the information about the frequency response of the corresponding system.

5.4.4 Feed-forward or Non-recirculating Configuration (Four-coupler System)

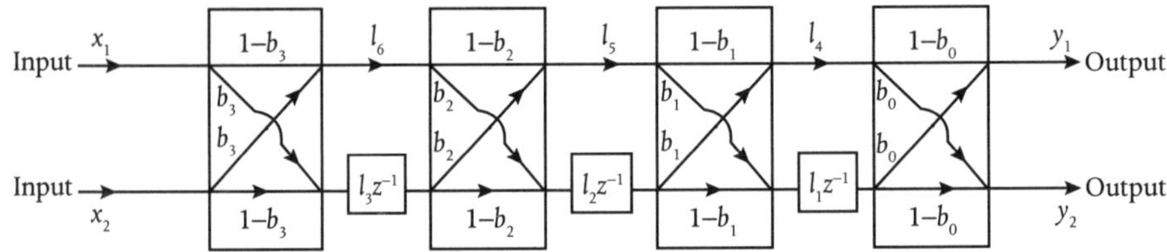

Fig. 5.18 Feed-forward or non-recirculating four-coupler system

Figure 5.18 shows the configuration for this case. The output Y_1 in terms of X_1 and X_2 can be written as:

$$
\begin{aligned}
Y_1 =\ &(1-b_3)(1-b_2)(1-b_0)l_1l_5l_4X_1 + b_3l_3z^{-1}b_2l_5(1-b_1)l_4(1-b_0)X_1 \\
&+(1-b_3)l_6l_2z^{-1}b_1l_4(1-b_0)X_1 + (1-b_3)l_6(1-b_2)l_5b_1l_1z^{-1}b_0X_1 \\
&+b_3l_3z^{-1}b_2l_5b_1l_1z^{-1}b_0X_1 + b_3l_3z^{-1}(1-b_2)l_2z^{-1}b_1l_4(1-b_0)X_1 \\
&+b_3l_3z^{-1}(1-b_2)l_2z^{-1}(1-b_1)l_1z^{-1}b_0X_1 + (1-b_3)l_3z^{-1}b_2l_5(1-b_1)l_4 \\
&(1-b_0)X_2 + b_3l_6(1-b_2)l_5(1-b_1)l_4(1-b_0)X_2 + (1-b_3)l_3z^{-1}(1-b_2)l_2 \\
&z^{-1}b_1l_4(1-b_0)X_2 + (1-b_3)l_3z^{-1}(1-b_2)l_2z^{-1}(1-b_1)l_1z^{-1}b_0X_2 + b_3l_6b_2
\end{aligned}
$$

$$l_2 z^{-1} b_1 l_4 (1-b_0) X_2 + b_3 l_6 b_2 l_2 z^{-1} b_1 l_4 (1-b_0) X_2 + b_3 l_6 (1-b_2) l_5 b_1 l_1 z^{-1} b_0$$
$$X_2 + b_3 l_6 b_2 l_2 z^{-1} l_4 (1-b_0) b_1 X_2$$

Now, we can calculate the value of H_{11} as:

$$H_{11} = \frac{Y_1}{X_1},$$

where $X_2 = 0$

$$\begin{aligned} Y_1 = {} & (1-b_0)(1-b_1)(1-b_2)(1-b_3) l_4 l_5 l_6 X_1 + (1-b_0)(1-b_1) b_2 b_3 l_3 l_4 l_5 z^{-1} X_1 \\ & + (1-b_0) b_1 b_2 (1-b_3) l_2 l_4 l_6 z^{-1} X_1 + b_0 b_1 (1-b_2)(1-b_3) l_1 l_5 l_6 z^{-1} X_1 + \\ & b_0 b_1 b_2 b_3 l_1 l_3 l_5 z^{-2} X_1 + (1-b_0) b_1 (1-b_2) b_3 l_2 l_3 l_4 z^{-2} X_1 + b_0 (1-b_1)(1-b_2) b_3 l_1 l_2 l_3 z^{-3} X_1 . \end{aligned}$$

Hence,

$$\begin{aligned} H_{11}(z) = {} & (1-b_1)(1-b_2)(1-b_3) l_4 l_5 l_6 + (1-b_0)(1-b_1) b_2 b_3 l_3 l_4 l_5 z^{-1} \\ & + (1-b_0) b_1 b_2 (1-b_3) l_2 l_4 l_6 z^{-1} + b_0 b_1 (1-b_2)(1-b_3) l_1 l_5 l_6 z^{-1} \\ & + b_0 b_1 b_2 b_3 l_1 l_3 l_5 z^{-2} + (1-b_0) b_1 (1-b_2) b_3 l_2 l_3 l_4 z^{-2} \\ & + b_0 (1-b_1)(1-b_2) b_3 l_1 l_2 l_3 z^{-3}. \end{aligned} \tag{5.12}$$

The magnitude and phase response of four-coupler system can be represented as shown in Fig. 5.19.

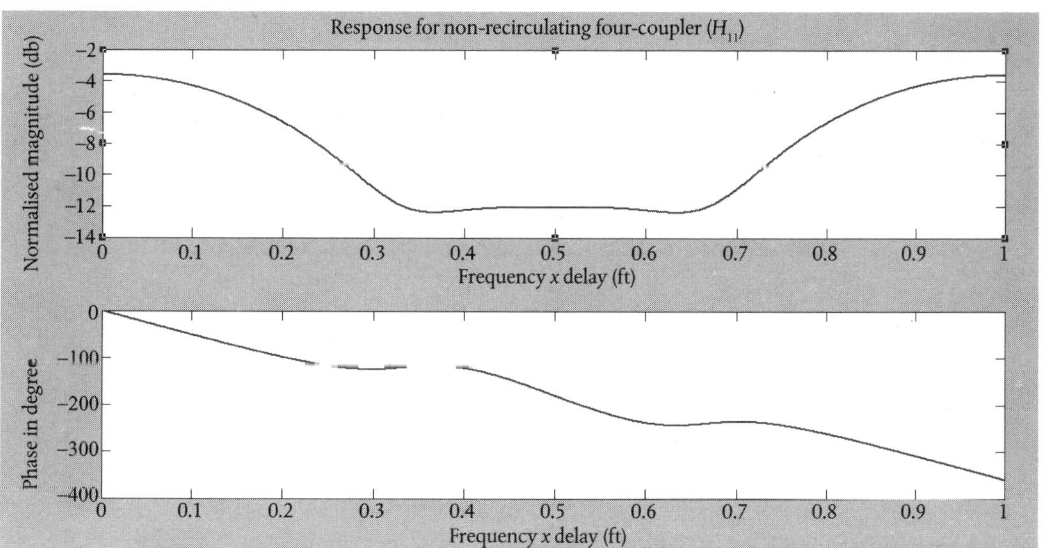

Fig. 5.19 Plot of normalised magnitude and phase response for non-recirculating four-coupler system (H_{11})

It is apparent from Fig. 5.19 that the bandwidth of band rejection increased substantially as compared with Fig. 5.13. Present case also shows the nonlinear phase response. However, no sharp cut-off frequency being achieved from this filter design.

Similarly,

$$H_{12} = \frac{Y_1}{X_2},$$

where $X_1 = 0$

$$\begin{aligned} Y_1 = {} & (1-b_0)(1-b_1)b_2(1-b_3)l_3l_4l_5z^{-1}X_2 + (1-b_0)(1-b_1)(1-b_2)b_3l_4l_5l_6X_2 + \\ & (1-b_0)b_1(1-b_2)(1-b_3)l_2l_3l_4z^{-2}X_2 + b_0(1-b_1)(1-b_2)(1-b_3)l_1l_2l_3z^{-3}X_2 + \\ & (1-b_0)b_1b_2b_3l_2l_4l_6z^{-1}X_2 + b_0b_1(1-b_2)b_3l_1l_5l_6z^{-1}X_2 + (1-b_0)b_1b_2b_3l_2l_4l_6z^{-1}X_2. \end{aligned}$$

Hence,

$$\begin{aligned} H_{12}(z) = {} & (1-b_0)(1-b_1)b_2(1-b_3)l_3l_4l_5z^{-1} + (1-b_0)(1-b_1)(1-b_2)b_3l_4l_5l_6 \\ & + (1-b_0)b_1(1-b_2)(1-b_3)l_2l_3l_4z^{-2} + b_0(1-b_1)(1-b_2)(1-b_3)l_1l_2l_3z^{-3} \\ & + (1-b_0)b_1b_2b_3l_2l_4l_6z^{-1} + b_0b_1(1-b_2)b_3l_1l_5l_6z^{-1} \\ & + (1-b_0)b_1b_2b_3l_2l_4l_6z^{-1}. \end{aligned} \tag{5.13}$$

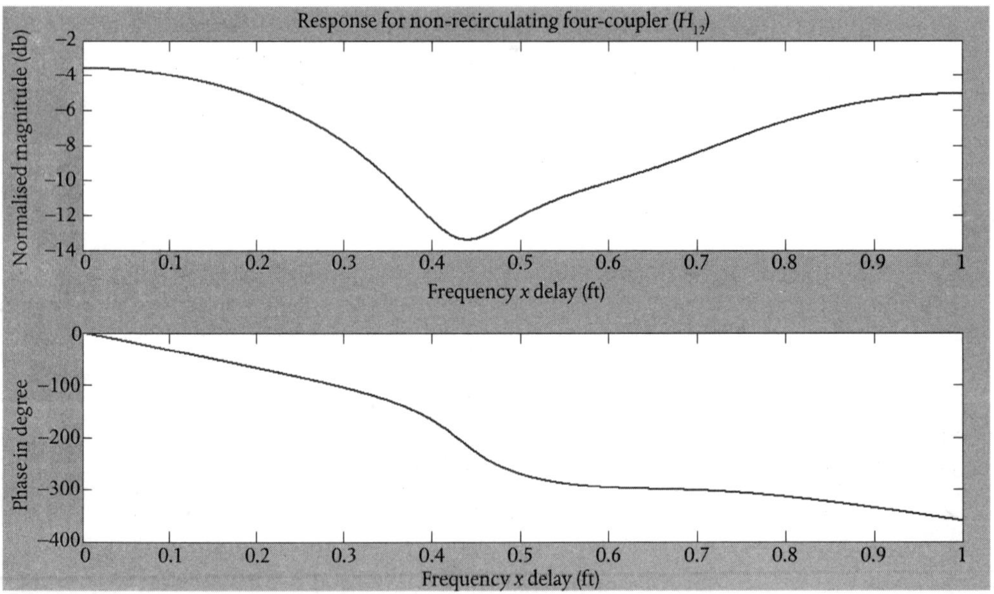

Fig. 5.20 Plot of normalised magnitude and phase response for non-recirculating four-coupler system (H_{12}).

Figure 5.20 shows the response of normalised magnitude and phase response of four-coupler non-recirculating system. Figure 5.20 shows the nonlinear response. There seem to be no advantages from this filter design.

Similarly, we can write the output Y_2 as:

$$\begin{aligned}Y_2 = {} & (1-b_3)l_6(1-b_2)l_5(1-b_1)l_4 b_0 X_1 + (1-b_3)l_6(1-b_2)l_5 b_1 l_1 z^{-1}(1-b_0)X_1 + \\ & (1-b_3)l_6 b_2 l_2 z^{-1}(1-b_1)l_1 z^{-1}(1-b_0)X_1 + b_3 l_3 z^{-1}(1-b_2)l_2 z^{-1}(1-b_1)l_1 z^{-1}(1-b_0)X_1 + \\ & b_3 l_3 z^{-1} b_2 l_5 b_1 l_1 z^{-1}(1-b_0)X_1 + (1-b_3)l_6 b_2 l_2 z^{-1} b_1 l_4 b_0 X_1 + (1-b_3)l_3 z^{-1}(1-b_2)l_2 z^{-1}(1-b_1)l_1 z^{-1}(1-b_0)X_2 + \\ & b_3 l_6(1-b_2)l_5(1-b_1)l_4 b_0 X_2 + b_3 l_6(1-b_2)l_5 b_1 l_1 z^{-1}(1-b_0)X_2 + (1-b_3)l_3 z^{-1}(1-b_2)l_2 z^{-1} b_1 l_4 b_0 X_2.\end{aligned}$$

$H_{21} = \dfrac{Y_2}{X_1}$, where $X_2 = 0$

Hence, we can write:

$$\begin{aligned}Y_2 = {} & b_0(1-b_1)(1-b_2)(1-b_3)l_4 l_5 l_6 X_1 + (1-b_0)b_1(1-b_2)(1-b_3)l_1 l_5 l_6 z^{-1} X_1 \\ & + (1-b_0)(1-b_1)b_2(1-b_3)l_1 l_2 l_6 z^{-2} X_1 \\ & + (1-b_0)(1-b_1)(1-b_2)b_3 l_1 l_2 l_3 z^{-3} X_1 + (1-b_0)b_1 b_2 b_3 l_1 l_3 l_5 z^{-2} X_1 \\ & + b_0 b_1 b_2 (1-b_3) l_2 l_4 l_6 z^{-1} X_1\end{aligned}$$

$$\begin{aligned}H_{21} = {} & b_0(1-b_1)(1-b_2)(1-b_3)l_4 l_5 l_6 + (1-b_0)b_1(1-b_2)(1-b_3)l_1 l_5 l_6 z^{-1} \\ & + (1-b_0)(1-b_1)b_2(1-b_3)l_1 l_2 l_6 z^{-2} + (1-b_0)(1-b_1)(1-b_2)b_3 l_1 l_2 l_3 z^{-3} \\ & + (1-b_0)b_1 b_2 b_3 l_1 l_3 l_5 z^{-2} + b_0 b_1 b_2 (1-b_3) l_2 l_4 l_6 z^{-1}.\end{aligned} \tag{5.14}$$

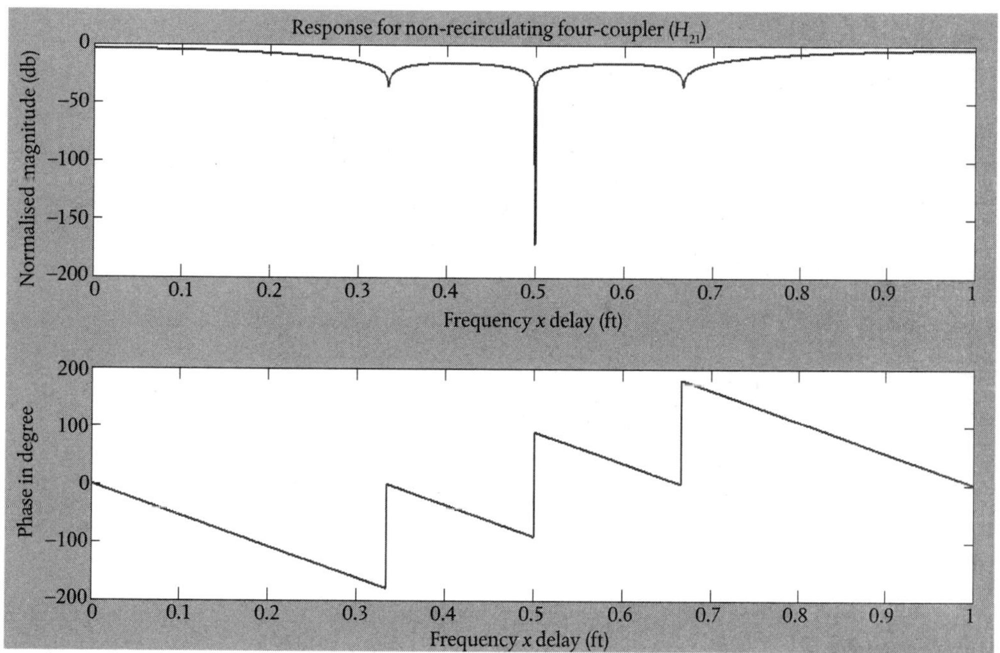

Fig. 5.21 Plot of normalised magnitude and phase response for non-recirculating four-coupler system (H_{21})

Figure 5.21 represents the normalised magnitude and phase response of the four-coupler non-recirculating system. The plot shows the decrement in normalised magnitude several times. The phase response changes several times at the corresponding point. It is apparent from the plot that frequency responses of filter have multiple ripples in its magnitude and phase response plot. To calculate of H_{22},

$$H_{22} = \frac{Y_2}{X_2}, \text{where } X_1 = 0$$

Hence,

$$\begin{aligned} Y_2 = {} & (1-b_0)(1-b_1)(1-b_2)(1-b_3)l_1l_2l_3z^{-3}X_2 + b_0(1-b_1)(1-b_2)b_3l_4l_5l_6X_2 + \\ & (1-b_0)b_1(1-b_2)b_3l_1l_5l_6z^{-1}X_2 + (1-b_0)(1-b_1)b_2b_3l_1l_2l_6z^{-2}X_2 + \\ & b_0(1-b_1)b_2(1-b_3)l_3l_4l_5z^{-1}X_2 + (1-b_0)b_1b_2(1-b_3)l_1l_3l_5z^{-2} + b_0b_1(1-b_2)l_2l_3l_4z^{-2}X_2. \end{aligned}$$

Therefore,

$$\begin{aligned} H_{22}(z) = {} & b_0(1-b_1)(1-b_2)b_3l_4l_5l_6 + (1-b_0)b_1(1-b_2)b_3l_1l_5l_6z^{-1} + b_0(1-b_1)b_2(1-b_3)l_3l_4l_5z^{-1} + \\ & (1-b_0)(1-b_1)b_2b_3l_1l_2l_6z^{-2} + (1-b_0)b_1b_2(1-b_3)l_1l_3l_5z^{-2} \\ & + b_0b_1(1-b_2)l_2l_3l_4z^{-2} + (1-b_0)(1-b_1)(1-b_2)(1-b_3)l_1l_2l_3z^{-3}. \end{aligned} \tag{5.15}$$

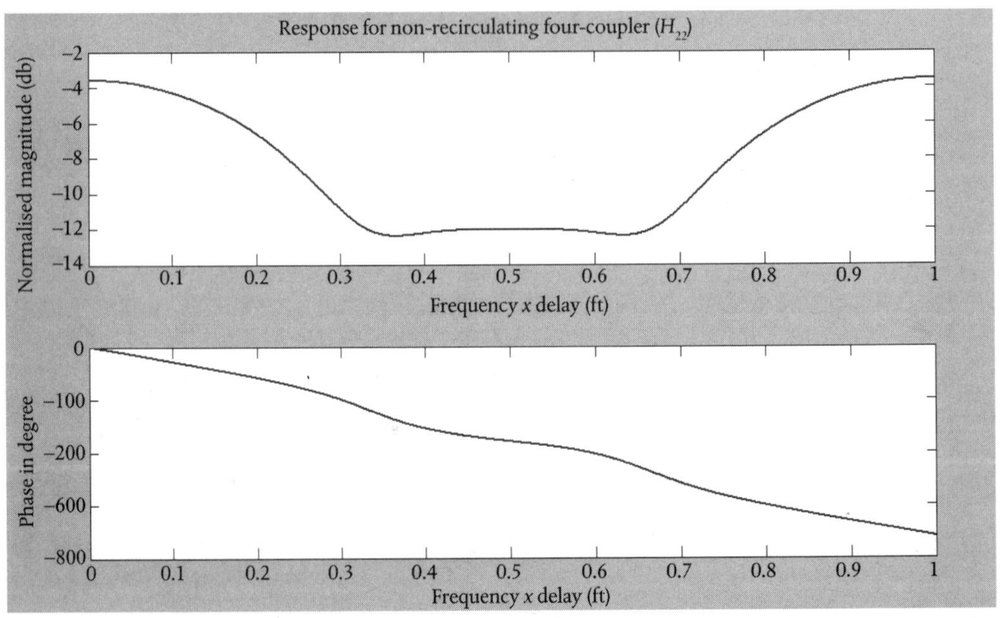

Fig. 5.22 Plot of normalised magnitude and phase response for non-recirculating four-coupler system (H_{22})

Figure 5.22 shows the variation of normalised magnitude and phase for the four-coupler non-recirculating system (H_{22}). The plot is very much similar to that of Fig. 5.19. This can be used as band stop filter.

5.4.5 Recirculating System (Four-coupler System)

The four-coupler system for this can be represented as shown in Fig. 5.23.

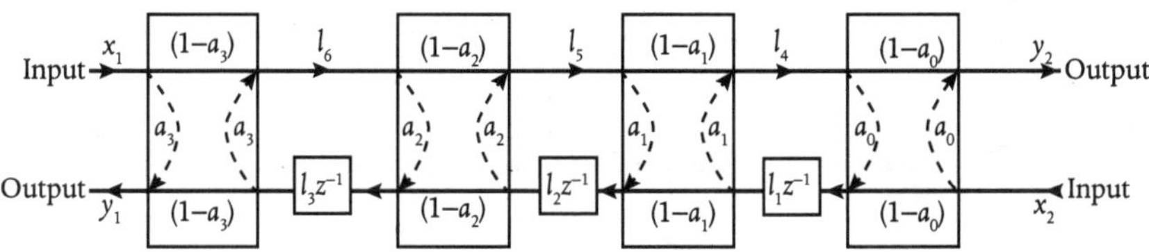

Fig. 5.23 Feed-backward or recirculating four-coupler systems

The output Y_1 can be represented as:

$$\begin{aligned}Y_1 = {}& a_3X_1 + (1-a_3)l_6a_2l_3z^{-1}(1-a_3)X_1 + (1-a_3)l_6(1-a_2)l_5a_1l_2z^{-1}(1-a_2)l_3z^{-1}X_1(1-a_3) + \\ & (1-a_3)l_6(1-a_2)l_5(1-a_1)l_4a_0l_1z^{-1}l_2z^{-1}l_3z^{-1}(1-a_3)X_1(1-a_1)(1-a_2) + \\ & (1-a_0)l_1z^{-1}(1-a_1)l_2z^{-1}(1-a_2)l_3z^{-1}(1-a_3)X_2.\end{aligned}$$

In order to calculate H_{11},

$$H_{11} = \frac{Y_1}{X_1},$$

where $X_2 = 0$; hence,

$$\begin{aligned}Y_1 = {}& a_3X_1 + a_2(1-a_3)^2l_3l_6z^{-1}X_1 + a_1(1-a_2)^2(1-a_3)^2l_2l_3l_5l_6z^{-2} + \\ & a_0(1-a_1)^2(1-a_2)^2(1-a_3)^2l_1l_2l_3l_4l_5l_6X_1z^{-3}\end{aligned}$$

or

$$\begin{aligned}Y_1 = {}& a_3X_1 + a_2(1-2a_3+a_3^2)l_3l_6z^{-1}X_1 + a_1(1-a_2)^2(1-2a_3+a_3^2)l_2l_3l_5l_6z^{-2}X_1 + \\ & a_0(1-a_1)^2(1-a_2)^2(1-2a_3+a_3^2)l_1l_2l_3l_4l_5l_6z^{-3}X_1\end{aligned}$$

$$\begin{aligned}Y_1 = {}& a_3X_1 + a_2(1-2a_3)l_3l_6z^{-1}X_1 + a_3^2l_3l_6z^{-1}X_1 \\ & + a_1(1-a_2)^2(1-2a_3)l_2l_3l_5l_6z^{-2}X_1 + a_1(1-a_2)^2a_3^2l_2l_3l_5l_6z^{-2}X_1 \\ & + a_0(1-a_1)^2(1-a_2)^2(1-2a_3)l_1l_2l_3l_4l_5l_6z^{-3}X_1 \\ & + a_0(1-a_1)^2(1-a_2)^2a_3^2l_1l_2l_3l_4l_5l_6z^{-3}X_1\end{aligned}$$

Substituting, $a_3X_1 = Y_1$, we get:

$$\begin{aligned}& Y_1\left[1 - a_2a_3l_3l_6z^{-1} - a_1a_3(1-a_2)^2l_2l_3l_5l_6z^{-2} - a_0(1-a_1)^2(1-a_2)^2a_3l_1l_2l_3l_4l_5l_6z^{-3}\right] \\ & \quad = [a_3 + a_2(1-2a_3)l_3l_6z^{-1} + a_1(1-a_2)^2(1-2a_3)l_2l_3l_5l_6z^{-2} \\ & \quad + a_0(1-a_1)^2(1-a_2)^2(1-2a_3)l_1l_2l_3l_4l_5l_6z^{-3}].\end{aligned}$$

Hence,

$$H_{11} = \frac{a_3 + a_2(1-2a_3)l_3 l_6 z^{-1} + a_1(1-a_2)^2(1-2a_3)l_2 l_3 l_5 l_6 z^{-2} + a_0(1-a_1)^2(1-a_2)^2(1-2a_3)l_1 l_2 l_3 l_4 l_5 l_6 z^{-3}}{\left[1 - a_2 a_3 l_3 l_6 z^{-1} - a_1 a_3 (1-a_2)^2 l_2 l_3 l_5 l_6 z^{-2} - a_0(1-a_1)^2(1-a_2)^2 a_3 l_1 l_2 l_3 l_4 l_5 l_6 z^{-3}\right]} \tag{5.16}$$

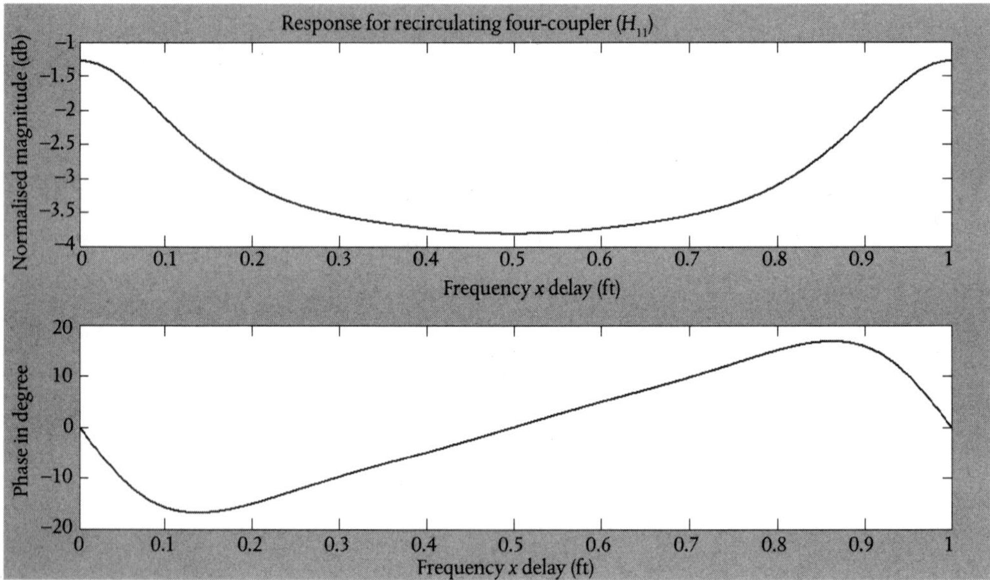

Fig. 5.24 Plot of normalised magnitude and phase response for recirculating four-coupler systems (H_{11})

Figure 5.24 shows the magnitude and phase response of the four-coupler system with X_1 as the input and Y_1 as the output. This combination can be used as the band-stop filter that rejects the large range of band of frequency. The phase response shows the symmetric variation. Hence, it will be more useful where low dispersion is needed.

Similarly; from Fig. 5.23,

$$\begin{aligned} Y_1 = {} & a_3 X_1 + (1-a_3) l_6 a_2 l_3 z^{-1} (1-a_3) X_1 \\ & + (1-a_3) l_6 (1-a_2) l_5 a_1 l_2 z^{-1} (1-a_2) l_3 z^{-1} (1-a_3) X_1 \\ & + (1-a_3) l_6 (1-a_2) l_5 (1-a_1) l_4 a_0 l_1 z^{-1} l_2 z^{-1} l_3 z^{-1} (1-a_3)(1-a_1)(1-a_2) X_1 \\ & + (1-a_0) l_1 z^{-1} (1-a_1) l_2 z^{-1} (1-a_2) l_3 z^{-1} (1-a_3) X_2 \end{aligned}$$

or

$$\begin{aligned} Y_1 = {} & a_3 X_1 + a_2 (1-a_3)^2 l_3 l_6 z^{-1} X_1 + a_1 (1-a_2)^2 (1-a_3)^2 l_2 l_3 l_5 l_6 z^{-2} X_1 \\ & + a_0 (1-a_1)^2 (1-a_2)^2 (1-a_3)^2 l_2 l_3 l_5 l_6 z^{-3} X_1 + (1-a_0)(1-a_1)(1-a_2)(1-a_3) l_1 l_2 l_3 z^{-3} X_2 \end{aligned}$$

or

$$Y_1 = a_3X_1 + a_2(1-2a_3)l_3l_6z^{-1}X_1 + a_2a_3^{2}l_3l_6z^{-1}X_1 + a_1(1-a_2)^2(1-2a_3)l_2l_3l_5l_6z^{-2}X_1 + a_1(1-a_2)^2a_3^{2}l_2l_3l_5l_6z^{-2}X_1 + (1-a_0)(1-a_1)(1-a_2)(1-a_3)l_1l_2l_3z^{-3}X_2$$

We can write:

$$Y_1 = a_2a_3l_3l_6z^{-1}Y_1 + a_1(1-a_2)^2a_3l_2l_3l_5l_6z^{-2}Y_1 + (1-a_0)(1-a_1)(1-a_2)(1-a_3)l_1l_2l_3z^{-3}X_2$$

Therefore,

$$H_{12} = \frac{Y_1}{X_2}.$$

Hence,

$$H_{12} = \frac{(1-a_0)(1-a_1)(1-a_2)(1-a_3)l_1l_2l_3z^{-3}}{1-a_2a_3l_3l_6z^{-1} - a_1(1-a_2)^2a_3l_2l_3l_5l_6z^{-2}}. \quad (5.17)$$

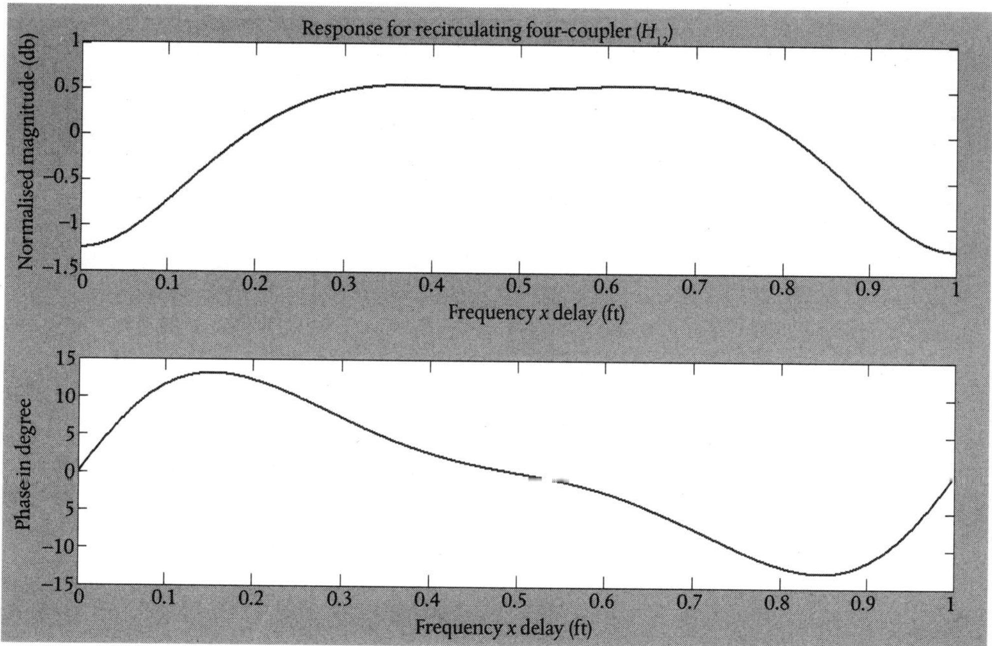

Fig. 5.25 Plot of normalised magnitude and phase response for recirculating four-coupler systems (H_{12})

Figure 5.25 shows the normalised magnitude and phase response of the four-coupler recirculating system (H_{12}). The response shows band-pass filtering action. The phase response is having the symmetric variation with some nonlinearity in its response. Similarly, the output in Fig. 5.23 can be expressed as to calculate H_{21}:

$$Y_2 = (1-a_3)l_6(1-a_2)l_5(1-a_1)l_4(1-a_0)X_1 + a_0X_2 + (1-a_0)^2a_1l_1l_4z^{-2}X_2 + (1-a_0)^2(1-a_1)^2a_2l_1l_2l_4l_5z^{-2}X_2 + (1-a_0)^2(1-a_1)^2(1-a_2)^2a_3l_1l_2l_3l_4l_5l_6z^{-3}X_2$$

or

$$
\begin{aligned}
Y_2 &= (1-a_3)l_6(1-a_2)l_5(1-a_1)l_4(1-a_0)X_1 + a_0X_2 + (1-2a_0)a_1l_1l_4z^{-1}X_2 + \\
&\quad a_0^2a_1l_1l_4z^{-1}X_2 + (1-2a_0)(1-a_1)^2a_2l_1l_2l_4l_5z^{-2}X_2 + a_0^2(1-a_1)^2a_2a_3l_1l_2l_4l_5z^{-2}X_2 + \\
&\quad (1-2a_0)(1-a_1)^2(1-a_2)^2a_3l_1l_2l_3l_4l_5l_6z^{-3}X_2 + a_0^2(1-a_1)^2(1-a_2)^2a_3l_1l_2l_3l_4l_5l_6z^{-3}X_2 \\
Y_2 &= (1-a_0)(1-a_1)(1-a_2)(1-a_3)l_4l_5l_6X_1 + a_1a_0l_1l_4z^{-1}Y_2 \\
&\quad + a_0(1-a_1)^2a_2a_3l_1l_2l_4l_5z^{-2}Y_2 + a_0(1-a_1)^2(1-a_2)^2a_3l_1l_2l_3l_4l_5l_6z^{-3}Y_2
\end{aligned}
$$

Since,

$$H_{21} = \frac{Y_2}{X_1}.$$

Hence,

$$H_{21} = \frac{(1-a_0)(1-a_1)(1-a_2)(1-a_3)l_4l_5l_6}{1-a_1a_0l_1l_4z^{-1} - a_0(1-a_1)^2a_2a_3l_1l_2l_4l_5z^{-2} - a_0(1-a_1)^2(1-a_2)^2a_3l_1l_2l_3l_4l_5l_6z^{-3}}. \tag{5.18}$$

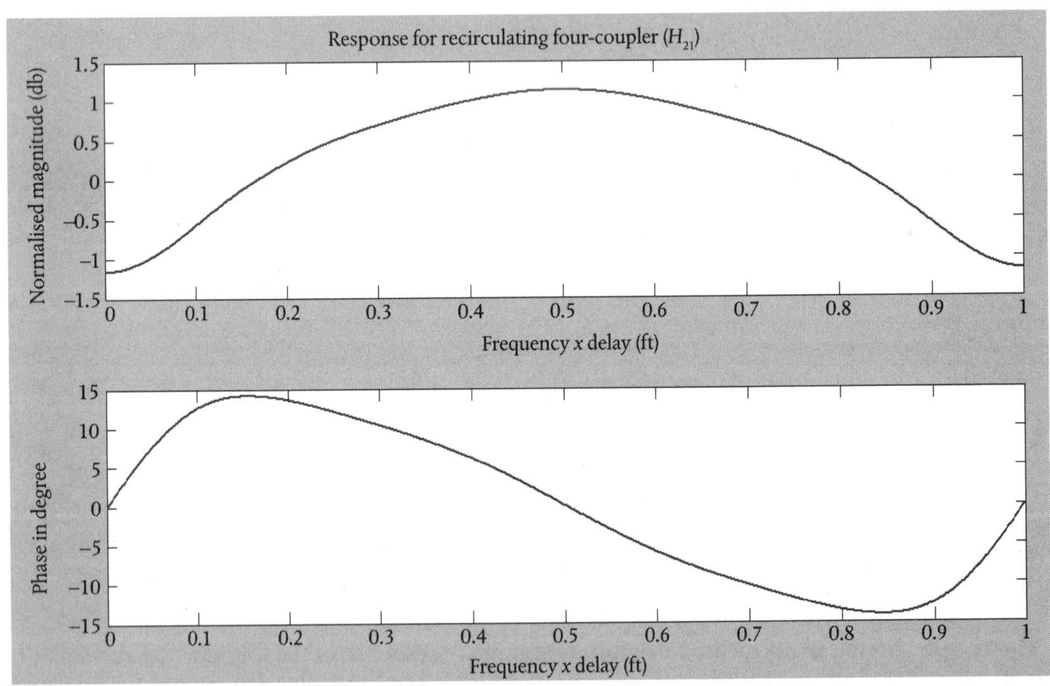

Fig. 5.26 Plot of normalised magnitude and phase response for recirculating four-coupler systems (H_{21})

Figure 5.26 shows the band-pass filtering action. However, phase response shows the symmetric variation, but it seems no sharp cut-off frequency response. Hence, this filter will not be useful where sharp cut-off is required in filtering action. Now, again from Fig. 5.23,

$$Y_2 = a_0 X_2 + (1-a_0)^2 a_1 l_1 l_4 z^{-1} X_2 + (1-a_0)^2 (1-a_1)^2 a_2 l_1 l_2 l_4 l_5 z^{-2} X_2$$
$$+ (1-a_0)^2 (1-a_1)^2 (1-a_2)^2 a_3 l_1 l_2 l_3 l_4 l_5 l_6 z^{-3} X_2,$$

or

$$Y_2 = a_0 X_2 + (1-2a_0) a_1 l_1 l_4 z^{-1} X_2 + a_0{}^2 a_1 l_1 l_4 z^{-1} X_2 + (1-2a_0)(1-a_1)^2 a_2 l_1 l_2 l_4 l_5 z^{-2} X_2$$
$$+ a_0{}^2 (1-a_1)^2 a_2 l_1 l_2 l_4 l_5 z^{-2} X_2 + (1-2a_0)(1-a_1)^2 (1-a_2)^2 a_3 l_1 l_2 l_3 l_4 l_5 l_6 z^{-3} X_2$$
$$+ a_0{}^2 (1-a_1)^2 (1-a_2)^2 a_3 l_1 l_2 l_3 l_4 l_5 l_6 z^{-3} X_2,$$

or

$$Y_2 = a_0 X_2 + (1-2a_0) a_1 l_1 l_4 z^{-1} X_2 + a_0 a_1 l_1 l_4 z^{-1} Y_2 + (1-2a_0)(1-a_1)^2 a_2 l_1 l_2 l_4 l_5 z^{-2} X_2 +$$
$$a_0 (1-a_1)^2 a_2 l_1 l_2 l_4 l_5 z^{-2} Y_2 + (1-2a_0)(1-a_1)^2 (1-a_2)^2 a_3 l_1 l_2 l_3 l_4 l_5 l_6 z^{-3} X_2 +$$
$$a_0 (1-a_1)^2 (1-a_2)^2 a_3 l_1 l_2 l_3 l_4 l_5 l_6 z^{-3} Y_2.$$

Since,

$$H_{22} = \frac{Y_2}{X_2}.$$

Hence,

$$H_{22} = \left[\frac{a_0 + (1-2a_0) a_1 l_1 l_4 z^{-1} + (1-2a_0)(1-a_1)^2 a_2 l_1 l_2 l_4 l_5 z^{-2} + (1-2a_0)(1-a_1)^2 (1-a_2)^2 a_3 l_1 l_2 l_3 l_4 l_5 l_6 z^{-3}}{1 - a_0 a_1 l_1 l_4 z^{-1} - a_0 (1-a_1)^2 a_2 l_1 l_2 l_4 l_5 z^{-2} - a_0 (1-a_1)^2 (1-a_2)^2 a_3 l_1 l_2 l_3 l_4 l_5 l_6 z^{-3}} \right] \quad (5.19)$$

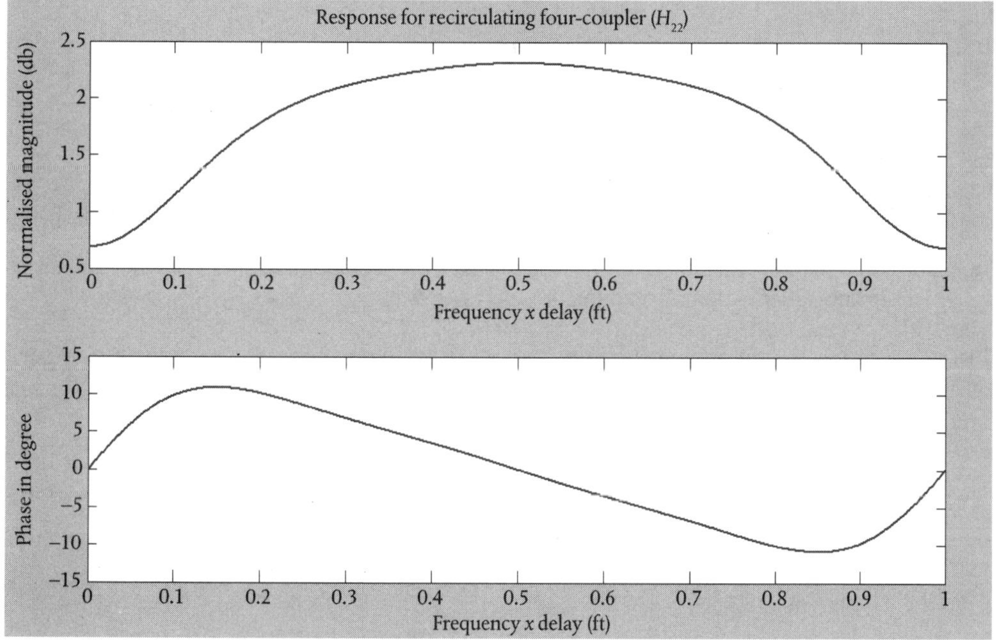

Fig. 5.27 Plot of normalised magnitude and phase response for recirculating four-coupler systems (H_{22}).

Figure 5.27 shows the variation in normalised magnitude and phase response of the four-coupler recirculating system (H_{22}). The plot is taken when X_2 acts as an input and Y_2 as an output. The plot shows the band-pass filtering action with somewhat improvement when compared to previous case.

5.4.6 Conversion from *Z*-domain to *S*-domain

Since we know that $s = j\omega$, $z = e^{sT}$ from (5.4). Hence, we can easily connect z-transform domain into s-domain. Here, s-domain is Laplace transform and we can easily apply control theory concepts on these transfer formulas.

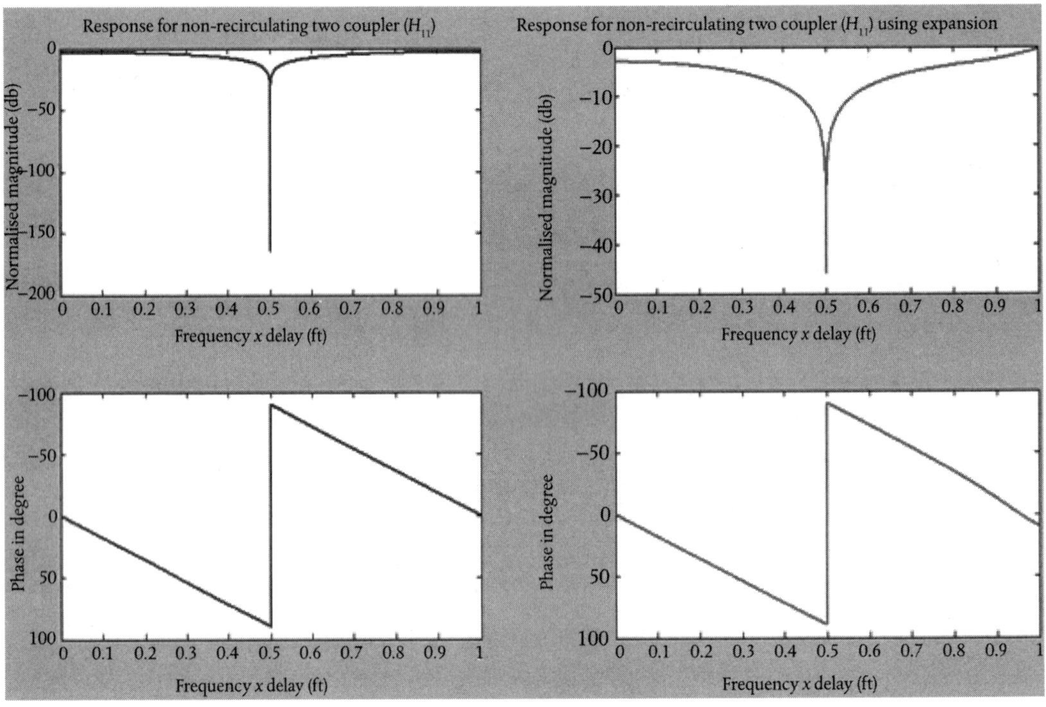

Fig. 5.28 Comparison of response for non-recirculating two-coupler system (*z*-domain and its approximation in *s*-domain expansion)

This $e^{j\omega T}$ describes a unit circle centred at the origin of the z-plane. The unit circle also plays an important role in problems regarding stability. Hence, for two-coupler system, e^{sT} is approximated up to 13 numbers of elements:

$$z = e^{sT} = 1 + sT + \frac{(sT)^2}{2} + \frac{(sT)^3}{3} + \frac{(sT)^4}{4} + \frac{(sT)^5}{5} + \frac{(sT)^6}{6} + \frac{(sT)^7}{7} + \frac{(sT)^8}{8} + \frac{(sT)^9}{9} + \frac{(sT)^{10}}{10} + \frac{(sT)^{11}}{11} + \frac{(sT)^{12}}{12} + \frac{(sT)^{13}}{13}. \tag{5.20}$$

Therefore, we can compare the response for non-recirculating and recirculating system in z-domain and s-domain with approximation of expansion as shown in Fig. 5.28.

Hence, Fig. 5.28 shows the similarity between z-domain and s-domain using the approximate expansion of e^{sT}. Now, applying this approximation, we can obtain the response of non-recirculating and recirculating system as follows:

As we know that for non-recirculating system,

$$H_{11} = (1-b_1)(1-b_0)\iota_{11} + b_1 b_0 \iota_{12} z^{-1}$$

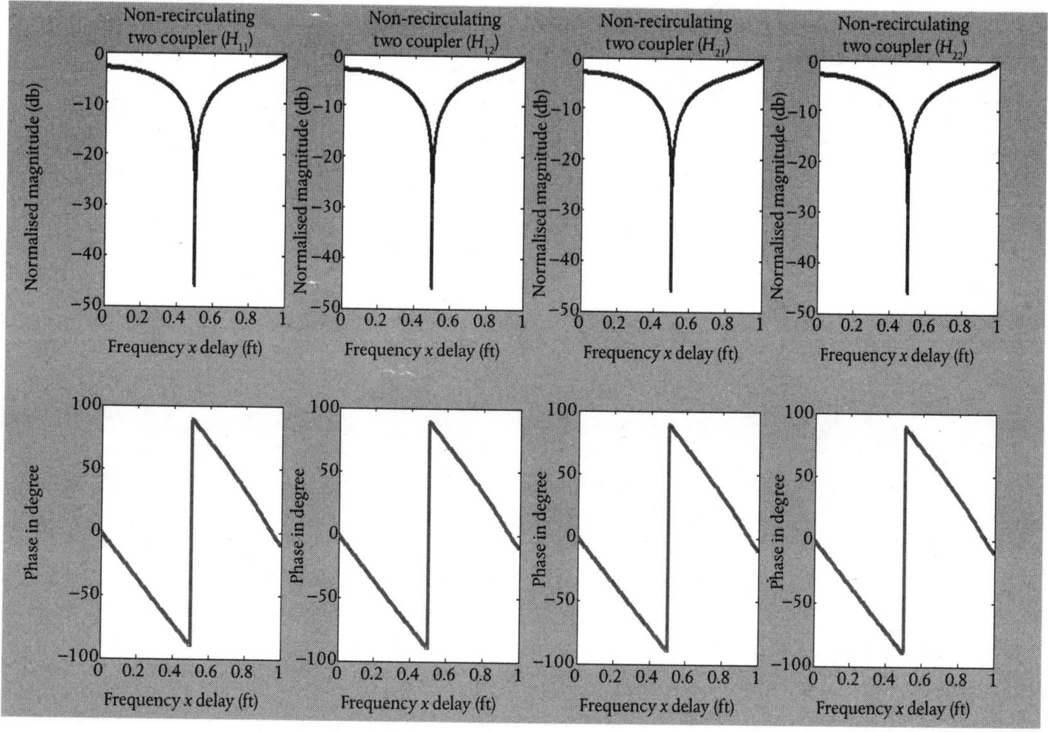

Fig. 5.29 Plot of normalised magnitude and phase responses for non-recirculating two-coupler system using expansion approximation

We can replace z-domain into s-domain by using (5.20). Hence, we can obtain the various response parameters. Figure 5.29 shows the plot of the normalised magnitude and phase response for the two-coupler non-recirculating system using the expansion of $z = e^{sT}$. All the parameters represent the notch filtering action. The response is very much similar to that of Fig. 5.16. For a recirculating system the similar type of graph can be obtained using the expansion approximation. As we know that, in the case of recirculating system, H_{11} is given in (5.5) as:

$$H_{11}(z) = \frac{a_1 + (1-2a_1) a_0 \iota_1 z^{-1}}{1 - a_1 a_0 \iota_1 z^{-1}}.$$

By using (5.20), the response of various parameters in s-domain of two-coupler recirculating system can be calculated.

Figure 5.30 represents the plot of normalised magnitude and phase response of the various parameters of the two-coupler recirculating system using the approximation of the $z = e^{sT}$ up to 13 numbers of the elements. The response is very much similar to that of Fig. 5.13 and Fig. 5.14. After the verification of our result in s-domain with z-domain, we try to compute the time domain response of the systems as discussed in the next section.

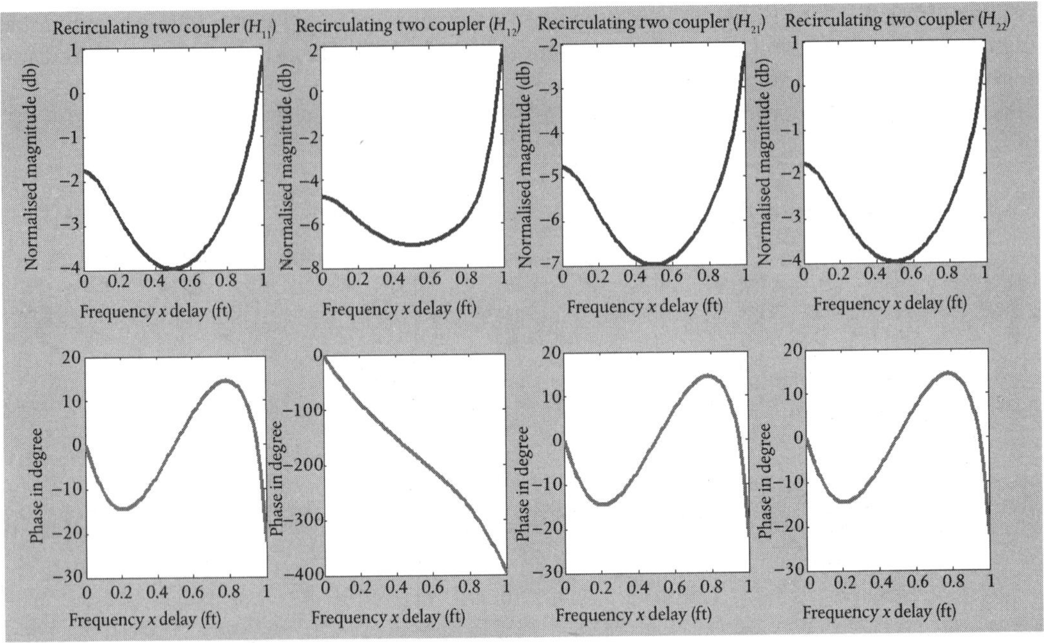

Fig. 5.30 Plot of normalised magnitude and phase response for recirculating two-coupler system (H_{11}) using expansion approximations

5.4.7 Time Response

For non-recirculating system, the time domain response can be computed by taking inverse Laplace transform of transfer function.

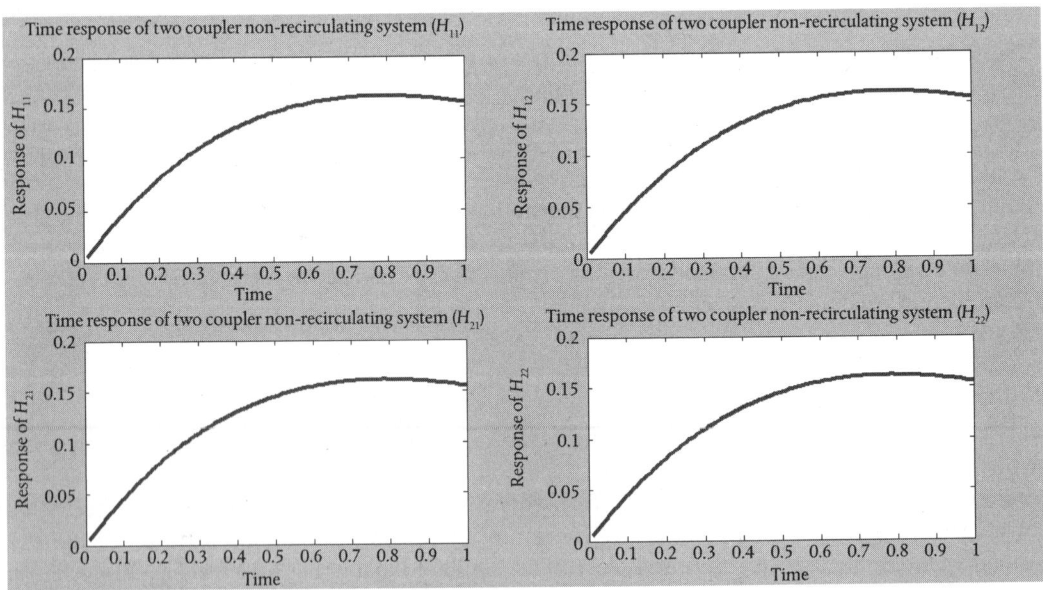

Fig. 5.31 Time domain response of various parameters of two-coupler non-recirculating system

Figure 5.31 shows the time response of the various parameters of two-coupler non-recirculating system. Hence, these waveforms are equivalent to that of the low-pass filtering action. Similarly, for recirculating system.

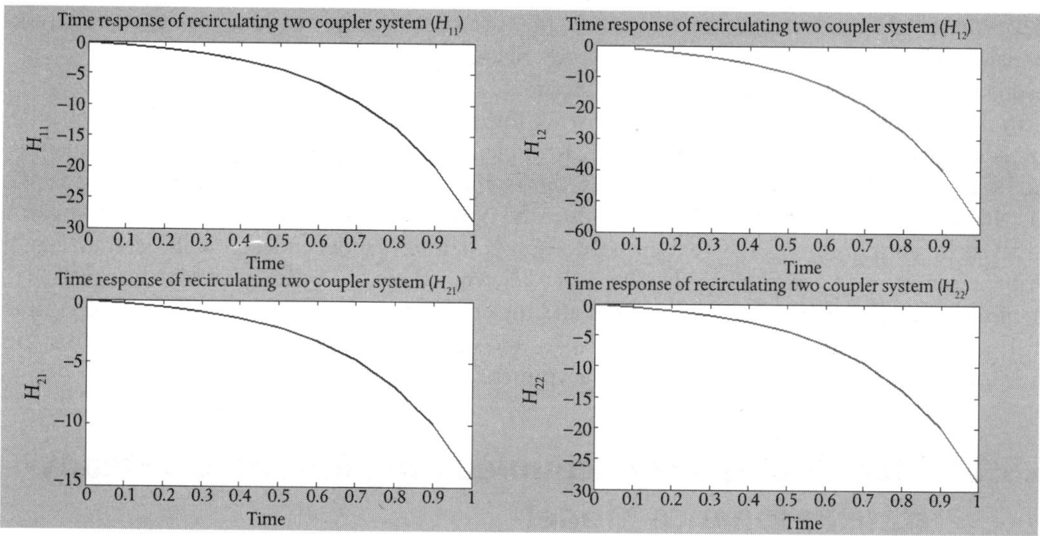

Fig. 5.32 Time domain response of various parameters of two-coupler recirculating system

Figure 5.32 represents the plot of the time response of various parameters of the two-coupler recirculating system. These waveforms represent the characteristics of the high pass filter. From this study, one can easily compute the group delay and group velocity dispersion of these various types of lattice structures.

In this section, the results have been represented up to four-coupler recirculating/non-recirculating lattice structure. However, the result can be easily extended to the higher order of the system. It is apparent from our study as we move from two-coupler recirculating/non-recirculating to four-coupler recirculating/non-recirculating systems, the response of the lattice structure improves substantially.

5.5 ANALYSIS OF OPTICAL RING RESONATOR INTEGRATED WITH ASYMMETRIC MACH-ZEHNDER INTERFEROMETER

Integrated optical ring resonators are useful components of optical switches, sensors, modulators, and wavelength-division multiplexing systems. In recent years, ring resonators integrated with thermo-optic or electro-optic phase shifters have been demonstrated to control the transmission. High-finesses resonators with a huge radius using gain medium or ultra low-loss waveguide have been realised. These technical progresses provide flexible methods for the design and utility of ring resonator-based optical device. In order to extend application of ring resonators, integrated devices constructed by the MZI and resonator (s) are considerable current interest. Micro-ring(s) coupled in one arm of the MZI has shown potential for the construction of optical time delay lines and nonlinear optical switches. Another noteworthy architecture, a ring resonator integrated with an MZI has shown extreme high-contrast switching characteristics and been expected to implement low-power optical switches and narrow bandwidth filters. By incorporating phase shifters into the circulation loop, such device can also serve as a chromatic dispersion compensator. Recently, a ring resonator that is asymmetric in path lengths of the waveguide in MZI has been demonstrated. This asymmetric configuration leads to an enhancement response to the phase shift in the long arm of the MZI when it works at optimal resonance. In order to achieve this optimal condition, however, a selected critical length for the feedback loop is indispensable. It is difficult to realise by practice micro-fabrication processes. In addition, the propagation losses and the asymmetric phase transfer operators of the MZI are usually neglected for conventional analytical models. Such models may be inapplicable to describe the transmission property of the MZI-coupled resonators with a highly asymmetrical configuration.

In this section, we present and characterise an improved configuration having two phase shifters embedded into different positions of the MZI-coupled ring resonators using an accurate analytical model by considering the propagation losses, the asymmetrical phase transfer operators of the MZI, and the radius of the ring cavity. One can show the optimal resonance of the ring resonator can be achieved by simply tuning the two embedded phase shifters, independent on the certain length. Distinct transmission and filtering characteristics determined by the structural parameters such as the asymmetrical path length of the waveguides and the coupling ratio of the directional couplers in the MZI are shown. We highlight the unique intensity and spectral response related to the phase variation of the outer waveguide of the MZI for highly asymmetric configurations.

5.5.1 Models of the MZI-coupled Ring Resonator: Steady-state Resonance Model

The scheme of the MZI-coupled ring resonator is shown in Fig. 5.33(a). It consists of two identical directional couplers with a coupling ratio k, a waveguide ring resonator (the radius is R), and a U-shaped waveguide with a length of L. Two phase shifters $(\Delta\varphi_6, \Delta\varphi_8)$ controlled thermo-optically or electro-optically are embedded into the ring resonator. The U-shaped waveguide is a key element in this structure serving as a high-sensitive sensor to detect small refractive index variation of the outer arm of the MZI. The induced phase variation of the U-shaped waveguide $\Delta\phi_5$ is directly proportional to the value of L. For simplicity, we assume that the two directional couplers are point devices with zero length, which is the phase shift of the cross-output port is $-\pi/2$ and the insertion loss is zero.

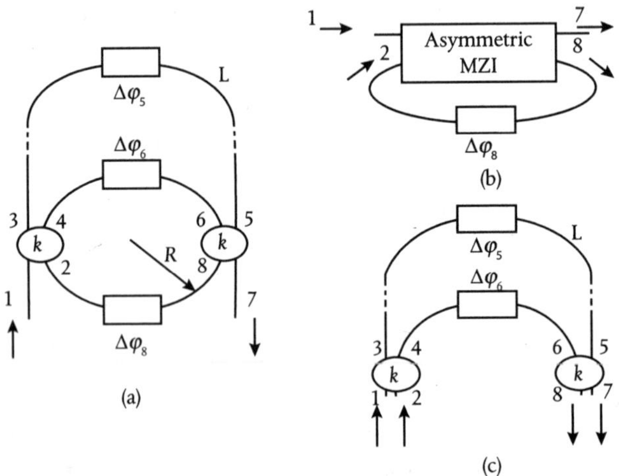

Fig. 5.33 (a) Schematic diagram of the MZI-coupled ring resonator. (b) An equivalent model corresponds to that of an asymmetrical MZI connected with a bent waveguide. (c) Schematic diagram of the asymmetric MZI

Signal launched into the circuit at port 1 is coupled into the circulation loop by the directional couplers, and comes from port 7. At steady-state resonance, the complex electric fields at different ports are given as:

$$\left.\begin{aligned} E_3 &= \sqrt{1-k}E_1 - i\sqrt{k}E_2 \\ E_4 &= -i\sqrt{k}E_1 + \sqrt{1-k}E_2 \\ E_5 &= \exp(-\alpha L)\exp(i\varphi_5)E_3 \\ E_6 &= \exp(-\alpha\pi R)\exp(i\varphi_6)E_5 \\ E_7 &= \sqrt{1-k}E_5 - i\sqrt{k}E_6 \\ E_8 &= -i\sqrt{k}E_5 + \sqrt{1-k}E_6 \\ E_2 &= \exp(-\alpha\pi R)\exp(i\varphi_3)E_8 \end{aligned}\right\}, \tag{5.21}$$

where $\varphi_5 = \beta L + \Delta\varphi_5$, $\varphi_8 = \beta\pi R + \Delta\varphi_8$ and $\varphi_6 = \beta\pi R + \Delta\varphi_6$ corresponding to the phase transfer operators through different waveguides and phase shifters, $L = V\pi R\,(V \geq 1)$, β is the mode propagation constant as a function of the mode effective index N_{eff}, α is the amplitude coefficient in the bent wave guide, and 2α is the propagation loss. From (5.21), we assume,

$$\left.\begin{aligned} A &= \sqrt{1-k} \\ B &= -j\sqrt{k} \\ C &= \exp(-\alpha L)\exp(j\varphi_5) \\ D &= \exp(-\alpha\pi R)\exp(j\varphi_6) \\ E &= \exp(-\alpha\pi R)\exp(j\varphi_8) \end{aligned}\right\}. \tag{5.22}$$

Therefore, (5.21) would transfer as:

$$\left.\begin{aligned} E_3 &= AE_1 + BE_2 \\ E_4 &= BE_1 + AE_2 \\ E_5 &= CE_3 \\ E_6 &= DE_4 \\ E_7 &= AE_5 + BE_6 \\ E_8 &= BE_5 + AE_6 \\ E_2 &= EE_8 \end{aligned}\right\}. \tag{5.23}$$

In (5.23) in to electric-field E_6, substituting value of E_3 in E_5, we get:

$$E_5 = ACE_1 + BCE_2\,. \tag{5.24}$$

Similarly, substituting value of E_4:

$$E_6 = BDE_1 + ADE_2. \tag{5.25}$$

Now, substituting the value of E_5 and E_6 from (5.24) and (5.25) in (5.23), we can obtain for E_7:

$$E_7 = A\left[ACE_1 + BCE_2\right] + B\left[BDE_1 + ADE_2\right]$$
$$E_7 = \left[A^2C + B^2D\right]E_1 + \left[ABC + ABD\right]E_2. \tag{5.26}$$

From (5.23), we get:

$$E_2 = E\left[BE_5 + AE_6\right] = BE\left[ACE_1 + BCE_2\right] + AE\left[BDE_1 + ADE_2\right],$$

or

$$E_2 = ABCEE_1 + B^2CEE_2 + ABDEE_1 + A^2DEE_2$$

There,

$$E_2\left[1 - B^2CE - A^2DE\right] = \left[ABCE + ABDE\right]E_1$$

or

$$E_2 = \frac{(ABCE + ABDE)E_1}{\left(1 - B^2CE - A^2DE\right)}. \tag{5.27}$$

Substituting the value of E_2 from (5.27) in (5.26)

$$E_7 = \left(A^2C + B^2D\right)E_1 + \frac{(ABC + ABD)(ABCE + ABDE)E_1}{\left(1 - B^2CE - A^2DE\right)}$$

or

$$\frac{E_7}{E_1} = \left(A^2C + B^2D\right) + \frac{(ABC + ABD)(ABCE + ABDE)}{\left(1 - B^2CE - A^2DE\right)}$$

or

$$\frac{E_7}{E_1} = \frac{\left(A^2C + B^2D\right)\left(1 - B^2CE - A^2DE\right) + (ABC + ABD)(ABCE + ABDE)}{\left(1 - B^2CE - A^2DE\right)}$$
$$= \frac{A^2C + B^2D + CDE\left[-A^4 - B^4 + 2A^2B^2\right]}{\left[\left(1 - B^2CE - A^2DE\right)\right]}$$

or

$$\frac{E_7}{E_1} = \frac{A^2C + B^2D - CDE\left(A^2 - B^2\right)^2}{\left[\left(1 - B^2CE - A^2DE\right)\right]}. \tag{5.28}$$

Finally, substituting value of A, B, C, D, and E in (5.28), we get:

$$\frac{E_7}{E_1} = \frac{(1-k)\exp\left(j\varphi_5 - \alpha L\right) + (-k)\exp\left(j\varphi_6 - \alpha\pi R\right) - \exp\left[j\left(\varphi_5 + \varphi_6 + \varphi_8\right) - \alpha L - 2\alpha\pi R\right]\left[1 - k + k\right]^2}{1 - \left(-j\sqrt{k}\right)^2 \exp\left(j\varphi_5 - \alpha L\right)\exp\left(-\alpha\pi R + j\varphi_8\right) - (1-k)\exp\left(j\varphi_6 - \alpha\pi R\right)\exp\left(j\varphi_8 - \alpha\pi R\right)}$$

or

$$\frac{E_7}{E_1}=\frac{g_1-\exp\left[j(\varphi_5+\varphi_8+\varphi_8)-\alpha L-2\alpha\pi R\right]}{g_2+k\exp\left[j(\varphi_5+\varphi_8)-\alpha L-\alpha\pi R\right]}, \tag{5.29}$$

where

$$\begin{cases} g_1=-k\exp(j\varphi_6-\alpha\pi R)+(1-k)\exp(j\varphi_5-\alpha L) \\ g_2=1-(1-k)\exp\left[j(\varphi_6+\varphi_8)-2\alpha\pi R\right] \end{cases}.$$

Now, from (5.23), (5.24), and (5.25), it can be shown:

$$E_8=BE_5+AE_6=B\left[ACE_1+BCE_2\right]+A\left[BDE_1+ADE_2\right],$$

or

$$E_8=\left[ABC+ABD\right]E_1+\left[B^2C+A^2D\right]E_{2.} \tag{5.30}$$

Substituting value of E_2 from (5.27a) in (5.30), we get:

$$E_8=\left[ABC+ABD\right]E_1+\frac{\left[B^2C+A^2D\right]\left[ABCE+ABDE\right]}{\left[1-B^2CE-A^2DE\right]}E_1$$

or

$$\frac{E_8}{E_1}=\frac{\left[ABC+ABD\right]\left[1-B^2CE-A^2DE\right]+\left[B^2C+A^2D\right]\left[ABCE+ABDE\right]}{\left[1-B^2CE-A^2DE\right]}$$

or

$$\frac{E_8}{E_1}=\frac{ABC+ABD}{\left[1-B^2CE-A^2DE\right]}.$$

Finally, substituting value of A, B, C, D, and E, we get:

$$\frac{E_8}{E_1}=\frac{\sqrt{1-k}\left(-j\sqrt{k}\right)\exp(j\varphi_5-\alpha L)+\sqrt{1-k}\left(-j\sqrt{k}\right)\exp(j\varphi_6-\alpha\pi R)}{1-\left(-j\sqrt{k}\right)^2\exp\left[j(\varphi_5+\varphi_8)-\alpha L-\alpha\pi R\right]-(1-k)\exp\left[j(\varphi_8+\varphi_6)-2\alpha\pi R\right]}$$

$$=\frac{-j\sqrt{k(1-k)}\exp(j\varphi_5-\alpha L)-j\sqrt{k(1-k)}\exp(j\varphi_6-\alpha\pi R)}{1+k\exp\left[j(\varphi_5+\varphi_8)-\alpha L-\alpha\pi R\right]-(1-k)\exp\left[j(\varphi_6+\varphi_8)-2\alpha\pi R\right]}.$$

Multiplying numerator and denominator by $\exp\left[j\varphi_8-\alpha\pi R\right]$

$$=\frac{-j\left[\sqrt{k(1-k)}\exp\left[j(\varphi_5+\varphi_8-\alpha L-\alpha\pi R)\right]+\sqrt{k(1-k)}\exp\left[j(\varphi_6+\varphi_8)-2\alpha\pi R\right]\right]}{\exp\left[j\varphi_8-\alpha\pi R\right]+k\exp\left[j(\varphi_5+2\varphi_8)-\alpha L-2\alpha\pi R\right]-(1-k)\exp\left[j(\varphi_6+2\varphi_8)-3\alpha\pi R\right]}$$

$$=\frac{\left[\sqrt{k(1-k)}\exp\left[j(\varphi_5+\varphi_8-\pi/2)-\alpha L-\alpha\pi R\right]+\sqrt{k(1-k)}\exp\left[j(\varphi_6+\varphi_8-\pi/2)-2\alpha\pi R\right]\right]}{\exp\left[j\varphi_8-\alpha\pi R\right]+k\exp\left[j(\varphi_5+2\varphi_8)-\alpha L-2\alpha\pi R\right]-(1-k)\exp\left[j(\varphi_6+2\varphi_8)-3\alpha\pi R\right]}$$

$$\frac{E_8}{E_1} = \frac{g_3 + \sqrt{k(1-k)}\exp\left[j(\varphi_6 + \varphi_8 - \pi/2) - 2\alpha\pi R\right]}{g_4 - (1-k)\exp\left[j(\varphi_6 + 2\varphi_8) - 3\alpha\pi R\right]}, \tag{5.31}$$

where

$$g_3 = \sqrt{k(1-k)}\exp\left[j(\varphi_5 + \varphi_8 - \pi/2) - \alpha L - \alpha\pi R\right]$$

$$g_4 = e^{[j\varphi_8 - \alpha\pi R]} + k\exp\left[j(\varphi_5 + 2\varphi_8) - \alpha L - 2\alpha\pi R\right].$$

The output intensity transmission operator is $\left|\frac{E_7}{E_1}\right|^2$. The comparisions of the normalised transmission through the MZI-coupled ring resonator with different structural parameters are shown in Fig. 5.34.

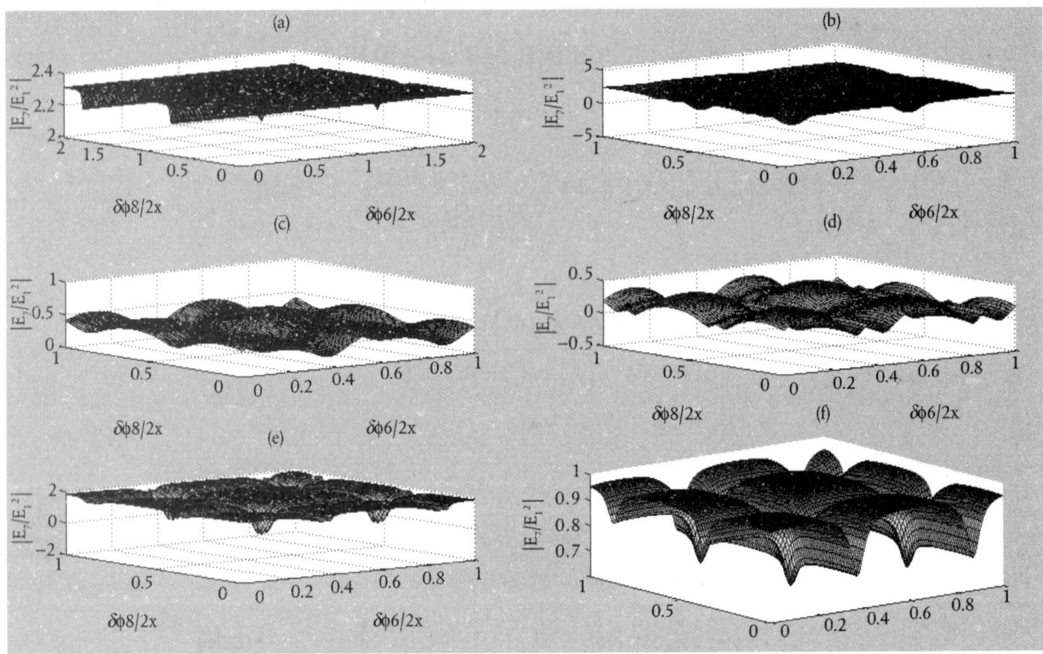

Fig. 5.34 Comparisons of the transmission through the MZI-coupled ring resonator with different structural parameters, where the wavelength $\lambda = 1.55$ μm, $N_{eff} = 1.45$ and $R = 0.5$ cm. (a) and (b) are transmission with parameters $k = 0.95$, $2\alpha = 0.05$ dB/cm, V is 1.10 and 10.18, respectively; (c) and (d) are transmission corresponds to $k = 0.5$, $2\alpha = 0.2$ dB/cm, V is 3.90 and 29.22, respectively; (e) is transmission with parameters $k = 0.1$, $2\alpha = 0.05$ dB/cm V is 2.90; (f) is transmission with parameter $k = 0.1$, $2\alpha = 0.2$ dB/cm, $V = 81.60$.

5.5.2 Resonance Condition

The desired high-sensitive transmission response to the phase change $\Delta\phi_5$ can be achieved at the optimal resonance corresponding to the maximum extinction ratio $(E_7/E_1 = 0)$. Here, we assume $\Delta\phi_5$

is zero at the given initial state. By setting the real and imaginary parts of the output field E_7 in(5.29) to zero, we achieved the phase condition at the optimal resonance:

$$\begin{cases} \beta L+\beta\pi R+\Delta\varphi_{8_r}=2m\pi+\arccos(\varepsilon) \\ 2\beta\pi R+\Delta\varphi_{6_r}+\Delta\varphi_{8_r}-\pi=(2n-1)\pi+\arccos(\eta) \end{cases} \tag{5.32}$$

or

$$\begin{cases} \beta L+\beta\pi R+\Delta\varphi_{8_r}=2(m+1)\pi-\arccos(\varepsilon) \\ 2\beta\pi R+\Delta\varphi_{6_r}+\Delta\varphi_{8_r}-\pi=(2n+1)\pi-\arccos(\eta). \end{cases} \tag{5.33}$$

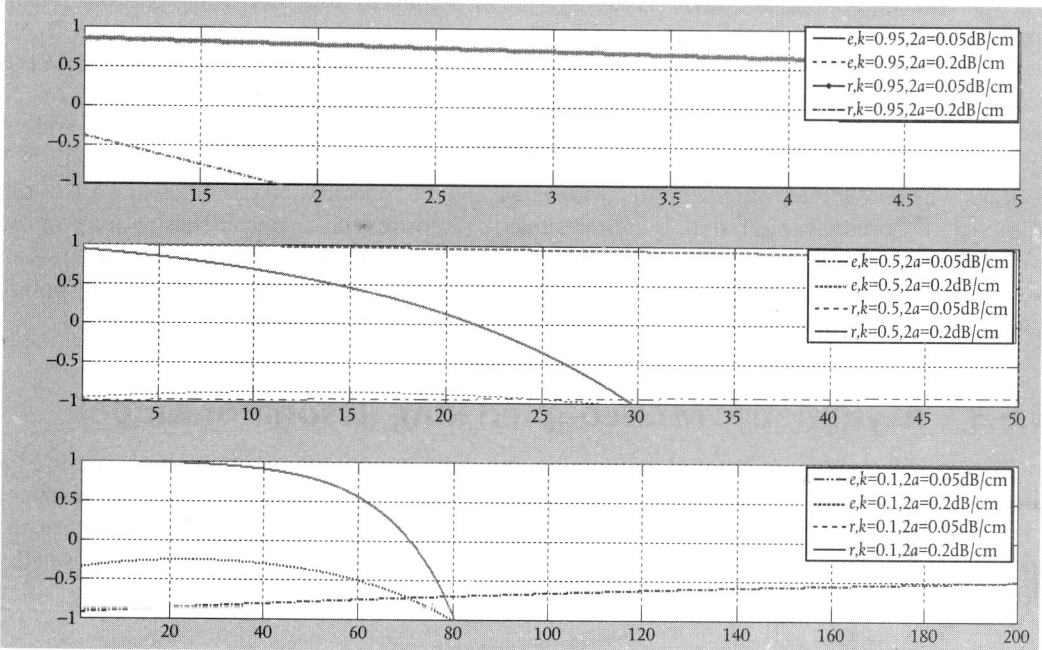

Fig. 5.35 The characteristics of the parameters ε and η as a function V with different 2α and k. We set $R=0.5$ cm, k = 0.95, 0.5, and 0.1, 2α = 0.2 dB/cm and 0.05 dB/cm, respectively.

where m and n are positive integer, $\Delta\varphi_{6_r}$ and $\Delta\varphi_{8_r}$ correspond to the phase factors of the two phase shifters under the optimal resonance condition for selected wavelength. ε and η are important parameters determining whether $\Delta\varphi_6$ and $\Delta\varphi_8$ can be turned to $\Delta\varphi_{6_r}$ and $\Delta\varphi_{8_r}$ simultaneously. This is given as:

$$\varepsilon=\frac{k^2\exp(2\alpha L+2\alpha\pi R)-(1-k)^2\exp(4\alpha\pi R)+1}{-2\kappa\exp(\alpha L+\alpha\pi R)} \tag{5.34}$$

$$\eta=\frac{-k^2\exp(2\alpha L+2\alpha\pi R)+(1-k)^2\exp(4\alpha\pi R)+1}{2(1-k)\exp(\alpha L+\alpha\pi R)} \tag{5.35}$$

By solving (5.32) and(5.33) with selected optimal resonant wavelength, the value of $\Delta\varphi_{6r}$ and $\Delta\varphi_{8r}$, can be achieved. Here, $\beta L+\beta\pi R+\Delta\varphi_{8r}$ and $2\beta\pi R+\Delta\varphi_{6r}+\Delta\varphi_{8r}-\pi$ in (5.32) and (5.33) correspond to the optimal resonance condition of the one-trip phase transfer operator in the outer and inner circulation loops determined by the values of ε and η. Only if $\varepsilon,\eta\epsilon[-1,1]$,it can simultaneously satisfy the optimal resonance condition.

Figure 5.35 shows the characteristics of the parameters ε and η as a function of V with different values of 2α and k. When $\Delta\varphi_6$ and $\Delta\varphi_8$ are turned over the range of 2π, optimised resonance condition can be achieved for arbitrary V with the condition of $\varepsilon,\eta\epsilon[-1,1]$. As shown, ε and η are determined by the coupling ratio k and the propagation losses 2α. By reducing the values of k ans 2α, the value of V can be increased significantly. For example, when $R=0.5\,\text{cm}, k=0.95$ and $2\alpha=0.2\,\text{dB/cm}$, the resonator cannot be turned to the optimal resonance once $V>2$. In contrast, when $R=0.5\,\text{cm}$, $k=0.1$, and $2\alpha=0.05\,\text{dB/cm}$, the MZI-coupled resonator with highly asymmetric configuration ($V>200$) can still be tuned to the optimal resonance.

Figure 5.35 shows clearly that the transmission characteristics depend on the structural parameters heavily. Typically, there are two resonant peaks when $\Delta\varphi_6$ and $\Delta\varphi_8$ are tuned over the range of 2π. The special case of single transmission peak occurs at $\varepsilon=\eta=-1$ as shown in Fig. 5.35 (b), 5.35 (d), and 5.35 (f). From the comparison, one can see that when $k=0.1$ and $2\alpha=0.2$ dB/cm, the maximum value of V is 81.60. The output intensity is zero in a wide range and the transmission characteristics are distinct from that of another configuration. It indicates that the transmission characteristics of our proposed MZI-coupled ring resonator can be useful in different applications by setting different parameters k and V. Figure 5.34 illustrate the intensity distribution of the ring resonator with different k at the optimal resonance.

5.5.3 Asymmetric MZI-coupled Ring Resonator Model

In order to reveal the underlying physics of the resonance characteristics more clearly, we used an alternate numerical model based on the multi-beam interference to calculate the optimal resonance conditions. It is illustrated in Figs. (5.33b) and (5.33c). A tunableMZI is connected with a bent waveguide and a closed resonant cavity is formed. When the signal is launched from port 2 $(E_1=0)$, the normalised electric field at ports 7 and 8 can be expressed as:

$$\frac{E_{72}}{E_2}=\sqrt{k(1-k)}e^{(-\alpha\pi R)}e^{[i(\varphi_6-2/\pi)]}+\sqrt{k(1-k)}e^{(-\alpha L)}e^{[i(\varphi_5-2/\pi)]}. \tag{5.36}$$

Byrewriting (5.21),we can prove (5.36) as follows:

$$E_3=\sqrt{1-k}E_1-i\sqrt{k}E_2 \tag{5.37}$$

$$E_4=-i\sqrt{k}E_1+\sqrt{1-k}E_2 \tag{5.38}$$

$$E_5=\exp(-\alpha L)\exp(i\varphi_5)E_3 \tag{5.39}$$

$$E_6=\exp(-\alpha\pi R)\exp(i\varphi_6)E_4 \tag{5.40}$$

$$E_7=\sqrt{1-k}E_5-i\sqrt{k}E_6 \tag{5.41}$$

$$E_8=-i\sqrt{k}E_5+\sqrt{1-k}E_6 \tag{5.42}$$

$$E_2=\exp(-\alpha\pi R)\exp(i\varphi_8)E_8 \tag{5.43}$$

As the signal is launched from port 2, it is assumed that E_1 is zero. Now,substituting the value of E_1 as zero in (5.37) and (5.38):

$$E_3 = -i\sqrt{k}E_2. \tag{5.44}$$

$$E_4 = \sqrt{1-k}E_2. \tag{5.45}$$

Substituting the value of E_3 in (5.39), we get:

$$E_5 = \exp(-\alpha L)\exp(i\varphi_5)\left(-i\sqrt{k}E_2\right). \tag{5.46}$$

Similarly,substituting the value of E_4 in (5.40):

$$E_6 = \exp(-\alpha\pi R)\exp(i\varphi_6)\sqrt{1-k}E_2. \tag{5.47}$$

Now, substituting the value of E_5 and E_6 from (5.46) and (5.47) in (5.41), we get:

$$E_{72} = \sqrt{(1-k)}\left(-i\sqrt{k}\right)E_2\exp(-\alpha L)\exp(i\varphi_5) - \left(-i\sqrt{k}\right)\sqrt{(1-k)}\exp(-\alpha\pi R)\exp(i\varphi_6)E_2$$

$$\frac{E_{72}}{E_2} = \sqrt{k(1-k)}\exp(-\alpha\pi R)\exp\left[i\left(\varphi_6 - \frac{\pi}{2}\right)\right] + \sqrt{k(1-k)}\exp(-\alpha L)\exp\left[i\left(\varphi_5 - \frac{\pi}{2}\right)\right]. \tag{5.48}$$

Again, substituting the value of E_5 and E_6 from (5.46) and (5.47) in (5.42), we get:

$$E_{82} = \left(-i\sqrt{k}\right)\exp(-\alpha L)\exp(i\varphi_5)\left(-i\sqrt{k}\right)E_2 + \sqrt{(1-k)}\exp(-\alpha\pi R)\exp(i\varphi_6)\sqrt{(1-k)}E_2$$

$$\frac{E_{82}}{E_2} = (1-k)\exp(-\alpha\pi R)\exp(i\varphi_6) + k\exp(-\alpha L)\exp(i\varphi_5) \tag{5.49}$$

We define k_l and $\gamma_l\,(l=1,2)$ as the effective coupling ratio and the insertion loss when signal launched from ports 1 and 2 of the MZI. Using (5.48) and (5.49), the effective coupling ratio K_2 and the insertion loss γ_2 are given as:

$$\begin{aligned} K_2 &= \frac{|E_{72}/E_2|^2}{|E_{72}/E_2|^2 + |E_{82}/E_2|^2} \\ &= \frac{(1-k)k.g_5}{(1-k)\exp(-2\alpha\pi R) + k\exp(-2\alpha L)}, \end{aligned} \tag{5.50}$$

where

$$g_5 = \exp(-2\alpha\pi R) + \exp(-2\alpha L) + 2\exp(-\alpha L - \alpha\pi R)\cos(\varphi_5 - \varphi_6).$$

$$\begin{aligned} \gamma_2 &= 1 - |E_{72}/E_2|^2 - |E_{82}/E_2|^2 \\ &= 1 - (1-k)\exp(-2\alpha\pi R) - k\exp(-2\alpha L). \end{aligned} \tag{5.51}$$

Now, we can prove (5.50) and (5.51) as follows, using (5.48),

$$\frac{E_{72}}{E_2} = \sqrt{k(1-k)}\left[\exp(-\alpha\pi R)\left\{\cos\left(\varphi_6 - \frac{\pi}{2}\right) + i\sin\left(\varphi_6 - \frac{\pi}{2}\right)\right\} + \exp(-\alpha L)\left\{\cos\left(\varphi_5 - \frac{\pi}{2}\right) + i\sin\left(\varphi_5 - \frac{\pi}{2}\right)\right\}\right]$$

or

$$\frac{E_{72}}{E_2} = \sqrt{k(1-k)}\left[\left\{\exp(-\alpha\pi R)\cos\left(\varphi_6 - \frac{\pi}{2}\right) + \exp(-\alpha L)\cos\left(\varphi_5 - \frac{\pi}{2}\right)\right\} + j\left\{\exp(-\alpha\pi R)\sin\left(\varphi_6 - \frac{\pi}{2}\right) + \exp(-\alpha L)\sin\left(\varphi_5 - \frac{\pi}{2}\right)\right\}\right].$$

Intensity can be calculated as:

$$\left|\frac{E_{72}}{E_2}\right|^2 = \{k(1-k)\}\left[\left\{\exp(-\alpha\pi R)\cos\left(\varphi_6 - \frac{\pi}{2}\right) + \exp(-\alpha L)\cos\left(\varphi_5 - \frac{\pi}{2}\right)\right\}^2 + \left\{\exp(-\alpha\pi R)\sin\left(\varphi_6 - \frac{\pi}{2}\right) + \exp(-\alpha L)\sin\left(\varphi_5 - \frac{\pi}{2}\right)\right\}^2\right]$$

or

$$\left|\frac{E_{72}}{E_2}\right|^2 = \{k(1-k)\}\left[\exp(-2\alpha\pi R)\cos^2\left(\varphi_6 - \frac{\pi}{2}\right) + \exp(-2\alpha L)\cos^2\left(\varphi_5 - \frac{\pi}{2}\right) + 2\exp(-\alpha L - \alpha\pi R)\cos\left(\varphi_6 - \frac{\pi}{2}\right)\cos\left(\varphi_5 - \frac{\pi}{2}\right) + \exp(-2\alpha\pi R)\sin^2\left(\varphi_6 - \frac{\pi}{2}\right) + \exp(-2\alpha L)\sin^2\left(\varphi_5 - \frac{\pi}{2}\right) + 2\exp(-\alpha L - \alpha\pi R)\sin\left(\varphi_6 - \frac{\pi}{2}\right)\sin\left(\varphi_5 - \frac{\pi}{2}\right)\right]$$

or

$$\left|\frac{E_{72}}{E_2}\right|^2 = k(1-k)\left[\exp(-2\alpha\pi R) + \exp(-2\alpha L) + 2\exp(-\alpha L - \alpha\pi R)\cos(\varphi_5 - \varphi_6)\right]. \tag{5.52}$$

Similarly, from (5.49)

$$\frac{E_{82}}{E_2} = (1-k)\exp(-\alpha\pi R)\{\cos\varphi_6 + i\sin\varphi_6\} - k\exp(-\alpha L)\{\cos\varphi_5 + i\sin\varphi_5\}$$

or

$$\frac{E_{82}}{E_2} = \{(1-k)\exp(-\alpha\pi R)\cos\varphi_6 - k\exp(-\alpha L)\cos\varphi_5\} + i\{(1-k)\exp(-\alpha\pi R)\sin\varphi_6 - k\exp(-\alpha L)\sin\varphi_5\}.$$

Intensity:

$$\left|\frac{E_{82}}{E_2}\right|^2 = \{(1-k)\exp(-\alpha\pi R)\cos\varphi_6 - k\exp(-\alpha L)\cos\varphi_5\}^2 + \{(1-k)\exp(-\alpha\pi R)\sin\varphi_6 - k\exp(-\alpha L)\sin\varphi_5\}^2$$

$$\left|\frac{E_{82}}{E_2}\right|^2 = (1-k)^2\exp(-2\alpha\pi R)\cos^2\varphi_6 + k^2\exp(-2\alpha L)\cos^2\varphi_5 - 2k(1-k)\exp(-(\alpha L + \alpha\pi R))\cos\varphi_6\cos\varphi_5 + (1-k)^2\exp(-2\alpha\pi R)\sin^2\varphi_6 + k^2\exp(-2\alpha L)\sin^2\varphi_5 - 2k(1-k)\exp(-(\alpha L + \alpha\pi R))\sin\varphi_6\sin\varphi_5$$

or

$$\left|\frac{E_{82}}{E_2}\right|^2 = (1-k)^2\exp(-2\alpha\pi R) + k^2\exp(-2\alpha L) - 2k(1-k)\exp(-(\alpha L + \alpha\pi R))\cos(\varphi_5 - \varphi_6) \quad (5.53)$$

Adding (5.52) and (5.53), we get:

$$\left|\frac{E_{72}}{E_2}\right|^2 + \left|\frac{E_{82}}{E_2}\right|^2 = k(1-k)\left[\exp(-2\alpha\pi R) + \exp(-2\alpha L) + 2\exp(-\alpha L - \alpha\pi R)\cos(\varphi_5 - \varphi_6)\right]$$
$$+(1-k)^2\exp(-2\alpha\pi R) + k^2\exp(-2\alpha L) - 2k(1-k)\exp(-(\alpha L + \alpha\pi R))\cos(\varphi_5 - \varphi_6)$$
$$\left|\frac{E_{72}}{E_2}\right|^2 + \left|\frac{E_{82}}{E_2}\right|^2 = (1-k)\exp(-2\alpha\pi R) + k\exp(-2\alpha L)$$

Therefore,

$$K_2 = \frac{\left|\frac{E_{72}}{E_2}\right|^2}{\left|\frac{E_{72}}{E_2}\right|^2 + \frac{E_{82}}{E_2}}$$

or

$$K_2 = \frac{k(1-k)\left[\exp(-2\alpha\pi R) + \exp(-2\alpha L) + 2\exp(-\alpha L - \alpha\pi R)\cos(\varphi_5 - \varphi_6)\right]}{(1-k)\exp((-2\alpha\pi R)) + k\exp(-2\alpha L)}$$

or

$$K_2 = \frac{k(1-k)g_5}{(1-k)\exp(-2\alpha\pi R) + k\exp(-2\alpha L)}. \quad (5.54)$$

Therefore,(5.54) is exactly same as (5.50), where

$$g_5 = \exp(-2\alpha\pi R) + \exp(-2\alpha L) + 2\exp(-\alpha L - \alpha\pi R)\cos(\varphi_5 - \varphi_6).$$

Now, since insertion loss in this case is given as:

$$\gamma_2 = 1 - \left|\frac{E_{72}}{E_2}\right|^2 - \left|\frac{E_{82}}{E_2}\right|^2.$$

Hence,from (5.52) and (5.53), we get

$$\gamma_2 = 1 - k(1-k)\exp(-2\alpha\pi R) - k(1-k)\exp(-2\alpha L) - 2k(1-k)\exp(-\alpha L$$
$$-\alpha\pi R)\cos(\varphi_5 - \varphi_6) - (1-k)^2\exp(-2\alpha\pi R) - k^2\exp(-2\alpha L)$$
$$+2k(1-k)\exp(-\alpha L - \alpha\pi R)\cos(\varphi_5 - \varphi_6)$$

$$\gamma_2 = 1-(1-k)\exp(-2\alpha\pi R)-k\exp(-2\alpha L) \tag{5.55}$$

When the signal is launched from port 1 $(E_2 = 0)$, the normalised output electric fields at port 7 and 8 are given as:

$$\frac{E_{71}}{E_1} = -k\exp(-\alpha\pi R)\exp(i\varphi_6)+(1-k)\exp(-\alpha L)\exp(i\varphi_5) \tag{5.56}$$

$$\frac{E_{81}}{E_1} = \sqrt{k(1-k)}\exp(-\alpha\pi R)\exp\left[i(\varphi_6-\pi/2)\right]+\sqrt{k(1-k)}\exp(-\alpha L)\exp\left[i(\varphi_5-\pi/2)\right]. \tag{5.57}$$

Since in this case $E_2 = 0$, we can prove (5.56) and (5.57) as follows from (5.37)–(5.43),

$$E_3 = \sqrt{(1-k)}E_1. \tag{5.58}$$

$$E_4 = -i\sqrt{k}E_1. \tag{5.59}$$

Now, substituting the value of E_3 and E_4 from (5.58) and (5.59) in(5.39)–(5.40),we get:

$$E_5 = \exp(-\alpha L)\exp(i\varphi_5)\sqrt{(1-k)}E_{1.} \tag{5.60}$$

$$E_6 = \exp(-\alpha\pi R)\exp(i\varphi_6)\left(-i\sqrt{k}\right)E_{1.} \tag{5.61}$$

Substituting the value of E_5 and E_6 in (5.41) from (5.60) and (5.61), we get

$$E_{71} = \sqrt{(1-k)}e^{-\alpha L}e^{i\varphi_5}\sqrt{(1-k)}E_1 - i\sqrt{k}e^{-\alpha\pi R}e^{i\varphi_6}\left(-i\sqrt{k}\right)E_1$$

or

$$\frac{E_{71}}{E_1} = -ke^{-\alpha\pi R}e^{i\varphi_6}+(1-k)e^{-\alpha L}e^{i\varphi_5}. \tag{5.62}$$

This equation is exactly same as (5.56); by substituting value of E_5 and E_6 from (5.60) and (5.61) in (5.42), we get

$$E_{81} = \left(-i\sqrt{k}\right)\exp(-\alpha L)\exp(i\varphi_5)\sqrt{(1-k)}E_1+\sqrt{(1-k)}\exp(-\alpha\pi R)\exp(i\varphi_6)\left(-i\sqrt{k}\right)E_1$$

$$\frac{E_{81}}{E_1} = \sqrt{k(1-k)}\exp(-\alpha\pi R)\exp i\left[\varphi_6-\frac{\pi}{2}\right]+\sqrt{k(1-k)}\exp(-\alpha L)\exp i\left(\varphi_5-\frac{\pi}{2}\right), \tag{5.63}$$

which is same as (5.57); in this case, the effective coupling ratio K_1 and the insertion loss γ_1 of the MZI are given as:

$$K_1 = \frac{|E_{81}/E_1|^2}{|E_{81}/E_1|^2+|E_{71}/E_1|^2}$$

$$= \frac{(1-k)kg_5}{k\exp(-2\alpha\pi R)+(1-k)\exp(-2\alpha L)}. \tag{5.64}$$

$$\begin{aligned}\gamma_1 &= 1-\left|E_{71}/E_1\right|^2-\left|E_{81}/E_1\right|^2\\ &= 1-k\exp(-2\alpha\pi R)-(1-k)\exp(-2\alpha L).\end{aligned} \tag{5.65}$$

From (5.62), we can prove (5.64) and (5.65) as follows:

$$\frac{E_{71}}{E_1} = -k\exp(-\alpha\pi R)\{\cos\varphi_6+i\sin\varphi_6\}+(1-k)\exp(-\alpha L)\{\cos\varphi_5+i\sin\varphi_5\}$$

or

$$\frac{E_{71}}{E_1} = \{-k\exp(-\alpha\pi R)\cos\varphi_6+(1-k)\exp(-\alpha L)\cos\varphi_5\}+j\{-k\exp(-\alpha\pi R)\sin\varphi_6+(1-k)\exp(-\alpha L)\sin\varphi_5\}$$

Hence, intensity would become as:

$$\begin{aligned}\left|\frac{E_{71}}{E_1}\right|^2 &= \{-k\exp(-\alpha\pi R)\cos\varphi_6+(1-k)\exp(-\alpha L)\cos\varphi_5\}^2\\ &\quad+\{-k\exp(-\alpha\pi R)\sin\varphi_6+(1-k)\exp(-\alpha L)\sin\varphi_5\}^2\end{aligned}$$

or

$$\begin{aligned}\left|\frac{E_{71}}{E_1}\right|^2 &= k^2\exp(-\alpha\pi R)\cos^2\varphi_6+(1-k)^2\exp(-2\alpha L)\cos^2\varphi_5-2k(1-k)\exp(-\alpha L\\ &\quad-\alpha\pi R)\cos\varphi_6\cos\varphi_5+k^2\exp(-2\alpha\pi R)\sin^2\varphi_6+(1-k)^2\exp(-2\alpha L)\sin^2\varphi_6\\ &\quad-2k(1-k)\exp(-\alpha L-\alpha\pi R)\sin\varphi_6\sin\varphi_5\end{aligned}$$

or

$$\left|\frac{E_{71}}{E_1}\right|^2 = k^2\exp(-2\alpha\pi R)+(1-k)^2\exp(-2\alpha L)-2k(1-k)\exp(-\alpha L-\alpha\pi R)\cos(\varphi_5-\varphi_6). \tag{5.66}$$

Similarly, from (5.63):

$$\frac{E_{81}}{E_1} = \sqrt{k(1-k)}\exp(-\alpha\pi R)\left\{\cos\left(\varphi_6-\frac{\pi}{2}\right)+i\sin\left(\varphi_6-\frac{\pi}{2}\right)\right\}+\sqrt{k(1-k)}\left\{\cos\left(\varphi_5-\frac{\pi}{2}\right)+i\sin\left(\varphi_5-\frac{\pi}{2}\right)\right\}$$

or

$$\begin{aligned}\frac{E_{81}}{E_1} &= \left\{\sqrt{k(1-k)}\exp(-\alpha\pi R)\cos\left(\varphi_6-\frac{\pi}{2}\right)+\sqrt{k(1-k)}\exp(-\alpha L)\cos\left(\varphi_5-\frac{\pi}{2}\right)\right\}\\ &\quad+i\left\{\sqrt{k(1-k)}\exp(-\alpha\pi R)\sin\left(\varphi_6-\frac{\pi}{2}\right)+\sqrt{k(1-k)}\exp(-\alpha L)\sin\left(\varphi_5-\frac{\pi}{2}\right)\right\}.\end{aligned}$$

Hence, intensity would become as:

$$\left|\frac{E_{81}}{E_1}\right|^2 = \left\{\sqrt{k(1-k)}\exp(-\alpha\pi R)\cos\left(\varphi_6 - \frac{\pi}{2}\right) + \sqrt{k(1-k)}\exp(-\alpha L)\cos\left(\varphi_5 - \frac{\pi}{2}\right)\right\}^2$$
$$+\left\{\sqrt{k(1-k)}\exp(-\alpha\pi R)\sin\left(\varphi_6 - \frac{\pi}{2}\right) + \sqrt{k(1-k)}\exp(-\alpha L)\sin\left(\varphi_5 - \frac{\pi}{2}\right)\right\}^2$$

or

$$\left|\frac{E_{81}}{E_1}\right|^2 = k(1-k)\exp(-2\alpha\pi R)\cos^2\left(\varphi_6 - \frac{\pi}{2}\right) + k(1-k)\exp(-2\alpha L)\cos^2\left(\varphi_5 - \frac{\pi}{2}\right)$$
$$+2k(1-k)\exp(-\alpha L - \alpha\pi R)\cos\left(\varphi_6 - \frac{\pi}{2}\right)\cos\left(\varphi_5 - \frac{\pi}{2}\right)$$
$$+k(1-k)\exp(-2\alpha\pi R)\sin^2\left(\varphi_6 - \frac{\pi}{2}\right) + k(1-k)\exp(-2\alpha L)\sin^2\left(\varphi_6 - \frac{\pi}{2}\right)$$
$$+2k(1-k)\exp(-\alpha L - \alpha\pi R)\sin\left(\varphi_6 - \frac{\pi}{2}\right)\sin\left(\varphi_5 - \frac{\pi}{2}\right)$$

$$\left|\frac{E_{81}}{E_1}\right|^2 = k(1-k)\exp(-2\alpha\pi R) + k(1-k)\exp(-2\alpha L) + 2k(1-k)\exp(-\alpha L - \alpha\pi R)\cos(\varphi_5 - \varphi_6). \quad (5.67)$$

Adding (5.66) and (5.67), we get:

$$\left|\frac{E_{71}}{E_1}\right|^2 + \left|\frac{E_{81}}{E_1}\right|^2 = k^2\exp(-2\alpha\pi R) + (1-k)^2\exp(-2\alpha L) - 2k(1-k)\exp(-\alpha L$$
$$-\alpha\pi R)\cos(\varphi_5 - \varphi_6) + k(1-k)\exp(-2\alpha\pi R) + k(1-k)\exp(-2\alpha L)$$
$$+2k(1-k)\exp(-\alpha L - \alpha\pi R)\cos(\varphi_5 - \varphi_6)$$

$$\left|\frac{E_{71}}{E_1}\right|^2 + \left|\frac{E_{81}}{E_1}\right|^2 = k\exp(-2\alpha\pi R) + (1-k)\exp(-2\alpha L).$$

Hence, the coupling coefficient for this case would become as:

$$K_1 = \frac{\left|\frac{E_{81}}{E_1}\right|^2}{\left|\frac{E_{71}}{E_1}\right|^2 + \left|\frac{E_{81}}{E_1}\right|^2}$$

or

$$K_1 = \frac{k(1-k)\exp(-2\alpha\pi R) + k(1-k)\exp(-2\alpha L) + 2k(1-k)\exp(-\alpha L - \alpha\pi R)\cos(\varphi_5 - \varphi_6)}{k\exp(-2\alpha\pi R) + (1-k)\exp(-2\alpha L)}$$

or

$$K_1 = \frac{k(1-k)g_5}{k\exp(-2\alpha\pi R)+(1-k)\exp(-2\alpha L)} \tag{5.68}$$

which is same as in (5.64), where

$$g_5 = \exp(-2\alpha\pi R)+\exp(-2\alpha L)+2\exp(-\alpha L-\alpha\pi R)\cos(\varphi_5-\varphi_6).$$

Similarly, the insertion loss term would become as:

$$\gamma_1 = 1-\left|\frac{E_{71}}{E_1}\right|^2-\left|\frac{E_{81}}{E_1}\right|^2$$

$$\begin{aligned}\gamma_1 = 1&-k^2\exp(-2\alpha\pi R)-(1-k)^2\exp(-2\alpha L)+2k(1-k)\exp(-\alpha L-\alpha\pi R)\cos(\varphi_5-\varphi_6)\\ &-k(1-k)\exp(-2\alpha\pi R)-k(1-k)\exp(-2\alpha L)-2k(1-k)\exp(-\alpha L-\alpha\pi R)\cos(\varphi_5-\varphi_6)\end{aligned}$$

$$\gamma_1 = 1-k\exp(-2\alpha\pi R)-(1-k)\exp(-2\alpha L) \tag{5.69}$$

which is same as (5.65). The phase transfer operator in the MZI can be obtained by the argument of (5.56), (5.57), (5.48), and (5.49) as follows:

$$\varphi_{jl} = arg\left(\frac{E_{jl}}{E_l}\right)(j=7,8\,l=1,2), \tag{5.70}$$

where subscript l and j denote the input port and the output port, respectively.

5.5.4 Transmission Performance of the Asymmetric MZI-coupled Ring Resonator

We deduce the field transmission operators as follows. The input light field E_1 is coupled in part into output port 7 by $E_{71}=\sqrt{(1-\gamma_1)(1-K_1)}\exp(i\varphi_{71})E_1$ and the rest field emerging from port 8 is $E_{8(1)}=\sqrt{(1-\gamma_1)K_1}exp(i\varphi_{81})E_1$. Then, the field circulates in the resonator and re-enters MZI from port 2, the field coupled to output port 7 is $E_{7(2)}=\sqrt{(1-\gamma_2)K_2}\exp(-\alpha\pi R)\exp[i(\varphi_{72}+\varphi_8)]E_{8(1)}$ and the rest field $E_{8(2)}=\sqrt{(1-\gamma)(1-K_2)}\exp(-\alpha\pi R)\exp[i(\varphi_{82}+\varphi_8)]E_{8(1)}$ continues circulating in the resonator until the field decreases to zero. When the signal circulates in the resonator by j_{th} turn, the field transmits to port 7 and port 8 are given as:

$$E_{7(j+1)} = \sqrt{(1-\gamma_2)K_2}\exp(-\alpha\pi R)\exp[i(\varphi_{72}+\varphi_8)]E_{8(j)} \tag{5.71}$$

$$E_{8(j+1)} = \sqrt{(1-\gamma)(1-K_2)}\exp(-\alpha\pi R)\exp[i(\varphi_{82}+\varphi_8)]E_{8(j)}. \tag{5.72}$$

Under the steady states, the resulting fields at port 7 and port 8 are given as:

$$E_{7\,final} = E_{7(1)}+E_{7(2)}+E_{7(3)}+\ldots\ldots.$$

$$= \frac{\left\{P_A \exp\left[i\left(\varphi_{81}+\varphi_{72}+\varphi_8\right)\right]+P_B \exp\left[i\left(\varphi_{82}+\varphi_{71}+\varphi_8\right)\right]-\sqrt{(1-K_1)(1-\gamma_1)}\exp\left(i\varphi_{71}\right)\right\}}{-P_c \exp\left[i\left(\varphi_{82}+\varphi_8\right)-1\right]}, \quad (5.73)$$

and similarly,

$$E_{8\,final} = E_{8(1)} + E_{8(2)} + E_{8(3)} + \ldots\ldots.$$
$$= \frac{K_1\left(1-\gamma_1\right)}{1+p_c^2+2p_c \exp\left[i\left(\varphi_{82}+\varphi_8\right)\right]}, \quad (5.74)$$

where

$$p_A = -\sqrt{K_1 K_2 \left(1-\gamma_1\right)\left(1-\gamma_2\right)}\exp(-\alpha\pi R),$$
$$P_B = \sqrt{(1-K_1)(1-K_2)(1-\gamma_1)(1-\gamma_2)}\exp(-\alpha\pi R),$$
$$p_C = -\sqrt{(1-\gamma_2)(1-K_2)}\exp(-\alpha\pi R).$$

Equations (5.73) and (5.74) are equivalent to (5.29) and (5.31). We can get the equivalent result for (5.71)–(5.74) as follows. The input light field E_1 is coupled in part intooutput port 7 by:

$$E_{7(1)} = \sqrt{(1-\gamma_1)(1-K_1)}\exp\left(i\varphi_{71}\right)E_{1.} \quad (5.75)$$

We can derive the equation as follows:

$$\gamma_1 = 1 - \left|\frac{E_{71}}{E_1}\right|^2 - \left|\frac{E_{81}}{E_1}\right|^2 \quad (5.76)$$

and

$$K_1 = \frac{\left|\frac{E_{81}}{E_1}\right|^2}{\left|\frac{E_{71}}{E_1}\right|^2 + \left|\frac{E_{81}}{E_1}\right|^2}. \quad (5.77)$$

Now, substitutingthe value of γ_1 and K_1 from (5.76) and (5.77) in (5.75), we get:

$$\frac{E_{7(1)}}{E_1} = \sqrt{\left\{1-\left(1-\left|\frac{E_{71}}{E_1}\right|^2 - \left|\frac{E_{81}}{E_1}\right|^2\right)\right\}\left\{1-\left(\frac{\left|\frac{E_{81}}{E_1}\right|^2}{\left|\frac{E_{71}}{E_1}\right|^2 + \left|\frac{E_{81}}{E_1}\right|^2}\right)\right\}}\, e^{\left(i\varphi_{71}\right)}$$

and this can be represented as general complex in polar form as:

$$\frac{E_{7(1)}}{E_1} = \left|\frac{E_{7(1)}}{E_1}\right| e^{\left(i\varphi_{71}\right)}, \quad (5.78)$$

$$\text{where } \varphi_{71} = arg\left(\frac{E_{7(1)}}{E_1}\right).$$

Equation (5.78) represents the proof of (5.75);in the same manner, we can write:

$$E_{8(1)} = \sqrt{(1-\gamma_1)(K_1)}\exp(i\varphi_{81})E_1. \quad (5.79)$$

Now, substitutingthe value of γ_1 and K_1 from (5.76) and (5.77), respectively, in (5.79), we get:

$$\frac{E_{8(1)}}{E_1} = \sqrt{\left\{1-\left(1-\left|\frac{E_{71}}{E_1}\right|^2 - \left|\frac{E_{81}}{E_1}\right|^2\right)\right\}\left(\frac{\left|\frac{E_{81}}{E_1}\right|^2}{\left|\frac{E_{71}}{E_1}\right|^2 + \left|\frac{E_{81}}{E_1}\right|^2}\right)}\exp(i\varphi_{81})$$

$$\frac{E_{8(1)}}{E_1} = \left|\frac{E_{8(1)}}{E_1}\right|\exp(i\varphi_{81}). \quad (5.80)$$

Equation (5.80) represents the proof of (5.79) by using (5.70):
Again,

$$E_{7(2)} = \sqrt{(1-\gamma_2)(K_2)}\exp(i\varphi_{72})E_2. \quad (5.81)$$

We know that:

$$\gamma_2 = 1-\left|\frac{E_{72}}{E_2}\right|^2 - \left|\frac{E_{82}}{E_2}\right|^2 \quad (5.82)$$

$$K_2 = \frac{\left|\frac{E_{72}}{E_2}\right|^2}{\left|\frac{E_{72}}{E_2}\right|^2 + \left|\frac{E_{82}}{E_2}\right|^2}. \quad (5.83)$$

Substituting the value of γ_2 and K_2 from (5.82) and (5.83), respectively, in (5.81):

$$\frac{E_{7(2)}}{E_2} = \sqrt{\left\{1-\left(1-\left|\frac{E_{72}}{E_2}\right|^2 - \left|\frac{E_{82}}{E_2}\right|^2\right)\right\}\left(\frac{\left|\frac{E_{72}}{E_2}\right|^2}{\left|\frac{E_{72}}{E_2}\right|^2 + \left|\frac{E_{82}}{E_2}\right|^2}\right)}\exp(i\varphi_{72})$$

$$\frac{E_{7(2)}}{E_2} = \left|\frac{E_{7(2)}}{E_2}\right|\exp(i\varphi_{72}). \quad (5.84)$$

Hence, (5.84) gives the proof of (5.81). Similarly,

$$E_{8(2)} = \sqrt{(1-\gamma_2)(1-K_2)}\exp(i\varphi_{82})E_2. \quad (5.85)$$

Substitutingthe value of the γ_2 and K_2 from (5.82) and (5.83), respectively, in (5.85):

$$\frac{E_{8(2)}}{E_2} = \sqrt{\left\{1-\left(1-\left|\frac{E_{72}}{E_2}\right|^2 - \left|\frac{E_{82}}{E_2}\right|^2\right)\right\}\left\{1-\left(\frac{\left|\frac{E_{72}}{E_2}\right|^2}{\left|\frac{E_{72}}{E_2}\right|^2 + \left|\frac{E_{82}}{E_2}\right|^2}\right)\right\}}\exp(i\varphi_{82})$$

$$\frac{E_{8(2)}}{E_2} = \left|\frac{E_{8(2)}}{E_2}\right|\exp(i\varphi_{82}). \tag{5.86}$$

Hence, (5.86) gives the proof of (5.85). Now, when the signal circulates in the resonator by the j_{th} turn, the transmitted to port 7 and port 8 are given as:

$$E_{7(j+1)} = \sqrt{(1-\gamma_2)(K_2)}\exp(-\alpha\pi R)\exp\left[i(\varphi_{72}+\varphi_8)\right]E_{8(j)} \tag{5.87}$$

$$E_{8(j+1)} = \sqrt{(1-\gamma_2)(1-K_2)}\exp(-\alpha\pi R)\exp\left[i(\varphi_{82}+\varphi_8)\right]E_{8(j)}. \tag{5.88}$$

Equations (5.87) and (5.88) can be represented as follows:
As we know that from (5.81):

$$E_{7(2)} = \sqrt{(1-\gamma_2)(K_2)}\exp(i\varphi_{72})E_2$$

Again, we know that:

$$E_2 = \exp(-\alpha\pi R)\exp(i\varphi_8)E_8. \tag{5.89}$$

Substitutingthe value of E_2 from (5.89) in (5.81), we get:

$$E_{7(2)} = \sqrt{(1-\gamma_2)(K_2)}\exp\left[i(\varphi_{72}+i\varphi_8)\right]\exp(-\alpha\pi R)E_{8(1)} \tag{5.90}$$

Similarly, substituting value E_2 from (5.89) in (5.85), we get:

$$E_{8(2)} = \sqrt{(1-\gamma_2)(1-K_2)}\exp\left[i(\varphi_{82}+\varphi_8)\right]\exp(-\alpha\pi R)E_{8(1)}. \tag{5.91}$$

Hence, (5.90) and (5.91) show the similar result to that of (5.87) and (5.88) for the value of $j=1$. Hence, using this method, we can find the similar results for other value such as $E_{7(3)}$ and $E_{8(3)}$. From (5.87), we can write:

$$E_{7(3)} = \sqrt{(1-\gamma_2)(K_2)}\exp(-\alpha\pi R)\exp\left[i(\varphi_{72}+\varphi_8)\right]E_{8(2)}. \tag{5.92}$$

Substituting the value of $E_{8(2)}$ from (5.91) in (5.92), we get:

$$E_{7(3)} = \sqrt{(1-\gamma_2)(K_2)}\exp(-\alpha\pi R)\exp\left[i(\varphi_{72}+\varphi_8)\right]\sqrt{(1-\gamma_2)(1-K_2)}\exp\left[i(\varphi_{82}+\varphi_8)\right]\exp(-\alpha\pi R)E_{8(1)}$$

$$E_{7(3)} = (1-\gamma_2)\sqrt{K_2(1-K_2)}\exp(-\alpha\pi R)\exp\left\{i(\varphi_{72}+\varphi_{82}+2\varphi_8)\right\}E_{8(1)}. \tag{5.93}$$

Similarly,

$$E_{8(3)} = \sqrt{(1-\gamma_2)(1-K_2)}\exp(-\alpha\pi R)\exp\left[i(\varphi_{82}+\varphi_8)\right]E_{8(2)}. \tag{5.94}$$

Substitutingvalue of $E_{8(2)}$ from (5.85)

$$E_{8(3)} = \sqrt{(1-\gamma_2)(1-K_2)}\exp(-\alpha\pi R)\exp\left[i(\varphi_{82}+\varphi_8)\right]\sqrt{(1-\gamma_2)(1-K_2)}\exp\left[i(\varphi_{82}+\varphi_8)\right]\exp(-\alpha\pi R)E_{8(1)}$$

$$E_{8(3)} = (1-\gamma_2)(1-K_2)\exp(-2\alpha\pi R)\exp\left[i(2\varphi_{82}+2\varphi_8)\right]E_{8(1)} \tag{5.95}$$

$$E_{8(final)} = E_{8(1)} + E_{8(2)} + E_{8(3)} + \ldots$$

$$E_{8(final)} = E_{8(1)} + \sqrt{(1-\gamma_2)(1-K_2)}\exp\left[i(\varphi_{82}+\varphi_8)\right]\exp(-\alpha\pi R)E_{8(1)}$$
$$+(1-\gamma_2)(1-K_2)\exp(-2\alpha\pi R)\exp\left[i(2\varphi_{82}+2\varphi_8)\right]E_{8(1)} + \ldots.$$

$$E_{8(final)} = \frac{E_{8(1)}}{1-\sqrt{(1-\gamma_2)(1-K_2)}\exp\left[i(\varphi_{82}+\varphi_8)\right]\exp(-\alpha\pi R)}$$

Let us assume that $p_c = -\sqrt{(1-\gamma_2)(1-K_2)}\exp(-\alpha\pi R)$. Substituting this value, we get:

$$E_{8(final)} = \frac{\sqrt{(1-\gamma_1)K_1}\exp(j\varphi_{81})E_1}{1+p_c\exp\left[i(\varphi_{82}+\varphi_8)\right]},$$

where we have used (5.79). Hence,

$$\left|\frac{E_{8(final)}}{E_1}\right|^2 = \frac{K_1(1-\gamma_1)}{1+p_c^2+2p_c\cos(\varphi_{82}+\varphi_8)}. \tag{5.96}$$

In the same manner, by using (5.75), (5.90), and (5.93), we get:

$$E_{7(final)} = E_{7(1)} + E_{7(2)} + E_{7(3)} + \ldots$$

$$\begin{aligned} E_{7(final)} = {} & \sqrt{(1-\gamma_1)(1-K_1)}\exp(i\varphi_{71})E_1 + \sqrt{(1-\gamma_2)(K_2)}\exp\left[i(\varphi_{72}+i\varphi_8)\right]\exp(-\alpha\pi R)E_{8(1)} \\ & +(1-\gamma_2)(1-K_2)\exp(-2\alpha\pi R)\exp\left[i(2\varphi_{82}+2\varphi_8)\right]E_{8(1)}. \end{aligned} \tag{5.97}$$

Now, substitutingthe value of $E_{8(1)}$ in (5.97) from (5.79), we get:

$$\begin{aligned} E_{7(final)} = {} & \sqrt{(1-\gamma_1)(1-K_1)}\exp(i\varphi_{71})E_1 \\ & +\sqrt{(1-\gamma_2)(K_2)}\exp\left[i(\varphi_{72}+i\varphi_8)\right]\exp(-\alpha\pi R)\sqrt{(1-\gamma_1)(K_1)}\exp(i\varphi_{81})E_1 \\ & +(1-\gamma_2)(1-K_2)\exp(-2\alpha\pi R)\exp\left[i(2\varphi_{82}+2\varphi_8)\right]\sqrt{(1-\gamma_1)(K_1)}\exp(i\varphi_{81})E_1. \end{aligned}$$

We can assume that:

$$p_A = -\sqrt{K_1K_2(1-\gamma_1)(1-\gamma_2)}\exp(-\alpha\pi R)$$

$$p_B = \sqrt{(1-K_1)(1-K_2)(1-\gamma_1)(1-\gamma_2)}\exp(-\alpha\pi R).$$

Then,

$$E_{7(final)} = \sqrt{(1-\gamma_1)(1-K_1)}\exp(i\varphi_{71})E_1 - p_A \exp i[\varphi_{81}+\varphi_{72}+\varphi_8]E_1 + \ldots\ldots \quad (5.98)$$

By solving $E_{7\,final} = 0$, we obtain the optimal resonance condition of the MZI-coupled ring resonator as shown in (5.99) and (5.100) at the end of this section, where $p_D = -\sqrt{K_1K_2(1-\gamma_2)/(1-K_1)}\exp(-\alpha\pi R)$, m and n are positive integers and $\Delta\phi_5 = 0$. Therefore, φ_{jl} is a function of $\Delta\varphi_6$, as shown in (5.48)–(5.70).

$$\begin{cases} \varphi_{82} + \beta L + \Delta\varphi_{8r} = 2m\pi + arc\cos\left[\left(p_D^2 + p_C^2 + 1\right)/2p_C\right] \\ \varphi_{81} + \varphi_{72} - \varphi_{71} + \beta L + \Delta\varphi_{8r} = 2n\pi + arc\cos\left[\left(-p_D^2 + p_C^2 + 1\right)/2p_D\right] \end{cases} \quad (5.99)$$

or

$$\begin{cases} \varphi_{82} + \beta L + \Delta\varphi_{8r} = 2(m+1)\pi - arc\cos\left[\left(p_D^2 + p_C^2 + 1\right)/2p_C\right] \\ \varphi_{81} + \varphi_{72} - \varphi_{71} + \beta L + \Delta\varphi_{8r} = 2(n+1)\pi - arc\cos\left[\left(-p_D^2 + p_C^2 + 1\right)/2p_D\right] \end{cases} \quad (5.100)$$

5.6 DSP APPROACH TO THE DESIGN OF OPTICAL DEVICES

In order to satisfy the ever-increasing demand for high bit rates, next generation optical communication networks can be made all-optical to overcome the electronics bottleneck and more efficiently utilise the intrinsic broad bandwidth of optical fibres. Currently, there are two possible technologies for achieving high transmission rate: optical time division multiplexing (OTDM) and dense wavelength division multiplexing (DWDM). However, neither the full potential of OTDM nor that of the DWDM technology has been realised due to lack of suitable nonlinear, all-optical devices that can perform signal regeneration, ultrafast switching, encoding/decoding, and/or wavelength conversion efficiently.

There are a number of problems with current nonlinear optical material and devices. There are two types nonlinear optical material from which devices can be made: non-resonant and resonant. Non-resonant materials have a weak nonlinear response, but the passage of light occurs with very low loss and the response is broadband, typically exceeding 10THz. However, because of the weak nonlinear response, these materials tend to be bulky and impose a long latency. Resonant materials have a very strong nonlinear response, but at the expense of reduced bandwidth and increased loss. Artificial resonances can be used in optical architectures to overcome the limitation of current nonlinear devices and materials.

The nonlinear optical response of many artificial resonant has been studied in literatures, but most of the studies have been limited to analysing the nonlinear properties of specific architecture and do not provide the synthesis approach to device design that can produce a specific nonlinear response. Discrete-time signal processing (DSP) provides an easy to use the mathematical framework, the z-transform, for the description of the discrete-time filters. The z-transform has already been used to emphasise numerous optical filter configurations in order to optimise their linear response.

This section shows that the DSP approach is a systematic and intuitive way to design nonlinear optical device using the DSP approach. First, a prototype linear frequency response (in the weak perturbation limit) is selected for the desired nonlinear optical devices. Next, the optical architecture's unit cell is selected and the multistage optical architecture is analysed using the z-transform.

5.6.1 Transfer Function of Optical Devices

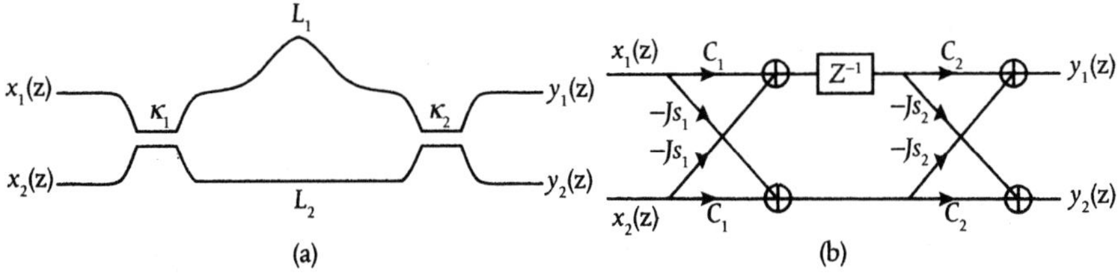

Fig. 5.36 MZI device (a) waveguide layout and (b) *z*-schematic

Consider anMZI device as shown in Fig. 5.36 (a).
Figure 5.36 (b) shows the equivalent lattice form of Fig 5.36 (a), where

$$\left.\begin{array}{l} s_1=\sqrt{\kappa_1}\ ,c_1=\sqrt{1-\kappa_1} \\ s_2=\sqrt{\kappa_2}\ ,c_2=\sqrt{1-\kappa_2} \end{array}\right\}. \tag{5.101}$$

From Fig. 5.36 (b), we can get:

$$Y_1(z)=c_1z^{-1}c_2X_1(z)+(-js_1)(-js_2)X_1(z)+(-js_1)z^{-1}c_2X_2(z)+c_1(-js_2)X_2(z). \tag{5.102}$$

Similarly,

$$Y_2(z)=c_1c_2X_2(z)+(-js_1)(-js_2)z^{-1}X_2(z)+c_1z^{-1}(-js_2)X_1(z)+(-js_1)c_2X_1(z) \tag{5.103}$$

Now, substituting $X_2=0$ in (5.102) to get $H_{11}(z)$ as $H_{11}(z)\big|_{X_2=0}=\dfrac{Y_1(z)}{X_1(z)}$

$$Y_1(z)=\left[c_1c_1z^{-1}-s_1s_2\right]X_1(z).$$

Hence, from (5.101):

$$H_{11}(z)=\sqrt{(1-\kappa_1)(1-\kappa_2)}z^{-1}-\sqrt{\kappa_1\kappa_2}. \tag{5.104}$$

Similarly, we can calculate $H_{12}(z)=\dfrac{Y_1(z)}{X_2(z)}\bigg|_{X_1(z)=0}$ from (5.102)as:

$$Y_1(z)=\left[-js_1c_2z^{-1}-jc_1s_2\right]X_2(z)$$

$$H_{12}(z)=-j\left[\sqrt{\kappa_1(1-\kappa_2)}z^{\ 1}+\sqrt{\kappa_2(1-\kappa_1)}\right]. \tag{5.105}$$

Similarly, from (5.103),

$$H_{21}(z)\big|_{X_2=0}=\frac{Y_2(z)}{X_1(z)}$$

$$H_{21}(z) = -j\left[\sqrt{\kappa_2(1-\kappa_1)}z^{-1} + \sqrt{(1-\kappa_2)\kappa_1}\right].$$

Similarly, $H_{22}(z)\Big|_{X_1=0} = \dfrac{Y_2(z)}{X_2(z)}$;

$$Y_2(z) = \left[c_1c_2 - s_1s_2z^{-1}\right]X_2(z)$$

Hence,

$$H_{22}(z) = \left[\sqrt{(1-\kappa_1)(1-\kappa_2)} - \sqrt{\kappa_1\kappa_2}z^{-1}\right]. \tag{5.106}$$

We can show the MATLAB plot for all these transfer functions, as shown in Fig. 5.37.

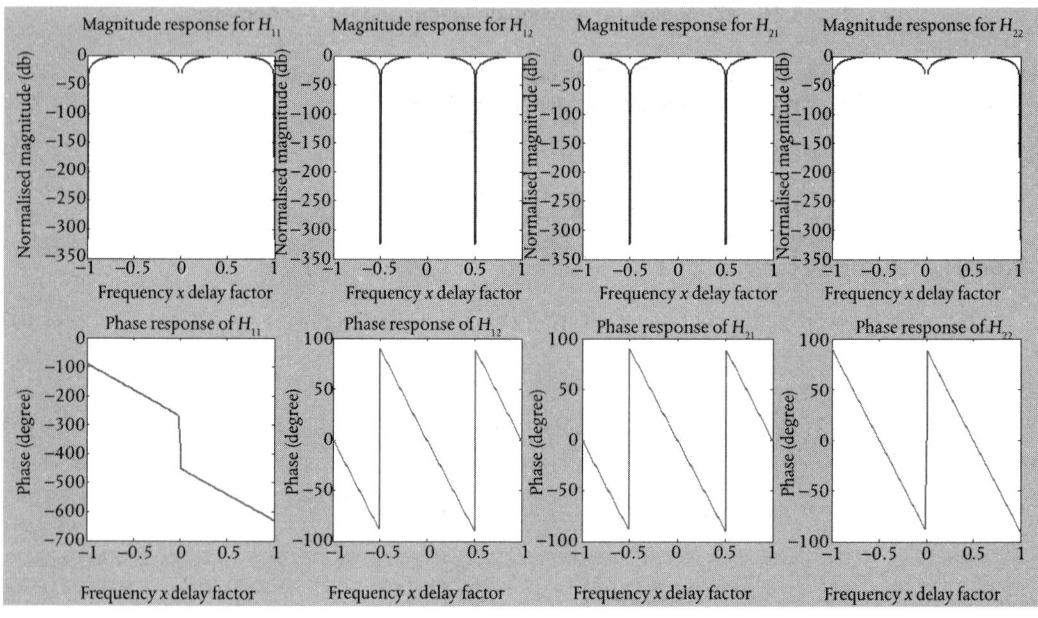

Fig. 5.37 The transmission plot of MZI as an optical filter

Figure 5.37 shows the variation in the magnitude and phase response when MZI is used as an optical filter. The figure shows the linear phase response. Hence, MZI can be used as an optical filter, especially as a notch filter.

This approach can be used to design the optical devices to obtain the various nonlinear responses;for example, all-optical switching, nonlinear phase shift, second harmonic generation, four-wave mixing (i.e., frequencies v_m and v_n mix to produce $2v_m - v_n$ and $2v_n - v_m$), soliton (which is a carrier of digital information), bi-stability (which results in twostable, switchable output states and can be used as a basis for logic operation and thresholding with restoration), and amplification (which can overcome loss). The nonlinear phase shift is a fundamental nonlinear process that enables many all-optical switching and logic devices, and it is the process used to demonstrate our approach. Artificial resonant structures are used in the devices to overcome the aforementioned traditional drawbacks.

5.6.2 Optical Filters and Z-transforms

In this section, we discuss the important aspects of optical filters in relation to discrete filters and explains how z-transform can be used to describe optical filters as well. This section borrows heavily from Madsen and Zhao's book on optical filters. Like discrete filters, optical filters are broadly into two categories: finite impulse response (FIR) and infinite impulse response (IIR). FIR filters have no feedback paths between the output and input and their transfer function has only zeros. These filters are also referred to as moving average (MA) filters. The IIR filters have feedback paths and their transfer function have poles and may or may not have zeros. When zeros are not present or all the zeros occur at the origin, IIR filters are referred as autoregressive (AR) filters. When both poles and non-origin zeros are present, they are referred to as autoregressive moving average (ARMA) filters.

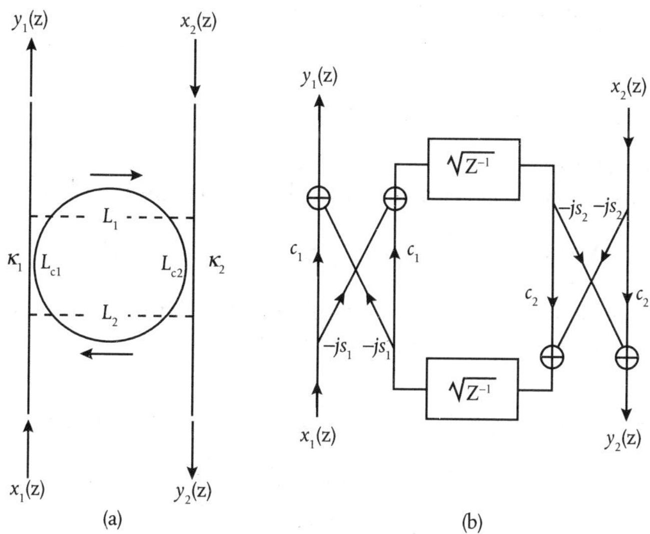

Fig 5.38 Ring resonator: (a) waveguide and (b) z-schematic

Optical architecture can be of restricted type or general type. With restricted architectures, we cannot obtain arbitrary frequency response, while general architectures, like discrete filters, allow arbitrary frequency response to be approximated over a frequency range of interest. To approximate any arbitrary function in discrete-time signal processing, a set of sinusoidal functions whose weighted sum yields a Fourier series approximation is used. The optical analogue is found in interferometers. Interferometers come in two general classes: (1) Mach–Zehnder interferometer (MZI) and (2) Fabry–Perot interferometer (FPI). The MZI has finite numbers of delay and no recirculating (or feedback) delay paths. Therefore, these are MA filters. The FPI consist of a cavity surrounded by two partial reflectors that are parallel to each other. The waveguide analogue of the FPI is the ring resonator shown in Fig. 5.38 (a). The output is the sum of delayed versions of the input signal weighted by the roundtrip cavity transmission. The transmission response is of the ARtype, while the reflection responses are of ARMA type. The ring resonator is an example of an artificial resonator.

The z-transform schematics for the MZI and FPI device are shown in Fig. 5.35 (b) and Fig. 5.38 (b), respectively. Here, κ is the power coupling ratio for each directional coupler, $c=\sqrt{1-\kappa}$ is the through-port transmission term, and $-js=-j\sqrt{\kappa}$ is the cross-port transmission term. Further, $z=e^{j\Omega T}$, and $\Omega T=\beta L_u$,where L_u is the smallest path length called the unit delay length, T is the unit delay and is equal to $2n\pi/\lambda$, n is the refractive index of the material, c is the speed of the light in vaccum, and λ is the wavelength of light. The propagation loss of a delay line is accounted for by multiplying z^{-1} by $\gamma=10^{-\alpha L/20}$, where α is the average loss per unit length in dB and L is the delay path length. Because delay is discrete values of the unit delay, the frequency response is periodic. One period is defined as the free spectral range (FSR) and is given by FSR = $1/T$. The normalised frequency, $f=\omega/2\pi$, is related to the optical frequency by $f=(v-v_c)T$ or $f=(\Omega-\Omega_c)T/2\pi$. The centre frequency $v_c=c/\lambda_c$ is defined so that the product of refractive index and unit length is equal to an integer number of wavelength, that is, $m\lambda_c=nL_u$, where m is an integer.

To analyse the frequency response of the MZI, the unit delay is set equal to the differences in path lengths, $L_u=L_1-L_2$. The overall transfer function matrix of the MZI is the product of the matrices:

$$\Phi_{MZI}=\Phi_{cplr}(\kappa_2)\Phi_{delay}\Phi_{cplr}(\kappa_1)$$

$$=\begin{bmatrix} c_2 & -js_2 \\ -js_2 & c_2 \end{bmatrix}\begin{bmatrix} z^{-1} & 0 \\ 0 & -1 \end{bmatrix}\begin{bmatrix} c_1 & -js_1 \\ -js_1 & c_1 \end{bmatrix}. \quad (5.107)$$

For the ring resonator, the unit delay is equal to $L_u=L_1+L_2+L_{c1}+L_{c2}$, where L_{c1} and L_{c2} are the coupling region lengths for each coupler. The sum of all-optical paths is given as:

$$Y_2(z)=-s_1s_2\sqrt{\gamma z^{-1}}\left(1+c_1c_2\gamma z^{-1}+c_1^2c_2^2\gamma^2z^{-2}+\ldots\right)X_1(z) \quad (5.108)$$

We can show that the detailed derivation of transfer function for various cases as follows. When we assume $X_2(z)=0$, we can calculate the value of H_{21} as follows:

$$Y_2(z)=(-js_1)(-js_2)\sqrt{\gamma z^{-1}}X_1(z)+c_2\sqrt{\gamma z^{-1}}c_1\sqrt{\gamma z^{-1}}(-js_1)(-js_2)\sqrt{\gamma z^{-1}}X_1(z)$$
$$+(-js_1)(-js_2)\sqrt{\gamma z^{-1}}\sqrt{\gamma z^{-1}}\sqrt{\gamma z^{-1}}\sqrt{\gamma z^{-1}}\sqrt{\gamma z^{-1}}c_1^2c_2^2X_1(z)$$
$$Y_2(z)=-s_1s_2\sqrt{\gamma z^{-1}}\left[1+c_1c_2\gamma z^{-1}+c_1^2c_2^2\left(\gamma z^{-1}\right)^2+\ldots\right]X_1(z).$$

Figure 5.38 (b) shows the equivalent form of Fig. 5.38 (a):

$$Y_2(z)=-s_1s_2\sqrt{\gamma z^{-1}}\left[\sum_{k=0}^{\infty}(c_1c_2\gamma)^k z^{-k}\right]X_1(z)$$

$$H_{21}(z)=\frac{-s_1s_2\sqrt{\gamma z^{-1}}}{\left(1-c_1c_2\gamma z^{-1}\right)}. \quad (5.109)$$

Substituting the value of $s_1, s_2, c_1,$ and c_2 from (5.101) in (5.109):

$$H_{21}(z)=\frac{-\sqrt{\kappa_1\kappa_2\gamma z^{-1}}}{\left(1-c_1c_2\gamma z^{-1}\right)}=\frac{\sigma\sqrt{\gamma z^{-1}}}{A(z)}, \quad (5.110)$$

where $\sigma=-\sqrt{\kappa_1\kappa_2}, A(z)=1-c_1c_2\gamma z^{-1}$.

Now, in the same manner, we can calculate the value of H_{11}. From Fig. 5.37 (b), we can write the expression for Y_1 as:

$$Y_1(z)=c_1X_1(z)+(-js_1)\sqrt{\gamma z^{-1}}c_2\sqrt{\gamma z^{-1}}(-js_1)X_1(z)$$
$$+(-js_1)\sqrt{\gamma z^{-1}}c_2\sqrt{\gamma z^{-1}}c_1\sqrt{\gamma z^{-1}}c_2\sqrt{\gamma z^{-1}}(-js_1)X_1(z)+\ldots.$$

or

$$Y_1(z)=c_1X_1(z)-s_1^2\left(\gamma z^{-1}\right)c_2X_1(z)-s_1^2\left(\gamma z^{-1}\right)^2 c_1c_2^2+X_1(z)....$$

or

$$Y_1(z)=X_1(z)\left[c_1-s_1^2\left(\gamma z^{-1}\right)c_2\left\{1+\left(\gamma z^{-1}\right)c_1c_2+\left(\gamma z^{-1}\right)^2 c_1^2c_2^2+....\right\}\right]$$

or

$$Y_1(z)=X_1(z)\left[c_1-\frac{s_1^2\gamma z^{-1}c_2}{\left(1-\gamma z^{-1}c_1c_2\right)}\right]$$

or

$$H_{11}(z)=\frac{c_1-c_2\gamma z^{-1}}{1-\gamma z^{-1}c_1c_2}=\frac{B(z)}{A(z)} \tag{5.111}$$

$$\left.\begin{array}{c} A(z)=1-\gamma z^{-1}c_1c_2,\sigma=-\sqrt{\kappa_1\kappa_2} \\ B(z)=c_1-c_2\gamma z^{-1} \end{array}\right\}.$$

Similarly,

$$\begin{aligned}Y_2(z)&=c_2X_2(z)+\left(-js_2\sqrt{\gamma z^{-1}}c_1\sqrt{\gamma z^{-1}}\left(-js_2\right)\right)X_2(z)\\&+\left(-js_2\right)\sqrt{\gamma z^{-1}}c_1\sqrt{\gamma z^{-1}}c_2\sqrt{\gamma z^{-1}}c_1\sqrt{\gamma z^{-1}}\left(-js_2\right)X_2(z)+....\end{aligned}$$

or

$$Y_2(z)=X_2(z)\left[c_2-s_2^2\gamma z^{-1}c_1-s_2^2\left(\gamma z^{-1}\right)^2 c_1^2c_2-....\right]$$

or

$$Y_2(z)=X_2(z)\left[c_2-s_2^2\gamma z^{-1}c_1\left\{1+\gamma z^{-1}c_1c_2+\left(\gamma z^{-1}\right)^2 c_1^2c_2^2+....\right\}\right]$$

or

$$H_{22}(z)=c_2-\frac{s_2^2\left(\gamma z^{-1}\right)c_1}{1-\gamma z^{-1}c_1c_2}$$

or

$$H_{22}(z)=\frac{c_2-\gamma z^{-1}c_1c_2^2-s_2^2\left(\gamma z^{-1}\right)c_1}{1-\gamma z^{-1}c_1c_2}=\frac{c_2-\gamma z^{-1}c_1\left(c_2^2+s_2^2\right)}{1-\gamma z^{-1}c_1c_2}=\frac{c_2-\gamma z^{-1}c_1\left(1-\kappa_2+\kappa_2\right)}{1-\gamma z^{-1}c_1c_2}$$

$$H_{22}(z)=\frac{c_2-\gamma z^{-1}c_1}{1-\gamma z^{-1}c_1c_2}=\frac{B^R(z)}{A(z)}, \tag{5.112}$$

where $B^R(z)=c_2-\gamma z^{-1}c_1$.
In the same manner, we can calculate the value of H_{12} when $X_1(z)=0$:

$$Y_1(z)=\Big[(-js_2)\sqrt{\gamma z^{-1}}(-js_1)+(-js_2)\sqrt{\gamma z^{-1}}c_1\sqrt{\gamma z^{-1}}c_2\sqrt{\gamma z^{-1}}(-js_1)$$
$$+(-js_2)\sqrt{\gamma z^{-1}}c_1\sqrt{\gamma z^{-1}}c_2\sqrt{\gamma z^{-1}}c_1\sqrt{\gamma z^{-1}}c_2\sqrt{\gamma z^{-1}}(-js_1)+\ldots\Big]X_2(z)$$

or

$$Y_1(z)=\Big[-s_1s_2\sqrt{\gamma z^{-1}}-s_1s_2\sqrt{\gamma z^{-1}}(\gamma z^{-1})c_1c_2-s_1s_2\sqrt{\gamma z^{-1}}(\gamma z^{-1})^2c_1^2c_2^2-\ldots\Big]X_2(z)$$

or

$$H_{12}(z)\big|_{X_1=0}=-s_1s_2\sqrt{\gamma z^{-1}}\Big[1+(\gamma z^{-1})c_1c_2+(\gamma z^{-1})^2c_1^2c_2^2+\ldots\Big]$$

$$H_{12}(z)=\frac{-s_1s_2\sqrt{\gamma z^{-1}}}{1-\gamma z^{-1}c_1c_2}=\frac{\sigma\sqrt{\gamma z^{-1}}}{A(z)}. \tag{5.113}$$

The overall transfer function can be written as a scattering matrix (because it relates the input and the outputs) like:

$$\begin{pmatrix}Y_1(z)\\Y_2(z)\end{pmatrix}=\begin{pmatrix}\dfrac{B(z)}{A(z)} & \dfrac{\sigma\sqrt{\gamma z^{-1}}}{A(z)}\\[2ex] \dfrac{\sigma\sqrt{\gamma z^{-1}}}{A(z)} & \dfrac{-B^R(z)}{A(z)}\end{pmatrix}\begin{pmatrix}X_1(z)\\X_2(z)\end{pmatrix}. \tag{5.114}$$

Hence, we see that optical resonances are represented by poles in a transfer function. Therefore, the filters built using artificial resonances are FIR filters.

We have used the MZI and micro-ring resonator as the building blocks to design the nonlinear optical devices for obtaining nonlinear phase shift in this section. Detailed description of using z-transform for analysing single-stage and multi-stage optical filter is provided.

5.6.3 Optical Architecture for Nonlinear Phase Shifter

The second step is to select the optical architecture's unit cell and analyse it using the z-transform. Artificial resonances produced by the ring resonators can be used to enhance the nonlinear phase shift response of the optical devices.

The presence of a ring resonator in the architecture implies the presence of a pole in the filter's transfer function. To select the optical architecture for obtaining a nonlinear phase shift response, we analyse two ring resonator configurations: (1) single pole and (2) single pole–zero with the pole and zero independent of each other.

1. Single-pole design shows single-pole architecture with zero at the origin. The transfer function for this architecture in the z-domain is given as:

$$\frac{E_{out}(z)}{E_{in}(z)}=\frac{\sqrt{\kappa_0\kappa_1}\sqrt{\gamma e^{-j\phi}z^{-1}}}{1-c_0c_1\gamma e^{-j\phi}z^{-1}}. \tag{5.115}$$

The total phase change in the fundamental range $-\pi \le \omega \le \pi$ for this unit cell is equal to π. By cascading N such cells, we can obtain a total phase change of $N\pi$ in the fundamental range.

1. Single pole–zero design with independent poles and zeros show a single pole–zero architecture with the poles and zeros independent of each other. The transfer function for this architecture in the z-domain is given as:

$$\frac{E_{out}(z)}{E_{in}(z)} = \frac{\left(c^2 c_r - s^2 e^{-j\phi_t}\right) - \left(c^2 e^{-j\phi_r} - s^2 c_r e^{-j(\phi_t+\phi_r)}\right) z^{-1}}{1 - c_r e^{-j\phi_r} z^{-1}}. \tag{5.116}$$

Equation (5.115) has been already derived and (5.116) can be derived as follows. Figure 3.49(a) and Fig. 3.49(b) shows a basic ARMA filter and its equivalent filter, respectively.

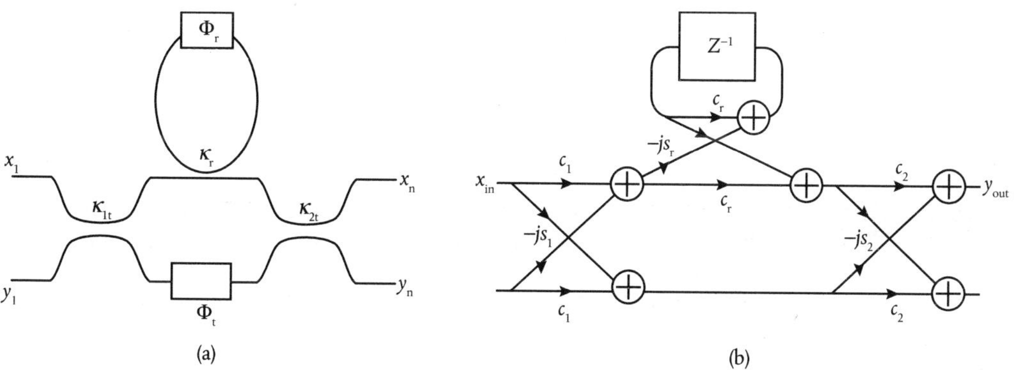

Fig. 5.39 (a) A basic ARMA Filter; (b) its equivalent lattice form

From Fig. 5.39 (b),

$$\begin{aligned} Y_{out}(z) = {} & c_1 c_2 c_r X_{in} + (-js_1)(-js_2) X_{in} + c_1(-js_r) z^{-1} c_2 (-js_r) X_{in} \\ & + c_1(-js_r) z^{-1} c_r z^{-1} c_2 (-js_r) X_{in} + c_1(-js_r) z^{-1} c_r z^{-1} c_r z^{-1} c_2 (-js_r) X_{in}(z) + \ldots \end{aligned}$$

$$Y_{out}(z) = X_{in}(z)\left[c_1 c_2 c_r - s_1 s_2 - c_1 c_2 s_r^2 z^{-1}\left\{1 + c_r z^{-1} + c_r^2 z^{-2} + \ldots\right\}\right]$$

$$Y_{out}(z) = X_{in}(z)\left[c_1 c_2 c_r - s_1 s_2 - \frac{c_1 c_2 s_r^2 z^{-1}}{1 - c_r z^{-1}}\right]$$

or the transfer function:

$$H(z) = \frac{c_1 c_2 c_r - s_1 s_2 + s_1 s_2 c_r z^{-1} - c_1 c_2 z^{-1}}{1 - c_r z^{-1}}. \tag{5.117}$$

Now, if $c_1 = c_2 = c$ And $s_1 = s_2 = s$,

$$H(z) = \frac{\left(c^2 c_r - s^2\right) - \left(c^2 - s^2 c_r\right) z^{-1}}{\left(1 - c_r z^{-1}\right)}, \tag{5.118}$$

where we assumed $\phi_r = 0$ and $\phi_t = 0$.

The output plot can be represented as shown in Fig. 5.40.

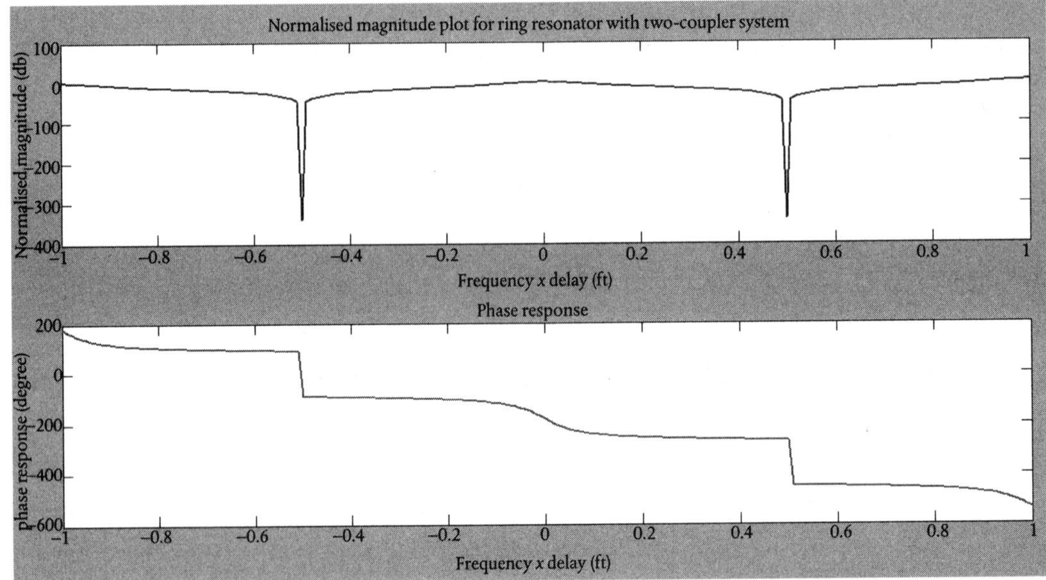

Fig. 5.40 Response of ring resonator with two-coupler system

The total phase change in the fundamental range $-\pi \leq \omega \leq \pi$ for this unit cell is equal to 2π if the filter is maximum phase and 0 if it is minimum phase. We are interested in low-pass maximum phase system $(|\text{zero}| > 1/|\text{pole}|)$. Since, they have the maximum net phase change and most of the phase change lies within the pass band. The architecture can be designed to be low-pass maximum phase system since the poles and zeros are independent of each other. By cascading N such unit cells, we can obtain a total phase change of $2N\pi$ in the fundamental range.

A third possible configuration is a ring resonator with a single coupler. However, this is a pole–zero architecture with dependent poles and zeros and is always high pass for a maximum phase system. The total phase change is equal to 2π but most of the phase change is present in the stop band, and hence, we cannot obtain the prototype response of Fig. 5.40 using the unit cell. Therefore, we decided to use the first and second configuration as the unit cell for our design. Joining the first configuration unit cell in a lattice structure gives us AR lattice filter architecture, as shown in Fig. 5.40, and joining the second configuration unit cell in a lattice gives us an ARMA lattice filter architecture; these are chosen since they have low-pass band loss and can operate at significantly higher components variation as compared to transversal or cascade structure.

The next step is to obtain a Z-transform description of the multistage architecture obtained by joining the unit cells. First, a DSP schematic is drawn for the architecture and then it is analysed to obtain a transfer function matrix. The AR and ARMA lattice architecture's DSP schematics and transfer function are given. The detailed derivation can be represented as follows.

5.6.4 Auto-regressive Lattice Filter

Figure 5.41 shows the waveguide layout and DSP schematic of auto-regressive (AR) lattice architecture. The detailed derivation of its transfer matrix can be represented as follows.

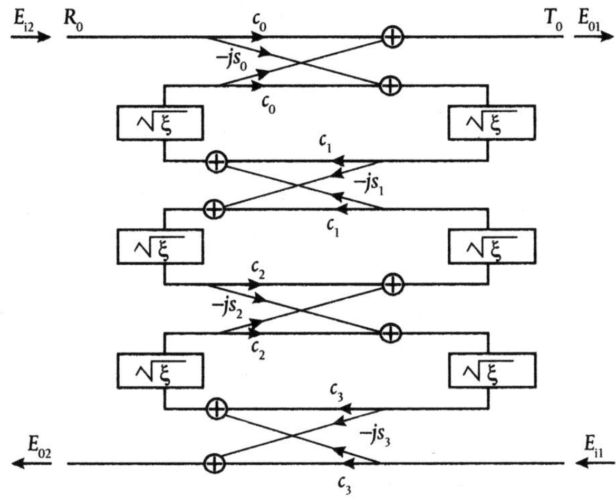

Fig. 5.41 Auto-regressive (AR) lattice filter having multistage

We can redraw the circuit in more understandable manner as shown in Fig. 5.42.

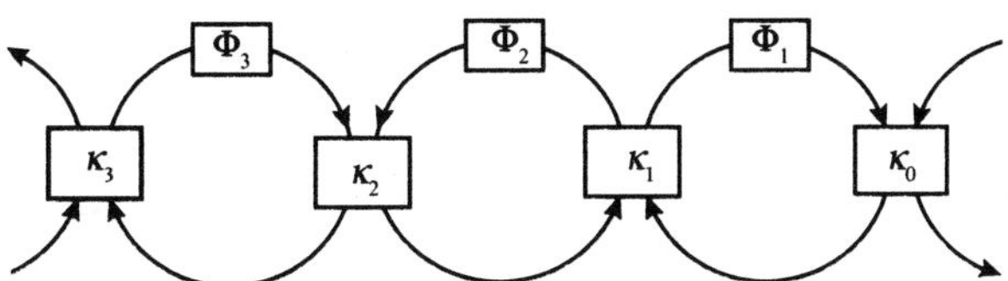

Fig. 5.42 The equivalent form of Fig. 5.41

The single stage can be represented as shown in Fig. 5.43.

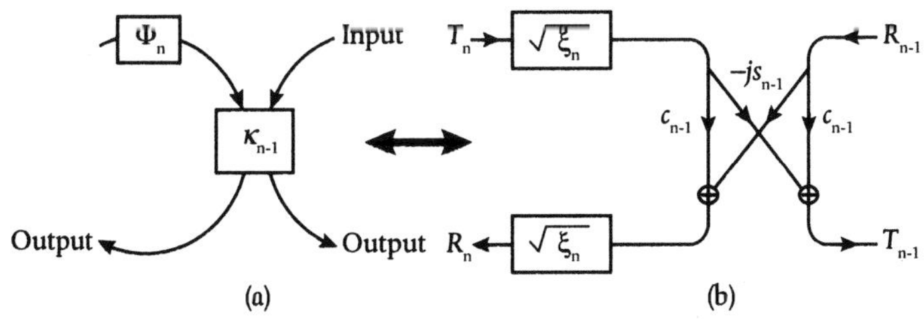

Fig. 5.43 (a) Single-stage AR filter;(b) Equivalent lattice diagram

Hence, $\xi_n = \gamma e^{-j\phi_n} z^{-1}$

and $c_n = \sqrt{1-\kappa_n}$

Now, for a single stage, we can write the equation as follows:

$$T_{n-1} = \sqrt{\xi_n}\left(-js_{n-1}\right)T_n + c_{n-1}R_{n-1} \tag{5.119}$$

$$R_n = \xi_n c_{n-1} T_n + \left(-js_{n-1}\right)\sqrt{\xi_n} R_{n-1} \tag{5.120}$$

In matrix form, we can write:

$$\begin{bmatrix} R_n \\ T_{n-1} \end{bmatrix} = S_n \begin{bmatrix} T_n \\ R_{n-1} \end{bmatrix},$$

where

$$S_n = \begin{bmatrix} \xi_n c_{n-1} & \sqrt{\xi_n}\left(-js_{n-1}\right) \\ \left(-js_{n-1}\right)\sqrt{\xi_n} & c_{n-1} \end{bmatrix}.$$

Now, from (5.119), we can write:

$$T_n\left(\sqrt{\xi_n}\left(js_{n-1}\right)\right) = -T_{n-1} + c_{n-1}R_{n-1}$$

$$T_n = \frac{1}{\left(\sqrt{\xi_n}\left(js_{n-1}\right)\right)}\left[-T_{n-1} + c_{n-1}R_{n-1}\right]$$

$$T_n = \frac{-1}{\left(\sqrt{\xi_n}\left(js_{n-1}\right)\right)}\left[T_{n-1} - c_{n-1}R_{n-1}\right]. \tag{5.121}$$

Substituting the value of T_n from (5.121) in (5.120), we get:

$$R_n = \xi_n c_{n-1}\left[\frac{-1}{\left(\sqrt{\xi_n}\left(js_{n-1}\right)\right)}\left[T_{n-1} - c_{n-1}R_{n-1}\right]\right] + \left(-js_{n-1}\right)\sqrt{\xi_n}R_{n-1}.$$

After solving this equation, we get:

$$R_n = \frac{-\xi_n c_{n-1}T_{n-1}}{\left(\sqrt{\xi_n}\left(js_{n-1}\right)\right)} + \frac{\xi_n R_{n-1}}{\left(\sqrt{\xi_n}\left(js_{n-1}\right)\right)}.$$

Therefore, simultaneously, we can write:

$$T_n = \frac{-1}{\left(\sqrt{\xi_n}\left(js_{n-1}\right)\right)}\left[T_{n-1} - c_{n-1}R_{n-1}\right] \tag{5.122}$$

$$R_n = \frac{-1}{\left(\sqrt{\xi_n}\left(js_{n-1}\right)\right)}\left[\xi_n c_{n-1}T_{n-1} - \xi_n R_{n-1}\right]. \tag{5.123}$$

In matrix form, we can write:

$$\begin{bmatrix} T_n \\ R_n \end{bmatrix} = \frac{1}{-js_{n-1}\sqrt{\xi_n}}\begin{bmatrix} 1 & -c_{n-1} \\ \xi_n c_{n-1} & -\xi_n \end{bmatrix}\begin{bmatrix} T_{n-1} \\ R_{n-1} \end{bmatrix}.$$

Replacing $n \to n+1$

$$\begin{bmatrix} T_{n+1} \\ R_{n+1} \end{bmatrix} = \frac{1}{-js_n\sqrt{\xi_{n+1}}}\begin{bmatrix} 1 & -c_n \\ \xi_{n+1}c_n & -\xi_{n+1} \end{bmatrix}\begin{bmatrix} T_n \\ R_n \end{bmatrix}$$

or

$$\begin{bmatrix} T_{n+1} \\ R_{n+1} \end{bmatrix} = \phi_n \begin{bmatrix} T_n \\ R_n \end{bmatrix}, \tag{5.124}$$

where $\phi_n = \dfrac{1}{-js_n\sqrt{\xi_{n+1}}} \begin{bmatrix} 1 & -c_n \\ \xi_{n+1}c_n & -\xi_{n+1} \end{bmatrix}$.

Substituting $\xi = \gamma e^{-j\phi_n} z^{-1}$, we get :

$$\phi_n = \frac{1}{-js_n\sqrt{\gamma e^{-j\phi_{n+1}} z^{-1}}} \begin{bmatrix} 1 & -c_n \\ \gamma e^{-j\phi_{n+1}} z^{-1} c_n & -\gamma e^{-j\phi_{n+1}} z^{-1} \end{bmatrix}. \tag{5.125}$$

In the same manner, we can show that:

$$\begin{bmatrix} T_{(n+1)}(z) \\ R_{(n+1)}(z) \end{bmatrix} = \phi_n \phi_{n-1} \phi_{n-2} \ldots.. \phi_1 \phi_0 \begin{bmatrix} T_0(z) \\ R_0(z) \end{bmatrix}. \tag{5.126}$$

5.6.5 ARMA Lattice Filter

Figure 5.44 shows the waveguide layout and DSP schematic of an ARMA lattice architecture. The transfer matrix for this architecture can be derived as follows. The ARMA lattice architecture can be represented as follows:

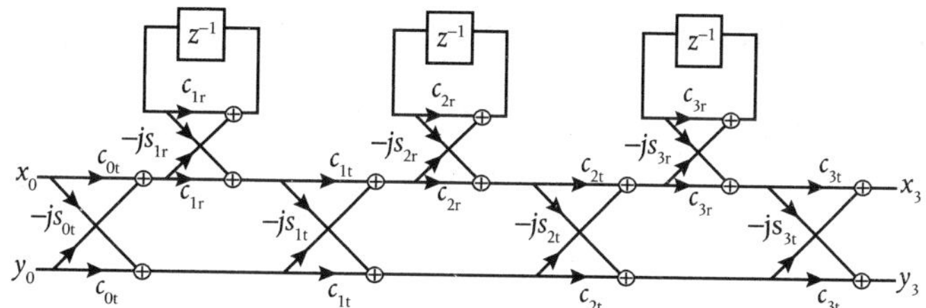

Fig. 5.44 ARMA lattice architecture

The single stage of structure can be represented as follows (Fig. 5.45):

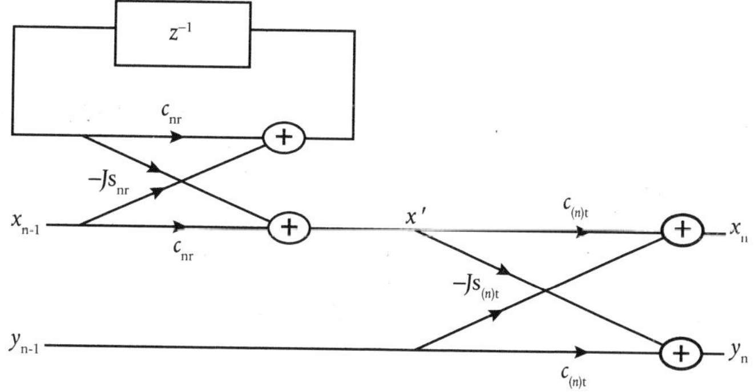

Fig. 5.45 Single stage of ARMA lattice structure

Now, from the above-mentioned single-structure architecture, we can write:

$$X'(z)=X_{(n-1)}(z)c_{nr}+(-js_{nr})^2 z^{-1}X_{(n-1)}(z)+(-js_{nr})^2 z^{-2}c_{nr}X_{(n-1)}(z)$$
$$+(-js_{nr})^2 z^{-3}c_{nr}^2 X_{(n-1)}(z)+\ldots\ldots \text{ up to}\infty$$

or

$$X'(z)=X_{(n-1)}(z)c_{nr}+z^{-1}X_{(n-1)}(z)(-js_{nr})^2\left[1+z^{-1}c_{nr}+z^{-3}c_{nr}^2+\ldots\ldots up\,to\,\infty\right]$$

or

$$X'(z)=X_{(n-1)}(z)c_{nr}+\frac{z^{-1}X_{(n-1)}(z)(-js_{nr})^2}{(1-z^{-1}c_{nr})}$$

$$X'(z)=X_{(n-1)}(z)\left[\frac{c_{nr}-z^{-1}}{1-z^{-1}c_{nr}}\right]. \quad (5.127)$$

From Fig.5.44, we can write

$$X_n(z)=X'(z)c_{nt}+Y_{(n-1)}(z)(-js_{nt})(5.128) \quad (5.128)$$

$$Y_n(z)=-js_{nt}X'(z)+Y_{(n-1)}(z)c_{nt}. \quad (5.129)$$

Substituting the value of $X'(z)$ in (5.128) from (5.127), we get:

$$X_n(z)=X_{(n-1)}(z)\left[\frac{c_{nr}-z^{-1}}{1-z^{-1}c_{nr}}\right]c_{nt}+Y_{(n-1)}(z)(-js_{nt})$$

or

$$X_n(z)=\frac{1}{1-z^{-1}c_{nr}}\left[c_{nt}(c_{nr}-z^{-1})X_{(n-1)}(z)+(-js_{nt})(1-z^{-1}c_{nr})Y_{(n-1)}(z)\right]. \quad (5.130)$$

As we have assumed that:

$$A_n(z)=1-z^{-1}c_{nr} \quad (5.131)$$

$$A_n^R(z)=-c_{nr}+z^{-1}. \quad (5.132)$$

Substituting the values of $1-z^{-1}c_{nr}$ and $-c_{nr}+z^{-1}$ from (5.131) and (5.132), respectively, in (5.130), we can obtain:

$$X_n(z)=\frac{1}{A_n(z)}\left[-c_{nt}A_n^R(z)X_{(n-1)}(z)+(-js_{nt})A_n(z)Y_{(n-1)}(z)\right]. \quad (5.133)$$

Similarly, substitutingthe value of $X'(z)$ from (5.127) in (5.129):

$$Y_n(z)=-js_{nt}X_{(n-1)}(z)\left[\frac{c_{nr}-z^{-1}}{1-z^{-1}c_{nr}}\right]+Y_{(n-1)}(z)c_{nt}$$

or

$$Y_n(z)=\frac{1}{1-z^{-1}c_{nr}}\left[\left(-js_{nt}\right)\left(c_{nr}-z^{-1}\right)X_{(n-1)}(z)+c_{nt}\left(1-z^{-1}c_{nr}\right)Y_{(n-1)}(z)\right]. \tag{5.134}$$

Substitutingvalues of $1-z^{-1}c_{nr}$ and $-c_{nr}+z^{-1}$ from (5.131) and (5.132), respectively, in (5.134), we can obtain:

$$Y_n(z)=\frac{1}{A_n(z)}\left[js_{nt}A_n^R(z)X_{(n-1)}(z)+c_{nt}A_n(z)Y_{(n-1)}(z)\right]. \tag{5.135}$$

By analysing(5.133) and (5.135), we can write in matrix form as follows:

$$\begin{bmatrix}X_n(z)\\Y_n(z)\end{bmatrix}=\frac{1}{A_n(z)}\begin{bmatrix}-c_{nt}A_n^R(z) & -js_{nt}A_n(z)\\ js_{nt}A_n^R(z) & c_{nt}A_n(z)\end{bmatrix}\begin{bmatrix}X_{n-1}(z)\\Y_{n-1}(z)\end{bmatrix} \tag{5.136}$$

or

$$\begin{bmatrix}X_n(z)\\Y_n(z)\end{bmatrix}=\phi_n\begin{bmatrix}X_{n-1}(z)\\Y_{n-1}(z)\end{bmatrix},$$

where

$$\phi_n=\frac{1}{A_n(z)}\begin{bmatrix}-c_{nt}A_n^R(z) & -js_{nt}A_n(z)\\ js_{nt}A_n^R(z) & c_{nt}A_n(z)\end{bmatrix}.$$

SUMMARY

- Optical communication is any form of telecommunication that uses light as the transmission medium. An optical communication system consists of a transmitter, which encodes a message into an optical signal, a channel, which carries the signal to its destination, and a receiver, which reproduces the message from the received optical signal.
- Optical fibre communication is the most common type of channel for optical communications; however, other types of optical waveguides are used within computers or communications gear, and they have even formed the channel of very short-distance (e.g., chip to chip, intra chip) links in laboratory trials. The transmitters in optical fibre links are generally light-emitting diodes (LEDs) or laser diodes.
- The function of optical communication system is to convey the signal from the informational source over the transmission medium to the destination. The communication system, therefore, consists of a transmitter or modulator linked to the informational source, the transmission medium, and a receiver or demodulator at the destination point.
- Optical fibre transmission has played a key role in increasing the bandwidth of telecommunications. The access mechanism is based on the processing of signals with a so-called optical signal processor, abbreviated as OSP.
- The OSP has the ability to modify the information content of data signals while, at the same time, preserving certain properties of the physical carrier. Applications are generally found in optical communication networks.

- Optical networks are high-capacity telecommunications networks based on optical technologies and components that provide routing, grooming, and restoration at the wavelength-level as well as wavelength-based services.
- Fibre optic signal processing devices can be constructed to perform various functions, such as convolution, correlation, matrix operation, and frequency filtering, pulse train generation, pulse compression, and matched filtering are also achievable.
- We can use various types of lattice ladder structure in order to perform the various types of filtering actions.
- The lattice structure allows a systematic approach to filter synthesis;it is easy to implement, more regular and modular, and has convenient numerical properties;these properties make it very popular for the implementation of various filtering actions.
- To perform the various signal processing operations described, we need to employ the elements such as time delay, tapping, time advance, summing elements, and branching elements.
- Recirculating (feed backward) and non-recirculating (feed forward) delay lines are two structures that are the basic to fibre optic signal processing.
- In the two-coupler recirculating delay line, a single-mode fibre loop (with loop delay *T*) is closed upon itself by two-directional coupler, with the second coupler feeding some portion of the light to the first coupler through the fibre loop.
- The two-coupler non-recirculating delay line also consist of two-directional coupler, but in this case, the output of the first coupler are both fed forward to the second coupler where they are recombined after a time delay T, which is the time delay difference between the two feed-forward fibre lines.
- The input–output relationship of the directional coupler can describe by a 2×2 complex transfer matrix.

$$\begin{pmatrix} E_3 \\ E_4 \end{pmatrix} = (\gamma)^{\frac{1}{2}} \begin{pmatrix} A & B \\ C & D \end{pmatrix} \cdot \begin{pmatrix} E_1 \\ E_2 \end{pmatrix}$$

- The input and output light intensities can be related to each other by the following expression:

$$\begin{pmatrix} I_3 \\ I_4 \end{pmatrix} = (\gamma) \begin{pmatrix} 1-\kappa & \kappa \\ \kappa & 1-\kappa \end{pmatrix} \begin{pmatrix} I_1 \\ I_2 \end{pmatrix},$$

  where (I_1, I_2) and (I_3, I_4) are the input and output intensities, respectively, *Y* is the overall intensity transmission factor, and κ is the intensity coupling coefficient.
- In order to increase our ability to handle more complex system problems, we utilise Z-transform techniques for describing the signal.
- The Z-transform, $F(z)$ of a signal, $f(k)$, is defined by the following expression:

$$F(z) \triangleq \sum_{k=-\infty}^{\infty} f(k).z^{-k},$$

  where k is the time index, which is an integer multiple of the basic time delay ($\kappa = nT$), and z is the transform variable thatrepresents a unit time advance (z^{-1} represents a unit time delay).
- The frequency response is obtained when we evaluate the transfer function at $z = e^{j\omega T}$. Since we know that $s = j\omega$, $z = e^{sT}$. Hence, we can easily connect Z-transform domain into s-domain. Here,*s*-domain is Laplace transform and we can easily apply control theory concepts on these transfer formulas.

- Using the definition of the Z-transform, we can find the expression for the system transfer functions, *H* (*z*), of the various types of coupler delay lines.
- The transfer function of the two-coupler recirculating delay lines in *z*-domain can be represented as:

$$H_{11}(z) = \frac{a_1 + (1-2a_1)a_0 l_1 z^{-1}}{1 - a_1 a_0 l_1 z^{-1}},$$

where a_1 and a_0 are the intensity coupling coefficients of the couplers, and l_1 is the loop intensity transmittance of the system.

- Similarly, the transfer function of two-coupler recirculating delay line, with X_1 as the input and Y_2 as the output, can be represented as follows:

$$H_{21}(z) = \frac{(1-a_1)(1-a_0) l_{11}}{1 - a_1 a_0 l_1 z^{-1}},$$

where l_{11} is the intensity transmission factor of the forward-fibre line.

- The transfer function of two-coupler non-recirculating delay lines, with X_1 as the input and Y_1 as the output can be derived and represented as follows:

$$H_{11} = (1-b_1)(1-b_0) l_{11} + b_1 b_0 l_{12} z^{-1}$$

- In the same manner, the transfer function for the four-coupler non-recirculating system can be shown as follows:

$$\begin{aligned} H_{11} = {} & (1-b_1)(1-b_2)(1-b_3) l_4 l_5 l_6 + (1-b_0)(1-b_1) b_2 b_3 l_3 l_4 l_5 z^{-1} + \\ & (1-b_0) b_1 b_2 (1-b_3) l_2 l_4 l_6 z^{-1} + b_0 b_1 (1-b_2)(1-b_3) l_1 l_5 l_6 z^{-1} + \\ & b_0 b_1 b_2 b_3 l_1 l_3 l_5 z^{-2} + (1-b_0) b_1 (1-b_2) b_3 l_2 l_3 l_4 z^{-2} + b_0 (1-b_1)(1-b_2) b_3 l_1 l_2 l_3 z^{-3}. \end{aligned}$$

- Similarly, the transfer function for the system H_{12} (non-recirculating system):

$$\begin{aligned} H_{12} = {} & (1-b_0)(1-b_1) b_2 (1-b_3) l_3 l_4 l_5 z^{-1} + (1-b_0)(1\ \ b_1)(1\ \ b_2) b_3 l_4 l_5 l_6 \\ & + (1-b_0) b_1 (1-b_2)(1-b_3) l_2 l_3 l_4 z^{-2} + b_0 (1-b_1)(1-b_2)(1-b_3) l_1 l_2 l_3 z^{-3} + (1-b_0) b_1 b_2 b_3 l_2 l_4 l_6 z^{-1} \\ & + b_0 b_1 (1-b_2) b_3 l_1 l_5 l_6 z^{-1} + (1-b_0) b_1 b_2 b_3 l_2 l_4 l_6 z^{-1}. \end{aligned}$$

- The transfer function for the four-coupler non-recirculating system (H_{21}) can be represented as follows:

$$\begin{aligned} H_{21} = {} & b_0 (1-b_1)(1-b_2)(1-b_3) l_4 l_5 l_6 + (1-b_0) b_1 (1-b_2)(1-b_3) l_1 l_5 l_6 z^{-1} \\ & + (1-b_0)(1-b_1) b_2 (1-b_3) l_1 l_2 l_6 z^{-2} + (1-b_0)(1-b_1)(1-b_2) b_3 l_1 l_2 l_3 z^{-3} \\ & + (1-b_0) b_1 b_2 b_3 l_1 l_3 l_5 z^{-2} + b_0 b_1 b_2 (1-b_3) l_2 l_4 l_6 z^{-1} \end{aligned}$$

- Similarly, we can derive the transfer function of the four-coupler recirculating system (H_{11}):

$$H_{11} = \frac{a_3 + a_2(1-2a_3) l_3 l_6 z^{-1} + a_1 (1-a_2)^2 (1-2a_3) l_2 l_3 l_5 l_6 z^{-2} + a_0 (1-a_1)^2 (1-a_2)^2 (1-2a_3) l_1 l_2 l_3 l_4 l_5 l_6 z^{-3}}{\left[1 - a_2 a_3 l_3 l_6 z^{-1} - a_1 a_3 (1-a_2)^2 l_2 l_3 l_5 l_6 z^{-2} - a_0 (1-a_1)^2 (1-a_2)^2 a_3 l_1 l_2 l_3 l_4 l_5 l_6 z^{-3}\right]}$$

- In the same manner, the transfer function of the four-coupler recirculating system (H_{12}) can be represented as follows:

$$H_{12} = \frac{(1-a_0)(1-a_1)(1-a_2)(1-a_3)l_1l_2l_3z^{-3}}{1-a_2a_3l_3l_6z^{-1}-a_1(1-a_2)^2a_3l_2l_3l_5l_6z^{-2}}$$

- The transfer function for the four-coupler recirculating system H_{21}:

$$H_{21} = \frac{(1-a_0)(1-a_1)(1-a_2)(1-a_3)l_4l_5l_6}{1-a_1a_0l_1l_4z^{-1}-a_0(1-a_1)^2a_2a_3l_1l_2l_4l_5z^{-2}-a_0(1-a_1)^2(1-a_2)^2a_3l_1l_2l_3l_4l_5l_6z^{-3}}$$

$$H_{22} = \left[\frac{a_0+(1-2a_0)a_1l_1l_4z^{-1}+(1-2a_0)(1-a_1)^2a_2l_1l_2l_4l_5z^{-2}+(1-2a_0)(1-a_1)^2(1-a_2)^2a_3l_1l_2l_3l_4l_5l_6z^{-3}}{1-a_0a_1l_1l_4z^{-1}-a_0(1-a_1)^2a_2l_1l_2l_4l_5z^{-2}-a_0(1-a_1)^2(1-a_2)^2a_3l_1l_2l_3l_4l_5l_6z^{-3}}\right]$$

- This $e^{j\omega T}$ describes a unit circle centred at the origin of the z-plane. The unit circle also plays an important role in problems regarding stability. Hence, for two-coupler system, e^{sT} is approximated up to 13 numbers of elements:

$$z=e^{sT}=1+sT+\frac{(sT)^2}{\lfloor\underline{2}}+\frac{(sT)^3}{\lfloor\underline{3}}+\frac{(sT)^4}{\lfloor\underline{4}}+\frac{(sT)^5}{\lfloor\underline{5}}+$$
$$\frac{(sT)^6}{\lfloor\underline{6}}+\frac{(sT)^7}{\lfloor\underline{7}}+\frac{(sT)^8}{\lfloor\underline{8}}+\frac{(sT)^9}{\lfloor\underline{9}}+\frac{(sT)^{10}}{\lfloor\underline{10}}+\frac{(sT)^{11}}{\lfloor\underline{11}}+\frac{(sT)^{12}}{\lfloor\underline{12}}+\frac{(sT)^{13}}{\lfloor\underline{13}}$$

- We can replace all z by above-mentioned equation and we can find the similar response to that of the z-transform function.
- Integrated optical ring resonators are useful components of optical switches, sensors, modulators, and wavelength-division multiplexing systems.
- In this chapter, we propose and characterise an improved configuration having two phase shifters embedded into different positions of the MZI-coupled ring resonators using an accurate analytical model by considering the propagation losses, the asymmetrical phase transfer operators of the MZI, and the radius of the ring cavity.
- In the case of steady-state resonance model, we can find the transfer function as follows:

$$\frac{E_7}{E_1}=\frac{(1-k)\exp(j\varphi_5-\alpha L)+(-k)\exp(j\varphi_6-\alpha\pi R)-\exp\left[j(\varphi_5+\varphi_6+\varphi_8)-\alpha L-2\alpha\pi R\right][1-k+k]^2}{1-\left(-j\sqrt{k}\right)^2\exp(j\varphi_5-\alpha L)\exp(-\alpha\pi R+j\varphi_8)-(1-k)\exp(j\varphi_6-\alpha\pi R)\exp(j\varphi_8-\alpha\pi R)}$$

- Similarly, we can obtain the function

$$\frac{E_8}{E_1}=\frac{\sqrt{1-k}\left(-j\sqrt{k}\right)\exp(j\varphi_5-\alpha L)+\sqrt{1-k}\left(-j\sqrt{k}\right)\exp(j\varphi_6-\alpha\pi R)}{1-\left(-j\sqrt{k}\right)^2\exp\left[j(\varphi_5+\varphi_8)-\alpha L-\alpha\pi R\right]-(1-k)\exp\left[j(\varphi_8+\varphi_6)-2\alpha\pi R\right]}.$$

- The desired high-sensitive transmission response to the phase change $\Delta\phi_5$ can be achieved at the optimal resonance corresponding to the maximum extinction ratio $(E_7/E_1=0)$. Here, we assume $\Delta\phi_5$ is zero at the given initial state. By setting the real and imaginary parts of the output field E_7, above-mentioned equation is set to zero.

- We can find the optimal condition as follows:

$$\beta L+\beta\pi R+\Delta\varphi_{8r}=2m\pi+arc\cos(\varepsilon)$$
$$2\beta\pi R+\Delta\varphi_{6r}+\Delta\varphi_{8r}-\pi=(2n-1)\pi+arc\cos(\eta),$$

where

$$\varepsilon=\frac{k^2\exp(2\alpha L+2\alpha\pi R)-(1-k)^2\exp(4\alpha\pi R)+1}{-2\kappa\exp(\alpha L+\alpha\pi R)}$$

$$\eta=\frac{-k^2\exp(2\alpha L+2\alpha\pi R)+(1-k)^2\exp(4\alpha\pi R)+1}{2(1-k)\exp(2\alpha\pi R)}$$

- In order to reveal the underlying physics of the resonance characteristics more clearly, we used an alternate numerical model based on the multi-beam interference to calculate the optimal resonance conditions.
- We can find the required transfer function as follows:

$$\frac{E_{72}}{E_2}=\sqrt{k(1-k)}\exp(-\alpha\pi R)\exp\left[i(\varphi_6-2/\pi)\right]+\sqrt{k(1-k)}\exp(-\alpha L)\exp\left[i(\varphi_5-2/\pi)\right]$$

$$\frac{E_{82}}{E_2}=(1-k)\exp(-\alpha\pi R)\exp(i\varphi_6)+k\exp(-\alpha L)\exp(i\varphi_5)$$

The effective coupling ratio K_2 and the insertion loss γ_2 are given as:

$$K_2=\frac{|E_{72}/E_2|^2}{|E_{72}/E_2|^2+|E_{82}/E_2|^2}=\frac{(1-k)k.g_5}{(1-k)\exp(-2\alpha\pi R)+k\exp(-2\alpha L)}$$

where

$$g_5=\exp(-2\alpha\pi R)+\exp(-2\alpha L)+2\exp(-\alpha L-\alpha\pi R)\cos(\varphi_5-\varphi_6).$$

Further,

$$\gamma_2=1-|E_{72}/E_2|^2-|E_{82}/E_2|^2$$
$$=1-(1-k)\exp(-2\alpha\pi R)-k\exp(-2\alpha L).$$

- Similarly, another set of equations can be obtained as follows:

$$\frac{E_{71}}{E_1}=-k\exp(-\alpha\pi R)\exp(i\varphi_6)+(1-k)\exp(-\alpha L)\exp(i\varphi_5)$$

$$\frac{E_{81}}{E_1}=\sqrt{k(1-k)}\exp(-\alpha\pi R)\exp\left[i(\varphi_6-\pi/2)\right]+\sqrt{k(1-k)}\exp(-\alpha L)\exp\left[i(\varphi_5-\pi/2)\right].$$

Further, in the same manner, we can obtain the remaining expression for coupling ratio and insertion loss as follows:

$$K_1 = \frac{|E_{81}/E_1|^2}{|E_{81}/E_1|^2 + |E_{71}/E_1|^2}$$

$$= \frac{(1-k)kg_5}{k\exp(-2\alpha\pi R)+(1-k)\exp(-2\alpha L)}$$

$$\gamma_1 = 1 - |E_{71}/E_1|^2 - |E_{81}/E_1|^2$$

$$= 1 - k\exp(-2\alpha\pi R) - (1-k)\exp(-2\alpha L)$$

➢ We can find the transmission performance as follows:

$$E_{7\,final} = E_{7(1)} + E_{7(2)} + E_{7(3)} + \ldots\ldots.$$

$$= \frac{\{P_A\exp[i(\varphi_{81}+\varphi_{72}+\varphi_8)] + P_B\exp[i(\varphi_{82}+\varphi_{71}+\varphi_8)] - \sqrt{(1-K_1)(1-\gamma_1)}\exp(i\varphi_{71})\}}{-P_c\exp[i(\varphi_{82}+\varphi_8)-1]}$$

$$E_{8\,final} = E_{8(1)} + E_{8(2)} + E_{8(3)} + \ldots\ldots.$$

$$= \frac{K_1(1-\gamma_1)}{1+p_c^2+2p_C\exp[i(\varphi_{82}+\varphi_8)]},$$

where

$$p_A = -\sqrt{K_1K_2(1-\gamma_1)(1-\gamma_2)}\exp(-\alpha\pi R)$$

$$P_B = \sqrt{(1-K_1)(1-K_2)(1-\gamma_1)(1-\gamma_2)}\exp(-\alpha\pi R),$$

$$p_C = -\sqrt{(1-\gamma_2)(1-K_2)}\exp(-\alpha\pi R).$$

THEORY QUESTIONS AND ANSWERS

Question 1: Describe the block diagram of optical communication system.

Answer: The block diagram of optical communication system can be shown as follows:

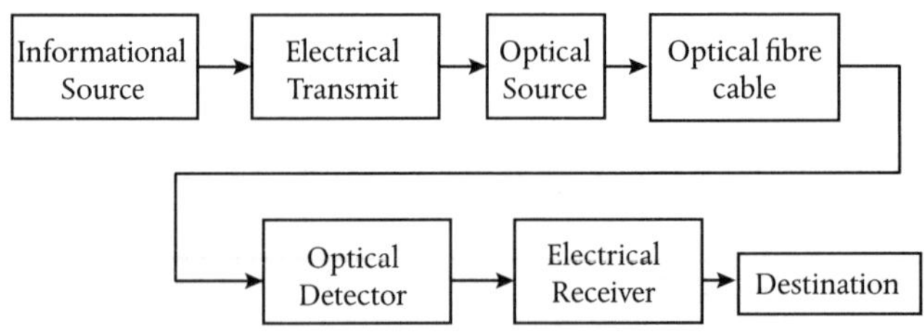

Fig.P1 Block diagram of optical communication system

An optical fibre communication system is similar in basic concept to any type of communication system. A block schematic of a general communication system is shown in Fig. P1. The function of which is to convey the signal from the informational source over the transmission medium to the destination. Therefore, the communication system consists of a transmitter or modulator linked to the informational source, the transmission medium, and a receiver or demodulator at the destination point. In electrical communication, the information source provides an electrical signal, usually derived from a message signal, which is not electrical (e.g., sound), to a transmitter comprising electrical and electronic component that converts the signal into a suitable form for propagating over the transmission medium. This is often achieved by modulating a carrier, which may be electromagnetic wave. The transmission medium can consist of a pair of wires, a coaxial cable, or a radio link through free space down which the signal is transmitted to the receiver, where it is transformed into the original electrical information signal (demodulated) before being passed to the destination.

Question 2: What are the advantages of optical fibre communication in signal processing application?

Answer: Communication using an optical carrier wave guided along a glass fibre has a number of extremely attractive features, several of which were apparent when the techniques were originally conceived. Furthermore, the advances in the technology to date have surpassed even the most optimistic prediction, creating additional advantages. The most powerful advantages in optical fibre communication system.

(a) **Enormous potential bandwidth**: The optical carrier frequency in the range 10^{13} to 10^{16} Hz yields a far greater potential transmission bandwidth than metallic cable systems.
(b) **Small size and weight:** Optical fibres have very small diameters that are often no greater than the diameter of a human hair. Hence, even when such fibres are covered with protective coating they are far smaller and much lighter than corresponding copper cable.
(c) **Electrical isolation:** Optical fibres that are fabricated from glass, or sometimes a plastic polymer, are electrical insulators; therefore, unlike their metallic counterpart, they do not exhibit earth loop and interface problem. Furthermore, this property makes optical fibres transmission ideally suited for communication in electrically hazardous environment as the fibres create no arcing or spark hazard at abrasion or short circuits.
(d) **Immunity to interference and cross talk:** Optical fibres form a dielectric waveguide and they are, therefore, free from electromagnetic interference (EMI), radio frequency interference (RFI), or switching transient giving electromagnetic pulses. Hence, the operation of an optical fibre communication system is unaffected by transmission through an electrically noisy environment and the fibre cable requires no shielding from EMI.
(e) **Signal security:** The light from optical fibres does not radiate significantly and they provide a high degree of signal security.
(f) **Low transmission loss:** The development of optical fibre over the last 20 years resulted in the production of optical fibre cables that exhibit very low attenuation or transmission loss in comparison with the best copper conductor. Fibres have been fabricated with the losses as low as 0.2 dB/km and this feature has the major advantages in the optical fibre communication.
(g) **Ruggedness and flexibility:** Although protective coatings are essential, optical fibres may be manufactured with a very high tensile strength. The fibres may also be bent to quite small radii or twisted without damage.
(h) **System reliability and ease of implementation:** These features primarily stem from the low-loss property of the optical fibre cables, which reduces the requirement for intermediate repeaters or line amplifiers to boost the transmitted signal strength.

Question 3: What are the advantages of lattice structures?

Answer: The lattice structures are having the following advantages:

(a) They allow a systematic approach to filter synthesis.
(b) They have a very straightforward implementation.
(c) They are both regular and modular. That is, the implementation is accomplished by interconnecting section having the same basic topology, making the assembly easily expandable.
(d) They have convenient numerical properties that make them suitable for application to digital signal processing.
(e) They provide a natural model for many important physical and mathematical phenomena like speech signal generation and they are useful in estimation problems.

Question 4: What is the difference between recirculating (feed backward) and non-recirculating (feed forward) delay line structures? Explain its operations.

Answer:

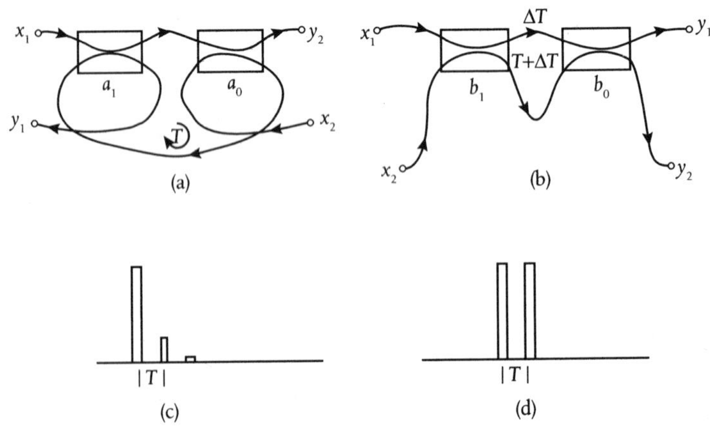

Fig. P2 Recirculating (feed backward) and non-recirculating (feed forward) delay line structures

Recirculating (feed backward) and non-recirculating (feed forward) delay lines are two structures that are the basic to fibre optic signal processing. In the two-coupler recirculating delay line, a single-modefibre loop (with loop delay *T*) is closed upon itself by two-directional coupler, with the second coupler feeding some portion of the light to the first coupler through the fibre loop. Therefore, optical signal sent into the input port, say at x_1, circulates repeatedly around the loop, sending a portion of the recirculating intensity to output port y_1 and y_2 (Fig. P2).

The two-coupler non-recirculating delay lines also consist of two-directional coupler; however, in this case, the output of the first coupler are both fed forward to the second coupler where they are recombined after a time delay *T*, which is the time delay difference between the two feed-forward fibre lines. The impulse response of the system is composed of just two impulses spaced by the amount Tin time.

SOLVED PROBLEMS

Problem 1: A causal LTI system is described by the following difference equation:

$$y(n)=0.81y(n-2)+x(n)-x(n-2)$$

Determine

(a) The system function $H(z)$
(b) The unit impulse response $h(n)$
(c) The unit step response $v(n)$, that is, the response to the unit step $u(n)$

Solution: Since, the system is causal, the Region of Convergence (ROC) will be outside of a circle with radius equal to the largest pole magnitude.

(a) Taking the z-transform of both sides of the difference equation, and then, solving for $Y(z)/X(z)$, we obtain:

$$H(z)=\frac{1-z^{-2}}{1-0.81z^{-2}}=\frac{1-z^{-2}}{\left(1-0.9z^{-1}\right)\left(1+0.9z^{-1}\right)}, \quad |z|>0.9$$

(a) using MATLAB for the partial expansion:

$$>>b=[1,0,-1];a=[1,0,-0.81];$$

$$>>[R,p,C]=\text{residuez}(b,a);$$

R =

0.1173
0.1173

P =

0.9000
0.9000

C = 1.2346

We have

$$H(z)=1.2346-0.1173\frac{1}{1+0.9z^{-1}}-0.1173\frac{1}{1-0.9z^{-1}},|z|>0.9\,.$$

Hence, we can write:

$$h(n)=1.2346\delta(n)-0.1173\left\{1+(-1)^n\right\}(0.9)^n u(n).$$

(b) As we know that, z-transform of unit step function is $u(n)=\dfrac{1}{1-z^{-1}}$:

$$V(z)=H(z)U(z)=\left[\frac{\left(1+z^{-1}\right)\left(1-z^{-1}\right)}{\left(1+0.9z^{-1}\right)\left(1-0.9z^{-1}\right)}\right]\frac{1}{1-z^{-1}}\cdot|z|>0.9\cap|z|>1$$

Hence,

$$=\left[\frac{\left(1+z^{-1}\right)}{\left(1+0.9z^{-1}\right)\left(1-0.9z^{-1}\right)}\right]|z|>0.9$$

or

$$V(z)=1.0556\frac{1}{1-0.9z^{-1}}-0.0556\frac{1}{1+0.9z^{-1}}.|z|>0.9$$

Finally,

$$v(n)=\left[1.0556(0.9)^n-0.0556(-0.9)^n\right]u(n).$$

Note that in the calculation of $V(z)$, there is a pole–zero cancellation at $z=1$. This has two implications. First, the ROC of $V(z)$ is still $\{|z|>0.9\}$ and not $\{|z|>0.9\cap|z|>1=|z|>1\}$. Second, the step response $v(n)$ contains no steady-state response.

Problem 2: If $H(z)=\dfrac{z+1}{z^2-0.9z+0.81}$ is a causal system, find:

(a) Its transfer function representation
(b) Its difference equation representation and
(c) Its impulse response representation.

Solution: The poles of the system function are at $z=0.9\angle\pm\pi/3$. Hence, the ROC of the above-mentioned causal system is $|z|>0.9$. Therefore, the unit circle is in the ROC, and the discrete-time Fourier transform $H(e^{j\omega})$ exists.

(a) Substituting $z=e^{j\omega}$ in $H(z)$

$$H(e^{j\omega})=\frac{e^{j\omega}+1}{e^{2j\omega}-0.9e^{j\omega}+0.81}=\frac{e^{j\omega}+1}{\left(e^{j\omega}-0.9e^{j\pi/3}\right)\left(e^{j\omega}-0.9e^{-j\pi/3}\right)}$$

(b) Using $H(z)=Y(z)/X(z)$

$$\frac{Y(z)}{X(z)}=\frac{z+1}{z^2-0.9z+0.81}\left(\frac{z^{-2}}{z^{-2}}\right)=\frac{z^{-1}+z^{-2}}{1-0.9z^{-1}+0.81z^{-2}}$$

Cross multiplying,

$$Y(z)-0.9z^{-1}Y(z)+0.81z^{-2}Y(z)=z^{-1}X(z)+z^{-2}X(z).$$

Now, taking the inverse z-transforms,

$$y(n)-0.9y(n-1)+0.81y(n-2)=x(n-1)+x(n-2)$$

or

$$y(n)=0.9y(n-1)-0.81y(n-2)+x(n-1)+x(n-2)$$

(c) Using MATLAB,

```
>>b = [0,1,1]; a = [1, 0.9,0,0.81];
>> [R,p,C] = residue z(b,a)
```

R =

  −0.6173+0.9979i
  −0.6713−0.9979i

P =

  0.4500−0.7794i
  0.4500+0.7794i

C= 1.2346

```
>> Mp = abs(p)
Mp = 0.9000      0.9000
>> Ap = angle(p)/pi
Ap = −0.3333     0.3333
```

We have:

$$H(z)=1.2346+\frac{-0.6173+j0.9979}{1-|0.9|e^{-j\pi/3}z^{-1}}+\frac{-0.6713-j0.9979}{1-|0.9|e^{-j\pi/3}z^{-1}},|z|>0.9$$

Hence, by taking inverse z-transform:

$$h(n)=1.2346\delta(n)+|0.9|^{n}\left[-1.2346\cos(n\pi/3)+1.9958\sin(n\pi/3)\right]u(n)$$

Problem 3: What are the lattice ladder structures?

Solution: Lattice structures are suitable forms for performing the various signal processing operations. It can be formed by joining the couplers in the required fashion in order to obtain the desired transfer function response. The lattice filter is extensively used in digital speech processing and in the implementation of adaptive filters.

Problem 4: Determine the lattice form of the following equation:

$$y(n)=2x(n)+\frac{13}{12}x(n-1)+\frac{5}{4}x(n-2)+\frac{2}{3}x(n-3).$$

Solution:

MATLAB Script

```
>>b = [2, 13/12, 5/4, 2/3];
>>K = dir2latc(b)
K = 2.000     0.2500     0.5000     0.3333
```

Hence,

$$K_0=2, K_1=\frac{1}{4}, K_2=\frac{1}{2}, K_3=\frac{1}{3}$$

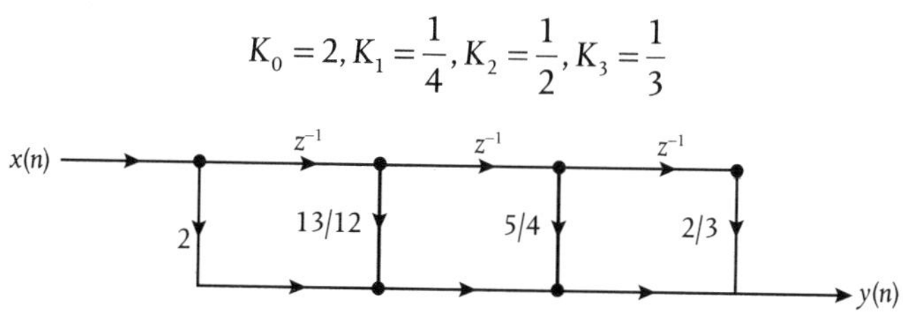

Fig. P3 Direct form

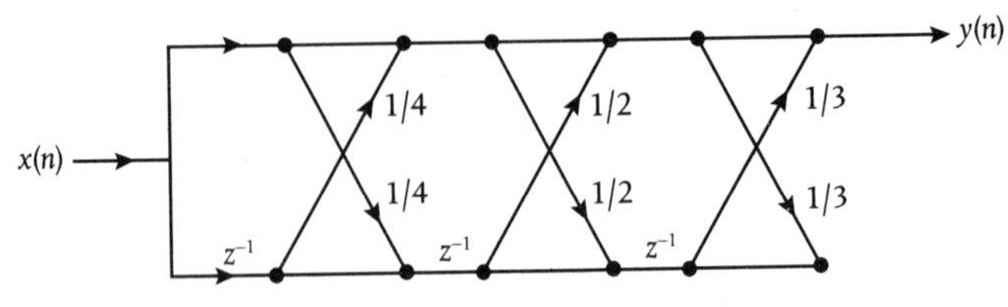

Fig. P4 Lattice form

Figures P3 and P4 show the desired results.

MULTIPLE CHOICE QUESTIONS

1. What makes optical fibres immune to EMI?
   (a) They transmit signals in as light rather than electric current.
   (b) They are too small for magnetic fields to introduce current in them.
   (c) Magnetic fields cannot penetrate the glass of the fibre.
   (d) They are readily shielded by outer conductors in cable.
2. What type of optical fibre would best be used in the multiple channel architecture (MCA)?
   (a) Refractive index
   (b) Step index
   (c) Graded index
   (d) None of these
3. Speeds of laboratory fibre optic Local Area Networks are now in the range of
   (a) 1 Mbits/s
   (b) 10 Mbits/s
   (c) Gigabits per second.
   (d) Hundreds of megabits per second
4. Photodiodes used as fibre optic directors are
   (a) Unbiased to generate a voltage same as a solar cell
   (b) Forward bias
   (c) Reversed bias
   (d) Thermoelectrically cooled
5. The wavelength of light has no role in
   (a) Diffraction
   (b) Interference
   (c) Polarisation
   (d) Reflection
6. What type of fibre has the highest modal dispersion?
   (a) Step-index multimode
   (b) Graded index multimode
   (c) Step-index single mode
   (d) Graded index mode
7. The wavelength of visible light extends from
   (a) 0.8 to 1.0 nm
   (b) 400 to 750 nm
   (c) 200 to 660 nm

8. Which of the following is the fastest light sensor?
   (a) PIN photodiode
   (b) Photovoltaic diode
   (c) Phototransistor
   (d) Avalanche photodiode
9. Splicing fibres mean
   (a) Fusion
   (b) Butt
   (c) Glue
   (d) Both fusion and butt

Answers to Multiple Choice Questions

1. (a)
2. (c)
3. (d)
4. (c)
5. (c)
6. (a)
7. (b)
8. (d)
9. (d)

QUESTIONS

1. Describe the principles of optical signal processor. Why an optical signal processor is called an optical computer?
2. What are the features of optical signal processing?
3. Draw the diagram of burst processing adaptive filter.
4. Comment on multidimensional processing.
5. What are the advantages of thin film optical design?
6. Explain optic lattice signal processing.
7. What is optical ring resonator?
8. Describe optical architecture of nonlinear phase shifter.

REFERENCES

1. Papageogiou D and Boucouvalas A C, Propagation constants of cylindrical dielectric waveguides with arbitrary refractive index profile using resonance techniques, *Electron. Lett.*, vol. 18, no. 18, pp. 786–788, 1982.
2. Raghuwanshi S K, Pandey R R, and Kumar V, Optimisation of substrate-radiation/substrate-cove radiation modes in planar slab optical waveguide structure, *Int. Nano Lett.*, vol. 1, no. 2, pp. 91–96, July 2011.
3. Raghuwanshi S K, Pandey R R, and Kumar V, Step discontinuity analysis in an asymmetric single mode planar slab taper optical waveguide, *I-managers J. on Electronics Engineering*, vol. 2, no. 2, Dec 2011–Feb 2012.
4. Shamir J, *Optical Systems and Processes*, Bellingham: SPIE Press, 1999.
5. Stokes L F, Chodorow M, and Shaw H J, All-single-mode fibre resonator, *Opt. Lett.*, vol. 7, pp. 288–290, 1982.

6. Taylor H F, Fibre and integrated optical devices for signal processing,*Proc. SPIE*, vol. 1976, pp. 17–27, 1979.
7. Taylor H F, Dispersion characteristics of diffused channel waveguide, *IEEE J. Quant. Electron.*, vol. QE-12, No. 12, December 1976.
8. Tur M, Goodman J W, Moslehi B, Bowser J E, and Shaw H J, Fibre-optic signal processor with applications to matrix–vector multiplication and lattice filtering, *Opt. Lett.*, vol. 7, pp. 463–465, 1982.
9. Turner J M, 'Use of a Digital Lattice Structure in Estimation and Filtering,' in *Signal Processing: Theories and Applications*, M. Kunt and F. de Coulon, Eds. Amsterdam, The Netherlands: North-Holland; also in *EURASIP*, p. 33, 1980.
10. Yano Y, Ono T, Fukuchi K, Ito T, Yamazaki H, Yamaguchi M, and Eumura K, "Terabits WDM transmission experiment using optical duobinary coding," *22nd European Conferences on Optical Communication*, Oslo, Norway, p. ThB.3.1, 1996.
11. Azaa J and Chen L R, Multi-wavelength optical signal processing using multistage ring resonators, *Photonics Tech. Lett.*, vol. 14, no. 5, pp. 654–656, May 2002.
12. Berginc G and Bourrely C, The small-slope approximation method applied to a three-dimensional slab with rough boundaries, *Prog. Electromagn. Res.*, PIER 73, pp. 131–211, 2007.
13. Boucouvalas A C and Papageogiou C D, Cut-off frequency in optical fibres of arbitrary refractive index profile using the resonance techniques, *IEEE J. Quant. Electron.*, vol. 18, no. 12, pp. 2027–2031, 1982.
14. Digonnet M J F and Shaw H J, Analysis of a tunable single mode optical fibre coupler, IEEE *J. Quant. Electron.*, vol. QE-18, pp. 746–754, 1982.
15. Heebnerand J E Boyd R W, Enhanced all-optical switching by use of a nonlinear fibre ring resonator, *Opt. Lett.*, vol. 24, no. 12, pp. 847–849, June 1999.
16. Ishigure T, Koike Y, and Fleming J W, Optimum index profile of the perfluorinated and polymerised gi polymer optical fibre and its dispersion properties, *IEEE J. Lightwave Technol.*, vol. 18, pp. 178–184, 2000.
17. Jain P, Raghuwanshi S K, and Talabattula S, "A novel approach to analyse Z-varying integrated optical waveguides,"*Proceedings of International conference on Optics & Optoelectronics*, 12–15 Dec. 2005, IRDE, Dehradun, India (PP-FIO-22).
18. Jeong D Y, Ye Y H, and Zhang Q M, Effective optical properties associated with wave propagation in photonic crystal of finite length along the propagation direction,*J. Appl. Phys.*, vol. 92, pp. 4194–4206, 2002.

6 Nonlinear Optics and Its Applications

Objectives

- This chapter deals with the application of nonlinear materials.
- Various types of fibres to mitigate the nonlinearities effect in optical communication system are considered.
- Pulse compression in optical fibres is demonstrated.
- The effect of self-phase modulation and nonlinear process on frequency is established.
- The effect of power losses in optical waveguides is expounded.

6.1 INTRODUCTION

The nonlinear effects in optical fibre occur either due to intensity dependence of refractive index of the medium or due to inelastic scattering phenomenon [1-3]. This chapter describes various types of nonlinear effects such as self-phase modulation (SPM), cross-phase modulation (CPM), and four-wave mixing (FWM). Their thresholds, managements, and applications are also discussed; further, comparative studies on these effects are presented.

The terms linear and nonlinear (Fig. 6.1), in optics, mean intensity-independent and intensity-dependent phenomena, respectively. Nonlinear effects in optical fibres occur due to (1) change in the refractive index of the medium with optical intensity and (2) inelastic scattering phenomenon. The power dependence of the refractive index is responsible for the Kerr effect [4-7]. Depending upon the type of input signal, the Kerr nonlinearity manifests itself in three different effects such as SPM, CPM, and FWM. At high-power level, the inelastic scattering phenomenon can induce stimulated effects such as stimulated Brillouin scattering (SBS) and stimulated Raman scattering (SRS). The intensity of scattered light grows exponentially if the incident power exceeds a certain threshold value. The difference between Brillouin and Raman scattering is that the Brillouin-generated phonons (acoustic) are coherent and give rise to a macroscopic acoustic wave in the fibre, while in the Raman scattering, the phonons (optical) are incoherent and no macroscopic wave is generated.

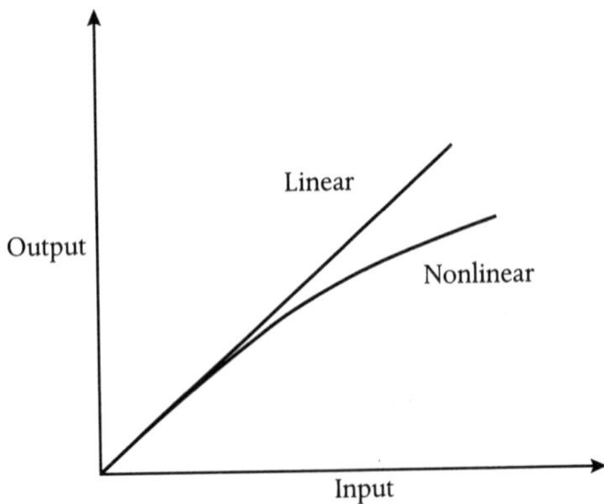

Fig. 6.1 Linear and nonlinear interactions

The description of optical communication system under the linearity assumption is adequate to understand the behaviour of these systems when they are operated at moderate power (a few milliwatts) and at bit rates up to about 2.5 Gb/s. However, at bit rates such as 10 Gb/s and above and/or at higher transmitted powers, it is important to consider the effect of nonlinearities. In the case of wavelength division multiplexing (WDM) systems, the effect can become important even at moderate power and bit rates [8-13].

There are two categories of nonlinear effects. The first arises due to the interaction of lightwaves with phonons (molecular vibrations) in the silica medium one of the several types of scattering effects, namely Rayleigh scattering. The two main effects in this category are SBS and SRS.

The second set of nonlinear effects in this category arises due to the dependence of the refractive index of the applied electric field, which in turn is proportional to the square of the field amplitude. The most important nonlinear effects in this category are SPM and FWM.

In the scattering effects, energy gets transferred from one lightwave to another wave at a longer wavelength or at lower energy. The lost energy is absorbed by the molecular vibrations, or phonons, in the medium. The type of phonon involved is different for SBS and SRS. This second wave is called the Stokes wave. The first wave can be thought of being a pump wave that causes amplification of the Stokes wave. As the pump wave propagates in the fibre, it loses power and the Stokes wave gains power. In the case of SBS, the pump wave is the signal wave and the Stokes wave is the unwanted wave that is generated due to the scattering process. In the case of SRS, the pump wave is a high-power wave and the Stokes wave is the signal wave that gets amplified at the expense of the pump wave.

In general, scattering effects are characterised by a gain coefficient "g", measured in meters per watt, and spectral width Δf over which the gain is present. The gain coefficient is a measure of the strength of the nonlinear effect.

In the case of SPM, the transmitted pulses undergo chirping. This induced chirp factor becomes significant at high-power levels. Pulse broadening effects of chromatic dispersion can be enhanced in the presence of chirp. Thus, the SPM-induced chirp can significantly increase the pulse spreading due to chromatic dispersion in these systems.

In a WDM system with multiple channels, the induced chirp in one channel depends on the variation of the refractive index with the intensity on the other channels. This effect is called CPM or XPM. When we discuss the induced chirp in a channel due to the variation of the refractive index with the intensity on the same channel, we call the effect as SPM.

In the case of WDM systems, another important nonlinear effect is that of FWM. If the WDM system consists of frequencies like $f_1 \ldots f_n$, FWM gives new signals at frequencies such as $2f_i - f_j$ and $f_i + f_j - f_k$. These signals appear as crosstalk to the existing signals in the system. These crosstalk effects are particularly severe when the channel spacing is tight. Reduced chromatic dispersion enhances the crosstalk induced by FWM effects than systems using standard single-mode fibre (SMF).

6.1.1 Effective Length and Area

The nonlinear interaction depends on the transmission length and the cross-sectional area of the fibre. The longer the link length, the more the interaction and the worse the effect of the nonlinearity. However, as signal propagates along the link, its power decreases because of fibre attenuation. Thus, most of the nonlinear effects occur early in the fibre span and diminish as the signal propagates, which is shown in Fig. 6.2 (a).

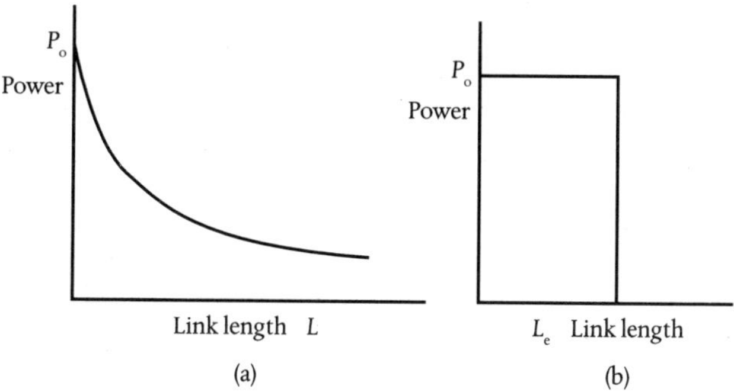

Fig. 6.2 Plot of power versus link length

Modelling this effect can be quite complicated, but in practice, it is a simple model that assumes that the power is constant over a certain effective length L_e. Further , it has proved to be quite sufficient in understanding the effect of nonlinearities. Let us suppose P_0 denotes the power transmitted into the fibre and $P(z) = P_0 e^{\alpha z}$ denotes the power at distance z along the link, with αbeing the fibre attenuation. Let L denote the actual link length. Then, the effective length is defined as the length L_e such that $P_0 L_e = P(z)dz$. This yield:

$$L_e = 1 \quad c^{\alpha L/\alpha}.$$

Typically, $\alpha = 0.22$ dB/km at 1.55 μm wavelength, and for long links where links are $L \gg 1/\alpha$, we have $L_e = 20$ km. Figure 6.2 (b) shows the power versus effective length.

In addition to the link length, the effect of nonlinearity also grows with the intensity in the fibre. For a given power, the intensity is inversely proportional to the area of the core, which is shown in Fig. 6.3(a). Since the power is not uniformly distributed within the cross-section of the fibre.

The effective area, as defined earlier, has the significance that the dependence of most nonlinear effects can be expressed in terms of the effective area for the fundamental mode propagating in the given type of fibre [14-17]. For example, the effective intensity of the pulse can be taken to be $I_e = P/A_e$, where P is the pulse power, in order to calculate the impact of certain nonlinear effects like SPM. The effective area of Single Mode Fibre (SMF) is around 85 μm^2. The dispersion compensating fibres even have smaller effective areas, and hence, they exhibit higher nonlinearities. Figure 6.3 (b) shows the intensity variation versus radius.

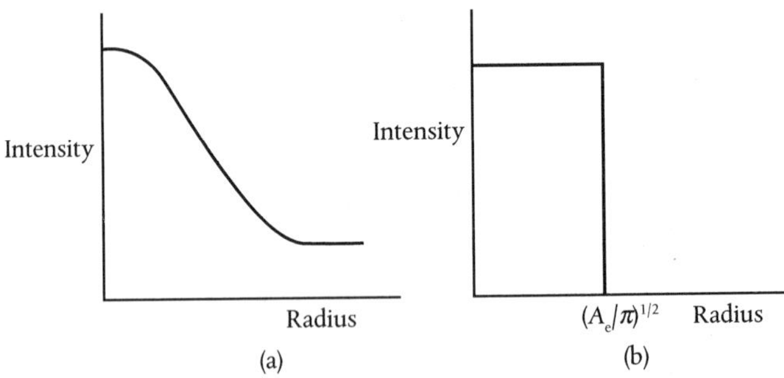

Fig. 6.3 Power distribution within the cross section of the fibre

6.2 STIMULATED BRILLOUIN SCATTERING

Brillouin scattering is an effect caused by the χ-nonlinearity of a medium, specifically by that part of the nonlinearity which is related to acoustic phonons. An incident photon can be converted into a scattered photon of slightly lower energy, usually propagating in the backward direction [18-19]. The coupling of optical fields and acoustic waves occurs via electrostriction. The effect can occur spontaneously even at low optical powers, and then reflecting the thermally generated phonon field. For higher optical powers, there can be a stimulated effect, where the optical fields substantially contribute to the phonon population. Above a certain threshold power of a light beam in a medium, stimulated Brillouin scattering can reflect most of the power of an incident beam. This process involves a strong nonlinear optical gain for the back reflected wave: an originally weak counter propagating wave at the suitable optical frequency can be strongly amplified. Here, the two counter propagating waves generate a travelling refractive index grating; the higher the reflected power, the stronger the index grating and the higher the effective reflectivity.

The frequency of the reflected beam is slightly lower than that of the incident beam; the frequency difference ν_B corresponds to the frequency of emitted phonons. This so-called Brillouin frequency shift is set by a phase-matching requirement. For pure backward Brillouin scattering, the Brillouin shift can be calculated from the refractive index n, the acoustic velocity v_a, and the vacuum wavelength λ: (For Brillouin scattering in fibres, the effective refractive index must be used.)

$$\nu_B = \frac{2nv_a}{\lambda}. \tag{6.1}$$

The Brillouin frequency shift depends on the material composition, and to some extent, the temperature and pressure of the medium. Such dependencies are exploited for fibre optic sensor.

Another important application of stimulated Brillouin scattering is optical phase conjugation. There are, for example, phase conjugate mirrors for high-power Q-switched lasers that make it possible that the thermal distortions occurring in forward and backward direction in the laser crystal compensate each other.

Stimulated Brillouin scattering (SBS) is frequently encountered when narrow-band optical signals (e.g., from a single-frequency laser) are amplified in a fibre amplifier or just propagated through a passive fibre. The material nonlinearity of silica is actually not very high, and therefore, the typically small effective mode area and long propagation length strongly favour nonlinear effects. For silica fibres, the Brillouin frequency shift is of the order of $10-20$ GHz, and the Brillouin gain has an intrinsic bandwidth of typically $50-100$ MHz, which is determined by the strong acoustic absorption (short phonon lifetime). However, the Brillouin gain spectrum may be strongly "smeared out" by various effects, such as transverse variations of the acoustic phase velocity or longitudinal temperature variations. Accordingly, the peak gain may be strongly reduced, leading to a substantially higher SBS threshold.

The Brillouin threshold of optical fibres for narrow-band continuous-wave light typically corresponds to a Brillouin gain of the order of 90 dB. With additional laser gain in an active fibre, the threshold can be lower. For trains of ultrashort pulses, the SBS threshold is determined not by a peak power, but rather by a power spectral density.

The SBS introduces the most stringent power limit for the amplification and the passive propagation of narrow-band optical signals in fibres. In order to raise the Brillouin threshold, it is possible to increase the bandwidth of the light beyond the Brillouin gain bandwidth, reduce the fibre length, concatenate fibres with slightly different Brillouin shift, or (in high-power active fibre devices) exploit the longitudinally varying temperature. There are also attempts to reduce the overlap of guided optical and acoustic waves or to introduce significant propagation losses for the acoustic wave. To some extent, SBS problems can be reduced via basic amplifier design modifications, concerning the doping concentration, effective mode area, and pump propagation direction.

On the other hand, the Brillouin gain can be used for operating a Brillouin fibre laser. Such devices are often made as fibre ring lasers. Due to low resonator loss, they can have a relatively low-pump threshold and a very small line width. The temperature dependence of the Brillouin shift can be used for temperature and pressure sensing.

6.3 STIMULATED RAMAN SCATTERING

Spontaneous Raman Scattering occurs in optical fibres when a pump wave is scattered by the silica molecules. Some pump photons give up their energy to create other photons of reduced energy at a lower frequency; the remaining energy is absorbed by silica molecules, which end up in an excited vibrational state. An important difference from Brillouin scattering is that the vibrational energy levels of silica dictate the value of the Raman shift $\Omega_R = \omega_p - \omega_s$. As an acoustic wave is not involved, spontaneous Raman scattering is an isotropic process and occurs in all directions.

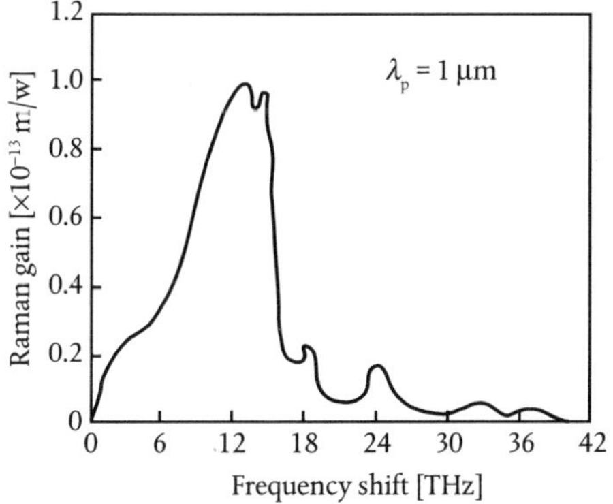

Fig. 6.4 Raman gain spectrum of silica fibres

Similar to the SBS case, the Raman scattering process becomes stimulated if the pump power exceeds a threshold value. The SRS can occur in both the forward and backward directions in optical fibres. Physically speaking, the beating of the pump and with the scattered light in these two directions creates a frequency component at the beat frequency $\omega_p - \omega_s$, which acts as a source that derives molecular oscillations. Since the amplitude of the scattered wave increases in response to these oscillations, a positive-feedback loop sets in the spectrum of the Raman gain depending on the decay time associated with the excited vibrational state. In the case of a molecular gas or liquid, the decay time is relatively long (1 ns), resulting in a Raman gain bandwidth of 1 GHz. In the case for optical fibres, the bandwidth exceeds 10 THz. Figure 6.4 shows the Raman gain spectrum of silica fibres. The broadband and multipeak nature of the spectrum is due to the amorphous nature of glass. More specifically, vibrational energy levels of silica molecules merge together to form a band. As a result, the stokes frequency ω_s can differ from the pump frequency ω_p over a wide range. The maximum gain occurs when the Raman shift $\Omega_R = \omega_p - \omega_s$ is about 13 THz. Another major peak occurs near 15 THz, while minor peaks persist for values of Ω_R as large as 35 THz. The peak value of the Raman gain g_R is about 1×10^{-13} m/W at a wavelength of 1 μm. This value scales linearly with ω_p (or inversely with the pump wavelength λ_p), resulting in $g_R \approx 6 \times 10^{-13}$ m/W at 1.55 μm. Similar to the case of SBS, the threshold power P_{th} is defined as the incident power at which half of the pump power is transferred to the Stokes field at the output end of a fibre of length L.

Both SRS and SBS can be seen as advantage while designing optical communication systems because they can amplify an optical signal by transferring energy to it from a pump beam whose wavelength is suitably chosen. The SRS is especially useful because of its extremely large bandwidth. Indeed, the Raman gain is used routinely for compensating fibre losses in modern light-wave systems.

6.4 PROPAGATION IN NONLINEAR MEDIUM

In order to understand SPM, CPM or XPM, and FWM, we need to understand how light propagates if we reduce linearity condition. However, we will assume isotropy, local responsitivity, homogeneity, losslessness on the silica medium. The losslessness assumption can be relaxed by carrying remaining discussion using complex variables for following fields and susceptibilities. For linear medium, we use the following relation between the Fourier transform of the induced dielectric polarisation and applied electric field, respectively.

$$P(r,\omega) = \varepsilon_0 \chi(r,\omega) E(r,\omega) \tag{6.2}$$

In discussing the effects of nonlinearities, we assume that the electric field of fundamental mode is assumed to be in x direction. Here, polarisation refers to the energy distribution of propagating mode and is different from the dielectric polarisation. The linear polarised modes here have no relation with the linear component of the dielectric polarisation. The following results can be generalised to this case. We consider linearly polarised electric field assumption because of isotropy assumption. Even in the presence of nonlinearities, the dielectric polarisation is in the same direction of the electric field, which is the $x-$ direction by assumption. Thus, vectors $E(r,t)$ and $P(r,t)$ can be represented by scalars as $E(x,t)$ and $P(x,t)$, respectively. With this assumption, in the presence of nonlinearities, we can write:

$$P(r,t) = P_L(r,t) + P_{NL}(r,t). \tag{6.3}$$

Here, $P_L(r,t)$ is the linear dielectric polarisation due to the linear polarisation assumption. The nonlinear dielectric polarisation $P_{NL}(r,t)$ is given as:

$$P_{NL}(r,t) = \varepsilon_0 \chi^{(3)} E^3(r,t), \tag{6.4}$$

where $\chi^{(3)}$ is the third-order nonlinear susceptibility and is assumed to be constant independent of t. Recall that the refractive index is related to the susceptibility. Thus, the nonlinear dielectric polarisation results in refractive index to be intensity dependent, which is root cause for nonlinearities. We will use above-mentioned equation as basic equation in analysing SPM, CPM, and FWM. For simplicity, we will assume the signals used are plane monochromatic waves that are of the form:

$$E(r,t)=E(z,t)=E\cos(\omega_0 t-\beta_0 z), \tag{6.5}$$

where E is a constant. The term monochromatic refers to single frequency signal, namely ω_0 and the term plane wave indicates that the electric field, which is constant in the plane perpendicular to the direction of propagation z. Hence, we can write $E(z,t)$ for $E(r,t)$ conveniently. In the case of WDM system, we assume that signals in every wavelength channel is a monochromatic plane wave. Thus, if there n wavelength channels at angular frequencies ω_0, ω_1 ... ω_n, with the corresponding propagating constants β_0, β_1,...... β_n, the electric field of composite WDM signal can be written as:

$$E(r,t)=E(z,t)=\sum_i E_i\cos(\omega_0 t-\beta_i z). \tag{6.6}$$

Since the signals in the WDM system need not necessarily in the phase, we need to add ϕ_I to each of the sinusoids, but we omit it here to keep the expression simple.

6.5 NONLINEAR PHASE MODULATION

The refractive index of silica was assumed to be power independent. In reality, all materials behave nonlinearly at high intensities and their refractive index increases with intensity. The physical origin of this effect lies in the harmonic response of electrons to optical fields, resulting in a nonlinear susceptibility. To include nonlinear refraction, we modify the core and cladding indices of a silica fibre as:

$$n'_j=n_j+n_2\left(\frac{P}{A_{eff}}\right), \quad j=1,2...., \tag{6.7}$$

where n_2 is the nonlinear index coefficient, P is the optical power, and A_{eff} is the effective mode area introduced earlier. The numerical value of n_2 is about $2.6\times10^{-20}\,m^2/W$ for silica fibres and varies somewhat with dopants used inside the core. Because of this relatively small value, the nonlinear part of the refractive index is quite small. Nevertheless, it affects modern lightwave systems considerably because of long fibre lengths. In particular, it leads to the phenomena of SPM and CPM.

6.5.1 Self-phase Modulation

If we use first-order perturbation theory to see how fibre modes are affected by the nonlinear term, we find that the mode shape does not change but the propagation constant becomes power dependent. The SPM-induced chirp affects the pulse shape through group velocity dispersion (GVD) and often leads to additional pulse broadening. In general, spectral broadening of the pulse induced by SPM increases the signal bandwidth considerably and limits the performance of lightwave systems.

If fibre losses α are compensated periodically using optical amplifiers, φ_{NL} should be multiplied by the number of amplifiers NA because the SPM-induced phase accumulates over multiple amplifiers. If we use $\varphi_{NL}=0.1$ as the maximum tolerable value and replace L_{eff} (effective length) by $1/\alpha$ for long fibres, this condition can be written as a limit on the input peak power as:

$$P_{\text{in}} < 0.1\alpha/(\gamma \text{NA}), \tag{6.8}$$

where γ are nonlinear parameter and NA stands for numerical aperture.

6.5.2 Cross-phase Modulation

The intensity dependence of the refractive index in above-mentioned equation can also lead to another nonlinear phenomenon known as CPM or XPM. It occurs when two or more optical channels are transmitted simultaneously inside an optical fibre using the WDM technique. In such systems, the nonlinear phase shift for a specific channel depends not only on the power of that channel but also on the power of other channels. The phase shift for the j_{th} channel becomes:

$$\varphi_{\text{NLj}} = \gamma L_{\text{eff}}\left(P_{\text{j}} + 2\sum_{\text{m}\neq\text{j}} P_{\text{m}}\right), \tag{6.9}$$

where the sum extends over the number of channels. The factor of 2 in the equation has its origin in the form of the nonlinear susceptibility and indicates that XPM is twice as effective as SPM for the same amount of power. The total phase shift depends on the powers in all channels and would vary from bit to bit depending on the bit pattern of the neighbouring channels. If we assume equal channel powers, the phase shift in the worst case in which all channels "M" simultaneously carry 1 bit and all pulses overlap is given as:

$$\varphi_{\text{NLj}} = 2\left(\frac{\gamma}{\alpha}\right)(M-1)P_{\text{j}}. \tag{6.10}$$

It is difficult to estimate the impact of XPM on the performance of multichannel lightwave systems. The reason is that the preceding discussion has implicitly assumed that XPM acts in isolation without dispersive effects and is valid only for continuous wave (CW) optical beams. In practice, pulses in different channels travel at different speeds. The XPM-induced phase shift can occur only when two pulses overlap in time. For widely separated channels, they overlap for such a short time that XPM effects are virtually negligible. On the other hand, pulses in neighbouring channels will overlap long enough for XPM effects to accumulate. A common method for studying the impact of SPM and XPM uses a numerical approach. The aforementioned equation can be generalised to include the SPM and XPM effects by adding a nonlinear term. The resulting equation is known as the nonlinear Schrödinger equation and has the form:

$$\frac{dA}{dz} + \beta_1\frac{dA}{dt} + \frac{i\beta_2}{2}\frac{d^2A}{dt^2} = \alpha A + i\gamma|A|^2 A, \tag{6.11}$$

where β_1 is group velocity term and β_2 is responsible for pulse broadening. Here, we neglected the third-order dispersion and added the term containing α to account for fibre losses. This equation is quite useful for designing lightwave systems. Since the nonlinear parameter γ depends inversely on the effective core area, the impact of fibre nonlinearities can be reduced considerably by enlarging A_{eff}. A_{eff} is about 80 μm^2 for standard fibres but reduces to 50 μm^2 for dispersion-shifted fibres (DSF). A new kind of fibre known as large effective area fibre (LEAF) has been developed for reducing the impact of fibre nonlinearities. The nonlinear effects are not always detrimental for lightwave systems. In fact, an optical pulse can propagate without distortion if the peak power of the pulse is chosen to correspond to a fundamental soliton.

6.5.3 Four-wave Mixing

The power dependence of the refractive index seen in (6.7) has its origin in the third-order nonlinear susceptibility denoted by $\chi^{(3)}$. The nonlinear phenomenon, known as FWM, also originates from $\chi^{(3)}$. If three optical fields with carrier frequencies ω_1, ω_2, and ω_3 copropagate inside the fibre simultaneously, $\chi^{(3)}$ generates a fourth field whose frequency ω_4 is related to other frequencies by a relation $\omega_4 = \omega_1 \pm \omega_2 \pm \omega_3$. Several frequencies corresponding to different plus or minus sign combinations are possible in principle. In practice, most of these combinations do not build-up because of a phase-matching requirement. Frequency combinations of the form $\omega_4 = \omega_1 + \omega_2 - \omega_3$ are often troublesome for multichannel communication systems since they can become nearly phase matched when channel wavelengths lie close to the zero-dispersion wavelengths. In fact, the degenerate FWM process for which $\omega_1 = \omega_2$ is often the dominant process and impacts the system performance most.

On a fundamental level, an FWM process can be viewed as a scattering process in which two photons of energies $h_{\omega 1}$ and $h_{\omega 2}$ are destroyed, and their energies appears in the form of two new photons of energies $h_{\omega 3}$ and $h_{\omega 4}$. The phase-matching condition then stems from the requirement of momentum conservation. Since all four waves propagate in the same direction, the phase mismatch can be written as:

$$\Delta = \beta(\omega_3) + \beta(\omega_4) - \beta(\omega_1) - \beta(\omega_2), \tag{6.12}$$

where $\beta(\omega)$ is the propagation constant for an optical field with frequency ω. In the degenerate case, $\omega_2 = \omega_1, \omega_3 = \omega_1 + \Omega$ and $\omega_4 = \omega_1 - \Omega$, where Ω represents the channel spacing. The FWM process is completely phase matched when $\beta_2 = 0$. When β_2 is small ($<1\,\text{ps/km}$) and channel spacing is also small ($\Omega < 100\,\text{GHz}$), this process can still occur and transfer power from each channel to its nearest neighbours. Such a power transfer not only results in the power loss for the channel but also induces interchannel crosstalk that degrades the system performance severely. Modern WDM systems avoid FWM by using the technique of dispersion management in which GVD is kept locally high in each fibre section even though it is low on average. Commercial dispersion-shifted fibres are designed with a dispersion of about 4 $\text{ps}/(\text{km}-\text{nm})$, a value found large enough to suppress FWM. The FWM can also be useful in designing lightwave systems. It is often used for demultiplexing individual channels when time division multiplexing is used in the optical domain. It can also be used for wavelength conversion. The FWM in optical fibres is sometimes used for generating a spectrally inverted signal through the process of optical phase conjugation.

6.6 NEW TYPES OF OPTICAL FIBRES

Just as DSF are developed to reduce the pulse spreading in 1.55-μm band, new types of fibres are also developed to mitigate the effects of the nonlinearities on optical communication system.

6.6.1 Non-zero Dispersion Fibre

Although dispersion-shifted fibre (DSF) overcomes the chromatic dispersion in the 1.55-μm wavelength window, unfortunately it is not useful to the WDM communication system because of severe penalties due to FWM and other nonlinearities. These penalties can be overcome if there is chromatic dispersion present in the fibre because different waves travel with different group velocities. This led to the development of nonzero dispersion optical fibres. Such fibres have a chromatic dispersion between 1 and 6 $\text{ps}/\text{nm}-\text{km}$, or between -1 and -6 $\text{ps}/\text{nm}-\text{km}$, in the 1.55-μm wavelength window. This reduces the penalties due to nonlinearities at the same time retaining the most of the advantages of DSF.

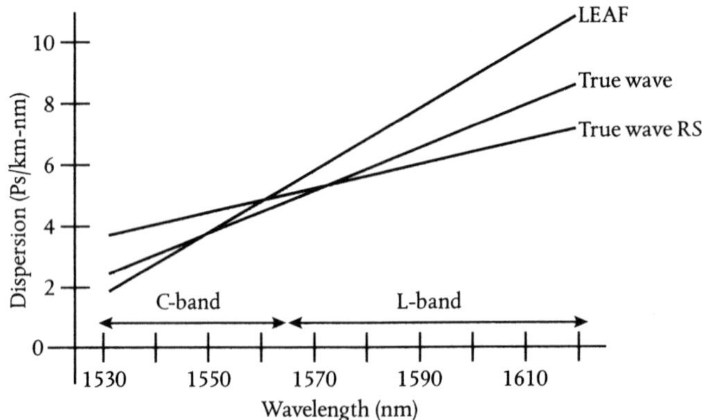

Fig. 6.5 Dispersion versus wavelength for various types of fibre

Since all NZ DSF are designed to have nonzero value of dispersion in the C-band, there zero dispersion lie outside the C-band but could lie in L-band or in the S-band. In such cases, a large portion of band around the zero dispersion becomes useless due to FWM. Figure 6.5 shows the dispersion for various types of fibres.

If the spread is small, that is, the accumulated chromatic dispersion in the different channels is close to being uniform; it may be possible to compensate the accumulated chromatic dispersion in all the channels with a single-chromatic dispersion compensator. This would be cheaper than using a chromatic dispersion compensator for each channel.

6.6.2 Large Effective Areas

The effects of nonlinearities can be reduced by designing a fibre with large effective areas. We have seen that non-zero dispersion fibres have a small value of chromatic dispersion in the 1.55-μm bandwidth to minimise the effects of chromatic dispersion. Unfortunately, such fibres have small effective areas. Fibres with large effective areas achieve better trade-offs between chromatic dispersion and nonlinearities than normal NZ DSF. However, these fibres tend to have large chromatic dispersion slopes. Another trade-off is that this reduces the effect of distributed Raman amplification.

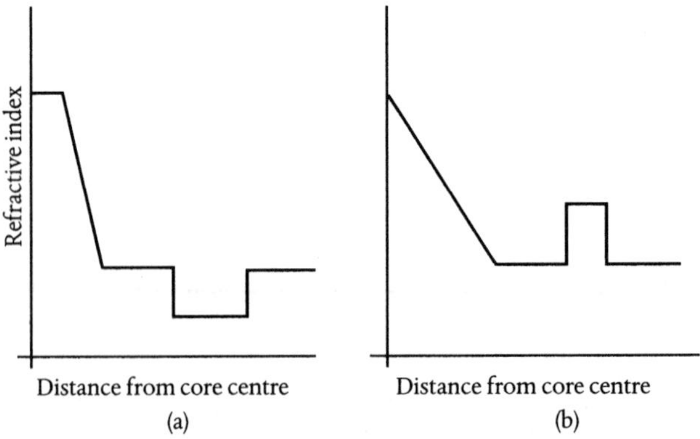

Fig. 6.6 Refractive index profile in fibre

Large effective area fibre (LEAF) consists of three regions. The core region of the LEAF has triangular variation of refractive index. The annular region has refractive index of cladding. This is surrounded by material of higher refractive index. Middle region being the region of lower refractive index does not confine the power, thus power gets distributed over a large area. This reduces the peak power in the core and increases the effective area. The refractive index profile of the LEAF is shown in Fig. 6.6.

6.6.3 Positive and Negative Dispersion Fibres

Fibres are designed to have either positive chromatic dispersion or negative chromatic dispersion in 1.55-μm wavelength window. Positive chromatic dispersion system is used for terrestrial systems, whereas negative chromatic dispersion fibres are used for submarine system [20-23]. Both positive and negative chromatic dispersion cause pulse spreading, and the amount of pulse spreading depends only on the magnitude of the chromatic dispersion and not on its sign (in the absence of chirping and nonlinearities).

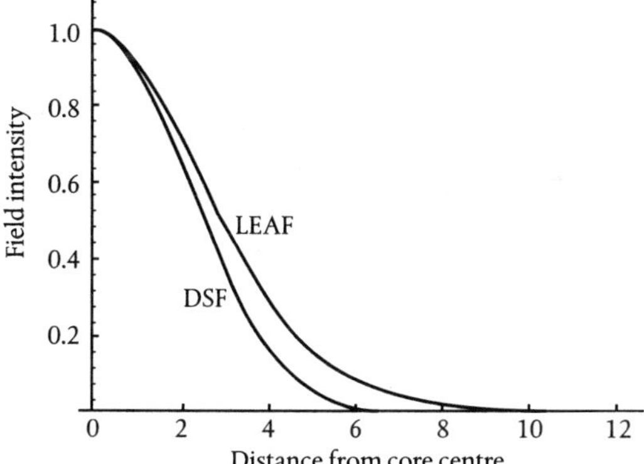

Fig. 6.7 Field intensity distribution in fibre

Then, why the need for fibres with different signs of chromatic dispersion, positive for terrestrial systems and negative for undersea links. To understand the motivation for this, we need to understand another nonlinear phenomenon: modulation instability. Figure 6.7 shows the electric field distribution for the core of DSF and LEAF.

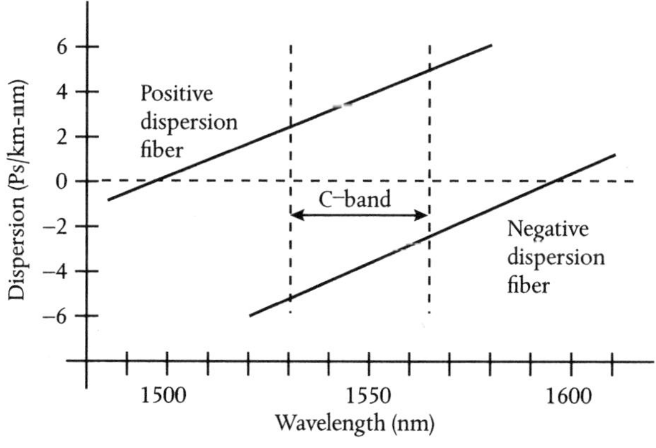

Fig. 6.8 Dispersion occurs at different wavelengths

We know that pulse compression occurs for a positively chirped pulse when the chromatic dispersion is positive. We also know that SPM causes positive chirping of pulses phenomena. Hence, chromatic dispersion and SPM-induced chirp leads to a breakup of relatively broad pulse of duration a few picoseconds. This phenomenon is referred to as modulation instability and leads to a significantly increased bit-error rate. Modulation instability occurs only in positive chromatic dispersion fibre, and thus, it can be avoided by the use of negative chromatic dispersion fibre. Its effects in positive chromatic dispersion fibre can be minimised by using lower power levels. The WDM systems cannot operate around the zero dispersion wavelength of the fibre due to the severity of FWM. For positive chromatic dispersion fibre, the dispersion zero lies below the $1.55-\mu m$ band and not in the L-band [24-26]. Hence, the systems using positive chromatic dispersion fibre can be upgraded to use the L-band. This upgradability is an important feature for terrestrial systems. Thus, positive chromatic dispersion fibre is preferred for terrestrial systems, and the power levels are controlled so that modulation instability is not significant. For undersea links, however, the use of higher power levels is very important due to the very long link lengths. These links are not capable of being upgraded anyway since they are buried on the ocean floor; therefore, the use of the L-band in these fibres at a later date is not possible. Hence, negative chromatic dispersion fibre is used for undersea links.

Since negative chromatic dispersion fibre is used for undersea links, the chromatic dispersion can be compensated using standard SMF, which has positive chromatic dispersion; that is, alternative lengths of negative chromatic dispersion fibre and (positive chromatic dispersion) SMF can be used to keep the total chromatic dispersion low. This is preferable to using dispersion compensating fibres since they are more susceptible to nonlinear effects because of their lower effective areas.

Note that all the fibres we have considered have positive chromatic dispersion slope; that is, the chromatic dispersion increases with increasing wavelength. This is mainly because material dispersion slope of silicon is positive and usually dominates the negative dispersion slope of waveguide dispersion. The negative chromatic dispersion slope fibre is useful in chromatic dispersion slope compensation, while it is possible a negative dispersion fibre with negative slope. It is considered difficult to design a positive chromatic dispersion fibre with negative slope.

6.7 NONLINEAR OPTICAL LOOP MIRRORS (NOLM)

The NOLM is the fibre implementation of a nonlinear anti-resonant ring interferometer. All optical switching with NOLM was first analysed by Nelson et al. 1991 [27],and was subsequently demonstrated by many groups.

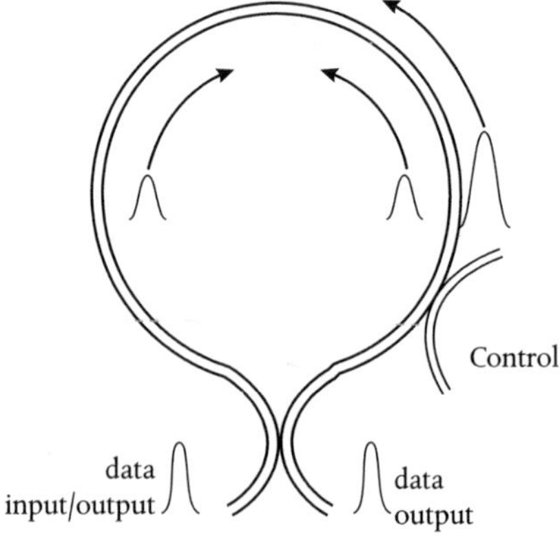

Fig. 6.9 Schematic of an NOLM

An NOLM employs a 3-dB directional coupler connected to a fibre loop acting as a nonlinear medium as shown in Fig. 6.9. A data pulse sent into the loop is split by the coupler into two equal intensity counter propagating pulses, which will recombine at the coupler and emerge out of the input port. Switching is achieved by injecting an intense control pulse into the loop, which travels in synchronism with one of the data pulses and induces phase shift to this data pulse via CPM. A phase shift will cause the recombined data pulse to emerge out from the output port.

6.7.1 Advantages of NOLMs

There are several advantages of NOLMs. We mention a few of them here.

1. Very fast switching time could be in principle achieved, since the Kerr nonlinearity in fibre is virtually instantaneous.
2. Stable operation can be maintained even for long fibre loops because thermal and vibrational disturbances, which typically act on a much longer time than the loop transit time, affect both interfering pulses equally.
3. Low switching power can be achieved by lengthening the fibre loop.

6.7.2 Major Challenges/Drawbacks for NOLMs

The following are the major challenges for NOLMs:

1. Polarisation dependence
2. Chromatic dispersion
3. Accomplishment of high contrast
4. Latency
5. Incompatibility with integrated optics.

Polarisation dependence is due to birefringence induced by linearly polarised control pulse. Elaborate polarisation diversity scheme employing two identical sections of birefringent fibre cross spliced at the centre of the fibre loop has been demonstrated to achieve polarisation insensitive switching as shown in Fig. 6.10. Another recently developed polarisation-insensitive NOLM uses a twisted fibre to emulate a circularly polarised fibre, but the resultant throughout is reduced by more than 3 dB.

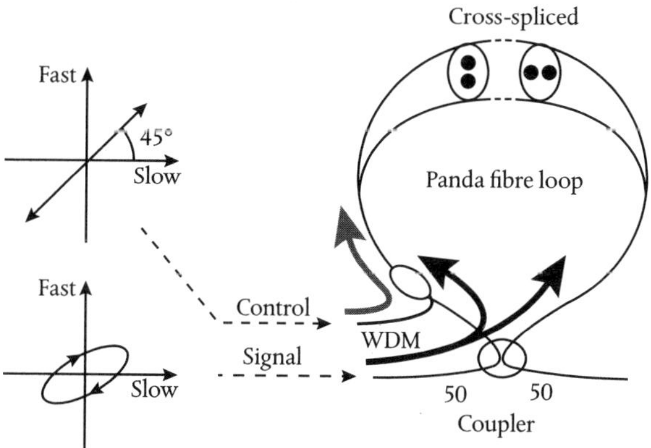

Fig. 6.10 Polarisation independent implementation of an NOLM using two identical, cross-spliced birefringent fibres

To facilitate separation of control from signal after switching, the control is usually set at a different wavelengths from the signal; hence, chromatic dispersion in fibre has a number of undesirable consequences:

1. It limits interaction length so that switching energy cannot always be reduced by lengthening the fibre.
2. It reduces the switching speed since chromatic dispersion causes a broadening of the switching window,
3. It limits the signal bandwidth.
4. Achieving high contrast ratio in many practical NOLMs represents another challenge because it requires a nearly perfect 50/50 split coupler;
5. The coupler is wavelength dependent; hence, high contrast ratio cannot be obtained over a large bandwidth and contrast ratio deteriorates with increasing switching repetition rate.

As the time interval between control pulses shortens, more control pulses are propagating in the loop at any given time, which introduce more unwanted phase shifts to the counter propagating signals, causing incomplete cancellation of the unwanted channel at the transmission port. Despite these drawbacks, NOLMs are promising for optical time-division multiplexed (OTDM) systems, and apart from performing signal demultiplexing, NOLMs have been shown to perform wavelength conversion, optical rate conversion, and signal regeneration.

6.7.3 Nonlinear Semiconductor Element in a Fibre Loop Mirror

To eliminate the long latency problem of NOLMs, semiconductors with large resonant nonlinearity have been used to replace fibre as the nonlinear medium. This also brings other advantages, such as improved stability, reduced switching energy, and the possibility of integration. Fast switching is achieved by using an asymmetric configuration as shown in Fig. 6.11 similar to the placing of nonlinear elements asymmetrically in a Mach–Zender (MZ) switch. Devices employing a semiconductor laser amplifier (SLA) as the nonlinear element in a fibre loop have been demonstrated to perform various functions, the most notable being the SLA in a Loop Mirror (SLALOM) 215 and the Terahertz Optical Asymmetric Demultiplexer (TOAD), which were used for demultiplexing high bit-rate OTDM signals. The major advantage of employing SLA in the loop, apart from low latency, is the very low switching energy required, typically about a picojoule; switching energy as low as 355 PJ was reported. This is due to: (a) the large nonlinearities in semiconductors and (b) the usage of current injected nonlinear elements (SLA) to provide gain as well as nonlinearity.

Although integration is possible with this device and a fully integrated version has been demonstrated, the integrated device is rather complex, consisting of passive loop waveguides, an active SLA, and a 3-dB directional coupler. When compared to passive integrated switches such as Nonlinear Directional Coupler (NLDC), this integrated device is bulky and its fabrication is considerably more complex.

The switching time of the TOAD or SLALOM is determined by the location x as well as the length l of the nonlinear element as shown in Fig. 6.11. The shortest switching time possible is twice the transit time of the nonlinear element. The length of the nonlinear element is determined by the SLA's nonlinear coefficient and the switching energy. It is typically several hundreds of microns, which limits the switching time to a few picoseconds. The demultiplexing repetition rate, on the other hand, is still primarily limited by the SLA's carrier lifetime.

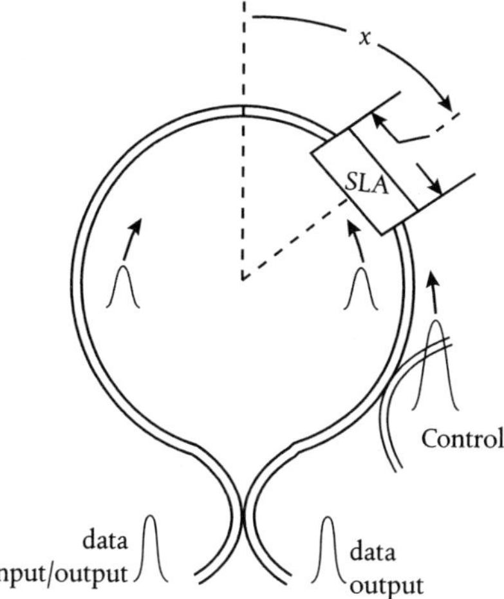

Fig. 6.11 Generic configuration for a nonlinear element in a loop mirror

The use of photonic technologies for data processing in all the optical domain is a field of growing importance with a strong potential for a variety of interesting applications in such diverse areas as ultrahigh speed optical telecommunications, ultrafast metrology, optical sensing, microwave engineering, image processing, and optical computing. The advantages of processing the information in all-optical domain include the tremendous available bandwidth and the parallelism intrinsic to the optical approach, which translate into ultrahigh processing speeds that are not possible otherwise. In the present optical communication systems, signal processing functions such as switching, routing, frequency conversion, and regeneration are typically performed electrically, which requires converting the optical signal to electrical and back to optical again for retransmission. While this traditional approach has the advantage of relying on mature technology, there is significant interest in bypassing the optoelectronic conversion and undertaking processing tasks entirely in the optical domain, since the all-optical approach provides a route to optical networks of increased capacity, scalability, and flexibility as well as reduced network management complexity and costs. Among the all-optical signal processing technologies, all-optical regeneration is considered a key enabling function for the deployment of future all optical communication systems as a means of restoring the quality of the signals, which is deteriorated by impairments such as noise accumulation from optical amplifiers, fibre chromatic dispersion, different types of crosstalk, jitter and nonlinearities, and other imperfections in transmission and cross connects. Although ultimately re-amplification, reshaping, and retiming (3R regeneration) of the signals is required for many applications, simple and cost-efficient 2R regenerators (devices with re-amplification and reshaping characteristics only) can be an attractive practical solution. The central component of most all-optical regenerators is a nonlinear optical gate, which perform the decision in the optical domain. This gate is most commonly realised through nonlinear phenomena in semiconductors and optical fibres. The semiconductor-based devices exploiting cross-gain/absorption/phase modulation in a semiconductor optical amplifier (SOA), a saturable absorber, and an electro-absorption modulator are compact in size and probably most suitable for mass production. However, the operating speed of such devices is often dominated by the carrier lifetime. Slow carrier recovery causes significant pattern effect, which can be a serious problem in regeneration applications since it rather degrades the quality of the recovered signal. On the other hand, for the fibre-based devices including those utilising FWM, SPM and CPM and nonlinear interferometry within an optical loop mirror, the operating speed is determined by ultrafast fibre nonlinearity, which provides response times in the femtosecond range. However, the fibre-based devices are typically larger in size compared with the SOA-based devices. This, however, is being improved with the recent developments of highly nonlinear fibres (HNLFs), which enable the implementation of short devices operated at reduced powers. The intrinsic simplicity and large speed potential of the fibre-based regenerators appear fascinating for the realisation of all optical regeneration in future photonic networks. Owing to its self-switching ability and flexibility, the fibre Sagnac interferometer or nonlinear optical loop mirror (NOLM), originally proposed applications were being recognised as a valuable tool in a large field of optical signal processing applications. These include optical switching, phase conjugation, wavelength demultiplexing, mode locking, pulse compression, and pedestal suppression; earlier studies have addressed the use of NOLMs to reshape and stabilise soliton pulses. More recent studies have established the NOLM as an effective 2R regenerative device in telecommunication systems. The NOLM is made of a coupler whose output ports are connected through a span of fibre; if a nonlinear phase shift difference appears between the two interfering beams, it offers a versatile way of obtaining a nonlinear transmission or switching characteristic through the optical Kerr effect. A variation of the differential phase by π will cause a change of the NOLM from high transmission to a high reflection. Most NOLM designs rely on SPM, which allows a nonlinear phase shift to accumulate if a power imbalance exists between the counter propagating beams in the loop. The breaking of the loop-power symmetry can be done with an asymmetric coupler or alternatively with a gain or loss element spliced asymmetrically within the loop or alternatively by making the dispersion of the fibre asymmetric. A variation of the NOLM design uses polarisation asymmetry between the counter propagating beams to achieve switching through nonlinear polarisation rotation. The NOLM typically exhibits a sinusoidal nonlinear transmission characteristic. Low transmission at low power enables suppression of noise and radiative background; whereas stable fixed points of the input–output power mapping can be used to reduce the amplitude jitter of the carrier pulses. In this section, we report on recent results on the development of the all-optical signal processing techniques utilising the NOLM for application in fibre optic communications. In this section, we describe the main results of the application of in-line NOLMs as a general technique for all-optical passive 2R regeneration of return to zero (RZ) on–off keyed (OOK) signals in high speed, ultralong distance transmission systems, and terrestrial photonic networks. A theoretical model enabling the description of the stable propagation of ultrashort optical pulses in fibre transmission systems in the quasi-linear regime with lumped nonlinear optical devices (NODs) placed periodically along the line and applied to a system with inline NOLMs. A novel transmission technique based on periodic inline all-optical format conversion between RZ and non-return to zero (NRZ) like is also demonstrated by the use of such NDF-enhanced NOLMs; further, we also present a concept for phase preserving all-optical 2R regeneration of signals using phase shift keying (PSK) modulation formats, which exploit a new design of NOLM based on distributed Raman amplification in the loop fibre (RANOLM). The regenerative scheme is applied to RZ differential PSK (DPSK) transmission. We review the basic properties of the NOLM.

6.7.4 Operation Principle of the NOLM

The principle of the regeneration function of the NOLM is as follows: low intensity noise and dispersive waves are filtered out from the higher power pulses through an effective saturable absorption action. This allows for restoration of the pulse amplitude and shape. Further, the operation of the NOLM just after the peak of its switching curve enables stabilisation of the pulse amplitude against fluctuations in the input power. Note that the loss- or gain-imbalanced NOLM provides higher discrimination against lower power radiation. On the contrary, the transfer functions of a coupler-imbalanced NOLM rises linearly at low powers. However, since the loss- or gain-imbalanced NOLM has sharper switching peaks, the portion of the switching curve on which operation is stable is smaller. However, this limitation can be overcome by placing the operation point only slightly past the switching peak. This device suppresses rather large pulse amplitude fluctuations at the expenses of a poor background noise reduction. In an NOLM setup, incorporating an additional phase shifter for extension of the low transmission region of the output characteristic was studied. In the case of an RZ-OOK pulse train, the noise and radiative background in the zero-bit time slots is suppressed by the loop-mirror saturable absorption action and the amplitude jitter in the one-bit pulses is also reduced. However, the NOLM is insensitive to pulse position in time, and thus, it does not aid problems of timing jitter in a system. Nevertheless, its intensity filtering action enables the use of strong guiding filters in the system without encountering pulse instability. Amplitude noise in the input signal is modelled by a Gaussian distribution of the power. It is seen that, although the loss- or gain-imbalanced NOLM mainly supports the zero bits, there is amplitude jitter reduction on the one bits too.

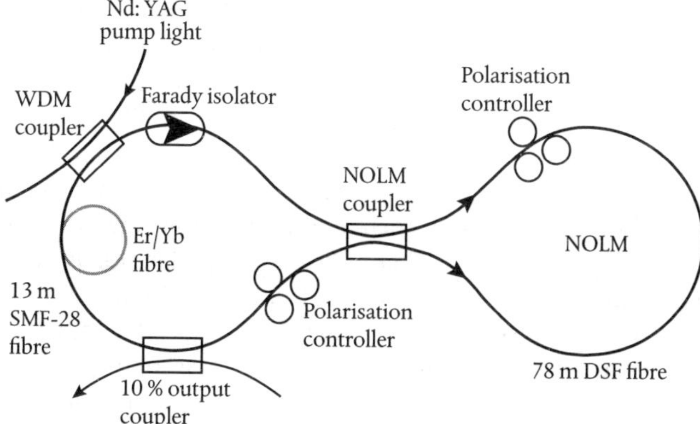

Fig. 6.12 Fibre laser diagram

6.8 EXPERIMENTAL STUDIES ON ALL-FIBRE LASER

Laboratory work utilises an all-fibre pulse laser as diagrammed in Fig. 6.12; it is a figure of eight formed entirely of optical fibres and fibre optic components. Rather than using mirrors to reflect the beam through the gain medium many times, fibre forms a closed loop around the gain medium with the output coupled directly out of the loop through a fibre coupler that removes 10% of the light from the cavity. Laser components are connected by splicing their fibre "pigtails" together. The primary lobe of the laser cavity as shown on the left side of Fig. 6.12 comprises approximately 15 m of fibre. One may be used, SMF 28, a standard silica glass communications fibre that supports only the TEM_{00} mode. The NOLM is a loop approximately 78 m in length and is composed of DSF. The DSF has been engineered to have near-zero dispersion at typical fibre communication wavelengths. This laser operates at 1.55 μm, a wavelength at which signal attenuation in silica glass fibres is at a minimum. The gain medium is a section of fibre doped with erbium and ytterbium ions. Pump light at 1.06 μm from a commercial Nd:YAG laser is injected into the fibre loop through a wavelength division multiplex (WDM) coupler that preferentially transmits 1.06 – μm light across the coupler while allowing very little 1.55 – μm laser light out of the fibre loop. The isolator is a fibre optic component that takes advantage of optical activity and a series of linear polariser to act as a one-way gate for light in the fibre loop. We can position the Faraday isolator in the laser cavity to limit 1.55 – μm light to one direction of travel around the loop. Since the isolator includes linear polariser, it introduces a loss into the cavity depending on the polarisation of the input light.

The transmission of a light beam through the loop mirror from the input pigtail to the output depends on the splitting ratio α of the coupler. If $\alpha = 0.5$, then the NOLM acts as a reflector. Light will be evenly split into two counter propagating beams, with one beam acquiring a $\pi/2$ phase shift when it crosses the coupler. After the beams have circulated around the loop, they interfere at the output pigtail. The beam that crosses the coupler again acquires a phase shift of $\pi/2$ so that when the beams combine; one has a total phase shift of zero, while the other has been shifted by π. No light is seen at the output, and by conservation of energy, all the light reflects to the input, and hence, the term "loop mirror". Here, NOLM incorporates a coupler with a splitting ratio of $\alpha = 0.513$. With slightly different intensities of light travelling in opposite directions around the loop, we must take into account the effects of the nonlinear index of refraction in optical fibres. In the case of a lightwave entering the input pigtail, the nonlinear index of refraction causes the more intense beam to propagate through the loop at a slower speed. Thus, the unequal splitting ratio introduces an intensity-dependent differential phase shift between the two beams. The fibre loop is long in order to allow this phase to build up. We can construct a power transfer curve (PTC) to capture the dependence of NOLM transmission on the input intensity. We use the input power-dependent transmission of the NOLM to mode lock the fibre laser. Passive mode locking is the result of interrupting any laser cavity with a nonlinear device that preferentially transmits high intensities. If the laser can minimise its loss by increasing beam intensity, but the population inversion is not sufficient to simply increase continuous wave (CW) intensity, then all the laser cavity modes will tend to "lock" together such that their Fourier sum is a pulse train. This process takes place as random intensity noise on a CW laser beam gets preferentially amplified at the expense of the less intense background. When the pump laser is activated, laser is initially in CW operation. The introduction of acoustic noise into the cavity by tapping the lab table initiates the process of mode locking. The fibre laser produces pulses on the order of one pico-second in duration, usually with one pulse per round trip of the laser cavity. A pulse travelling through an optical fibre is subject to dispersion and nonlinearity. We know that at 1.55 μm is in the anomalous dispersion regime for silica glass, and therefore, the fibres in our laser can support soliton pulses. Indeed, if any pulse with energy equal to or exceeding that of a fundamental soliton propagates in an optical fibre, it should evolve into a soliton due to the actions of dispersion and SPM. Therefore, we believe that the fibre laser should produce soliton like pulses when it mode locks. The influence of other pulse-shaping mechanisms in the fibre cavity causes the laser to produce pulses that are close to, but not quite, solitons.

6.8.1 NOLM Power Transfer Curves (PTC)

Aubryn Murray did significant work to characterise power transfer curves, their effect on the NOLM, and their role in causing the laser to mode lock. The fibre laser does not have a single PTC; the polarisation of light inside the NOLM complicates the system. Since the PTC depends on interference at the NOLM coupler, we can manipulate the PTC by changing the relative polarisations of the counter propagating beams in the loop. Figure 6.13 shows the PTC for the present case.

The horizontal axis displays the power of a light beam coupled into the NOLM input pigtail; the vertical axis shows the fractional power observed on the output pigtail. In particular, a polarisation mismatch between the recombining beams can alter the depth of modulation of the PTC since waves of identical polarisation interfere while waves of crossed polarisation do not. Additionally, changing the polarisation of beams in the NOLM can alter the phase of the PTC because, if the counter propagating beams travel along different birefringent axes in the fibre, their phases will differ when they recombine at the coupler. The PTC for NOLM is always sinusoidal.

1. PTC neglecting polarisation in the NOLM.
2. Example of two possible PTCs when polarisation is included in the model.

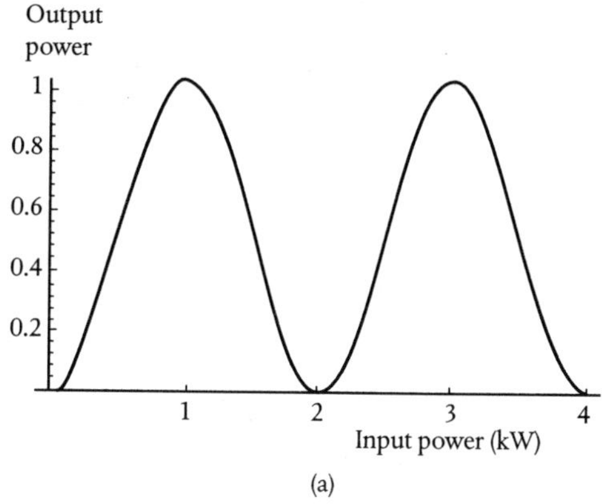

(a)

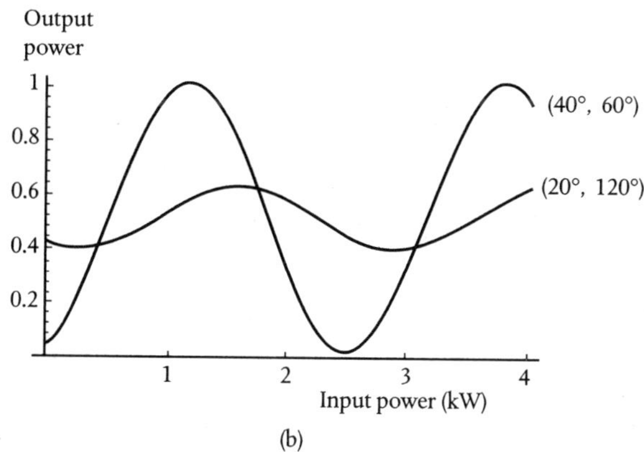

(b)

Fig. 6.13 Example of plots of power transfer through the NOLM

Measurements of average laser power output indicate that the peak power of a pulse in the laser cavity is close to the vertical axis of the NOLM power transfer curve. Our current understanding is that mode locking only occurs when the PTC has a positive slope at low input powers and allows intermediate transmission at low input powers. Since NOLM, PTCs are sinusoidal; an intermediate fractional transmission when the PTC depth of modulation is high indicates that the PTC slope is near maximum. It makes sense for the laser to mode lock under these conditions because a positive PTC slope at low input power means that the laser cavity discriminates against low intensity light, which creates an "incentive" for the laser to pack its energy into short pulses and maximise their transmission. A larger slope increases this discrimination and thus increases the likelihood of mode locking. Conversely, a negative PTC slope would discourage pulse formation because pulses, with high peak power compared to a CW beam, would be discriminated against instead.

6.8.2 OTDM Demultiplexing using Highly Nonlinear Fibre (HNLF) in an NOLM at 160 Gb/s

An NOLM has been studied extensively and is one of the most promising methods for fast all-optical switching. Recently Tbit/s switching speeds are possible due to the ultrafast nonlinearities in the fibre. Demultiplexing from 320 and 640 Gb/s to a 10 Gb/s base rate has already been reported. Several experiments that use orthogonally polarised signal and control pulse streams in a fibre with randomly varying birefringence have been reported. The advantage of this type of NOLM is the simple design and lower loss than an NOLM based on splicing of polarisation-maintaining fibres. In this chapter, we present demultiplexing from 160 Gb/s to a 10 and 40 Gb/s base rate with a commercially available highly nonlinear fibre (HNLF). In the NOLM, a single 160 Gb/s OTDM signal input is split into two counter propagating signals. One part travels clockwise (CW) and the other part travels counter clockwise (CCW) through the loop. The optical path length is exactly the same for CW and CCW, since they follow the same path but in opposite direction. In CCW direction, a 40-GHz clock signal is coupled into the loop through the first polarisation beam splitter (PBS1) and leaves the loop through PBS2. The CCW signal acquires an extra nonlinear phase shift, due to XPM between the CCW signal and the clock, when the data coincides with the clock. Thus, when the CW and CCW signal interfere at the coupler, the signal pulse can be switched to the drop port.

6.8.2.1 NOLM

The clock signal and the OTDM signal are obtained from the same mode locked laser (MLL) source operating at 1550 nm; however, to avoid crosstalk, the clock is converted to 1540 nm based on super continuum (SC) generation in a 2.2-km DSF. The schematic for the wavelength conversion is shown in Fig. 6.14.

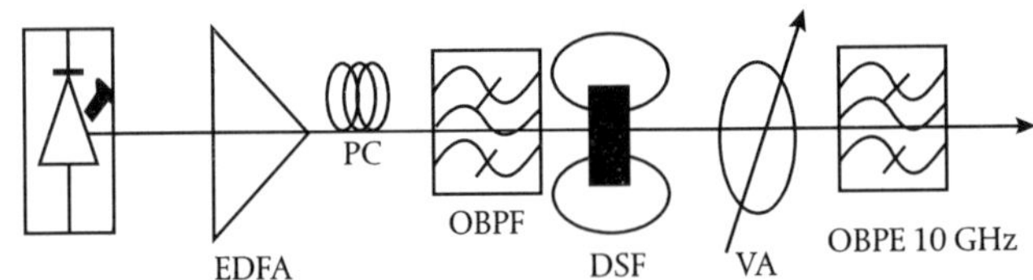

Fig. 6.14 Schematic of the experimental setup for wavelength conversion on SC generation in 2.25-km DSF

The MLL generates pulses of $1.9-\text{ps}$ duration with a repetition rate of 10 GHz. The 10-GHz pulse stream is then amplified to an average power of $+20$ dBm for SC generation. The output of the DSF is filtered with a 3-nm bandpass filter with centre frequency clock of 1540 nm. The time trace of the converted clock pulse is shown in Fig. 6.15. For the 40-Gb/s base rate, we also need a 40-GHz clock signal. To get the 40-GHz clock signal, the 10-GHz pulse from the MLL is multiplexed to 40 GHz first, and then, the above SC generation procedure is carried out to convert the clock to another wavelength. The pump signal average power for SC generation is $+25$ dBm. At this power level, a clean converted 40-GHz clock signal is found at 1560 nm.

6.8.2.2 Demultiplexing

The all-optical demultiplexer was first tested for the $10-\text{Gb/s}$ base rate and later for the $40-\text{Gb/s}$ base rate. For the $10-\text{Gb/s}$ base rate, we used the wavelengths: data $=1551$ nm and clock $=1539$ nm. For the base rate of 10 Gb/s, the $10-\text{GHz}$ pulse stream from the MLL is modulated by an MZI intensity modulator driven by a pattern generator producing pseudo-random binary sequence (PRBS) 2^7-1. Then, the 10 Gb/s signal is multiplexed to a 80 Gb/s and 160 Gb/s OTDM signal. For the 40 Gb/s base rate, the 40-GHz optical pulse is obtained by multiplexing the $10-\text{GHz}$ pulse from MLL first, and then the PRBS pattern is modulated onto the pulse streams by an electron-absorption modulator (EAM) to produce the $40-\text{Gb/s}$ base signal. Then, the $40-\text{Gb/s}$ signal is multiplexed to 80 Gb/s and 160 Gb/s. Pattern effect is observed in the multiplexed signal, which is introduced by the EAM. The clock power injected into the HNLF is $+10-\text{dBm}$ averages, corresponding to $665-\text{mW}$ peak power. The clock pulse measured on the autocorrelator is $1.5-\text{ps}$ full width at half maximum (FWHM). The BER for demultiplexing from $80\ \text{Gb/s}$ and $160\ \text{Gb/s}$ to the $10\ \text{Gb/s}$ base rate is measured. The power penalty to obtain a BER of 10^{-9} is varying between 1.2 and 5 dB. The measurements were unstable because of wandering of the clock signal due to temperature fluctuations in the DSF. The FWHM is measured to be 2.2 ps. The BER measurements with a $40\ \text{Gb/s}$ base rate are performed by connecting a 50-GHz photodiode output directly to the input of the BER test set. The sensitivity penalty for demultiplexing from $80\ \text{Gb/s}$ to $40\ \text{Gb/s}$ is 3.7 dB. The power of the clock signal injected in the HNLF is $+16\ \text{dBm}$ average. This corresponds to $665-\text{mW}$ peak power. The loss between the EDFA and the HNLF is 4 dB. Demultiplexing from $80\ \text{Gb/s}$ to a $10-\text{Gb/s}$ or $40-\text{Gb/s}$ base rate, and $160\ \text{Gb/s}$ to a $10-\text{Gb/s}$ base rate with a BER lower than 10^{-9} is demonstrated using an NOLM with a 500-m HNLF as nonlinear element. A better and more stable performance is expected using a better clock signal and good OTDM signal for the all-optical demultiplexing.

6.9 PULSE COMPRESSION USING OPTICAL FIBRES

Pulse compression is an established technique for generating optical pulses shorter than those produced by lasers or amplifiers. Most commonly, additional bandwidth is generated by SPM in an optical fibre. The group velocity dispersion (GVD) required for compression of the pulse is typically provided by gratings or prisms. Compressors based on SFMs are limited to nano-joule pulse energies by higher order nonlinear effects and ultimately by damage to fibre.

6.9.1 Need for Optical Pulse Compression

1. Conventional methods generate optical pulses having widths in the picoseconds region.
2. Femto-second pulses find applications in OTDM communication system.
3. An OTDM is a high-capacity, high bit-rate communication system.
4. High bit rate requires high repetition rate of the pulse.
5. To generate high repetition-rate pulses, we need ultrashort widths of optical pulses.

6.9.2 Pulse Compression (Linear Fibres)

When pulses are chirped, their duration can be reduced by removing (or at least reducing) this chirp, that is, by flattening the spectral phase. De-chirping can be accomplished by sending the pulses through an optical element with a suitable amount of chromatic dispersion such as a pair of diffraction gratings (a grating compressor), a prism pair, an optical fibre, a chirped mirror, a chirped fibre Bragg grating, or a volume Bragg grating. The smallest possible pulse duration is then set by the optical bandwidth of the pulses, which is not modified by dispersive (linear) compression. In the ideal case, bandwidth-limited pulses are obtained.

6.9.2.1 Chirp

The chirp of an optical pulse is usually understood as the time dependence of its instantaneous frequency. Specifically, an up-chirp (down chirp) means that the instantaneous frequency raises (decreases) with time as shown in Fig. 6.15. As an example, consider a pulse with a Gaussian envelope and a quadratic temporal phase:

$$E(t) = E_0 \exp\left(-at^2 + ibt^2\right) \tag{6.13}$$

This is associated with a linear chirp, that is, with a linear variation of the instantaneous frequency.

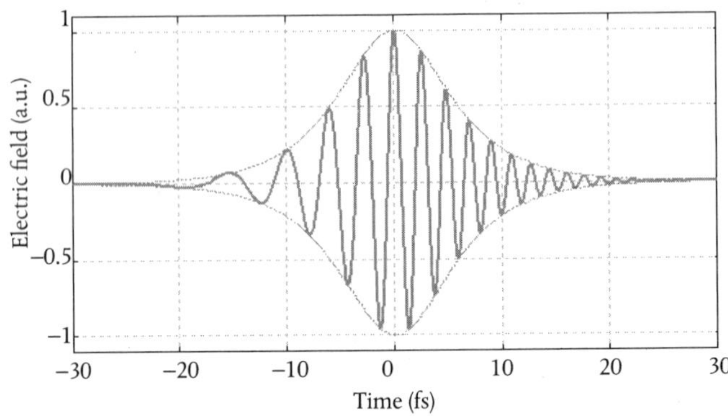

Fig. 6.15 Electric field of a strongly up-chirped pulse, where the instantaneous frequency grows with time

A pulse can acquire a chirp, for example, during propagation in a transparent medium due to the effects of chromatic dispersion and nonlinearities (e.g., SPM arising from the Kerr effect). In semiconductor lasers or amplifiers, chirps can also result from refractive index changes associated with changes in the carrier density. The chirp of a pulse can be removed or reversed by propagating it through optical components with suitable chromatic dispersion. For a given pulse spectrum, the minimum pulse duration is obtained when there is no chirp ("unchirped pulses"). This condition is equivalent to a constant instantaneous frequency and (nearly) equivalent to a constant spectral phase.

6.9.2.2 Quantifying pulse chirp

There are different ways of quantifying the chirp of a pulse:

1. The magnitude of a chirp can be considered to be the rate of change in the instantaneous frequency, for example, in units of hertz per second. In the case of nonlinear chirps, this quantity is time dependent, that is, not a constant number for one pulse.
2. It is also possible to specify the amount of group delay dispersion (in units of seconds squared) required compressing a pulse dispersively. More precisely, this may be the second-order dispersion that leads to the smallest pulse duration or to the highest peak power.
3. Such definitions are not related to each other in a simple way. For example, an increasing amount of normal chromatic dispersion applied to an initially unchirped (transform limited) pulse will obviously increase the amount of dispersion required for compression. On the other hand, the chirp in terms of the rate of change of the instantaneous frequency will initially increase but later will decrease. The latter results from the fact that the same change in instantaneous frequency will occur within a larger and larger time interval as the pulse duration increases.
4. As a second example, an initially unchirped pulse subject to self-phase modulation via Kerr nonlinearity will exhibit an increasing optical bandwidth. For increasingly strong spectral broadening, the rate of change of the instantaneous frequency increases, whereas the amount of dispersion as required for maximum compression can decrease.

6.9.2.3 Prism pairs

The pairs of (typically Brewster angled) prisms can be used for introducing anomalous chromatic dispersion, for example, into a laser resonator without introducing significant power losses. A first prism refracts different wavelength components to slightly different angles. A second prism then refracts all components again to let them propagate in parallel directions as shown in Fig. 6.16, but with a wavelength-dependent position (which is sometimes called a spatial chirp).

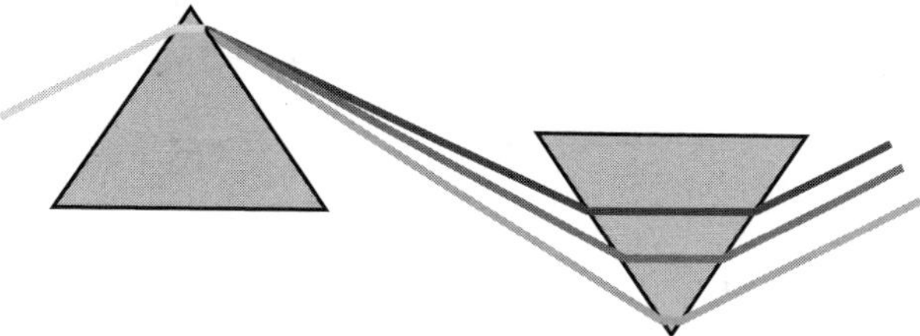

Fig. 6.16 A prism pair for spatially dispersing different wavelength components, and thus, also introducing wavelength-dependent phase changes and chromatic dispersion

With a second prism pair, or simply by reflecting the beams back through the original prism pair, all wavelength components can later be spatially recombined. The spatial separation of different wavelength (or frequency) components can be utilised in different ways:

1. An optical filter can be realised, for example, by inserting a knife edge from one side, attenuating primarily the short or the long wavelength components. This can be used, for example, for wavelength tuning of lasers by placing such a prism pair within the laser resonator.
2. The wavelength-dependent optical path lengths of such a dispersive delay line lead to anomalous dispersion, which may be partly offset by material dispersion in the prisms. The overall dispersion can be adjusted by varying the insertion of one or both crystals into the beam. This technique is often used to provide adjustable dispersion compensation in mode locked lasers as shown in Fig. 6.17 (e.g., for mode locking of solitons) and for dispersive compression (or stretching) of optical pulses.

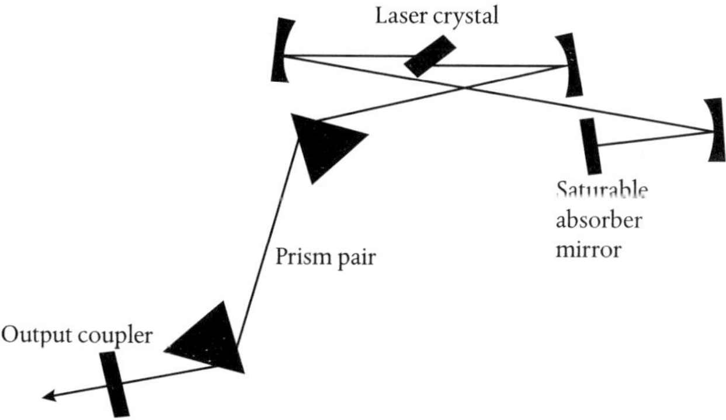

Fig. 6.17 Resonator setup of a mode-locked laser. A prism pair is used for dispersion compensation. The overall anomalous chromatic dispersion allows for soliton mode locking and can be adjusted via the prism insertion

Typical amounts of anomalous dispersion from prism pairs are a few thousand. For larger amounts of dispersion, a pair of diffraction gratings may be required. The attraction of using a prism pair, however, is that anomalous dispersion can be provided without introducing significant losses into a laser resonator. For the compression of ultrashort pulses in the few cycle regions, prisms with a fairly small apex angle (and anti-reflection coatings) are sometimes used. Such configurations can achieve a lower residual chirp from higher order dispersion. However, it is often necessary to compensate the higher order dispersion with other means (e.g. with additional chirped mirrors). Figure 6.18 shows a group delay dispersion of a prism pairs.

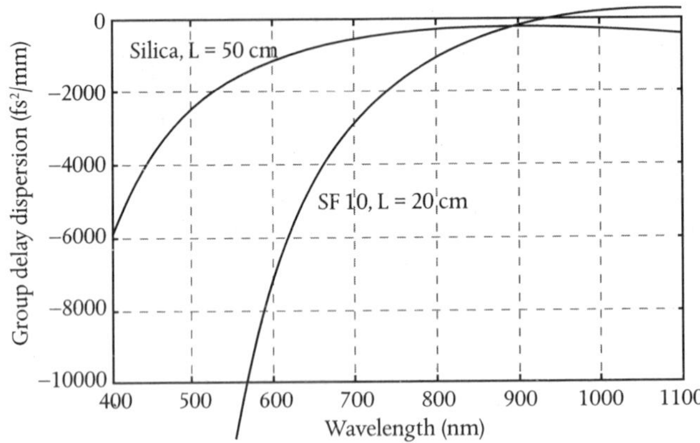

Fig. 6.18 Group delay dispersion of prism pairs: comparison of a setup with silica prisms and 50-cm spacing, and another one with SF10 prisms and 20-cm spacing. The insertion of the prisms at 800 nm is 2 mm in both cases. The SF10 prisms can generate more dispersion, but the higher order dispersion is significantly higher

6.9.2.4 Chirped mirrors

A chirped mirror is a kind of dielectric mirror, which is usually used for dispersion compensation in mode locked lasers. Another interesting feature of chirped mirrors is that they make it possible to achieve a broader reflection bandwidth than ordinary Bragg mirrors. The basic idea of chirped mirror designs is that the Bragg wavelength is not constant but varies within the structure so that light at different wavelengths penetrates to a different extent into the mirror structure, thus experiencing a different group delay. Figure 6.19 reveals the operation of a chirped mirror.

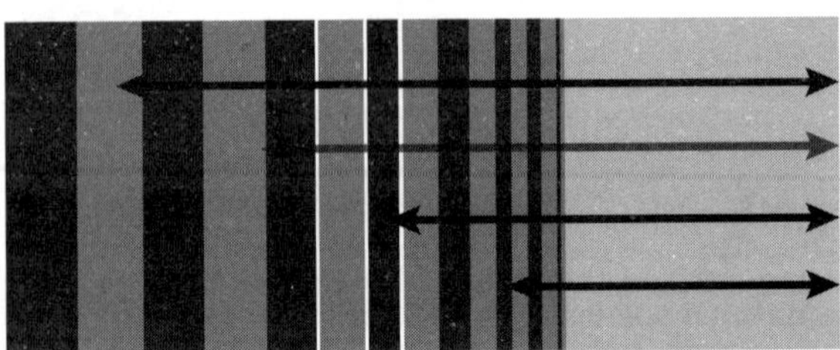

Fig. 6.19 Operation principle of a chirped mirror. Light with a long wavelength penetrates deeper into the mirror structure and thus experiences a larger group delay. This leads to anomalous chromatic dispersion

However, a native design directly based on this idea would not work: it would exhibit strong oscillations of the group delay and even more so of the group delay dispersion. This disturbing effect can to some extent be mitigated by numerical optimisation of the layer structure, but this is difficult because the optimisation has to be done in a multi-dimensional space (resulting from the large number of layers) where a huge number of local optima exist, most of which do not correspond to satisfactory designs.

Later, it was found that the disturbing oscillations have two origins:

1. There is a Fresnel reflection at the front face (the interface to air), which leads to strong additional dispersion as in a Gires–Tournois interferometer.
2. The sudden "switching" of the coupling of counter propagating waves from zero in air to a finite value in the structure causes a kind of impedance mismatch.

Both problems can be eliminated with a double-chirped design, which has two additional features:

1. The Fresnel reflection is removed with an additional anti-reflection structure on top of the double-chirped section.
2. The coupling of counter propagating waves is turned on smoothly by also varying the duty cycle, that is, the ratio of optical thickness of high- and low-index layers.

Even without numerical optimisation, double-chirped designs can have a dispersion profile that relatively matches the design goal. Further refinement is then achieved with numerical optimisation, that is, with fine tuning of the layer thickness values.

6.9.3 Pulse Compression (Nonlinear Fibres)

In the first step, the optical bandwidth is increased, typically with a nonlinear interaction like SPM. In most cases, this leads to chirped pulses, often with a duration, which is even larger than the original pulse duration. Thereafter, the pulse duration can be strongly reduced by linear compression, which removes or at least decreases the chirp.

6.9.3.1 Methods for nonlinear pulse compression

Nonlinear pulse compression can be done with different configurations of optical elements, and with methods that are based on different physical principles. Some examples are discussed in the following sections.

6.9.3.2 Soliton pulse compression

A fundamental soliton is an optical pulse that can propagate in a dispersive medium (e.g., an optical fibre) with a constant shape of the temporal intensity profile, that is, without any temporal broadening as is usually caused by dispersion. This can happen when the pulse has a certain shape and an energy, which is determined by the parameters of the medium (in particular, by the dispersion and nonlinearity) and the pulse duration. The conditions for (fundamental) soliton pulse propagation in a lossless medium are: if a positive value of the nonlinear coefficient n_2 (as occur for most media), the chromatic dispersion needs to be anomalous.

The temporal shape of the pulse has to be that of an unchirped sech^2 pulse (assuming that the group delay dispersion is constant, that is, there is no higher order dispersion):

$$P(t) = P_p \operatorname{sech}^2(t/\tau) = \frac{P_p}{\cosh^2(t/\tau)} \tag{6.14}$$

The pulse energy E_p and soliton pulse duration τ have to meet the following conditions:

$$E_p = \frac{2|\beta_2|}{|\gamma|\tau} \tag{6.15}$$

Here, the FWHM pulse duration is $1.7627 \times \tau$, γ is the SPM coefficient in $\text{rad}/(\text{W m})$, and β_2 is the group velocity dispersion defined as a derivative with respect to angular frequency, that is, the group delay dispersion per unit length (fs²/mm).

A higher order soliton is a soliton pulse the energy of which is higher than that of a fundamental soliton by a factor, which is the square of an integer number (i.e., 4, 9, 16, etc.). The temporal shape of such a pulse is not constant, but rather varies periodically during propagation. The period of their evolution is the so-called soliton period.

Higher order solitons can be used for nonlinear pulse compression: a sech^2-shaped pulse with a suitable energy, injected into an optical fibre with anomalous dispersion, can evolve as a higher order soliton, and after certain propagation distance, the pulse duration can be substantially decreased. High soliton orders allow for strong compression, but also lead to a critical choice of the pump wavelength.

On the other hand, fundamental solitons are usually fairly stable and higher order solitons can break up into fundamental solitons under the influence of various effects such as higher order dispersion, Raman scattering, or two-photon absorption. Such soliton breakup sometimes plays an essential role in the process of SC generation in photonic crystal fibres.

Fig. 6.20 Setup for pulse compression with a fibre only. The compression mechanism could be higher order soliton compression or adiabatic soliton compression

Soliton effect compression is the term coined for the compression of higher order solitons in an optical fibre with constant anomalous group velocity dispersion (GVD) characteristics along the length as shown in Fig. 6.20. Through the interplay of the GVD and nonlinear SPM effects, optical soliton propagation is possible. In the case of a fundamental or first-order soliton, the effects of SPM cancel the effects of anomalous GVD perfectly and the soliton propagates while preserving sech^2 shape in a lossless fibre. On the other hand, a higher order soliton changes its shape periodically as it propagates along the fibre. Such a soliton always experiences an initial pulse narrowing phase before recovering its original shape at integral multiples of the soliton period. The soliton period equals $(\pi/2)Z_0$ where Z_0 is the dispersion length defined in terms of the pulse width, FWHM, and GVD parameter $\beta_2(\text{ps}^2/\text{km})$ as:

$$Z_0 = T_2\text{FWHM}/3.11|\beta_2| \tag{6.16}$$

Optimum pulse compression is achieved when the soliton leaves the fibre just as it attains its narrowest width. The optimum fibre length equals a fraction of Z_0 and decreases as the soliton order, N, is increased. Thus, the required fibre lengths for soliton-effect compression of picoseconds pulses are typically short. On the other hand, the required pulse peak power is typically very high and increases in proportion to the square of the soliton order. The peak power required for the formation of an nth order soliton is given as:

$$PN = N_2 / \gamma Z_0, \tag{6.17}$$

where γ is the fibre nonlinearity coefficient in units of $W^{-1}km^{-1}$.

An inherent feature of the soliton effect compression technique is the optimally compressed output pulse consisting of a narrow spike accompanied by a broad pedestal component. For a $N=3$ soliton input, a maximum compression factor of 8 can be achieved with approximately 28% of the output energy contained in the pedestal component. To realise high-quality compression, the pedestal must be minimised since the interaction between the pedestal and the compressed spike occurs upon further propagation, leading to undesirable periodic pulse reshaping. Such behaviour is detrimental for optical communication applications, and therefore, pedestal suppression techniques are necessary. In general, increasing the input soliton order not only enhances the compression factor and reduces the required fibre length but also increases the power requirements and reduces the pulse quality. Optimum compression of a $N=15$ soliton is capable of a much higher compression factor of 60, with the drawback, however, of approximately 80% of the output energy contained in the pedestal.

6.9.3.3 Adiabatic soliton compression

Adiabatic soliton compression is a technique for the temporal compression of ultrashort pulses in a fibre. The principle of operation is described in the following. For a fundamental soliton pulse in a fibre, the product of pulse energy and pulse duration is proportional to the group velocity dispersion divided by the nonlinearity of the fibre. Thus, the pulse duration must be reduced if the dispersion is reduced while keeping constant the pulse energy. Significant pulse compression can, therefore, be obtained by propagating the pulses through a dispersion decreasing fibre.

However, the following conditions must be satisfied:

1. The initial pulses must fulfil the soliton condition at the input fibre end.
2. The fibre dispersion must be varied sufficiently slowly to allow adiabatic adaptation of the pulses to the fibre parameters (otherwise, the pulses can become distorted). More precisely stated, the dispersion must not vary significantly over a length scale of a soliton period. As the latter scales with the square of the pulse duration, rather long fibres are required if the initial pulses are longer than 1 ps.
3. The fibre dispersion must stay sufficiently constant over the whole spectral range of the compressed pulses. In other words, higher order dispersion must be sufficiently weak. However, it has been shown that slightly normal dispersion in the wings of the generated pulse spectrum can be beneficial.
4. Interestingly, there are situations where Raman scattering and higher order dispersion combine in such a way that the pulse compression stays adiabatic, even though each of the mentioned effects separately would lead to severe pulse distortion.
5. Even though the method is elegant and powerful, it suffers from the need to use a dispersion decreasing fibre. The latter requirement is eliminated by a variant of the method, where the fibre

has constant dispersion but contains a laser active dopant that allows the amplification of the pulses. Here, increasing pulse energy for constant dispersion also results in temporal compression.

6. Instead of using a dispersion decreasing fibre, it is also possible to concatenate (fusion splice) different fibres with different dispersion values. This may lead to more reproducible results, but as the dispersion does not vary continuously, the compression factor and/or the pulse quality can be compromised.
7. Generally, adiabatic soliton compression is limited to fairly low-pulse energies, since the soliton energies of pulses in fibres cannot be made very high. Therefore, the technique is mainly applied to high repetition rate pulse trains, for example, in the context of optical fibre communications.

6.9.4 Pulse Compression with Similar on Pulse Propagation

A parabolic pulse is an optical pulse (usually an ultrashort pulse) that has a temporal intensity profile with a parabolic shape not only near the pulse centre but also well towards the wings up to the point where the intensity goes to zero. In the ideal case, the pulse would have an intensity profile according to:

$$P(t)=\begin{cases} P_0\left(1-(t/\tau)^2\right) & \text{for} |t|<\tau \\ 0 & \text{otherwise} \end{cases} \tag{6.18}$$

Figure 6.21 shows an approximately parabolic pulse as it can result from pulse propagation in a fibre amplifier under certain circumstances. Parabolic pulses have a number of remarkable properties, which have turned out to be particularly important in the context of fibre amplifiers for generating high-energy ultrashort pulses.

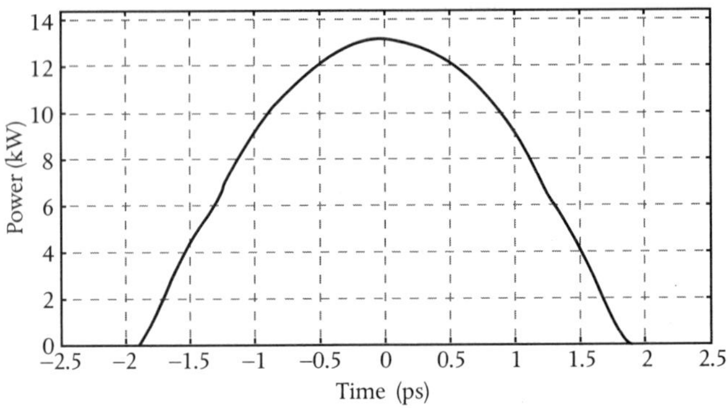

Fig. 6.21 Temporal profile of a nearly parabolic pulse

Temporal profile of a nearly parabolic pulse can be generated in a fibre amplifier fed with Gaussian pulses. Up-chirped parabolic pulses can be generated in a fibre amplifier where the amplified pulses not only experience optical gain but also the Kerr nonlinearity and normal chromatic dispersion. In this situation, a chirped parabolic pulse is the asymptotic solution for long propagation lengths. After a sufficiently long propagation length associated with a sufficiently large amplification factor, the shape of the resulting pulses no longer depends on the shape or duration of the seed pulses and the peak power and duration after a long propagation length depends only on the seed pulse energy. In that regime, the time bandwidth product of the pulses is large.

With increasing propagation length, the peak power rises. This is a case of self-similar pulse propagation, since the parabolic pulse shape, although not the pulse duration, is preserved. Therefore, such pulses are sometimes called similaritons. The pulses have a very nearly linear up-chirp, that is, a linearly rising instantaneous frequency. The chirp rate (in THz/ps) approaches a constant value, which is determined only by the fibre dispersion and gain per unit length.

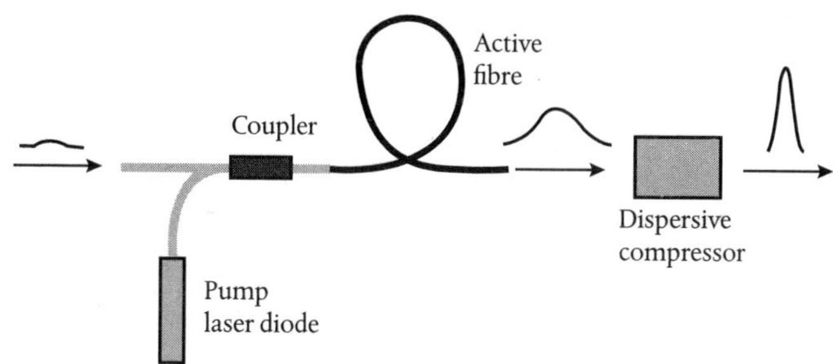

Fig. 6.22 Setup for pulse compression with similariton pulse propagation

While the pulse is amplified in a rare earth-doped fibre, its duration and spectral width both increase. A dispersive compressor can subsequently reduce the pulse duration strongly as revealed in Fig. 6.22. In a fibre amplifier with normal dispersion, self-similar parabolic pulses experience spectral broadening while a high-pulse quality is preserved. The parameters of the input signal pulses are fairly uncritical and high pulse energies are possible. The resulting linear chirp makes it relatively easy to obtain strong temporal compression in a subsequent dispersive optical element.

6.9.5 Nonlinear Fibre Bragg Gratings

The two widely used techniques to compress an optical pulse, namely using higher order soliton effect and adiabatic pulse compression. The former can have large compression ratio, but it suffers from significant pedestal generation. Adiabatic pulse compression can be achieved by the use of distributed gain or dispersion decreasing fibre. However, adiabatic condition must be satisfied, that is, the change must be gradual. A better solution that consists of a transmission-based pulse compressor employing a nonlinear fibre Bragg grating (FBG) with exponentially decreasing dispersion. The large dispersion in the spectral vicinity of the grating stop band leads to very short devices (centimetres as opposed to kilometres). Numerical examples confirm that compressed pulses are of large compression factor, little chirp, and almost free of pedestal. A figure of merit is defined for the optimisation of the grating parameters to generate compressed pulse with good quality.

Nonlinear pulse propagation in FBGs can be described by the nonlinear coupled-mode equations (NLCMEs) that describe the coupling between the forward propagating wave (u) and the backward propagating wave (v)

$$i\frac{\partial u}{\partial z}+i\beta_1\frac{\partial u}{\partial T}+\delta u+\kappa v+\gamma\left(|u|^2+2|v|^2\right)u=0 \tag{6.19}$$

$$-i\frac{\partial v}{\partial z}+i\beta_1\frac{\partial v}{\partial T}+\delta v+\kappa u+\gamma\left(|v|^2+2|u|^2\right)v=0 \tag{6.20}$$

where β_1 is the group delay per unit length, $\delta = n(\omega_c - \omega_b)/c$ is the frequency detuning from the Bragg frequency ω_e with n being the average refractive index and c being the speed of light in vacuum; the parameter $\kappa = \eta.\Delta n.\pi/\Lambda/n/2$ is the coupling co-efficient with η being the fraction of the fibre mode that overlaps with the grating, Δn being the refraction index modulation, and Λ being the period of FBG. The parameter γ is the nonlinear coefficient, z is distance, and T is the time.

Under certain conditions, the NLCMEs can be reduced to the well-known nonlinear Schrodinger-type equation (NLSE). Self-similar pulse compression near the photonic bandgap of gratings in the NLSE limit has been described.

A figure of merit (M) is defined to quantify the relative importance of the second-order and third-order dispersions. M is defined to be the ratio of second-order dispersion length (LD) to the third-order dispersion length (LD').

6.9.6 Compression due to Self-phase Modulation

Self-phase modulation (SPM) in optical fibre is ordinarily associated with spectral broadening of an ultrashort optical pulse. However, for appropriate initial conditions of the input pulse, SPM can result in significant spectral compression. Indeed, SPM causes spectral compression or broadening depending on the initial frequency modulation (chirp) of the pulse electric field, $E(t)$. Specifically, a negatively chirped pulse is compressed by the effects of SPM. We use the term spectral compression to stress that for negatively chirped pulse SPM transfers pulse energy to spectral components that are closer to the centre wavelength without energy loss. Applications of spectral compression include the low-loss generation of 1 ps transform limited pulses from $100-\text{fs}$ pulses and the use of these spectrally compressed pulses in fibre links.

The effect of SPM is to redistribute the pulse energy to different spectral components. The origin of this redistribution is the intensity-dependent nonlinear phase shift $\varphi_{NL}(t) \propto |E(t)|^2$. The result of this nonlinear phase shift, in silica, is a frequency downshift $(\delta\omega < 0)$ in the leading edge of the pulse and an up-shift $(\delta\omega > 0)$ in the trailing edge. Specifically,

$$\delta\omega(t) = -\frac{\partial \varphi_{NL}}{\partial t} \propto \frac{\partial}{\partial t}|E(t)|^2. \tag{6.21}$$

Thus, the effect of SPM is dependent on the sign of the initial chirp. Negatively chirped pulses, where the long and the short wavelengths are in the trailing and the leading edges, respectively, are spectrally compressed, since both the long and the short wavelengths are shifted toward the centre wavelength (λ_0). Spectral broadening occurs for pulses that are positively chirped or transform limited. Therefore, it is important to know the initial chirp if the effects of SPM are to be predicted. Second harmonic generation is employed for frequency resolved optical gating to determine the pulse intensity and phase. The initial such pulse is generated by a mode-locked Ti: sapphire laser at $\lambda_0 = 810$ nm. The pulse duration is 110 fs (FWHM), with a time bandwidth product of 0.34. The negative chirp can be created by first propagating the pulse through double-pass SF 10 prism-pair geometry. The negatively chirped pulse is then coupled into 0.5 m of SMF that served as the source of SPM. An amplitude and phase assessment is made by use of frequency resolved optical gating (FROG) on (i) the initial laser pulse (location 1), (ii) the negatively chirped pulse (location 2), and (iii) the spectrally compressed pulse (location 3) as shown in Fig. 6.23.

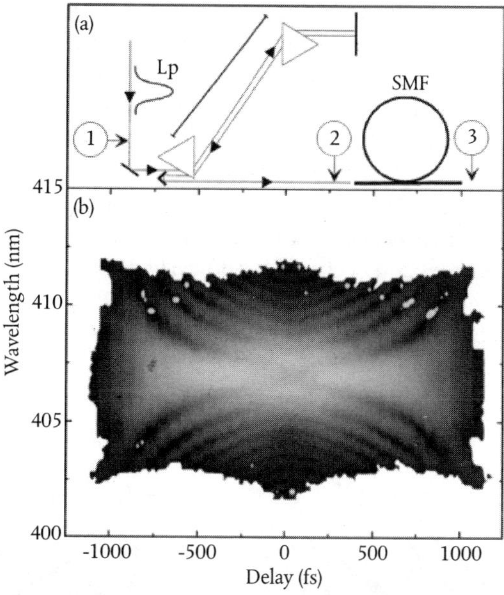

Fig. 6.23 (a) Experimental spectral compression setup showing the prism separation of $L_p = 3\,\text{m}$. Pulse characterisation performed at locations 1, 2, and 3. (b) Measured FROG spectrogram after the prism pair and the SMF (location 3). Spectral compression increases near-zero delay resulting from larger SPM at higher intensities. Average power in the fibre is $70\ \text{mW}\left(8400\text{W peak power}\right)$.

The negative chirp is characterised before the fibre and determined that $L_p = 3\,\text{m}$ the pulse was strongly chirped with a temporal FWHM of $665\,\text{fs}$. The spectral phase is retrieved by the FROG algorithm and fitted to a Taylor series expansion including quadratic:

$$\left(\varphi_2 \equiv d^2\varphi/d\omega^2\Big|_{\omega_0}\right) \tag{6.22}$$

and cubic

$$\left(\varphi_3 \equiv d^3\varphi/d\omega^3\Big|_{\omega_0}\right) \tag{6.23}$$

phase distortion terms. The rms error G between the normalised experimental spectrogram and the retrieved FROG spectrogram is very small, typically $G = 0.0022$. Although the sign of the observed chirp is ambiguous for second harmonic generation FROG (SHG FROG), it is also observed that the expected linear dependence of the magnitude of the linear chirp on the prism separation. This dependence, together with the observation of enhanced compression with increasing prism separation, confirms the existence of the expected negative chirp. The phase distortion measured after the prism pair is $\varphi_2 = -30{,}600\ \text{fs}^2$ and $\varphi_3 = -79{,}100\ \text{fs}^3$. These values compare well with the calculated phase distortion imposed by the isolated prism-pair geometry, which is $\varphi_{2,\text{cal}} = -45{,}000\,\text{fs}^2$ and $\varphi_{3,\text{cal}} = -102{,}000\,\text{fs}^3$. The smaller measured phase distortion results mostly from the positive φ_2 and φ_3 of the experimental optics. The near-transform limited initial pulse exhibited a small phase distortion, typically $\varphi_2 = +120\ \text{fs}^2$ and $\varphi_3 = +2000\ \text{fs}^3$.

The negatively chirped pulse is then spectrally compressed in 0.5 m of SMF as shown in Fig. 6.23(b). The pulse power dependence of the compression is investigated by use of both FROG and separate spectral measurements. The average power at the SMF output is varied from 5 to 70 mW by attenuation at the fibre input. An FROG spectrogram for 70-mW average power is shown in Fig. 6.23(b). The SHG spectral width can be seen to narrow near-zero delay resulting from the enhanced SPM at higher intensities. It is noted that the retrieved phase revealed significant positive chirp. Thus, at 70-mW average power, the total SPM is in excess of that needed to compensate for the initial negative chirp. At 40 mW, the resultant phase distortion is near minimum, producing a near-transform limited pulse. Indeed, significant spectral compression occurred for powers greater than 25 mW as shown in Fig. 6.24. Spectra at these powers also show structure and a slight asymmetry, which can be associated with the cubic phase distortion produced by the prism pair.

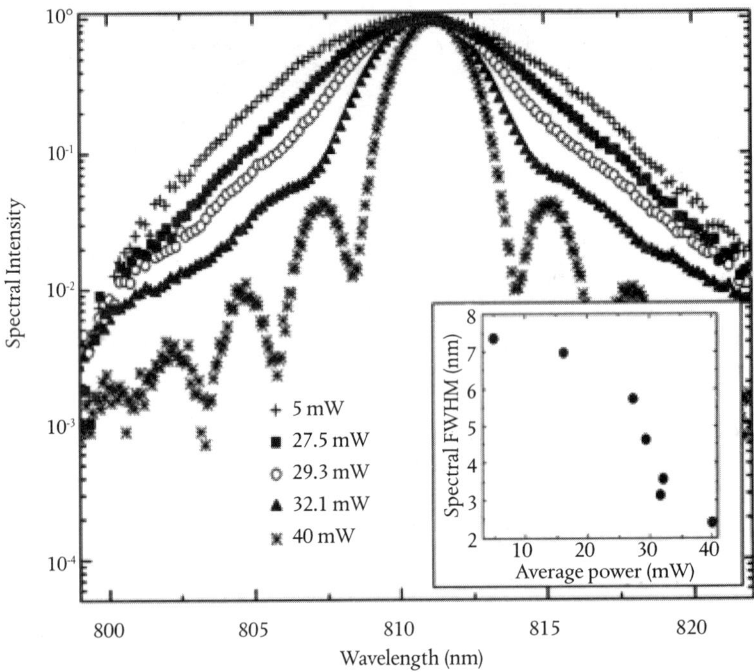

Fig. 6.24 Observed spectra measured at location 3 for increasing average power. Inset, spectral FWHM versus average power

The retrieved spectrum, temporal intensity, and corresponding phase are shown in Fig. 6.24 for 40-mW average power. The initial spectral FWHM of 8.4 nm is compressed to 2.4 nm, and the temporal FWHM broadens from 110 to 600 fs. Importantly, the phase is constant over the FWHM of the spectrum, which is indicative of a transform limited pulse. It is noted that the temporal FWHM of the negatively chirped pulse (measured at location 1) remains essentially unchanged after spectral compression, although the pulse shape becomes somewhat asymmetric owing to the residual phase distortion. The propagation of an ultrashort pulse through SMF is theoretically described by the nonlinear Schrodinger equation (NLSE). The NLSE accounts for the effects of material absorption, group velocity dispersion, and optical nonlinearity owing to propagation through the fibre. To verify these results, the NLSE is solved for propagation through the prism pair and the SMF using the split-step Fourier method.

$$E_p(\omega) = F\left[E_0 \operatorname{sech}(t/t_0)\right] \exp\left[i\left(\varphi_2\omega^2/2 + \varphi_3\omega^3/6\right)\right], \tag{6.24}$$

where $E_0 \operatorname{sech}(t/t_0)$ is the initial electric field and F denotes the Fourier transform.

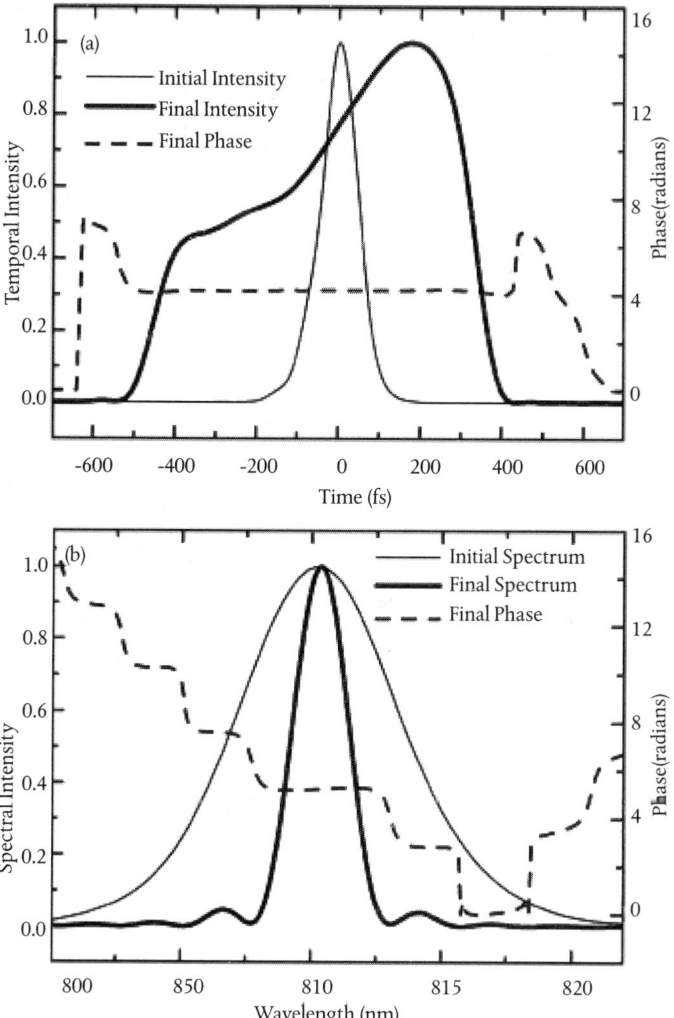

Fig. 6.25 Retrieved intensity and phase: (a) temporal and (b) spectral for an average power of 40 mW measured at location C (FROG error of G = 0.0027)

The phase over the temporal and spectral FWHM is free of significant distortion. Such initial pulse temporal and spectral intensity (location 1) are shown in comparison (thin curves) as shown in Fig. 6.25. The inverse Fourier transform is then used to generate $E_p(t)$, which is used as the input for the split-step Fourier method. The calculated intensity $|E_p(t)|$ compares well with the experimental intensity at location 2. The NLSE is solved for pulse propagation through 0.5 m of SMF with a spatial step size of

100 mm. The NLSE includes quadratic (β_2) and cubic (β_3) group velocity dispersion calculated from the manufacturer's dispersion parameters. The effect of third-order nonlinearity (SPM) is contained in the nonlinear refractive index $(n_2 = 3.2\times10^{-20}\,\mathrm{m}^2/\mathrm{W})$; higher order nonlinearities and material absorption are assumed to be negligible owing to the short length of fibre. The spatial step size is chosen to ensure conservation of pulse energy. The final pulse energy is within 5×10^{-5} % of the initial energy for the chosen spatial step. A comparison of the calculated and the measured spectra for an average power of 40 mW (4800 W peak power) is shown in Fig. 6.26.

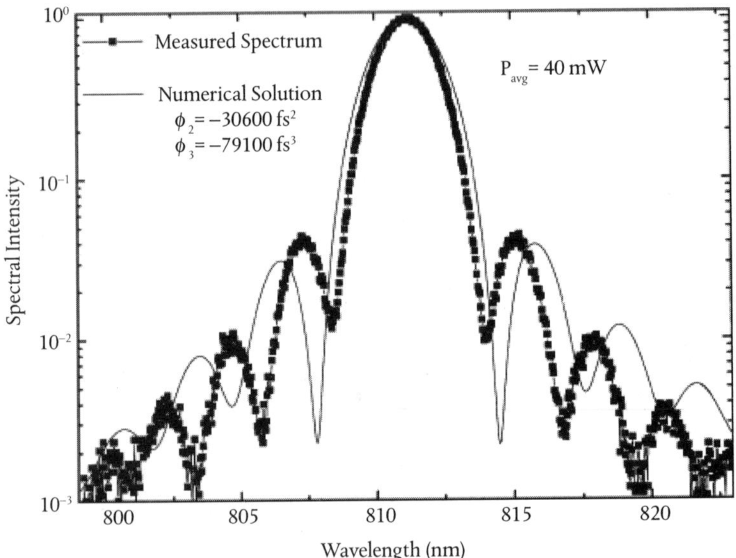

Fig. 6.26 Numerical solution where $Ep(\omega)$ is calculated with the measured phase distortions ϕ_2 and ϕ_3.

The experimentally measured spectrum is shown to be very similar to it. Note that the measured spectrum is nearly identical to the FROG-retrieved spectrum of Fig. 6.25. Importantly, the directly measured spectrum of Fig. 6.24 is nearly identical to the retrieved spectrum of Fig. 6.26. The spectral FWHM calculated by the split-step Fourier method was 2.89 nm when compared with the measured FWHM of 2.4 nm. The numerical phase compared well with the retrieved phase and is also constant over the pulse FWHM. It is to be noted that if $E_p(\omega)$ is calculated with $\varphi_{2,\mathrm{cal}}$ and $\varphi_{3,\mathrm{cal}}$; instead, the numerical spectrum does not compare well with the experimental spectrum. This illustrates the sensitivity of the spectral compression to the degree of initial negative chirp. The results are not altered by the dispersion of the short fibre. However, the compression performance is ultimately limited by the dispersion and its interplay with SPM.

In conclusion of this section, the ability to spectrally compress ultrashort pulses has been demonstrated with commensurate temporal broadening to produce longer pulses with constant phase over the field envelope. This ability is clearly illustrated by the results from the spectral and the FROG measurements. The numerical solution to the NLSE is also in good agreement with the aforementioned results.

6.9.7 Soliton Pulse Propagation in Optical Fibre

A soliton pulse is a self-reinforcing solitary pulse that maintains its shape while it travels at constant speed. Soliton pulses are caused by a cancellation of nonlinear and dispersion effect of the medium.

1. Nonlinear effect
2. Dispersion effect

6.9.7.1 Nonlinear effect in optical fibre (Kerr effect)

The Kerr effect, also called the quadratic electro-optic effect (QEO effect), is a change in the refractive index of a material in response to an applied electric field. The Kerr effect is the induced index change directly proportional to the square of the electric field.

6.9.7.2 Dispersion effect in optical fibre

Dispersion is the spreading of a light pulse in time as it propagates down the fibre. Dispersion is the phenomenon in which the phase velocity of a wave depends on its frequency or alternatively when the group velocity depends on the frequency. There are generally two sources of dispersion: material dispersion and waveguide dispersion.

Dispersion and nonlinearity can interact to produce permanent and localised waveforms. Consider a pulse of light travelling in glass. This pulse can be thought of as consisting of light of several different frequencies. Since glass shows dispersion, these different frequencies will travel at different speeds and the shape of the pulse will, therefore, change overtime.

However, there is also the nonlinear Kerr effect: the refractive index of a material at a given frequency depends on the light's amplitude or strength. If the pulse has just the right shape, the Kerr effect will exactly cancel the dispersion effect and the pulse's shape will not change overtime: a soliton.

There are mainly two types of solitons:

6.9.7.2.1 Spatial Solitons

The nonlinear effect can balance the diffraction. The electromagnetic field can change the refractive index of the medium while propagating, thus creating a structure similar to a graded index fibre. If the field is also a propagating mode of the guide it has created, then it will remain confined and it will propagate without changing its shape.

6.9.7.2.2 Temporal Solitons

If the electromagnetic field is already spatially confined, it is possible to send pulses that will not change their shape because the nonlinear effects will balance the dispersion. Those solitons were discovered first and they are often simply referred as "solitons" in optics.

6.10 SPATIAL SOLITONS

In order to understand how a spatial soliton can exist, we have to make some considerations about a simple convex lens. As shown in Fig. 6.27, an optical field approaches the lens and then it is focused. The effect of the lens is to introduce a non-uniform phase change that causes focusing. This phase change is a function of the space and can be represented with (x) , whose shape is approximately represented in the picture.

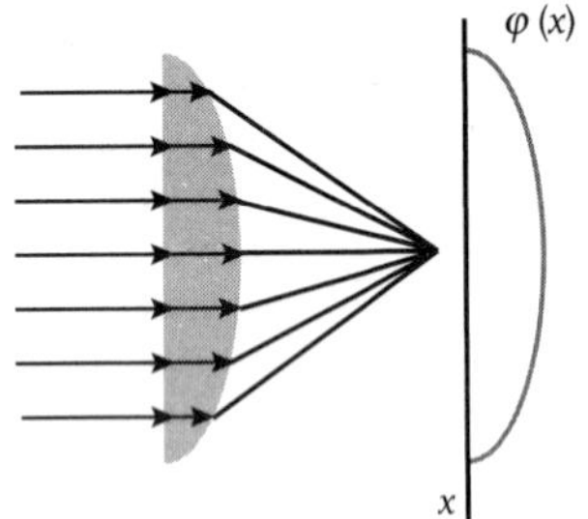

Fig. 6.27 Soliton lens equivalent

The phase change can be expressed as the product of the phase constant and the width of the path the field has covered. We can write it as:

$$\varphi(x) = k_0 n L(x), \tag{6.25}$$

where $L(x)$ is the width of the lens, changing in each point with a shape that is the same of $\varphi(x)$ because k_0 and n are constants. In other words, in order to get a focusing effect, we just have to introduce a phase change of such a shape, but we are not obliged to change the width. If we keep the width L fixed in each point, but change the value of the refractive index $n(x)$, we will get exactly the same effect but with a completely different approach.

In other words, this is the way graded index fibres work: the change in the refractive index introduces a focusing effect that can balance the natural diffraction of the field. If the two effects balance each other perfectly, then we have a confined field propagating within the fibre.

Spatial solitons are based on the same principle: the Kerr effect introduces an SPM that changes the refractive index according to the intensity:

$$\varphi(x) = k_0 n(x) L = k_0 L\left[n + n_2 I(x)\right] \tag{6.26}$$

If $I(x)$ has a shape similar to the one shown in Fig. 6.25, then we have created the phase behaviour we wanted and the field will show a self-focusing effect. In other words, the field creates a fibre-like guiding structure while propagating. If the field creates a fibre and it is the mode of such a fibre at the same time, it means that the focusing nonlinear and diffractive linear effects are perfectly balanced and the field will propagate forever without changing its shape (as long as the medium does not change and if we can neglect losses, obviously). In order to have a self-focusing effect, we must have a positive n_2, otherwise we will get the opposite effect and we will not notice any nonlinear behaviour.

The optical waveguide using the soliton creates while propagating is not only a mathematical model but also actually exists and can be used to guide other waves at different frequencies. This way it is possible to let light interact with light at different frequencies (this is impossible in linear media).

An electric field is propagating in a medium showing optical Kerr effect, and therefore, the refractive index is given as:

$$n(I) = n + n_2 I. \tag{6.27}$$

We remember that the relationship between intensity and electric field is (in the complex representation):

$$I = \frac{[E]^2}{2\eta}, \tag{6.28}$$

where $\eta = \eta_0 / n$ and η_0 is the impedance of free space given as:

$$\eta_0 = \sqrt{\frac{\mu_0}{\varepsilon_0}} \approx 377\Omega \tag{6.29}$$

The field is propagating in the z-direction with a phase constant $k_0 n$. About now, we will ignore any dependence on the y-axis, assuming that it is infinite in that direction. Then, the field can be expressed as:

$$E(x,z,t) = A_m a(x,z) e^{i(k_0 nz - \omega t)}, \tag{6.30}$$

where A_m is the maximum amplitude of the field and $a(x,z)$ is a dimensionless normalised function (so that its maximum value is 1) that represents the shape of the electric field among the x-axis. In general, it depends on z because fields change their shape while propagating. Now, we have to solve Helmholtz equation:

$$\nabla^2 E + k_0^2 n^2 (I) E = 0, \tag{6.31}$$

where it was pointed out clearly that the refractive index (thus the phase constant) depends on intensity. If we replace the expression of the electric field in the equation, assuming that the envelope $a(x,z)$ changes slowly while propagating, that is,

$$\left| \frac{\partial^2 a(x,z)}{\partial z^2} \right| \ll \left| k_0 \frac{\partial a(x,z)}{\partial z} \right|. \tag{6.32}$$

Then (6.31) becomes:

$$\frac{\partial^2 a}{\partial x^2} + i2k_0 n \frac{\partial a}{\partial z} + k_0^2 \left[n^2 (I) - n^2 \right] a = 0. \tag{6.33}$$

Let us introduce an approximation that is valid because the nonlinear effects are always much smaller than the linear ones:

$$\left[n^2 (I) - n^2 \right] = \left[n(I) - n \right] \left[n(I) + n \right] = n_2 I (2n + n_2 I) \approx 2nn_2 I. \tag{6.34}$$

Now, we express the intensity in terms of the electric field:

$$\left[n^2 (I) - n^2 \right] \approx 2nn_2 \frac{|A_m|^2 |a(x,z)|^2}{2\eta_0 / n} = n^2 n_2 \frac{|A_m|^2 |a(x,z)|^2}{\eta_0}. \tag{6.35}$$

Equation (6.33) becomes:

$$\frac{1}{2\kappa_0 n} \frac{\partial^2 a}{\partial x^2} + i \frac{\partial a}{\partial z} + \frac{k_0 nn_2 |A_m|^2}{2\eta_0} |a|^2 a = 0\,. \tag{6.36}$$

We will now assume $n_2 > 0$ so that the nonlinear effect will cause self-focusing. In order to make this evident, we will write in the equation $n_2 = | n_2 |$. Let us now define some parameters and replace them in the equation:

1. $\xi = \frac{x}{X_0}$, we can express the dependence on the x-axis with a dimensionless parameter; X_0 is a length, whose physical meaning will become clear later.

2. $L_d = X_0^2 k_0 n$; after the electric field has propagated across z for this length, the linear effects of diffraction cannot be neglected anymore.
3. $\zeta = \frac{z}{L_d}$, for studying the z-dependence with a dimensionless variable.
4. $L_{nl} = \frac{2\eta_o}{k_0 n |n_2| . |A_m|^2}$, after the electric field has propagated across z for this length, the nonlinear effects cannot be neglected anymore. This parameter depends upon the intensity of the electric field, that is, typical for nonlinear parameters.

Then, (6.36) becomes:

$$\frac{1}{2}\frac{\partial^2 a}{\partial \xi^2} + i\frac{\partial a}{\partial \zeta} + N^2 |a|^2 a = 0, \tag{6.37}$$

where

$$N = \frac{k_0 \eta x_0^2}{L_{nl}} = \frac{L_d}{L_{nl}}.$$

This is a common equation known as nonlinear Schrödinger equation. From this form, we can understand the physical meaning of the parameter N:

1. If $N \ll 1$, then we can neglect the nonlinear part of the equation. It means $L_d \ll L_{nl}$, then the field will be affected by the linear effect (diffraction) much earlier than the nonlinear effect, it will just diffract without any nonlinear behaviour.
2. If $N \gg 1$, then the nonlinear effect will be more evident than diffraction and, because of SPM, the field will tend to focus.
3. If $N \approx 1$, then the two effects balance each other and we have to solve the equation.

For $N = 1$, the solution of the equation is simple and it is the fundamental soliton:

$$a(\xi, \zeta) = \text{sech}(\xi) e^{i\zeta/2}, \tag{6.38}$$

where sech is the hyperbolic secant. It still depends on z, but only in phase, and therefore, the shape of the field will not change during propagation. Figure 6.28 shows the first-order soliton $(N = 1)$.

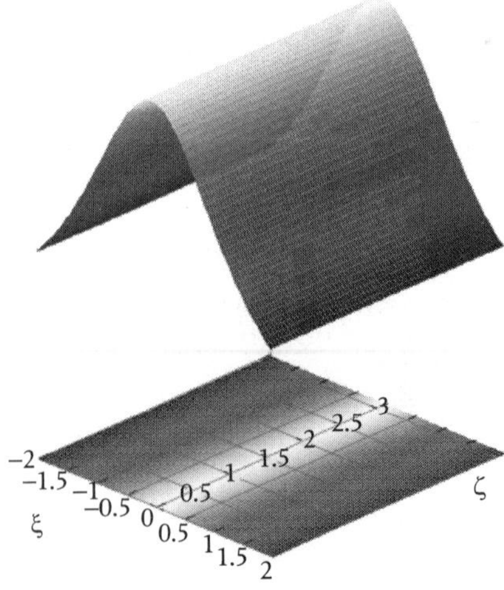

Fig. 6.28 Soliton's shape while propagating with $N = 1$ and it does not change its shape

For $N = 2$, it is still possible to express the solution in a closed form, but it has a more complicated form:

$$a(\xi,\zeta) = \frac{4\left[\cosh(3\xi) + 3e^{4i\zeta}\cosh(\xi)\right]e^{i\zeta/2}}{\cosh(4\xi) + 4\cosh(2\xi) + 3\cos(4\zeta)}. \tag{6.39}$$

It does change its shape during propagation, but it is a periodic function of z with period $\zeta = \pi/2$. Figure 6.29 shows the second-order soliton $(N = 2)$.

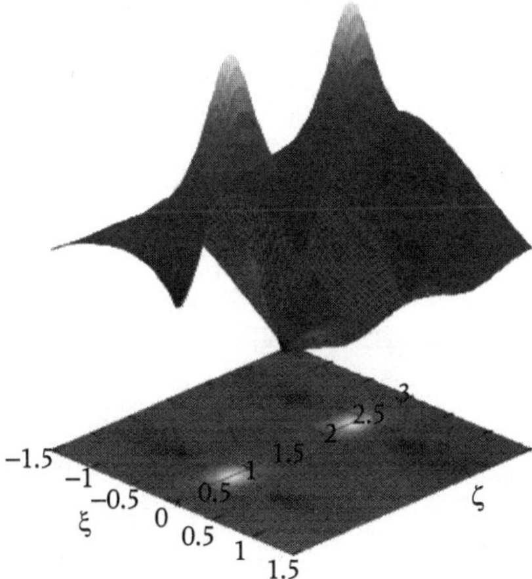

Fig. 6.29 Soliton's shape while propagating with $N = 2$, and it changes its shape periodically

For soliton solutions, N must be an integer and it is said to be the order of the soliton. For higher values of N, there are no closed-form expressions, but the solitons exist and they are all periodic with different periods. Their shape can easily be expressed only immediately after generation:

$$a(\xi,\zeta = 0) = \text{N}\operatorname{sech}(\xi). \tag{6.40}$$

On the right side of Fig. 6.29 is the second-order soliton: at the beginning, it has a shape of a sech, then the maximum amplitude increases and then returns to the sech shape. Since high intensity is necessary to generate solitons, if the field increases its intensity even further, the medium could be damaged.

The condition to be solved if we want to generate a fundamental soliton, which is obtained by expressing N in terms of all the known parameters and then substituting $N = 1$:

$$1 = \text{N} = \frac{L_d}{L_{nl}} = \frac{X_0^2 k_0^2 n^2 |n_2| |A_m|^2}{2\eta_0}. \tag{6.41}$$

In other words, in terms of maximum intensity, value becomes:

$$I_{\max} = \frac{|A_m|^2}{2\eta_0 / n} = \frac{1}{X_0^2 k_0^2 n |n_2|} \tag{6.42}$$

In most of the cases, the two variables that can be changed are the maximum intensity $I_{\max}$ and the pulse width X_0.

6.11 TEMPORAL SOLITONS

If the electromagnetic field is already spatially confined, it is possible to send pulses that will not change their shape because the nonlinear effects will balance the dispersion. Those solitons were discovered first and they are often simply referred as "solitons" in optics.

The main problem that limits transmission bit rate in optical fibres is group velocity dispersion. It is because generated impulses have a nonzero bandwidth and the medium they are propagating through has a refractive index that depends on frequency (or wavelength). This effect is represented by the group delay dispersion parameter D; using it, it is possible to calculate exactly how much the pulse will widen.

$$\Delta\tau \approx DL\Delta\lambda , \tag{6.43}$$

where L is the length of the fibre and $\Delta\lambda$ is the bandwidth in terms of wavelength.

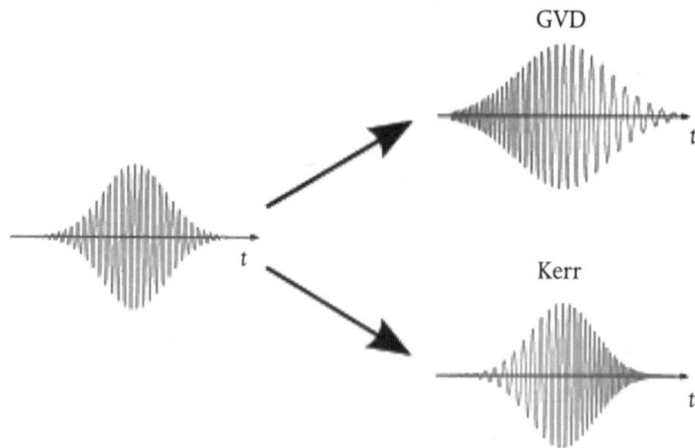

Fig. 6.30 Envelope of the field oscillating at a defined frequency

The approach in modern communication systems is to balance such dispersion with other fibres having D with different signs in different parts of the fibre; this way the pulses keep on broadening and shrinking while propagating. With temporal solitons, it is possible to remove such a problem completely.

Consider the picture shown in Fig. 6.30, there is a standard Gaussian pulse, that is, the envelope of the field oscillating at a defined frequency. We assume that the frequency remains perfectly constant during the pulse.

6.12 LINEAR AND NONLINEAR EFFECTS ON GAUSSIAN PULSES

Now, we let this pulse propagate through a fibre with $D > 0$, it will be affected by group velocity dispersion. For this sign of D, the dispersion is anomalous so that the higher frequency components will propagate a little bit faster than the lower frequencies, thus arriving before at the end of the fibre. The overall signal we get is a wider chirped pulse, as shown in Fig. 6.31.

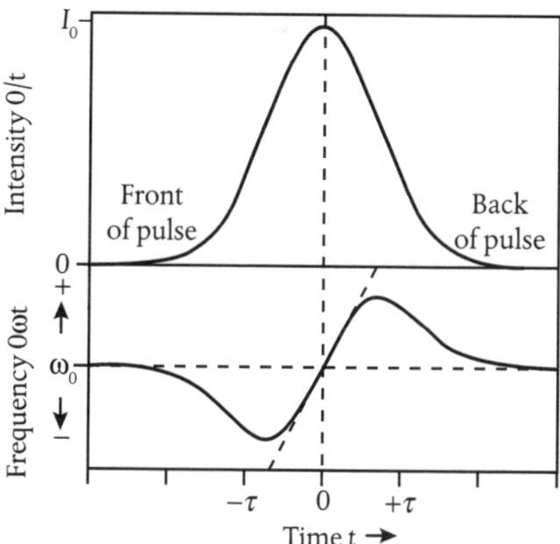

Fig. 6.31 Linear and nonlinear effects on Gaussian pulses

6.13 EFFECT OF SPM ON FREQUENCY

Now, let us assume we have a medium that shows only nonlinear Kerr effect but its refractive index does not depend on frequency; however, such a medium does not exist, but it is worth considering it to understand the different effects. The phase of the field is given as:

$$\varphi(t)=\omega_o t-\kappa_0 z=\omega_o t-\kappa z\left[n+n_2 I(t)\right]. \tag{6.44}$$

The frequency (according to its definition) is given as:

$$\omega(t)=\frac{\partial\varphi(t)}{\partial t}=\omega_o-\kappa_0 z n_2\frac{\partial I(t)}{\partial t}. \tag{6.45}$$

At the beginning of the pulse, the frequency is lower; at the end, it is higher. After the propagation through ideal medium, we will get a chirped pulse with no broadening because we have neglected dispersion. From the first picture in Fig. 6.31, we see that the two effects introduce a change in frequency in two different opposite directions. It is possible to make a pulse so that the two effects will balance each other. Considering higher frequencies, linear dispersion will tend to let them propagate faster, while nonlinear Kerr effect will slow them down. The overall effect will be that the pulse does not change while propagating and such pulses are called temporal solitons.

An electric field is propagating in a medium showing optical Kerr effect through a guiding structure (such as an optical fibre) that limits the power on the xy-plane. If the field is propagating towards z with a phase constant β_0, then it can be expressed in the following form;

$$E(r,t)=A_m a(t,z)f(x,y)e^{i(\beta_0 z-\omega_0 t)}, \tag{6.46}$$

where A_m is the maximum amplitude of the field and $a(t,z)$ is the envelope that shapes the impulse in the time domain; in general, it depends on z because the impulse can change its shape while propagating; $f(x,y)$ represents the shape of the field on the xy-plane and it does not change during propagation because we have assumed the field is guided. Both a and f are normalised dimensionless functions whose maximum value is 1 so that A_m really represents the field amplitude. The relationship between the electric field and its polarisation is given by a convolution integral. Anyway, using a representation in the Fourier domain, we can replace the convolution with a simple product, thus using standard relationships that are valid in simpler media. We Fourier transform the electric field using the following definition:

$$\tilde{E}(r,\omega-\omega_0)=\int_{-\infty}^{\infty}E(r,t)e^{-i(\omega-\omega_0)}tdt. \tag{6.47}$$

Using this definition, a derivative in the time domain corresponds to a product in the Fourier domain:

$$\frac{\partial}{\partial t}\tilde{E}\Leftrightarrow -i(\omega-\omega_0)\tilde{E}. \tag{6.48}$$

The complete expression of the field in the frequency domain is:

$$\tilde{E}(r,\omega-\omega_0)=A_m\tilde{a}(\omega-\omega_0,z)f(x,y)e^{i\beta_0 z}. \tag{6.49}$$

Now, we can solve Helmholtz equation in the frequency domain:

$$\nabla^2\tilde{E}+n^2(\omega)k_0^2\tilde{E}=0. \tag{6.50}$$

We decide to express the phase constant with the following notation:

$$n(\omega)k_0=\beta(\omega)=\beta_0+\beta_l(\omega)+\beta_{nl}=\beta_0+\Delta\beta(\omega), \tag{6.51}$$

where we assume that $\Delta\beta$ (the sum of the linear dispersive component and the nonlinear part) is a small perturbation, that is, $|\beta_0|\gg|\Delta\beta(\omega)|$. The phase constant can have any complicated behaviour, but we can represent it with a Taylor series centred on ω_0:

$$\beta(\omega)\approx\beta_0+(\omega-\omega_0)\beta_1+\frac{(\omega-\omega_0)^2}{2}\beta_2+\beta_{nl}, \tag{6.52}$$

where:

$$\beta_u=\left.\frac{d^u\beta(\omega)}{d\omega^u}\right|_{\omega=\omega_0}. \tag{6.53}$$

We substitute the expression of the electric field in the equation and make some calculations. If we assume the slowly varying envelope approximation:

$$\frac{\partial^2\tilde{a}}{\partial z^2}\ll\left|\beta_0\frac{\partial\tilde{a}}{\partial z}\right|. \tag{6.54}$$

We get, for (6.50):

$$2i\beta_0\frac{\partial\tilde{a}}{\partial z}+\left[\beta^2(\omega)-\beta_0^2\right]\tilde{a}=0. \tag{6.55}$$

We are ignoring the behaviour in the *xy*-plane because it is already known and given by $f(x,y)$. We make a small approximation, as we did for the spatial soliton:

$$\beta^2(\omega)-\beta_0^2=\left[\beta(\omega)-\beta_0\right]\left[\beta(\omega)+\beta_0\right]=\left[\beta_0+\Delta\beta(\omega)-\beta_0\right]\left[2\beta_0+\Delta\beta(\omega)\right]\approx 2\beta_0\Delta\beta(\omega).$$

Replacing this in (6.55), we get simply:

$$i\frac{\partial\tilde{a}}{\partial z}+\Delta\beta(\omega)\tilde{a}=0. \tag{6.56}$$

Now, we want to return in the time domain. Expressing the products by derivatives, we get the duality:

$$\Delta\beta(\omega) \Leftrightarrow i\beta_1 \frac{\partial}{\partial t} - \frac{\beta_2}{2}\frac{\partial^2}{\partial t^2} + \beta_{nl.} \tag{6.57}$$

We can write the nonlinear component in terms of the amplitude of the field:

$$\beta_{nl} = k_0 n_2 I = k_0 n_2 \frac{|E|^2}{2\eta_0 / n} = k_0 n_2 n \frac{|A_m|^2}{2\eta_0}|a|^2. \tag{6.58}$$

For duality with the spatial soliton, we define:

$$L_{nl} = \frac{2\eta_0}{k_0 n n_2 |A_m|^2} \tag{6.59}$$

and this symbol has the same meaning of the previous case, even if the context is different. Equation (6.56) becomes

$$i\frac{\partial a}{\partial z} + i\beta_1 \frac{\partial a}{\partial t} - \frac{\beta_2}{2}\frac{\partial^2 a}{\partial t^2} + \frac{1}{L_{nl}}|a|^2 a = 0. \tag{6.60}$$

We know that the impulse is propagating among the *z*-axis with a group velocity given by $v_g = 1/\beta_1$, and therefore, we are not interested in it because we just want to know how the pulse changes its shape while propagating. We decide to study the impulse shape, that is, the envelope function $a(.)$ using a reference that is moving with the field at the same velocity. Thus, we make the substitution:

$$T = t - \beta_1 z\ , \tag{6.61}$$

and (6.60) becomes:

$$i\frac{\partial a}{\partial z} - \frac{\beta_2}{2}\frac{\partial^2 a}{\partial T^2} + \frac{1}{L_{nl}}|a|^2 a = 0. \tag{6.62}$$

We assume the medium where the field is propagating to show anomalous dispersion, that is, $\beta_2 < 0$ or in terms of the group delay dispersion parameter $= \frac{-2\pi c}{\lambda^2};\ \beta_2 > 0$. We make this more evident by replacing $\beta_2 = -|\beta_2|$ in the equation . Now, let us define the following parameters (the duality with the previous case is evident):

$$L_d = \frac{T_0^2}{|\beta_2|};\ \tau = \frac{T}{T_0}\ ; \zeta = \frac{z}{L_d};\ N^2 = \frac{L_d}{L_{nl}}. \tag{6.63}$$

Replacing those in (6.62), we get:

$$\frac{1}{2}\frac{\partial^2 a}{\partial \tau^2} + i\frac{\partial a}{\partial \zeta} + N^2 |a|^2 a = 0. \tag{6.64}$$

In other words, it is exactly the same equation we have obtained in the previous case. The first-order soliton is given as:

$$a(\tau, \zeta) = \operatorname{sech}(\tau) e^{i\zeta/2}. \tag{6.65}$$

The same considerations we have made are valid in this case. The condition $N = 1$ becomes a condition on the amplitude of the electric field:

$$|A_m|^2 = \frac{2\eta_0 |\beta_2|}{T_0^2 n_2 k_0 n}. \tag{6.66}$$

or in terms of intensity:

$$I_{max} = \frac{|A_m|^2}{2\eta_0 / n} = \frac{|\beta_2|}{T_0^2 n_2 k_0}. \tag{6.67}$$

or we can express it in terms of power if we introduce an effective area A_{eff} defined so that $P = IA_{eff}$.

$$P = \frac{|\beta_2| A_{eff}}{T_0^2 n_2 k_0}. \tag{6.68}$$

6.14 STABILITY OF SOLITONS

We have described what optical solitons are, and using mathematics, we have seen that if we want to create them, we have to create a field with a particular shape (just such for the first order) with a particular power related to the duration of the impulse. However, what if we are slightly wrong in creating such impulses. Adding small perturbations to the equations and solving them numerically, it is possible to show that mono-dimensional solitons are stable. They are often referred as (1 + 1) *D* solitons, meaning that they are limited in one dimension (*x* or *t*, as we have seen) and propagate in another one (*z*).

If we create such a soliton using slightly wrong power or shape, then it will adjust itself until it reaches the standard such shape with the right power. Unfortunately, this is achieved at the expense of some power loss, which can cause problems because it can generate another non-soliton field propagating together with the field we want. Mono-dimensional solitons are very stable; for example, if $0.5 < N < 1.5$, we will generate a first-order soliton anyway; if N is greater, we will generate a higher order soliton, but the focusing it does while propagating may cause high-power peaks damaging the media. The only way to create a $(1+1)D$ spatial soliton is to limit the field on the *y*-axis using a dielectric slab, then limiting the field on *x* using the soliton.

On the other hand, $(2+1)D$ spatial solitons are unstable; therefore, any small perturbation (e.g., due to noise) can cause the soliton to diffract as a field in a linear medium or to collapse, thus damaging the material. It is possible to create stable $(2+1)D$ spatial solitons using saturating nonlinear media, where the Kerr relationship $n(I) = n + n_2 I$ is valid until it reaches a maximum value. Working close to this saturation level makes it possible to create a stable soliton in a three-dimensional space.

If we consider the propagation of shorter (temporal) light pulses or over a longer distance, we need to consider higher order corrections, and therefore, the pulse carrier envelope is governed by the higher order nonlinear Schrödinger equation (HONSE) for which there are some specialised (analytical) soliton solutions.

6.15 EFFECT OF POWER LOSSES

As we have seen, in order to create a soliton, it is necessary to have the right power when it is generated. If there are no losses in the medium, then we know that the soliton will keep on propagating forever without changing shape (first order) or changing its shape periodically (higher orders). Unfortunately, any medium introduces losses, and therefore, the actual behaviour of power will be in the form:

$$P(z) = P_0 e^{-\alpha z}. \tag{6.69}$$

This is a serious problem for temporal solitons propagating in fibres for several kilometres. Let us consider what happens for the temporal soliton; generalisation to the spatial ones is immediate. We have proved that the relationship between power P_0 and impulse length T_0 is:

$$P = \frac{|\beta_2| A_{\text{eff}}}{T_0^2 n_2 k_0}. \tag{6.70}$$

If the power changes, the only thing that can change in the second part of the relationship is T_0. If we add losses to the power and solve the relationship in terms of T_0, we get:

$$T(z) = T_0 e^{(\alpha/2)z}. \tag{6.71}$$

The width of the impulse grows exponentially to balance the losses. This relationship is true as long as the soliton exists, that is, until this perturbation is small; therefore, it must be $\alpha z \ll 1$, otherwise we cannot use the equations for solitons and we have to study standard linear dispersion. If we want to create a transmission system using optical fibres and solitons, we have to add optical amplifiers in order to limit the loss of power.

6.15.1 Information Transmission with Solitons

The basic idea is to use a soliton in each bit slot representing 1 in a bit stream as shown in Fig. 6.32. The neighbouring solitons in this scheme should be well separated, and therefore, the spacing between two solitons exceeds a few times their FWHM. The soliton width T_0 is related to the bit rate B as:

$$B = \frac{1}{T_B} = \frac{1}{2q_0 T_0}. \tag{6.72}$$

T_B is the duration of the bit slot and $2q_0 = T_B / T_0$ is the distance between neighbouring solitons in normalised units.

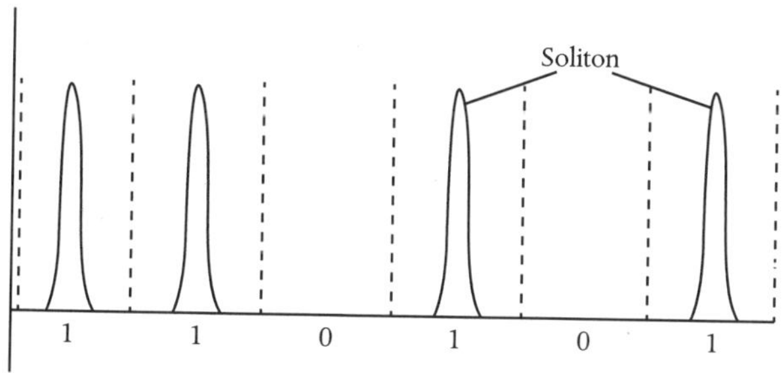

Fig. 6.32 Information Transmission with Solitons

6.16 SOLITON BIT STREAM IN RZ FORMAT SOLITON INTERACTION

The presence of solitons in the neighbouring bits perturbs a soliton simply because the combined optical field is not a solution to the NLSE. Neighbouring solitons either come closer together or move apart because of nonlinear interaction between them, which introduces error in the data. However, relatively large soliton spacing, necessary to avoid soliton interaction, limits the bit rate of soliton communication system. The phenomenon of soliton interaction has been studied extensively. It has been found that this interaction can be reduced by using unequal amplitudes for neighbouring solitons. This interaction can be modified by other factors like initial frequency chirp imposed on input pulse. Loss-managed solitons SPM effect is power dependent. As the pulse propagates through the fibre, it loses power due to fibre losses, thereby weakening the SPM effect. As a result, the SPM effect is not strong enough to counteract dispersion, which results in broadening of the pulse. This brings the need for amplifiers in the system. There are two kinds of amplification techniques: lumped and distributed amplification techniques.

6.16.1 Lumped Amplification Technique

In the lumped amplification scheme, optical amplifiers are placed periodically along the fibre link such that the fibre losses between two amplifiers are exactly compensated by the amplifier gain. An important design parameter is the amplifier spacing. It should be as large as possible to minimise the cost. For non-soliton systems, it is typically 80–100 km. For soliton systems using Lumped amplification scheme, amplifier spacing is restricted to much smaller values.

The reason is that optical amplifiers can boost soliton energy to the input level over a length of few metres without allowing for gradual recovery of fundamental soliton. The amplified soliton adjusts its width dynamically in the fibre section following the amplifier. However, it also sheds a part of its energy as dispersive waves during this adjustment phase, which can accumulate to significant levels over a large number of amplifier stages and must be avoided. It can be shown that this effect can be minimised by keeping the amplifier spacing "LA" much smaller than the dispersion length "LD" of the pulse.

6.16.2 Distributed Amplification

The condition $L_A << L_D$, imposed on lumped amplified soliton systems, becomes increasingly difficult to satisfy in practice as bit rates exceed 10 Gbps. This condition can be relaxed considerably when distributed amplification scheme is used. In this technique, signal is amplified locally at every point along the fibre length.

6.17 GENERATION OF SOLITON PULSE

Experiments has been carried out to analyse the effect of high-frequency (20 MHz to 1 GHz) external magnetic field-induced nonlinear Kerr effect on single-mode optical fibre of considerable length (50–100 m) to compensate group velocity dispersion (GVD) and subsequent evolution of soliton pulse (peak energy, narrow, secant hyperbolic pulse). The generation of soliton pulse in fibre is an obvious conclusion as SPM due to high energy of pulse offset GVD, whereas the evolution length is 2000 km (the laser wavelength chosen is greater than $1.3\ \mu m$).

Moreover, peak soliton pulse is of period $1-3 ps$ so that it is safely accommodated in the optical bandwidth. Once soliton pulse is generated, it is least dispersed over thousands of kilometres length of fibre limiting the number of repeater station.

6.17.1 Dark Solitons

In the analysis of both types of solitons, we have assumed particular conditions about its medium:

1. In spatial solitons, $n_2 > 0$; this means the SPM causes self focusing.
2. In temporal solitons, $\beta_2 \langle 0 \text{ or } D \rangle 0$; this means anomalous dispersion.

Is it possible to obtain solitons if those conditions are not verified. If we assume $n_2 \langle 0 \text{ or } \beta_2 \rangle 0$, we get the following differential equation (it has the same form in both cases; we will use only the notation of the temporal soliton):

$$-\frac{1}{2}\frac{\partial^2 a}{\partial \tau^2}+\frac{\partial a}{\partial \zeta}+N^2|a|^2 a=0\,. \tag{6.73}$$

This equation has soliton-like solutions. For the first-order $(N=1)$:

$$a(\tau,\zeta)=\tanh(\tau)e^{i\zeta}. \tag{6.74}$$

6.17.2 Power of a Dark Soliton

For higher order solitons $(N>1)$, we can use the following closed form expression:

$$a(\tau,\zeta=0)=N\tanh(\tau)e^{i\zeta}. \tag{6.75}$$

It is a soliton, in the sense that it propagates without changing its shape, but it is not made by a normal pulse; rather, it is a lack of energy in a continuous time beam. The intensity is constant, but for a short time during which it jumps to zero and back again, thus generating a dark pulse. Those solitons can actually be generated introducing short dark pulses in much longer standard pulses. Dark solitons are more difficult to handle than standard solitons, but they have shown to be more stable and robust to losses.

Soliton has promised a lot of potentiality since its theory was put forward. Seven years elapsed between the discovery of optical solitons in fibres and its experimental confirmation. However, attempts for its practical implementation have been a real challenge for the soliton community for more than 20 years because of the ever-increasing demands for higher bit rate and longer distance transmission as well as the successful improvement of linear transmission quality owing to the continuous improvements in fibres and amplifiers. The potential still exists and poses a challenge to the soliton community.

SUMMARY

- There are two categories of nonlinear effects. The first arises due to the interaction lightwaves with phonons (molecular vibrations) in the silica medium from one of the several types of scattering effects, namely Rayleigh scattering. The two main effects in this category are Stimulated Brillouin Scattering (SBS) and Stimulated Raman Scattering (SRS).
- The second set of nonlinear effects in this category arises due to the dependence of the refractive index of the applied electric field, which in turn is proportional to the square of the field amplitude. The most important nonlinear effects in this category are SPM and FWM.
- In the case of SBS, the pump wave is the signal wave, and the Stokes wave is the unwanted wave that is generated due to the scattering process. In the case of SRS, the pump wave is a high-power wave, and the Stokes wave is the signal wave that gets amplified at the expense of the pump wave.
- In a WDM system with multiple channels, the induced chirp in one channel depends on the variation of the refractive index with the intensity on the other channels. This effect is called cross-phase modulation (CPM or XPM).
- Brillouin scattering is an effect caused by the χ-nonlinearity of a medium, specifically by that part of the nonlinearity, which is related to acoustic phonons. An incident photon can be converted

into a scattered photon of slightly lower energy, usually propagating in the backward direction and a phonon.

- For pure backward Brillouin scattering, the Brillouin shift can be calculated from the refractive index n, the acoustic velocity v_a, and the vacuum wavelength λ. (For Brillouin scattering in fibres, the effective refractive index must be used.)

$$v_B = \frac{2nv_a}{\lambda}$$

- Both SRS and SBS can be used to advantage while designing optical communication systems because they can amplify an optical signal by transferring energy to it from a pump beam whose wavelength is suitably chosen.
- The SRS is especially useful because of its extremely large bandwidth. Indeed, the Raman gain is used routinely for compensating fibre losses in modern lightwave systems.
- For linear medium, the following shows the relation between the Fourier transform of the induced dielectric polarisation and applied electric field, respectively.

$$P(r,\omega) = \varepsilon_0 \chi(r,\omega) E(r,\omega)$$

- The FWM can also be useful in designing lightwave systems. It is often used for demultiplexing individual channels when time division multiplexing is used in the optical domain. It can also be used for wavelength conversion. The FWM in optical fibres is sometimes used for generating a spectrally inverted signal through the process of optical phase conjugation.
- Although DSF overcomes the chromatic dispersion in the 1.55-μm window; unfortunately, it is not useful to the WDM communication system because of severe penalties due to FWM and other nonlinearities. These penalties can be overcome if there is little chromatic dispersion present in the fibre because different waves travel with different group velocities. This led to the development of nonzero dispersion optical fibres.
- The effects of nonlinearities can be reduced by designing a fibre with large effective areas. We have seen that nonzero dispersion fibres have a small value of chromatic dispersion in the 1.55-μm band to minimise the effects of chromatic dispersion.
- Fibres are designed to have either positive chromatic dispersion or negative chromatic dispersion in 1.55-μm window. Positive chromatic dispersion system is used for terrestrial systems, while negative chromatic dispersion fibres are used for submarine system.
- An NOLM employs a 3-dB directional coupler connected to a fibre loop acting as a nonlinear medium. A data pulse sent into the loop is split by the coupler into two equal intensity counter propagating pulses, which will recombine at the coupler and emerge out of the input port.
- To eliminate the long latency problem of NOLMs, semiconductors with large resonant nonlinearity have been used to replace fibre as the nonlinear medium. This also brings other advantages, such as improved stability, reduced switching energy, and the possibility of integration.
- The chirp of an optical pulse is usually understood as the time dependence of its instantaneous frequency. Specifically, an up-chirp (down-chirp) means that the instantaneous frequency raises (decreases) with time.
- Pairs of (typically Brewster angled) prisms can be used for introducing anomalous chromatic dispersion into a laser resonator without introducing significant power losses.
- A chirped mirror is a kind of dielectric mirror, which is usually used for dispersion compensation in mode-locked lasers. Another interesting feature of chirped mirrors is that they make it possible to achieve a broader reflection bandwidth than ordinary Bragg mirrors.

- A fundamental soliton is an optical pulse that can propagate in a dispersive medium (e.g., an optical fibre) with a constant shape of the temporal intensity profile, that is, without any temporal broadening as is usually caused by dispersion. This can happen when the pulse has a certain shape and an energy, which is determined by the parameters of the medium (in particular, by the dispersion and nonlinearity) and the pulse duration.
- The soliton period equals $(\pi/2)Z_0$ where Z_0 is the dispersion length defined in terms of the pulse width, FWHM, and GVD parameter $\beta_2\left(\text{ps}^2/\text{km}\right)$ as

$$Z_0 = T_2\,\text{FWHM}/3.11|\beta_2|$$

- Adiabatic soliton compression is a technique for the temporal compression of ultrashort pulses in a fibre.
- A parabolic pulse is an optical pulse (usually an ultrashort pulse) that has a temporal intensity profile with a parabolic shape not only near the pulse centre but also well towards the wings up to the point where the intensity goes to zero. In the ideal case, the pulse would have an intensity profile according to

$$P(t)=\begin{cases} P_0\left(1-(t/\tau)^2\right) & \text{for } |t|<\tau \\ 0 & \text{otherwise} \end{cases}$$

- The Kerr effect, also called the quadratic electro-optic effect (QEO effect), is a change in the refractive index of a material in response to an applied electric field. The Kerr effect is the induced index change directly proportional to the square of the electric field.
- Dispersion is the spreading of a light pulse in time as it propagates down the fibre. Dispersion is the phenomenon in which the phase velocity of a wave depends on its frequency or alternatively when the group velocity depends on the frequency.
- The Kerr effect introduces an SPM that changes the refractive index according to the intensity:

$$\varphi(x)=k_o n(x)L=k_0 L\left[n+n_2 I(x)\right]$$

- The relationship between intensity and electric field is (in the complex representation):

$$I=\frac{[E]^2}{2\eta},$$

where $\eta=\eta_0/n$ and η_0 is the impedance of free space, and it is given as:

$$\eta_0=\sqrt{\frac{\mu_0}{\varepsilon_0}}\approx 377\Omega.$$

- In the analysis of both types of solitons, we have assumed particular conditions about its medium: In spatial solitons, $n_2>0$ means the SPM causes self focusing.
In temporal solitons, $\beta_2\langle 0 \text{ or } D\rangle 0$ means anomalous dispersion.
- Optical fibres exhibit a variety of nonlinear effects. Fibre nonlinearities are feared by telecom system designers because they can affect system performance adversely. Fibre nonlinearities can be managed thorough proper system design.
- Nonlinear effects are useful for many devices and system applications: optical switching, soliton formation, wavelength conversion, broadband amplification, channel demultiplexing, etc.
- New kinds of fibres have been developed for enhancing nonlinear effects (micro-structured fibres with air holes). Nonlinear effects in such fibres are finding new applications.

THEORY QUESTIONS AND ANSWERS

Question 1: Write down short notes on nonlinear scattering in fibres.

Answer: Optical waveguides do not always behave as complete linear channels whose increase in output optical power is directly proportional to the input optical power. Several nonlinear effects occur; due to this scattering causes disproportionate attenuation usually at high optical power levels. This nonlinear scattering causes the transfer of optical power from one mode, either in the forward or backward direction, to the same or other modes at a different frequency. It depends critically upon the optical power density within the fibre, and hence, only becomes significant above the threshold power.

The most important types of nonlinear scattering within optical fibres are:

1. Stimulated Brillouin scattering
2. Stimulated Raman scattering

Both are usually observed at high optical power densities in long SMFs. These scattering mechanisms in fact give optical gain but with a shift in frequency, thus contributing to attenuation for light transmission at a specific wavelength. However, it may be noted that such nonlinear phenomena can also be used to give optical amplification in the context of integrated optical techniques.

1. **Stimulated Brillouin scattering (SBS)**
   Stimulated Brillouin scattering (SBS) may be regarded as the modulation of light through thermal molecular vibrations within the fibre. The scattered light appears as thermal molecular vibrations within the fibre. The scattered light appears as upper and lower sidebands that are separated from the incident light by the modulation frequency. The incident photon in this scattering process produces a phonon of acoustic frequency as well as a scattered photon. This produces an optical frequency shift that varies with the scattering angle because the frequency of the sound wave varies with acoustic wavelength. The frequency shift is maximum in the backward direction reducing to zero in the forward direction making SBS a mainly backward process.

   Brillouin scattering is only significant above a threshold power density. Assuming that the polarisation state of the transmitted light is not maintained, it may be shown that the threshold power P_B is given as:

   $$P_B = 4.4 \times 10^{-3} d^2 \lambda^2 A v,$$

   where

   $d =$ fibre core diameter in micrometre

   $\lambda =$ operating wavelength in micrometer

   $A(\text{dB}) =$ fibre attenuation in decibels per kilometre

   $v =$ source bandwidth (i.e., injection laser) in GHz

   The above-mentioned expression allows the determination of the threshold optical power that must be launched into a single-mode optical fibre before SBS occurs.

2. **Stimulated Raman scattering (SRS)**
   It is similar to stimulated Brillouin scattering except that a high-frequency optical phonon rather than an acoustic phonon is generated in the scattering process. Further, SRS can occur in both the forward and backward directions in an optical fibre and may have an optical power threshold of up to three orders of magnitude higher than that of the Brillouin threshold in a particular fibre.

Using the same criteria as those specified for the Brillouin scattering threshold given in the following equation, it may be shown that the threshold optical power for SRS, P_R, in a long SMF is given as:

$$P_R = 5.9 \times 10^{-2} d^2 \lambda A,$$

where d, λ, and A (dB) are same as specified earlier.

The Brillouin threshold occurs at an optical power level of around 65 MW, whilst the Raman threshold is approximately 15 times larger. It is, therefore, apparent that the losses introduced by nonlinear scattering may be avoided by the use of a suitable optical signal level (i.e., working below the threshold optical powers). However, it must be noted that the Brillouin threshold can occur at optical power as low as 10 mW in SMFs. Nevertheless, this is still a high-power level for optical communication and may be easily avoided. The SBS and SRS are not usually observed in multimode fibres because their relatively large-core diameters make the threshold optical power levels extremely high. Moreover, it should be noted that the threshold optical powers for both these scattering mechanisms may be increased by the suitable adjustment of the other parameters. Operation at the longest possible wavelength is advantageous although this may be offset by the reduced fibre attenuation (from Rayleigh scattering and material absorption) normally obtained.

Question 2: Discuss in detail: nonlinear effects in optical fibre communication system.

Answer: In an optical system, obtaining the maximum transmission rate is possible by merging TDM and WDM concepts as it depends on temporal and spectral characteristics of light. However, TDM has limiting factors in terms of chromatic and polarisation mode dispersion, whereas WDM has limiting factors in terms of the nonlinear effects for the transmission performances in the fibre. Optimised maximum transmission capacity depends on a few factors such as dispersion, signal power, and fibre length.

Nonlinearity in optical fibre is caused due to high intensity of light in the core. There are two possibilities for the occurrence of nonlinear effects in optical fibre. They are due to inelastic scattering phenomenon or due to change in the refractive index of the medium related with intensity of light. We summarise the nonlinear effects in optical fibre communication system as follows:

Fibre nonlinearities

1. Studied during the 1970s.
2. Ignored during the 1980s.
3. Feared during the 1990s.
4. Are being used in this decade

Major Nonlinear Effects

1. Self-phase modulation (SPM)
2. Cross-phase modulation (XPM)
3. Four-wave mixing (FWM)
4. Stimulated Raman scattering (SRS)
5. Stimulated Brillouin Scattering (SBS)

Origin of Nonlinear Effects in Optical Fibres

1. Ultrafast third order susceptibility $\chi^{(3)}$.
2. Real part leads to SPM, XPM, and FWM.
3. Imaginary part leads to SBS and SRS.

Kerr nonlinearity further gives rise to three different effects based on input signal power, as shown in Fig. P1.

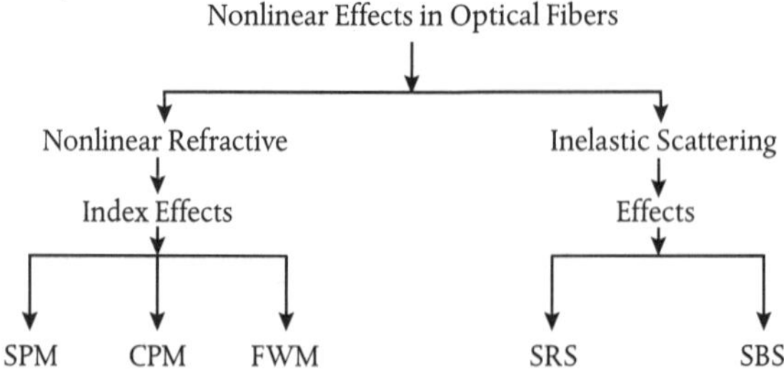

Fig. P1 Types of nonlinear effects

Nonlinear effects have their own advantages and disadvantages. These effects limit transmission capacity and can be overcome by using large core fibres, reverse dispersion fibres, etc. On the other side, these effects can be used to improve the performance of the system by wavelength conversion. Nonlinearity in the refractive index is known as Kerr nonlinearity, producing a carrier-induced phase modulation of the propagating signal, which is called the Kerr effect. Refractive index is a function of electric field E denoted by $n(E)$. The function of refractive index is evaluated by Taylor's series, which is given as:

$$n(E)=n+\frac{dn}{dE}+\frac{1}{2}\frac{d^2n}{dE^2}E^2+\ldots$$

For symmetry material, first-order term of E is zero as refractive index is an even function. Second order is considered for studying the Kerr effect and given by an expression:

$$n(E)=n-\frac{n^3\xi}{2}E^2,$$

where Kerr coefficient is given as $\xi=\frac{1}{n^3}\frac{d^2n}{dE^2}$; $n=$ refractive index of the medium. The typical value of Kerr coefficient is 10^{-18} to $10^{-14}\frac{m^2}{V^2}$.

Question 3: Discuss in detail the SPM in nonlinear optics.

Answer: Nonlinear effect depends on intensity of light and refractive index. Input pulse travels through fibre, which has high intensity of light inside core resulting in higher refractive index. Signal intensity changes with respect to time leads to variations in refractive index with time, which is similar to intensity-dependent refractive index. These variations in refractive index resulting in time-dependent phase changes. These phase changes are the same as optical signal change with time, and therefore, the name self-phase modulation (SPM). Different parts of the input pulse change the phase of the signal randomly, resulting a frequency chirp, which is defined as a signal whose frequency varies with time (increases up-chirp or decreases down-chirp). These variations in frequencies cause pulse broadening, which can be seen significantly high in the systems with high transmission power because the transmission power is directly dependent on frequency chirp. Phase shift by field over fibre length is given as:

$$\phi=\frac{2\pi nL}{\lambda},$$

where $n=$ refractive index of the medium, $L=$ length of the fibre, $\lambda=$ wavelength of the optical pulse, and $nL=$ optical path length.

In SPM, pulse broadening is seen in the time domain and spectral characteristics are unaltered. Chirp produced by SPM is used to reduce the effects of dispersion caused by pulse broadening. These effects depend mostly on input power of the signal transmitted, which can be used as a threshold condition for the frequency chirp to occur. The SPM is major limitations in single-channel systems.

SPM: Good or Bad

1. The SPM-induced spectral broadening can degrade performance of a lightwave system.
2. Modulation instability often enhances system noise.
3. Modulation instability can be used to produce ultrashort pulses at high repetition rates.
4. The SPM often used for fast optical switching (NOLM or MZI).
5. Formation of standard and dispersion managed optical solitons.
6. Useful for all-optical regeneration of WDM channels.
7. Other applications (pulse compression, chirped pulse amplification, passive mode locking, etc.).

Applications of SPM

Applications of SPM are solitons and pulse compression. A soliton is a short duration pulse that does not change during the propagation because of cancelling linear and nonlinear effects. In SPM, a frequency chirp generated from linear dispersion in which the leading edge has lower frequencies and trailing edge has higher frequencies and vice versa for chirp generated from nonlinear dispersion. There is a possibility to compensate these effects by varying shape and power of the pulse.

Pulse compression is seen when chromatic dispersion is positive and the leading edge of the pulse travels slowly and narrows down the gap between the leading edge and centre. It is the same with the trailing edge; but here, instead of moving slowly it moves faster and comes closer to centre of the pulse.

Question 4: What do you understand by optical solitons and its stability?

Answer: Combination of SPM and anomalous GVD produces solitons. Solitons preserve their shape in spite of the dispersive and nonlinear effects occurring inside fibres. They are useful for optical communications systems. Dispersive and nonlinear effects are balanced when two lengths become equal. $L_{NL} = L_D$.

Stability of Optical Solitons

Solitons are remarkably stable. Fundamental solitons can be excited with any pulse shape.

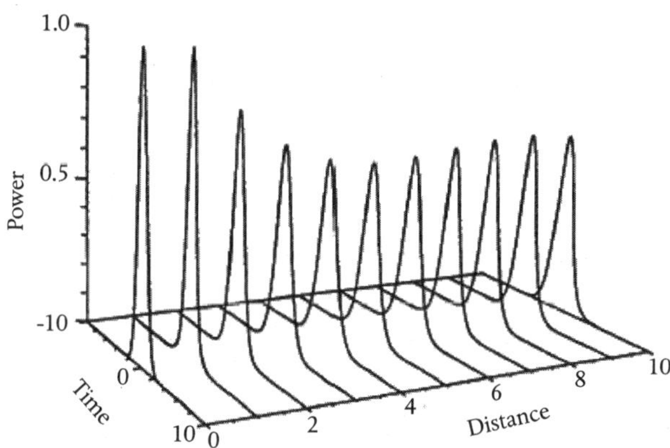

Fig. P2 Soliton pulse generation

It can be interpreted as temporal modes of an SPM-induced waveguide. Some pulse energy is lost through dispersive waves. Figure P2 shows the soliton pulse generation.

Question 5: Discuss the cross-phase modulation (CPM) in nonlinear optics.

Answer: Cross-phase modulation is similar nonlinear effect to SPM except it occurs when there are two or more optical signals propagate through fibre. The distortion of signal and pulse broadening will be asymmetric because of more than one signal is propagating and depends on intensity of the propagating pulse and co-propagating pulses. "CPM converts power fluctuations in a particular wavelength channel to phase fluctuations in other co propagating channels".

Expression for phase shift caused by nonlinear effect is given as:

$$\varnothing_{nl}^{i} = k_{nl} L_{eff}\left(P_i + 2\sum_{n \neq i}^{N} P_n\right)$$

where N = N-channel transmission system, $n = 1,2,3......N$, L_{eff} = effective length of link, and k_{nl} = propagation constant.

However, CPM is better than SPM as its propagating transmission capacity is twice; however, it is advantageous only when all the propagating signal are superimposed with each other for every time slice. If the signals are not superimposed, it can lead to severe damage to the system performance when compared to SPM.

CPM-induced Chirp

1. Fibre dispersion affects the CPM considerably.
2. Pulses belonging to different WDM channels travel at different speeds.
3. XPM occurs only when pulses overlap.
4. Asymmetric XPM-induced chirp and spectral broadening.

CPM: Good or Bad

1. CPM leads to interchannel crosstalk in WDM systems.
2. It can produce amplitude and timing jitter.

Beneficial uses of CPM

1. Nonlinear pulse compression
2. Passive mode locking
3. Ultrafast optical switching
4. Demultiplexing of OTDM channels
5. Wavelength conversion of WDM channels

Applications of CPM

One of the applications of CPM is "Pulse Compression". It also produces frequency chirp, which is used for pulse compression, which is same as SPM; however, for SPM, the signal has to be stronger to generate frequency chirp, while in CPM, because of co-propagating pulses, it can produce the frequency chirp from a weak signal also. Phase shift due to CPM is used for optical switching. Optical switching is used to block the transmission if the phase shift between the pulses is too large and if the phases of the signal are identical, then it will vary the phases of signal using CPM.

Question 6: Discuss the four-wave mixing (FWM) effect in context of nonlinear optics.

Answer: In this nonlinear effect, three optical fields when propagated through fibre will give rise to a new optical field, which depends on the three optical fields and it was originated from. The frequency of the new optical field is given as:

$$\omega_4 = \omega_1 \pm \omega_2 \pm \omega_3 ,$$

where $\omega_1, \omega_2, \omega_3$ are the frequencies of three original optical fields. Figure P3 shows the FWM effect.

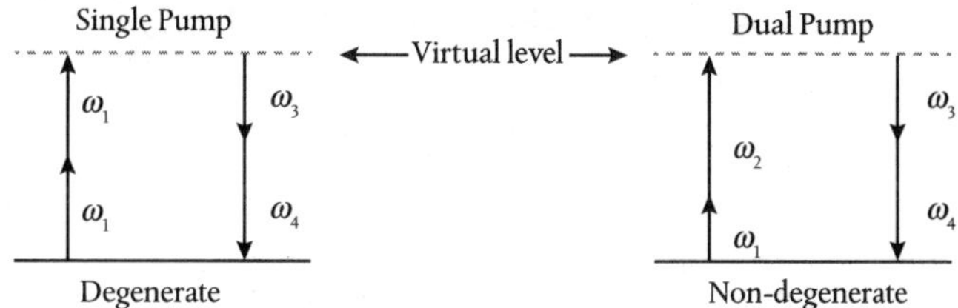

Fig. P3 Four-wave mixing (FWM)

The FWM is not dependent on bit rate like the other two nonlinear effects; instead they depend on channel spacing and dispersion of fibre. Dispersion depends on wavelength, and therefore, newly generated optical wave and reference signal wave have different group velocities. With different group velocities, phase matching is not possible, which decrease the power transfer to new optical wave. At the same time, if the newly generated wave and original wave have same wavelength, results are in interference. The interference of the signals decreases signal to noise ratio. Whenever we observe different group velocities, then FWM effect decreases, channel spacing increases, and dispersion increases.

FWM: Good or Bad

1. FWM leads to interchannel crosstalk in WDM systems.
2. It generates additional noise and degrades system performance.

Beneficial uses of FWM

1. Optical amplification and wavelength conversion
2. Phase conjugation and dispersion compensation
3. Ultrafast optical switching and signal processing
4. Generation of correlated photon pairs

Applications of FWM

Applications of FWM are wavelength conversion and squeezing. Wavelength conversion is important because when an incoming signal wavelength is already utilised by some other wave and is residing at the destination, then we can change the wavelength by using wavelength conversation technique and allow both signals to travel through fibre at the same time. The FWM is used to squeeze the signal, that is, reducing the quantum noise.

Question 7: Discuss the stimulated Raman scattering in context of nonlinear optics.

Answer: The following are few points on stimulated Raman scattering.

1. Scattering of light from vibrating silica molecules as shown in Fig. P4.

2. Amorphous nature of silica turns vibrational state into a band.
3. Raman gain spectrum extends over 40 THz.

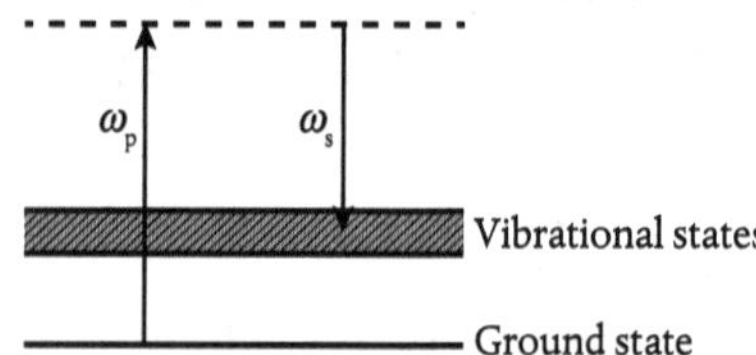

Fig. P4: Stimulated Raman scattering

4. Raman gain is maximum near 13 THz.
5. Scattered light red shifted by 100 nm in the 1.5-mm region.

SRS: Good or Bad

1. Raman gain introduces interchannel crosstalk in WDM systems.
2. Crosstalk can be reduced by lowering channel powers but it limits the number of channels.
3. Raman amplifiers are a boon for WDM systems.
4. It can be used in the entire 1300 to 1650 nm range.
5. EDFA bandwidth limited to ~ 40 nm near 1550 nm.
6. The distributed nature of Raman amplification lowers noise.
7. It is required for opening new transmission bands in telecom systems.

Question 8: List out the comparison of nonlinear refractive effects.

Answer: Nonlinear effects degrade the performance of fibre optic systems. They provide gain to some channels by increasing transmission rate but at an expense of depleting power. The SPM and CPM affect only phase of the signals and can cause spectral broadening, which leads to increased dispersion. The nonlinear effects depend on transmission length and intensity. Table P1 shows the comparison of nonlinear refractive effects.

Table P1 Comparison of nonlinear refractive effects

Nonlinear Phenomenon	SPM	CPM	FWM
Bit rate	Dependent	Dependent	Independent
Origin	Nonlinear susceptibility	Nonlinear susceptibility	Nonlinear susceptibility
Effects	Phase shift due to pulse itself only	Phase shift is due to co-propagating pulses	New waves are generated
Shape of broadening	Symmetric	May be symmetric or asymmetric	Symmetric
Channel Spacing	No effect	Increases when the spacing is decreased	Increases when the spacing is decreased

Question 9: What is pulse compression?

Answer: Pulse compression is a signal processing technique (linear/nonlinear) for reducing the pulse duration. It is a generic term that is used to describe a wave-shaping process that is produced as a propagating waveform, which is modified by the electrical network properties of the transmission line. The pulse is frequency modulated, and it provides a method to further resolve targets that may have overlapping returns.

There are various methods for temporally compressing (shortening) optical pulses, that is, reducing the pulse duration. Typically, such methods start in the picoseconds or femtosecond region, that is, already in the regime of ultrashort pulses. Pulse duration means time difference between FWHM points of the optical power versus time curve.

Question 10: Why compress the pulse?

Answer: Pulse compression originated with the desire to amplify the transmitted impulse (peak) power by temporal compression. It is a method that combines the high energy of a long pulse width with the high resolution of a short pulse width. Pulse compression technique is used in radar, sonar, and echography to increase the range resolution as well as signal to noise ratio.

Question 11: Discuss the key terms in nonlinear fibre optics.

Answer:

Chromatic Dispersion: The chromatic dispersion of an optical medium is the phenomenon that the phase velocity and group velocity of light propagating in a transparent medium depend on the optical frequency. A related quantitative measure is the group velocity dispersion.

Pulse duration: In a pulse waveform, the interval between the time, during the first transition, that the pulse amplitude reaches a specified fraction (level) of its final amplitude, and the time the pulse amplitude drops, on the last transition, to the same level. The most frequently used definition is based on the FWHM of the optical power versus time.

Chirp: The chirp of an optical pulse is usually understood as the time dependence of its instantaneous frequency. Specifically, an up-chirp (down-chirp) means that the instantaneous frequency raises (decreases) with time.

Kerr Effect: The Kerr effect is a nonlinear optical effect occurring when intense light propagates in crystals and glasses, but also in other media like gases. Its physical origin is a nonlinear polarisation generated in the medium, which itself modifies the propagation properties of the light. The Kerr effect is the effect of an instantaneously occurring nonlinear response, which can be described as modifying the refractive index.

Self-phase Modulation (SPM): Due to the Kerr effect, high optical intensity in a medium (e.g. an optical fibre) causes a nonlinear phase delay that has the same temporal shape as the optical intensity. This nonlinear phase modulation of a beam caused by its own intensity via the Kerr effect is called self-phase modulation.

Photonic Crystal Fibres (PCF): Optical fibres with a built-in macrostructure in most cases consisting of small air holes in glass.

Solitons: In general, the temporal and spectral shape of a short optical pulse changes during propagation in a transparent medium due to the Kerr effect and chromatic dispersion. Under certain circumstances, however, the effects of Kerr nonlinearity and dispersion can exactly cancel each other, apart from a constant phase delay per unit propagation distance, so that the temporal and spectral shape of the pulses is preserved even over long propagation distances. These types of pulses with a certain balance of nonlinear and dispersive effects are called solitons.

Question 12: Discuss the pulse compression technique in optical fibre.

Answer: There are two methods of pulse compression:

Linear pulse compression: Using linear frequency modulation or chirping, how can one have a large enough pulse (to still have a good SNR at the receiver) without poor resolution? This is where pulse compression enters the picture. Following is the basic principle: A signal is transmitted, with a long enough length so that the energy budget is correct. This signal is designed so that after matched filtering, the width of the inter-correlated signals is smaller than the width obtained by the standard sinusoidal pulse, as explained earlier (hence, the name of the technique: pulse compression). Linear chirps are the most typically used signals to achieve pulse compression.

Nonlinear pulse compression:

Here, we consider the nonlinear compression of pulses. First, the pulses are spectrally broadened by the nonlinearity of a fibre; at the same time, the pulse duration increases somewhat due to chromatic dispersion. After the fibre, a dispersive compressor is used for obtaining near bandwidth-limited pulses with accordingly shorter duration.

The dispersive compressor is numerically optimized for maximum peak power of the pulses. Interestingly, for a longer fiber piece one requires less group delay dispersion (GDD). This is due to the increased pulse bandwidth.

SOLVED PROBLEMS

Problem 1: An SMF working at 1300 nm has an attenuation of 0.5 dB/km. The fibre core diameter is $5\,\mu\text{m}$ and bandwidth of laser diode is 700 MHz. Calculate the threshold power of stimulated Brillouin and Raman scattering.

Solution:

Given: $A = 0.5\ \text{dB/km},\ d = 5\ \mu\text{m},\ \lambda = 1300\ \text{nm},\ v = 700\ \text{MHz}$ Then

(a) $P_B = 4.4\times10^{-3} d^2\lambda^2 Av = 65.06\ \text{mW}$

(b) $P_R = 5.9\times10^{-2} d^2\lambda A = 0.96\ \text{W}$

Problem 2: Calculate mode parameter V for a step index optical fibre having refractive indices 1.503 and 1.500 for core and cladding, respectively. The core radius is $4\,\mu\text{m}$ at a light wavelength of $1\,\mu\text{m}$.

Solution:

Given: $n_1 = 1.503,\ n_2 = 1.500,\ a = 4\ \mu\text{m},\ \lambda = 1\ \mu\text{m}$

$$V = \frac{2\pi.a.\left(n_1^2 - n_2^2\right)^{1/2}}{\lambda} = 2.385.$$

If wavelength is changed to $0.5\,\mu\text{m}$, then the value of V becomes 4.768. The value of $V = 2.385$ means that optical fibre is an SMF.

Poblem 3: Calculate the numerical aperture (NA) if the value of mode parameter is 2.111 for an SMF, $a = 4.01\ \mu m$, and the wavelength of laser light is $1.3\ \mu m$.

Solution:

Given

$$V = 2.111,\ a = 4.01\ \mu m,\ \lambda = 1.3\ \mu m$$

Then,

$$V = \frac{2\pi}{\lambda}.a.NA$$

$$NA = \frac{V\lambda}{2\pi a}$$

$$\Rightarrow \frac{2.111 \times 1.3 \times 10^{-6}}{2 \times 3.143 \times 4.01 \times 10^{-6}} = 0.11.$$

The value of numerical aperture is 0.11.

MULTIPLE CHOICE QUESTIONS

1. What are the consequences of SPM in nonlinear optics?
   (a) Modification in pulse spectrum
   (b) Limited transmission rate
   (c) Dispersion effect
   (d) All the above
2. Which reasons are responsible for the occurrence of nonlinear CPM?
   (a) Difference in transmission phase of peak pulse and leading or trailing edges of pulse
   (b) Third-order optical nonlinearity
   (c) Intensity dependence of refractive index
   (d) All the above
3. Which type of scattering occurs due to interaction of light in a medium with time-dependent optical density variations, thereby resulting into the change of energy (frequency) and path?
   (a) Stimulated Brillouin Scattering (SBS)
   (b) Stimulated Raman Scattering (SRS)
   (c) Mie Scattering
   (d) Rayleigh Scattering
4. In Rayleigh scattering of light in glass, at which type of temperature does the glass attain the state of thermal equilibrium and exhibits its relativity to annealing temperature?
   (a) Junction
   (b) Fictive
   (c) Breakdown
   (d) Decomposition
5. In Kerr effect, induced index change has its proportionality with respect to _________
   (a) square of electric field
   (b) cube of electric field

(c) cube root of electric field
(d) one-fourth power of electric field

6. Sky looks blue because the sun light is subjected to __________.
(a) Rayleigh scattering
(b) Compton scattering
(c) Both
(d) None

7. Assuming no ISI, the maximum possible bandwidth of a multimode graded index fibre with 5 MHz shows the total pulse broadening of 0.1s for the distance of about 12km. What would be the value of bandwidth length product?
(a) 40 MHz
(b) 60 MHz
(c) 90 MHz
(d) 120 MHz

8. Which among the following is regarded as an inelastic scattering of a photon?
(a) Kerr Effect
(b) Raman Effect
(c) Hall Effect
(d) Miller Effect

Answers to Multiple Choice Questions

1. (d)
2. (c)
3. (a)
4. (b)
5. (a)
6. (a)
7. (b)
8. (b)

QUESTIONS

1. Compare stimulated Brillouin and stimulated Raman scattering in optical fibres, and indicate the way in which they may be avoided in optical fibre communications.
2. Calculate the propagation distance over which the injected optical power is reduced by a factor of two for three fibres with losses of 0.2, 20, and 2000 dB/km. Further, calculate the attenuation constant $\alpha(\text{ cm}^{-1})$ for the three fibres.
3. An SMF is measured to have $\lambda^2\left(dn/d\lambda^2\right)=0.02$ at $0.8\,\mu\text{m}$. Calculate the dispersion parameters β_2 and D. Calculate the numerical values of β_2 (ps^2/km) and D $\left[\text{ps}/(\text{km}-\text{nm})\right]$ at $1.5\,\mu\text{m}$ when the modal index varies with wavelength as $n((\lambda)=1.45-s(\lambda-1.3\,\mu\text{m})^3$, where $s=0.003\,\mu\text{m}^3$.
4. A 1-km-long SMF with the zero-dispersion wavelength at $1.4\,\mu\text{m}$ is measured to have $D=10\text{ps}/(\text{km}-\text{nm})$ at $1.55\,\mu\text{m}$. Two pulses from Nd:YAG lasers operating at 1.06 and $1.32\,\mu\text{m}$ are launched simultaneously into the fibre. Calculate the delay in the arrival time of the two pulses at the fibre output assuming that β_2 varies linearly with wavelength over the range $1.0-1.6\,\mu\text{m}$.
5. Use Maxwell's equations to express the field components $E_\rho, E_\varphi, H_\rho$, and H_φ inside the fibre core in terms of E_z and H_z.
6. Derive eigenvalue equation by matching the boundary conditions at the core–cladding interface of a stepindex fibre.
7. Use the eigenvalue equation to derive the single-mode condition in optical fibres.

8. An SMF has an index step of 0.005. Calculate the core radius if the fibre has a cut-off wavelength of 1 μm. Assume a core index of 1.45.
9. Derive an expression for the confinement factor Γ of SMFs defined as the fraction of the total mode power contained inside the fibre core. Use the Gaussian approximation used for the fundamental fibre mode.
10. Estimate the FWHM of the spot size associated with the fibre mode and the fraction of the mode power inside the core when the fibre of Problem 8 is used to transmit 1.3- μm light.
11. Consider the Fourier transform of the Raman response function and plot the real and imaginary parts as a function of frequency. What is the physical meaning of the resulting curves?
12. The Raman-gain spectrum of a fibre is approximated by a Lorentzian profile whose FWHM is 5 THz. The gain peak is located at 15 THz from the carrier frequency of the pulse. Derive an expression for the Raman response function of this fibre.
13. A dispersion-shifted fibre is measured to have $D = 2\,\text{ps}/(\text{km}-\text{nm})$ at $1.55\,\mu\text{m}$. It has an effective core area of $40\,\mu\text{m}^2$. Calculate the dispersion and nonlinear lengths when (i) 10-ps pulses with 100-mW peak power and (ii) 1-ps pulses with 1-W peak power are launched into the fibre. Are nonlinear effects important in both cases?
14. Determine the temporal and spectral widths (FWHM) of this pulse for a chirped Gaussian pulse with $C = 5$ and $T_0 = 50\,\text{ps}$.
15. Prove that for an unchirped Gaussian pulse of arbitrary width, the product $\Delta\nu\Delta t$ approximately equals 0.44, where Δt and $\Delta\nu$ are the temporal and spectral widths (both measured as FWHM), respectively.
16. Repeat problem 15 for an unchirped "sech" pulse and prove that $\Delta\nu\Delta t$ approximately equals 0.315.
17. Show that a chirped Gaussian pulse is compressed initially inside anSMF when $\beta_2 C < 0$. Derive expressions for the minimum width and the fibre length at which the minimum occurs.
18. Estimate the limiting bit rate for a 60-km SMF link at 1.3- and 1.55-μm wavelengths assuming transform-limited 50-ps (FWHM) input pulses. Assume $\beta_2 = 0$ and $-20\,\text{ps}^2/\text{km}$ and $\beta_3 = 0.1$ and $0\,\text{ps}^3/\text{km}$ at 1.3- and 1.55-μm wavelengths, respectively.
19. An optical communication system is operating with chirped Gaussian input pulses. Assume $\beta_3 = 0$ and $V\omega\_1$ and obtain a condition on the bit rate in terms of the parameters C, β_2, and L.

REFERENCES

1. Miller S E, Integrated Optics: An Introduction, *Bell Syst. Tech. J.*, vol. 48, pp. 2059–2068, 1999.
2. Dong L, Archambault J L, Reekie L, Russell P S J and Payne D N, Single-pulse Bragg gratings written during fibre drawing, *Electron. Lett.*, vol. 29, no. 17 pp. 1577–1578, 1993.
3. Durmus A, Yasuk F and Boztosun I, Exact analytical solution of the Kline Gordan equation for the pionic atom by asymptotic iteration method, *Int. J. Mol. Phys. E (IJMPE)*, vol. 15, 1243, 2006.
4. Erdogan T, Fibre grating spectra, *IEEE J. Lightwave Technol.*, vol. 15, no. 8, August 1997.
5. Nishihara H, Haruna M and Suhara T, *Optical Integrated Circuits*, New York: McGraw-Hill, 1989.
6. Nicolae C, Osgood R M, Zhang S and Brueck S R T, Zero average refractive index band gap in PC superlattices, *J. Opt. Soc. Am. B*, vol. 23, pp. 506–512, 2006.
7. Chaudhuri P R and Roy S, Analysis of arbitrary index profile planar optical waveguides and Multilayer Nonlinear Structures: A Simple Finite Difference Algorithm, *Opt. Quant. Electron.*, vol. 39, pp. 221–237, 2007.
8. Poulsen M R, Borel P I, Pedersen J F, Hubner J M, Kristensen J H P, Rotwitt K, Svalgaard M and Svendsen W, Advances in silica-based integrated optics, *Opt. Eng.*, vol. 42, no. 10, pp. 2821–2834, 2003.

9. Raghuwanshi S K, An algorithm to design the optimal flat gain fibre Raman amplifier, *National Conference on Microwave and Optoelectronics* (NCMO), Aurangabad, Dec. 2004.
10. Raghuwanshi S K, Gupta V, Dinesh V K and Talabattula S, Bidirectional optical fibre transmission scheme through Raman amplification: Effect of pump depletion, *J. Indian Inst. Sci.*, vol. 86, no. 6, pp. 655–665, Nov–Dec 2006.
11. She S X, Propagation characteristics and loss of metal-clad graded index optical waveguides, *Opt. Commun.*, vol. 103, pp. 365–369, 1993.
12. Shumpert J D, Modeling of periodic dielectric structures (electromagnetic crystals), Ph.D. dissertation, University of Michigan, 2001.
13. Smith G D, Numerical solution of partial differential equations, in *Finite Difference Methods*, New York: Oxford University Press, 1985.
14. Stegeman G I and Seaton C T, Nonlinear integrated optics, *J. Appl. Phys.*, vol. 58, no. 12, pp. R57–R78, Dec., 1985.
15. Stubbe R, Sahlgren B, Sandgren S and Asseh A, in *Photosensitivity and Quadratic Nonlinearity in Glass Waveguides* (Fundamentals and Applications) p. PD1, Post deadline papers, Portland, OR, 1995.
16. Taranenko V B and Weiss C O, Spatial solitons in semiconductor micro-resonator,*IEEE J. Sel. Top. Quantum Electron.*, vol. 8, no.3, pp. 488–496, May/June 2002.
17. Wang M -Y, Xu J, Wu J, Wei B, Li H -L, Xu T and Ge D -B, FDTD study on wave propagation in layered structures with biaxial anisotropic metamaterials, *Prog. Electromagn. Res.*, vol. 81, pp. 253–265, 2008.
18. Yeh P, *Optical Waves in Layered Media*, New York: John Wiley and Sons, 1988.
19. Yevick D, A guide to electric field propagation technique for guided-wave optics, *Opt. Quantum Electron.*, vol. 26, pp. 185–197, 1994.
20. Agrawal G P, *Nonlinear Fibre Optics*, San Diego, CA: Academic Press Inc., 1989.
21. Aktas M and Sever R, Exact supersymmetric solution of Schrodinger equation for central confining potentials by using the Nikiforov-Uvarov method, *J. Mol. Struct. (Theo chem)*, vol. 710, 223, 2004.
22. Berkhoer A L and Zakharaov V, Self excitation of waves with different polarisations in nonlinear media, *Sov. Phys. JETP*, vol. 31, pp. 486–490, 1970.
23. Birks T A, Knight J C and Russell P S J, Endlessly single-mode photonic crystal fibre, *Opt. Lett.*, vol. 22, no. 13, p. 961–963, 1997.
24. Blow K J and Wood D, Theoretical description of transient Stimulated Raman scattering in optical fibres, *IEEE J Quantum Electron.*, vol. 25, pp. 2665–2673, 1989.
25. Groothoff N, Canning J and Buckley E, Bragg gratings in air-silica structured fibres, *Opt. Lett.*, vol. 28, pp. 233–235, 2003.
26. Jiang H, Chen H, Li H, Zhang Y, Zi J and Zhu S, Properties of one dimensional photonics crystals containing single negative material, *Phys. Rev. E*, vol. 69, pp. 066607–066610, 2004.
27. Nelson B P, Blow K J, Constantine P D, Doran N J, Lucek J K, Marshall I W and Smith K, All-optical Gbit/s switching using nonlinear optical loop mirror, *Electron. Lett.*, vol. 27, issue 9, pp. 704 – 705, 1991.

7 Study of Electro-optic Modulators

Objectives

- This chapter discusses the propagation of wave in anisotropic media.
- Various devices using electro-optic effect are considered.
- The use of Mach–Zehnder interferometer for controlling the amplitude of an optical wave is analysed.
- Coordinate system for various geometries of E- and H-type mode is discussed.

7.1 INTRODUCTION

In this chapter, we will discuss wave propagation in anisotropic media in which the velocity of propagation depends on the propagation direction and also on the state of polarisation. Here, one can observe the phenomenon of double refraction. Anisotropic media form the basis of a large number of polarisation devices such as quarter-wave and half-wave plates and Soleil–Babinet compensator. The studies on them are also very important for understanding various light modulators based on the electro-optic effect. We will introduce the index ellipsoid and show how from the index ellipsoid, one can obtain the velocities of propagation and the polarisations of the two waves that can propagate along any given direction [1-9].

7.2 WAVE REFRACTIVE INDEX

In anisotropic media, the direction of displacement (*D*) and electric field (*E*) are not parallel for all possible directions of *E*. Hence, electric field applied along the *x*-direction generates a *D*, which has all the three components [10-13]. Thus,

$$D_x = \epsilon_{xx} E_x, D_y = \epsilon_{yx} E_x, D_z = \epsilon_{zx} E_x,$$

where $E_y = 0 = E_z$; $\epsilon_{xx}, \epsilon_{yx}$ and ϵ_{zx} are the corresponding permittivity components. *D* may be written in the form:

$$D_x = \epsilon_{xx} E_x + \epsilon_{xy} E_y + \epsilon_{xz} E_z$$

$$D_y = \epsilon_{yx} E_x + \epsilon_{yy} E_y + \epsilon_{yz} E_z$$

$$D_z = \epsilon_{zx} E_x + \epsilon_{zy} E_y + \epsilon_{zz} E_z .$$

In matrix form:

$$\begin{pmatrix} D_x \\ D_y \\ D_z \end{pmatrix} = \begin{pmatrix} \epsilon_{xx} & \epsilon_{xy} & \epsilon_{xz} \\ \epsilon_{yx} & \epsilon_{yy} & \epsilon_{yz} \\ \epsilon_{zx} & \epsilon_{zy} & \epsilon_{zz} \end{pmatrix} \begin{pmatrix} E_x \\ E_y \\ E_z \end{pmatrix}. \tag{7.1}$$

Equation (7.1) can be written as $D = \overset{=}{\epsilon}\, E$, where $\overset{=}{\epsilon}$ is the dielectric tensor. Consider a plane wave propagating through an anisotropic medium. The time and space dependence of the electric and magnetic fields will be given as:

$$E = E_0 e^{i(\omega t - \boldsymbol{k}r)} \tag{7.2}$$

$$H = H_0 e^{i(\omega t - \boldsymbol{k}r)}, \tag{7.3}$$

where E_0 and H_0 are constants, $\boldsymbol{k}$ is the propagation vector of the plane wave, and ω is the frequency of the electromagnetic wave. The wave refractive index n_w is defined by the equation:

$$|\boldsymbol{k}| = \mathrm{k} = \left(\frac{\omega}{C}\right) n_w = \left(\frac{\omega}{C}\right) n_r \hat{s}. \tag{7.4}$$

Consider following Maxwell equations:

$$\nabla \times E = -\frac{\partial B}{\partial t} = -i\omega\mu_0 H$$

$$\nabla \times H = \frac{\partial D}{\partial t} = i\omega B \text{ using } (B = \mu_0 H)$$

$$\nabla \times E = \begin{bmatrix} i & j & k \\ \frac{\partial}{\partial x} & \frac{\partial}{\partial y} & \frac{\partial}{\partial z} \\ E_x & E_y & E_z \end{bmatrix}$$

$$(\nabla \times E)_x = \left(\frac{\partial E_z}{\partial y} - \frac{\partial E_y}{\partial z}\right).$$

Substituting $\frac{\partial}{\partial y} = -ik_y$ and $\frac{\partial}{\partial z} = -ik_z$ and using (7.1)

$$= \left(-ik_y E_{0z} + ik_z E_{0y}\right) e^{i(\omega t - \boldsymbol{k}\tau)} = -i(\boldsymbol{k} \times E)_x$$

Hence, $\quad (\nabla \times E) = -i(\boldsymbol{k} \times E) = -i\omega\mu_0 H$

$$H = \frac{1}{\omega\mu_0}(\boldsymbol{k} \times E). \tag{7.5}$$

Similarly,

$$D = -\frac{1}{\omega}(\boldsymbol{k} \times H). \tag{7.6}$$

From (7.5) and (7.6),

$$D = -\frac{1}{\omega}\boldsymbol{k} \times \left(\frac{1}{\omega\mu_0}\boldsymbol{k} \times E\right) = -\frac{1}{\omega^2\mu_0}\boldsymbol{k} \times (\boldsymbol{k} \times E). \tag{7.7}$$

$$(\boldsymbol{k}.E)\boldsymbol{k} - \boldsymbol{k}^2 E = -\omega^2\mu_0 \overline{\overline{\epsilon}} E, \tag{7.8}$$

where $\overline{\overline{\epsilon}}$ is the dielectric constant tensor.

$$(\boldsymbol{k}.E)\boldsymbol{k} - \boldsymbol{k}^2 E = -\omega^2\mu_0 \overline{\overline{\epsilon}} E = -\omega^2\mu_0 D, D = -\frac{1}{\omega^2\mu_0}\left[(\boldsymbol{k}.E)\boldsymbol{k} - \boldsymbol{k}^2 E\right]$$

Hence, $$D.E = -\frac{1}{\omega^2\mu_0}\left[(\boldsymbol{k}.E)^2 - \boldsymbol{k}^2 E^2\right] = -\frac{1}{\omega^2\mu_0}\left[\left(|\boldsymbol{k}|.|E|\cos\left(90^0 + \varphi\right)\right)^2 - \boldsymbol{k}^2 E^2\right]$$

$$= -\frac{k^2E^2}{\omega^2\mu_0}\left[1 - sin^2\varphi\right] = \frac{k^2E^2}{\omega^2\mu_0}cos^2\varphi = D.E / E^2 = \frac{k^2}{\omega^2\mu_0}cos^2\varphi. \tag{7.9}$$

Let $\hat{k}$ represent the unit vector along k, that is, along the normal to the wavefront. Thus,

$$\boldsymbol{k} = \left(\omega/c\right)n_w\hat{\boldsymbol{k}}. \tag{7.10}$$

Further, $$\overline{\overline{\mathbf{K}}} = \frac{\overline{\overline{\epsilon}}}{\epsilon_0}, \tag{7.11}$$

where $\overline{\overline{\mathbf{K}}}$ is the dielectric constant tensor and ϵ_0 is the free-space permittivity. Using (7.10) and (7.11), (7.8) becomes:

$$(\hat{\boldsymbol{k}}.E)\hat{\boldsymbol{k}}\left(\omega^2/c^2\right)n_w^2 - \left(\omega^2/c^2\right)n_w^2 E = -\omega^2 \epsilon_0 \mu_0 \overline{\overline{\mathbf{K}}}\ E = -\left(\omega^2/c^2\right)\overline{\overline{\mathbf{K}}}\ E$$

$$(\hat{\boldsymbol{k}}.E)\hat{\boldsymbol{k}} - E = -\left(1/n_w^2\right)\overline{\overline{\mathbf{K}}}E. \tag{7.12}$$

As $\hat{\boldsymbol{k}} = k_x a_x + k_y a_y + k_z a_z$ and $E = E_x a_x + E_y a_y + E_z a_z$. Solving (7.12)

$$\left[\left(k_x a_x + k_y a_y + k_z a_z\right).\left(E_x a_x + E_y a_y + E_z a_z\right)\right]\left(k_x a_x + k_y a_y + k_z a_z\right) - \left(E_x a_x + E_y a_y + E_z a_z\right)$$
$$= -\left(1/n_w^2\right)\overline{\overline{\mathbf{K}}}\left(E_x a_x + E_y a_y + E_z a_z\right)$$

$$\left(k_x E_x + k_y E_y + k_z E_z\right)\left(k_x a_x + k_y a_y + k_z a_z\right) - \left(E_x a_x + E_y a_y + E_z a_z\right) = -\left(1/n_w^2\right)\overline{\overline{\mathbf{K}}}\left(E_x a_x + E_y a_y + E_z a_z\right).$$

$$\left(k_x^2E_xa_x + k_xk_yE_xa_y + k_xk_zE_xa_z + k_yk_xE_ya_x + k_y^2E_ya_y + k_yk_zE_ya_z + k_zk_xE_za_x + k_zk_yE_za_y + k_z^2E_za_z\right) - \left(E_xa_x + E_ya_y + E_za_z\right) = -\left(1/n_w^2\right)\overline{\overline{\mathbf{K}}}\left(E_xa_x + E_ya_y + E_za_z\right)$$

$$\left[\left(k_x^2E_x + k_yk_xE_y + k_zk_xE_z\right) - E_x\right]a_x + \left[\left(k_xk_yE_x + k_y^2E_y + k_zk_yE_z\right) - E_y\right]a_y + \left[\left(k_xk_zE_x + k_yk_zE_y + k_z^2E_z\right) - E_z\right]a_z = -\left(1/n_w^2\right)\overline{\overline{\mathbf{K}}}\left(E_xa_x + E_ya_y + E_za_z\right).$$

In vector form:

$$\begin{bmatrix} \left(k_x^2E_x + k_yk_xE_y + k_zk_xE_z\right) - E_x \\ \left(k_xk_yE_x + k_y^2E_y + k_zk_yE_z\right) - E_y \\ \left(k_xk_zE_x + k_yk_zE_y + k_z^2E_z\right) - E_z \end{bmatrix} = -\frac{1}{n_w^2}\begin{bmatrix} K_x & 0 & 0 \\ 0 & K_y & 0 \\ 0 & 0 & K_z \end{bmatrix}\begin{bmatrix} E_x \\ E_y \\ E_z \end{bmatrix}$$

$$\begin{bmatrix} \left(k_x^2E_x + k_yk_xE_y + k_zk_xE_z\right) - E_x \\ \left(k_xk_yE_x + k_y^2E_y + k_zk_yE_z\right) - E_y \\ \left(k_xk_zE_x + k_yk_zE_y + k_z^2E_z\right) - E_z \end{bmatrix} = -\frac{1}{n_w^2}\begin{bmatrix} K_xE_x \\ K_yE_y \\ K_zE_z \end{bmatrix}.$$

Equating both sides:

$$\left(k_x^2E_x + k_yk_xE_y + k_zk_xE_z\right) - E_x = -\frac{1}{n_w^2}K_xE_x \tag{7.13}$$

$$\left(k_xk_yE_x + k_y^2E_y + k_zk_yE_z\right) - E_y = -\frac{1}{n_w^2}K_yE_y \tag{7.14}$$

$$\left(k_xk_zE_x + k_yk_zE_y + k_z^2E_z\right) - E_z = -\frac{1}{n_w^2}K_zE_z. \tag{7.15}$$

From (7.13),

$$E_x\left[k_x^2 - 1 + \frac{1}{n_w^2}K_x\right] + k_yk_xE_y + k_zk_xE_z = 0.$$

Using

$$k_x^2 + k_y^2 + k_z^2 = 1, \tag{7.16}$$

$$E_x\left[1 - k_y^2 - k_z^2 - 1 + \frac{1}{n_w^2}K_x\right] + k_yk_xE_y + k_zk_xE_z = 0$$

$$\left[\left(1/n_w^2\right)K_x - k_y^2 - k_z^2\right]E_x + k_xk_yE_y + k_xk_zE_z = 0. \tag{7.17}$$

Similarly, from (7.14)–(7.16)

$$k_xk_yE_x + \left[\left(1/n_w^2\right)K_y - k_x^2 - k_z^2\right]E_y + k_yk_zE_z = 0 \tag{7.18}$$

$$k_x k_z E_x + k_y k_z E_y + \left[\left(1/n_w^2\right)K_z - k_x^2 - k_y^2\right]E_z = 0. \tag{7.19}$$

For a nontrivial solution of (7.17)–(7.19), the determinant of the coefficients of $E_x, E_y,$ and E_z must disappear in order to give the following equation:

$$\begin{vmatrix} K_x/n_w^2 - k_y^2 - k_z^2 & k_x k_y & k_x k_z \\ k_y k_x & K_y/n_w^2 - k_x^2 - k_z^2 & k_y k_z \\ k_x k_z & k_y k_z & K_z/n_w^2 - k_x^2 - k_y^2 \end{vmatrix} = 0. \tag{7.20}$$

The above-mentioned determinant is an eigenvalue equation; for a given direction of propagation of the plane wave, that is, for a set $k_x, k_y,$ and k_z, solving the above-mentioned determinant gives us the possible values of the refractive index n_w, which are the eigenvalues.

7.2.1 Wave Propagating Along the Z-direction

Consider wave propagating along the z-direction. In such a case, we will have:

$$k_x = 0, k_y = 0, k_z = 1. \tag{7.21}$$

Further, (7.20) becomes:

$$\begin{vmatrix} K_x/n_w^2 - 1 & 0 & 0 \\ 0 & K_y/n_w^2 - 1 & 0 \\ 0 & 0 & K_z/n_w^2 \end{vmatrix} = 0, \tag{7.22}$$

which gives

$$\left(\frac{K_x}{n_w^2} - 1\right)\left(\frac{K_y}{n_w^2} - 1\right)\frac{K_z}{n_w^2} = 0$$

$$\left(\frac{K_x}{n_w^2} - 1\right) = 0, \left(\frac{K_y}{n_w^2} - 1\right) = 0, \frac{K_z}{n_w^2} = 0.$$

that is,

$$K_x = n_w^2, K_y = n_w^2, K_z = 0$$

$$n_w = n_{w1} = K_x^{\frac{1}{2}}, n_w = n_{w2} = K_y^{\frac{1}{2}}.$$

These are two eigenvalues. The corresponding polarisation directions can be determined by substituting $n_w = n_{w1}$ and $n_w = n_{w2}$ in (7.17)–(7.19) and using (7.21). Thus, for a wave that has a refractive index n_{w1}, (7.18) gives us:

$$\left(K_y/K_x - 1\right)E_y = 0.$$

For $K_x \neq K_y$, we have $E_y = 0$. Similarly, (7.19) gives us $E_z = 0$. Thus, a wave with its E along x and propagating along z propagates with a phase velocity $c/K_x^{\frac{1}{2}}$. Further, other root $n_w = n_{w2} = K_y^{\frac{1}{2}}$ corresponds to a wave with its E lying along the y-direction.

Thus, for a wave propagating along the z-direction, the eigen waves are:

1. A wave with its E along *x* (and hence, its D also along x since x, y, z is a principal axis system) that travels with a phase velocity of $c/K_x^{1/2}$.
2. A wave with its E along *y* (and hence D also along y) that travels with a phase velocity $c/K_y^{1/2}$.
3. A wave with its E along z and propagating along either *x* or *y* direction travels with a phase velocity $c/K_z^{1/2}$.

7.3 RAY REFRACTIVE INDEX

Since *D* and E are not along the same direction, the component of *D* along E is calculated as: consider two vectors *D* and E, as shown in Fig. 7.1. Dot product of two vectors *D* and E as:

$$D.E = |D||E|\cos\varphi,$$

where φ is the angle between vectors *D* and E. Using the definition, the length of vector E can be expressed as:

$$|E|^2 = E.E$$

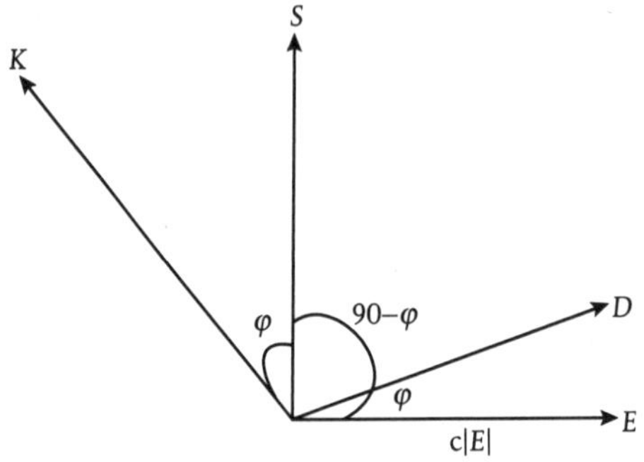

Fig. 7.1 The relative direction of *E*, *D*, *k*, and *S* in an anisotropic medium

Now, the length of the component of *D* along E is $|D|cos\varphi$, but it can be $c|E|$. Therefore,

$$c|E| = |D|cos\varphi$$

Multiplying both sides by |E| yields:

$$c|E|^2 = |D||E|cos\varphi = D.E$$

and $c = \dfrac{D.E}{|E|^2}$.

This can be written as:

$$D = \epsilon\left[E - \left(E.\hat{s}\right)\hat{s}\right] = \frac{(D.E)E}{E^2}.$$

Using $D = \epsilon\, E$,

$$\Rightarrow D - \left(D.\hat{s}\right)\hat{s} = \frac{(D.E)E}{E^2}, \tag{7.23}$$

where $\hat{s}$ is the unit vector along the ray direction (i.e., along S).

Now, from (7.23) and (7.9):

$$\Rightarrow D-\left(D.\hat{\mathbf{s}}\right)\hat{\mathbf{s}}=\left(D.E/E^2\right)E=\left(\frac{k^2}{\omega^2\mu_0}\cos^2\varphi\right)E.$$

7.3.1 Relation Between Ray Velocity and Wave Velocity

Let ray velocity be represented by $v_r=|S|/W$,

where S is ray direction and W is the energy density. The propagation direction is along the pointing vector, which is defined as $S=E\times H$.

$$\text{Energy density} = W=\frac{1}{2}(E.D+B.H)=\frac{1}{2}(E.D+\mu_0 H.H).$$

Substituting H and D from (7.5) and (7.6):

$$W=\frac{1}{2}\left[-\frac{1}{\omega}E.(\mathbf{k}\times H)+\mu_0\left(1/\omega\mu_0\right)(\mathbf{k}\times E).H\right]$$

$$W=\frac{1}{2}\left[-\frac{1}{\omega}E.(\mathbf{k}\times H)+\frac{1}{\omega}(\mathbf{k}\times E).H\right].$$

Using vector relation a · (b × c) = (a × b) · c,

$$W=\frac{1}{2}\left[-\frac{1}{\omega}(E\times\mathbf{k}).H+\frac{1}{\omega}(\mathbf{k}\times E).H\right].$$

Using a × b = −b × a,

$$W=\frac{1}{2}\left[\frac{1}{\omega}(\mathbf{k}\times E).H+\frac{1}{\omega}(\mathbf{k}\times E).H\right]$$

$$W=\frac{1}{\omega}(\mathbf{k}\times E).H.$$

Using vector relation a · (b × c) = (a × b) · c:

$$=\left(\frac{1}{\omega}\right)\mathbf{k}.(E\times H)=\left(\frac{1}{\omega}\right)\mathbf{k}.S$$

Thus,

$$v_r=\frac{\omega|S|}{\mathbf{k}.S}=\frac{\omega}{\mathbf{k}cos\varphi}=\frac{v_w}{cos\varphi}. \tag{7.24}$$

From (7.9) and (7.24),

$$\frac{k^2}{\omega^2\mu_0}cos^2\varphi=cos^2\varphi/v_w^2\mu_0=1/v_r^2\mu_0.$$

Hence, $$D-\left(D.\hat{s}\right)\hat{s}=\left(1/v_r^2\mu_0\right)E.$$

Using $c=\dfrac{1}{\sqrt{\mu_0\epsilon_0}}$ and substituting value of μ_0 in above-mentioned equation, we get:

$$\left(D.\hat{s}\right)\hat{s}-D=-\left(\epsilon_0 c^2/v_r^2\right)E$$

As $n_r=c/v_r$,

$$\left(D.\hat{s}\right)\hat{s}-D=-n_r^{\,2}\epsilon_0 E$$

$$E=E_0 e^{i(\omega t-\mathbf{k}r)} \tag{7.25}$$

$$|\mathbf{k}|=k=\left(\frac{\omega}{C}\right)n_w=\left(\frac{\omega}{C}\right)n_r\hat{\mathbf{s}}. \tag{7.26}$$

From (7.5),

$$D=-\frac{1}{\omega^2\mu_0}\mathbf{k}\times\left(\mathbf{k}\times E\right).$$

Converting k to s by using (7.26)

$$D=-\frac{1}{\omega^2\mu_0}\left[\left(\frac{\omega}{C}\right)n_r\hat{\mathbf{s}}\times\left(\left(\frac{\omega}{C}\right)n_r\hat{\mathbf{s}}\times E\right)\right]$$

$$D=-\frac{1}{\omega^2\mu_0}.\left(\left(\frac{\omega}{C}\right)n_r\right)^2\left[\hat{\mathbf{s}}\times\left(\hat{\mathbf{s}}\times E\right)\right]$$

$$D=-\frac{n_r^{\,2}}{c^2\mu_0}\left[(\hat{\mathbf{s}}.E)\hat{\mathbf{s}}-(\hat{\mathbf{s}}.\hat{\mathbf{s}})E\right]\ \text{(Using } \hat{\mathbf{s}}.\hat{\mathbf{s}}=1\text{)}$$

$$D=-\frac{n_r^{\,2}}{c^2\mu_0}\left[(\hat{\mathbf{s}}.E)\hat{\mathbf{s}}-E\right]=\frac{n_r^{\,2}}{c^2\mu_0}\left[E-(\hat{\mathbf{s}}.E)\hat{\mathbf{s}}\right]$$

$$D=n_r^{\,2}\in_0\left[E-(\hat{\mathbf{s}}.E)\hat{\mathbf{s}}\right]=n_r^{\,2}\in_0 E$$

Using the dielectric tensor constant $\overline{\overline{\mathbf{K}}}=\dfrac{\overline{\overline{\in}}}{\in_0}$,

$$D=n_r^{\,2}\in_0\left[E-(\hat{\mathbf{s}}.E)\hat{\mathbf{s}}\right]=n_r^{\,2}\frac{\overline{\overline{\in}}}{\overline{\overline{\mathbf{K}}}}E. \tag{7.27}$$

From (7.27),

$$D-\left(D.\hat{\mathbf{s}}\right)\hat{\mathbf{s}}=n_r^{\,2}\frac{\overline{\overline{\in}}}{\overline{\overline{\mathbf{K}}}}E.$$

$$\Rightarrow\left(D.\hat{\mathbf{s}}\right)\hat{\mathbf{s}}-D=-\frac{n_r^{\;2}}{\underline{\underline{\mathbf{K}}}}D. \tag{7.28}$$

As $\hat{\mathbf{s}}=s_xa_x+s_ya_y+s_za_z$ and D = $D_xa_x+D_ya_y+D_za_z$ and solving (7.28), we get

$$\left[\left(s_xa_x+s_ya_y+s_za_z\right).\left(D_xa_x+D_ya_y+Da_z\right)\right]\left(s_xa_x+s_ya_y+s_za_z\right)-\left(D_xa_x+D_ya_y+Da_z\right)$$
$$=-\frac{n_r^{\;2}}{\underline{\underline{\mathbf{K}}}}\left(D_xa_x+D_ya_y+D_za_z\right).$$

$$\left(s_xD_x+s_yD_y+s_zD_z\right)\left(s_xa_x+s_ya_y+s_za_z\right)-\left(D_xa_x+D_ya_y+D_Za_z\right)=-\frac{n_r^{\;2}}{\underline{\underline{\mathbf{K}}}}\left(D_xa_x+D_ya_y+D_za_z\right)..$$

$$\left(s_x^2D_xa_x+s_xs_yD_xa_y+s_xs_zD_xa_z+s_ys_xD_ya_x+s_y^2D_ya_y+s_ys_zD_ya_z+s_zs_xD_za_x+s_zs_yD_za_y+s_z^2D_za_z\right)-$$
$$\left(D_xa_x+D_ya_y+D_za_z\right)=-\frac{n_r^{\;2}}{\underline{\underline{\mathbf{K}}}}\left(D_xa_x+D_ya_y+D_za_z\right).$$

$$\left[\left(s_x^2D_x+s_ys_xD_y+s_zs_xD_z\right)-D_x\right]a_x+\left[\left(s_xs_yD_x+s_y^2D_y+s_zs_yD_z\right)-D_y\right]a_y+\left[\left(s_xs_zD_x+s_ys_zD_y+s_z^2D_z\right)-D_z\right]a_z$$
$$=-\frac{n_r^{\;2}}{\underline{\underline{\mathbf{K}}}}\left(D_xa_x+D_ya_y+D_za_z\right)$$

In vector form,

$$\begin{bmatrix}\left(s_x^2D_x+s_ys_xD_y+s_zs_xD_z\right)-D_x\\\left(s_xs_yD_x+s_y^2D_y+s_zs_yD_z\right)-D_y\\\left(s_xs_zD_x+s_ys_zD_y+s_z^2D_z\right)-D_z\end{bmatrix}=-n_r^2\begin{bmatrix}\frac{1}{K_x}&0&0\\0&\frac{1}{K_y}&0\\0&0&\frac{1}{K_z}\end{bmatrix}\begin{bmatrix}D_x\\D_y\\D_z\end{bmatrix}$$

$$\begin{bmatrix}\left(s_x^2D_x+s_ys_xD_y+s_zs_xD_z\right)-D_x\\\left(s_xs_yD_x+s_y^2D_y+s_zs_yD_z\right)-D_y\\\left(s_xs_zD_x+s_ys_zD_y+s_z^2D_z\right)-D_z\end{bmatrix}=-n_r^2\begin{bmatrix}\frac{D_x}{K_x}\\\frac{D_y}{K_y}\\\frac{D_z}{K_z}\end{bmatrix}. \tag{7.29}$$

By equating both sides,

$$\left(s_x^2D_x+s_ys_xD_y+s_zs_xD_z\right)-D_x=-n_r^2\frac{D_x}{K_x} \tag{7.30}$$

$$\left(s_x s_y D_x + s_y^2 D_y + s_z s_y D_z\right) - D_y = -n_r^2 \frac{D_y}{K_y} \tag{7.31}$$

$$\left(s_x s_z D_x + s_y s_z D_y + s_z^2 D_z\right) - D_z = -n_r^2 \frac{D_z}{K_z}. \tag{7.32}$$

From (7.30),

$$D_x\left[s_x^2 - 1 + \frac{n_r^2}{K_x}\right] + s_y s_x D_y + s_z s_x D_z = 0.$$

Using

$$s_x^2 + s_y^2 + s_z^2 = 1, \tag{7.33}$$

$$\left[\frac{n_r^2}{K_x} - s_y^2 - s_z^2\right] D_x + s_x s_y D_y + s_x s_z D_z = 0. \tag{7.34}$$

Similarly, from (7.31)–(7.33), we get

$$s_x s_y D_x + \left[\frac{n_r^2}{K_y} - s_x^2 - s_z^2\right] D_y + s_y s_z D_z = 0 \tag{7.35}$$

$$s_x s_z D_x + s_y s_z D_y + \left[\frac{n_r^2}{K_z} - s_x^2 - s_y^2\right] D_z = 0. \tag{7.36}$$

For nontrivial solution, the determinant of the coefficients of D_x, D_y, and D_z must disappear and we get:

$$\begin{vmatrix} \frac{n_r^2}{K_x} - s_y^2 - s_z^2 & s_x s_y & s_x s_z \\ s_x s_y & \frac{n_r^2}{K_y} - s_x^2 - s_z^2 & s_y s_z \\ s_x s_z & s_y s_z & \frac{n_r^2}{K_z} - s_x^2 - s_y^2 \end{vmatrix} = 0 \tag{7.36a}$$

Consider the wave propagating along the z-direction. In such a case, we have $s_x = 0, s_y = 0,$ and $s_z = 1$. Equation (7.36a) reduces to

$$\begin{vmatrix} \frac{n_r^2}{K_x} - 1 & 0 & 0 \\ 0 & \frac{n_r^2}{K_y} - 1 & 0 \\ 0 & 0 & \frac{n_r^2}{K_z} - 1 \end{vmatrix} = 0$$

The wave refractive index for a ray propagating along the z-direction, there are two values of n_r :

1. For a ray with E along x,

$$\frac{n_r^2}{K_x}-1=0,$$

$n_r = K_x^{1/2}$ and corresponding ray velocity is $c/K_x^{1/2}$.

2. For a ray with E along y,

$$\frac{n_r^2}{K_y}-1=0,$$

$n_r = K_y^{1/2}$ and corresponding ray velocity is $c/K_y^{1/2}$.

3. For a ray with E along z,

$$\frac{n_r^2}{K_z}-1=0$$

$n_r = K_z^{1/2}$ and corresponding ray velocity is $c/K_z^{1/2}$.

Consider a ray propagating in the (x–z) plane at an angle θ with the z-axis as shown in the Fig. 7.2:

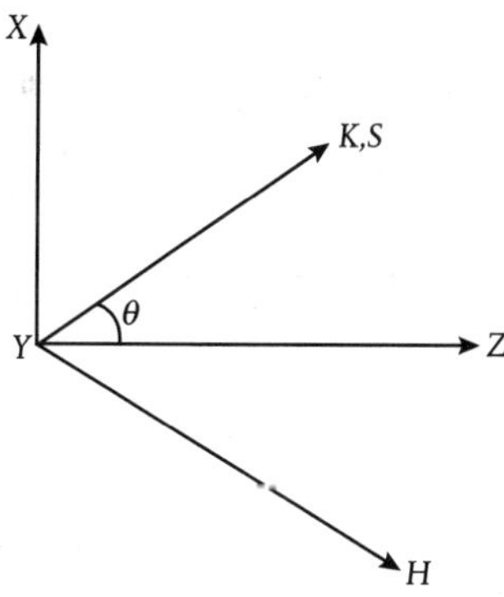

Fig. 7.2 Wave propagates in the x–z plane at an angle θ with the z-direction

For such case, we get:

$$s_x = sin\theta, s_y = 0, s_z = cos\theta.$$

Further, the determinant from (7.34)–(7.36) gives us two roots:

$$n_{ro} = K_y^{\frac{1}{2}}. \tag{7.37}$$

$$n_{re}^2 = K_x cos^2\theta + K_z sin^2\theta. \tag{7.38}$$

Since $c/n_r = v_r$ and defining $v_r s_x = v_{rx}, v_r s_y = v_{ry}$, and $v_r s_z = v_{rz}$, the three Cartesian components of the ray velocity and the determinant equation obtained from (7.34)–(7.36) can be written as:

$$\begin{vmatrix} c^2/K_x - v_{ry}^2 - v_{rz}^2 & v_{rx}v_{ry} & v_{rx}v_{rz} \\ v_{rx}v_{ry} & c^2/K_y - v_{rx}^2 - v_{rz}^2 & v_{ry}v_{rz} \\ v_{rx}v_{rz} & v_{ry}v_{rz} & c^2/K_z - v_{rx}^2 - v_{ry}^2 \end{vmatrix} = 0. \quad (7.39)$$

This equation can be represented schematically as shown in Fig. 7.3. For a given direction of propagation (i.e., for given $s_x, s_y,$ and s_z), one can obtain two ray velocities as a solution to the determinant (7.39). These two solutions are plotted as vectors in the same direction $\hat{s}(s_x, s_y, s_z)$ and of lengths equal to the two-ray velocities.

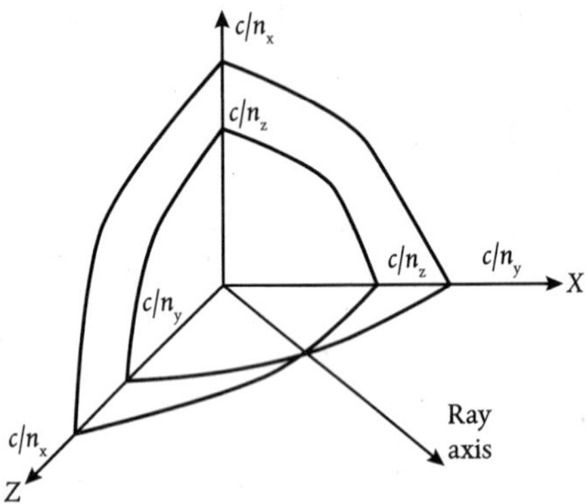

Fig. 7.3 The ray velocity surface on a biaxial medium

7.3.2 Method of Generating of Ray Velocity Surface

Consider ray propagation in *x*–*z* plane. Let us assume that

$$K_x < K_y < K_z.$$

As in the *x*–*z* plane, the two-ray refractive index are given as (7.37) and (7.38). Thus, in the *x*–*z* plane we would get two curves:

(a) $v_{rx}^2 + v_{rz}^2 = v_r^2 = c^2/n_r^2 = c^2/K_y$, which is a circle of radius $c/K_y^{1/2}$?
(b) By replacing $cos\theta = v_{rz}/v_r = v_{rz}n_r/c$ and $\sin\theta = v_{rx}/v_r = v_{rx}n_r/c$ in (7.38) leads to

$$\frac{v_{rz}^2}{c^2/K_x} + \frac{v_{rx}^2}{c^2/K_z} = 1,$$

which represents an ellipse with axes $c/K_y^{\frac{1}{2}}$ and $c/K_y^{\frac{1}{2}}$ along *x* and *z*-directions, respectively.

7.4 INDEX ELLIPSOID

When light propagates in a crystal, **D** vectors of the two eigenwaves and the wave normal direction ***s*** form a mutually perpendicular triad. It is, therefore, more convenient to present the field vectors in the **D** space. The electric energy density U_e is given as:

$$U_e = \frac{1}{2}E.D = \frac{1}{2}\left(\frac{D_x^2}{\epsilon_x} + \frac{D_y^2}{\epsilon_y} + \frac{D_z^2}{\epsilon_z}\right) = \frac{1}{2\epsilon_0}\left(\frac{D_x^2}{n_x^2} + \frac{D_y^2}{n_y^2} + \frac{D_z^2}{n_z^2}\right).$$

The constant energy density surface in **D** space is then:

$$\left(\frac{D_x^2}{n_x^2} + \frac{D_y^2}{n_y^2} + \frac{D_z^2}{n_z^2}\right) = 2\epsilon_0 U_e$$

If we denote $D/\sqrt{2\epsilon_0 U_e} = r = (x, y, z)$, then we have

$$\frac{x^2}{n_x^2} + \frac{y^2}{n_y^2} + \frac{z^2}{n_z^2} = 1$$

This surface is called the index ellipsoid. The role of the index ellipsoid is to:

1. Find the indices of refraction of the two eigenwaves.
2. Find the corresponding directions of the D vectors of the two eigenwaves.

The prescription is as follows:

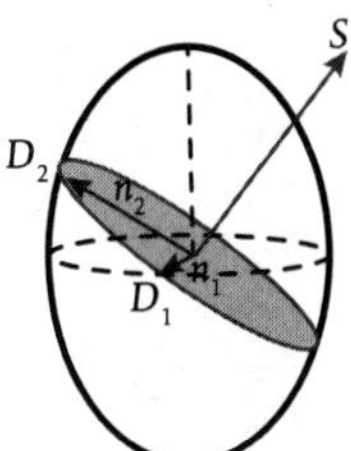

Fig. 7.4 Index ellipsoid

For a given arbitrary wave normal direction **s**, the index ellipsoid can be used to:

1. Draw a plane that is through the origin and is perpendicular to **s**. This plane intersects the index ellipsoid surface with a particular intersection ellipse.
2. The lengths of the two semi-axes of the intersection ellipse, n_1 and n_2, are the two indices of refraction of the eigenwaves.
3. The two axes of the intersection ellipse are each parallel to the allowed D vectors of the eigenwaves.

7.4.1. Index Ellipsoid in the Presence of an External Electric Field

If we refer to the principal axis system of the crystal, a general biaxial crystal can be described by an equation:

$$\frac{x^2}{n_x^2} + \frac{y^2}{n_y^2} + \frac{z^2}{n_z^2} = 1, \tag{7.40}$$

where $n_x, n_y,$ and n_z are the three principal indices of the crystal and (*x*, *y*, *z*) represents the principal axis co-ordinate system. For a uniaxial crystal with its *c*-axis parallel to the *z*-direction $n_x = n_y = n_0$ and

$n_z = n_e$; n_0 and n_e represent the ordinary and extraordinary refractive indices of the crystal. In such a case, (7.40) reduces to:

$$\frac{x^2+y^2}{n_0^2}+\frac{z^2}{n_e^2}=1.$$

The above-mentioned form of equation for index ellipsoid refers to the principal axis system. After the electric field is applied, both the size and the orientation of the index ellipsoid are changed. The equation of the index ellipsoid is generally modified:

$$E_i=\sum_j \frac{1}{\epsilon_{ij}} D_j.$$

7.4.2 Energy Density

$$U=\frac{1}{2}D.E=\frac{1}{2}\sum_{ij}\frac{1}{\epsilon_{ij}}\times D_i D_j$$

$$U=\frac{1}{2}\sum_{ij}\frac{D_i D_j}{\varepsilon_{ij}}$$

$$1=\frac{1}{2U}\left[\frac{D_x D_x}{\varepsilon_{xx}}+\frac{D_x D_y}{\varepsilon_{xy}}+\frac{D_x D_z}{\varepsilon_{xz}}+\frac{D_y D_x}{\varepsilon_{yx}}+\frac{D_y D_y}{\varepsilon_{yy}}+\frac{D_y D_z}{\varepsilon_{yz}}+\frac{D_z D_x}{\varepsilon_{zx}}+\frac{D_z D_y}{\varepsilon_{zy}}+\frac{D_z D_z}{\varepsilon_{zz}}\right]$$

The dielectric permeability matrix is real and symmetric, and therefore, the off-axis terms are equal: $(\varepsilon_{xy}=\varepsilon_{yx},\ \varepsilon_{xz}=\varepsilon_{zx},\ \varepsilon_{yz}=\varepsilon_{zy})$.

$$1=\frac{1}{2U}\left[\frac{D_x D_x}{\varepsilon_{xx}}+\frac{D_x D_y}{\varepsilon_{xy}}+\frac{D_x D_z}{\varepsilon_{xz}}+\frac{D_y D_x}{\varepsilon_{xy}}+\frac{D_y D_y}{\varepsilon_{yy}}+\frac{D_y D_z}{\varepsilon_{yz}}+\frac{D_z D_x}{\varepsilon_{xz}}+\frac{D_z D_y}{\varepsilon_{yz}}+\frac{D_z D_z}{\varepsilon_{zz}}\right]$$

$$1=\frac{1}{2U}\left[\frac{D_x D_x}{\varepsilon_{xx}}+\frac{D_y D_y}{\varepsilon_{yy}}+\frac{D_z D_z}{\varepsilon_{zz}}+2\frac{D_x D_y}{\varepsilon_{xy}}+2\frac{D_y D_z}{\varepsilon_{yz}}+2\frac{D_x D_z}{\varepsilon_{xz}}\right]$$

Using $x=\frac{D_x}{\sqrt{2U}}, y=\frac{D_y}{\sqrt{2U}}, z=\frac{D_z}{\sqrt{2U}}$.

$$1=\frac{1}{2U}\left[\frac{x.x.\sqrt{2U}\sqrt{2U}}{\varepsilon_{xx}}+\frac{y.y\sqrt{2U}\sqrt{2U}}{\varepsilon_{yy}}+\frac{z.z\sqrt{2U}\sqrt{2U}}{\varepsilon_{zz}}+2\frac{x.y\sqrt{2U}\sqrt{2U}}{\varepsilon_{xy}}+2\frac{y.z\sqrt{2U}\sqrt{2U}}{\varepsilon_{yz}}+2\frac{x.z\sqrt{2U}\sqrt{2U}}{\varepsilon_{xz}}\right]$$

$$1=\left[\frac{x^2}{\varepsilon_{xx}}+\frac{y^2}{\varepsilon_{yy}}+\frac{z^2}{\varepsilon_{zz}}+2\frac{x.y}{\varepsilon_{xy}}+2\frac{y.z}{\varepsilon_{yz}}+2\frac{x.z}{\varepsilon_{xz}}\right]$$

Using $\eta_{ij}=\left(\varepsilon^{-1}\right)_{ij}$,

$$1=\left[\eta_{xx}x^2+\eta_{yy}y^2+\eta_{zz}z^2+2\eta_{xy}x.y+2\eta_{yz}y.z+2\eta_{zx}z.x\right]$$

Substitute

$$\left(\frac{1}{n^2}\right)_1\equiv\eta_{xx},\left(\frac{1}{n^2}\right)_2\equiv\eta_{yy},\left(\frac{1}{n^2}\right)_3\equiv\eta_{zz},\left(\frac{1}{n^2}\right)_4\equiv\eta_{yz}=\eta_{zy},\left(\frac{1}{n^2}\right)_5\equiv\eta_{xz}=\eta_{zx},\left(\frac{1}{n^2}\right)_6\equiv\eta_{xy}=\eta_{yx}.$$

$$1=\left[\left(\frac{1}{n^2}\right)_1x^2+\left(\frac{1}{n^2}\right)_2y^2+\left(\frac{1}{n^2}\right)_3z^2+2\left(\frac{1}{n^2}\right)_4y.z+2\left(\frac{1}{n^2}\right)_5z.x+2\left(\frac{1}{n^2}\right)_6x.y\right] \tag{7.41}$$

Thus, after applying electric field equation of the index ellipsoid is generally modified as (7.41) and the six terms may be changed.

7.5 ELECTRO-OPTIC EFFECT

The electro-optic effect refers to changes in the refractive indices of the crystal due to an applied external electric field. The Pockels effect refers to changes in the refractive indices proportional to the first power of the applied electric field [14-19]. If the change is proportional to the square of the applied field, then the effect is referred as quadratic electro-optic effect or the Kerr effect. The change in the crystal refractive indices can be described in terms of the changes in the shape, dimensions, and orientation of the crystal index ellipsoid. The Pockels effect can exist only in those crystals that do not possess an inversion symmetry. A crystal is said to possess an inversion symmetry if inversion about any of its lattice points (i.e., replacing *r* by – *r*, *r* being the position vector of any lattice point measured from the point about which the inversion is being performed leaves the crystal structure invariant.)

Instead of specifying the changes in the refractive indices, it is more convenient to consider the changes in $(1/n^2)$ due to external electric field. The changes in $1/n^2$ due to Pockels effect are described by the following equation:

$$\Delta\left(1/n^2\right)_i=\sum_{k=x,y,z}r_{ik}E_k,$$

where $i=(1,2\ldots.6)$ and $k=x,y,z$

When only the linear electro-optic effect is considered, the change induced by the electric field E = (E_x, E_y, E_z) can be expressed as:

$$\begin{bmatrix}\Delta\left(\frac{1}{n^2}\right)_1\\ \Delta\left(\frac{1}{n^2}\right)_2\\ \Delta\left(\frac{1}{n^2}\right)_3\\ \Delta\left(\frac{1}{n^2}\right)_4\\ \Delta\left(\frac{1}{n^2}\right)_5\\ \Delta\left(\frac{1}{n^2}\right)_6\end{bmatrix}=\begin{bmatrix}r_{11}&r_{12}&r_{13}\\ r_{21}&r_{22}&r_{23}\\ r_{31}&r_{32}&r_{33}\\ r_{41}&r_{42}&r_{43}\\ r_{51}&r_{52}&r_{53}\\ r_{61}&r_{62}&r_{63}\end{bmatrix}\begin{bmatrix}E_x\\ E_y\\ E_z\end{bmatrix}. \tag{7.42}$$

where the 6×3 matrix $[r_{ij}]$ is called linear electro-optic tensor and its elements are called electro-optic coefficients. For instance,

$$(1/n^2)_1=(1/n_x^2)+\Delta(1/n^2)_1,(1/n^2)_2=(1/n_y^2)+\Delta(1/n^2)_2,(1/n^2)_3=(1/n_z^2)+\Delta(1/n^2)_3,$$

$$\Delta(1/n^2)_4=(1/n^2)_4,\ \Delta(1/n^2)_5=(1/n^2)_5,\ \Delta(1/n^2)_6=(1/n^2)_6 \tag{7.43}$$

7.5.1 Electro-optic Effect in Lithium Niobate Crystal

$LiNbO_3$ is the most widely used material for the manufacture of electro-optics devices, including phase modulators, polarisation modulators, and Mach–Zehnder intensity modulators. $LiNbO_3$ is intrinsically a birefringent crystal. Its extraordinary optical axis is usually chosen as the z-axis of the principal coordinate system of the index ellipsoid. Therefore, the principal indexes in (7.40) are

$$n_x=n_0,n_y=n_0,n_z=n_e, \tag{7.44}$$

where the ordinary index $n_0=2.29$ and the extraordinary index $n_e=2.2$. Due to the symmetry properties of $LiNbO_3$, the corresponding EO tensor takes the form:

$$\begin{bmatrix} 0 & -r_{22} & r_{13} \\ 0 & r_{22} & r_{13} \\ 0 & 0 & r_{33} \\ 0 & r_{51} & 0 \\ r_{51} & 0 & 0 \\ -r_{22} & 0 & 0 \end{bmatrix}. \tag{7.45}$$

In this tensor, $r_{61}=-r_{22}, r_{42}=r_{51}, r_{12}=-r_{22}$ and $r_{23}=r_{13}$. For an $LiNbO_3$ crystal, the EO-coefficient are $r_{13}=8.6$ pm/V, $r_{33}=30.8$ pm/V, $r_{22}=3.4$ pm/V, and $r_{51}=28.0$ pm/V. Combining (7.42) and (7.45), the index ellipsoid for $LiNbO_3$ in the presence of the electric field is represented as:

$$\begin{bmatrix} \Delta\left(\frac{1}{n^2}\right)_1 \\ \Delta\left(\frac{1}{n^2}\right)_2 \\ \Delta\left(\frac{1}{n^2}\right)_3 \\ \Delta\left(\frac{1}{n^2}\right)_4 \\ \Delta\left(\frac{1}{n^2}\right)_5 \\ \Delta\left(\frac{1}{n^2}\right)_6 \end{bmatrix} = \begin{bmatrix} 0 & -r_{22} & r_{13} \\ 0 & r_{22} & r_{13} \\ 0 & 0 & r_{33} \\ 0 & r_{51} & 0 \\ r_{51} & 0 & 0 \\ -r_{22} & 0 & 0 \end{bmatrix} \begin{bmatrix} E_x \\ E_y \\ E_z \end{bmatrix}$$

$$\Delta\left(\frac{1}{n^2}\right)_1 = -r_{22}E_y + r_{13}E_z \tag{7.46}$$

Using (7.43), that is, $\left(\frac{1}{n^2}\right)_1 - \left(\frac{1}{n_x^2}\right) = \Delta\left(\frac{1}{n^2}\right)_1$

From (7.46) and $\left(\frac{1}{n^2}\right)_1 - \left(\frac{1}{n_x^2}\right) = \Delta\left(\frac{1}{n^2}\right)_1$

$$\left(\frac{1}{n^2}\right)_1 = \left(\frac{1}{n_x^2}\right) - r_{22}E_y + r_{13}E_z$$

Using $n_x = n_0$, we get:

$$\left(\frac{1}{n^2}\right)_1 = \left[\frac{1}{n_0^2} - r_{22}E_y + r_{13}E_z\right]. \tag{7.47}$$

$$\Delta\left(\frac{1}{n^2}\right)_2 = r_{22}E_y + r_{13}E_z. \tag{7.48}$$

From (7.43), (7.44), and (7.48)

$$\left(\frac{1}{n^2}\right)_2 = \left[\frac{1}{n_0^2} + r_{22}E_y + r_{13}E_z\right]. \tag{7.49}$$

$$\Delta\left(\frac{1}{n^2}\right)_3 = r_{33}E_z,\ \Delta\left(\frac{1}{n^2}\right)_4 = r_{51}E_y,\ \Delta\left(\frac{1}{n^2}\right)_5 = r_{51}E_y,\ \Delta\left(\frac{1}{n^2}\right)_6 = -r_{22}E_y. \tag{7.50}$$

From (7.41), (7.48), (7.49), and (7.50)

$$x^2\left(\frac{1}{n_o^2} - r_{22}E_y + r_{13}E_z\right) + y^2\left(\frac{1}{n_o^2} + r_{22}E_y + r_{13}E_z\right) + z^2\left(\frac{1}{n_e^2} + r_{33}E_z\right) + 2yzr_{51}E_y + 2zxr_{51}E_x - 2xyr_{22}E_x = 1.$$

Since r_{33} is the largest EO coefficient, it is desirable to exploit it for EO modulation. This means electric field is applied along the z-direction and the incident optical wave is polarised along the z-axis. With the electric field along the z-direction, $E_x = E_y = 0$, the index ellipsoid becomes

$$x^2\left(\frac{1}{n_o^2} + r_{13}E_z\right) + y^2\left(\frac{1}{n_o^2} + r_{13}E_z\right) + z^2\left(\frac{1}{n_e^2} + r_{33}E_z\right) = 1.$$

Let us assume $r_{33}E_z << n_e$, $r_{13}E_z << n_o$, the refractive indices for *x*, *y*, and *z*-polarised waves will be given as:

$$n_x = \left(\frac{1}{n_o^2} + r_{13}E_z\right)^{-(1/2)} \approx n_0 - \frac{1}{2}n_0^3 r_{13}E_z$$

$$n_y = \left(\frac{1}{n_o^2} + r_{13}E_z\right)^{-(1/2)} \approx n_0 - \frac{1}{2}n_0^3 r_{13}E_z.$$

$$n_z = \left(\frac{1}{n_o^2} + r_{13}E_z\right)^{-(1/2)} \approx n_e - \frac{1}{2}n_e^3 r_{13}E_z \ldots$$

Hence, $$\Delta n = \frac{1}{2} n_0^3 r_{ij} E = \frac{1}{2} n_0^3 r_{ij} \frac{V}{d}$$

When only one arm is modulated, the phase difference becomes

$$\Delta\phi = \Delta n \frac{2\pi}{\lambda} L,$$

where Δn is the optical index change in the waveguide active layer, λ is the optical wavelength, L is the modulation length. Hence, $\Delta\phi = \frac{1}{2} n_0^3 r_{ij} \frac{V}{d} \frac{2\pi}{\lambda} L$

$$\Delta\phi = n_0^3 r_{ij} \frac{V}{d} \frac{\pi}{\lambda} L = \pi \frac{V}{V_\pi}$$

where $V_\pi = \frac{\lambda}{r_{ij} n_0^3} \frac{d}{L}$. V_π is the voltage value at which the voltage-induced phase difference reaches 180°.

7.6 MACH–ZEHNDER MODULATOR

The operating principle of MZM is based on the electro-optic effect, which is characterised by the variation of an applied electrical field causing refractive index changes in the modulator arms. The variation of the refractive index in the modulator arms induces a change of material propagation constant β, resulting in different phases in both modulator arms. The input optical signal E_0 is divided by a 3-dB coupler into two equal parts E_1 and E_2, which propagate in lower and upper arm of the modulator.

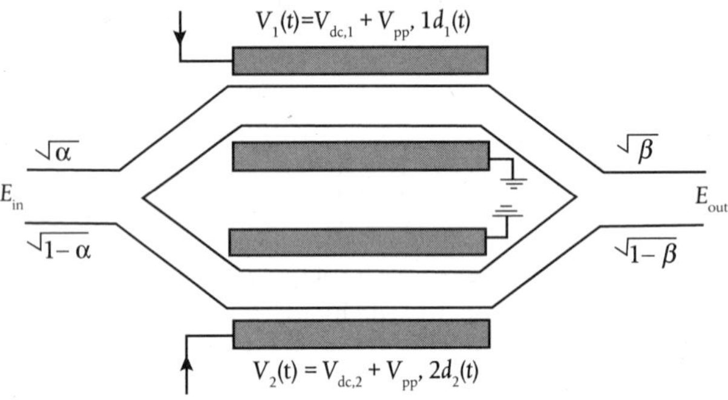

Fig. 7.5 Mach–Zehnder modulator (MZM) structure

In order to illustrate the operation principle of such a device, the simple interferometric structure represented in Fig. 7.5 is considered. It is based on a Mach–Zehnder interferometer including one electro-optic material into one of the arms. When used as a data modulator, such a structure is generally integrated by diffusing some waveguides in the electro-optic material and depositing the electrodes on top or around the waveguides. Assuming the power is split or combined equally in the input and output splitter and couplers of the Mach–Zehnder interferometer, the power at the output of the interferometer depends on the difference between the phase shifts experienced by the light propagating in the upper and lower arm of the structure, $\phi\omega$ and ϕ_0 respectively, according to

$$P_{out} = P_{in} \cos^2\left(\frac{\Delta\phi}{2}\right),$$

where $\Delta\phi = \phi(t) - \phi_0$.

The phase shift induced in the upper arm of the interferometer depends on its refractive index, which itself depends on the applied external electric field through the electro-optic effect [20-23]. If a time-dependent voltage $V(t)$ is applied across the upper waveguide of the modulator, its refractive index will become time dependent and, in turn, the transmission P_{out}/P_{in} of the Mach–Zehnder interferometer will also depend on time. If a continuous optical wave is applied to the input of the modulator, the output power will thus be modulated according to the electrical data $V(t)$. The value of the phase shift created by an applied external voltage depends upon many parameters, including the choice of the electro-optic material, the orientation of the external electric field with respect to the principal axis of the crystal, the polarisation of the incoming lightwave as well as the geometry and dimensions of the waveguide. Lithium-niobate ($LiNbO_3$) is customarily used as a suitable electro-optic material for high-speed modulation in optical communication systems, owing to its relatively large electro-optic coefficient and well-controlled waveguide fabrication processes. In any case, it is possible to make abstraction of the actual physical implementation of the modulator and describe the ability of the material and chosen configuration to respond to an applied voltage by introducing a quantity known as the half-wave voltage (V_π).

By applying a voltage of V_π to the electrode of an electro-optic waveguide will result in a voltage-induced phase shift of π. The electro-optically induced phase shift $\varphi(t)$ can, therefore, be related to the applied voltage $V(t)$ according to:

$$\phi(t) = \frac{V(t)}{V_\pi}.$$

7.6.1 Derivation of Output Electric Field of Mach-Zehnder Modulator

The field at the output of the modulator is the sum of the contributions propagating through the upper and lower paths. The loss experienced by the optical field in the upper and lower arm of the structure, including excess loss in the input and output Y-junctions, is described by ρ_1 and ρ_2, respectively.

$$E_{out}(t) = \left[\rho_1\sqrt{\alpha}\sqrt{\beta}e^{-j\phi_1} + \rho_2\sqrt{1-\alpha}\sqrt{1-\beta}e^{-j\phi_2}\right].E_{in}(t)$$

Considering,

$$a = \rho_1\sqrt{\alpha}\sqrt{\beta} \text{ and } b = \rho_2\sqrt{1-\alpha}\sqrt{1-\beta}$$

$$E_{out}(t) = \left[ae^{-j\left[\frac{(\phi_1+\phi_2)+(\phi_1-\phi_2)}{2}\right]} + be^{-j\left[\frac{(\phi_1+\phi_2)-(\phi_1-\phi_2)}{2}\right]}\right].E_{in}(t)$$

$$E_{out}(t) = e^{-j\phi}\left[ae^{-j\left(\frac{\Delta\phi}{2}\right)} + be^{j\left(\frac{\Delta\phi}{2}\right)}\right].E_{in}(t),$$

where $\Delta\phi = (\phi_1 - \phi_2)$ and $\phi = \dfrac{(\phi_1 + \phi_2)}{2}$

$$E_{out}(t)=e^{-j\phi}\left[a\cos\left(\frac{\Delta\phi}{2}\right)-ja\sin\left(\frac{\Delta\phi}{2}\right)+b\cos\left(\frac{\Delta\phi}{2}\right)+jb\sin\left(\frac{\Delta\phi}{2}\right)\right]E_{in}(t)$$

$$E_{out}(t)=e^{-j\phi}\left[(a+b)\cos\left(\frac{\Delta\phi}{2}\right)+j(b-a)\sin\left(\frac{\Delta\phi}{2}\right)\right]E_{in}(t),$$

For $a=b$

$$E_{out}(t)=2ae^{-j\phi}\left[\cos\left(\frac{\Delta\phi}{2}\right)\right]E_{in}(t),$$

The phase shifts in the upper and lower arms of the structure, ϕ_1 and ϕ_2, can be expressed as a function of the applied voltages V_1 and V_2 , respectively, as well as the half-wave voltage V_π, which is assumed to be equal in both arms.

$$\phi_1(t)=\frac{V_1(t)}{V} \text{ and } \phi_2(t)=\frac{V_2(t)}{V}$$

$$e^{-j\phi}=e^{-j\left(\frac{\phi_1+\phi_2}{2}\right)}=e^{-j\frac{\pi}{V_\pi}\left\{\frac{V_1(t)+V_2(t)}{2}\right\}+\phi_0}.$$

In a particular case,

$$V_1(t)=V_e\cos(\phi_e t) \text{ and } V_2(t)=0$$

$$E_{out}(t)=2ae^{-j\frac{\pi}{V_\pi}\left\{\frac{V_e\cos(\phi_e t)+\phi_0}{2}\right\}}.\cos\left\{\frac{1}{2}\left(\phi_0+\frac{\pi}{V_\pi}V_e\cos(\phi_e t)\right)\right\}E_0\cos(\phi_0 t)$$

$$|E_{out}(t)|\propto E_0\cos\left\{\frac{1}{2}\left(\phi_0+\frac{\pi}{V_\pi}\cos(\phi_0 t)\right)\right\}.\cos(\phi_0 t).$$

7.7 ELECTRO-OPTIC EFFECT IN KDP CRYSTAL: PHASE MODULATION (LONGITUDINAL CONFIGURATION)

When a light propagates through the crystal in the presence of an externally applied electric field, the refractive index of the crystal get altered, and thus, it could induce birefringence in isotropic crystals. This effect is known as the electro-optic effect. Consider a linearly polarised plane wave propagating along z-direction in a KDP crystal and assume that it is polarised along the x-axis as shown in Fig. 7.6. An electric field is applied along the z-direction. Since the x -direction corresponds to a principal axis, the linearly polarised lightwave will propagate without any change in the state of polarisation. If the crystal is of length 'l' along the z-direction, then the wave emerging from the crystal will be:

$$E_{x'}(z=l)=E_{x'}(0)e^{i(\omega t-kl)}$$

$$E_{x'}(z=l)=E_{x'}(0)\left[\cos(\omega t-kl)+i\sin(\omega t-kl)\right].$$

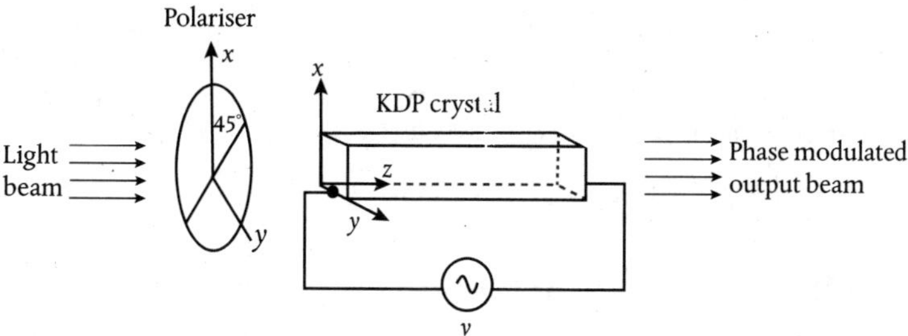

Fig. 7.6 Light beam polarised at 45° to the *x*-axis passes through a KDP crystal

Considering real part only,

$$E_{x'}(z=l)=E_{x'}(0)\cos(\omega t-kl).$$

Using $k=k_0 n_{x'}$,

$$E_{x'}(z=l)=E_{x'}(0)\cos(\omega t-k_0 n_{x'}l).$$

Using $n_{x'}=n_0-\frac{1}{2}n_0^3 r_{63}E_z$,

$$E_{x'}(z=l)=E_{x'}(0)\cos\left[\omega t-k_0\left(n_0-\frac{1}{2}n_0^3 r_{63}E_z\right)l\right].$$

Using $E_z l=V$,

$$E_{x'}(z=l)=E_{x'}(0)\cos\left[\omega t-k_0 n_0 l+\frac{1}{2}n_0^3 r_{63}k_0 V\right].$$

Substituting $V=V_0\sin\omega_m t$ and $k_0=\omega/c$, we get :

$$E_{x'}(z=l)=E_{x'}(0)\cos\left[\omega t-k_0 n_0 l+\frac{1}{2}n_0^3 r_{63}\frac{\omega}{c}V_0\sin\omega_m t\right].$$

Substituting $\xi=\frac{1}{2}n_0^3 r_{63}\omega/cV_0$, we get:

$$E_{x'}(z=l)=E_{x'}(0)\cos\left[(\omega t-k_0 n_0 l)+\xi\sin\omega_m t\right].$$

$$E_{x'}(z=l)=E_{x'}(0)\left[\cos(\omega t-k_0 n_0 l)\cos(\xi\sin\omega_m t)-\sin(\omega t-k_0 n_0 l)\sin(\xi\sin\omega_m t)\right] \quad (7.51)$$

Using Bessel function conditions, we get:

$$\cos(\xi\sin\omega_m t)=J_0(\xi)+2J_2(\xi)\cos 2\omega_m t+2J_4(\xi)\cos 4\omega_m t+\ldots$$

$$\sin(\xi\sin\omega_m t)=2J_1(\xi)\sin\omega_m t+2J_3(\xi)\sin 3\omega_m t+\ldots$$

From (7.51),

$$E_{x'}(z=l)=E_{x'}(0)\{\cos(\omega t-k_0n_0l)[J_0(\xi)+2J_2(\xi)\cos 2\omega_m t+2J_4(\xi)\cos 4\omega_m t+\ldots]$$
$$-\sin(\omega t-k_0n_0l)[2J_1(\xi)\sin\omega_m t+2J_3(\xi)\sin 3\omega_m t+\ldots]\}.$$

$$E_{x'}(z=l)=E_{x'}(0)\{J_0(\xi)\cos(\omega t-k_0n_0l)+2J_2(\xi)\cos(\omega t-k_0n_0l)\cos 2\omega_m t+2J_4(\xi)\cos(\omega t-k_0n_0l)\cos 4\omega_m t-$$
$$2J_1(\xi)\sin(\omega t-k_0n_0l)\sin\omega_m t-2J_3(\xi)\sin(\omega t-k_0n_0l)\sin 3\omega_m t+\ldots\}.$$

$$E_{x'}(z=l)=E_{x'}(0)\{J_0(\xi)\cos(\omega t-k_0n_0l)+2J_2(\xi)\cos(\omega t-k_0n_0l)\cos 2\omega_m t+2J_4(\xi)\cos(\omega t-k_0n_0l)\cos 4\omega_m t-$$
$$2J_1(\xi)\sin(\omega t-k_0n_0l)\sin\omega_m t-2J_3(\xi)\sin(\omega t-k_0n_0l)\sin 3\omega_m t+\ldots\}.$$

Using $2\sin A\sin B=\cos(A+B)-\cos(A-B)$ and $2\cos A\cos B=\cos(A+B)+\cos(A-B)$

$$E_{x'}(z=l)=E_{x'}(0)\{J_0(\xi)\cos(\omega t-k_0n_0l)-J_1(\xi)[\cos((\omega+\omega_m)t-k_0n_0l)-\cos((\omega-\omega_m)t-k_0n_0l)]+$$
$$J_2(\xi)[\cos((\omega+2\omega_m)t-k_0n_0l)+\cos((\omega-2\omega_m)t-k_0n_0l)]+\ldots\}.$$

Thus, the optical beam contains in addition to the wave at the fundamental frequency with an amplitude $J_0(\xi)$, and also various sidebands at frequency $\omega\pm\omega_m$, $\omega\pm 2\omega_m\ldots$ with amplitude $J_1(\xi), J_2(\xi)$.

7.7.1 Disadvantages of Phase Modulation

1. The electrode should be transparent, which will not create any hurdle in propagation.
2. We apply high voltage along the direction of propagation of the beam.
3. Retardation is independent of the length of the crystal and depends only on the applied voltage.

7.8 ELECTRO-OPTIC EFFECT IN KDP CRYSTAL: AMPLITUDE MODULATION

Consider a linearly polarised wave along the x' - direction, as shown in Fig 7.7, and travelling along the z-direction in a KDP crystal in which we have applied an external electric field E_z along the z-direction. Then the output wave at $z=l$:

$$E_{x'}(z=l)=E_{x'}(0)\exp\{i(\omega t-k_0n_{x'}l\}.$$

Using $n_{x'}=n_0-\frac{1}{2}n_0^3r_{63}E_z$ & $n_{y'}=n_0+\frac{1}{2}n_0^3r_{63}E_z$,

$$E_{x'}(z=l)=E_{x'}(0)\exp\left\{i\left[\omega t-k_0\left(n_0-\frac{1}{2}n_0^3r_{63}E_z\right)l\right]\right\}$$

$$E_{x'}(z=l)=E_{x'}(0)\exp\left\{i\left[\omega t-k_0n_0l+\frac{1}{2}n_0^3r_{63}k_0E_zl\right]\right\}.$$

Substituting $k_0 = \omega/c$,

$$E_{x'}(z=l) = E_{x'}(0)\exp\left\{i\left[\omega t - \frac{\omega}{c}(n_0 l) + \frac{\omega}{2c} n_0^3 r_{63} E_z l\right]\right\}. \tag{7.52}$$

If instead of a beam polarised along the x' -direction, we have a beam polarised along the y' -direction, the output wave at $z=l$ will be given as:

$$E_{y'}(z=l) = E_{y'}(0)\exp\left\{i(\omega t - k_0 n_{y'} l)\right\}.$$

Using $n_{y'} = n_0 + \frac{1}{2} n_0^3 r_{63} E_z$:

$$E_{y'}(z=l) = E_{y'}(0)\exp\left\{i\left[\omega t - \frac{\omega}{c}(n_0 l) - \frac{\omega}{2c} n_0^3 r_{63} E_z l\right]\right\}. \tag{7.53}$$

Electro-optic retardation between two components $E_{x'}(z=l)$ and $E_{y'}(z=l)$, $\gamma =$ phase of $E_{x'}(z=l) -$ phase of $E_{y'}(z=l)$:

$$\gamma = \frac{\omega}{c} n_0^3 r_{63} E_z l = \frac{\omega}{c} n_0^3 r_{63} V.$$

Hence, electro-optic retardation is independent of length and depends only on the externally applied voltage. The following are the cases:

1. When the applied voltage is 0.

$$\gamma = 0$$

2. When V = half-wave voltage = V_π

$$\gamma = \pi = \frac{\omega}{c} n_0^3 r_{63} V_\pi$$

$$V_\pi = \frac{\pi c}{\omega n_0^3 r_{63}} = \frac{\lambda_0}{2 n_0^3 r_{63}}.$$

The superposition of two retarded linearly polarised waves leads to an elliptically polarised wave.

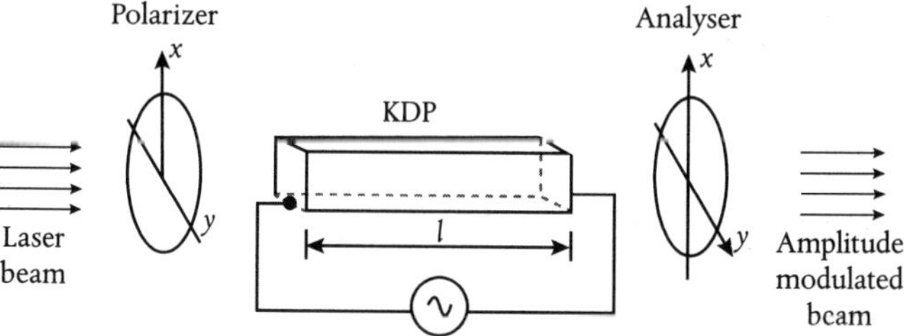

Fig. 7.7 An electro-optic amplitude modulator

On passing this elliptically polarised beam through an analyser will be modulated. Therefore, light emerging from the analyser will be given as:

$$E = E_{x'}\cos\frac{\pi}{4} + E_{y'}\sin\frac{\pi}{4} \tag{7.54}$$

$$E = \frac{1}{\sqrt{2}}\left[E_{x'} + E_{y'}\right], \tag{7.55}$$

If input is linearly polarised,

$$E_{x'}(0) = E_{y'}(0) = \frac{A}{\sqrt{2}}.$$

From (7.52)–(7.55). we get:

$$E = \frac{A}{2}\exp\left\{i\left[\omega t - \frac{\omega}{c}(n_0 l) + \frac{\omega}{2c}n_0^3 r_{63}E_z l\right]\right\} + \frac{A}{2}\exp\left\{i\left[\omega t - \frac{\omega}{c}(n_0 l) - \frac{\omega}{2c}n_0^3 r_{63}E_z l\right]\right\}.$$

$$E = \frac{A}{2}\exp\left\{i\left[\omega t - \frac{\omega}{c}(n_0 l) + \frac{\omega}{2c}n_0^3 r_{63}E_z l\right]\right\} + \frac{A}{2}\exp\left\{i\left[\omega t - \frac{\omega}{c}(n_0 l) - \frac{\omega}{2c}n_0^3 r_{63}E_z l\right]\right\} \times$$
$$\exp\left\{i\left[\omega t - \frac{\omega}{c}(n_0 l) + \frac{\omega}{2c}n_0^3 r_{63}E_z l\right]\right\} \times \exp\left\{-i\left[\omega t - \frac{\omega}{c}(n_0 l) + \frac{\omega}{2c}n_0^3 r_{63}E_z l\right]\right\}.$$

$$E = \frac{A}{2}\exp\left\{i\left[\omega t - \frac{\omega}{c}(n_0 l) + \frac{\omega}{2c}n_0^3 r_{63}E_z l\right]\right\}\left[1 - \exp\left\{i\left[\omega t - \frac{\omega}{c}(n_0 l) - \frac{\omega}{2c}n_0^3 r_{63}E_z l\right] - i\left[\omega t - \frac{\omega}{c}(n_0 l) + \frac{\omega}{2c}n_0^3 r_{63}E_z l\right]\right\}\right].$$

$$E = \frac{A}{2}\exp\left\{i\left[\omega t - \frac{\omega}{c}(n_0 l) + \frac{\omega}{2c}n_0^3 r_{63}E_z l\right]\right\}\left[1 - \exp\left\{-i\frac{\omega}{c}n_0^3 r_{63}E_z l\right\}\right],$$

where
$$\gamma = \frac{\omega}{c}n_0^3 r_{63}V = \frac{\pi V}{V_\pi} \tag{7.56}$$

$$E = \frac{A}{2}\exp\left\{i\left[\omega t - \frac{\omega}{c}(n_0 l) + \frac{\omega}{2c}n_0^3 r_{63}E_z l\right]\right\}\left[1 - \exp(-i\gamma)\right].$$

Let $\varphi = \left[\omega t - \frac{\omega}{c}(n_0 l) + \frac{\omega}{2c}n_0^3 r_{63}E_z l\right]$.

Intensity of the output beam is given as:

$$I_0 = \frac{1}{2}Re\left[E.E^*\right] = \frac{1}{2}\left\{\frac{A^2}{2}\exp(i\varphi)\left[1 - \exp(-i\gamma)\right] \times \exp(-i\varphi)\left[1 - \exp(i\gamma)\right]\right\}.$$

$$= \frac{A^2}{4}\left[1 - \exp(-i\gamma)\right]\left[1 - \exp(i\gamma)\right].$$

$$= \frac{A^2}{4}\left[1 - \exp(-i\gamma) - \exp(i\gamma) + 1\right].$$

$$=\frac{A^2}{4}[1-\cos\ddagger-i\sin\ddagger-\cos\ddagger+i\sin\ddagger+1]$$

$$=\frac{A^2}{4}[2-2\cos\gamma]$$

$$=\frac{A^2}{2}\left(\sin^2\frac{\gamma}{2}\right).$$

Intensity of the input beam $I_i=\frac{1}{2}A^2$,

$$\frac{I_0}{I_i}=\sin^2\left(\frac{\gamma}{2}\right)$$

From (7.56),

$$\frac{I_0}{I_i}=\sin^2\left(\frac{1}{2}\pi\frac{V}{V_\pi}\right).$$

Hence, the output intensity depends on sinusoidal applied voltage, as shown in Fig.7.8.

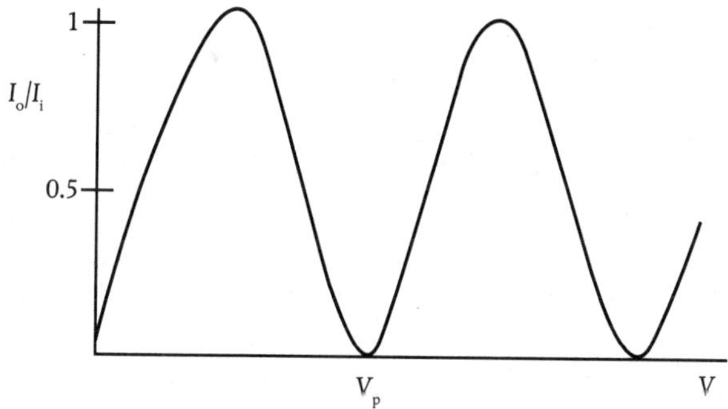

Fig. 7.8 Variation of the transmitted intensity of the amplitude modulator versus externally applied voltage

Let $V=V_0\cos t$

Therefore,
$$\frac{I_0}{I_i}=\sin^2\left(\frac{1}{2}\pi\frac{V}{V_\pi}\right)=\sin^2\left(\frac{1}{2}\pi\frac{V_0\cos\varepsilon t}{V_\pi}\right).$$

If we assume $V_0\ll V_\pi$, then

$$\frac{I_0}{I_i}\approx\left(\frac{1}{2}\pi\frac{V_0\cos t}{V_\pi}\right)^2\approx\frac{\pi^2}{4}\frac{V_0^2}{V_\pi^2}\cos^2 t\approx\frac{\pi^2}{8}\frac{V_0^2}{V_\pi^2}(1+\cos 2t).$$

Frequency is doubled, as shown in Fig.7.9. To overcome this problem, we place a $\frac{\lambda}{4}$ plate of the output as shown in Fig.7.10.

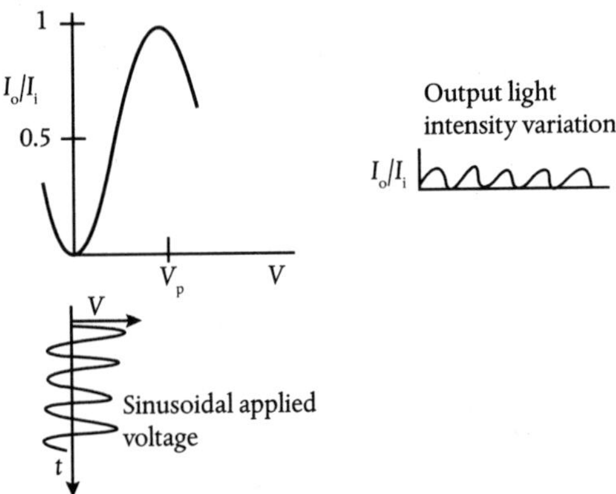

Fig. 7.9 For the amplitude modulator, the output intensity variation is not linearly related to the applied voltage

7.8.1 Remedy for Amplitude Modulation

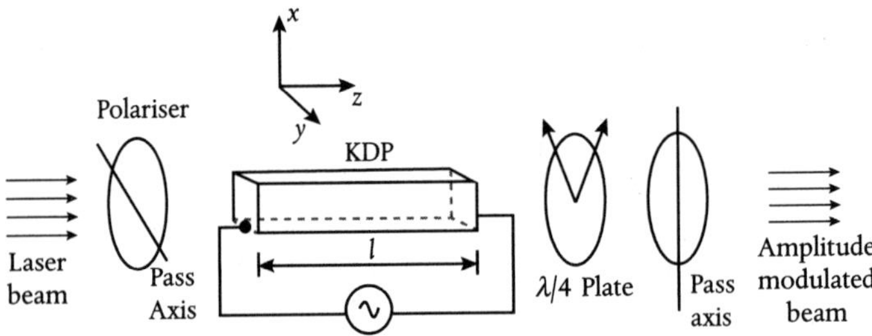

Fig. 7.10 An electro-optic amplitude modulator using KDP

$\frac{\lambda}{4}$ plate introduces an additional phase shift of $\frac{\pi}{2}$ between $E_{x'}$ and $E_{y'}$. The output of the modulator for an applied voltage V will be:

$$E=\frac{A}{2}\exp\left\{i\left[\omega t-\frac{\omega}{c}(n_0 l)+\frac{\omega}{2c}n_0^3 r_{63}E_z l\right]\right\}\left[1-\exp(-i\gamma)\exp\left(-i\frac{\pi}{2}\right)\right].$$

$$E=\frac{A}{2}\exp\left\{i\left[\omega t-\frac{\omega}{c}(n_0 l)+\frac{\omega}{2c}n_0^3 r_{63}E_z l\right]\right\}\left[1+i\exp(-i\gamma)\right].$$

Let $\varphi=\left[\omega t-\frac{\omega}{c}(n_0 l)+\frac{\omega}{2c}n_0^3 r_{63}E_z l\right].$

Output intensity = I_0:

$$I_0=\frac{1}{2}Re\left[E.E^*\right]=\frac{1}{2}\left\{\frac{A^2}{4}\exp(i\varphi)\left[1+i\exp(-i\gamma)\right]\times\exp(-i\varphi)\left[1-i\exp(i\gamma)\right]\right\}.$$

$$=\frac{A^2}{8}\left[1+i\exp(-i\gamma)\right]\left[1-i\exp(i\gamma)\right].$$

$$=\frac{A^2}{8}\left(1-i\exp(i\gamma)+i\exp(-i\gamma)+1\right).$$

$$=\frac{A^2}{8}\left(1-i\cos\gamma+\sin\gamma+i\cos\gamma+\sin\gamma+1\right).$$

$$=\frac{A^2}{8}\left(2+2\sin\gamma\right).$$

$$=\frac{A^2}{4}\left(1+\sin\gamma\right).$$

Intensity of input beam $I_i=\frac{1}{2}A^2$:

$$\frac{I_0}{I_i}=\frac{1}{2}\left(1+\sin\gamma\right)\approx\frac{1}{2}\left(1+\gamma\right).$$

From (7.56),

$$\frac{I_0}{I_i}=\frac{1}{2}\left(1+\pi\frac{V}{V_\pi}\right).$$

Hence, the output intensity depends on sinusoidal applied voltage.

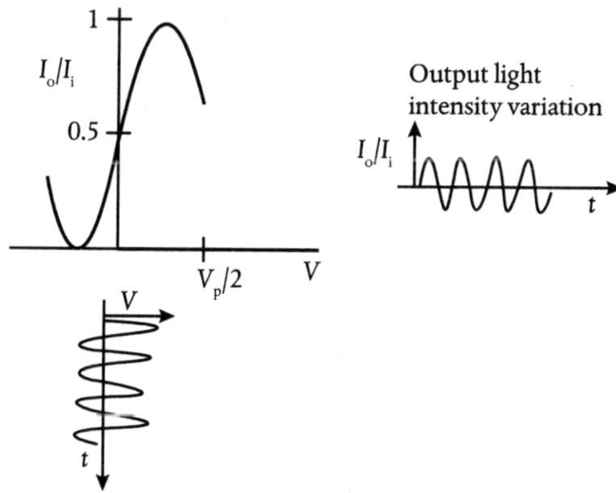

Fig. 7.11 Variation of the transmitted intensity of the amplitude modulator versus externally applied voltage

Let $V=V_0\cos t$.

Therefore,
$$\frac{I_0}{I_i}=\frac{1}{2}\left(1+\pi\frac{V}{V_\pi}\right)=\frac{1}{2}\left(1+\pi\frac{V_0\cos t}{V_\pi}\right).$$

Therefore, transmitted intensity is linearly related to the applied voltage as shown in Fig. 7.11.

7.9 ELECTRO-OPTIC EFFECT IN KDP CRYSTALS: TRANSVERSE MODE

The optical path is perpendicular to the direction of applied electric field. In transverse configuration, the light beam propagates along the y'-direction and the field is again applied along the z-direction as shown in Fig. 7.12. In the presence of a z-directed electric field, the refractive indices for a wave propagating along the y'-direction and polarised along the x'-direction and z'-direction are given as:

$$n_{x'} = n_0 - \frac{1}{2} n_0^3 r_{63} E_z \text{ and } n_{z'} = n_e$$

If the light beam incident on the crystal is linearly polarised along 45^0 to the x'-direction, then the emergent field components along x'-direction and z'-direction after travelling a length of l of the crystal will be:

$$E_{x'}(y=l) = \frac{A}{\sqrt{2}} \exp[i(\omega t - kl)].$$

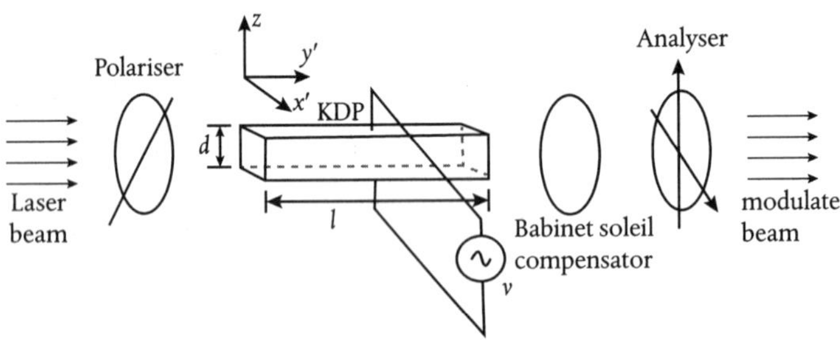

Fig. 7.12 Electro-optic KDP modulator in the transverse mode of operation

Using $k = k_0 n_{x'}$,

$$E_{x'}(y=l) = \frac{A}{\sqrt{2}} \exp(i(\omega t - k_0 n_{x'} l)).$$

Using $n_{x'} = n_0 - \frac{1}{2} n_0^3 r_{63} E_z$,

$$E_{x'}(y=l) = \frac{A}{\sqrt{2}} \exp\left\{i\left[\omega t - k_0\left(n_0 - \frac{1}{2} n_0^3 r_{63} E_z\right)l\right]\right\}$$

$$E_{z'}(y=l) = \frac{A}{\sqrt{2}} \exp\{i[\omega t - k_0 n_e l]\}.$$

At the input end,

$$E_{x'}(y'=0) = \frac{A}{\sqrt{2}} \exp(i\omega t), E_{z'}(y'=0) = \frac{A}{\sqrt{2}} \exp(i\omega t).$$

Hence, the retardation between the two linearly polarised components when the beam emerges from the crystal would be:

$$\gamma = \text{phase of } E_{x'} - \text{phase of } E_{z'}$$

$$= \left[(n_0 - n_e) - \frac{1}{2} n_0^3 r_{63} E_z \right] \frac{\omega}{c} l.$$

Using $E_z d = V$,

$$= \left[(n_0 - n_e) - \frac{1}{2} n_0^3 r_{63} \frac{V}{d} \right] \frac{\omega}{c} l.$$

Cases:

1. When V = 0, then the applied voltage is zero and there is a finite retardation due to the intrinsic birefringence of the material:

$$\gamma = (n_0 - n_e) \frac{\omega}{c} l.$$

2. Even in the absence of a field, a light wave that is propagating along the y'-direction will be an ordinary wave if its polarisation is along the x'-direction and an electromagnetic wave if its polarisation is along the z'-direction.
   Phase shift induced by external modulation voltage is

$$\gamma = \frac{1}{2c} n_0^3 r_{63} \frac{l}{d} V\omega$$

3. When $V = V_\pi$ = half-wave voltage

$$(n_0 - n_e) \frac{\omega}{c} l + \pi = \left[(n_0 - n_e) - \frac{1}{2} n_0^3 r_{63} \frac{V_\pi}{d} \right] \frac{\omega}{c} l$$

$$\pi = \frac{1}{2} n_0^3 r_{63} \frac{V_\pi}{d} \frac{\omega}{c} l$$

$$\frac{2\pi}{n_0^3 r_{63}} \frac{c}{\omega} \frac{d}{l} = V_\pi$$

7.9.1 Advantages of Transverse Mode

1) No external obstacle
2) Low-driving voltage
3) Half-wave voltage is not independent of the modulator length, but dependent on the ratio $\frac{d}{l}$

7.10 TRANSIT TIME LIMITATIONS

As the optical beam passes through the length of the modulator, there are limitations on the frequencies of the modulating signal because of finite transit time [24-29]. Time (τ_d) taken by the optical beam to pass through the crystal is given as:

$$\tau_d = \frac{nl}{c},$$

where l is length of modulator crystal and n is either n_e or n_0.

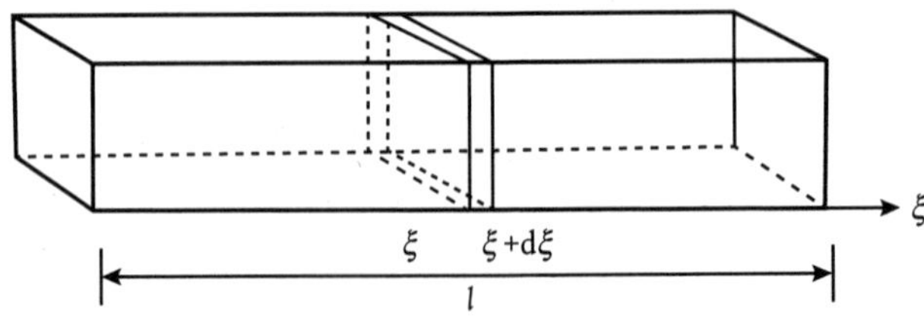

Fig. 7.13 Light beam propagation along the ξ-direction in a modulator crystal

If the applied electric field is $E = E_0 \cos\omega_m t$. Let l be the length of the crystal and $\xi = 0$ be the input plane of the crystal and $\xi = l$ be the output plane. $\xi -$ axis be along the length of the crystal. If the beam of light is emerging from the crystal at a time t, this beam will have been at ξ at a time $= t - \Delta t = t - \frac{(l-\xi)n}{c}$, where $\Delta t = \frac{(l-\xi)n}{c}$, c is the velocity of light, and n is the refractive index. Hence, $E = E_0 \cos\omega_m \left[t - \frac{(l-\xi)n}{c}\right]$. Let us consider the longitudinal configuration of a KDP crystal modulator, as shown in Fig. 7.13. The phase shift suffered by the x' -polarised wave as t travels from the plane ξ to the plane $\xi + d\xi$ is given by (7.52):

$$\Phi = \frac{\omega}{c}(n_0 l) - \frac{\omega}{2c} n_0^3 r_{63} E l$$

$$d\Phi_0 = \frac{\omega}{c}(n_0 d\xi) - \frac{\omega}{2c} n_0^3 r_{63} E_0 \cos\left\{\omega_m \left[t - \frac{(l-\xi)n'_x}{c}\right]\right\} d\xi$$

Thus, the total phase shift suffered by the x' -polarised wave as it propagates from the input plane $\xi = 0$ to the output plane $\xi = l$ will be:

$$\Phi_0 = \int_{\xi=0}^{\xi=l} d\Phi_0 = \frac{\omega}{c}\left(n_0 \int_{\xi=0}^{\xi=l} d\xi\right) - \frac{\omega}{2c} n_0^3 r_{63} E_0 \int_{\xi=0}^{\xi=l} \cos\left\{\omega_m \left[t - \frac{(l-\xi)n'_x}{c}\right]\right\} d\xi.$$

Let $A = \int_{\xi=0}^{\xi=l} \cos\left\{\omega_m \left[t - \frac{(l-\xi)n'_x}{c}\right]\right\} d\xi$

$$\Phi_0 = \frac{\omega}{c}(n_0 l) - \frac{\omega}{2c} n_0^3 r_{63} E_0 A \tag{7.57}$$

$$\Rightarrow A = \int_{\xi=0}^{\xi=l} \cos\left\{\left[\omega_m t - \frac{\omega_m (l-\xi)n'_x}{c}\right]\right\} d\xi$$

$$\Rightarrow A=\int_{\xi=0}^{\xi=l}\left\{\cos\omega_m t.\cos\left[\frac{\omega_m(l-\xi)n'_x}{c}\right]+\sin\omega_m t\sin\left[\frac{\omega_m(l-\xi)n'_x}{c}\right]\right\}d\xi.$$

$$\Rightarrow A=\cos\omega_m t\int_{\xi=0}^{\xi=l}\cos\left[\frac{\omega_m(l-\xi)n'_x}{c}\right]d\xi+\sin\omega_m t\int_{\xi=0}^{\xi=l}\sin\left[\frac{\omega_m(l-\xi)n'_x}{c}\right]d\xi.$$

$$\Rightarrow A=\cos\omega_m t\left|\frac{\sin\left[\frac{\omega_m(l-\xi)n'_x}{c}\right]}{-\frac{\omega_m n'_x}{c}}\right|_{\xi=0}^{\xi=l}+\sin\omega_m t\left|\frac{\sin\left[\frac{\omega_m(l-\xi)n'_x}{c}\right]}{\frac{\omega_m n'_x}{c}}\right|_{\xi=0}^{\xi=l}.$$

$$\Rightarrow A=\cos\omega_m t\left(-\frac{\sin\left[\frac{\omega_m l n'_x}{c}\right]}{-\frac{\omega_m n'_x}{c}}\right)+\sin\omega_m t\left(\frac{1-\cos\left[\frac{\omega_m l n'_x}{c}\right]}{\frac{\omega_m n'_x}{c}}\right).$$

$$\Rightarrow A=\left[1/\left(\frac{\omega_m n'_x}{c}\right)\right]\left[\cos\omega_m t 2\sin\left(\frac{\omega_m l n'_x}{2c}\right)\cos\left(\frac{\omega_m l n'_x}{2c}\right)+\sin\omega_m t\, 2\sin^2\left(\frac{\omega_m l n'_x}{2c}\right)\right].$$

$$\Rightarrow A=\left[2\sin\left(\frac{\omega_m l n'_x}{2c}\right)/\left(\frac{\omega_m n'_x}{c}\right)\right]\left[\cos\omega_m t\cos\left(\frac{\omega_m l n'_x}{2c}\right)+\sin\omega_m t\sin\left(\frac{\omega_m l n'_x}{2c}\right)\right].$$

By multiplying and dividing by l, we get:

$$A=\left[l\sin\left(\frac{\omega_m l n'_x}{2c}\right)/\left(\frac{\omega_m l n'_x}{2c}\right)\right]\left[\cos\left(\omega_m t-\frac{\omega_m l n'_x}{2c}\right)\right]. \tag{7.58}$$

From (7.57) and (7.58),

$$\Phi_0=\frac{\omega}{c}(n_0 l)-\frac{\omega}{2c}n_0^3 r_{63} l E_0\left\{\left[\sin\left(\frac{\omega_m l n'_x}{2c}\right)/\left(\frac{\omega_m l n'_x}{2c}\right)\right]\left[\cos\left(\omega_m t-\frac{\omega_m l n'_x}{2c}\right)\right]\right\}.$$

Similarly, the total phase shift suffered by the (y'-polarised) wave as it propagates from the input plane $\xi=0$ to the output plane $\xi=l$ will be:

$$\Phi_e=\frac{\omega}{c}(n_0 l)+\frac{\omega}{2c}n_0^3 r_{63} l E_0\left\{\left[\sin\left(\frac{\omega_m l n'_y}{2c}\right)/\left(\frac{\omega_m l n'_y}{2c}\right)\right]\left[\cos\left(\omega_m t-\frac{\omega_m l n'_y}{2c}\right)\right]\right\}.$$

Considering the retardation using
Assume $n_0=n'_x=n'_y$

$$\Phi = \Phi_e - \Phi_0$$

$$\Phi = \frac{\omega}{c} n_0^3 r_{63} l E_0 \left\{ \left[\sin\left(\frac{\omega_m n_0 l}{2c} \right) / \left(\frac{\omega_m n_0 l}{2c} \right) \right] \left[\cos\left(\omega_m t - \frac{\omega_m n_0 l}{2c} \right) \right] \right\}$$

Let $\Phi^0 = \frac{\omega}{c} n_0^3 r_{63} l E_0$ and $\tau_d = \frac{n_0 l}{c}$

$$\Phi = \Phi^0 \left\{ \left[\sin\left(\frac{1}{2} \omega_m \tau_d \right) / \left(\frac{1}{2} \omega_m \tau_d \right) \right] \left[\cos\left(\omega_m t - \frac{1}{2} \omega_m \tau_d \right) \right] \right\} \tag{7.59}$$

The factor
$$\sin\left(\frac{1}{2} \omega_m \tau_d \right) / \left(\frac{1}{2} \omega_m \tau_d \right). \tag{7.60}$$

It gives the reduction in the peak retardation due to the finite transit time. If $\omega_m \tau_d = 2\pi$ or $\tau_d = 2\pi / \omega_m$ i.e the time taken by the light beam to traverse the crystal is exactly equal to the time period of the applied electric field, then there is no retardation.

7.11 TRAVELLING WAVE MODULATORS

The transit time limitation on the maximum modulating frequency due to transit time effects can be overcome by the following methods:

(a) By having the applied electric field also propagating along the length of the crystal. Such modulators are referred to as travelling wave modulators.
(b) It is important for the phase velocity of the modulating signal to be equal to the phase velocity of the light wave propagating through the crystal.

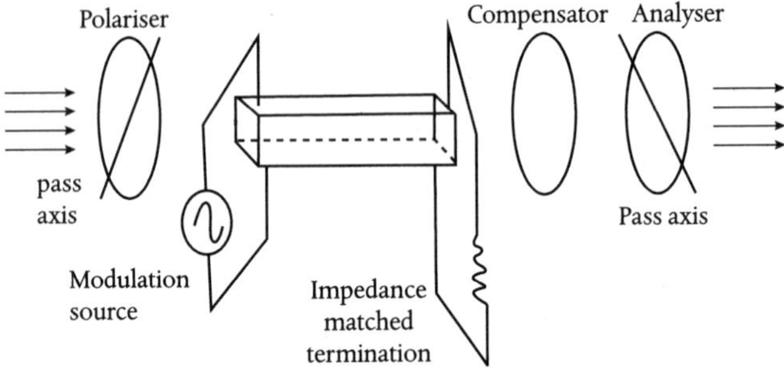

Fig. 7.14 Travelling wave electro-optic modulator

In travelling wave modulator, the modulator crystal is placed between a pair of electrodes that forms a transmission line as shown in Fig. 7.14. The modulating field also propagates along the crystal and the two velocities are matched [30–36]. For a travelling wave, the applied electric field is no longer given by $E = E_0 \cos \omega_m t$. However, by the following equation:

$$E = E_0 \cos\left(\omega_m t - k_m \xi \right), \tag{7.61}$$

where ω_m is the modulation frequency and k_m represents the wave vector of the applied electric field. The velocity of propagation of the applied field will be:

$$v_m = \frac{\omega_m}{k_m}.$$

The time taken by the optical beam to pass through the crystal is given by:

$$= t - \Delta t = t - \frac{(l-\xi)n}{c}$$

The applied electric field (7.61) becomes

$$E = E_0 \cos\left\{\omega_m\left[t - \frac{(l-\xi)n}{c}\right] - k_m\xi\right\}.$$

Thus, phase shift in travelling from the plane ξ to $\xi + d\xi$ will be:

$$\Delta\Phi = (\omega/c)\Delta n d\xi,$$

where Δn is the difference in the refractive indices of the two polarisation states induced by the applied electric field. Since Δn is proportional to the applied field:

$$d\Phi = \alpha\cos[\omega_m(t - ln/c) + \xi((\omega_m n/c) - k_m)]d\xi,$$

where α is the retardation per unit length due to a constant applied electric field of strength E_0 and in the longitudinal configuration, it is given by $\alpha = \frac{\omega}{2c} n_0^3 r_{63} E_0$. Thus, the total retardation in the wave as it propagates through the length l of the crystal will be:

$$\Phi = \int_{\xi=0}^{\xi=l} d\Phi = \alpha\int_{\xi=0}^{\xi=l} \cos[\omega_m(t - ln/c) + \xi((\omega_m n/c) - k_m)]d\xi$$

$$= \alpha\int_{\xi=0}^{\xi=l} \{\cos[\omega_m(t - ln/c)]\cos[\xi((\omega_m n/c) - k_m)] - \sin[\omega_m(t - ln/c)]\sin[\xi((\omega_m n/c) - k_m)]\}d\xi$$

$$= \alpha\left\{\cos[\omega_m(t - ln/c)]\int_{\xi=0}^{\xi=l} \cos[\xi((\omega_m n/c) - k_m)]d\xi - \sin[\omega_m(t - ln/c)]\int_0^l \sin[\xi((\omega_m n/c) - k_m)]d\xi\right\}$$

$$= \alpha\left\{\cos[\omega_m(t - ln/c)]\left|\frac{\sin[\xi((\omega_m n/c) - k_m)]}{(\omega_m n/c) - k_m}\right|_0^l + \sin[\omega_m(t - ln/c)]\left|\frac{\cos[\xi((\omega_m n/c) - k_m)]}{(\omega_m n/c) - k_m}\right|_0^l\right\}$$

$$= \alpha[1/((\omega_m n/c) - k_m)]\{\cos[\omega_m(t - ln/c)]\sin[l((\omega_m n/c) - k_m)] + \sin[\omega_m(t - ln/c)]\{\cos[l((\omega_m n/c) - k_m)] - 1\}\}$$

$$= \alpha[1/((\omega_m n/c) - k_m)]\left\{\cos[\omega_m(t - ln/c)]\sin[l((\omega_m n/c) - k_m)] - \sin[\omega_m(t - ln/c)]2\sin^2\left[\frac{l}{2}((\omega_m n/c) - k_m)\right]\right\}$$

$$= \alpha[1/((\omega_m n/c) - k_m)]\left\{\cos[\omega_m(t - ln/c)]2\sin\left[\frac{l}{2}((\omega_m n/c) - k_m)\right]\cos\left[\frac{l}{2}((\omega_m n/c) - k_m)\right]\right.$$
$$\left. - \sin[\omega_m(t - ln/c)]2\sin^2\left[\frac{l}{2}((\omega_m n/c) - k_m)\right]\right\}$$

$$= 2\alpha \sin\left[\frac{l}{2}((\omega_m n/c) - k_m)\right]\left[1/((\omega_m n/c) - k_m)\right]\left\{\cos\left[\omega_m (t - ln/c)\right] 2\cos\left[\frac{l}{2}((\omega_m n/c) - k_m)\right]\right.$$

$$\left. - \sin\left[\omega_m (t - ln/c)\right]\sin\left[\frac{l}{2}((\omega_m n/c) - k_m)\right]\right\}$$

$$= 2\alpha \sin\left[\frac{l}{2}((\omega_m n/c) - k_m)\right]\left[1/((\omega_m n/c) - k_m)\right]\left\{\cos\left[\omega_m t - l\omega_m n/c - l\omega_m n/2c - k_m\right]\right\}$$

By multiplying and dividing by l:

$$= \alpha l \sin\left[\frac{l}{2}((\omega_m n/c) - k_m)\right]\left[1/2l((\omega_m n/c) - k_m)\right]\left\{\cos\left[\omega_m t - \frac{1}{2} l((\omega_m n/c) + k_m)\right]\right\}$$

$$\Phi = \Phi_0 \sin\left[\frac{l}{2}((\omega_m n/c) - k_m)\right]\left[1/2l((\omega_m n/c) - k_m)\right]\left\{\cos\left[\omega_m t - \frac{1}{2} l((\omega_m n/c) + k_m)\right]\right\} \tag{7.62}$$

where $\Phi_0 = \alpha l$ is the peak-induced retardation over length l of the crystal.

Let $v = c/n =$ phase velocity of the light beam and $v_m = \omega_m / k_m =$ phase velocity of the modulation field. Then, (7.62) reduces to:

$$\Phi = \Phi_0 \frac{\sin\left[\frac{1}{2}\omega_m \tau_d \left(1 - (v/v_m)\right)\right]}{\frac{1}{2}\omega_m \tau_d \left(1 - (v/v_m)\right)} \left\{\cos\left[\omega_m t - \frac{\omega_m ln}{2c}\left(1 + \frac{v}{v_m}\right)\right]\right\}$$

where $\tau_d = ln/c$. The factor $\dfrac{\sin\left[\frac{1}{2}\omega_m \tau_d \left(1 - (v/v_m)\right)\right]}{\frac{1}{2}\omega_m \tau_d \left(1 - (v/v_m)\right)}$ gives the reduction in the peak retardation, which is similar to (7.60) except for the factor $\left(1 - (v/v_m)\right)$. Thus, if $v = v_m$, then the finite τ_d has no restriction on the modulation and maximum retardation is achieved irrespective of the length of the crystal.

7.12 CORE FIELD IN GRADED INDEX FIBRES WITH NON-CIRCULAR INDEX CONTOURS

7.12.1 Analysis of Core Mode Fields in Elliptical-Cylindrical Coordinate System

Let us consider an ellipse in the plane (x, y) whose equation reads $(x/a)^2 + (y/b)^2 = 1$, where $a > b$. The semi-focal distance f is given as $f^2 = a^2 - b^2$, and the eccentricity e reads as $e = f/a$, $e \in [0,1]$. The ellipse is defined by any pair of parameters selected from $\{a, b, f, e\}$.

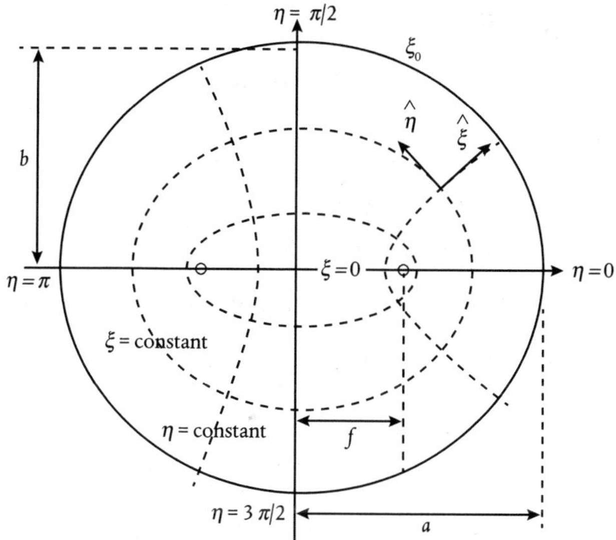

Fig. 7.15 Elliptical–cylindrical coordinate system.

The elliptic coordinates (ξ, η, z) are defined according to the transformation $(x+iy)=f\cosh(\xi+i\eta)$. Expanding

$$x+iy = f\cosh(\xi)\cosh(i\eta)+f\sinh(\xi)\sinh(i\eta)\text{, and}$$

using $\cosh(i\theta)=\cos(\theta)$ and $\sinh(i\theta)=i\sin(\theta)$,

$$x+iy = f\cosh(\xi)\cos(\eta)+if\sinh(\xi)\sin(\eta)$$

Now, by equating the real and the imaginary parts of each side, we obtain:

$$x=f\cosh\xi\cos\eta;\ y=f\sinh\xi\sin\eta;\ z=z,$$

where ξ works as a radial coordinate and takes the values $\xi\in(0,\infty)$; the coordinate η is an angular coordinate taking the range $\eta\in(0,2\pi)$; and $z\in(-\infty,\infty)$. The surface $\xi=\xi_0$ reduces to the elliptic cylinder:

$$\left[\frac{x}{f\cosh\xi_0}\right]^2+\left[\frac{y}{f\sinh\xi_0}\right]^2=\cos^2\eta+\sin^2\eta=1$$

with semi-major axis $a=f\cosh\xi_0$, and semi-minor axis $b=f\sinh\xi_0$, as shown in Fig.7.15. Consider fibres invariant in the z-direction, and thus, mode fields vary as $exp(i\beta z)$. The exact fields satisfy Maxwell's equations, but in practice, the refractive index changes very little in a distance of one wavelength. Therefore, the modes obtained by using the scalar wave approximation will be quite accurate. Each mode $E(x,y,z)$ satisfies the scalar wave equation,

$$\left[\nabla^2+k^2n^2(x,y)\right]E=0$$

where $n(x,y)$ is the refractive index and $k=2\pi/\lambda$, with λ being the free-space wavelength.

Taking $E(x,y,z)=\phi(x,y)\exp(i\beta z)$,

$$\left[\nabla_t^2+k^2n^2-\beta^2\right]\phi=0 \tag{7.63}$$

where $\nabla_t^2=\nabla^2-\dfrac{\partial^2}{\partial z^2}$

Derivation of $\nabla_t^2\phi=\dfrac{1}{f^2\left(\sinh^2\xi+\sin^2\eta\right)}\left(\dfrac{\partial^2\phi}{\partial\xi^2}+\dfrac{\partial^2\phi}{\partial\eta^2}\right)$

Let $z=x+iy=f\cosh(\xi+i\eta),\bar{z}=x-iy=f\cosh(\xi-i\eta)$, then $z\bar{z}=x^2+y^2$ and

$$\frac{4\partial^2 z\bar{z}}{\partial z\partial\bar{z}}=\left(\frac{\partial^2}{\partial x^2}+\frac{\partial^2}{\partial y^2}\right)z\bar{z}$$

$$\frac{4\partial^2}{\partial z\partial\bar{z}}=\left(\frac{\partial^2}{\partial x^2}+\frac{\partial^2}{\partial y^2}\right) \tag{7.64}$$

We transform to elliptical coordinates substituting $\zeta=\xi+i\eta,\bar{\zeta}=\xi-i\eta$. Therefore, we get $z=f\cosh\zeta,\bar{z}=f\cosh\bar{\zeta}$.

Then,

$$\zeta\bar{\zeta}=\xi^2+\eta^2 \tag{7.65}$$

Using $\dfrac{d\cosh\theta}{d\theta}=\sinh\theta$ and $\dfrac{d\sinh\theta}{d\theta}=\cosh\theta$

As

$$z=f\cosh\zeta,\ \frac{\partial z}{\partial\zeta}=f\sinh\zeta \tag{7.66}$$

Hence,

$$\frac{\partial\zeta}{\partial z}=\frac{1}{f\sinh\zeta} \tag{7.67}$$

Similarly,

$$\frac{\partial\bar{\zeta}}{\partial\bar{z}}=\frac{1}{f\sinh\bar{\zeta}} \tag{7.68}$$

as $\zeta\bar{\zeta}=\xi^2+\eta^2$

From (7.64),

$$\frac{4\partial^2\zeta\bar{\zeta}}{\partial\zeta\partial\bar{\zeta}}=\left(\frac{\partial^2}{\partial\xi^2}+\frac{\partial^2}{\partial\eta^2}\right)\zeta\bar{\zeta} \tag{7.69}$$

From (7.67)–(7.69),

$$\frac{\partial}{\partial z}=\frac{1}{f\sinh\zeta}\frac{\partial}{\partial\zeta} \tag{7.70}$$

$$\frac{\partial}{\partial \bar{z}}=\frac{1}{f\sinh\bar{\zeta}}\frac{\partial}{\partial\bar{\zeta}} \tag{7.71}$$

$$\frac{4\partial^2}{\partial\zeta\partial\bar{\zeta}}=\left(\frac{\partial^2}{\partial\xi^2}+\frac{\partial^2}{\partial\eta^2}\right) \tag{7.72}$$

Multiplying (7.70) and (7.71), we get:

$$\frac{\partial}{\partial z}\times\frac{\partial}{\partial\bar{z}}=\frac{1}{f\sinh\zeta}\frac{1}{f\sinh\bar{\zeta}}\frac{\partial}{\partial\zeta}\frac{\partial}{\partial\bar{\zeta}}$$

Hence,

$$\frac{4\partial^2}{\partial z\partial\bar{z}}=\frac{4}{f^2\sinh\zeta\sinh\bar{\zeta}}\frac{\partial^2}{\partial\zeta\partial\bar{\zeta}} \tag{7.73}$$

Using $2\sinh(A+B)\sinh(A-B)=\cosh(A+B)+\cosh(A-B)$,

$$\sinh\zeta\sinh\bar{\zeta}=\sinh(\xi+i\eta)\sinh(\xi-i\eta)=\frac{1}{2}\{\cosh 2\xi-\cosh 2i\eta\}$$

Using $\cosh i\theta=\cos\theta$,

$$\frac{1}{2}\{\cosh 2\xi-\cosh 2i\eta\}=\frac{1}{2}\{\cosh 2\xi-\cos 2\eta\} \tag{7.74}$$

From (7.73) and (7.74),

$$\frac{4\partial^2}{\partial z\partial\bar{z}}=\frac{2}{f^2(\cosh 2\xi-\cos 2\eta)}\frac{4\partial^2}{\partial\zeta\partial\bar{\zeta}}$$

Using $\cosh 2\theta=1+2\sinh^2\theta$ and $\cosh 2\theta=1-2\sin^2\theta$,

$$\frac{4\partial^2}{\partial z\partial\bar{z}}=\frac{2}{f^2\left[\left(\left(1+2\sinh^2\xi\right)\right)-\left(1-2\sin^2\eta\right)\right]}\frac{4\partial^2}{\partial\zeta\partial\bar{\zeta}}$$

$$\frac{4\partial^2}{\partial z\partial\bar{z}}=\frac{2}{f^2\left[2\left(\sinh^2\xi+\sin^2\eta\right)\right]}\frac{4\partial^2}{\partial\zeta\partial\bar{\zeta}}$$

Using $\frac{4\partial^2}{\partial\zeta\partial\bar{\zeta}}=\left(\frac{\partial^2}{\partial\xi^2}+\frac{\partial^2}{\partial\eta^2}\right)$

$$\frac{4\partial^2}{\partial z\partial\bar{z}}=\frac{2}{f^2\left[2\left(\sinh^2\xi+\sin^2\eta\right)\right]}\left(\frac{\partial^2}{\partial\xi^2}+\frac{\partial^2}{\partial\eta^2}\right)$$

$$\frac{4\partial^2}{\partial z\partial\bar{z}}=\left(\frac{\partial^2}{\partial x^2}+\frac{\partial^2}{\partial y^2}\right)=\frac{1}{f^2\left(\sinh^2\xi+\sin^2\eta\right)}\left(\frac{\partial^2}{\partial\xi^2}+\frac{\partial^2}{\partial\eta^2}\right) \tag{7.75}$$

From (7.75),

$$\nabla_t^2\phi=\frac{1}{f^2\left(\sinh^2\xi+\sin^2\eta\right)}\left(\frac{\partial^2\phi}{\partial\xi^2}+\frac{\partial^2\phi}{\partial\eta^2}\right) \tag{7.76}$$

Solving by method of "separable of variables"
From (7.63), we get

$$\left[\nabla_t^2+k^2n^2-\beta^2\right]\phi=0$$

$$\nabla_t^2\phi+\left(k^2n^2-\beta^2\right)\phi=0$$

From (7.76), we get:

$$\frac{1}{f^2\left(\sinh^2\xi+\sin^2\eta\right)}\left(\frac{\partial^2\phi}{\partial\xi^2}+\frac{\partial^2\phi}{\partial\eta^2}\right)+\left(k^2n^2-\beta^2\right)\phi=0$$

Let $\phi=Q(\xi)S(\eta)$,

$$\frac{\partial\phi}{\partial\xi}=S(\eta)\frac{\partial Q}{\partial\xi}\text{ and }\frac{\partial\phi}{\partial\eta}=Q(\xi)\frac{\partial S(\eta)}{\partial\eta}$$

$$\frac{\partial^2\phi}{\partial\xi^2}=S(\eta)\frac{\partial^2 Q}{\partial\xi^2}\text{ and }\frac{\partial^2\phi}{\partial\eta^2}=Q(\xi)\frac{\partial^2 S}{\partial\eta^2}$$

$$\frac{1}{f^2\left(\sinh^2\xi+\sin^2\eta\right)}\left(S(\eta)\frac{\partial^2 Q}{\partial\xi^2}+Q(\xi)\frac{\partial^2 S}{\partial\eta^2}\right)+\left(k^2n^2-\beta^2\right)Q(\xi)S(\eta)=0.$$

Dividing both sides by $Q(\xi)S(\eta)$, we get:

$$\frac{1}{f^2\left(\sinh^2\xi+\sin^2\eta\right)}\left(\frac{1}{Q(\xi)}\frac{\partial^2 Q}{\partial\xi^2}+\frac{1}{S(\eta)}\frac{\partial^2 S}{\partial\eta^2}\right)+\left(k^2n^2(\xi,\eta)-\beta^2\right)=0 \tag{7.77}$$

This equation can be separated if we can find a value of f for which:

$$n^2(\xi,\eta)=n_0^2-\frac{g(\xi)+h(\eta)}{\sinh^2\xi+\sin^2\eta} \tag{7.78}$$

From (7.77) and (7.78),

$$\frac{1}{f^2\left(\sinh^2\xi+\sin^2\eta\right)}\left(\frac{1}{Q(\xi)}\frac{\partial^2 Q}{\partial\xi^2}+\frac{1}{S(\eta)}\frac{\partial^2 S}{\partial\eta^2}\right)+\left[k^2\left(n_0^2-\frac{g(\xi)+h(\eta)}{\sinh^2\xi+\sin^2\eta}\right)-\beta^2\right]=0$$

$$\frac{1}{Q(\xi)}\frac{\partial^2 Q}{\partial\xi^2}+\frac{1}{S(\eta)}\frac{\partial^2 S}{\partial\eta^2}+f^2\left(k^2n_0^2-\beta^2\right)\left(\sinh^2\xi+\sin^2\eta\right)-k^2f^2\left(g(\xi)+h(\eta)\right)=0$$

By separating $Q(\xi)$ and $S(\eta)$ terms, we get:

$$\frac{1}{Q(\xi)}\frac{\partial^2 Q}{\partial \xi^2}+f^2\left(k^2n_0^2-\beta^2\right)\sinh^2\xi-k^2f^2g(\xi)$$
$$=-\left[\frac{1}{S(\eta)}\frac{\partial^2 S}{\partial \eta^2}+f^2\left(k^2n_0^2-\beta^2\right)\sin^2\eta-k^2f^2h(\eta)\right]$$

Equating each term equal to a constant μ:

$$\frac{1}{Q(\xi)}\frac{\partial^2 Q}{\partial \xi^2}+f^2\left(k^2n_0^2-\beta^2\right)\sinh^2\xi-k^2f^2g(\xi)=\mu$$

$$\frac{1}{Q(\xi)}\frac{\partial^2 Q}{\partial \xi^2}+f^2\left(k^2n_0^2-\beta^2\right)\sinh^2\xi-k^2f^2g(\xi)-\mu=0$$

$$\frac{1}{Q(\xi)}\frac{\partial^2 Q}{\partial \xi^2}+k^2f^2\left[\left(n_0^2-\frac{\beta^2}{k^2}\right)\sinh^2\xi-g(\xi)\right]-\mu=0.$$

$$\frac{\partial^2 Q}{\partial \xi^2}+k^2\left\{f^2\left[\left(n_0^2-\frac{\beta^2}{k^2}\right)\sinh^2\xi-g(\xi)\right]-\frac{\mu}{k^2}\right\}Q(\xi)=0.$$

Let $G(\xi)=f^2\left[\left(n_0^2-\frac{\beta^2}{k^2}\right)\sinh^2\xi-g(\xi)\right]-\frac{\mu}{k^2}$,

$$\frac{\partial^2 Q}{\partial \xi^2}+k^2G(\xi)Q(\xi)=0 \tag{7.79}$$

Similarly,

$$\left[\frac{1}{S(\eta)}\frac{\partial^2 S}{\partial \eta^2}+f^2\left(k^2n_0^2-\beta^2\right)\sin^2\eta-k^2f^2h(\eta)\right]=-\mu$$

$$\frac{1}{S(\eta)}\frac{\partial^2 S}{\partial \eta^2}+f^2\left(k^2n_0^2-\beta^2\right)\sin^2\eta-k^2f^2h(\eta)=\mu$$

$$\frac{1}{S(\eta)}\frac{\partial^2 S}{\partial \eta^2}+f^2\left(k^2n_0^2-\beta^2\right)\sin^2\eta-k^2f^2h(\eta)-\mu=0$$

$$\frac{1}{S(\eta)}\frac{\partial^2 S}{\partial \eta^2}+k^2f^2\left[\left(n_0^2-\frac{\beta^2}{k^2}\right)\sin^2\eta-h(\eta)\right]-\mu=0$$

$$\frac{\partial^2 S}{\partial \eta^2}+k^2\left\{f^2\left[\left(n_0^2-\frac{\beta^2}{k^2}\right)\sin^2\eta-h(\eta)\right]-\frac{\mu}{k^2}\right\}S(\eta)=0$$

Let $H(\eta)=f^2\left[\left(n_0^2-\frac{\beta^2}{k^2}\right)\sin^2\eta-h(\eta)\right]+\frac{\mu}{k^2}$

$$\frac{\partial^2 S}{\partial\eta^2}+k^2H(\eta)S(\eta)=0 \tag{7.80}$$

Solution for $\frac{\partial^2 Q}{\partial\xi^2}+k^2G(\xi)Q(\xi)=0$

Equation (7.79) can be solved using WKB approximation. Consider general second-order differential equation:

$$\frac{\partial^2 y}{\partial x^2}+\lambda^2u(x)y=0 \tag{7.81}$$

where λ is a large parameter and we take x to be real.

The asymptotic solution of (7.81) can be written as:

$$y(x)=e^{\lambda\phi(x)}\left[(z_0(x)+\frac{1}{\lambda}z_1(x)\ldots\right] \tag{7.82}$$

$$y'(x)=e^{\lambda\phi(x)}\lambda\phi'(x)\left[z_0(x)+\frac{1}{\lambda}z_1(x)\ldots\right]+e^{\lambda\phi(x)}\left[z_0'(x)+\frac{1}{\lambda}z_1'(x)\ldots\right] \tag{7.83}$$

$$\begin{aligned}y''^{(x)}=e^{\lambda\phi(x)}\lambda^2\phi'^2(x)\left[z_0(x)+\frac{1}{\lambda}z_1(x)\ldots\right]+e^{\lambda\phi(x)}\lambda\phi''(x)\left[z_0(x)+\frac{1}{\lambda}z_1(x)\ldots\right]+\\ e^{\lambda\phi(x)}\lambda\phi'(x)\left[z_0'(x)+\frac{1}{\lambda}z_1'(x)\ldots\right]+e^{\lambda\phi(x)}\lambda\phi'(x)\left[z_0'(x)+\frac{1}{\lambda}z_1'(x)\ldots\right]+e^{\lambda\phi(x)}\left[z_0''(x)+\frac{1}{\lambda}z_1''(x)\ldots\right]\end{aligned} \tag{7.84}$$

From (7.81), (7.82), and (7.84)

$$\begin{aligned}&\left\{e^{\lambda\phi(x)}\lambda^2\phi'^2(x)\left[z_0(x)+\frac{1}{\lambda}z_1(x)\ldots\right]+e^{\lambda\phi(x)}\lambda\phi''(x)\left[z_0(x)+\frac{1}{\lambda}z_1(x)\ldots\right]\right.\\ &+e^{\lambda\phi(x)}\lambda\phi'(x)\left[z_0'(x)+\frac{1}{\lambda}z_1'(x)\ldots\right]+e^{\lambda\phi(x)}\lambda\phi'(x)\left[z_0'(x)+\frac{1}{\lambda}z_1'(x)\ldots\right]\\ &\left.+e^{\lambda\phi(x)}\left[z_0''(x)+\frac{1}{\lambda}z_1''(x)\ldots\right]\right\}+\lambda^2u(x)\left\{e^{\lambda\phi(x)}\lambda\phi'(x)\left[z_0(x)+\frac{1}{\lambda}z_1(x)\ldots\right]\right\}=0\end{aligned}$$

$$\begin{aligned}&\lambda^2\phi'^2(x)z_0(x)+\lambda\phi'^2(x)z_1(x)+\lambda\phi'^2(x)z_0(x)+\phi(x)''z_1(x)+\lambda\phi'(x)z_0'(x)+\phi'(x)z_1'(x)\\ &+\lambda.\phi'(x)z_0'(x)+\phi'(x)z_1'(x)+z_0''(x)+\frac{1}{\lambda}z_1''(x)+\lambda^2u(x)z_0(x)+\lambda u(x)z_1(x)\quad+\ldots=0\end{aligned}$$

$$\lambda^2\left(\phi'^2(x)+u(x)\right)z_0(x)+\lambda\left[2\phi'(x)z_0'(x)+\phi''(x)z_0(x)\right]+2\phi'(x)z'_1(x)++z''_0(x)+\ldots=0$$

Equating coefficients of $\lambda^2=0$,

$$\phi'^2(x)+u(x)=0$$

$$\phi'^2(x)=-u(x)$$

Integrating with respect to x,

$$\phi'(x)=-u^{-\frac{1}{2}}(x)$$

Integrating with respect to x

$$\phi(x)=\pm\int u^{-\frac{1}{2}}(x)dx \tag{7.85}$$

Equating coefficients of λ:

$$2\phi'(x)z_0'(x)+\phi''(x)z_0(x)=0$$

$$\frac{z_0'(x)}{z_0(x)}=-\frac{\phi''(x)}{2\phi'(x)}$$

Integrating both sides:

$$\int\frac{z_0'(x)}{z_0(x)}dx=-\int\frac{\phi''(x)}{2\phi'(x)}dx$$

$$ln\left(z_0(x)\right)=-\frac{1}{2}\ln\left(\phi'(x)\right)+ln(k1)$$

$$ln\left(z_0(x)\right)=-\frac{1}{2}\ln\left(k1\phi'(x)\right)$$

$$ln\left(z_0(x)\right)=\ln\left(k1\phi'(x)\right)^{-\frac{1}{2}}$$

$$z_0(x)=\frac{\text{constant}}{\sqrt{\phi'(\text{x})}} \tag{7.86}$$

From (7.85) and (7.86),

$$z_0(x)=\frac{c_\pm}{u^{\frac{1}{4}}(x)}$$

From (7.82), (7.85), and (7.86)
Asymptotic solution for large λ and assuming $u(x)\neq 0$:

$$y(x) \sim \frac{1}{u^{\frac{1}{4}}(x)}\left(c_{+}e^{i\lambda\phi(x)} + c_{-}e^{-i\lambda\phi(x)}\right)$$

$$y(x) \sim \frac{1}{u^{\frac{1}{4}}(x)}\left(c_{+}e^{i\lambda\int u^{\frac{1}{2}}(x)dx} + c_{-}e^{-i\lambda\int u^{\frac{1}{2}}(x)dx}\right) \tag{7.87}$$

Hence, from (7.79) and (7.87), we get:

$$Q(\xi) \sim \frac{1}{G(\xi)^{\frac{1}{4}}} exp\left\{\pm ik\int\sqrt{G(\xi)}d\xi\right\} \tag{7.88}$$

Cases:
When $G(\xi)>0$, the field is oscillatory.
When $G(\xi)<0$, the field is evanescent.
Similarly,

$$S(\eta) \sim \frac{1}{H(\eta)^{\frac{1}{4}}} exp\left\{\pm ik\int\sqrt{H(\eta)}d\eta\right\} \tag{7.89}$$

7.12.2 Hyperbolic Type Mode (H-type)

Condition $\mu < 0$:

Calculating $G(0), H(0)$ and $H\left(\frac{\pi}{2}\right)$

1. $G(0)$

$$G(\xi) = \left[f^2\left(n_0^2 - \frac{\beta^2}{k^2}\right)\sinh^2\xi - g(\xi)\right] - \frac{\mu}{k^2}$$

Assume $g(\xi)=0$ using $\sinh^2\theta = \dfrac{\cosh 2\theta - 1}{2}$

$$G(\xi) = \left[f^2\left(n_0^2 - \frac{\beta^2}{k^2}\right)\frac{\cosh 2\xi - 1}{2}\right] - \frac{\mu}{k^2}$$

$$G(0) = \left[f^2\left(n_0^2 - \frac{\beta^2}{k^2}\right)\frac{\cosh 0 - 1}{2}\right] - \frac{\mu}{k^2}$$

$$G(0) = -\frac{\mu}{k^2}$$

As $\mu < 0$, $G(0) > 0$

2. $H(0)$

$$H(\eta) = \left[f^2\left(n_0^2 - \frac{\beta^2}{k^2}\right)\sin^2\eta - h(\eta)\right] + \frac{\mu}{k^2}$$

Assume $h(\eta)=0$ and using $\sin^2\theta = \dfrac{1-\cos 2\theta}{2}$.

$$H(\eta)=\left[f^2\left(n_0^2-\frac{\beta^2}{k^2}\right)\frac{1-\cos 2\theta}{2}\right]+\frac{\mu}{k^2}$$

$$H(0)=\left[f^2\left(n_0^2-\frac{\beta^2}{k^2}\right)\frac{1-\cos 0}{2}\right]+\frac{\mu}{k^2}$$

$$H(0)=\frac{\mu}{k^2}$$

As $\mu<0$, $H(0)<0$

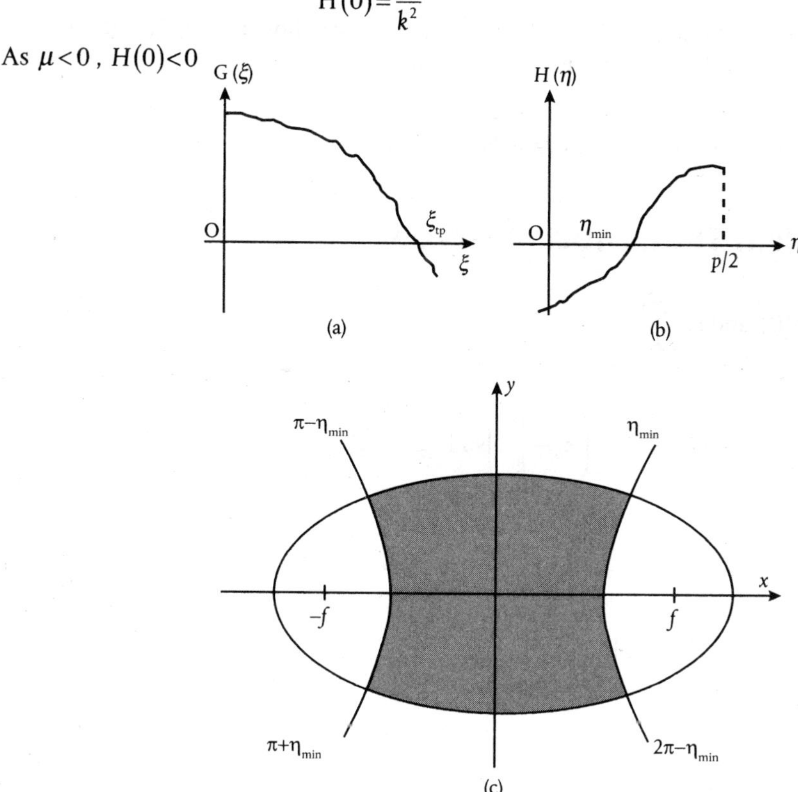

Figure 7.16 G(ξ) and $H(\eta)$ and the mode pattern for a 'hyperbolic' (*H*-type) mode

$$H(\eta)=\left[f^2\left(n_0^2-\frac{\beta^2}{k^2}\right)\sin^2\eta-h(\eta)\right]+\frac{\mu}{k^2}$$

Assume $h(\eta)=0$ and Using, $\sin^2\theta=\dfrac{1-\cos 2\theta}{2}$

$$H\left(\frac{\pi}{2}\right)=\left[f^2\left(n_0^2-\frac{\beta^2}{k^2}\right)\frac{1-\cos\pi}{2}\right]+\frac{\mu}{k^2}$$

$$H\left(\frac{\pi}{2}\right)=\left[f^2\left(n_0^2-\frac{\beta^2}{k^2}\right)\right]+\frac{\mu}{k^2}$$

Condition for propagation is $H\left(\frac{\pi}{2}\right)>0$.

$$\left[f^2\left(n_0^2-\frac{\beta^2}{k^2}\right)\right]+\frac{\mu}{k^2}>0$$

$$\mu>-f^2\left(k^2n_0^2-\beta^2\right)$$

Then, $H(\eta)>0$ for $\eta_{min}<\eta<\pi/2$, where $H(\eta_{min})=0$. G and H are shown in Fig. 7.16 (a) and 7.16(b), respectively. Thus, the mode is largely confined to the region.

$\left\{\begin{matrix} 0<\xi<\xi_{tp} \\ \eta_{min}<\eta<\pi/2 \end{matrix}\right\}$ and typical pattern is shown in Fig. 7.16(c).

7.12.3 Elliptical Type Mode (E-type)

Condition $\mu>0$

Calculating $G(0), H(0)$, and $H\left(\frac{\pi}{2}\right)$:

1. $G(0)$

$$G(\xi)=\left[f^2\left(n_0^2-\frac{\beta^2}{k^2}\right)\sinh^2\xi-g(\xi)\right]-\frac{\mu}{k^2}$$

Assume $g(\xi)=0$

Using $\sinh^2\xi=\frac{\cosh 2\xi-1}{2}$:

$$G(\xi)=\left[f^2\left(n_0^2-\frac{\beta^2}{k^2}\right)\frac{\cosh 2\xi-1}{2}\right]-\frac{\mu}{k^2}$$

$$G(0)=\left[f^2\left(n_0^2-\frac{\beta^2}{k^2}\right)\frac{\cosh 0-1}{2}\right]-\frac{\mu}{k^2}$$

$$G(0)=-\frac{\mu}{k^2}$$

As $\mu>0$,

$$G(0)<0 \tag{7.90}$$

2. $H(0)$

$$H(\eta)=\left[f^2\left(n_0^2-\frac{\beta^2}{k^2}\right)\sin^2\eta-h(\eta)\right]+\frac{\mu}{k^2}$$

Assume $h(\eta)=0$ and using $\sin^2\theta=\frac{1-\cos 2\theta}{2}$:

$$H(\eta)=\left[f^2\left(n_0^2-\frac{\beta^2}{k^2}\right)\frac{1-\cos 2\theta}{2}\right]+\frac{\mu}{k^2}$$

$$H(0)=\left[f^2\left(n_0^2-\frac{\beta^2}{k^2}\right)\frac{1-\cos 0}{2}\right]+\frac{\mu}{k^2}$$

$$H(0)=\frac{\mu}{k^2}$$

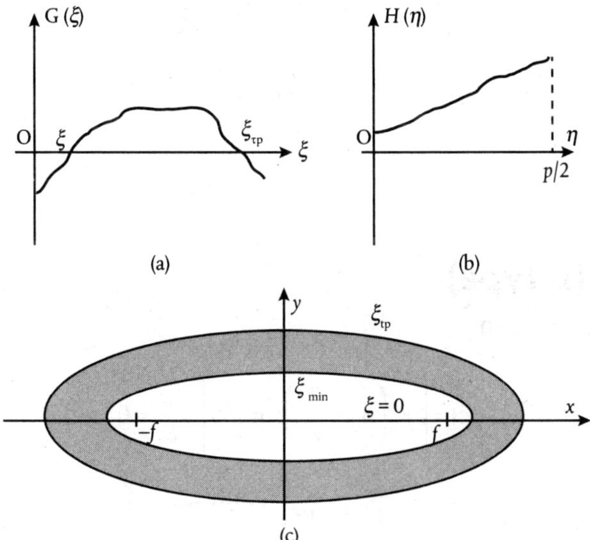

Fig. 7.17 **G**(ξ) and $H(\eta)$ and the mode pattern for a 'hyperbolic' (*H*-type) mode

As $\mu>0$, $H(0)>0$. Generally, $G(\xi)$ will have zeros in the core: ξ_{min} and ξ_{tp}. Fig 7.17(a) and Fig.7.17 (b) show typical forms of G and H, and Fig (c) shows mode pattern of this type.

7.13 STEP-INDEX FIBRE WITH ELLIPTICAL BOUNDARY

It is obtained by assuming $h(\eta)=0$ for all η and $g(\xi)=0; \xi<\xi_b$

7.13.1 $\mu<0$ (H-type)

$$G(\xi)=\left[f^2\left(n_0^2-\frac{\beta^2}{k^2}\right)\sinh^2\xi-g(\xi)\right]-\frac{\mu}{k^2}$$

As $g(\xi)=0; \xi<\xi_b$ and $\mu<0$,

$$G(\xi)>0 \text{ for } \xi<\xi_b \text{ and } G(\xi)<0 \text{ for } \xi>\xi_b$$

Condition for propagation: $H\left(\frac{\pi}{2}\right)>0$

$$\left[f^2\left(n_0^2-\frac{\beta^2}{k^2}\right)\right]+\frac{\mu}{k^2}>0$$

$$\mu>-f^2\left(k^2n_0^2-\beta^2\right)$$

Hence, for $H(\eta)>0$

$$\left[f^2\left(n_0^2-\frac{\beta^2}{k^2}\right)\sin^2\eta\right]+\frac{\mu}{k^2}>0$$

$$\left[f^2\left(k^2n_0^2-\beta^2\right)\sin^2\eta\right]+\mu>0$$

$$\sin^2\eta>\frac{-\mu}{f^2\left(k^2n_0^2-\beta^2\right)}=\sin^2\eta_{min}$$

7.13.2 $\mu>0$ (E-type)

Now, from (7.90), we get $G(0)<0$

$$G(\xi)=\left[f^2\left(n_0^2-\frac{\beta^2}{k^2}\right)\sinh^2\xi\right]-\frac{\mu}{k^2}<0$$

$$\sinh^2\xi<\frac{\mu}{f^2\left(k^2n_0^2-\beta^2\right)}=\sinh^2\xi_{min}\ at\ \xi=\xi_{min}$$

Otherwise, $G(\xi)>0$ for $\xi_{min}<\xi<\xi_b$. These are elliptic-type modes as in Fig. 7.17(c) with ξ_b replacing ξ_{tp}.

7.14 LIMITING CASES

Optical fibres are usually analysed using either rectangular or circular polar coordinates. We now study limiting cases $f\to\infty$ and $f\to 0$, respectively.

7.14.1 $f\to\infty\left(\xi\to 0\ \text{and}\ \eta\to\frac{\pi}{2}\right)$

x and y in this limit becomes:

$$x=f\cosh 0\cos\eta\sim f\cos\eta\to x \text{ and } y=f\sinh\xi\sin\frac{\pi}{2}\sim f\sinh\xi\to y$$

The value of $n^2(\xi,\eta)$ under these limits becomes:

$$n^2(\xi,\eta)=n_0^2-\frac{g(\xi)+h(\eta)}{\sinh^2\xi+\sin^2\eta}$$

$$=n_0^2-\frac{g(\xi)+h(\eta)}{\frac{\cosh 2\xi-1}{2}+\frac{1-\cos 2\eta}{2}}.$$

$$n^2(\xi,\eta) = n_0^2 - \frac{g(\xi)+h(\eta)}{\frac{\cosh 0 - 1}{2} + \frac{1-\cos\pi}{2}} \left(\text{using } \xi \to 0 \text{ and } \eta \to \frac{\pi}{2}\right)$$

$$n^2(\xi,\eta) = n_0^2 - \frac{g(\xi)+h(\eta)}{\frac{1-1}{2} + \frac{1+1}{2}}$$

$$n^2(\xi,\eta) = n_0^2 - g(\xi) - h(\eta).$$

$x-$ axis read as azimuthal angle φ of any point (ξ,η) with respect to $\varphi = \text{arc}\tanh\left(\frac{y}{x}\right) = \text{arc}\tanh\left(\frac{y/f}{x/f}\right)$

$$= \frac{\text{arc}\sinh(y/f)}{\text{arc}\cosh(x/f)} = \frac{g_1(y)}{h_1(x)}$$

$$n^2(\xi,\eta) = n_0^2 - g_1(y) - h_1(x),$$

where $g_1(y) = \text{arc}\sinh(y/f)$ and $h_1(x) = \text{arc}\cosh\left(\frac{x}{f}\right)$

The value of $G(\xi)$ and $H(\eta)$ under this limit reduces to:

$$G(\xi) = f^2\left[\left(n_0^2 - \frac{\beta^2}{k^2}\right)\sinh^2\xi - g(\xi)\right] - \frac{\mu}{k^2}$$

$$G(\xi) = f^2\left[\left(n_0^2 - \frac{\beta^2}{k^2}\right)\frac{\cosh 2\xi - 1}{2} - g(\xi)\right] - \frac{\mu}{k^2}$$

Using limits $f \to \infty \left(\text{hence, } \xi \to 0 \text{ and } \eta \to \frac{\pi}{2}\right)$:

$$G(\xi) \to f^2[-g(\xi)] - \frac{\mu}{k^2}$$

$$G(\xi) \to -g_1(y) f^2 - \frac{\mu}{k^2} \tag{7.91}$$

Similarly,

$$H(\eta) = f^2\left[\left(n_0^2 - \frac{\beta^2}{k^2}\right)\sin^2\eta - h(\eta)\right] + \frac{\mu}{k^2}$$

$$H(\eta) = f^2\left[\left(n_0^2 - \frac{\beta^2}{k^2}\right)\frac{1-\cos 2\eta}{2} - h(\eta)\right] + \frac{\mu}{k^2}$$

Using limits $f \to \infty \left(\text{hence, } \xi \to 0 \text{ and } \eta \to \frac{\pi}{2}\right)$,

$$H(\eta)\rightarrow f^{2}\left[\left(n_{0}^{2}-\frac{\beta^{2}}{k^{2}}\right)-h(\eta)\right]+\frac{\mu}{k^{2}}$$

$$H(\eta)\rightarrow f^{2}\left[\left(n_{0}^{2}-\frac{\beta^{2}}{k^{2}}\right)-h_{1}(x)\right]+\frac{\mu}{k^{2}} \tag{7.92}$$

The oscillatory part is contained within a rectangle and the caustics x_m and y_m are found from zeros of the above-mentioned functions:

From (7.91), we get:

$$-g_{1}(y_{m})f^{2}-\frac{\mu}{k^{2}}=0$$

$$g_{1}(y_{m})=-\frac{\mu}{f^{2}k^{2}} \tag{7.93}$$

From (7.92)

$$f^{2}\left[\left(n_{0}^{2}-\frac{\beta^{2}}{k^{2}}\right)-h_{1}(x_{m})\right]+\frac{\mu}{k^{2}}=0$$

$$\left(n_{0}^{2}-\frac{\beta^{2}}{k^{2}}\right)-h_{1}(x_{m})=-\frac{\mu}{f^{2}k^{2}} \tag{7.94}$$

From (7.93) and (7.94), we get:

$$\left(n_{0}^{2}-\frac{\beta^{2}}{k^{2}}\right)-h_{1}(x_{m})=-\frac{\mu}{f^{2}k^{2}}=g_{1}(y_{m}) \tag{7.95}$$

As $n^{2}(x_{m},y_{m})=n_{0}^{2}-g_{1}(y_{m})-h_{1}(x_{m})$

From (7.94),

$$n^{2}(x_{m},y_{m})=n_{0}^{2}-\left[\left(n_{0}^{2}-\frac{\beta^{2}}{k^{2}}\right)-h_{1}(x_{m})\right]-h_{1}(x_{m})$$

$n^{2}(x_{m},y_{m})=\frac{\beta^{2}}{k^{2}}>n_{cl}^{2}$ (as x_m, y_m must be in the core).

Now, as $f\cos\eta\rightarrow x$ and $f\sinh\rightarrow y$

Taking derivative from the aforementioned equation:

$$-f\sin\eta d\eta\rightarrow dx \text{ and } f\cosh\xi d\xi\rightarrow dy$$

Using limits $f\rightarrow\infty\left(\xi\rightarrow 0 \text{ and } \eta\rightarrow\frac{\pi}{2}\right)$,

$$-f\sin\frac{\pi}{2}d\eta\rightarrow dx \text{ and } f\cosh 0 d\xi\rightarrow dy$$

$$-fd\eta \rightarrow dx \text{ and } fd\xi \rightarrow dy \tag{7.96}$$

From (7.88)

$$Q(\xi) \sim \frac{1}{G(\xi)^{\frac{1}{4}}} exp\left\{\pm ik\int\sqrt{G(\xi)}d\xi\right\}$$

Using limits and (7.91),

$$Q \rightarrow \frac{1}{G(\xi)^{\frac{1}{4}}} exp\left\{\pm ik\int\left[-g_1(y)f^2 - \frac{\mu}{k^2}\right]^{\frac{1}{2}} d\xi\right\}$$

From (7.96) and changing to y-coordinate, we get:

$$Q \rightarrow \frac{1}{G(\xi)^{\frac{1}{4}}} exp\left\{\pm ik\int\left[-g_1(y)f^2 - \frac{\mu}{k^2}\right]^{\frac{1}{2}} \frac{dy}{f}\right\}$$

$$Q \rightarrow \frac{1}{G(\xi)^{\frac{1}{4}}} exp\left\{\pm ik\int\left[-g_1(y) - \frac{\mu}{f^2k^2}\right]^{\frac{1}{2}} dy\right\} \tag{7.97}$$

From (7.89),

$$S(\eta) \sim \frac{1}{H(\eta)^{\frac{1}{4}}} exp\left\{\pm ik\int\sqrt{H(\eta)}d\eta\right\}$$

Using limits and (7.92), we get:

$$S(\eta) \rightarrow \frac{1}{H(\eta)^{\frac{1}{4}}} exp\left\{\pm ik\int\left[f^2\left(n_0^2 - \frac{\beta^2}{k^2} - h_1(x)\right) + \frac{\mu}{k^2}\right]^{\frac{1}{2}} \frac{dx}{f}\right\}$$

$$S(\eta) \rightarrow \frac{1}{H(\eta)^{\frac{1}{4}}} exp\left\{\pm ik\int\left[n_0^2 - \frac{\beta^2}{k^2} - h_1(x) + \frac{\mu}{f^2k^2}\right]^{\frac{1}{2}} dx\right\} \tag{7.98}$$

The analysis of rectangular coordinates can be done using method of "separation of variables". Therefore, (7.63) can be written in rectangular coordinates:

$$\frac{\partial^2\phi}{\partial x^2} + \frac{\partial^2\phi}{\partial y^2} + \left(k^2n^2 - \beta^2\right)\phi = 0$$

Using $\phi = X(x)Y(y)$,

$$Y\frac{\partial^2 X}{\partial x^2} + X\frac{\partial^2 Y}{\partial y^2} + \left(k^2n^2 - \beta^2\right)XY = 0$$

Dividing both sides by XY, we get

$$\frac{1}{X}\frac{\partial^2 X}{\partial x^2}+\frac{1}{Y}\frac{\partial^2 Y}{\partial y^2}+\left(k^2n^2-\beta^2\right)=0$$

As $n^2\left(x_m, y_m\right)=n_0^2-g_1\left(y_m\right)-h_1\left(x_m\right)$

$$\frac{1}{X}\frac{\partial^2 X}{\partial x^2}+\frac{1}{Y}\frac{\partial^2 Y}{\partial y^2}+k^2\left[n_0^2-g_1(y)-h_1(x)\right]-\beta^2=0$$

Separating the X variables:

$$\frac{1}{X}\frac{\partial^2 X}{\partial x^2}+k^2\left(n_0^2-h_1(x)\right)-\beta^2=\delta$$

$$\frac{\partial^2 X}{\partial x^2}+\left[k^2\left(n_0^2-h_1(x)\right)-\beta^2\right]X=\delta X$$

$$\frac{\partial^2 X}{\partial x^2}+k^2\left[\left(n_0^2-h_1(x)\right)-\frac{\left(\beta^2+\delta\right)}{k^2}\right]X=0$$

$$\frac{\partial^2 X}{\partial x^2}+k^2H_1(x)X=0 \qquad (7.99)$$

where $H_1(x)=\left[\left(n_0^2-h_1(x)\right)-\frac{\left(\beta^2+\delta\right)}{k^2}\right]$

Separating the Y variable:

$$-\left[\frac{1}{Y}\frac{\partial^2 Y}{\partial y^2}+k^2\left(-g_1(y)\right)\right]=\delta$$

$$\frac{\partial^2 Y}{\partial y^2}+\left[k^2\left(-g_1(y)\right)\right]Y=\delta Y$$

$$\frac{\partial^2 Y}{\partial y^2}+k^2\left[\left(-g_1(y)\right)+\frac{(\delta)}{k^2}\right]Y=0$$

$$\frac{\partial^2 Y}{\partial y^2}+k^2G_1(y)Y=0 \qquad (7.100)$$

where $G_1(y)=\left[\left(-g_1(y)\right)+\frac{\delta}{k^2}\right]$. Using (7.87) and the solution of (7.99) and (7.100) Thus,

$$X\sim\frac{1}{H_1^{\frac{1}{4}}}\exp\left(\pm ik\int H_1^{\frac{1}{2}}dx\right)$$

$$Y \sim \frac{1}{G_1^{\frac{1}{4}}} exp\left(\pm ik \int G_1^{\frac{1}{2}} dy \right)$$

7.14.2 $f \to 0$

This case is for cylindrical polar coordinates (r, ψ). As in cylindrical fibres, the refractive index varies radially $(f\sinh\xi \to r)$ and angular coordinate $(\eta \to \psi)$ remain unchanged. Therefore, (7.78) reduces to:

$$n^2(r) = n_0^2 - \frac{g(\xi)}{\sinh^2 \xi}$$

Let,

$$P(r) = n^2(r) - \frac{\left(\frac{\mu}{r^2} + \beta^2 \right)}{k^2}$$

$$G(\xi) = f^2 \left[\left(n_0^2 - \frac{\beta^2}{k^2} \right) \sinh^2 \xi - g(\xi) \right] - \frac{\mu}{k^2}$$

Dividing by $\sinh^2 \xi$, we get:

$$\frac{G(\xi)}{\sinh^2 \xi} = f^2 \left[\left(n_0^2 - \frac{\beta^2}{k^2} \right) - \frac{g(\xi)}{sinh^2 \xi} \right] - \frac{\mu}{sinh^2 \xi k^2}$$

Dividing by f^2:

$$\frac{G(\xi)}{f^2 sinh^2 \xi} = \left[\left(n_0^2 - \frac{\beta^2}{k^2} \right) - \frac{g(\xi)}{sinh^2 \xi} \right] - \frac{\mu}{f^2 sinh^2 \xi k^2}$$

$$\frac{G(\xi)}{r^2} = n^2(r) - \left(\frac{\beta^2}{k^2} + \frac{\mu}{r^2 k^2} \right) = P(r)$$

$$G(\xi) \to r^2 P(r)$$

Hence, $Q(\xi)$ reduces to:

$$Q \sim \frac{1}{G(\xi)^{\frac{1}{4}}} exp\left\{ \pm ik \int \sqrt{G(\xi)} d\xi \right\} \to \frac{1}{\left| r^2 P(r) \right|^{\frac{1}{4}}} exp\left\{ \pm ik \int \sqrt{r^2 P(r)} \frac{dr}{r} \right\}$$

$$Q \to \frac{1}{\left| r^2 P(r) \right|^{\frac{1}{4}}} exp\left\{ \pm ik \int \sqrt{P(r)} dr \right\} \quad (7.101)$$

Similarly,

$$H(\eta)=f^2\left[\left(n_0^2-\frac{\beta^2}{k^2}\right)\sin^2\eta-h(\eta)\right]+\frac{\mu}{k^2}$$

Neglecting η terms as refractive index in cylindrical fibre changes radially.

$$H(\eta)\to\frac{\mu}{k^2}$$

From (7.89),

$$S(\eta)=\frac{1}{H(\eta)^{\frac{1}{4}}}exp\left\{\pm ik\int\sqrt{H(\eta)}d\eta\right\}\sim exp\left\{\pm ik\int\sqrt{H(\eta)}d\eta\right\}$$

$$S(\eta)\to exp\left\{\pm ik\int\sqrt{\frac{\mu}{k^2}}d\eta\right\}\to exp\left\{\pm i\sqrt{\mu}\eta\right\}$$

Let $\sqrt{\mu}$ must be an integer $=\nu$.
Using $\phi=R(r)\exp(i\nu\psi)$ and $G(\xi)\to r^2P(r)$ (7.79) reduces to:

$$\frac{d^2R}{dr^2}+k^2r^2P(r)R(r)=0 \tag{7.102}$$

Using $r=e^w$ and taking derivative on both sides,

$$dr=e^w dw=rdw$$

$$\frac{dr}{r}=dw$$

$$\frac{\partial^2R}{\partial w^2}+k^2e^{2w}P(r)R(r)=0$$

Using WKB approximation method, we get:

$$R\sim\frac{1}{e^{2w}P(r)^{\frac{1}{4}}}exp\left\{\pm ik\int\left(e^{2w}P(r)\right)^{\frac{1}{2}}dw\right\}.$$

$$R\sim\frac{1}{r^2P(r)^{\frac{1}{4}}}exp\left\{\pm ik\int\left(r^2P(r)\right)^{\frac{1}{2}}\frac{dr}{r}=\right\}\sim\frac{1}{r^2P(r)^{\frac{1}{4}}}exp\left\{\pm ik\int\left(P(r)\right)^{\frac{1}{2}}dr\right\}$$

7.15 MODE REPRESENTATION FOR PARABOLIC PROFILE WITH CIRCULAR REFRACTIVE INDEX CONTOURS

The circularly symmetric fibre with a parabolic profile:

$$n^2(r)=n_0^2-\gamma^2\left(\frac{r}{\rho}\right)^2;r\le\rho \tag{7.103}$$

where $\gamma^2 = 2\Delta n_0^2$ and ρ is the radius of the core

$$n^2(r) = n_0^2 - 2\Delta n_0^2\left(\frac{r}{\rho}\right)^2 = n_0^2\left[1 - 2\Delta\left(\frac{r}{\rho}\right)^2\right]; r \le \rho$$

Equation (7.103) can be represented in the form of (7.78) using:

$$g(\xi) = f^2(\gamma^2/\rho^2)\sinh^2\xi\cosh^2\xi$$

$$h(\eta) = f^2(\gamma^2/\rho^2)\sin^2\eta\cos^2\eta$$

Derivation of $g(\xi)$and $h(\eta)$ can be obtained by using (7.103) and (7.78):

$$n_0^2 - \gamma^2\left(\frac{r}{\rho}\right)^2 = n_0^2 - \frac{g(\xi)+h(\eta)}{\sinh^2\xi+\sin^2\eta}$$

$$\gamma^2\left(\frac{x^2+y^2}{\rho^2}\right) = \frac{g(\xi)+h(\eta)}{\sinh^2\xi+\sin^2\eta}$$

Taking $x = f\cosh\eta$ and $y = f\sinh\xi$,

$$\gamma^2\left[\frac{(f\cos\eta)^2 + (f\sinh\xi)^2}{\rho^2}\right] = \frac{g(\xi)+h(\eta)}{(\sinh^2\xi+\sin^2\eta)}$$

$$\left(\frac{\gamma^2}{\rho^2}\right)f^2\left[\cos^2\eta+\sinh^2\xi\right]\left[\sinh^2\xi+\sin^2\eta\right] = g(\xi)+h(\eta)$$

Using $\cosh^2 \; - \sinh^2 \; = 1$,

$$\left(\frac{\gamma^2}{\rho^2}\right)f^2\left[(\cos^2\eta+\sinh^2\xi)(1+\cosh^2\xi+\sin^2\eta)\right] = g(\xi)+h(\eta)$$

$$\left(\frac{\gamma^2}{\rho^2}\right)f^2\left(\cos^2\eta+\cos^2\eta\cosh^2\xi+\cos^2\eta\sin^2\eta+\sinh^2\xi+\sinh^2\xi\cosh^2\xi+\sinh^2\xi\sin^2\eta\right) = g(\xi)+h(\eta)$$

Equating ξ and η terms on both sides:

$$g(\xi) = \left(\frac{\gamma^2}{\rho^2}\right)f^2\sinh^2\xi\cosh^2\xi \text{ and } h(\eta) = \left(\frac{\gamma^2}{\rho^2}\right)f^2\cos^2\eta\cosh^2\eta$$

Thus, $G(\xi)$ becomes:

$$G(\xi) = f^2\left[\left(n_0^2 - \frac{\beta^2}{k^2}\right)\sinh^2\xi - g(\xi)\right] - \frac{\mu}{k^2}$$

$$= f^2\left[\left(n_0^2-\frac{\beta^2}{k^2}\right)\sinh^2\xi-\left[f^2(\gamma^2/\rho^2)\sinh^2\xi\cosh^2\xi\right]\right]-\frac{\mu}{k^2}$$

Using $\cosh^2\xi-\sinh^2\xi=1$,

$$= f^2\left[\left(n_0^2-\frac{\beta^2}{k^2}\right)\sinh^2\xi-\left[f^2(\gamma^2/\rho^2)\sinh^2\xi(1-\sinh^2\xi)\right]\right]-\frac{\mu}{k^2}$$

$$= f^2\left(n_0^2-\frac{\beta^2}{k^2}-\gamma^2\frac{f^2}{\rho^2}\right)\sinh^2\xi-f^4\frac{\gamma^2}{\rho^2}\sinh^4\xi-\frac{\mu}{k^2}$$

Similarly,

$$H(\eta)= f^2\left[\left(n_0^2-\frac{\beta^2}{k^2}\right)\sin^2\eta-h(\eta)\right]+\frac{\mu}{k^2}$$

$$= f^2\left(n_0^2-\frac{\beta^2}{k^2}-\gamma^2\frac{f^2}{\rho^2}\right)\sin^2\eta+f^4\frac{\gamma^2}{\rho^2}\sin^4\eta+\frac{\mu}{k^2}$$

SUMMARY

- In anisotropic media, velocity of propagation depends on the propagation direction and also on the state of polarisation; further, one can observe the phenomenon of double refraction.
- Anisotropic media form the basis of a large number of polarisation devices such as quarter-wave and half-wave plates.
- In anisotropic media, the direction of displacement (*D*) and electric field (*E*) are not parallel for all possible directions of E. Hence, electric field applied along the *x*-direction generates a *D* that has all the three components.
- A wave with its E along *x* (and hence, its *D* also along *x* since *x*, *y*, and *z* is a principal axis system), that travels with a phase velocity of $c/K_x^{1/2}$. A wave with its E along *y* (hence, *D* also along *y*), which travels with a phase velocity $c/K_y^{1/2}$. A wave with its E along *z* and propagating along either *x*- or *y*- direction travels with a phase velocity $c/K_z^{1/2}$.
- Relation between ray velocity and wave velocity $v_r = |S|/W$, where *S* is ray direction and *W* is energy density. The propagation direction is along pointing vector defined by $S = E\times H$.
- When light propagates in a crystal, the *D* vectors of the two eigenwaves and the wave normal direction **s** form a mutually perpendicular triad. It is, therefore, more convenient to present the field vectors in the *D*-space. The electric energy density *Ue* is given as:

$$U_e=\frac{1}{2}E.D=\frac{1}{2}\left(\frac{D_x^2}{\epsilon_x}+\frac{D_y^2}{\epsilon_y}+\frac{D_z^2}{\epsilon_z}\right)=\frac{1}{2\epsilon_0}\left(\frac{D_x^2}{n_x^2}+\frac{D_y^2}{n_y^2}+\frac{D_z^2}{n_z^2}\right).$$

- For a given arbitrary wave normal direction **s**, the index ellipsoid can be used to: (a) find the indices of refraction of the two eigenwaves; (b) find the corresponding directions of the *D* vectors of the two eigenwaves.
- The electro-optic effect refers to changes in the refractive indices of the crystal due to an applied external electric field. The Pockels effect refers to changes in the refractive indices proportional to

the first power of the applied electric field. If the change is proportional to the square of the applied field, then the effect is referred as quadratic electro-optic effect or the Kerr effect.

- The change in the crystal refractive indices can be described in terms of the changes in the shape, dimensions, and orientation of the crystal index ellipsoid.
- The Pockels effect can exist only in those crystals that do not possess an inversion symmetry. A crystal is said to possess an inversion symmetry if inversion about any of its lattice points (i.e., replacing *r* by – *r*, *r* being the position vector of any lattice point measured from the point about which the inversion is being performed) leaves the crystal structure invariant.
- Instead of specifying the changes in the refractive indices, it is more convenient to consider the changes in $(1/n^2)$ due to external electric field. The changes in $1/n^2$ due to Pockels effect are described by the following equation:

$$\Delta\left(1/n^2\right)_i = \sum_{k=x,y,z} r_{ik} E_k ,$$

where $i=(1, 2\ldots.6)$ and $k=x, y, z$

- $LiNbO_3$ is the most widely used material for the manufacture of electro-optics devices, including phase modulators, polarisation modulators, and Mach–Zehnder intensity modulators. $LiNbO_3$ is intrinsically a birefringent crystal. Its extraordinary optical axis is usually chosen as the *z*-axis of the principal coordinate system of the index ellipsoid.
- The operation principle of MZM is based on the electro-optic effect, which is characterised by the variation of an applied electrical field causing refractive index changes in the modulator arms.
- The variation of the refractive index in the modulator arms induces a change of material propagation constant β, resulting in different phases in both modulator arms. The input optical signal E_0 is divided by a 3-dB coupler into two equal parts E_1 and E_2, which propagate in lower and upper arm of the modulator.
- When we assume that the power is split or combined equally in the input and output couplers of the Mach–Zehnder interferometer, the power at the output of the interferometer depends on the difference between the phase shifts experienced by the light propagating in the upper and lower arm of the structure, $\phi(t)$ and ϕ_0, respectively. According to:

$$P_{out} = P_{in}\cos^2\left(\frac{\Delta\phi}{2}\right),$$

where $\Delta\phi = \phi(t) - \phi_0$.

- Lithium niobate ($LiNbO_3$) is customarily used as a suitable electro-optic material for high-speed modulation in optical communication systems, owing to its relatively large electro-optic coefficient and well-controlled waveguide fabrication processes.
- As the optical beam passes through the length of the modulator, there are limitations on the frequencies of the modulating signal because of finite transit time. Time τ_d taken by the optical beam to pass through the crystal is given as:

$$\tau_d = \frac{nl}{c},$$

where l is length of modulator crystal and *n* is either n_e or n_0.

- The transit time limitation on the maximum modulating frequency due to transit time effects can be overcome by: (a) Having the applied electric field propagating along the length of the crystal: such modulator is referred to as travelling wave modulators. (b) The phase velocity of the modulating signal should be equal to the phase velocity of the lightwave propagating through the crystal.

THEORY QUESTIONS AND ANSWERS

Question 1: What is the Pockels effect?

Answer: Both phase and amplitude modulators are based upon the Pockels effect: The electro-optic effect where the refractive index along one or more axes is proportional to an externally applied electric field. Therefore, by applying a voltage across the electrodes of an electro-optic crystal, we can change the phase of light as it passes through the crystal. By placing the crystal between crossed polarisers, this phase modulation can be converted into amplitude modulation.

Question 2: What is the Kerr effect?

Answer: The electro-optic effect where the refractive index along one or more axes is proportional to a square of the externally applied electric field. Therefore, by applying a voltage across the electrodes of an electro-optic crystal, phase of the signal is changes.

Question 3: What is electro-optic effect?

Answer: The electro-optic effect refers to change in the refractive indices of the crystal due to an applied external electric field. The Pockels effect refers to changes in the refractive indices proportional to the first power of the applied electric field. If the change is proportional to the square of the applied field, then the effect is referred as quadratic electro-optic effect or the Kerr effect. The change in the crystal refractive indices can be described in terms of the changes in the shape, dimensions, and orientation of the crystal-index ellipsoid.

Question 4: Can phase modulators be used in interferometric applications?

Answer: No, the electro-optic crystals that we use in phase modulators contain refractive index inhomogeneities. This spatially non-uniform refractive index imparts a significant wavefront distortion to any beam that passes through the modulator. When the beams are recombined, the wavefront distortion will cause the on: off contrast ratio to be very low, perhaps 3:1 or 5:1.

Question 5: What is the half-wave voltage or V_π?

Answer: This is the voltage (specified either at DC or at the maximum operating frequency) required to achieve a phase shift equal to π (3.14) radians. It is proportionately smaller at shorter wavelengths and it is much smaller for resonant devices due to the voltage enhancement provided by the resonant circuit or cavity on resonance. For amplitude modulators, V_π is the voltage necessary to change from maximum transmission to maximum extinction.

Question 6: What is the modulation depth at a given wavelength?

Answer: The modulation depth is the resulting phase change when 1 V (peak) is applied to the modulator, and is inversely proportional to the V_p. The modulation depth in most of cases be 1.06 μm. Since it is proportionately larger at shorter wavelengths, the modulation depth at 532 nm is twice than that at 1.06 μm. The modulation depth is much larger for resonant phase modulation because the resonant circuit provides voltage enhancement when operating on resonance.

Question 7: Can we use phase modulator as an amplitude modulator?

Answer: No, we cannot use phase modulator as an amplitude modulator. First, the phase modulators use just one electro-optic crystal. Since these crystals exhibit thermal birefringence, a phase modulator will have significant problems with thermal drift if it is used as an amplitude modulator. Second, most of phase modulators are made with MgO-doped lithium niobate crystals. This material works well for phase modulation and the MgO doping helps to increase the material's resistance to photorefractive damage. However, one drawback to this material is that it exhibits increased wavefront distortion (as compared with lithium tantalate or undoped lithium niobate). For most phase modulation applications, wavefront distortion of the optical beam does not present any problems, but when used as an amplitude modulator, the wavefront distortion causes poor on:off contrast ratios.

Question 8: How does a piezo resonance affect a modulator's performance?

Answer: Because all electro-optic crystals are also piezoelectric, sometimes phase modulation can be accompanied by unwanted amplitude modulation and beam deflection. This effect is most dramatic when the crystal is driven at a frequency that coincides with an acoustic resonance. By avoiding these piezoelectric resonances, the most problems are in the range of 1−100 MHz and modulator performance will be unaffected by piezo effects.

Question 9: How do the amplitude modulators work?

Answer: Amplitude modulators basically consist of an electro-optic modulator followed by a polariser. If the input polarisation is oriented at 45° to the crystal axes, the applied voltage will produce a variable phase delay between the ordinary and extraordinary field components, simulating a voltage-tunable wave plate.

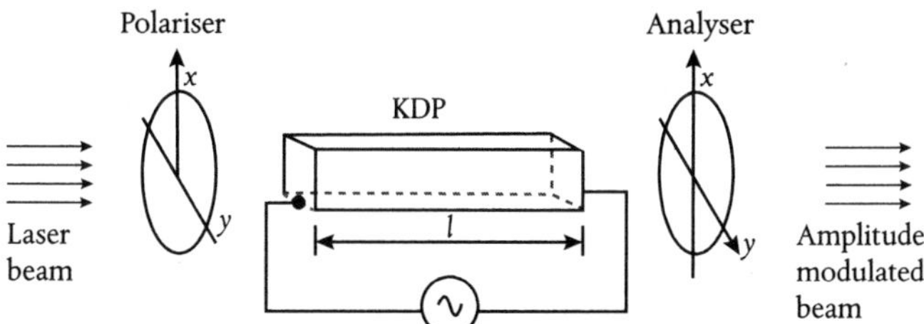

Fig. P1 An electro-optic amplitude modulator

Thus, modulation of intensity is a function of the square of sine. New focus simplifies optical setup by mounting the crystal at 45°. Thus, the input polarisation can be either vertical or horizontal. In order to suppress birefringence variations due to temperature changes, we use two matched crystals arranged in series with their applied electric fields oriented at 90° relative to each other. We cancel thermal birefringence while doubling the electro-optically induced polarisation rotations by reversing the crystal axes such that both polarisation components travel equal optical paths in the ordinary and extraordinary orientations.

SOLVED PROBLEMS

Problem 1: Consider a channel waveguide with a transverse dimension of d = 10 μm and an interaction length of $l = 5$ cm. Take $n_e = 2.2$ and $r_{33} = 30$ pm/V. Calculate half-wave voltage at 1.5 μm.

Solution:

Given: Transverse dimension of $d = 10\ \mu m$

An interaction length of $l = 5$ cm

$n_e = 2.2$ and $r_{33} = 30\ pm/V$

Using relation;

$$V = \frac{h}{r_{33}n_e^3}\cdot\frac{d}{l} = \frac{1.5\times10^{-6}}{30\times10^{-12}\times(2.2)^3}\cdot\frac{10\times10^{-6}}{5\times10^{-2}} = 0.94V$$

Problem 2: Show that the principal axes of the index ellipsoid for KH_2PO_4 in the presence of an electric field along the z-direction are at 45^0 to the $x-$ and y-directions?

Solution:

For KDP crystal,

$$\Delta(1/n^2)_i = \sum_{k=x,y,z} r_{ik}E_k,$$

where i =(1,2….6) and k = x,y,z

For KDP, the corresponding EO tensor takes the form:

$$\begin{bmatrix} 0 & 0 & 0 \\ 0 & 0 & 0 \\ 0 & 0 & 0 \\ r_{41} & 0 & 0 \\ 0 & r_{41} & 0 \\ 0 & 0 & r_{63} \end{bmatrix}.$$

When only the linear electro-optic effect is considered, the change induced by the electric field E = (E_x, E_y, E_z) can be expressed as:

$$\begin{bmatrix} \Delta\left(\frac{1}{n^2}\right)_1 \\ \Delta\left(\frac{1}{n^2}\right)_2 \\ \Delta\left(\frac{1}{n^2}\right)_3 \\ \Delta\left(\frac{1}{n^2}\right)_4 \\ \Delta\left(\frac{1}{n^2}\right)_5 \\ \Delta\left(\frac{1}{n^2}\right)_6 \end{bmatrix} = \begin{bmatrix} r_{11} & r_{12} & r_{13} \\ r_{21} & r_{22} & r_{23} \\ r_{31} & r_{32} & r_{33} \\ r_{41} & r_{42} & r_{43} \\ r_{51} & r_{52} & r_{53} \\ r_{61} & r_{62} & r_{63} \end{bmatrix} \begin{bmatrix} E_x \\ E_y \\ E_z \end{bmatrix}$$

$$\begin{bmatrix} \Delta\left(\frac{1}{n^2}\right)_1 \\ \Delta\left(\frac{1}{n^2}\right)_2 \\ \Delta\left(\frac{1}{n^2}\right)_3 \\ \Delta\left(\frac{1}{n^2}\right)_4 \\ \Delta\left(\frac{1}{n^2}\right)_5 \\ \Delta\left(\frac{1}{n^2}\right)_6 \end{bmatrix} = \begin{bmatrix} 0 & 0 & 0 \\ 0 & 0 & 0 \\ 0 & 0 & 0 \\ r_{41} & 0 & 0 \\ 0 & r_{41} & 0 \\ 0 & 0 & r_{63} \end{bmatrix} \begin{bmatrix} E_x \\ E_y \\ E_z \end{bmatrix}$$

$$\Delta\left(\frac{1}{n^2}\right)_1 = 0, \Delta\left(\frac{1}{n^2}\right)_2 = 0, \Delta\left(\frac{1}{n^2}\right)_3 = 0, \Delta\left(\frac{1}{n^2}\right)_4 = r_{41}E_x, \Delta\left(\frac{1}{n^2}\right)_5 = r_{41}E_y, \Delta\left(\frac{1}{n^2}\right)_6 = r_{63}E_z.$$

From (7.43.1) and above-mentioned condition:

$$\Big\{\left[\left(1/n_x^2\right)+\Delta\left(1/n^2\right)_1\right]x^2 + \left[\left(1/n_y^2\right)+\Delta\left(1/n^2\right)_2\right]y^2 + \left[\left(1/n_z^2\right)+\Delta\left(1/n^2\right)_3\right]z^2$$
$$+2\Delta\left(1/n^2\right)_4 y.z + 2\Delta\left(1/n^2\right)_5 zx + 2\Delta\left(1/n^2\right)_6 x.y\Big\} = 1$$

$$\left\{\left(1/n_x^2\right)x^2 + \left(1/n_y^2\right)y^2 + \left(1/n_z^2\right)z^2 + 2r_{41}E_x yz + 2\Delta r_{41}E_y zx + 2r_{63}E_z xy\right\} = 1$$

As $n_x = n_y = n_0$ and $n_z = n_e$,

$$\left\{\left(\frac{x^2+y^2}{n_0^2}\right) + \frac{1}{n_e^2}z^2 + 2r_{41}E_x yz + 2\Delta r_{41}E_y zx + 2r_{63}E_z xy\right\} = 1$$

Let us assume that the external electric field is applied along the *c*-axis, that is, along the z-axis. in such a case, the index ellipsoid is described as:

$$\left\{\left(\frac{x^2+y^2}{n_0^2}\right) + \frac{1}{n_e^2}z^2 + 2r_{63}E_z x.y\right\} = 1$$

Let us consider a rotated coordinate system (x', y', z'), which is obtained from the crystallographic axes (x, y, z) by a rotation by an angle ψ about the z-axis. Therefore, we can write:

$$x' = x\cos\psi + y\sin\psi$$

$$y' = -x\sin\psi + y\cos\psi$$

$$z' = z$$

Calculating *x*, *y*, and *z* from the above-mentioned equation:

$$x = x'\cos\psi + y'\sin\psi$$

$$y = -x'\sin\psi + y'\cos\psi$$

$$z = z'$$

Substituting x, y, and z in terms of x', y', and z':

$$\left\{\left(\frac{x^2+y^2}{n_0^2}\right)+\frac{z^2}{n_e^2}+2r_{63}E_z xy\right\}=1$$

$$\left\{\left[\frac{(x'\cos\psi + y'\sin\psi)^2+(-x'\sin\psi + y'\cos\psi)^2}{n_0^2}\right]+\frac{z^2}{n_e^2}+2r_{63}E_z(x'\cos\psi + y'\sin\psi)(-x'\sin\psi + y'\cos\psi)\right\}=1$$

$$\left\{\left[\frac{x'^2\cos^2\psi+y'^2\sin^2\psi-2x'y'\cos\psi\sin\psi+x'^2\sin^2\psi+y'^2\cos^2\psi+2x'y'\cos\psi\sin\psi}{n_0^2}\right]+\frac{z^2}{n_e^2}\right.$$
$$\left.+2r_{63}E_z\left(x'^2\cos\psi\sin\psi+x'y'\cos^2\psi-x'y'\sin^2\psi-y'^2\cos\psi\sin\psi\right)\right\}=1$$

$$\left\{\left[\frac{x'^2+y'^2}{n_0^2}\right]+\frac{z^2}{n_e^2}+2r_{63}E_z\left[x'^2\cos\psi.\sin\psi+x'y'\left(\cos^2\psi-\sin^2\psi\right)-y'^2\cos\psi\sin\psi\right]\right\}=1$$

In order to find the principal axes, the angle ψ must be so chosen to make cross-product term $x'y'$ term disappear.

This happens if $\left(\cos^2\psi-\sin^2\psi\right)=0$,

$$\cos 2\psi = 0$$

$$\psi = \frac{\pi}{4} = 45^\circ.$$

Hence, the principal axes of the index ellipsoid for KH_2PO_4 in the presence of an electric field along the z-direction are at 45° to the x- and y- directions.

Problem 3: Consider a transverse configuration of a KDP modulator where the light wave propagated along the y'-direction and the field was applied along the z-direction. Consider the light propagates along the z-direction and assume that the external field is applied along the x-direction. Calculate the direction of the axes of the index ellipsoid in the presence of a field and the principal indices of refraction?

Solution: If the applied electric field is along the x-direction, then the equation of the index ellipsoid

$$\left\{\left(\frac{x^2+y^2}{n_0^2}\right)+\frac{1}{n_e^2}z^2+2r_{41}E_x yz+2\,\Delta r_{41}E_y zx+2r_{63}E_z xy\right\}=1$$

reduces to

$$\frac{x^2+y^2}{n_0^2}+\frac{z^2}{n_e^2}+2r_{41}E_x yz=1$$

In order to determine the new y- and z-axes, we rotate the coordinate axes by an angle α about the x-axis. In such a case, we will have:

$$y = y'\cos\alpha - z'\sin\alpha \text{ and } z = y'\sin\alpha + z'\cos\alpha$$

Substituting y and z in the above-mentioned equation:

$$\frac{x^2 + (y'\cos\alpha - z'\sin\alpha)^2}{n_0^2} + \frac{(y'\sin\alpha + z'\cos\alpha)^2}{n_e^2} + 2r_{41}E_x(y'\cos\alpha - z'\sin\alpha)(y'\sin\alpha + z'\cos\alpha) = 1$$

$$\frac{x^2 + (y'\cos\alpha - z'\sin\alpha)^2}{n_0^2} + \frac{(y'\sin\alpha + z'\cos\alpha)^2}{n_e^2} + 2r_{41}E_x(y'\cos\alpha - z'\sin\alpha)(y'\sin\alpha + z'\cos\alpha) = 1$$

$$\left\{\left[\frac{x^2 + y'^2\cos^2\alpha + z'^2\sin^2\alpha - 2y'z'\cos\alpha.\sin\alpha}{n_0^2}\right] + \frac{y'^2\cos^2\alpha + z'^2\sin^2\alpha + 2y'z'\cos\alpha.\sin\alpha}{n_e^2} + 2r_{41}E_x\left(y'^2\cos\alpha.\sin\alpha + y'z'\cos^2\alpha - y'z'\sin^2\alpha - z'^2\cos\alpha.\sin\alpha\right)\right\} = 1$$

$$\left\{\left[\frac{x^2 + y'^2\cos^2\alpha + z'^2\sin^2\alpha}{n_0^2}\right] + \frac{y'^2\cos^2\alpha + z'^2\sin^2\alpha + 2y'z'\cos\alpha.\sin\alpha}{n_e^2} + 2r_{41}E_x\left(y'^2\cos\alpha.\sin\alpha - z'^2\cos\alpha.\sin\alpha\right) + y'z'\left[\frac{-2\cos\alpha.\sin\alpha}{n_0^2} + \frac{2\cos\alpha.\sin\alpha}{n_e^2} + 2r_{41}E_x\left(\cos^2\alpha - \sin^2\alpha\right)\right]\right\} = 1.$$

Equating cross-product term $y'z' = 0$ in order to find principal axis:

$$\left[\sin 2\alpha\left(\frac{1}{n_e^2} - \frac{1}{n_0^2}\right) + 2r_{41}E_x\cos 2\alpha\right] = 0$$

$$2r_{63}E_x\cos 2\alpha = \sin 2\alpha\left(\frac{1}{n_e^2} - \frac{1}{n_0^2}\right)$$

$$2r_{63}E_x \Big/ \left(\frac{1}{n_e^2} - \frac{1}{n_0^2}\right) = \tan 2\alpha$$

Thus, in the presence of a field in the x-direction, the principal y' and z'-axes are rotated through an angle α in the y–z plane where α is given by the above-mentioned equation.

Now, calculating new principal indices and

neglecting cross-product term $y'z'$, we get

$$\left\{\left[\frac{x^2+y'^2\cos^2\alpha+z'^2\sin^2\alpha}{n_0^2}\right]+\frac{y'^2\cos^2\alpha+z'^2\sin^2\alpha+2y'z'\cos\alpha.\sin\alpha}{n_e^2}\right.$$
$$\left.+2r_{41}E_x\left(y'^2\cos\alpha.\sin\alpha-z'^2\cos\alpha.\sin\alpha\right)\right\}=1$$

$$\left\{\frac{x^2}{n_0^2}+y'^2\left(\frac{\cos^2\alpha}{n_0^2}+\frac{\sin^2\alpha}{n_e^2}+2r_{41}E_x\cos\alpha.\sin\alpha\right)+\right.$$
$$\left.z'^2\left(\frac{\sin^2\alpha}{n_0^2}+\frac{\cos^2\alpha}{n_e^2}-2r_{41}E_x\cos\alpha.\sin\alpha\right)\right\}=1$$

By comparing with standard equation, we get:

$$\frac{x^2}{n_x^2}+\frac{y^2}{n_y^2}+\frac{z^2}{n_z^2}=1$$

$$n_x=n_0,\ n_y=\left(\frac{\cos^2\alpha}{n_0^2}+\frac{\sin^2\alpha}{n_e^2}+2r_{41}E_x\cos\alpha.\sin\alpha\right)^{-\frac{1}{2}},$$

$$n_z=\left(\frac{\sin^2\alpha}{n_0^2}+\frac{\cos^2\alpha}{n_e^2}-2r_{41}E_x\cos\alpha.\sin\alpha\right)^{-\frac{1}{2}}$$

MULTIPLE CHOICE QUESTIONS

1. Which of the following is not a common type of fibre-optic cable?
   (a) Single-mode step-index
   (b) Multimode graded-index
   (c) Single-mode graded-index
   (d) Multimode step-index
2. Which type of fibre optic cable is most widely used?
   (a) Single-mode step-index
   (b) Multimode step-index
   (c) Single-mode graded-index
   (d) Multimode graded-index
3. Which type of fibre-optic cable is the best for very high speed data?
   (a) Single-mode step-index
   (b) Multimode step-index
   (c) Single-mode graded-index
   (d) Multimode graded-index

4. Which type of fibre-optic cable has the least modal dispersion?
   (a) Single-mode step-index
   (b) Multimode step-index
   (c) Single-mode graded-index
   (d) Multimode graded-index
5. A step-index multimode optical fibre has a core diameter of ______ μm.
   (a) 62.5
   (b) 130
   (c) 35
   (d) 100
6. The graded-index multimode optical fibre has a core diameter of ______ μm.
   (a) 0.5
   (b) 0.05
   (c) 0.0005
   (d) 5
7. A single-mode optical fibre has a core diameter of ______ μm.
   (a) 0.1
   (b) 0.01
   (c) 0.2
   (d) 0.05
8. How many modes possible with a multimode step-index with a core diameter of 50 μm, a core refractive index of 1.6, a cladding refractive index of 1.584, and a wavelength of 1300 nm.
   (a) 456
   (b) 213
   (c) 145
   (d) 372
9. A type of index profile of an optical fibre that has a central core and outside cladding with a uniform refractive index:
   (a) multimode
   (b) graded index
   (c) step-index
   (d) single mode

Answers to Multiple Choice Questions

1. (c)
2. (a)
3. (a)
4. (a)
5. (a)
6. (b)
7. (b)
8. (d)
9. (c)

QUESTIONS

1. Define Pockels effect and Kerr effect?
2. Can phase modulators be used in interferometric applications?
3. What is the half-wave voltage or V_π?
4. Explain modulation depth at a given wavelength.

5. Define Q or quality factor, of the modulators.
6. What is VSWR?
7. What is the extinction ratio?
8. What is practical limitations of electro-optic modulator?
9. What are transit time limitations of crystal?
10. What is optical insertion loss?
11. Suppose that we generate 250 MHz acoustic wave on an $LiNbO_3$ substrate. The acoustic velocity in $LiNbO_3$ is 6.57 km/s and refractive index is about 2.2. Consider modulating a red laser beam from He–Ne laser, λ =632.8 nm. Calculate the acoustic wavelength, and hence, the Bragg deflection angle. What is the Doppler shift in wavelength?
12. Suppose that two optical guides embedded in a substrate like $LiNbO_3$ are coupled and the transmission length is $L_0 = 10\,\text{nm}$. If the coupling separation d is $\sim 10\,\mu\text{m}$, $n \approx 2.2$, the operating wavelength is $1.3\,\mu\text{m}$, taking a rough Pockels co-efficient $r \approx 10 \times 10^{-12}\,m/v$. What is the switching voltage?
13. Suppose that an $LiNbO_3$ crystal is used. $LiNbO_3$ is uniaxial and $n_1 = n_2 = n_0$ (polarisations parallel to x and y) and $n_3 = n_e$ (polarisations parallel to z). Neglecting the rotation of axes, if E_0 is the field along z, then the new refractive index is:

$$n_0' \approx n_0 - \frac{1}{2} n_0^3 r_{13} E_a$$

14. Calculate the half-wave voltage required to induce a retardation of π between the emerging and incident waves if the free space wavelength is $1\,\mu\text{m}$. What are their polarisations?
15. State the principle of operation of Mach–Zehnder interferometer.
16. Give an account of application of coupler in optical network. What is 3-dB coupler?
17. Calculate the rms pulse width σ of the rectangular pulse shown in Fig. UP1 in terms of its FWHM. τ.

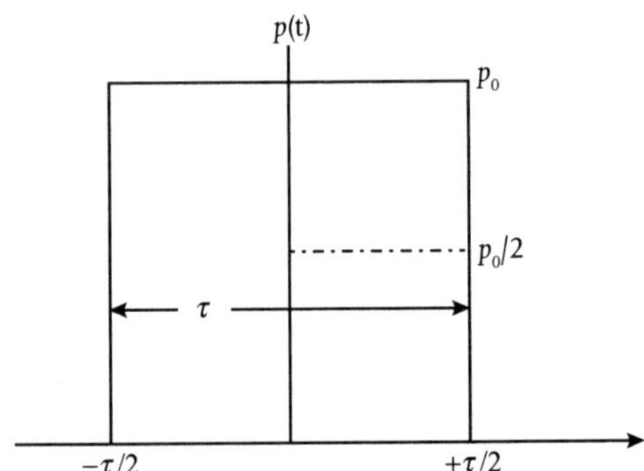

Fig. UP1 Rectangular pulse

18. Explain the working and construction of a fused biconical tapered coupler. A 2 × 2 biconical tapered fibre coupler has input power level of $P_1 = 400\,\mu\text{W}$. The output power at the other three ports is: $P_2 = 100\,\mu\text{W}$, $P_3 = 95\,\mu\text{W}$, and $P_4 = 8.3\,\mu\text{W}$ as shown in Fig. UP2. Calculate splitting ratio, excess loss, insertion loss, and cross talk.

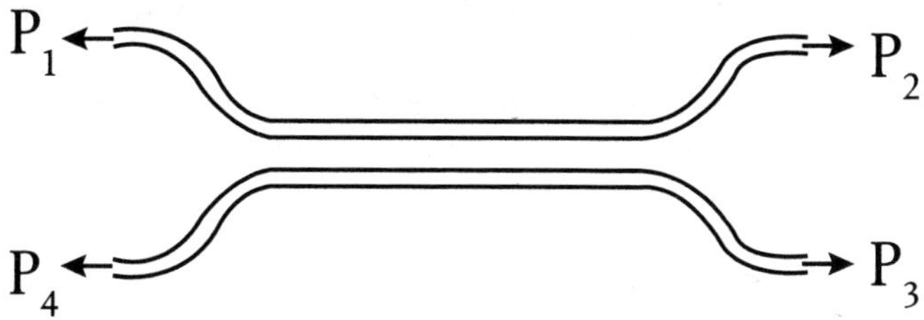

Fig. UP2

REFERENCES

1. Billings A, *Optics, Optoelectronics and Photonics Engineering Principles and Applications,* New York: Prentice Hall , 1993.
2. Guida G, Lustrac A D and Priou A, An introduction to photonic band gap (PBG) materials, *Progress In Electromagnetics Research,* PIER, vol. 41, pp. 1–20, 2003.
3. Yevick D and Hermansson B, Efficient beam propagation techniques, *IEEE J. Quantum Electron.* vol. 29, pp. 109–112, 1990.
4. Bates R J, *Optical Switching and Networking Handbook,* New York: McGraw-Hill, 2001.
5. Hill K O, Photosensitivity in optical fiber waveguides: Application to reflection fibre fabrication *Appl. Phys. Lett.,* vol. 32, pp. 647, 1978.
6. Ho K M, Chan C T and Soukoulis C M, Existence of a photonic gap in periodic dielectric structures, *Phys. Rev. Lett.,* vol. 65, pp. 3152, 1990.
7. Pollack C R, *Fundamental of Optoelectronics,* Irwin, 1995.
8. Pollock C R and Lipson M, *Integrated Photonics,* Boston: Kluwer Academic Publishers, 2003.
9. Raghuwanshi S K and Srinivas T, International conference on Microwave and Optoelectronics (ICMO), pp. 3, Aurangabad, India, 2007.
10. Yariv A and Yeh P, *Photonics,* 6th ed., New York: Oxford University Press, 2007.
11. Yasuk F, Durmus A and Boztosun I, Exact analytical solution to the relativistic Klein Gordon equation with non-central equal scalar and scalar and vector potentials, *J. Math. Phys.,* vol. 47, 082302, 2006.
12. Yeh P, *Optical Waves in Layered Media,* New York: John Wiley and Sons, 1988.
13. Yokouchi N, Danner A J and Choquette K D, Effective index model of 2D photonic crystal confined VCSELs, presented at *LEOS* VCSEL Summer Topical, Mont Tremblant, Quebec, 2002.
14. Zhuangqi C, Jiang Y, and Yingli C, Analytical Investigations of Planar Optical Waveguides with Arbitrary Index Profiles, *Opt. Quant. Electron.,* vol. 31, pp. 637–644, 1999.
15. Ziolkowski R W and Heyman E, Wave propagation in media having negative permittivity and permeability, *Phys. Rev. E,* vol. 64, pp. 056625, 2001.
16. Raghuwanshi S K and Talabattula S, Analytical approximation solutions for 3-D Optical Waveguides: Review, *Indian J. Phys.,* vol. 83, No. 2, 2009.
17. Raghuwanshi S K, Study of 1-D photonic crystal fibre by using plane wave expansion method, N. workshop on Quant. confined system and nano scale devices, Kerala, 3–5, Dec. 2009.
18. Raghuwanshi S K, Talabattula S and Selvarajan A, Fourier decomposition of the transverse field for analysing optical waveguides using beam propagation method, *J. Indian Inst. Sci.,* vol. 86, no. 6, pp. 667–680, Nov.–Dec. 2006.

19. Ramaswami R and Kumar N S, Optical networks: a practical prospective, 2nd ed., San Diego, CA: Academic Press, 2002.
20. Sakoda K, *Optical Properties of Photonic Crystals.* Berlin, Germany: Springer, 2001.
21. Rojas J A M, Alpuente J, Pineiro J and Sanchez R, Rigorous full vectorial analysis of electromagnetic wave propagation in 1d Inhomogeneous media, *Progress In Electromagnetics Research,* PIER, vol. 63, pp. 89–105, 2006.
22. Rostami A and Moayedi S K, Exact solution for light propagation through inhomogeneous media, *Indian J. Phys.* B, vol. 75, 357, 2001.
23. Smith G D, Numerical solution of partial differential equations, in *Finite Difference Methods,* New York: Oxford University Press, 1985.
24. Snyder A W and Love J D, *Optical Waveguide Theory,* London: Chapman and Hall, 1983.
25. Søndergaard T, Spontaneous emission in two-dimensional photonic crystal microcavities, *IEEE J. Quant. Electron.,* vol. 36, no. 4, pp. 450–457, 2000.
26. Sözüer H S and Haus J W, Photonic bands: Convergence problems with the plane-wave method, *Phys. Rev. B,* vol. 45, no. 24, pp. 13962–13972, 1992.
27. Special issue on Micro-opto-electro-mechanical Systems, *Opt. Eng.,* vol. 36, no.5, May 1997.
28. Survaiya S P and Shevgaonkar R K, Dispersion Characteristics of an Optical Fibre Having Linear Chirp Refractive Index Profile, *IEEE J. Lightwave Technol.,* vol. 17, No. 10, pp. 1797–1805, 1999.
29. Tamir T, *Guided Wave Optoelectronics,* Springer Series in Electronic and Photonics, 1982.
30. Utika H, Micromechanical photonics, *Opt. Rev.,* vol. 4, no. 6, p 623–633, 1997.
31. Villeneuve P R and Piché M, Photonic band gaps in two-dimensional square and hexagonal lattices, *Phys. Rev. B,* vol. 46, no. 8, pp. 4969–4972, 1992.
32. Xu W, Wang Z H and Huang Z M, Propagation constant of a planar dielectric waveguide with arbitrary refractive index variation, *Opt. Lett.,* vol. 18, pp. 805–807, 1993.
33. Barra A, Cassagne D and Jouanin C, Existence of two-dimensional absolute photonic band gaps in the visible region, *Appl. Phys. Lett.,* vol. 72, no. 6, pp. 627–629, 1998.
34. Gerken M and Miller D A B, Multilayer Thin-Film Stacks with Step like Spatial Beam Shifting, *IEEE J. Lightwave Technol.,* vol. 22, no. 2, pp. 612–618, Feb. 2004.
35. Gloge D, Dispersion in weakly guiding fibres, *App. Opt.* 10, pp. 2442–2445, 1971.
36. Inoue J K and Ohtaka K, *Photonic Crystals: Physics, Fabrication and Application,* New York: Springer Verlag, 2004.

8 Introduction to Microwave Photonics

Objectives

- This chapter describes the various methods for optical generation of microwave signals.
- Optical microwave signal processing is explained.
- Frequency modulation is delineated.
- Generation of millimetre-wave signal is explicated.

8.1 INTRODUCTION

With advantageous features of photonics, such as high time–bandwidth product, low loss capacity, immunity to electromagnetic interference, have led to the increasing interest in its use for generation, processing, control, and distribution of microwave signals, an area called microwave photonics. Further, the frequency response of an incoherent photonics microwave filter is more robust than a filter that is operating in the coherent region. In this chapter, a review on microwave photonics is presented with an emphasis on optical generation of microwave signals and all-optical microwave signal processing.

Microwave photonics is an interdisciplinary area that studies the interaction between the microwave and the optical signals for application such as radar, communication, sensor networks, warfare systems, and instrumentation [1-4]. In the past few years, there has been an increasing effort in researching new microwave photonics solutions for these applications. In general, the microwave photonics techniques cover the following topics: photonic generation of microwave signals, photonic processing of microwave signals, photonic distribution of microwave signals, and photonic analogue-to-digital conversion. In this chapter, techniques developed in the last few years are reviewed with an emphasis on the system architectures to implement the functions for the photonic generation and processing of microwave signals [5-9].

8.2 OPTICAL GENERATION OF MICROWAVE SIGNALS

A low-phase noise and frequency-tunable microwave or mm-wave source is desirable for many applications. For many applications, the generated microwave or mm-wave signal should be distributed to a remote site. The distribution of a microwave or mm-wave signal in the electrical domain is not practical due to the high loss associated with electrical distribution lines, for example, coaxial cable. The solution is the use of broad bandwidth and low-loss optical fibre.

8.2.1 Optical Heterodyning

A microwave or mm-wave signal can be generated in optical domain based on optical heterodyning in which two optical waves of different wavelengths beat at a photodetector [10-11]. An electrical beat note is then generated at the output of the photodetector with a frequency corresponding to the wavelength spacing of the two optical waves, as shown in Fig. 8.1.

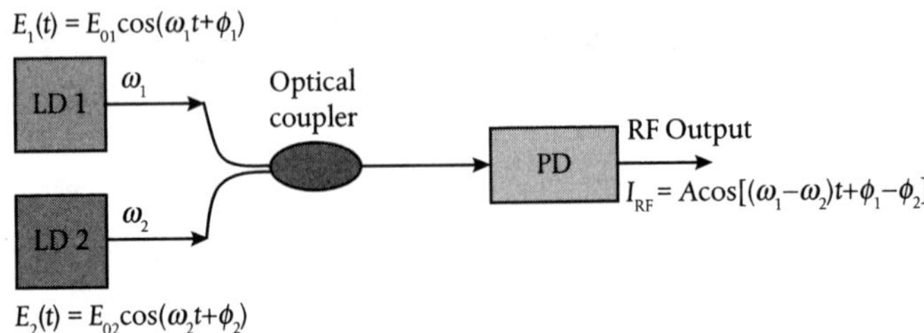

Fig. 8.1 Beating of two optical waves at a PD for the generation of a microwave signal

Let us assume that we have two optical waves, which is given as:

$$E_1(t) = E_{01}\cos(\omega_1 t + \phi_1) \tag{8.1}$$

and

$$E_2(t) = E_{02}\cos(\omega_2 t + \phi_2), \tag{8.2}$$

where E_{01}, E_{02} are the amplitude terms and ϕ_1, ϕ_2 are the phase terms of the two optical waves. The signal at the output of the PD is given as:

$$I_{PD}(t) \propto R_{PD}2\left[E_{01}\cos(\omega_1 t + \phi_1) + E_{02}\cos(\omega_2 t + \phi_2)\right]^2$$

$$I_{PD}(t) \propto R_{PD}\left[2E_{01}{}^2\cos^2(\omega_1 t + \phi_1) + 2E_{02}{}^2\cos^2(\omega_2 t + \phi_2) + 2E_{01}.E_{02}.2\cos(\omega_1 t + \phi_1)\cos(\omega_2 t + \phi_2)\right].$$

$$I_{PD}(t) \propto R_{PD}\left\{E_{01}{}^2\left[1 + \cos 2(\omega_1 t + \phi_1)\right] + E_{02}{}^2[1 + \cos 2(\omega_2 t + \phi_2)] + 2E_{01}.E_{02}.\left[\cos(\omega_1 t + \omega_2 t + \phi_1 + \phi_2) + \cos(\omega_1 t - \omega_2 t + \phi_1 - \phi_2)\right]\right\}.$$

$$I_{PD}(t) \propto R_{PD}\left\{E_{01}{}^2 + E_{02}{}^2 + 2E_{01}.E_{02}.\cos\left[(\omega_1 - \omega_2)t + \phi_1 - \phi_2\right]\right\}, \tag{8.3}$$

where R_{PD} is the responsivity of the PD. The above-mentioned calculation is referred to a coherent beating of the input optical signals. Terms $2(\omega_2 t + \phi_2)$, $2(\omega_2 t + \phi_2)$, and $\omega_1 t + \omega_2 t + \phi_1 + \phi_2$ disappear due to averaging.

Considering the limited bandwidth of the PD, the current at the output of the PD is given as:

$$I_{RF} = 2R_{PD}\sqrt{P_1 P_2}\cos\left[(\omega_1 - \omega_2)t + (\phi_1 - \phi_2)\right]. \tag{8.4}$$

As can be seen from (8.4), a microwave signal with a frequency equal to the frequency difference between the two optical waves is generated. This technique is capable of generating an electrical signal with a frequency up to the THz band, limited only by the bandwidth of the PD. However, by beating two optical waves from two free-running LDs would lead to a microwave signal having high-phase noise since the phase terms of the two optical waves are not correlated and the beating process will transfer the phase noise of the two optical waves to generate microwave signals with the two optical waves being locked in phase. These techniques can be classified into three categories: optical injection locking, optical phase-lock loop (OPLL), and microwave generation using external modulation. In addition, a low-phase-noise microwave signal can also be generated using an opto-electronic oscillator (OEO).

8.2.2 Optical Injection Locking

To generate a high-quality microwave signal via heterodyning, the phase terms of the two optical waves used for heterodyning must be highly correlated. A solution is to lock the phase term of two optical waves via optical injection locking [12-13]. Figure 8.2 shows a scheme to achieve optical injection locking, in which a master laser diode (LD) is frequency modulated by a microwave at f_m, and the output from the master LD, consisting of an optical carrier at λ_0, two first-order sidebands, and other higher order sidebands, is injected into the slave LDs. The two slave LDs have a free-running wavelength close to wavelengths of two sidebands, say -2nd- and $+2^{nd}$-order sidebands; thus, the two slave LDs are injection locked and the optical waves at the outputs of the two slave LDs are phase correlated and beating of two waves would generate a microwave signal with its phase noise determined by the reference RF source. Note that the system can achieve frequency multiplying operation. Since the two wavelengths of the two slave lasers are phase correlated, the beating of the two wavelengths at a photodetector would generate a beat note with low-phase noise. For the configuration shown in Fig. 8.2, the generated frequency is four times the frequency of the reference RF sources, and a frequency quadrupling operation is thus achieved. Note also that microwave signal can generate remotely. This is the key motivation of using such a technique for the optical distribution of microwave signals.

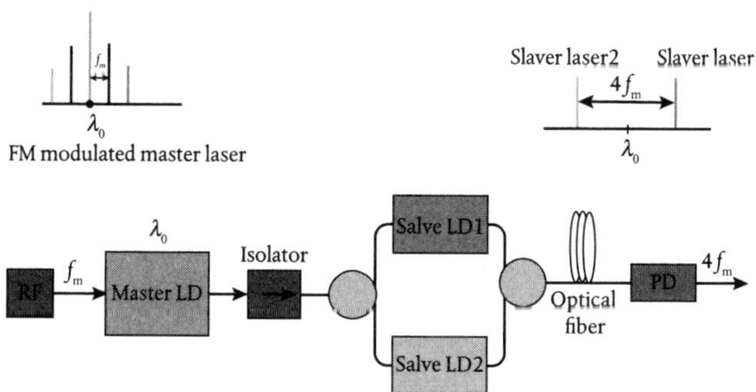

Fig. 8.2 Optical injection locking of two slave LDs

8.2.3 Optical Phase-lock Loop (OPLL)

In this technique, phase of one laser is actively locked to the phase that of the second laser by an OPLL. Two free-running laser sources can also be phase locked using a phase lock loop.

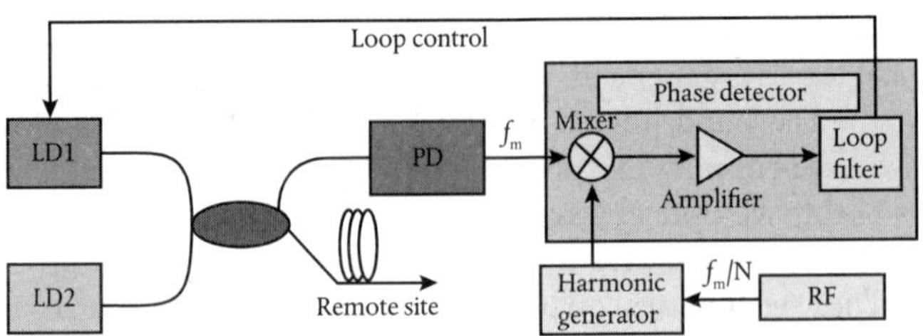

Fig. 8.3 Optical phase locking of two LDs using a phase-lock loop

The mixer, the amplifier, and the loop filter form a phase detection module. To achieve effective phase locking, the two lasers should be selected to have a narrow line width, and therefore, they have phase fluctuation one at low frequencies, which would ease significantly the requirement for a very short feedback loop. Figure 8.3 shows a scheme to achieve optical phase locking of two LDs. As can be seen, two optical waves from two LDs are applied to a PD. At the output of the PD, a beat note with the frequency corresponding to the wavelength spacing is generated. The generated microwave signal is then sent to a phase detector consisting of an electronic mixer and a low-pass filter. The output voltage, which is proportional to the phase difference between the generated microwave signal and the reference signal, is sent to one LD to control the injection current, and thus, its phase is accordingly changed, leading to the lock of the phase terms of the two optical waves.

The locking range is determined by the linewidth of the LDs and the loop length. To achieve effective phase locking, we need to use narrow linewidth LDs and also to make the loop length shorter. Again, similar to the injection locking scheme, the optical signal can be distributed over optical fibre for remote distribution. Note that it is not necessary that the frequency of the RF reference source is identical to the frequency of the generated microwave signal. In fact, if a harmonic generator is used, then an RF signal with different orders of harmonics can be sent to the mixer, and thus, a relatively low-frequency RF reference signal can be used. For example, to generate a 60-GHz microwave signal, one may use a 15-GHz reference source, and the fourth harmonic will be mixed with the beat note to produce a voltage signal to control the phase of one LD.

8.2.4 Optical Injection Phase Locking

In optical injection phase locking, the two techniques of optical injection locking and optical phase locking can be combined in a single optical locking system [14-18]. The system consists of a master laser, a slave laser, an external modulator, a photodetector, and a phase detection module (dotted box). Figure 8.4 shows the light from the master laser is divided into two channels, with one channel coupled into the modulator before being injected into the slave laser. The slave laser is locked to one sideband of the modulated signal for optical injection locking of the system. The other channel of the light from the master laser is combined with the output of the slave laser and the beat of the photodetector.

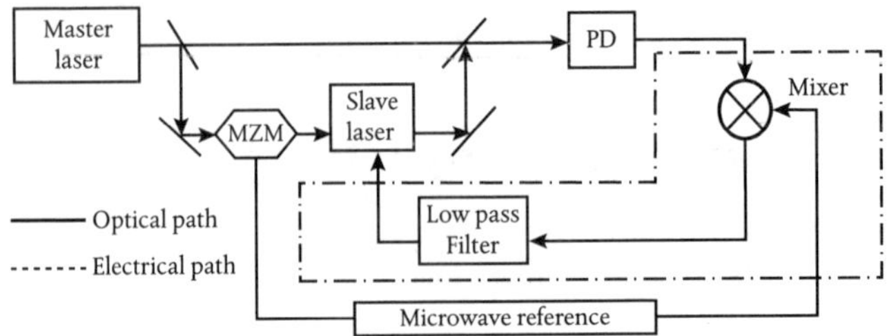

Fig. 8.4 Optical injection phase locking

The beat note is mixed with a microwave reference, with the mixing output filtered by a low-pass filter to achieve phase locking of the wavelength of the master laser.

8.2.5 Microwave Generation Based on External Modulation

Microwave signal generation can also be implemented based on external modulation by which the frequency of a low-frequency microwave signal can be increased to a high frequency through frequency multiplication [15-18]. Figure 8.5(a, b) shows a scheme to achieve frequency doubling using a Mach–Zehnder modulator (MZM). An RF signal is applied to the MZM via the RF port. The MZM is based on the minimum transmission point (MITP) to suppress the optical carrier. At the output of the MZM, only two first-order sidebands are generated. By beating the first-order sideband at a PD, a frequency-doubled microwave signal is generated. If the MZM is biased at the maximum transmission point (MATP), at the output of the MZM, an optical signal with an optical carrier and two second-order sidebands are generated. Key significance is the system simplicity, frequency tunability, and operation stability.

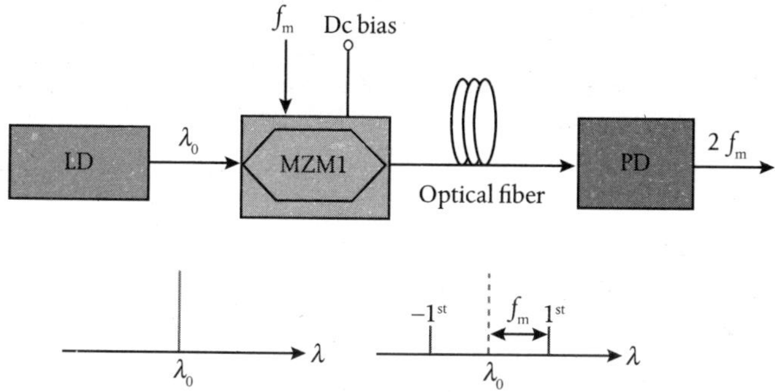

Fig. 8.5 (a) Frequency doubling by biasing the MZM at the minimum transmission point

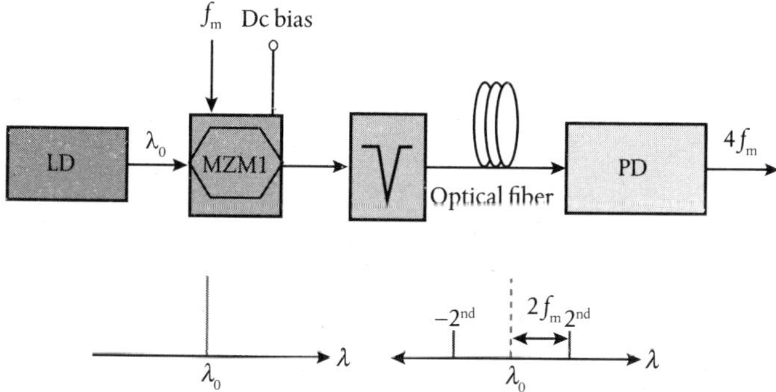

Fig. 8.5 (b) Frequency doubling by biasing the MZM at the maximum transmission point. The optical carrier is removed by an optical notch filter

By using a notch filter to remove the optical carrier and beat the two sidebands at the PD, a frequency quadrupled microwave signal is generated. To generate a microwave signal with a higher multiplication factor, a configuration that employs two cascaded MZMs may be used. The multiplication factor can be as high as 12, which enable the generation of a high-frequency microwave or terahertz signal using a low-frequency microwave source. Different techniques have been proposed for the generation of a frequency- quadrupled, sextupled, octupled, and twelvetupled microwave signal.

All these approaches were implemented by biasing the MZMs at the MATP or MITP in conjunction with the use of an optical or microwave phase shifter and an optical filter. For two cascaded MZMs, there are four possible bias combinations: (i) MATP, MATP, (ii) MITP, MITP, (iii) MATP, MITP, and (iv) MITP, MATP, with each combination corresponding to a specific multiplication factor under different operation conditions.

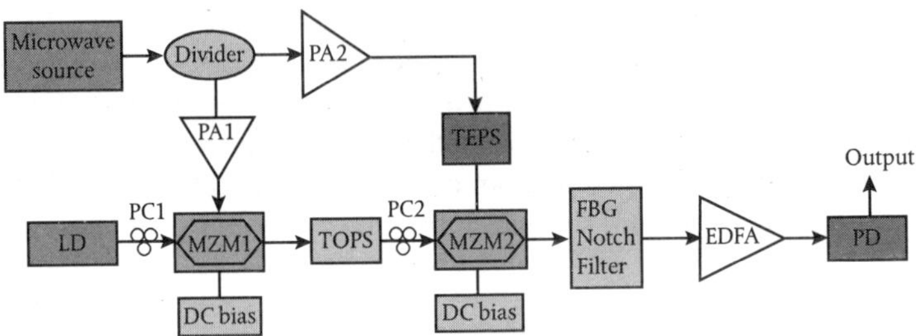

Fig. 8.6 A generalised microwave frequency multiplication system using two cascaded MZMs. **TOPS**: tunable optical phase shifter, **FBG**: fibre Bragg grating, **EDFA**: erbium doped fibre amplifier, **TEPS**: tunable electrical phase shifter, **PA**: power amplifier

Figure 8.6 shows a microwave generation system using two cascade MZMs. A theoretical analysis leading to the operating condition to achieve frequency quadrupling, sextupling, and octupling was developed and verified experimentally.

8.2.6 Microwave Generation Based on an Opto-electronic Oscillator

In addition to the beating of two optical waves at a PD to generate a microwave signal, a high spectral purity microwave signal can also be generated using an opto-electronic oscillator (OEO). When compared with the techniques using injection locking, phase-lock loop or external modulation, the use of an OEO can generate a microwave signal without the need for a reference microwave source [19-22].

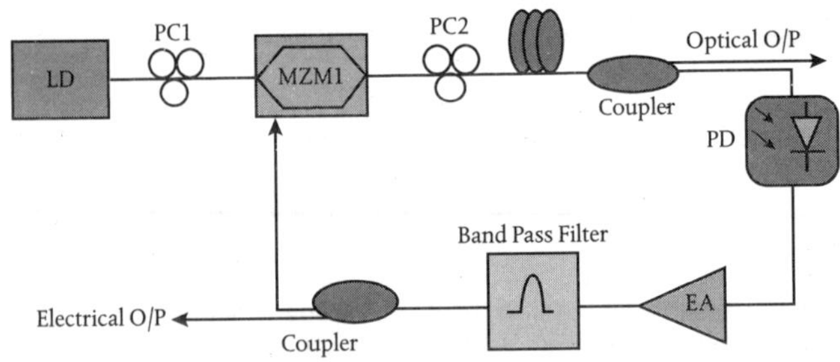

Fig. 8.7 An opto-electronic oscillator

Figure 8.7 shows a typical OEO. If the gain in the loop is greater than the loss, the OEO will oscillate. The phase noise performance is determined by the Q-factor of the OEO loop. To generate a microwave signal with low-phase noise, the loop length should be long, that is from a few to tens of km. The use of a long loop length will cause the OEO to generate a large number of closely spaced eigen modes. To ensure a single-frequency oscillation, a high-Q microwave filter with an ultra-narrow bandwidth must be used or high level of side modes will be generated. The use of multiple loops in an OEO would make the mode spacing greater, which would ease the requirement for a high-Q microwave filter. However, an OEO with multiple loops would have a poorer system stability and higher system cost. More importantly, the frequency tunability would be complicated if a multiple-loop structure is employed. Recently, a technique to achieve an optically tunable OEO with a wide frequency tunable range incorporating a tunable microwave photonic filter implemented based on phase-modulation of intensity-modulated (PM-IM) conversion using a phase-shifted fibre Bragg grating (PS-FBG) was proposed, with the configuration shown in Fig. 8.8. When a microwave signal is applied to the PM, a phase-modulated optical signal consisting of an optical carrier and two first-order sidebands are generated. When the phase-modulated light wave is sent to the PS-FBG that has an notch in the reflection spectrum, if the optical wavelength is tuned such that one sideband of the modulated optical signal falls into the notch and is removed by the PS-FBG; the phase modulated signal becomes a single-sideband intensity modulated signal, thus PM-IM conversion is achieved, and the joint operation of the phase modulator and the PS-FBG corresponds to a microwave photonic filter, with the filter bandwidth determined by the bandwidth of the PS-FBG.

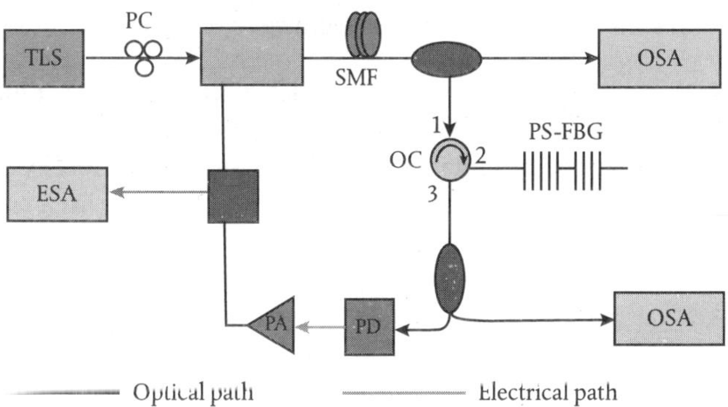

Fig. 8.8 An optically tunable OEO

It is known that a PS-FBG can have a notch as narrow as a few MHz, and thus, the microwave photonic filter can effectively select one eigen mode, making the OEO operate in single frequency. The frequency tuning can be simply done by tuning the wavelength of the lightwave from the tunable laser source (TLS). Thus, an OEO with a large frequency tunable range is achieved.

8.3 ALL-OPTICAL MICROWAVE SIGNAL PROCESSING: INTRODUCTION TO PHOTONIC MICROWAVE DELAY-LINE FILTERS

High-frequency and wideband microwave signals can also be processed in the optical domain. The key signal processing function is microwave filtering. In this section, we will review the techniques to implement a photonic microwave delay-line filters, with an emphasis on the generation of negative and complex coefficients to achieve a bandpass filtering [23-26]. The use of non-uniform time delays to generate equivalent complex coefficients to achieve photonic microwave delay-line filters with an arbitrary spectral response will also be discussed.

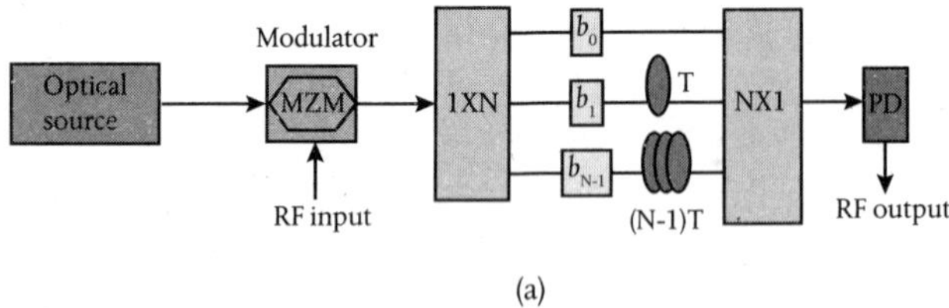

Fig. 8.9 (a) A generic photonic microwave delay-line filter with a finite impulse response

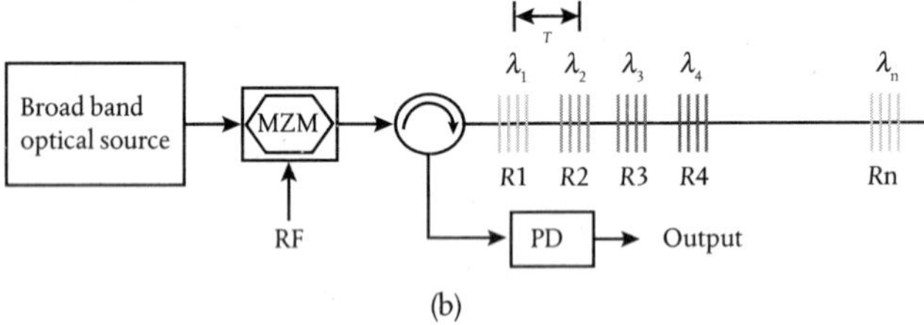

Fig. 8.9 (b) The implementation of a photonic microwave delay-line filter using an FBG array

Figure 8.9(a) shows a generic photonic microwave delay-line filter with a finite impulse response (FIR). As can be seen, the filter consists of a light source, a modulator, a delay-line module, and a PD, provided that the nonlinearity in the system is small and negligible. The entire system can be considered as a linear, time-invariant (LTI) system, in which the output $y(t)$ can be written as:

$$y(t)=\sum_{k=0}^{N-1}a_k x(t-kt), \tag{8.5}$$

where $x(t)$ is the microwave input signal, T is the time delay difference between two adjacent taps, and a_k is the coefficient of the k-th tap. By applying the Fourier transforms to both sides of (8.5), the system transfer function is then obtained:

$$H(\omega)=\sum_{k=0}^{N-1}a_k \exp(-j\omega T). \tag{8.6}$$

The time delay unit T determines the free spectral range (FSR) of the microwave filter. The coefficient a_k is proportional to the optical intensity of the k-th tap and cannot be negative. Therefore, the frequency response described by (8.6) is a typical transfer function of an FIR delay-line filter with all-positive coefficients. Figure 8.9(b) shows an incoherent photonic microwave delay-line filter using an FBG array. The microwave signal to be processed is modulated on an incoherent optical wave, such as a lightwave from a light emitting diode (LED) or an amplified spontaneous emission (ASE) source, which is then sent to an FBG array. Due to the reflection of the FBGs in the FBG array, lightwaves with different time delays are detected at the PD. Since the optical source is incoherent, no optical interference will be generated at the PD. The time-delayed optical signals are then converted to time-delayed microwave signals. The overall function of the system is equivalent to a delay-line filter with the time delay difference determined by the physical separation of adjacent FBGs. The tap coefficients are set by controlling the reflectivities of the FBGs.

8.3.1 Photonic Microwave Delay-line Filters with Negative Coefficients

It is known a photonic microwave delay-line filter operating in the incoherent regime would have all-positive tap coefficients. Based on signal processing theory, an all-positive-coefficient microwave delay-line filter can only operate as a low-pass filter. To overcome this limitation, considerable efforts have been taken to design and implement photonic microwave delay-line filters with negative or complex coefficients, and to achieve bandpass filtering functionality in the incoherent regime. A straightforward solution to generate negative coefficients is to use differential detection.

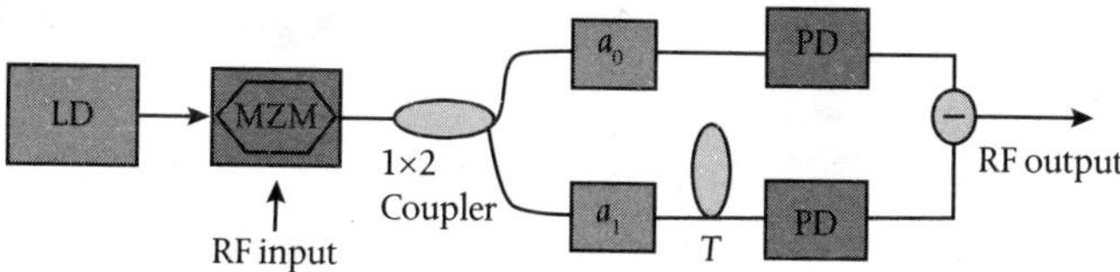

Fig. 8.10 A photonic microwave delay-line filter with a negative coefficient using differential detection

As shown in Fig. 8.10, a lightwave from an LD is modulated by a microwave signal, which is then time delayed by optical fibre delay lines with a time delay difference of T. The output signals from the fibre delay lines are fed to a differential PD, which consists of two matched PDs with the detected microwave signals combined and subtracted electrically, leading to the generation of a positive and a negative coefficient. The two-tap photonic microwave delay-line filter shown in Fig. 8.10 can be extended to have N taps if the single-wavelength source is replaced by an N wavelength source and the 3-dB coupler is replaced by a $1\times N$ WDM demultiplexer. In Fig. 8.10, the negative coefficient is not generated directly in the optical domain.

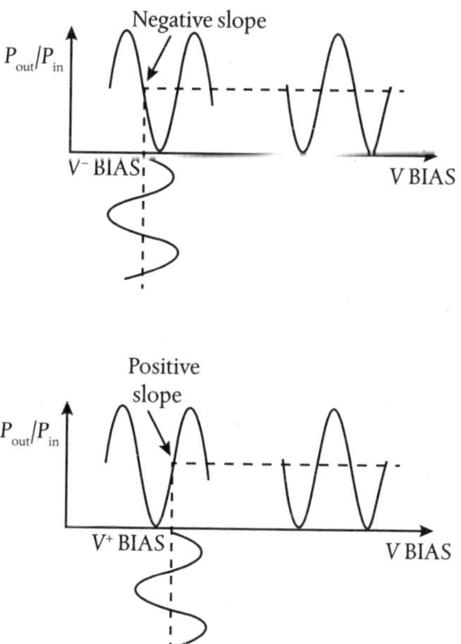

Fig. 8.11 Photonic microwave delay-line filter with negative coefficients based on phase inversion using complementarily biased intensity modulators. (a) The operation of phase inversion

The limitation of this technique is that for a filter with N taps, N PDs are needed; this makes the filter complicated and costly. A few techniques have been proposed to implement an all-optical photonic microwave delay-line filter with negative coefficients using a single PD. As an example, negative coefficients can be generated using two MZMs that are biased at the quadrature points of the complementary transmission slopes, as shown in Fig. 8.11(a).

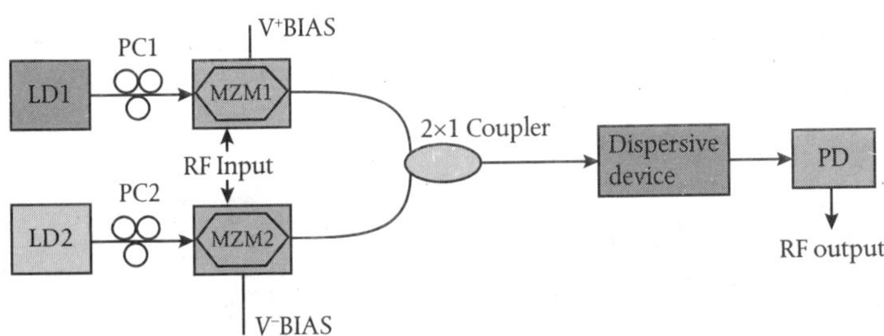

Fig. 8.11 (b) The schematic of the filter.

The schematic of the filter is shown in Fig. 8.11(b). As can be seen, the two MZMs are biased at the linear regions of the left and the right slopes of the transfer functions. When a microwave signal is applied to the two MZMs, the envelopes of the modulated optical signals are complementary. At the output of a PD, two complementary microwave signals are generated, leading to the generation of a negative coefficient. A similar technique using only a single MZM was proposed. Considering the wavelength dependence nature of the transfer function of the MZM, a proper DC bias would make the modulator operate on the complementary slopes of the transfer when the optical wavelengths are at 1550 nm and 1310 nm windows. All the filters discussed earlier are implemented based on the use of one or two MZMs. A photonic microwave bandpass with negative coefficients can also be implemented based on an optical phase modulator [39]. The negative coefficients are generated based on phase modulation (PM) and PM-IM conversion in two dispersive elements with complementary dispersion profiles, such as linearly chirped FBGs (LCFBGs). With reflection of the phase-modulated optical signals from the LCFBGs with complementary chirp profiles, microwave signals without or with a phase inversion is generated at the PD. An added advantage of using an optical phase modulator is that a phase modulator is not biased, which eliminates the bias drifting problem existing in an MZM-based system. The fundamental concept of the filter operation is shown in Fig. 8.12(a). An RF signal is applied to the optical phase modulator via the RF port to phase modulate the multiple wavelengths at the phase modulator. Since a PD functions as an envelope detector, if a phase-modulated signal is directly applied to a PD, no modulating signal will be detected except a DC. This conclusion can also be explained based on the spectrum of a phase-modulated signal.

As shown in Fig. 8.12(a), a small-signal phase-modulated signal has a spectrum with the $+1^{st}$- and -1st- order sidebands of phase. The beating between the optical carrier and the $+1^{st}$-order sideband exactly cancels the beating between optical carrier and the -1st-order sideband. However, if the phase-modulated optical signal passes through a dispersive element, the phase relationship between the two sidebands and the optical carrier will be changed, leading to the conversion from phase modulation to intensity modulation.

In addition, depending on the sign of the chromatic dispersion, a detected RF signal with or without a phase inversion would be obtained, leading to the generation of negative coefficients. The system configuration of the filter is shown in Fig. 8.12(b).

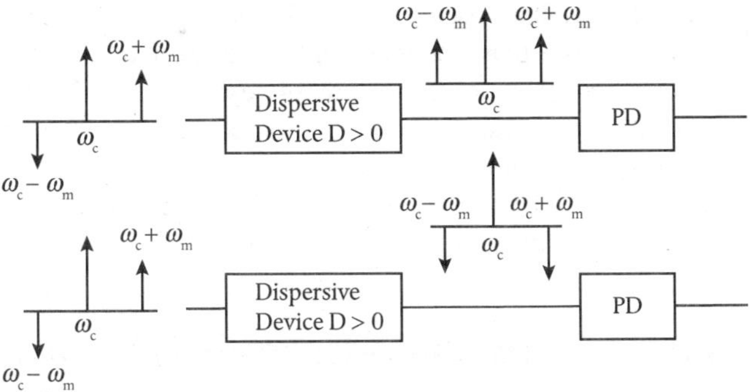

Fig. 8.12 (a) RF phase inversion based on phase modulation and PM-IM conversion through opposite dispersions

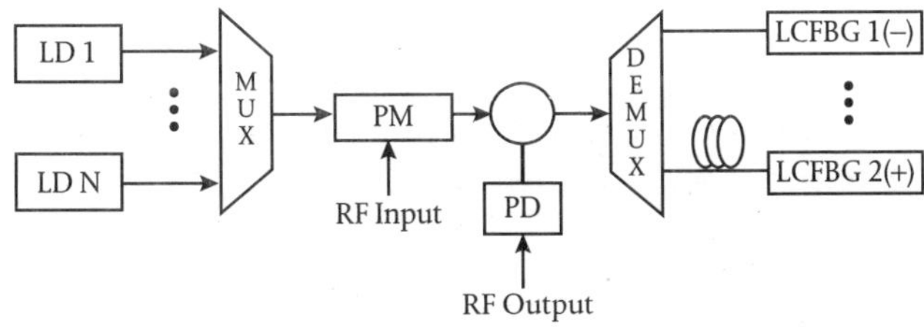

Fig. 8.12 (b) A photonic microwave delay-line filter with a negative coefficient based on PM-IM conversion in two LCFBGs with opposite dispersions

8.3.2 Photonic Microwave Delay-line Filters with Complex Coefficients

The tunability of a photonic microwave delay-line filter is usually achieved by adjusting the time-delay difference. However, the change of the time-delay difference would lead to the change of the FSR, which results in the change of the 3-dB bandwidth as well as the entire shape of the frequency response. For many applications, it is highly desirable that only the centre frequency of the passband or stop-band is changed while maintaining the shape of the frequency response unchanged. A solution to this problem is to design a photonic microwave delay-line filter with complex. An N-tap microwave delay-line filter with complex coefficients should have a transfer function given as:

$$H(\omega)=a_0+a_1e^{-j\theta}e^{-j\omega T}+\ldots+a_{N-1}e^{-j(N-1)\theta}e^{-j\omega(N-1)T}$$
$$=\sum_{n=0}^{N-1}a_n e^{-jn\theta}e^{-j\omega nT}, \quad (8.7)$$

where T is again the time delay difference between two adjacent taps. To tune the filter while maintaining the shape of the frequency response, the phase shifts of all the taps should maintain a fixed relationship, as can be seen from (8.7). Therefore, the phase shift of each tap should be tuned independently. A photonic microwave delay-line filter with a complex coefficient using a wideband tunable optical RF-phase shifter was demonstrated. The RF-phase shift is generated using two electro-optic MZMs by simply adjusting the bias voltages applied to the two MZMs, and the phase shift remains constant over the microwave frequency band of interest. A wideband phase shifter can also be realised based on a combined use of optical single-sideband modulation (SSB) and stimulated Brillouin scattering (SBS). It was demonstrated that the phase of a microwave signal carried by an optical carrier will experience microwave phase shift if the spectrum of the optical carrier or the side band is falling in the SBS gain spectrum when passing through an optical fibre in which an SBS is resulted.

8.3.3 Non-uniformly Spaced Photonic Microwave Delay-line Filters

The filter with complex coefficients is having the potential to be extended to have multiple taps, but the filter complexity would be significantly increased since each tap needs a wideband phase shifter. To design a photonic microwave delay-line filter with complex coefficients having a simple structure, a new concept was proposed by generating complex coefficients based on nonuniformly spaced taps. It was demonstrated that the complex coefficients can be equivalently generated by introducing additional time delays to the taps. As given in (8.6), the frequency response of a uniformly spaced microwave delay-line filter has a frequency response, which is given as:

$$H(\omega)=\sum_{k=0}^{N-1} a_k \exp(-j\omega kT).$$

As can be seen, $H(\omega)$ has a multi-channel frequency response with adjacent channels separated by an FSR, with the mth-channel located at $\omega = m$. In a regular photonic microwave delay-line filter based on detection, the coefficients are usually all positive or special designs have to be employed to generate negative or complex coefficients. However, a phase term can be introduced to a specific coefficient by adding an additional time delay at the specific tap, which is termed time delay-based phase shift. For example, at $\omega = m$ a time delay shift of $\Delta\tau$ will generate a phase shift given by $\Delta\phi = -\Delta\tau \times m$. Note that such a phase shift is frequency dependent, which is accurate only for the frequency at $m\Omega$, but approximately accurate for a narrow frequency band at around $m\Omega$. For most applications, the filter is designed to have a very narrow frequency band, and therefore, for the frequency band of interest, the phase shift can be considered constant over the entire bandwidth. As a result, if the mth bandpass response, where $m \neq 0$, is considered, one can then achieve the desired phase shift at the kth tap by adjusting the time delay shift by $\Delta\tau k$. Considering the time delay shift of $\Delta\tau k$, one can get the frequency response of the non-uniformly spaced delay-line filter at around $\omega = m\Omega$:

$$H_{Non}(\omega)=\sum_{k=0}^{N-1} a_k \exp\left[-j\left(\frac{2\pi k}{\Omega}+\Delta\tau_k\right)\omega\right]$$

$$=\sum_{k=0}^{N-1}\left[a_k \exp(-jm\Omega\Delta\tau_k)\right]\exp\left(-jk\frac{2\pi}{\Omega}\omega\right). \quad (8.8)$$

As can be seen from (8.8), an equivalent phase shift is generated for each tap coefficient. Specifically, if the desired phase shift for the kth tap is ϕ_k, the total time delay $\Delta\tau k$ for the kth tap is $\tau_k = kT - \phi_k / m\Omega\tau$ k. As a result, if the time delay of each tap is adjusted, the filter coefficients would have the required phase shifts to generate the required passband with the desired bandpass characteristics. The design of a non-uniformly spaced photonic microwave delay-line filter to achieve a bandpass filter with a flat top was demonstrated. The technique is particularly useful for advanced microwave signal processing. For example, the use of a non-uniformly spaced photonic microwave delay-line filter to achieve microwave phase coding, chirped microwave pulse generation, microwave pulse compression, microwave differentiation, and microwave Hilbert transformation has been reported.

8.4 FREQUENCY MODULATION: REVIEW

In frequency modulation, the frequency of the carrier is varied according to the amplitude of the modulating signal. The carrier frequency $\omega(t)$ would be varied with time so that $\omega(t)=\omega_c+km(t)$, where k is an arbitrary constant. If the peak amplitude of $m(t)$ is m_p, then the maximum and minimum values of the carrier frequency would be ω_c+km_p and ω_c-km_p, respectively. Hence, the spectral components would remain within this band with a bandwidth $2\,km_p$ centered at ω_c. The bandwidth is controlled by the arbitrarily constant k. By using an arbitrarily small k, we could make the information bandwidth arbitrarily small.

8.4.1 Concept of Instantaneous Frequency

By definition, a sinusoidal signal has a constant frequency, and hence, the variation of frequency with time appears to be contradictory to the conventional definition of a sinusoidal signal frequency. In FM, the carrier frequency is changing continuously every instant. We must have a sinusoidal signal at least over one cycle with the same frequency. Let us consider a generalised sinusoidal signal $\varphi(t)$ given by:

$$\varphi(t)=A\cos\theta(t) \tag{8.9}$$

Over the small interval $\Delta t \to 0$, the signal $\varphi(t)=A\cos\theta(t)$ and the sinusoidal $\varphi(t)=A\cos(\omega_c t+\theta_0)$ are identical:

$$\varphi(t)=A\cos(\omega_c t+\theta_0)t \text{ where } t_1<t<t_2$$

Over this small interval Δt, the frequency of $\varphi(t)$ is ω_c. Because $(\omega_c t+\theta_0)$ is tangential to $\theta(t)$, the frequency of $\varphi(t)$ is the slope of its angle $\theta(t)$ over this small interval. Generalise this concept at every instant and say that the instantaneous frequency ω_i at any instant t is the slope of $\theta(t)$ at t. Thus, for $\varphi(t)$ in equation:

$$\omega_i(t)=\frac{d\theta}{dt} \tag{8.10a}$$

$$\theta(t)=\int_{-\infty}^{t}\omega_i(\alpha)d\alpha \tag{8.10b}$$

Hence, there is a possibility of transmitting the information of $m(t)$ by varying angle θ of a carrier. Such techniques of modulation, where the angle of the carrier is varied in some manner with a modulating signal $m(t)$, are known as angle modulation. Two possibilities are phase modulation (PM) and frequency modulation (FM). In FM, the instantaneous frequency ω_i is:

$$\omega_i(t)=\omega_c+k_f m(t),$$

where k_f is a constant. The angle $\theta(t)$ is now:

$$\theta(t) = \int_{-\infty}^{t} \left[\omega_c + k_f m(\alpha)\right] d\alpha \tag{8.11a}$$

$$= \omega_c t + k_f \int_{-\infty}^{t} m(\alpha) d\alpha \tag{8.11b}$$

Assuming the constant term in $\theta(t)$ to be zero without loss of generality. The FM wave is:

$$\varphi_{FM}(t) = A\cos\left[\omega_c t + k_f \int_{-\infty}^{t} m(\alpha) d\alpha\right] \tag{8.11c}$$

8.4.2 Power of an Angle-modulated Wave

Although the instantaneous frequency and phase of an angle-modulated wave can vary with time, the amplitude A always remains constant. Hence, the power of a frequency-modulated wave is always $A^2/2$, regardless of the value of k_p or k_f.

8.4.3 Bandwidth of Angle-modulated Waves

In order to determine the bandwidth of an FM wave, let us define:

$$a(t) = \int_{-\infty}^{t} m(\alpha) d\alpha \tag{8.12}$$

and

$$\hat{\varphi}FM(t) = Ae^{j\left[\omega_c t + k_f a(t)\right]} = Ae^{jk_f a(t)} e^{j\omega_c t} \tag{8.13a}$$

Now,

$$\varphi_{FM(t)} = \text{Re } \hat{\varphi}FM(t) \tag{8.13b}$$

Expanding the exponential $e^{jk_f a(t)}$ in (8.13a) in power series yields:

$$\hat{\varphi}FM(t) = A\left[1 + jk_f a(t) - \frac{k_f^2}{2!}a^2(t) + \ldots + j^n \frac{k_f^n}{n!} a^n(t)\right] e^{j\omega_c t} \tag{8.14a}$$

$$\varphi\,FM(t) = \text{Re } \hat{\varphi}\,FM(t)$$

$$= A\left[\cos\omega_c t - k_f a(t)\sin\omega_c t - \frac{k_f^2}{2!}a^2(t)\cos\omega_c t + \frac{k_f^2}{3!}a^3(t)\sin\omega_c t \ldots\ldots\right] \tag{8.14b}$$

The modulated wave consists of an unmodulated carrier and various amplitude-modulated terms, such as $a(t)\sin\omega_c t$, $a^2(t)\cos\omega_c t$, $a^3(t)\sin\omega_c t$ .... and the signal $a(t)$, which is an integral of $m(t)$. If $M(\omega)$ is band limited to B, $A(\omega)$ is also band limited to B. The spectrum of $a^2(t)$ is simply $A(\omega) \times A(\omega)/2\pi$ and is band limited to 2B. Similarly, the spectrum of $a^n(t)$ is band limited to nB.

Hence, the spectrum consists of an unmodulated carrier and spectra of $a(t)$, $a^2(t)$, ..., $a^n(t)$, ... centred at ω_c. Clearly, the modulated wave is not band limited. It has an infinite bandwidth and is not related to the modulating-signal spectrum in any case as in AM.

Although the theoretical bandwidth of an FM wave is infinite, most of the modulated-signal power resides in a finite bandwidth. There are two distinct possibilities in term of bandwidths: narrow-band FM and wide-band FM.

8.4.4 Narrow-band Angle Modulation

Unlike AM, angle modulation is nonlinear. The principle of superposition does not apply. This may be verified from the fact that:

$$\mathrm{A}\cos\left\{\omega_c t + k_f\left[a_1(t) + a_2(t)\right]\right\} \neq \mathrm{A}\cos\left[\omega_c t + k_f a_1(t)\right] + \mathrm{A}\cos\left[\omega_c t + k_f a_2(t)\right].$$

The principle of superposition does not hold. If, however, k_f is very small (i.e., if $\left|k_f a(t)\right| \ll 1$), then all but the first two terms in (8.14) are negligible, and we have:

$$\varphi_{\mathrm{FM}}(t) = \mathrm{A}\left[\cos\omega_c t - k_f a(t)\sin\omega_c t\right]. \tag{8.15}$$

This is a linear modulation. This expression is similar to that of the AM wave. Because the bandwidth of $a(t)$ is B, the bandwidth of $\varphi_{\mathrm{FM}}(t)$ in (8.15) is only $2B$.

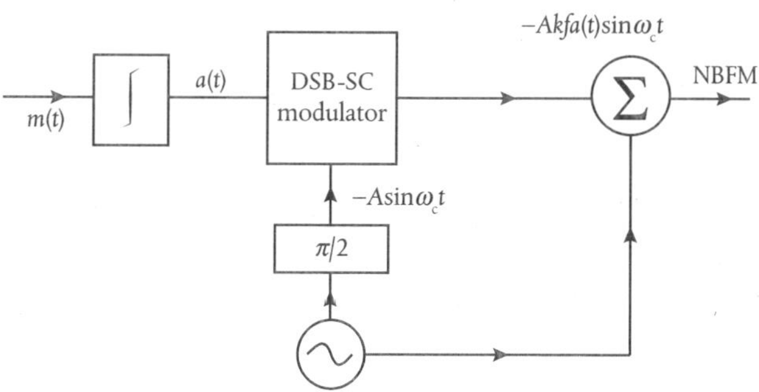

Fig. 8.13 Narrow-band FM wave generation

For this reason, the case $\left|k_f a(t)\right| \ll 1$ is called narrow-band FM (NBFM). A comparison of NBFM with AM shows that both cases have a carrier term and sidebands centred at $\pm\omega_c$. The modulated-signal bandwidth is identical ($2B$). The sideband spectrum for FM has a phase shift of $\pi/2$ with respect to the carrier, whereas that of AM is in phase with the carrier. Equation (8.15) suggest a possible method of generating narrow-band FM as shown in block in Fig. 8.13.

8.4.5 Wide-band FM (WBFM)

If the deviation in the carrier frequency is large (i.e., if the constant *kf* is chosen large enough so that the condition $\left|k_f a(t)\right| \ll 1$) is not satisfied), we cannot ignore the higher order terms in (8.14). The maximum and minimum carrier frequencies are $\omega_c + k_f m_p$ and $\omega_c - k_f m_p$, respectively. Hence, it was reasoned that the spectral component must lie in this range, resulting the FM bandwidth of $2\,k_f m_p$. The deviation of the carrier frequency is $\pm k_f m_p$. Let the carrier frequency deviation by $\Delta\omega$. Thus,

$$\Delta\omega = k_f m_p \tag{8.16}$$

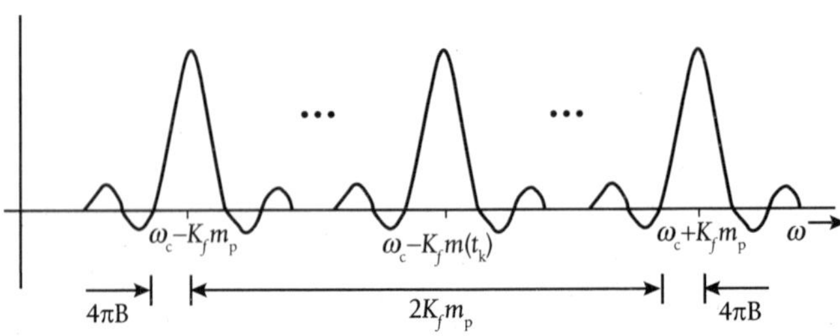

Fig. 8.14 Estimation of FM bandwidth

The carrier frequency deviation in hertz will be denoted by Δf. Thus,

$$\Delta f = \frac{k_f m_p}{2\pi}$$

The estimated FM bandwidth (in hertz) can be expressed as:

$$BFM = \frac{1}{2\Pi}\left(2k_f m_p + 8\pi B\right) = 2\left(\Delta f + 2B\right) \tag{8.17}$$

The estimated bandwidth thus obtained is somewhat higher than the actual value. For the narrow-band case, k_f is very small. Hence, Δf is very small (when compared to B). In this case, we can ignore the Δf term: BFM = $4B$

However, we have shown that for narrow band, the FM bandwidth is $2B$ Hz. This indicates that a better bandwidth is:

$$\text{BFM} = 2\left(\Delta f + 2B\right) = 2\left(\frac{k_f m_p}{2\pi} + B\right)$$

This result is obtained by Carson. Observe that for a truly wide-band case, where $\Delta f << B$. Let $\beta = \dfrac{\Delta f}{B}$. Carson's rule can be expressed in terms of the deviation ratio as BFM = $2B(\beta+1)$. The deviation ratio controls the amount of modulation and play a role similar to the modulation index in AM. The deviation ratio β is called the modulation index.

8.4.6 Verification of FM Bandwidth Relationship

We can verify the bandwidth relations for a specific case of tone modulation when $m(t)$ is a sinusoid. Let $m(t) = \alpha\cos\omega_m t$. From (8.12),

$$a(t) = \frac{\alpha}{\omega_m}\sin\omega_m t$$

Thus, from (8.13),

$$\hat{\varphi}_{FM}(t) = Ae^{\left(j\omega_c t + \frac{k_f a}{\omega_m}\sin\omega_m t\right)}$$

$\Delta\omega = k_f m_p = \alpha k_f$ and the bandwidth of $m(t)$ is $B = f_m \text{Hz}$. The deviation ratio is:

$$\beta = \frac{\Delta f}{f_m} = \frac{\Delta\omega}{\omega_m} = \frac{\alpha k_f}{\omega_m}$$

Hence, $\hat{\varphi}_{FM}(t) = \text{A}e^{j[\omega_c t + \beta\sin\omega_m t]}$

$$= \text{A}e^{j\omega_c t}e^{j\beta\sin\omega_m t} \tag{8.18}$$

The exponential term in parenthesis is a periodic signal with period $2\pi/\omega_m$ and can be expanded by the exponential Fourier series:

$$e^{j\beta\sin\omega_m t} = \sum_{n=-\infty}^{\infty} C_n e^{j\omega_m t}$$

where $C_n = \frac{\omega_m}{2\pi}\int_{-\pi/\omega_m}^{\pi/\omega_m} e^{j\beta\sin\omega_m t}e^{-n\omega_m t}dt$. Let $\omega_m t = x$, we get:

$$C_n = \frac{1}{2\pi}\int_{-\pi}^{\pi} e^{j\beta\sin x - nx}dx$$

The integral on the right-hand side cannot be evaluated in a closed form but must be integrated by expanding the integrand in infinite series. This integral is denoted by $Jn(\beta)$, the Bessel function of the first kind and n^{th}-order. Thus,

$$e^{j\beta\sin\omega_m t} = \sum_{n=-\infty}^{\infty} J_n(\beta)e^{jn\omega_m t} \tag{8.19}$$

Substituting (8.19) into (8.18), we get:

$$\hat{\varphi}_{FM}(t) = \text{A}\sum_{n=-\infty}^{\infty} J_n(\beta)e^{j[\omega_c t + n\omega_m t]}$$

$$= \text{A}\sum_{n=-\infty}^{\infty} J_n(\beta)\cos(\omega_c t + n\omega_m t)$$

The modulated signal has a carrier component and an infinite number of sidebands of frequencies $\omega_c \pm \omega_m$, $\omega_c \pm 2\omega_m$, $\omega_c \pm n\omega_m$, …. The strength of the nth sideband at $\omega = \omega_c + n\omega_m$ is $J_n(\beta)$. From the plots of $J_n(\beta)$ Bessel function, it can be seen that for a given β, $J_n(\beta)$ decreases with n. For a sufficiently large n, $J_n(\beta)$ is negligible, and there are only a finite number of significant sidebands. It can be seen from Bessel function that $J_n(\beta)$ is negligible for $n > \beta + 1$. Hence, the number of significant sidebands is $\beta + 1$. The bandwidth of the FM carrier is given by BFM = $2nf_m = 2(\beta+1)f_m$ = $2(\Delta f + B)$. When $\beta << 1$ (NBFM), there is only one significant sideband and the bandwidth BFM = $2f_m$ = 2B.

8.5 GENERATION AND DISTRIBUTION OF A WIDEBAND CONTINUOUSLY TUNABLE MILLIMETRE-WAVE SIGNAL WITH OPTICAL EXTERNAL MODULATION TECHNIQUE

Generation and transmission of microwave and millimetre-wave signals over optical fibres are of great interest for applications such as broad-band wireless access networks operating at millimetre-wave bands, antenna remoting, phased-array antennas, optical sensors, and radars. Optical generation and transmission of electrical signals have been extensively investigated with the most work to date focused on high-frequency signals, especially millimetre-wave signals. This is because there are fewer difficulties with conventional electronic techniques or optical techniques in generating and distributing lower-frequency electrical signals. Furthermore, the low-frequency modulated double-sideband (DSB) optical signal suffers less from the chromatic dispersion of the fibre than that of the high-frequency modulated signal when transmitting over standard single-mode fibre (SSMF).

Therefore, the highest frequency of the signal produced by these DSB techniques is limited by the bandwidth of the laser or the external modulator and the fibre chromatic dispersion. To overcome these limitations, an optical generation scheme that uses narrow bandwidth optical components to generate high frequency electrical signals becomes attractive. Optical microwave or millimetre-wave signal generation is usually based on heterodyne techniques by beating two optical carriers separated by the desired frequency in a square-law photodetector (PD). If the offset frequency of the two optical carriers is stable and their phases are correlated, a high-quality electrical signal will be generated. However, after the transmission of the two optical carriers over an SSMF, the generated microwave or millimetre-wave signal quality will be deteriorated because of the fibre chromatic dispersion.

Electrical signal generation based on optical heterodyning can be achieved by using either two stabilised lasers or one laser with an external optical modulator. The quality of the electrical signal produced by beating two free-running lasers rarely meets application specifications. Methods to further improve the signal quality, such as optical injection locking and optical phase-locked loop (OPLL), have been proposed. The OPLL techniques allow the suppression of the low-frequency components of the phase noise that deteriorate the generated signal. However, it is difficult to suppress the high-frequency components of the phase noise unless very narrow line-width (in kilohertz) optical sources are used.

Methods using a laser with an external optical modulator, such as an optical intensity modulator or optical phase modulator, have shown great potential for producing high-purity high-frequency millimetre-wave signals. These approaches are based on the inherent nonlinearity of the response of the optical modulator for generating high-order optical sidebands. Taking advantage of this property can dramatically lower the bandwidth requirements for the optical modulator and allows the use of a much lower frequency electrical drive signal. This can greatly reduce the cost of the system and makes it more practical to use.

A method to generate millimetre-wave signals using an external optical modulation technique was proposed by O'Reilly *et al.* in 1992 [27]. A frequency-doubled electrical signal was optically generated by biasing the Mach–Zehnder modulator (MZM) to suppress even-order optical sidebands. A 36-GHz millimetre-wave signal was generated when the MZM was driven by an 18-GHz microwave signal. Such a system was employed for a remote delivery of video services.

In 1994, O'Reilly and Lane [28] proposed another method to generate a frequency-quadrupled electrical signal. Instead of biasing the MZM to suppress the even-order optical sidebands, the method was based on the quadratic response of an optical intensity modulator. The optical carrier and the first and third-order optical sidebands were suppressed by adjusting the drive signal level. A 60-GHz millimetre-wave signal was generated when a 15-GHz drive signal was applied to the MZM.

However, to ensure a clean spectrum at the output of a PD, an imbalanced Mach–Zehnder filter with a free spectral range (FSR) equal to the spacing of the two second-order optical sidebands are used to suppress the unwanted optical components. Recently, an approach using an optical phase modulator to generate a frequency-quadrupled electrical signal was proposed. In this approach, a Fabry–Perot filter was used to select the two second-order optical sidebands. An electrical signal that has four times the frequency of the electrical drive signal was generated by beating the two second-order sidebands at a PD. A key advantage of these approaches is that an optical modulator with a maximum operating frequency of 15 GHz can generate a millimetre-wave signal up to 60 GHz. However, since both approaches rely on the optical filter to select the two optical sidebands to generate tunable millimetre-wave signals, a tunable optical filter must be used, which significantly increases the complexity and the cost of the system. For system applications with frequency reconfigurability, such as a wide-band surveillance radar, spread-spectrum or software-defined radio (SDR), continuously tunable millimetre-wave signals, are highly desired. The prospect of generating a wideband continuously tunable single-frequency millimetre-wave signal using fixed optical filters and narrow bandwidth optical modulators becomes very attractive. In this report, we propose a new approach that can optically generate a wide-band continuously tunable millimetre-wave signal without using a tunable optical filter. The system employs an optical intensity modulator, which is biased to suppress the odd-order optical sidebands. A fibre Bragg grating (FBG) serving as a wavelength fixed notch filter is then used to filter the optical carrier. A stable low-phase noise millimetre-wave signal at four times the frequency of the electrical drive signal is generated at the output of a PD. A 32–50-GHz millimetre-wave signal is observed on an electrical spectrum analyser (ESA) when the electrical drive signal is tuned from 8 to 12.5 GHz. The quality of the generated millimetre-wave signal is maintained after transmission over a 25-km SSMF.

8.5.1 Principle of the Proposed Approach

The proposed millimetre-wave signal generation system is shown in Fig. 8.15. An electrical drive signal is applied to an MZM. The MZM is biased to suppress the odd-order optical sidebands. An optical notch filter is connected at the output of the MZM to remove the optical carrier. Two second-order optical sidebands are obtained at the output of the notch filter. A beat signal with four times the frequency of the electrical drive signal is generated at a PD. It is known from our previous discussion on $LiNbO_3$ that the electric field at the output of a lithium niobate MZM, that is, $E_{out1}(t)$, can be approximately expressed as:

$$E_{out1}(t) = E_o \cos\left[\frac{\Phi[V(t)]}{2}\right]\cos(\omega_o t) \tag{8.20}$$

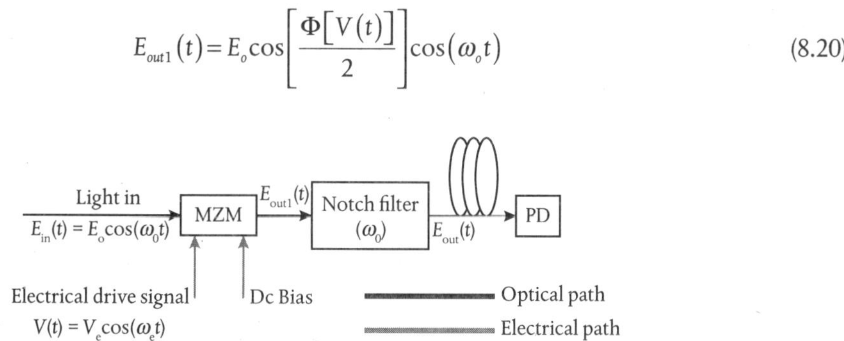

Fig. 8.15 Diagram of proposed microwave signal generation system

where E_o and ω_o are, respectively, the electric field amplitude and the angular frequency of the input optical carrier, $V(t)$ is the applied electrical drive voltage, and $\Phi[V(t)]$ is the optical phase difference caused by between the two arms of the MZM. If the MZM is driven by a sinusoidal electrical signal and biased with a constant dc voltage, $\Phi[V(t)]$ is expressed as:

$$\Phi[V(t)] = \phi_0 + \frac{\pi}{V_\pi} V_e \cos(\omega_e t) \tag{8.21}$$

where ϕ_0 is a constant phase shift determined by the constant DC-bias voltage, V_π is the half-wave voltage at high frequency, and V_e and ω_e are the amplitude and angular frequency of the electrical drive signal, respectively. Substituting (8.21) in (8.20), the electric field at the output of modulator

$$E_{out1}(t) = E_0 \cos\left[\frac{\phi V(t)}{2}\right] \cos((\omega_o t)) = E_0 \cos\left[\frac{\phi_0}{2} + \frac{}{2v_\pi} v_e \cos(\omega_e t)\right] \cos\omega_0 t$$

Let $\frac{\pi}{v_\pi} v_e = \beta,$

$$= E_0 \cos\omega_0 t \left\{\cos\left(\frac{\phi_0}{2}\right)\cos\left[\frac{\beta}{2}\cos(\omega_e t)\right] - \sin\left(\frac{\phi_0}{2}\right)\sin[\beta\cos(\omega_e t)]\right\}$$

$$= \left\{E_0 \cos\omega_0 t \cos\left(\frac{\phi_0}{2}\right)\cos\left[\frac{\beta}{2}\cos(\omega_e t)\right]\right\} - \left\{E_0 \cos\omega_0 t \sin\left(\frac{\phi_0}{2}\right)\sin[\beta\cos(\omega_e t)]\right\}$$

$$= E_0 \cos\left(\frac{\phi_0}{2}\right)\{\cos\omega_0 t \cos[\beta\cos(\omega_e t)]\} - E_0 \sin\left(\frac{\phi_0}{2}\right)\{\sin[\beta\cos(\omega_e t)]\cos\omega_0 t\}$$

Using,

1. $\cos(z\cos\eta) = J_0(z) + 2\sum_{n=1}^{\infty}(-1)^n J_{2n}(z)\cos 2n\eta$

2. $\sin(z\cos\eta) = 2\sum_{n=1}^{\infty}(-1)^n J_{2n-1}(z)\cos(2n-1)\eta$

$$= E_0 \cos\left(\frac{\phi_0}{2}\right)\left\{\cos\omega_0 t\, J_0(\beta) + 2\sum_{n=1}^{\infty}(-1)^n J_{2n}(\beta)\cos 2n\omega_e t \cos\omega_0 t\right\}$$

$$- E_0 \sin\left(\frac{\phi_0}{2}\right)\left\{\cos\omega_0 t\, 2\sum_{n=1}^{\infty}(-1)^n J_{2n-1}(\beta)\cos(2n-1)\omega_e t\right\}$$

$$\Rightarrow \left\{E_0 \cos\left(\frac{\phi_0}{2}\right)\cos\omega_0 t\, J_0(\beta)\right\} + \left\{\frac{E_0}{2}\cos\left[\frac{\phi_0}{2}\right] 2\sum_{n=1}^{\infty}(-1)^n J_{2n}(\beta)\cos 2n\omega_e t \cos\omega_0 t\right\}$$

$$- \left\{E_0 \sin\left(\frac{\phi_0}{2}\right)\cos\omega_0 t\, 2\sum_{n=1}^{\infty}(-1)^n J_{2n-1}(\beta)\cos(2n-1)\omega_e t\right\}$$

$$\Rightarrow\left\{E_0\cos\left(\frac{\phi_0}{2}\right)\cos\omega_0 t\,J_0(\beta)\right.$$

$$+E_0\cos\left(\frac{\phi_0}{2}\right)\left\{\sum_{n=1}^{\infty}(-1)^n J_{2n}(\beta)\left[\cos(\omega_0 t+2n\omega_e t)+\cos\left(\omega_0 t-2n\omega_e t\right)\right]\right\}$$

$$\left.-E_0\sin\left(\frac{\phi_0}{2}\right)\left\{\sum_{n=1}^{\infty}(-1)^n J_{2n-1}(\beta)\left[\cos(\omega_0 t+(2n-1)\omega_e t)+\cos\left(\omega_0 t-(2n-1)\omega_e t\right)\right]\right\}\right\}$$

$$\Rightarrow E_0\cos\left(\frac{\phi_0}{2}\right)\cos\omega_0 t\,J_0(\beta)$$

$$+E_0\cos\left(\frac{\phi_0}{2}\right)\left\{\sum_{n=1}^{\infty}(-1)^n J_{2n}(\beta)\left[\cos(\omega_0 t+2n\omega_e t)+\cos\left(\omega_0 t-2n\omega_e t\right)\right]\right\}$$

$$-E_0\sin\left(\frac{\phi_0}{2}\right)\left\{\sum_{n=1}^{\infty}(-1)^n J_{2n-1}(\beta)\left[\sin(\omega_0 t+(2n-1)\omega_e t-n\pi+\frac{\pi}{2})+\sin\left(\omega_0 t-(2n-1)\omega_e t\right)+n\pi-\frac{\pi}{2}\right]\right\}$$

$$E_{out1}(t)=E_0\cos\left(\frac{\phi_0}{2}\right)\cos\omega_0 t\;J_0(\beta)+E_0\cos\left(\frac{\phi_0}{2}\right)\left\{\sum_{n=1}^{\infty}J_{2n}(\beta)\left[\cos(\omega_0 t+2n\omega_e t-n\pi)\right.\right.$$

$$\left.\left.+\cos\left(\omega_0 t-2n\omega_e t+n\pi\right)\right]\right\}$$

$$-E_0\sin\left(\frac{\phi_0}{2}\right)\left\{\sum_{n=1}^{\infty}J_{2n-1}(\beta)\left[\sin(\omega_0 t-(2n-1)\omega_e t+n\pi-\frac{\pi}{2})-\sin\left(\omega_0 t+(2n-1)\omega_e t\right)-n\pi+\frac{\pi}{2}\right]\right\} \quad (8.22)$$

where J_n is the Bessel function of the first kind of order and $\beta=(V_e/V_\pi)(\pi/2)$ is the phase modulation index (PMI). When the MZM is driven by an electrical signal with adequate power, a large value of β is obtained. In this case, (8.22) shows that the power in the input optical carrier will be spread out among the first-order, second-order, third-order, and higher order optical sidebands. The amplitude distribution of these sidebands is governed by the variation of Bessel functions parameterised by β. Their amplitude is also affected by ϕ_0. If all these optical sidebands are fed to a square-law PD, harmonics of the electrical drive signal will be generated. The parameters β and ϕ_0 can be optimised, from the point-of-view of obtaining higher order electrical harmonics and maximising its conversion efficiency. For example, ϕ_0 is tuned to suppress all even-order optical sidebands so that the power of all even-order sidebands is transferred to the odd-order ones. An efficiency-improved frequency-doubled electrical signal is then obtained. In our proposed method, the DC bias of the MZM is tuned to have $\phi_0=0, 2\ \ , 4\ \ , \ldots\ldots$ All the odd-order optical sidebands associated with the term $\sin(\phi_0/2)$ then vanish, as indicated in 8.22. Only the even-order optical sidebands are kept. The power in the odd-order optical sidebands is transferred to the even-order sidebands improving the signal generation efficiency. If the electrical drive signal is applied to the MZM with an appropriate power level, optical sidebands up to the second order are generated and all optical sidebands above the second order have an amplitude low enough to be ignored, but the optical carrier represented by the term with the zeroth-order Bessel function (J_0) in 8.22 is still part of the spectrum. The optical signal can then be approximately expressed as:

$$E_{out1}(t)\cong E_0 J_0(\beta)\cos(\omega_0 t)-E_0 J_2(\beta)\cos(\omega_0 t-2\omega_e t)-E_0 J_2(\beta)\cos(\omega_0 t+2\omega_e t) \quad (8.23)$$

When this optical signal is fed to a PD, a strong frequency-doubled electrical signal and a weaker frequency-quadrupled electrical signal will be generated. However, when this optical signal that is composed of two second-order sidebands and one optical carrier is transmitted over a long span of optical fibre, the frequency-doubled electrical signal suffers from the chromatic-dispersion-induced power penalty, which limits its applications. In addition, the presence of a frequency-doubled electrical signal will cause interference to the operation of the frequency-quadrupled electrical signal in a wide-band system application. To eliminate the frequency-doubled electrical signal, we propose the use of a wavelength-fixed optical notch filter to filter out the optical carrier, as shown in Fig. 8.15. The electric field of the optical signal at the output of the optical notch filter can then be approximately expressed as:

$$E_{out}(t) \cong -E_0 J_2(\beta)\cos(\omega_0 t - 2\omega_e t) - E_0 J_2(\beta)\cos(\omega_0 t + 2\omega_e t) \tag{8.24}$$

Therefore, at the output of the optical notch filter, only two optical sidebands separated by four times the frequency of the drive signals are present. By applying this optical signal to a PD, an electrical signal that has four times the frequency of the electrical drive signal will be generated. The generated electrical signal can be written as:

$$V_{out}(t) = CJ_2^2(\beta)\cos(4\omega_e t) \tag{8.25}$$

where C is a constant that is related to the responsivity of the PD. Since the two optical sidebands originate from the same optical source, the frequency stability and phase noise of the generated signal are predominately determined by the electrical drive signal. Equation (8.25) also shows that the amplitude of the generated electrical signal can be maximised by optimising the value $J_2(\beta)$.

It is important to note that the DC-bias level for which the odd-order optical sidebands are eliminated is not dependent on the frequency of the electrical drive signal. In addition, the optical carrier has a fixed wavelength; therefore, the optical notch filter does not need to be tunable. These characteristics ensure that the proposed approach can generate a frequency-tunable electrical signal by simply tuning the frequency of the electrical drive signal at a low-frequency band.

8.6 ELECTRICAL HARMONIC SUPPRESSION ANALYSIS

Usually on a wide-band electrical heterodyne system, especially when the system operates over an octave bandwidth, an electrically tuned bandpass filter is inserted between the local oscillator and the mixer to suppress unwanted harmonics. This prevents potential in-band interference from corrupting the received signal. However, the use of an electrically tunable filter makes the system very complicated. A direct solution to this problem is to use a highly harmonic-suppressed local oscillator. In the following, we will analyse the harmonic-suppression characteristics of the proposed approach.

Let us assume that all odd-order optical sidebands generated by the modulation of the MZM by a sinusoidal signal can be completely suppressed by using an appropriate DC-bias voltage. This means that the condition of $\phi_0 = 0, 2\pi, 4\pi, \ldots\ldots$ is satisfied with a constant DC bias. Let us assume also that the attenuation of the optical notch filter at its centre notch wavelength is in α dB. Based on the above-mentioned assumptions, from (8.22), the optical signal at the output of the optical notch filter can be written as:

$$E_{out}(t) = E_0 kJ_0(\beta)\cos(\omega_0 t) + E_0\left\{\sum_{n=1}^{\infty} J_{2n}(\beta)\left[\cos(\omega_0 t - 2n\omega_e t + n\pi) + \cos(\omega_0 t + 2n\omega_e t - n\pi)\right]\right\}$$

where k is the optical electrical field attenuation factor, which is related to α by $\alpha = -20\log_{10} k$. Usually, for a commercially available MZM, the maximum available PMI β_{max} is 2. When $0 \leq \beta \leq 2$, Bessel functions for all are monotonically increasing with respect to β and monotonically decreasing with respect to the order of Bessel function n, and $J_2(2) = 0.35283$, $J_4(2) = 0.033996$, and $J_6(2) = 0.0012024$.

Thus, it is reasonable to ignore the optical sidebands with a Bessel coefficient higher. Therefore, it can be further simplified to:

$$\begin{aligned} E_{out} &\cong E_0 k J_0(\beta)\cos(\omega_0 t) - E_0 J_2(\beta)\cos(\omega_0 t - 2\omega_e t) \\ &- E_0 J_2(\beta)\cos(\omega_0 t + 2\omega_e t) + E_0 J_4(\beta)\cos(\omega_0 t - 4\omega_e t) \\ &+ E_0 J_4(\beta)\cos(\omega_0 t + 4\omega_e t) \end{aligned} \tag{8.27}$$

Equation (8.27) shows that the optical signal consists of an attenuated optical carrier and four optical sidebands. The spectrum of this optical signal is illustrated as shown in Fig. 8.16. The arrow direction shows their initial phase with respect to the phase of the optical carrier before transmission.

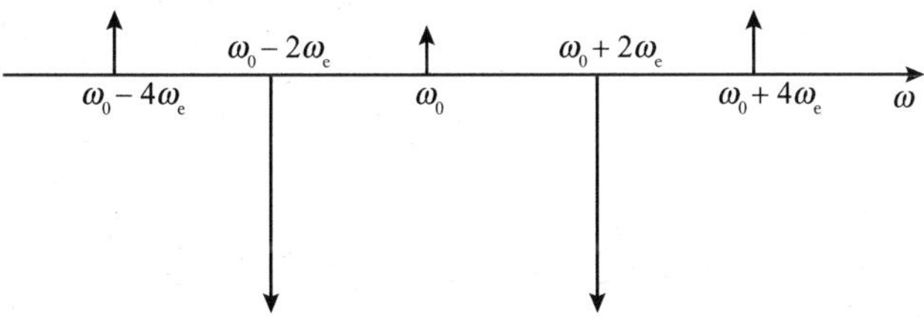

Fig. 8.16 Illustration of the optical spectrum at the output of the optical notch filter

When the optical signal shown in Fig. 8.16 is transmitted over a single-mode fibre, the chromatic dispersion of the fibre will cause an extra phase shift to each optical sideband when compared to the optical carrier. By expanding the propagation constant $\beta(\omega)$ of the fibre for each optical sideband to a Taylor series around the angular frequency of the optical carrier, that is,

$$\beta(\omega_0 \pm 2n\omega_e) = \beta(\omega_0) + \beta'(\omega_0)(\pm 2n\omega_e) + \frac{1}{2}\beta''(\omega_0)(\pm 2n\omega_e)^2 + \frac{1}{6}\beta'''(\omega_0)(\pm 2n\omega_e)^3 + \ldots \tag{8.28}$$

For $n = 1$,

$$\beta(\omega_0 \pm 2\omega_e) = \beta(\omega_0) + \beta'(\omega_0)(\pm 2\omega_e) + \frac{1}{2}\beta''(\omega_0)(\pm 2\omega_e)^2 + \frac{1}{6}\beta'''(\omega_0)(\pm 2\omega_e)^3 + \ldots \tag{8.29}$$

For $n = 2$,

$$\beta(\omega_0 \pm 4\omega_e) = \beta(\omega_0) + \beta'(\omega_0)(\pm 4\omega_e) + \frac{1}{2}\beta''(\omega_0)(\pm 4\omega_e)^2 + \frac{1}{6}\beta'''(\omega_0)(\pm 4\omega_e)^3 + \ldots \tag{8.30}$$

where $\beta'(\omega_0)$ and $\beta''(\omega_0)$ are the first-order and second-order derivative of the propagation constant at the angular frequency ω_0, respectively. The effect of higher order dispersion is neglected for the single-mode fibre at 1550-nm band, and $\beta''(\omega_0)$ can be expressed by the chromatic dispersion parameter as:

$$\beta''(\omega_0) = -\frac{c}{2\pi f_0^2} D \tag{8.31}$$

where c is the speed of light in free space and f_0 is the frequency of the optical carrier. The electric field representing the optical signal at the end of the transmission over a single-mode fibre of length can be obtained by adding the transmission phase delay $\beta(\omega_0 \pm 2n\omega_e)L$ to the corresponding optical sideband shown in (8.27). Electrical harmonics will be generated by applying this optical signal to a PD. The output voltage of the generated high-frequency electrical signal is:

$$\begin{aligned}E_{out} = {} & E_0 k J_0(\beta)\cos(\omega_0 t + \beta(\omega_0)L) - E_0 J_2(\beta)\cos(\omega_0 t - 2\omega_e t + \beta(\omega_0 - 2\omega_e)L) - \\ & E_0 J_2(\beta)\cos(\omega_0 t + 2\omega_e t + \beta(\omega_0 + 2\omega_e)L) + E_0 J_4(\beta)\cos(\omega_0 t - 4\omega_e t + \beta(\omega_0 - 4\omega_e)L) + \\ & E_0 J_4(\beta)\cos(\omega_0 t + 4\omega_e t + \beta(\omega_0 + 4\omega_e)L)\end{aligned} \tag{8.32}$$

From (8.29), (8.30), and (8.32),

$$\begin{aligned}E_{out} \cong {} & E_0 k J_0(\beta)\cos(\omega_0 t) - E_0 J_2(\beta)\Big[\cos\Big(\omega_0 t - 2\omega_e t + \beta(\omega_0)L - \beta'(\omega_0)(2\omega_e L) + \frac{1}{2}\beta''(\omega_0)4\omega_e^2 L\Big) + \\ & \cos\Big(\omega_0 t + 2\omega_e t + \beta(\omega_0)L + \beta'(\omega_0)(2\omega_e L) + \frac{1}{2}\beta''(\omega_0)4\omega_e^2 L\Big)\Big] + \\ & E_0 J_4(\beta)\Big[\cos\Big(\omega_0 t - 4\omega_e t + \beta(\omega_0)L - \beta'(\omega_0)(4\omega_e L) + \frac{1}{2}\beta''(\omega_0)16\omega_e^2 L\Big) + \\ & \cos\Big(\omega_0 t + 4\omega_e t + \beta(\omega_0)L + \beta'(\omega_0)(4\omega_e L) + \frac{1}{2}\beta''(\omega_0)16\omega_e^2 L\Big)\Big]\end{aligned} \tag{8.33}$$

Using $\cos(A+B)+\cos(A-B) = 2\cos A\cos B$ (8.34)

Let $A = \omega_0 t + \beta(\omega_0)L + \frac{1}{2}\beta''(\omega_0)4\omega_e^2 L$ and $B = 2\omega_e t + \beta'(\omega_0)(2\omega_e L)$

From (8.33) and (8.34),

$$\begin{aligned}E_{out} \cong {} & E_0 k J_0(\beta)\cos(\omega_0 t) - E_0 J_2(\beta)2\Big[\cos\Big(\omega_0 t + \beta(\omega_0)L + \frac{1}{2}\beta''(\omega_0)4\omega_e^2 L\Big)\cos(2\omega_e t + \beta'(\omega_0)(2\omega_e L))\Big] \\ & + E_0 J_4(\beta)2\Big[\cos\Big(\omega_0 t + \beta(\omega_0)L + \frac{1}{2}\beta''(\omega_0)16\omega_e^2 L\Big)\cos(4\omega_e t + \beta'(\omega_0)(4\omega_e L))\Big]\end{aligned}$$

Considering $\beta(\omega_0) \approx 0$

$$\begin{aligned}E_{out} \cong {} & E_0 k J_0(\beta) - E_0 J_2(\beta)2\Big[\cos\Big(\frac{1}{2}\beta''(\omega_0)4\omega_e^2 L\Big)\cos(2\omega_e t + \beta'(\omega_0)(2\omega_e L))\Big] \\ & + E_0 J_4(\beta)2\Big[\cos\Big(\frac{1}{2}\beta''(\omega_0)16\omega_e^2 L\Big)\cos(4\omega_e t + \beta'(\omega_0)(4\omega_e L))\Big]\end{aligned}$$

$$V_{out} = (E_{out})^2$$

Using $(a-b+c)^2 = a^2 + b^2 + c^2 - 2ab - 2bc + 2ca$

$$
\begin{aligned}
V_{out} = {} & E_0^2k^2J_0^2(\beta)+4E_0^2J_2^2(\beta)\cos^2\left(2\beta''(\omega_0)\omega_e{}^2L\right)\cos^2\left(2\omega_e t+\beta'(\omega_0)2\omega_e L\right)+ \\
& 4E_0^2J_4^2(\beta)\cos^2\left(8\beta''(\omega_0)\omega_e{}^2L\right)\cos^2\left(4\omega_e t+\beta'(\omega_0)4\omega_e L\right)- \\
& 4E_0^2kJ_0(\beta)J_2(\beta)\left[\cos\left(2\beta''(\omega_0)\omega_e{}^2L\right)\cos\left(2\omega_e t+\beta'(\omega_0)2\omega_e L\right)\right]- \\
& 8E_0^2J_2(\beta)J_4(\beta)\left[\cos\left(2\beta''(\omega_0)\omega_e{}^2L\right)\cos\left(2\omega_e t+\beta'(\omega_0)2\omega_e L\right)\right]\left[\cos\left(8\beta''(\omega_0)\omega_e{}^2L\right)\cos\left(4\omega_e t+\beta'(\omega_0).4\omega_e L\right)\right] \\
& +4E_0^2kJ_0(\beta)J_4(\beta)\left[\cos\left(8\beta''(\omega_0)\omega_e{}^2L\right)\cos\left(4\omega_e t+\beta'(\omega_0)4\omega_e L\right)\right]
\end{aligned}
$$

Using $2\cos^2\theta=1+\cos 2\theta$,

$$
\begin{aligned}
V_{out} = {} & E_0^2k^2J_0^2(\beta)+E_0^2J_2^2(\beta)\left[1+\cos\left(4\beta''(\omega_0)\omega_e{}^2L\right)\right]\left[1+\cos\left(4\omega_e t+\beta'(\omega_0)4\omega_e L\right)\right]+ \\
& 4E_0^2J_4^2(\beta)\left[1+\cos\left(16\beta''(\omega_0)\omega_e{}^2L\right)\right]\left[1+\cos\left(8\omega_e t+\beta'(\omega_0)8\omega_e L\right)\right]- \\
& 4E_0^2kJ_0(\beta)J_2(\beta)\left[\cos\left(2\beta''(\omega_0)\omega_e{}^2L\right)\cos\left(2\omega_e t+\beta'(\omega_0)2\omega_e L\right)\right]- \\
& 8E_0^2J_2(\beta)J_4(\beta)\left[\cos\left(2\beta''(\omega_0)\omega_e{}^2L\right)\cos\left(2\omega_e t+\beta'(\omega_0)2\omega_e L\right)\right]\left[\cos\left(8\beta''(\omega_0)\omega_e{}^2L\right)\cos\left(4\omega_e t+\beta'(\omega_0)4\omega_e L\right)\right] \\
& +4E_0^2kJ_0(\beta)J_{41}(\beta)\left[\cos\left(8\beta''(\omega_0)\omega_e{}^2L\right)\cos\left(4\omega_e t+\beta'(\omega_0)4\omega_e L\right)\right]
\end{aligned}
$$

Neglecting higher order term: $4E_0^2J_4^2(\beta)\left[1+\cos\left(16\beta''(\omega_0)\omega_e{}^2L\right)\right]\left[1+\cos\left(8\omega_e t+\beta'(\omega_0)8\omega_e L\right)\right]$

$$
\begin{aligned}
V_{out} = {} & E_0^2k^2J_0^2(\beta)+E_0^2J_2^2(\beta)\left[1+\cos\left(4\beta''(\omega_0)\omega_e{}^2L\right)\right]\left[1+\cos\left(4\omega_e t+\beta'(\omega_0)4\omega_e L\right)\right]- \\
& 4E_0^2kJ_0(\beta)J_2(\beta)\left[\cos\left(2\beta''(\omega_0)\omega_e{}^2L\right)\cos\left(2\omega_e t+\beta'(\omega_0)2\omega_e L\right)\right]- \\
& 8E_0^2J_2(\beta)J_4(\beta)\left[\cos\left(2\beta''(\omega_0)\omega_e{}^2L\right)\cos\left(2\omega_e t+\beta'(\omega_0)2\omega_e L\right)\right]\left[\cos\left(8\beta''(\omega_0)\omega_e{}^2L\right)\cos\left(4\omega_e t+\beta'(\omega_0).4\omega_e L\right)\right] \\
& +4E_0^2kJ_0(\beta)J_{41}(\beta)\left[\cos\left(8\beta''(\omega_0)\omega_e{}^2L\right)\cos\left(4\omega_e t+\beta'(\omega_0)4\omega_e L\right)\right]
\end{aligned}
$$

Using $1+\cos\theta\approx\cos\theta$,

$$
\begin{aligned}
V_{out} = {} & E_0^2k^2J_0^2(\beta)+E_0^2J_2^2(\beta)\cos\left(4\beta''(\omega_0)\omega_e{}^2L\right)\cos\left(4\omega_e t+\beta'(\omega_0)4\omega_e L\right)- \\
& 4E_0^2kJ_0(\beta)J_2(\beta)\left[\cos\left(2\beta''(\omega_0)\omega_e{}^2L\right)\cos\left(2\omega_e t+\beta'(\omega_0)2\omega_e L\right)\right]- \\
& 8E_0^2J_2(\beta)J_4(\beta)\left[\cos\left(2\beta''(\omega_0)\omega_e{}^2L\right)\cos\left(2\omega_e t+\beta'(\omega_0)2\omega_e L\right)\right]\left[\cos\left(8\beta''(\omega_0)\omega_e{}^2L\right)\cos\left(4\omega_e t+\beta'(\omega_0)4\omega_e L\right)\right] \\
& +4E_0^2kJ_0(\beta)J_{41}(\beta)\left[\cos\left(8\beta''(\omega_0)\omega_e{}^2L\right)\cos\left(4\omega_e t+\beta'(\omega_0)4\omega_e L\right)\right]
\end{aligned}
\tag{8.35}
$$

Solving term from (8.35),

$$
8E_0^2J_2(\beta)J_4(\beta)\left[\cos\left(2\beta''(\omega_0)\omega_e{}^2L\right)\cos\left(2\omega_e t+\beta'(\omega_0)2\omega_e L\right)\right]\left[\cos\left(8\beta''(\omega_0)\omega_e{}^2L\right)\cos\left(4\omega_e t+\beta'(\omega_0)4\omega_e L\right)\right]
$$

$$
2E_0^2J_2(\beta)J_4(\beta)\left[\cos\left(10\beta''(\omega_0)\omega_e{}^2L\right)+\cos\left(6\beta''(\omega_0)\omega_e{}^2L\right)\right]\left[\cos\left(6\omega_e t+\beta'(\omega_0)6\omega_e L\right)+\cos\left(2\omega_e t+\beta'(\omega_0)2\omega_e L\right)\right]
$$

Considering $\cos\left(10\beta''(\omega_0)\omega_e^{\,2}L\right)+\cos\left(6\beta''(\omega_0)\omega_e^{\,2}L\right)\approx\cos\left(6\beta''(\omega_0)\omega_e^{\,2}L\right)$

$$=2E_0^2J_2(\beta)J_4(\beta)\left[\cos\left(6\beta''(\omega_0)\omega_e^{\,2}L\right)\right]\left[\cos\left(6\omega_et+\beta'(\omega_0)6\omega_eL\right)+\cos\left(2\omega_et+\beta'(\omega_0)2\omega_eL\right)\right]$$

$$\begin{aligned}&=2E_0^2J_2(\beta)J_4(\beta)\left[\cos\left(6\beta''(\omega_0)\omega_e^{\,2}L\right)\cos\left(6\omega_et+\beta'(\omega_0)6\omega_eL\right)\right]+\\&\left[\cos\left(6\beta''(\omega_0)\omega_e^{\,2}L\right)\cos\left(2\omega_et+\beta'(\omega_0)2\omega_eL\right)\right]\end{aligned}\tag{8.36}$$

Using (8.35) and (8.36),

$$\begin{aligned}V_{out}=\;&E_0^2k^2J_0^2(\beta)+E_0^2J_2^2(\beta)\left\{\cos\left[\left(4\beta''(\omega_0)\right)\omega_e^{\,2}L\right]\cos\left[4\omega_et+\beta'(\omega_0)4\omega_eL\right]\right\}-\\&4E_0^2kJ_0(\beta)J_2(\beta)\left\{\cos\left[\left(2\beta''(\omega_0)\omega_e^{\,2}L\right)\right]\cos\left[\left(2\omega_et+\beta'(\omega_0)2\omega_eL\right)\right]\right\}-\\&2E_0^2J_2(\beta)J_4(\beta)\left\{\cos\left[\left(6\beta''(\omega_0)\right)\omega_e^{\,2}L\right]\cos\left[\left(6\omega_et+\beta'(\omega_0)6\omega_eL\right)\right]+\right.\\&\left.\cos\left[\left(6\beta''(\omega_0)\right)\omega_e^{\,2}L\right]\cos\left[\left(2\omega_et+\beta'(\omega_0)2\omega_eL\right)\right]\right\}+\\&4E_0^2kJ_0(\beta)J_{4126}(\beta)\left\{\cos\left[\left(8\beta''(\omega_0)\right)\omega_e^{\,2}L\right]\cos\left[\left(4\omega_et+\beta'(\omega_0)4\omega_eL\right)\right]\right\}\end{aligned}$$

$$\begin{aligned}V_{out}\propto\;&-2E_0^2\left\{2kJ_0(\beta)J_2(\beta).cos\left[\left(2\beta''(\omega_0)\right)\omega_e^{\,2}L\right]+\right.\\&\left.J_2(\beta)J_4(\beta)\cos\left[\left(6\beta''(\omega_0)\right)\omega_e^{\,2}L\right]\right\}\cos\left(2\omega_et+\beta'(\omega_0)2\omega_eL\right)+\\&E_0^2\left\{J_2^2(\beta)cos\left[4\beta''(\omega_0)\omega_e^{\,2}L\right]+4kJ_0(\beta)J_{4126}(\beta)\cos\left[8\beta''(\omega_0)\omega_e^{\,2}L\right]\right\}\cos\left[4\omega_et+\beta'(\omega_0)4\omega_eL\right]-\\&2E_0^2J_2(\beta)J_4(\beta)\left\{\cos\left[6\beta''(\omega_0)\omega_e^{\,2}L\right]\cos\left[6\omega_et+\beta'(\omega_0)6\omega_eL\right]\right\}\end{aligned}$$

Using $\beta''(\omega_0)=-\dfrac{c}{2\pi f_0^2}D$,

$$\begin{aligned}V_{out}\propto\;&-2E_0^2\left\{2kJ_0(\beta)J_2(\beta)\cos\left(-\frac{c}{2\pi f_0^2}2D\omega_e^{\,2}L\right)+\right.\\&\left.J_2(\beta)J_4(\beta)\cos\left(-\frac{c}{2\pi f_0^2}D6\omega_e^{\,2}L\right)\right\}\cos\left(2\omega_et+\beta'(\omega_0)2\omega_eL\right)+\\&E_0^2\left\{J_2^2(\beta)cos\left(-\frac{c}{2\pi f_0^2}4D(\omega_0)\omega_e^{\,2}L\right)+\right.\\&\left.4E_0^2kJ_0(\beta)J_{4126}(\beta)\left[\cos\left(-\frac{c}{2\pi f_0^2}D8\omega_e^{\,2}L\right)\right]\right\}\cos\left(4\omega_et+\beta'(\omega_0)4\omega_eL\right)-\\&2E_0^2J_2(\beta)J_4(\beta)\left\{\left[\cos\left(-\frac{c}{2\pi f_0^2}D6\omega_e^{\,2}L\right)\cos\left(6\omega_et+\beta'(\omega_0)6\omega_eL\right)\right]\right\}\end{aligned}$$

$$V_{out} \propto -2E_0^2\left\{2kJ_0(\beta)J_2(\beta)\left[\cos\left(-\frac{c}{2\pi f_0^2}2D\omega_e{}^2L\right)\right]\right.$$

$$\left.+J_2(\beta)J_4(\beta)\cos\left(-\frac{c}{2\pi f_0^2}D6\omega_e{}^2L\right)\right\}\cos(2\omega_e t+\beta'(\omega_0)2\omega_e L)$$

$$+E_0^2\left\{J_2^2(\beta)cos\left(-\frac{c}{2\pi f_0^2}4D(\omega_0)\omega_e{}^2L\right)\right.$$

$$\left.+4E_0^2kJ_0(\beta)J_{4126}(\beta)\left[\cos\left(-\frac{c}{2\pi f_0^2}D8\omega_e{}^2L\right)\right]\right\}\cos(4\omega_e t+\beta'(\omega_0)4\omega_e L)$$

$$-2E_0^2J_2(\beta)J_4(\beta)\left\{\left[\cos\left(-\frac{c}{2\pi f_0^2}D6\omega_e{}^2L\right)\cos(6\omega_e t+\beta'(\omega_0)6\omega_e L)\right]\right\}$$

Using $\omega_e = 2\pi f_e$,

$$V_{out} \propto -2E_0^2\left\{2kJ_0(\beta)J_2(\beta)\left[\cos\left(4\pi cDL\frac{f_e^2}{f_o^2}\right)\right]+\right.$$

$$\left.J_2(\beta)J_4(\beta)\cos\left(12\pi cDL\frac{f_e^2}{f_o^2}\right)\right\}\cos(2\omega_e t+\beta'(\omega_0)2\omega_e L)+$$

$$E_0^2\left\{J_2^2(\beta)\cos\left(8\pi cDL\frac{f_e^2}{f_o^2}\right)+4E_0^2kJ_0(\beta)J_{4126}(\beta)\left[\cos\left(16\pi cDL\frac{f_e^2}{f_o^2}\right)\right]\right\}\cos(4\omega_e t+\beta'(\omega_0)4\omega_e L)-$$

$$2E_0^2J_2(\beta)J_4(\beta)\left\{\left[\cos\left(12\pi cDL\frac{f_e^2}{f_o^2}\right)\cos(6\omega_e t+\beta'(\omega_0)6\omega_e L)\right]\right\}$$

$$V_{out} \propto -2E_0^2\left\{2kJ_0(\beta)J_2(\beta)\left[\cos\left(4\pi cDL\frac{f_e^2}{f_o^2}\right)\right]+J_2(\beta)J_4(\beta).\cos\left(12\pi cDL\frac{f_e^2}{f_o^2}\right)\right\}\cos(2\omega_e t+\beta'(\omega_0)2\omega_e L)+$$

$$\left\{J_2^2(\beta)cos\left(8\pi cDL\frac{f_e^2}{f_o^2}\right)+4E_0^2kJ_0(\beta)J_{4126}(\beta)\left[\cos\left(16\pi cDL\frac{f_e^2}{f_o^2}\right)\right]\right\}\cos(4\omega_e t+\beta'(\omega_0)4\omega_e L)-$$

$$2E_0^2J_2(\beta)J_4(\beta)\left\{\left[\cos\left(12\pi cDL\frac{f_e^2}{f_o^2}\right)\cos(6\omega_e t+\beta'(\omega_0)6\omega_e L)\right]\right\} \quad (8.37)$$

where f_e is the frequency of the electrical drive signal. The electrical spectrum of the generated signal expressed by (8.37) is illustrated as shown in Fig. 8.17.

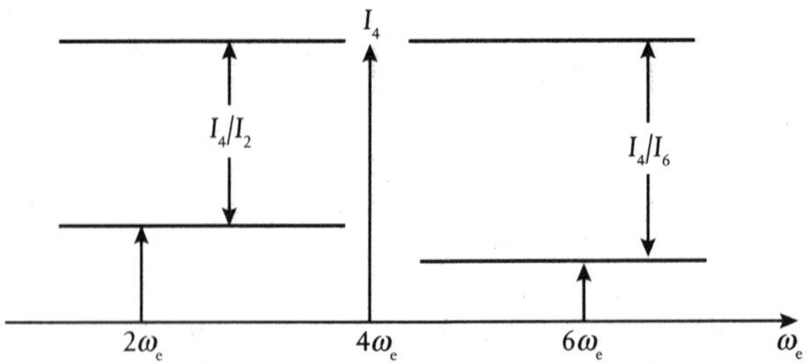

Fig. 8.17 Illustration of the electrical spectrum at the output of a PD

From (8.37), the power intensity of the fourth-order electrical harmonic I_4 is proportional to the coefficients of optical sidebands:

$$I_4 \propto \frac{E_0^4}{2}\left\{J_2^2(\beta)\cos\left(8\pi cDL\frac{f_e^2}{f_o^2}\right)+2kJ_0(\beta)J_{4126}(\beta)\left[\cos\left(16\pi cDL\frac{f_e^2}{f_o^2}\right)\right]\right\}^2 \tag{8.38}$$

The power intensities of the second- and sixth-order electrical harmonics are:

$$I_2 \propto 2E_0^4\left\{2kJ_0(\beta)J_2(\beta)\left[\cos\left(4\pi cDL\frac{f_e^2}{f_o^2}\right)\right]+\cos\left(12\pi cDL\frac{f_e^2}{f_o^2}\right)\right\}^2 \tag{8.39}$$

$$I_6 \propto 2E_0^4\left\{J_2(\beta)J_4(\beta)\left[\cos\left(12\pi cDL\frac{f_e^2}{f_o^2}\right)\right]\right\}^2 \tag{8.40}$$

For a distribution system that operates at 1550 nm with a transmission distance of 25 km over a standard single-mode fibre with $D=17\,\text{ps/nm-km}$, the power intensity I_4 and harmonic suppressions of I_4/I_2 and I_4/I_6 versus the modulation depth $\beta(0\le\beta\le 2)$ are plotted in Fig. 8.18.

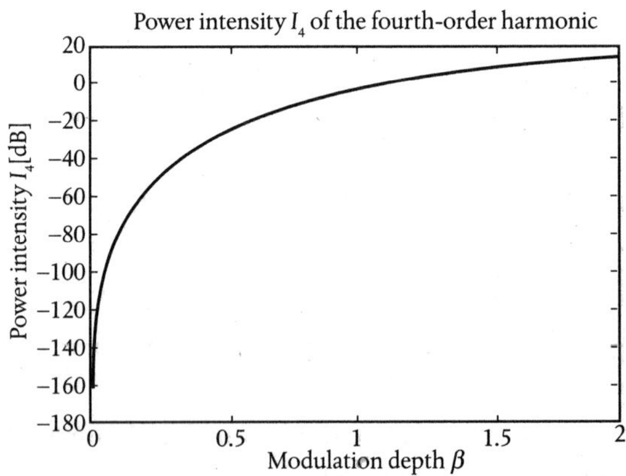

Fig. 8.18 Power intensity I_4 of the fourth-order harmonic

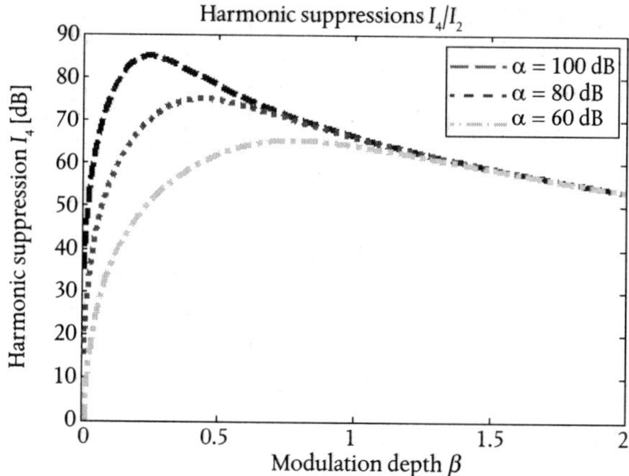

Fig. 8.19 Harmonic suppression (I_4/I_2) of the fourth-order harmonic (frequency of the electrical drive f_e = 12.5 GHz)

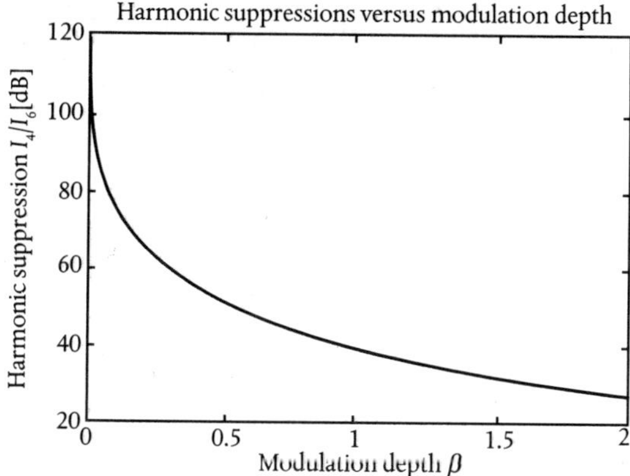

Fig. 8.20 Harmonic suppression (I_4/I_6) of the fourth-order harmonic (frequency of the electrical drive f_e = 12.5 GHz)

Figure 8.18 shows that the power intensity I_4 is monotonically increasing for $0 \le \beta \le 2$, and Fig. 8.19 shows that the harmonic suppression I_4/I_2 is monotonically decreasing for $0.5 \le \beta \le 2$ and $\alpha \ge 80$ dB, and for $0.25 \le \beta \le 2$ and $\alpha \ge 100$ dB. Figure 8.20 shows that the harmonic suppression I_4/I_6 is monotonically decreasing for $0 \le \beta \le 2$, which is independent of the attenuation of the optical notch filter.With a large attenuation of the optical notch filter, a lower modulation depth corresponds to an improved harmonic suppression. As can be seen from Fig. 8.19 for $\alpha = 100$ and $0.25 \le \beta \le 2$, the lower the modulation depth, the higher the harmonic suppression. However, lower modulation depth leads to a lower output power of the fourth-order electrical harmonic, as shown in Fig. 8.18. This problem can be solved at a low cost by using erbium-doped fibre amplifiers (EDFAs) in the 1550-nm band. Note that a lower modulation depth means a less power requirement for the electrical drive signal. Figure 8.21(a) shows that when the modulation depth $\beta = 0.6$, the minimum value of $I_4/I_2$is around 2 dB with $\alpha = 40$ dB, greater than 18.5 dB with $\alpha = 60$ dB, 28.3 dB with $\alpha = 80$ dB, and 30 dB with $\alpha = 100$ dB;

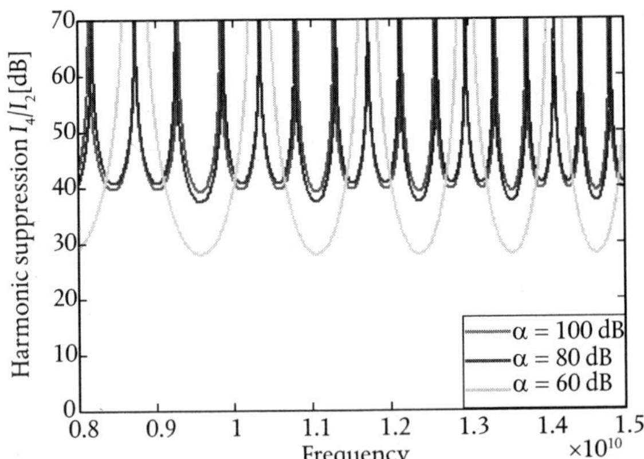

Fig. 8.21 (a) Harmonic suppression (I_4/I_2) versus frequency of the electrical drive signal with modulation depth $\beta=0.6$.

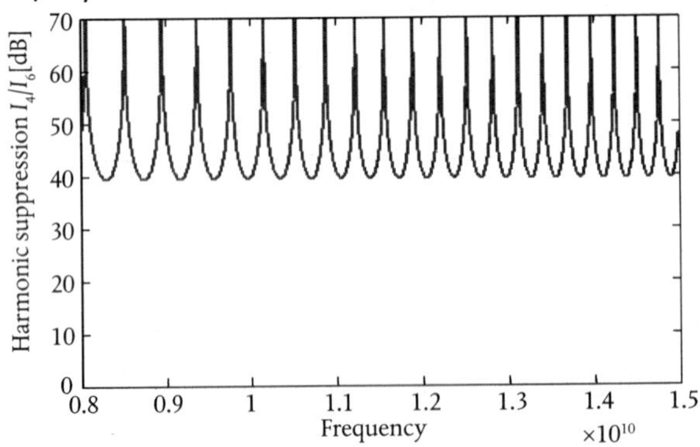

Fig. 8.21 (b) Harmonic suppression (I_4/I_6) versus frequency of the electrical drive signal with modulation depth $\beta=0.6$.

Figure 8.21(b) shows that the minimum value of I_4/I_6 is greater than 30 dB and is independent of α. This means that with a small value of modulation depth, the harmonic suppression is mostly affected by the attenuation of the optical notch filter.

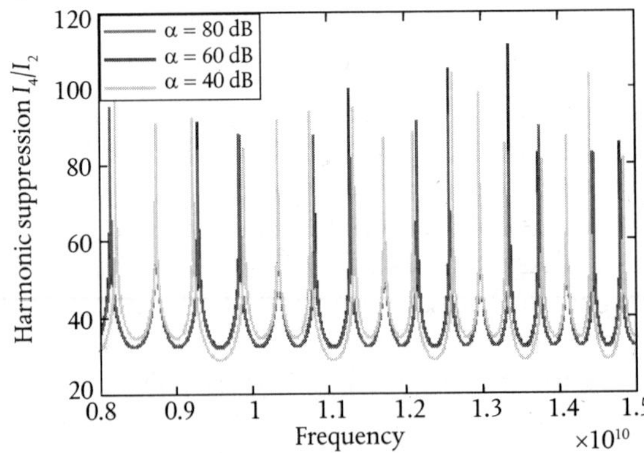

Fig. 8.22 (a) Harmonic suppression (I_4/I_2) versus frequency of the electrical drive signal with modulation depth $\beta=0.9$.

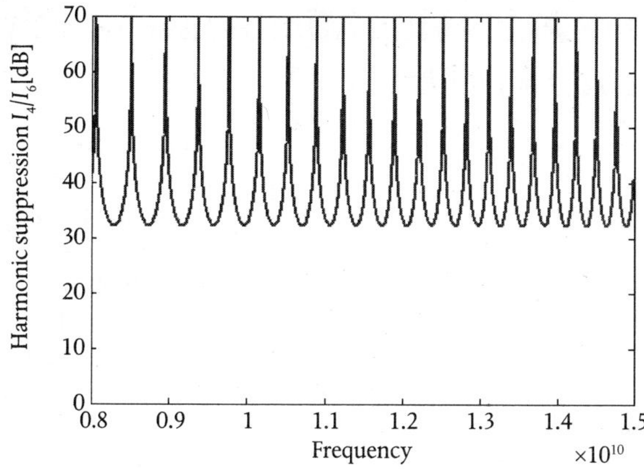

Fig. 8.22 (b) Harmonic suppression (I_4 / I_6) versus frequency of the electrical drive signal with modulation depth $\beta = 0.6$.

Figure 8.22(a) and (b) indicate that with a small value of the attenuation of the optical notch filter, increasing the modulation depth of the MZM from 0.6 to 0.9 improves the harmonic suppression. The minimum value of I_4/I_2 has increased from 2 to 7.7 dB for $\alpha = 40$ dB case, and 18.5 to 19.7 dB for $\alpha = 60$ dB case.

8.7 INVESTIGATION OF PHOTONICALLY ASSISTED MICROWAVE FREQUENCY MULTIPLICATION BASED ON EXTERNAL MODULATION

Microwave frequency multiplication based on external modulation using two cascaded Mach–Zehnder modulators (MZMs) has been considered an effective solution for high-frequency and frequency-tunable microwave signal generation. In this section, a generalised approach to achieving microwave frequency multiplication using two cascaded MZMs is presented. A theoretical analysis leading to the operating conditions to achieve frequency quadrupling, sextupling, or octupling is developed. The generation of a microwave signal can also be done using external modulation. The key significance of using external modulation is the system simplicity, frequency tunability, and operation stability. Mach–Zehnder modulator (MZM) is biased at the minimum transmission point (MITP) to eliminate all the even-order sidebands. If the modulation signal is small, then the higher order sidebands are small and are ignored. Therefore, only the two first-order sidebands would be present at the output of the MZM. The beating of the two sidebands at a PD would generate a microwave signal with a frequency that is twice that of the sinusoidal drive signal.

The key advantage of using a phase modulator is the better long-term stability, since a phase modulator is not biased, which makes the system free from DC bias drift. To generate a microwave signal with a higher multiplication factor, a configuration that employs two cascaded MZMs may be used. The multiplication factor can be as high as eight, which enables the generation of a high-frequency microwave or terahertz signal using a low-frequency microwave drive signal. Different techniques have been proposed for the generation of a frequency-quadrupled, sextupled, or octupled microwave signal. For two cascaded MZMs, there are four possible bias combinations: (i) MATP, MATP, (ii) MITP, MITP, (iv) MATP, MITP, and (iv) MITP, MATP, with each combination corresponding to a specific multiplication factor under different operating conditions.

8.7.1 Principle of Achieving Microwave Multiplication Using Two Cascaded MZMs

A generalised system using two cascaded MZMs to achieve microwave multiplication is shown in Fig. 8.23. The system consists of a continuous-wave (CW) laser source, two MZMs, a tunable optical phase shifter (TOPS), two polarisation controllers (PCs), an erbium-doped fibre amplifier (EDFA), and a photomixer. A low-frequency microwave drive signal from a microwave source is divided into two paths by a power divider, amplified by two microwave power amplifiers (PAs), and applied to the two MZMs. A tunable electrical phase shifter (TEPS) can be used between one PA and the corresponding MZM to replace the TOPS. The functions of the TOPS and the TEPS are the same, which will be explained further in this section. The two PCs are used to minimise the polarisation-dependent losses of the two MZMs. A wavelength-fixed FBG notch filter may also be incorporated in the system to remove the optical carrier, which will be explained later. A harmonic mixer may also be incorporated to down convert the generated microwave frequency to make it measurable by an electrical spectrum analyser (ESA). In the system, each of the two MZMs is biased at either the MATP or the MITP to suppress the odd- or the even-order sidebands. The cascade of two MZMs would lead to four bias combinations: (i)MATP, MATP, (ii) MITP, MITP, (iii) MATP, MITP, and (iv) MITP, MATP. By properly controlling the powers of the microwave drive signals applied to the two MZMs and the phase shift introduced by the TOPS or the TEPS, only two dominant sidebands would be generated at the output of the second (MZM2). By properly controlling the powers of the microwave drive signals applied to the two MZMs and the phase shift introduced by the TOPS or the TEPS, only two dominant sidebands would be generated at the output of the second (MZM2).

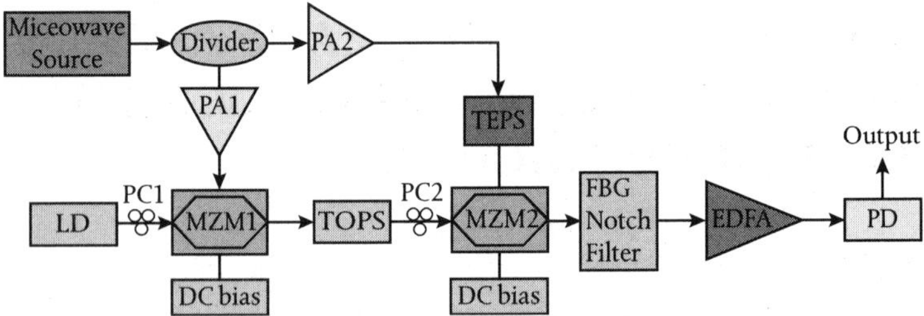

Fig. 8.23 A generalised microwave frequency multiplication system using two cascaded MZMs

By properly controlling the powers of the microwave drive signals applied to the two MZMs and the phase shift introduced by the TOPS or the TEPS, only two dominant sidebands would be generated at the output of the second (MZM2). Depending on the bias combinations, the frequency spacing of the two sidebands could be four, six, or eight times the frequency of the microwave drive signal. By beating the two sidebands at the photomixer, a frequency quadrupled, sextupled, or octupled microwave signal is generated.

When the incident lightwave is $E_0\cos(\omega_0 t)$, the electrical field $E(t)$ at the output of the MZM can be expressed as:

$$
\begin{aligned}
E(t) \propto\ & E_0\cos(\Phi/2)\Big\{J_0(\beta)\cos(\omega_0 t+\Phi/2)+ \\
& \sum_{n=1}^{\infty}(-1)^n J_{2n}(\beta)\Big[\cos\big(\omega_0 t+2n(\omega_m t+\phi)+\Phi/2\big)+\cos\big(\omega_0 t-2n(\omega_m t+\phi)+\Phi/2\big)\Big]\Big\}+ \\
& E_0\sin(\Phi/2)\Big\{\sum_{n=1}^{\infty}(-1)^n J_{2n-1}(\beta)\Big[\cos\big(\omega_0 t+(2n-1)(\omega_m t+\phi)+\Phi/2\big)+ \\
& \cos\big(\omega_0 t-(2n-1)(\omega_m t+\phi)+\Phi/2\big)\Big]\Big\}
\end{aligned}
\tag{8.41}
$$

where E_0 is the electrical amplitude of the incident lightwave, ω_0 and ω_m are, respectively, the angular frequencies of the lightwave and the microwave drive signal, $J_i(\beta)$ is the i th-order Bessel function of the first kind, ϕ is the initial phase of the microwave drive signal applied to the MZM, Φ is an additional phase difference between the two optical signals from the two arms of the MZM, which is introduced by the bias voltage, β is the PMI, which is equal to $\pi V/2V_\pi$, where V is the amplitude of the microwave drive signal, and V_π is the half-wave voltage of the MZM. From (8.41), if Φ is 0, the MZM is biased at the MATP and all the odd-order sidebands would be suppressed; and if Φ is π, the MZM is biased at the MITP and all the even-order sidebands would be suppressed.

8.7.2 Four Different Bias Combinations for Two MZMs

MATP, MATP

The two MZMs are both biased at the MATP, which is, the additional phase difference between the two optical signals from the two arms of the ith MZM is zero, $\Phi_i = 0$ for $i = 1,2$. According to (8.41), the signal at the output of the first MZM (MZM1) can be written as:

$$E_1(t) \propto E_0 J_0(\beta_1)\cos(\omega_0 t) - E_0 J_2(\beta_1)\cos(\omega_0 t - 2\omega_m t - 2\phi_1) - E_0 J_2(\beta_1)\cos(\omega_0 t + 2\omega_m t + 2\phi_1) \quad (8.42)$$

where β_1 is the PMI of MZM1 and ϕ_1 is the initial phase of the microwave drive signal applied to MZM. When the lightwave is travelling through the TOPS, a phase difference between the two adjacent even-order sidebands would be introduced and the electrical field $E_{11}(t)$ at the output of the TOPS can be written as:

$$\begin{aligned} E_{11}(t) \propto\ & E_0 J_0(\beta_1)\cos(\omega_0 t + 2\varphi) - E_0 J_2(\beta_1)\cos(\omega_0 t - 2\omega_m t - 2\phi_1 + \varphi) - \\ & E_0 J_2(\beta_1)\cos(\omega_0 t + 2\omega_m t + 2\phi_1 + 3\varphi) \end{aligned} \quad (8.43)$$

Let $E_0' = E_0 J_0(\beta_1)$, $E_0'' = -E_0 J_2(\beta_1) = E_0'''$ and

$\phi' = \omega_0 t + 2\varphi$, $\phi'' = \omega_0 t - 2\omega_m t - 2\phi_1 + \varphi$, $\phi''' = \omega_0 t + 2\omega_m t + 2\phi_1 + 3\varphi$

$E_{11}(t)$ is then sent to MZM2. The electrical field $E_2(t)$ at the output of MZM_2 is given by $E_2(t)$. Substituting (8.43) in (8.42), and replacing β_1 to β_2 and ϕ_1 to ϕ_2 ,

$$\begin{aligned} E_2(t) =\ & E_0' J_0(\beta_2)\cos\phi' - E_0' J_2(\beta_2)\cos(\phi' - 2\omega_m t - 2\phi_2) - E_0' J_2(\beta_2)\cos(\phi' + 2\omega_m t + 2\phi_2) + \\ & E_0'' J_0(\beta_2)\cos\phi'' - E_0'' J_2(\beta_2)\cos(\phi'' - 2\omega_m t - 2\phi_2) - \\ & E_0'' J_2(\beta_2)\cos(\phi'' + 2\omega_m t + 2\phi_2) + \\ & E_0''' J_0(\beta_2)\cos\phi''' - E_0''' J_2(\beta_2)\cos(\phi''' - 2\omega_m t - 2\phi_2) - \\ & E_0''' J_2(\beta_2)\cos(\phi''' + 2\omega_m t + 2\phi_2) \end{aligned}$$

$$\begin{aligned} E_2(t) =\ & E_0 J_0(\beta_1) J_0(\beta_2)\cos(\omega_0 t + 2\varphi) - E_0 J_0(\beta_1) J_2(\beta_2)\cos(\omega_0 t + 2\varphi - 2\omega_m t - 2\phi_2) - \\ & E_0 J_0(\beta_1) J_2(\beta_2)\cos(\omega_0 t + 2\varphi + 2\omega_m t + 2\phi_2) - E_0 J_2(\beta_1) J_0(\beta_2)\cos(\omega_0 t - 2\omega_m t - 2\phi_1 + \varphi) + \\ & E_0 J_2(\beta_1) J_2(\beta_2)\cos(\omega_0 t - 2\omega_m t - 2\phi_1 + \varphi - 2\omega_m t - 2\phi_2) + \\ & E_0 J_2(\beta_1) J_2(\beta_2)\cos(\omega_0 t - 2\omega_m t - 2\phi_1 + \varphi + 2\omega_m t + 2\phi_2) - \\ & E_0 J_2(\beta_1) J_0(\beta_2)\cos(\omega_0 t + 2\omega_m t + 2\phi_1 + 3\varphi) + \\ & E_0 J_2(\beta_1) J_2(\beta_2)\cos(\omega_0 t + 2\omega_m t + 2\phi_1 + 3\varphi - 2\omega_m t - 2\phi_2) + \\ & E_0 J_2(\beta_1) J_2(\beta_2)\cos(\omega_0 t + 2\omega_m t + 2\phi_1 + 3\varphi + 2\omega_m t + 2\phi_2) \end{aligned}$$

$$\begin{aligned}E_2(t) = E_0 \{ & J_0(\beta_1)J_0(\beta_2)\cos(\omega_0 t + 2\varphi) - J_0(\beta_1)J_2(\beta_2)\cos((\omega_0 - 2\omega_m)t - 2\phi_2 + 2\varphi) - \\ & J_0(\beta_1)J_2(\beta_2)\cos((\omega_0 + 2\omega_m)t + 2\phi_2 + 2\varphi) - J_2(\beta_1)J_0(\beta_2)\cos((\omega_0 - 2\omega_m)t - 2\phi_1 + \varphi) + \\ & J_2(\beta_1)J_2(\beta_2)\cos((\omega_0 - 4\omega_m)t - 2\phi_1 + \varphi - 2\phi_2) + J_2(\beta_1)J_2(\beta_2)\cos(\omega_0 t - 2\phi_1 + \varphi + 2\phi_2) - \\ & J_2(\beta_1)J_0(\beta_2)\cos((\omega_0 + 2\omega_m)t + 2\phi_1 + 3\varphi) + J_2(\beta_1)J_2(\beta_2)\cos(\omega_0 t + 2\phi_1 + 3\varphi - 2\phi_2) - \\ & J_2(\beta_1)J_2(\beta_2)\cos((\omega_0 + 4\omega_m)t + 2\phi_1 + 3\varphi + 2\phi_2)\}\end{aligned} \tag{8.44}$$

where β_2 is the PMI for MZM_2 and ϕ_2 is the initial phase of the drive signal applied to MZM_2. Consider two conditions:

1. $\beta_1 = \beta_2 = \beta$, we have $J_0(\beta_1)J_2(\beta_2) = J_0(\beta_2)J_2(\beta_1)$
2. $\varphi + 2\phi_1 - 2\phi_2 = (2k+1)\pi, k$ is an integer,

Using the first condition, (8.44) can be simplified into

$$\begin{aligned}E_2(t) = E_0 \{ & J_0^2(\beta)\cos(\omega_0 t + 2\varphi) - J_0(\beta)J_2(\beta)\cos((\omega_0 - 2\omega_m)t - 2\phi_2 + 2\varphi) - \\ & J_0(\beta)J_2(\beta)\cos((\omega_0 + 2\omega_m)t + 2\phi_2 + 2\varphi) - J_0(\beta)J_2(\beta)\cos((\omega_0 - 2\omega_m)t - 2\phi_1 + \varphi) + \\ & J_2^2(\beta)\cos((\omega_0 - 4\omega_m)t - 2\phi_1 + \varphi - 2\phi_2) + J_2^2(\beta)\cos(\omega_0 t - 2\phi_1 + \varphi + 2\phi_2) - \\ & J_0(\beta)J_2(\beta)\cos((\omega_0 + 2\omega_m)t + 2\phi_1 + 3\varphi) + J_2^2(\beta)\cos(\omega_0 t + 2\phi_1 + 3\varphi - 2\phi_2) - \\ & J_2^2(\beta)\cos((\omega_0 + 4\omega_m)t + 2\phi_1 + 3\varphi + 2\phi_2)\}\end{aligned}$$

Using $\varphi + 2\phi_1 - 2\phi_2 = (2k+1)\pi, k$ is an integer, that is, $\cos(\theta + (2k+1)\pi) = -\cos\theta$.

Simplifying $J_0(\beta)J_2(\beta)$ terms, we get

$$\begin{aligned}E_2(t) = E_0 \{ & J_0^2(\beta)\cos(\omega_0 t + 2\varphi) - J_0(\beta)J_2(\beta)\cos((\omega_0 - 2\omega_m)t - 2\phi_2 + \varphi + \varphi + 2\phi_1 - 2\phi_{1 J_0(\beta) J_2(\beta)}) - \\ & J_0(\beta)J_2(\beta)\cos((\omega_0 + 2\omega_m)t + 2\phi_2 + 2\varphi) - J_0(\beta)J_2(\beta)\cos((\omega_0 - 2\omega_m)t - 2\phi_1 + \varphi) + \\ & J_2^2(\beta)\cos((\omega_0 - 4\omega_m)t - 2\phi_1 + \varphi - 2\phi_2) + J_2^2(\beta)\cos(\omega_0 t - 2\phi_1 + \varphi + 2\phi_2) - \\ & J_0(\beta)J_2(\beta)\cos((\omega_0 + 2\omega_m)t + 2\phi_1 + 2\varphi + \varphi - 2\phi_2 + 2\phi_2) + \\ & J_2^2(\beta)\cos(\omega_0 t + 2\phi_1 + 3\varphi - 2\phi_2) - J_2^2(\beta)\cos((\omega_0 + 4\omega_m)t + 2\phi_1 + 3\varphi + 2\phi_2)\}\end{aligned}$$

$$\begin{aligned}E_2(t) = E_0 \{ & J_0^2(\beta)\cos(\omega_0 t + 2\varphi) + J_0(\beta)J_2(\beta)\cos((\omega_0 - 2\omega_m)t - 2\phi_1 + \varphi) - \\ & J_0(\beta)J_2(\beta)\cos((\omega_0 + 2\omega_m)t + 2\phi_2 + 2\varphi) - J_0(\beta)J_2(\beta)\cos((\omega_0 - 2\omega_m)t - 2\phi_1 + \varphi) + \\ & J_2^2(\beta)\cos((\omega_0 - 4\omega_m)t - 2\phi_1 + \varphi - 2\phi_2) + J_2^2(\beta)\cos(\omega_0 t - 2\phi_1 + \varphi + 2\phi_2) + \\ & J_0(\beta)J_2(\beta)\cos((\omega_0 + 2\omega_m)t + 2\phi_2 + 2\varphi) + J_2^2(\beta)\cos(\omega_0 t + 2\phi_1 + 3\varphi - 2\phi_2) - \\ & J_2^2(\beta)\cos((\omega_0 + 4\omega_m)t + 2\phi_1 + 3\varphi + 2\phi_2)\}\end{aligned}$$

Simplifying $J_2^2(\beta)$ terms,

$$E_2(t)=E_0\{J_0^2(\beta)\cos(\omega_0 t+2\varphi)+J_2^2(\beta)\cos((\omega_0-4\omega_m)t-2\phi_1+\varphi-2\phi_2)$$
$$+J_2^2(\beta)\cos(\omega_0 t-2\phi_1+\varphi+\varphi-\varphi+2\phi_2)+J_2^2(\beta)\cos(\omega_0 t+2\phi_1+2\varphi+\varphi-2\phi_2)$$
$$-J_2^2(\beta)\cos((\omega_0+4\omega_m)t+2\phi_1+3\varphi+2\phi_2)\}$$

$$E_2(t)=E_0\{J_0^2(\beta)\cos(\omega_0 t+2\varphi)+J_2^2(\beta)\cos((\omega_0-4\omega_m)t-2\phi_1+\varphi-2\phi_2)$$
$$+J_2^2(\beta)\cos(\omega_0 t+2\varphi-(2\phi_1+\varphi-2\phi_2))$$
$$+J_2^2(\beta)\cos(\omega_0 t+2\varphi+2\phi_1+\varphi-2\phi_2)$$
$$-J_2^2(\beta)\cos((\omega_0+4\omega_m)t+2\phi_1+3\varphi+2\phi_2)\}$$

Using $\varphi+2\phi_1-2\phi_2=(2k+1)\pi, k$ is an integer, that is, $\cos(\theta+(2k+1)\pi)=-\cos\theta$:

$$E_2(t)\propto E_0\{J_2{}^2(\beta)\cos((\omega_0-4\omega_m)t+\varphi-2\phi_1-2\phi_2)+[J_0{}^2(\beta)-2J_2{}^2(\beta)]\cos(\omega_0 t+2\varphi)$$
$$+J_2{}^2(\beta)\cos((\omega_0+4\omega_m)t+3\varphi+2\phi_1+2\phi_2)\} \tag{8.45}$$

As can be seen, only the carrier and the fourth-order sidebands are present. If $J_0{}^2(\beta)=2J_2{}^2(\beta)$, then the optical carrier will be suppressed and only the fourth-order sidebands will be present at the output of second MZM (MZM2). Note that to meet the condition, β should be equal to 1.6995. By beating the two fourth-order sidebands at the photomixer, an electrical signal with a frequency that is eight times the frequency of the microwave drive signal will be generated. Thus, frequency octupling is achieved. The frequency multiplication factor (FMF) is eight. In addition to the dominant fourth-order sidebands, the second largest sidebands are the sixth-order sidebands. From (8.44), the power of the fourth-order sidebands is proportional to $J_2(\beta_1)J_2(\beta_2)$. Similarly, we can easily obtain the power of the sixth-order sidebands, which is proportional to $J_4(\beta)J_2(\beta)$. Therefore, in the optical domain, the optical sideband suppression ratio, OSSR, using $\beta_1=\beta_2$ is given as:

$$\mathrm{OSSR}_{\mathrm{MATP,MATP}}=20\log_{10}\frac{J_2(\beta)J_2(\beta)}{J_4(\beta)J_2(\beta)}=23.52\,\mathrm{d}$$

To satisfy the condition $J_0{}^2(\beta)=2J_2{}^2(\beta)$, the PMI should be very large, which is not easy to realise since a large microwave drive signal is needed. When $\beta=1.6995$, for an MZM with a V_π of 7.5 V, the voltage of the microwave drive signal should be 8.11 V, which corresponds to a microwave power of 29 dBm. As shown in Fig. (8.23), the notch filter is an FBG. Let us assume an FBG notch filter with a notch depth of D dB is used to remove the residual optical carrier, then the optical sideband suppression ratio, $\mathrm{OSSR}_{\mathrm{MATP,\,MATP}}$, would be given by

$$\mathrm{OSSR}_{\mathrm{MATP,MATP}}=20\log_{10}[J_2(\beta_1)J_2(\beta_2)]-\mathrm{Max}\{20\log_{10}[J_4(\beta)J_2(\beta)],20\log_{10}[J_0{}^2(\beta)-2J_2{}^2(\beta)]-D\} \tag{8.46}$$

According to (8.46), the relationship between the $\mathrm{OSSR}_{\mathrm{MATP,MATP}}$ and the PMI is evaluated by numerical computation. As can be seen, three different values of D are provided. The $\mathrm{OSSR}_{\mathrm{MATP,MATP}}$ will increase the first, and then, decreases as the PMI increases. As the value of D increases, the value of β needed to achieve a maximum $OSSR_{MATP,MATP}$ decreases. In the electrical domain, undesired harmonics would be generated due to the beating between the two fourth-order sidebands with the optical carrier or with the two sixth-order sidebands.

MITP, MITP

The MZMs are both biased at the MITP, that is, $\Phi_i = \pi$ for $i=1,2$. According to (8.41), the signal at the output of MZM1 can be written as:

$$E_1(t) \propto E_0 \cos(\pi/2)\{J_0(\beta)\cos(\omega_0 t+\pi/2)+$$

$$\sum_{n=1}^{\infty}(-1)^n J_{2n}(\beta)\left[\cos(\omega_0 t+2n(\omega_m t+\phi)+\pi/2)+\cos(\omega_0 t-2n(\omega_m t+\phi)+\pi/2)\right]\Big\}+$$

$$E_0 \sin(\pi/2)\left\{\sum_{n=1}^{\infty}(-1)^n J_{2n-1}(\beta)\left[\cos(\omega_0 t+(2n-1)(\omega_m t+\phi)+\pi/2)+\right.\right.$$

$$\left.\left.\cos(\omega_0 t-(2n-1)(\omega_m t+\phi)+\pi/2)\right]\right\}$$

$$E_1(t) \propto E_0 J_1(\beta_1)\sin((\omega_0-\omega_m)t-\phi_1)+E_0 J_1(\beta_1)\sin((\omega_0+\omega_m)t+\phi_1) \tag{8.47}$$

If a phase difference φ between the two sidebands is introduced by the TOPS, we have

$$E_{11}(t) \propto E_0 J_1(\beta_1)\sin((\omega_0-\omega_m)t-\phi_1+\varphi)+E_0 J_1(\beta_1)\sin((\omega_0+\omega_m)t+\phi_1+2\varphi) \tag{8.48}$$

Let $E_0' = E_0 J_1(\beta_1)$, $E_0'' = E_0 J_1(\beta_1)$ and $\phi' = ((\omega_0-\omega_m)t-\phi_1+\varphi)$, $\phi'' = ((\omega_0+\omega_m)t+\phi_1+2\varphi)$.

The optical signal is then sent to second MZM (MZM2). The electrical filed at the output of second MZM (MZM2) is $E_2(t)$.

Substituting (8.48) into (8.47), and changing β_1 to β_2 and ϕ_1 to ϕ_2

$$E_2(t) = E_0' J_1(\beta_2)\sin(\phi'-\omega_m t-\phi_2+\pi/2)+E_0' J_1(\beta_2)\sin(\phi'+\omega_m t+\phi_2+\pi/2)+$$
$$E_0'' J_1(\beta_2)\sin(\phi''-\omega_m t-\phi_2+\pi/2)+E_0'' J_1(\beta_2)\sin(\phi''+\omega_m t+\phi_2+\pi/2)$$

$$E_2(t) = E_0 J_1(\beta_1)J_1(\beta_2)\sin((\omega_0-\omega_m)t-\phi_1+\varphi-\omega_m t-\phi_2+\pi/2)+$$
$$E_0 J_1(\beta_1)J_1(\beta_2)\sin((\omega_0+\omega_m)t-\phi_1+\varphi+\omega_m t+\phi_2+\pi/2)+$$
$$E_0 J_1(\beta_1)J_1(\beta_2)\sin((\omega_0+\omega_m)t+\phi_1+2\varphi-\omega_m t-\phi_2+\pi/2)+$$
$$E_0 J_1(\beta_1)J_1(\beta_2)\sin((\omega_0+\omega_m)t+\phi_1+2\varphi+\omega_m t+\phi_2+\pi/2)$$

$$\begin{aligned} E_2(t) \propto E_0 J_1(\beta_1)J_1(\beta_2)\{&\cos((\omega_0-2\omega_m)t-\phi_1-\phi_2+\varphi)+\cos(\omega_0 t+\varphi-\phi_1+\phi_2)+ \\ &\cos(\omega_0 t+2\varphi+\phi_1-\phi_2)+\cos((\omega_0+2\omega_m)t+\phi_1+\phi_2+2\varphi)\} \end{aligned} \tag{8.49}$$

If the phase condition $\varphi+2\phi_1-2\phi_2=(2k+1)\pi, k$ is an integer. $\varphi=(2k+1)\pi+2\phi_2-2\phi_1$. Consider $\cos(\omega_0 t+\varphi-\phi_1+\phi_2)$,

$$= \cos(\omega_0 t+(2k+1)\pi+2\phi_2-2\phi_1-\phi_1+\phi_2)$$

$$= \cos(\omega_0 t+(2k+1)\pi+3\phi_2-3\phi_1)$$

$$= -\cos\left(\omega_0 t + 3\phi_2 - 3\phi_1\right)$$

Similarly, $\cos\left(\omega_0 t + 2\varphi + \phi_1 - \phi_2\right)$

$$= \cos\left(\omega_0 t + 2\left((2k+1)\pi + 2\phi_2 - 2\phi_1\right) + \phi_1 - \phi_2\right)$$

$$= \cos\left(\omega_0 t + (2k+1)2\pi + 3\phi_2 - 3\phi_1\right)$$

$$= \cos\left(\omega_0 t + 3\phi_2 - 3\phi_1\right)$$

Hence, (8.49) reduces to

$$E_2(t) \propto E_0 J_1(\beta_1) J_1(\beta_2)\left\{\cos\left((\omega_0 - 2\omega_m)t - \phi_1 - \phi_2 + \varphi\right) + \cos\left((\omega_0 + 2\omega_m)t + \phi_1 + \phi_2 + 2\varphi\right)\right\} \quad (8.50)$$

As can be seen, only the second-order sidebands are generated at the output of second MZM (MZM2). By beating the second-order sidebands at the photomixer, a frequency-quadrupled signal would be generated; thus, frequency quadrupling is achieved. The FMF is four. In addition to the dominant second-order sidebands, the second largest sidebands are the fourth-order sidebands. From (8.50), the power of the second-order sidebands is proportional to $J_1(\beta_1)J_1(\beta_2)$. Similarly, we can easily obtain the power of the fourth-order sidebands, which is proportional to $J_1(\beta_1)J_3(\beta_2)$. Therefore, the optical sideband suppression ratio, $OSSR_{MITP,MITP}$, would be given as:

$$OSSR_{MITP,\,MITP} = 20\log_{10}\frac{J_1(\beta_1)J_1(\beta_2)}{J_1(\beta_1)J_3(\beta_2)}\,\text{dB} \quad (8.51)$$

If the phase condition is not fully satisfied, the optical carrier would appear, which can be removed by a wavelength-fixed notch filter.

MATP, MITP

If MZM1 is biased at the MATP with $\Phi_1 = 0$, and MZM2 is biased at the MITP with $\Phi_2 = \pi$, only the third-order side bands would be generated and frequency sextupling would be achieved. From (8.42), the electrical fields at the outputs of MZM1:

$$E_1(t) = E_0 J_0(\beta_1)\cos(\omega_0 t) - E_0 J_2(\beta_1)\cos\left(\omega_0 t - 2\omega_m t - 2\phi_1\right) - E_0 J_2(\beta_1)\cos\left(\omega_0 t + 2\omega_m t + 2\phi_1\right)$$

If a phase difference φ between the two sideband is introduced by the TOPS,

$$E_{11}(t) \propto E_0 J_0(\beta_1)\cos\left(\omega_0 t + 2\varphi\right) - E_0 J_2(\beta_1)\cos\left(\omega_0 t - 2\omega_m t - 2\phi_1 + \varphi\right) E_0 J_2(\beta_1)\cos\left(\omega_0 t + 2\omega_m t + 2\phi_1 + 3\varphi\right)$$

Let $E_0' = E_0 J_0(\beta_1),\ E_0'' = E_0''' = E_0 J_2(\beta_1)$ and

$$\phi' = \left(\omega_0 t + 2\varphi\right),\ \phi'' = \left(\omega_0 t - 2\omega_m t - 2\phi_1 + \varphi\right),\ \phi''' = \left(\omega_0 t + 2\omega_m t + 2\phi_1 + 3\varphi\right) \quad (8.52)$$

The electrical filed at the output of MZM2 can be expressed using (8.47) and (8.52):

$$E_2(t) \propto \left\{E_0' J_1(\beta_2)\sin\left(\phi' - \omega_m t - \phi_2\right) + E_0' J_1(\beta_2)\sin\left(\phi' + \omega_m t + \phi_2\right) - E_0'' J_1(\beta_2)\sin\left(\phi'' - \omega_m t - \phi_2\right) - E_0'' J_1(\beta_2)\sin\left(\phi'' + \omega_m t + \phi_2\right) - E_0''' J_1(\beta_2)\sin\left(\phi''' - \omega_m t - \phi_2\right) - E_0''' J_1(\beta_2)\sin\left(\phi''' + \omega_m t + \phi_2\right)\right\}$$

$$E_2(t) \propto \{E_0 J_0(\beta_1) J_1(\beta_2)\sin(\omega_0 t + 2\varphi - \omega_m t - \phi_2) + E_0 J_0(\beta_1) J_1(\beta_2)\sin(\omega_0 t + 2\varphi + \omega_m t + \phi_2)$$
$$- E_0 J_2(\beta_1) J_1(\beta_2)\sin(\omega_0 t - 2\omega_m t - 2\phi_1 + \varphi - \omega_m t - \phi_2)$$
$$- E_0 J_2(\beta_1) J_1(\beta_2)\sin(\omega_0 t - 2\omega_m t - 2\phi_1 + \varphi + \omega_m t + \phi_2)$$
$$- E_0 J_2(\beta_1) J_1(\beta_2)\sin(\omega_0 t + 2\omega_m t + 2\phi_1 + 3\varphi - \omega_m t - \phi_2)$$
$$- E_0 J_2(\beta_1) J_1(\beta_2)\sin(\omega_0 t + 2\omega_m t + 2\phi_1 + 3\varphi + \omega_m t + \phi_2)\}$$

$$E_2(t) \propto E_0 J_1(\beta_2)\{J_0(\beta_1)\sin((\omega_0 - \omega_m)t + 2\varphi - \phi_2) + J_0(\beta_1)\sin((\omega_0 + \omega_m)t + 2\varphi + \phi_2) -$$
$$J_2(\beta_1)\sin((\omega_0 - 3\omega_m)t - 2\phi_1 + \varphi - \phi_2) - J_2(\beta_1)\sin((\omega_0 - \omega_m)t - 2\phi_1 + \varphi + \phi_2) -$$
$$J_2(\beta_1)\sin((\omega_0 + \omega_m)t + 2\phi_1 + 3\varphi - \phi_2) - J_2(\beta_1)\sin(\omega_0 t + 3\omega_m t + 2\phi_1 + 3\varphi + \phi_2)\}$$

$$E_2(t) \propto E_0 J_1(\beta_2)\{J_2(\beta_1)\sin((\omega_0 - \omega_m)t + 2\varphi - \phi_2) + J_2(\beta_1)\sin((\omega_0 + \omega_m)t + 2\varphi + \phi_2) -$$
$$J_2(\beta_1)\sin((\omega_0 - 3\omega_m)t - 2\phi_1 + \varphi - \phi_2) - J_2(\beta_1)\sin((\omega_0 - \omega_m)t - 2\phi_1 + \varphi + \phi_2) -$$
$$J_2(\beta_1)\sin((\omega_0 + \omega_m)t + 2\phi_1 + 3\varphi - \phi_2) - J_2(\beta_1)\sin(\omega_0 t + 3\omega_m t + 2\phi_1 + 3\varphi + \phi_2)\} \quad (8.53)$$

Using 2 PMI conditions,

1. $J_0(\beta_1) = J_2(\beta_1)$
2. phase condition $\varphi - 2\phi_1 + \phi_2 + 2k\pi = 2\varphi - \phi_2$, $3\varphi + 2\phi_1 - \phi_2 = 2\varphi + \phi_2 + 2k\pi$ or $\varphi + 2\phi_1 - 2\phi_2 = 2k\pi$

Using conditions (1) and (2),

$$E_2(t) \propto E_0 J_1(\beta_2)\{J_2(\beta_1)\sin((\omega_0 - \omega_m)t + \varphi - 2\phi_1 + \phi_2 + 2k\pi) +$$
$$J_2(\beta_1)\sin((\omega_0 + \omega_m)t + 2\varphi + \phi_2) - J_2(\beta_1)\sin((\omega_0 - 3\omega_m)t - 2\phi_1 + \varphi - \phi_2) -$$
$$J_2(\beta_1)\sin((\omega_0 - \omega_m)t - 2\phi_1 + \varphi + \phi_2) - J_2(\beta_1)\sin((\omega_0 + \omega_m)t + 2\varphi + \phi_2 + 2k\pi) -$$
$$J_2(\beta_1)\sin(\omega_0 t + 3\omega_m t + 2\phi_1 + 3\varphi + \phi_2)\}$$

$$E_2(t) \propto E_0 J_1(\beta_2)\{J_2(\beta_1)\sin((\omega_0 - \omega_m)t + \varphi - 2\phi_1 + \phi_2) +$$
$$J_2(\beta_1)\sin((\omega_0 + \omega_m)t + 2\varphi + \phi_2) - J_2(\beta_1)\sin((\omega_0 - 3\omega_m)t - 2\phi_1 + \varphi - \phi_2) -$$
$$J_2(\beta_1)\sin((\omega_0 - \omega_m)t - 2\phi_1 + \varphi + \phi_2) - J_2(\beta_1)\sin((\omega_0 + \omega_m)t + 2\varphi + \phi_2) -$$
$$J_2(\beta_1)\sin(\omega_0 t + 3\omega_m t + 2\phi_1 + 3\varphi + \phi_2)\}$$

The above-mentioned equation reduces to:

$$E_2(t) \propto E_0 J_1(\beta_2)\{J_2(\beta_1)\sin((\omega_0 - 3\omega_m)t + \varphi - 2\phi_1 - \phi_2) + J_2(\beta_1)\sin((\omega_0 + 3\omega_m)t + 3\varphi + 2\phi_1 + \phi_2)\} \quad (8.54)$$

As from (8.54), only the third-order sidebands are generated. By beating the third-order sidebands at the photomixer, a frequency-sextupled signal is generated; thus, frequency sextupling is achieved. The FMF is six. To satisfy the PMI condition, the value of β_1 should be 1.8412. Considering the fact that is very large, the fourth-order sidebands at the output of MZM1 would have a relatively high power and would generate the fifth-order sidebands at the output of MZM2, which would contribute to the generation of a frequency-doubled and a frequency-octupled signals at the output of the photomixer. It is theoretically calculated that the optical sideband suppression ratio, $OSSR_{MATP,MITP}$, would be approximately equal to $20\log_{10}\left[J_2(\beta_1)/J_4(\beta_1)\right] \approx 21.97$ dB.

MITP, MATP

If MZM1 is biased at the MITP with $\Phi_1 = \pi$, and MZM2 is biased at the MATP with $\Phi_2 = 0$, only the third-order sidebands would be generated and frequency sextupling would be achieved. The electrical fields at the outputs of MZM1:

From (8.47),

$$E_1(t) \propto E_0 J_1(\beta_1)\sin((\omega_0 - \omega_m)t - \phi_1) + E_0 J_1(\beta_1)\sin((\omega_0 + \omega_m)t + \phi_1)$$

The electrical fields at the outputs the TOPS is:

$$E_{11}(t) \propto E_0 J_1(\beta_1)\sin((\omega_0 - \omega_m)t - \phi_1 + \varphi) + E_0 J_1(\beta_1)\sin((\omega_0 + \omega_m)t + \phi_1 + 2\varphi).$$

Let $E_0' = E_0'' = E_0''' = E_0 J_1(\beta_1)$ and $\phi' = ((\omega_0 - \omega_m)t - \phi_1 + \varphi)$, $\phi'' = ((\omega_0 + \omega_m)t + \phi_1 + 2\varphi)$ (8.55)

The electrical field at the output of MZM2 can expressed using (8.42):

$$\begin{aligned} E_2(t) \propto \{ & E_0' J_0(\beta_2)\cos(\phi' + \pi/2) - E_0' J_2(\beta_2)\cos(\phi' - 2\omega_m t - 2\phi_2 + \pi/2) \\ & - E_0' J_2(\beta_2)\cos(\phi' + 2\omega_m t + 2\phi_2 + \pi/2) + E_0'' J_0(\beta_2)\cos(\phi'' + \pi/2) \\ & - E_0'' J_2(\beta_2)\cos(\phi'' - 2\omega_m t - 2\phi_2 + \pi/2) \\ & - E_0'' J_2(\beta_2)\cos(\phi'' + 2\omega_m t + 2\phi_2 + \pi/2)\} \end{aligned}$$

$$\begin{aligned} E_2(t) \propto \{ & E_0 J_1(\beta_1) J_0(\beta_2)\cos((\omega_0 - \omega_m)t - \phi_1 + \varphi + \pi/2) \\ & - E_0 J_1(\beta_1) J_2(\beta_2)\cos((\omega_0 - \omega_m)t - \phi_1 + \varphi - 2\omega_m t - 2\phi_2 + \pi/2) \\ & - E_0 J_1(\beta_1) J_2(\beta_2)\cos((\omega_0 - \omega_m)t - \phi_1 + \varphi + 2\omega_m t + 2\phi_2 + \pi/2) \\ & + E_0 J_1(\beta_1) J_0(\beta_2)\cos((\omega_0 + \omega_m)t + \phi_1 + 2\varphi + \pi/2) \\ & - E_0 J_1(\beta_1) J_2(\beta_2)\cos((\omega_0 + \omega_m)t + \phi_1 + 2\varphi - 2\omega_m t - 2\phi_2 + \pi/2) \\ & - E_0 J_1(\beta_1) J_2(\beta_2)\cos((\omega_0 + \omega_m)t + \phi_1 + 2\varphi + 2\omega_m t + 2\phi_2 + \pi/2)\} \end{aligned}$$

$$\begin{aligned} E_2(t) \propto E_0 J_1(\beta_1)\{ & J_2(\beta_2)\sin((\omega_0 - 3\omega_m)t + \varphi - \phi_1 - 2\phi_2) - \\ & J_0(\beta_2)\sin((\omega_0 - \omega_m)t + \varphi - \phi_1) + J_2(\beta_2)\sin((\omega_0 - \omega_m)t + 2\varphi + \phi_1 - 2\phi_2) - \\ & J_2(\beta_2)\sin((\omega_0 + \omega_m)t + \varphi - \phi_1 + 2\phi_2) + J_0(\beta_2)\sin((\omega_0 + \omega_m)t + 2\varphi + \phi_1) + \\ & J_2(\beta_2)\sin((\omega_0 + 3\omega_m)t + 2\varphi + \phi_1 + 2\phi_2)\} \end{aligned} \quad (8.56)$$

Using two conditions,

1) PMI condition $J_0(\beta_2) = J_2(\beta_2)$, i.e., $\beta_2 \approx 1.8412$,
2) phasecondition $2\varphi + \phi_1 - 2\phi_2 = \varphi - \phi_1 + 2k\pi$, $\varphi - \phi_1 + 2\phi_2 + 2k\pi = 2\varphi + \phi_1$,thatis, $\varphi + 2\phi_1 - 2\phi_2 = 2k\pi$,

$$\begin{aligned} E_2(t) \propto E_0 J_1(\beta_1)\{ & J_2(\beta_2)\sin((\omega_0 - 3\omega_m)t + \varphi - \phi_1 - 2\phi_2) \\ & - J_2(\beta_2)\sin((\omega_0 - \omega_m)t + \varphi - \phi_1) + J_2(\beta_2)\sin((\omega_0 - \omega_m)t + 2\varphi + \phi_1 - 2\phi_2) \\ & - J_2(\beta_2)\sin((\omega_0 + \omega_m)t + \varphi - \phi_1 + 2\phi_2) + J_2(\beta_2)\sin((\omega_0 + \omega_m)t + 2\varphi + \phi_1) \\ & + J_2(\beta_2)\sin((\omega_0 + 3\omega_m)t + 2\varphi + \phi_1 + 2\phi_2)\} \end{aligned}$$

$$E_2(t) \propto E_0 J_1(\beta_1)\{J_2(\beta_2)\sin((\omega_0 - 3\omega_m)t + \varphi - \phi_1 - 2\phi_2) - J_2(\beta_2)\sin((\omega_0 - \omega_m)t + \varphi - \phi_1)$$
$$+ J_2(\beta_2)\sin((\omega_0 - \omega_m)t + \varphi - \phi_1 + 2k\pi) - J_2(\beta_2)\sin((\omega_0 + \omega_m)t + \varphi - \phi_1 + 2\phi_2)$$
$$+ J_2(\beta_2)\sin((\omega_0 + \omega_m)t + \varphi - \phi_1 + 2\phi_2 + 2k\pi) + J_2(\beta_2)\sin((\omega_0 + 3\omega_m)t + 2\varphi + \phi_1 + 2\phi_2)\}$$

$$E_2(t) \propto E_0 J_1(\beta_1)\{J_2(\beta_2)\sin((\omega_0 - 3\omega_m)t + \varphi - \phi_1 - 2\phi_2) - J_2(\beta_2)\sin((\omega_0 - \omega_m)t + \varphi - \phi_1)$$
$$+ J_2(\beta_2)\sin((\omega_0 - \omega_m)t + \varphi - \phi_1) - J_2(\beta_2)\sin((\omega_0 + \omega_m)t + \varphi - \phi_1 + 2\phi_2)$$
$$+ J_2(\beta_2)\sin((\omega_0 + \omega_m)t + \varphi - \phi_1 + 2\phi_2) + J_2(\beta_2)\sin((\omega_0 + 3\omega_m)t + 2\varphi + \phi_1 + 2\phi_2)\}.$$

Equation (8.56) can be simplified to:

$$E_2(t) \propto E_0 J_1(\beta_1)\{J_2(\beta_2)\sin((\omega_0 - 3\omega_m)t + \varphi - \phi_1 - 2\phi_2) + J_2(\beta_2)\sin((\omega_0 + 3\omega_m)t + 2\varphi + \phi_1 + 2\phi_2)\} \quad (8.57)$$

As from (8.57), only the third-order sidebands are kept and the other sidebands are suppressed. By beating the two-third order sidebands at the photomixer, a frequency-sextupled signal would be generated. Again, the FMF is six. Similar to the case where MZM1 and MZM2 are, respectively, biased at the MATP and the MITP, a frequency-doubled signal and a frequency-octupled signal at the output of the photomixer would be also generated due to the existence of the fifth-order sidebands. The theoretically calculated optical sideband suppression ratio $OSSR_{MITP,MITP}$ would also be as same as those in the previous case.

SUMMARY

- An overview about the optical generation of microwave signals and all optical microwave signal processing was presented. The key significance of using photonics to generate and process microwave signals is that high-frequency microwave signals can be generated and processed in the optical domain that may not be possible using the state-of-the-art electronics.
- In addition, the generated and processed microwave signals can be distributed over optical fibre for long distance distribution to take advantage of the extremely low loss of the state-of-the art fibre, which is an added advantage of photonic techniques for the generation and processing of microwave signals.
- With advantageous features of photonics such as high-time bandwidth product, low loss capacity, and immunity to electromagnetic interference has led to increasing interest in its use for generation, processing, control and distribution of microwave signals, an area called microwave photonics.
- The frequency response of an incoherent photonics microwave filter is more robust than a filter that is operating in the coherent region.
- Microwave photonics is an interdisciplinary area that studies the interaction between microwave and optical signals for application such as radar, communication, sensor networks, warfare systems, and instrumentation.
- A microwave or mm-wave signal can be generated in optical domain based on optical heterodyning in which two optical waves of different wavelengths beat at a photodetector. An electrical beat note is then generated at the output of the photodetector with a frequency corresponding to the wavelength spacing of the two optical waves.

- In optical injection phase locking, the two techniques of optical injection locking and optical phase locking can be combined in a single optical locking system. The system consists of a master laser, a slave laser, an external modulator, a photodetector, and a phase detection module.
- Microwave signal generation can also be implemented based on external modulation by which the frequency of a low-frequency microwave signal can be increased to a high frequency through frequency multiplication.
- The tunability of a photonic microwave delay-line filter is usually achieved by adjusting the time-delay difference. However, the change of the time-delay difference would lead to the change of the free-spectral range (FSR), which results in the change of the 3-dB bandwidth as well as the entire shape of the frequency response.
- The theoretical bandwidth of an FM wave is infinite; most of the modulated-signal power resides in a finite bandwidth. There are two distinct possibilities in term of bandwidths: narrow-band FM and wide-band FM.
- Generation and transmission of microwave and millimetre-wave signals over optical fibres are of great interest for applications such as broadband wireless access networks operating at millimetre-wave bands, antenna remoting, phased-array antennas, optical sensors, and radars.
- Usually on a wide-band electrical heterodyne system, especially when the system operates over an octave bandwidth, an electrically tuned bandpass filter is inserted between the local oscillator and the mixer to suppress unwanted harmonics. This prevents potential in-band interference from corrupting the received signal.

THEORY QUESTIONS AND ANSWERS

Question 1: What do you understand by microwave photonics?

Answer: Microwave photonics (MWP) has been defined as the study of photonic devices operating at microwave frequencies and their application to microwave and optical systems. Initially, the advantages of photonic technologies provided the functions in microwave systems that are very complex or even impossible to carry out directly in the radio frequency (RF) domain. However, MWP has also succeeded in incorporating a variety of techniques used in microwave engineering to improve the performance of photonic communication networks and systems.

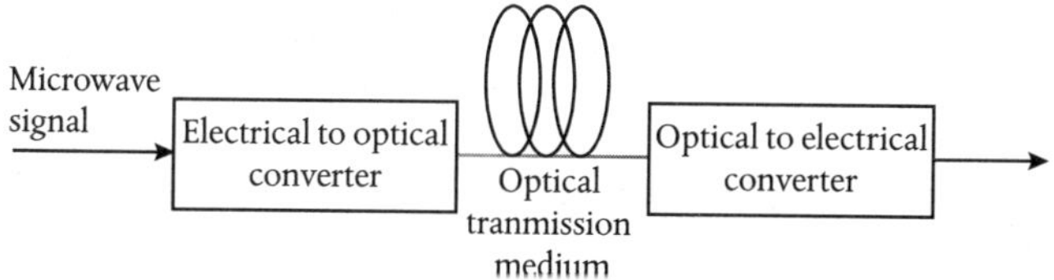

Fig. P1 Schematic highlighting the fundamental concept of an analogue microwave photonic link. One or more microwave signals are converted into the optical domain before transmission through a photonic or optical fibre link. At the other end of the photonic link, the signal is converted back into the electrical domain

Figure P1 shows a schematic highlighting the fundamental concept of a simple microwave photonic link. One or more analogue electrical signal at a microwave frequency are transported over an optical fibre or photonic link, with electrical-to-optical and optical-to-electrical conversion at the transmitting and receiving side, respectively, of the photonic link. The key advantages of microwave photonic links over conventional electrical-transmission systems, such as coaxial cables or waveguides, include reduced size, weight and cost, low and constant attenuation over the entire microwave and millimetre-wave modulation frequency range, immunity to electromagnetic interference, low dispersion, and high data-transfer capacity. The weight and attenuation benefits of microwave photonic links over coaxial cables are particularly compelling: typically, 1.7 kgkm^{-1} and 0.5 dBkm^{-1} for fibre with 567 kgkm^{-1} and 360 dBkm^{-1} at 2 GHz for a coaxial cable.

Question 2: Write down short notes on microwave photonic filters.

Answer: By microwave photonic filter, we understand a photonic subsystem designed with the aim of carrying equivalent tasks to those of an ordinary microwave filter within a radio frequency (RF) system or link, bringing supplementary advantages inherent to photonics such as low loss, high bandwidth, immunity to electromagnetic interference (EMI), tunability, and reconfigurability.

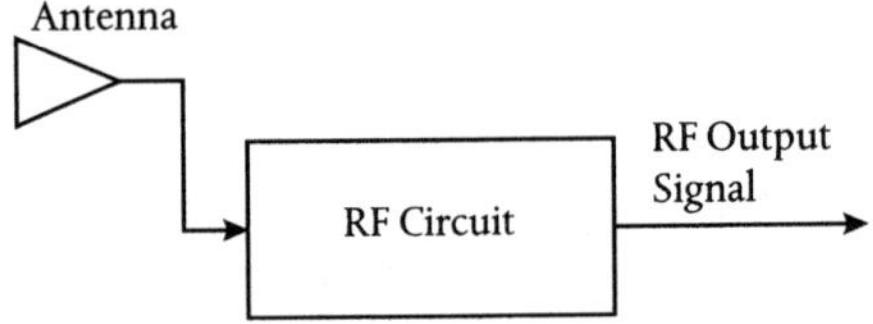

Fig. P2 Traditional configuration

The general concept behind microwave photonic filters is to replace the traditional approach (as shown in Fig. P2) toward RF-signal processing, where an RF signal originating at an RF source or coming from an antenna is fed to an RF circuit that performs the signal processing tasks (usually at an intermediate frequency band after a down conversion operation) by a novel technique. In this approach, the RF signal that was previously made to modulate an optical carrier that is directly processed in the optical domain by a photonic filter based on fibre and integrated photonic devices and circuits.

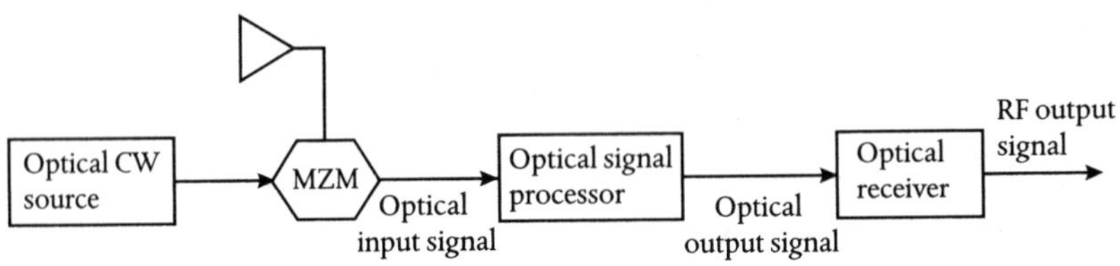

Fig. P3 Replacement of RF filter by a microwave photonic filter

The addition of extra photonic components (as shown in Fig. P3) implies increased filter complexity on one hand, and on the other hand, they bring other several advantages: Optical delay lines have very low loss (independent of the RF-signal frequency), provide very high time bandwidth products, are immune to EMI, are lightweight, and can provide very short delays that result in very high-speed sampling frequencies (over 100 GHz in comparison with a few gigahertz with the available electronic technology). It provides the possibility of spatial and wavelength parallelism using wavelength division multiplexing (WDM) techniques.

Question 3: What do you understand by optical heterodyning method of generating microwave?

Answer: A microwave or mm-wave signal can be generated in optical domain based on optical heterodyning in which two optical waves of different wavelengths beat at a photodetector.

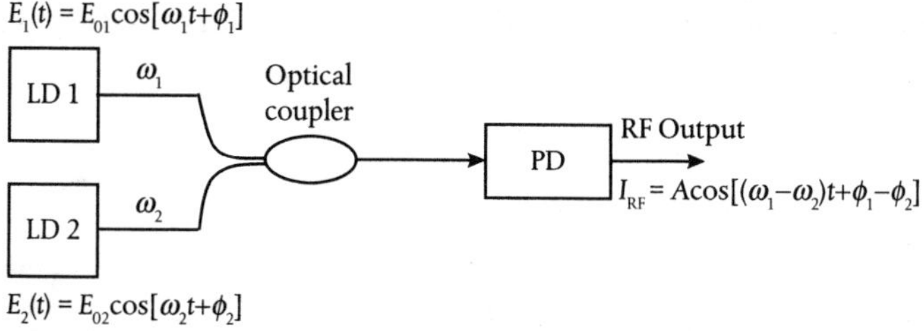

Fig. P4 Beating of two optical waves at a PD for the generation of a microwave signal

An electrical beat note is then generated at the output of the photodetector with a frequency corresponding to the wavelength spacing of the two optical waves, as shown in Fig. P4.

Let us assume we have two optical waves, which are given as:

$$E_1(t) = E_{01}\cos(\omega_1 t + \phi_1)$$

and

$$E_2(t) = E_{02}\cos(\omega_2 t + \phi_2),$$

where E_{01}, E_{02} are the amplitude terms and ϕ_1, ϕ_2 are the phase terms of the two optical waves. The signal at the output of the PD is given as:

$$I_{PD}(t) \propto R_{PD} 2\left[E_{01}\cos(\omega_1 t+\phi_1) + E_{02}\cos(\omega_2 t+\phi_2)\right]^2$$

$$I_{PD}(t) \propto R_{PD}\left[2E_{01}^2\cos^2(\omega_1 t+\phi_1) + 2E_{02}^2\cos^2(\omega_2 t+\phi_2) + 2E_{01}.E_{02}.2\cos(\omega_1 t+\phi_1)\cos(\omega_2 t+\phi_2)\right].$$

$$I_{PD}(t) \propto R_{PD}\left\{E_{01}^2\left[1+\cos 2(\omega_1 t+\phi_1)\right] + E_{02}^2[1+\cos 2(\omega_2 t+\phi_2)] + 2E_{01}.E_{02}.\left[\cos(\omega_1 t+\omega_2 t+\phi_1+\phi_2) + \cos(\omega_1 t-\omega_2 t+\phi_1-\phi_2)\right]\right\}.$$

$$I_{PD}(t) \propto R_{PD}\left\{E_{01}^2 + E_{02}^2 + 2E_{01}.E_{02}.\cos\left[(\omega_1-\omega_2)t+\phi_1-\phi_2\right]\right\},$$

where R is the responsivity of the PD. The above-mentioned calculation is referred to a coherent beating of the input optical signals. Terms $2(\omega_2 t+\phi_2)$, $2(\omega_2 t+\phi_2)$ and $\omega_1 t+\omega_2 t+\phi_1+\phi_2$ disappear due to averaging.

Considering the limited bandwidth of the PD, the current at the output of the PD is given as:

$$I_{RF} = 2R\sqrt{P_1 P_2}\cos\left[(\omega_1-\omega_2)t+(\phi_1-\phi_2)\right]$$

Question 4: What are the applications of microwave photonic filters?

Answer:

(1) A typical application of microwave photonic filter is in a moving target identification (MTI) ground radar system. The MTI radar uses the Doppler effect to separate the targets of interest from clutter (land, sea water, rain, etc.). To do this, the radar sends a pulse sequence with pulse width τ and inter-pulse period $PRI = 1/PRF$, where PRF identifies the pulse repetition frequency. Any moving object will generate a Doppler frequency shift Δv of the radar central frequency f_0 according to its speed

(dR/dt), where $R(t)$ designates the time varying distance from the target to the radar.

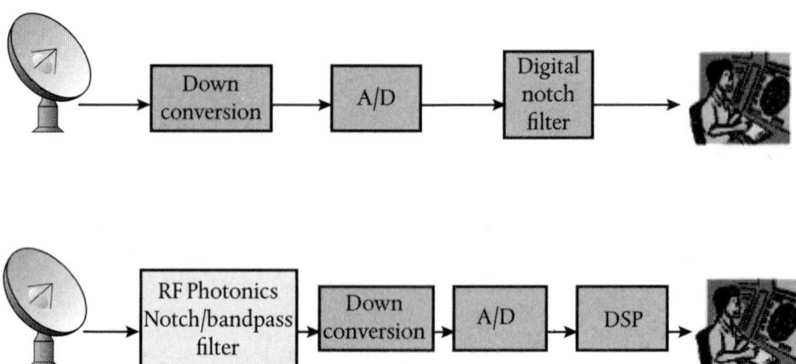

Fig. P5 Typical signal processing configuration in an MTI radar system. The second part of the figure shows the modified version including a microwave photonic filter prior to down conversion

The spectral signature of each object repeats in the spectrum periodically with a period given by the PRF, which obviously sets the limit on determining an unambiguous Doppler shift. This is usually performed as shown in the upper part of Fig. P5 by using a digital notch filter placed after frequency down conversion to baseband and using analogue to digital conversion (ADC). In order to distinguish the small echo from the target and the large echo from the fixed objects, high performance (14- to 18-bit resolution) ADCs are required, which represents a major bottleneck in the system. If the clutter or noise can be removed before down conversion, then the high-resolution requirements on the ADCs can be relaxed. For example, with 30-dB clutter attenuation, the required ADC resolution is reduced by 5 bits. This operation is difficult and costly in the microwave domain, but it is simple if the RF signal is modulated onto an optical carrier and the whole signal is processed directly in the optical domain by means of a photonic filter, as shown in the lower part of Fig. P5.

(2) In the emerging broadband wireless access networks and standards spanning from universal mobile telecommunications system (UMTS) to fixed access pico-cellular networks and including wireless local area network (WLAN), World Interoperability for Microwave Access, Inc. (WIMAX), local multipoint distribution service (LMDS) etc., there is a need to increase the capacity by reducing the coverage area. An enabling technology to obtain this objective is radio over fibre (RoF) systems, where radio signals are distributed from a central location to remote antenna units (RAUs) using fibre optic transmission as shown in the upper part of Fig. P6.

RoF makes it possible to centralise the RF-signal processing functions in one shared location (headend). By doing so, RAUs are simplified significantly as they only need to perform optoelectronic conversion and amplification functions. The centralisation of RF-signal processing functions enable equipment sharing, dynamic allocation of resources, and simplified system operation and maintenance. The processing at the head end involves a prior frequency down conversion, ADC, and baseband processing using a DSP as shown in the intermediate part of Fig. P6. This illustrates a direct fibre link joining a given RAU and the head end. The down conversion operation can be eliminated or divided into two steps, putting less stringent requirements on the ADC and DSP operations if a microwave photonic filter is placed prior to optical detection as shown in the lower part of Fig. P6.

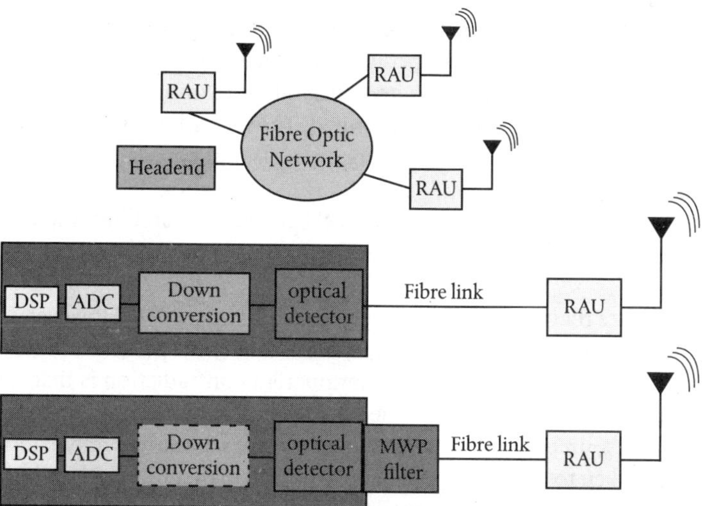

Fig. P6 RoF access network (upper). Potential application of microwave photonic filters at the head end on the centralised station (lower) replacing the RF filter of standard configuration (intermediate)

The microwave photonic filter can be employed either for channel rejection or for channel selection applications. In the first case, we deal with an optical link where not only the desired signal is carried by the fibre but also unwanted interfering signals that are also chosen by the antenna. A paradigmatic example can be found in radio astronomy applications, where signal transmission from several stations to a central site requires removing strong manmade interfering signals from astronomy bands. The ability to reject these interfering RF signals directly in the optical domain is a unique characteristic of these photonic filters.

(3) Another important application of microwave photonic filters is in the field of true time delay beam steering of antenna arrays. A photonic true-time-delay system for feeding an array of antennas is based on the use of broadband photonic delay lines. The feeder network for an array of N antennas is essentially equivalent to an N tap microwave photonic tunable FIR filter where the basic filter delay T can be altered; the only difference is that each filter sample is detected by a different optical receiver that is placed before each antenna unit in the array.

Question 5: What are the limitations and performance parameters of microwave photonic filters?

Answer: Microwave photonic filters must overcome a series of potential limitations prior to their practical realisation. We classify these limitations into two groups according to whether these limitations appear mainly in the optical domain or whether they manifest in the electrical domain.

1. **Optical Sources of Performance Limitation**
   Limitations arising in the optical domain include nonlinear optical effects, polarisation, and positive nature of the filter coefficients due to the incoherent operation, the limited range of attainable spectral periods, spectral periodicity, filter re-configurability, and tunability.
2. **Spectral Periodicity**
   The spectral periodicity of microwave photonic filters limits the bandwidth of the RF signals to

be processed to a fraction of the FSR in order to avoid spectral overlapping. Single resonance (i.e., non-periodic filters) is, therefore, desired for certain applications.

3. **Positive Coefficients**
   Filters working under the incoherent regime are linear in optical intensity, thus the coefficients of their impulse responses are always positive. This has two important implications as derived from the theory of positive systems. The first one and more important is that the range of transfer functions that can be implemented is quite limited. The second one is that regardless of its spectral period, the transfer function always has a resonance place at baseband.
4. **Fibre Nonlinearities**
   Filter linearity can be compromised if the optical carriers used in filter implementation to deliver enough power to stimulate fibre nonlinearities.
5. **Polarisation**
   Polarisation effects are mainly important under coherent operation. However, it has been outlined and experimentally demonstrated that even under incoherent operation, the filter can be sensitive to signal polarisation. The main cause for this apparent contradiction is that some signal samples experience exactly the same delay within the filter leading to a coherent interference between them even if a broadband source is employed. Further, when laser sources and external modulators are used, care must be taken to adjust the source polarisation to that required by the modulator. The use of polarisation preserving fibre pigtails at the modulator input helps to overcome this limitation.
6. **Re-configurability**
   This property refers to the possibility to dynamically change the values of operating wavelength. Passive structures are incapable of this feature. Several solutions have been proposed to overcome this limitation including the use of optical amplifiers (OAs), modulators, fibre gratings, and laser arrays.
7. **Tunability**
   This property refers to the possibility to dynamically change the position of filter resonances or notches. To provide tunability, it is necessary to alter the value of the sampling period T. Solutions that include the use of switched fibre delay lines, high dispersion fibres, and FBGs have been proposed.

Question 6: How to generate quadrupled frequency millimetre wave using external modulation technique?

Answer: An electrical drive signal is applied to an optical intensity modulator, which is biased to completely suppress the even-order optical sidebands including the carrier. Thus, two coherent carriers are obtained.

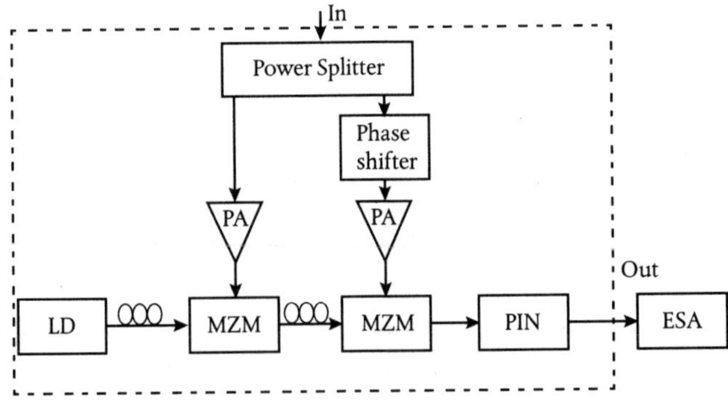

Fig. P7 Diagram of mm-wave generation. (ESA: Electrical spectrum analyser, LD: Laser diode, MZM: optical modulator, PA: Power amplifier, PC: polarisation controller, PIN: p-i-n diode, PS: phase shifter)

They are then modulated by the same microwave signal with phase deviation of ϕ introduced by a phase shifter in the next optical intensity modulator, which is also biased to completely suppress the even-order optical sidebands including the carrier (Fig. P7). As a result, four optical components are obtained. Two of them have the frequency difference with four times the frequency of the drive signal, while the other two have the same frequency but with a phase difference of 2ϕ. Therefore, the two components with same frequency will be counteracted when the phase deviation of ϕ is $\pi/2$.

From (8.50),

$$E_2(t) \propto E_0 J_1(\beta_1) J_1(\beta_2) \times \left\{\cos\left((\omega_0 - 2\omega_m)t - \phi_1 - \phi_2 + \varphi\right) + \cos\left((\omega_0 + 2\omega_m)t + \phi_1 + \phi_2 + 2\varphi\right)\right\}.$$

Therefore, only two optical sidebands separated by four times the frequency of the microwave signal are present. Then, a beat signal with four times the frequency of the electrical drive signal is generated at the output of a p-i-n diode.

Question 7: How to generate octupled frequency millimetre wave using external modulation technique?

Answer: The proposed system is shown in Fig. P8, which consists of a continuous-wave (CW) laser source, two MZMs that are connected by a tunable optical shifter (TOPS), and a PD. A low-frequency microwave signal from a microwave source is amplified by a microwave power amplifier (PA), and then divided into two signals by a power divider and applied to the two MZMs. Both of the MZMs are biased at the MATP to suppress the odd-order sidebands. By properly adjusting the phase shift introduced by the TOPS and the powers of the drive signals, only the 4th-order sidebands are generated. By beating the ± 4th-order sidebands at the PD, a frequency-octupled microwave signal is generated.

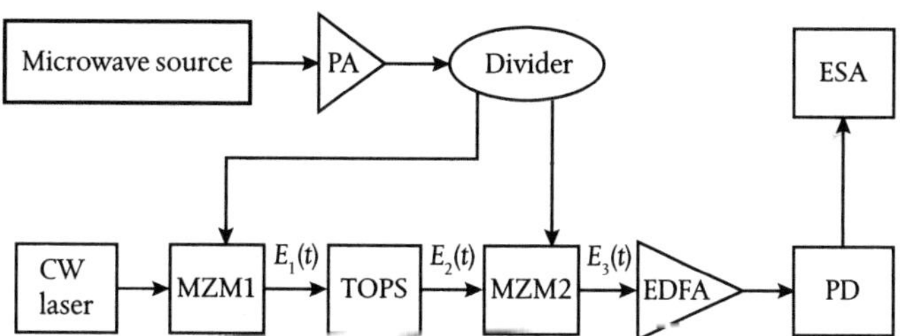

Fig. P8 Schematic diagram of the proposed microwave frequency octupling system using two cascaded MZMs (EDFA: erbium-doped fibre amplifier; ESA: electrical spectrum analyser, TOPS: tunable optical phase shifter)

Mathematically calculation:

From (8.42), the electrical field at the output of the first MZM (MZM1) can be expressed mathematically as

$$E_1(t) \propto E_0 J_0(\beta_1)\cos(\omega_0 t) - E_0 J_2(\beta_1)\cos(\omega_0 t - 2\omega_m t - 2\phi_1) - E_0 J_2(\beta_1)\cos(\omega_0 t + 2\omega_m t + 2\phi_1),$$

where E_0 is the amplitude of the incident lightwave to each arm of the MZM, ω_0 and ω_m are, respectively, the angular frequencies of the lightwave and the input microwave signal, J_n is the nth-order Bessel function of the first kind, β_1 and ϕ_1 are, respectively, the phase modulation index (PMI) and the initial phase of the microwave signal applied to MZM1. When the lightwave is travelling through the

TOPS, a phase difference φ between two adjacent even-order side bands would be introduced and the electrical field $E_2(t)$ at the output of the TOPS can be written from (8.43) as:

$$E_2(t) \propto E_0 J_0(\beta_1)\cos(\omega_0 t+2\varphi) - E_0 J_2(\beta_1)\cos(\omega_0 t-2\omega_m t-2\phi_1+\varphi) - E_0 J_2(\beta_1)\cos(\omega_0 t+2\omega_m t+2\phi_1+3\varphi).$$

The optical signal is then sent to MZM2. The electrical field $E_3(t)$ at the output of MZM is given by (8.44):

$$\begin{aligned}E_3(t)=E_0\{&J_0(\beta_1)J_0(\beta_2)\cos(\omega_0 t+2\varphi)-J_0(\beta_1)J_2(\beta_2)\cos((\omega_0-2\omega_m)t-2\phi_2+2\varphi)-\\&J_0(\beta_1)J_2(\beta_2)\cos((\omega_0+2\omega_m)t+2\phi_2+2\varphi)-J_2(\beta_1)J_0(\beta_2)\cos((\omega_0-2\omega_m)t-2\phi_1+\varphi)+\\&J_2(\beta_1)J_2(\beta_2)\cos((\omega_0-4\omega_m)t-2\phi_1+\varphi-2\phi_2)+J_2(\beta_1)J_2(\beta_2)\cos(\omega_0 t-2\phi_1+\varphi+2\phi_2)-\\&J_2(\beta_1)J_0(\beta_2)\cos((\omega_0+2\omega_m)t+2\phi_1+3\varphi)+J_2(\beta_1)J_2(\beta_2)\cos(\omega_0 t+2\phi_1+3\varphi-2\phi_2)-\\&J_2(\beta_1)J_2(\beta_2)\cos((\omega_0+4\omega_m)t+2\phi_1+3\varphi+2\phi_2)\}.\end{aligned}$$

Let us consider the following two conditions:

1. $\beta_1=\beta_2=\beta$, we have $J_0(\beta_1)J_2(\beta_2)=J_0(\beta_2)J_2(\beta_1)$
2. $\varphi+2\phi_1-2\phi_2=(2k+1)\pi$, k is an integer,

$$E_3(t) \propto E_0\{J_2^2(\beta)\cos((\omega_0-4\omega_m)t+\varphi-2\phi_1-2\phi_2)+[J_0^2(\beta)-2J_2^2(\beta)]\cos(\omega_0 t+2\varphi)+ J_2^2(\beta)\cos((\omega_0+4\omega_m)t+3\varphi+2\phi_1+2\phi_2)\}.$$

Since the output from MZM2 consists of only the $\pm$ 4th-order sidebands, the beating of the two sidebands at the PD would generate a frequency-octupled microwave signal.

Question 8: What is the condition for generating only two optical sidebands using two MZMs biased at different modes?

Answer: The conditions for generating only two optical sidebands can be summarised as given in Table P1.

Table P1 Condition for generating only two optical sidebands where maximum transmission point (MATP) means MZM arm is biased at $\phi=0$, (and minimum transmission point (MITP) means MZM arm biased at $\phi=\pi$.

Bias modes	Phase modulation index	Phase	Frequency multiplication factor
MATP + MATP	$\beta_1=\beta_1=\beta=1.6995$	$\varphi+2\phi_1-2\phi_2=(2k+1)\pi$	8
MITP + MITP	NONE	$\varphi+2\phi_1-2\phi_2=(2k+1)\pi$	4
MATP + MITP	$J_0(\beta_1)=J_2(\beta_1)$	$\varphi+2\phi_1-2\phi_2=2k\pi$	6
MITP + MATP	$J_0(\beta_2)=J_2(\beta_2)$	$\varphi+2\phi_1-2\phi_2=2k\pi$	6

Question 9: Show that the spectrum of a microwave photonic filter is periodic.

Answer:

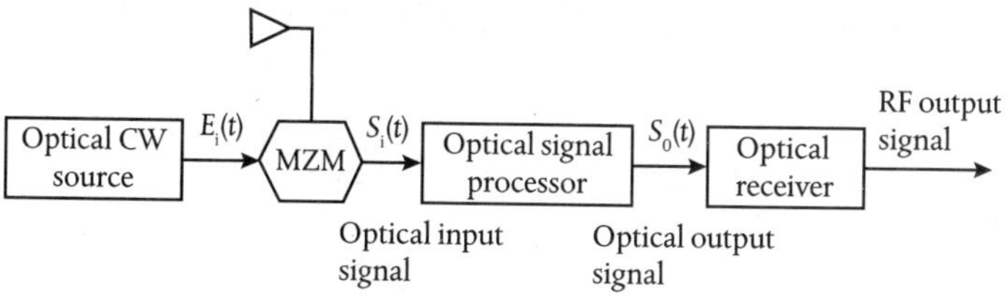

Fig. P9 Layout of microwave photonic filter

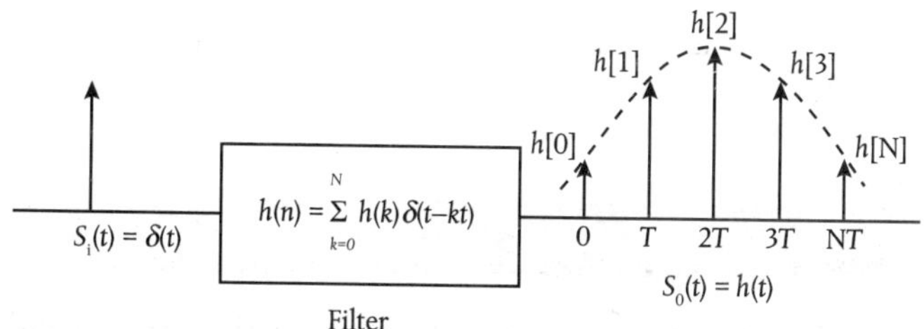

Fig. P10 Black-box representation of microwave photonic filter

Linear relationship between input optical field and output optical field $E_0(t)$ to the optical subsystem is established through the optical field transfer function $H_0(\omega)$, and hence,

$$\tilde{E}_0(\omega) = \tilde{E}_i(\omega) H_0(\omega)$$

Let for linear operation:

$$s_0(t) = \sum_{n=-N}^{N} a_n s_i(t - nT)$$

$$s_0(t) - s_i(t) * h(t),$$

where

$$h(t) = \sum_{n=-N}^{N} a_n \delta(t - nT) \tag{P1}$$

According to the number of samples N in the impulse response sequence, the filter can be classified as either a finite impulse response (FIR) filter or infinite impulse response (IIR) filter:

FIR filter if $N < \infty$ and IIR filter if $N = \infty$

From (P1), *h*(*t*) can be regarded as discrete-time signal, and thus, usual *z*-transform and discrete-time Fourier (DTF) can be fully employed for its analysis:

Therefore,

$$H(z) = \sum_{n=-\infty}^{n=\infty} h(n) z^{-n} \text{ and } H(\Omega) = \sum_{n=-\infty}^{n=\infty} h(n) e^{-jn\Omega T} \tag{P2}$$

Operation of microwave photonic filter can be described in terms of a system difference equation:

$$s_0(t - nT) = b_0 s_i(t) + b_1 s_i(t - T) + \ldots + b_m s_i(t - MT) - c_1 s_0(t - T) - \ldots - c_N s_0(t - NT).$$

Taking Z-transform on both sides

$$s_0(z)z^{-n} = b_0 s_i(z) + b_1 s_i(z)z^{-1} + \ldots + b_m s_i(z)e^{-MT} - c_1 s_0(z)z^{-n} - \ldots - c_N s_0(z)z^{-N}$$

$$s_0(z)z^{-n} + c_1 s_0(z)z^{-1} + \ldots + c_N s_0(z)z^{-N} = b_0 s_i(z) + b_1 s_i(z)z^{-1} + \ldots + b_m s_i(z)e^{-MT}.$$

In summation form:

$$\left[1 + \sum_{n=1}^{N} c_N z^{-n}\right] s_0(z) = s_i(z) \sum_{m=0}^{M} b_m z^{-m}$$

$$\frac{s_0(z)}{s_i(z)} = \frac{\sum_{m=0}^{M} b_m z^{-m}}{\left[1 + \sum_{n=1}^{N} c_N z^{-n}\right]}.$$

The location of the filter zeros and poles depend on the values of the filter coefficients b_m and c_N. Equation (P9.2) reveals that it is spectrally periodic with a period given by $2\pi/T$ in angular frequency units. This period is known as the filter-free spectral range (FSR).

MULTIPLE CHOICE QUESTIONS

1. ....is the progressive decrease of signal strength with increasing distance.
   (a) Radiation
   (b) Attenuation
   (c) Modulation
   (d) Propagation
2. Calculate the effective earth's radius if the surface refractivity is 301.
   (a) 8493 km
   (b) 8491 km
   (c) 6370 km
   (d) 6371 km
3. If *k*-factor is greater than 1, the array beam is bent
   (a) Away from the earth
   (b) Towards the ionosphere
   (c) Towards the earth
   (d) Towards the outer space
4. Rainfall is an important factor for fading of radio waves at frequencies above
   (a) 10 GHz
   (b) 100 GHz
   (c) 1 MHz
   (d) 1 GHz
5. What do you call an attenuation that occurs over many different wavelengths of the carrier?
   (a) Rayleigh Fading
   (b) Rician Fading
   (c) Wavelength Fading
   (d) Slow Fading
6. In microwave transmission using digital radio, what causes most inter-symbol interference?
   (a) Delayed Spreading
   (b) Rayleigh Fading

(c) Random Doppler Fading
(d) Slow Fading

7. At what position, is the input signal inserted into a travelling-wave tube?
   (a) At the cathode end of Helix
   (b) At the collector
   (c) At the collector end of Helix
   (d) At the control grid of electron gun
8. A high-power microwave pulse of the order of megawatts can be generated by a
   (a) Travelling-wave tube
   (b) Magnetron
   (c) Reflex klystron
   (d) Gunn diode
9. Microwave frequencies are normally regarded as those in the range of
   (a) 1 to 500 MHz
   (b) 1000 to 10,000 GHz
   (c) 1 to 100 GHz
   (d) 10 to 1000 GHz
10. What ferrite device can be used instead of duplexer of isolate microwave transmitter and receiver when both are connected to the same antenna?
   (a) Isolator
   (b) Magnetron
   (c) Simplex
   (d) Circulator

Answers to Multiple Choice Questions

1	(b)	6	(a)
2	(a)	7	(c)
3	(c)	8	(b)
4	(a)	9	(c)
5	(d)	10	(d)

QUESTIONS

1. What is microwave photonics?
2. What is the need of microwave photonics?
3. Explain the block diagram of a basic microwave photonics link.
4. Draw and explain the microwave photonic link architectures for direct and external modulation, and for direct and coherent detection.
5. Describe neatly with diagram the device technology based on microwave photonics.
6. Discuss the various applications of microwave photonics.
7. Explain microwave photonics filter.
8. Briefly explain the following microwave photonic components:
   (a) Electro-absorption Modulator
   (b) Photodetector and Mixer

   (c) Electro-absorption Transceiver
   (d) Microwave laser diodes
9. Explain briefly photonic microwave phase shifter/time delay control.
10. Draw a neat diagram of broadband fibre optic links and explain it in detail.
11. Explain microwave photonic systems in brief:
   (a) Photonic Local Oscillators
   (b) EMC sensor
   (c) Hybrid Fibre Coax (HFC) Systems
   (d) Fibre-radio systems

REFERENCES

1. Seeds A J, Lee C H, Funk E and Nagamura M, Guest editorial: Microwave photonics, *IEEE Photon. Technol. Lett.*, vol. 21, no. 12, pp. 2959–2961, 2003.
2. Seeds A J, Microwave photonics, *IEEE Trans. Microwave Theory Tech.*, vol. 50, no. 3, pp. 877–887, Mar. 2002.
3. Seeds A J and Williams K, Microwave photonics, *J. Lightwave Technol.*, vol. 24, no. 12, pp. 4628–4641, Dec. 2006.
4. Capmany J and Novak D, Microwave photonics combines two worlds, *Nat. Photon.*, vol.1, pp. 319–330, Sep. 2007.
5. Goldberg L, Taylor H F, Weller J F and Bloom D M, Microwave signal generation with injection locked laser diodes, *Electron. Lett.*, vol. 19, no. 13, pp. 491–493, Jun. 1983.
6. Goldberg L, Yurek A, Taylor H F and Weller J F, 35 GHz microwave signal generation with injection locked laser diode, *Electron Lett.*, vol. 21, no. 18, pp. 714–715, Aug. 1985.
7. Gliese U, Nielsen T N, Bruun M, Lintz Christensen E, Stubkjaer K E, Lindgren S and Broberg B, A wideband heterodyne optical phase-locked loop for generation of 3–18 GHz microwave carriers, *IEEE Photon. Technol. Lett.*, vol. 4, no. 8, pp. 936–938, Aug. 1992.
8. Yao X S and Maleki L, Multiloop optoelectronic oscillator, *IEEE J. Quan. Electron.*, vol. 36, no. 1, pp. 79–84, Jan. 2000.
9. Braun R P, Grosskopf G, Rohde D and Schmidt F, Optical millimeter-wave generation and transmission experiments for mobile 60 GHz band communications, *Electron. Lett.*, vol. 32, no. 7, pp. 626–628, 1996.
10. Li W and Yao J P, Microwave and terahertz generation based on photonically assisted microwave frequency twelvetupling with large tunability, *IEEE Photon. J.*, vol. 2, no. 6, pp. 954–959, Dec. 2010.
11. Li W and Yao J P, "Microwave frequency multiplication using two cascaded Mach-Zehnder modulators," *Proc. 2009 Asia-Pacific Microwave Photonics Conf.*, Beijing, China, Apr. 2009.
12. Li W and Yao J P, Investigation of photonically assisted microwave frequency multiplication based on external modulation, *IEEE Trans. Microw. Theory Tech.*, vol. 58, no. 11, pp. 3259–3268, Nov. 2010.
13. Li W and Yao J P, Microwave generation based on optical-domain microwave frequency octupling, *IEEE Photon. Technol. Lett.* vol. 22, no. 1, pp. 24–26, Jan. 2010.
14. Lin C, Shih P, Jiang W, Chen J, Peng P and Chi S, A continuously tunable and filter less optical millimeter wave generation via frequency octupling, *Opt. Express*, vol. 17, no. 22, pp. 19749–19756, Oct. 2009.
15. Ma J, Xin X, Yu J, Yu C, Wang K, Huang H and Rao L, Optical millimeter wave generated by octupling the frequency of the local oscillator, *J. Opt. Netw.* vol. 7, no. 10, pp. 837–845, Oct. 2008.
16. Zhang J, Chen H, Chen M, Wang T and Xie S, Photonic generation of a millimeter-wave signal

based on sextuple-frequency multiplication, *Opt. Lett.*, vol. 32, no. 9, pp. 1020–1022, Apr. 2007.

17. Fan Z F and Dagenais M, Optical Generation of a mHz-Linewidth microwave signal using semiconductor lasers and a discriminator-aided phase-locked loop, *IEEE Trans. Microwave Theory Tech.*, vol. 45, no. 8, pp. 1296–1300, Aug. 1997.
18. Mohamed M, Zhang X, Hraimel B and Wu K, Frequency sextupler for millimeter-wave over fiber systems, *Opt. Express*, vol. 16, no. 14, pp. 10141–10151, Jun. 2008.
19. Goldberg L, Taylor H F, Weller J F and Bloom D M, Microwave signal generation with injection-locked laser diodes, *Electron. Lett.*, vol. 19, no. 13, pp. 491–493, 1983.
20. Zhao Y, Zheng X, Wen H and Zhang H, Simplified optical millimeter-wave generation configuration by frequency quadrupling using two cascaded Mach–Zehnder modulators, *Opt. Lett.*, vol. 34, no. 21, pp. 3250–3252, Nov. 2009.
21. Jung T., Shen J -L, Tong D T K, Murthy S, Wu M C, Tanbun-Ek T, Wang W, Lodenkamper R, Davis R, Lembo L J and Brock J C, CW injection locking of a mode-locked semiconductor laser as a local oscillator comb for channelizing broad-band RF signals, *IEEE Trans. Microw. Theory Tech.*, vol. 47, no. 7, pp. 1225–1232, Jul. 1999.
22. Nöel L, Wake D, Moodie D G, Marcenac D D, Westbrook L D and Nesset D, Novel techniques for high capacity 60 GHz fiber-radio transmission systems, *IEEE Trans. Microw. Theory Tech.*, vol. 45, no. 8, pp. 1146–1423, Aug. 1997.
23. Ogawa H, Polifko D and Bamba S, Millimeter-wave fiber optics systems for personal radio communications, *IEEE Trans. Microw. Theory Tech.*, vol. 40, no. 12, pp. 2285–2293, Dec. 1992.
24. Qi G, Yao J P, Seregelyi J, Bélisle C and Paquet S, Generation and distribution of a wide-band continuously tunable mm-wave signal with an optical external modulation technique, *IEEE Trans. Microw. Theory Tech.*, vol. 53, no. 10, pp. 3090–3097, Oct. 2005.
25. Zhang J, Chen H, Chen M, Wang T, Xie S, A photonic microwave frequency quadrupler using two cascaded intensity modulators with repetitious optical carrier suppression, *IEEE Photon. Technol. Lett.*, vol. 19, no. 14, pp.1057–1059, Jul. 2007.
26. Kawanishi T, Kiuchi H, Yamada M, Sakamoto T, Tsuchiya M, Amagai J and Izutsu M, Quadruple frequency double sideband carrier suppressed modulation using high extinction ratio optical modulators for photonic local oscillators, in *Proc. IEEE Microwave Photonics Conf.*, pp. 1–4, Oct. 2005.
27. O'Reilly J J, Lane P M, Heidemann R and Hofstetter R, Optical generation of very narrow linewidth millimeter wave signals, *Electron. Lett.*, vol. 28, no. 25, pp. 2309–2311, 1992.
28. O'Reilly J J and Lane P M, Remote delivery of video services using mm-wave and optics, *J. Lightw. Technol.*, vol. 12, no. 2, pp. 369–375, Feb. 1994.

9 Application of Fibre Optics

Objectives

- This chapter probes the various applications of fibre optics.
- Use of optical fibre in telecommunication and network analysis is explored.
- Broadband networking architecture for optical fibre is reviewed.
- The application of optical fibre in medical science and engineering are emphasised.

9.1 INTRODUCTION

Optical fibres are widely used in fibre optic communications, which permits transmission over longer distances and at higher bandwidths (data rates) than other forms of communication. Fibres are used instead of metal wires because signals travel along them with less loss and are also immune to electromagnetic interference. Fibres are also used for illumination and are wrapped in bundles so they can be used to carry images, thus allowing viewing in tight spaces. Specially designed fibres are used for a variety of other applications including sensors and fibre lasers [1-10].

Optical fibre typically consists of a transparent core surrounded by a transparent cladding material with a lower index of refraction [11-19]. Light is kept in the core by total internal reflection. This causes the fibre to act as a waveguide. Fibres that support many propagation paths or transverse modes are called multimode fibres (MMF), while those that can only support a single mode are called single-mode fibres (SMF). Multimode fibres generally have a larger core diameter and are used for short-distance communication links and for applications where high power must be transmitted. Single-mode fibres are used for most communication links longer than 1050 m.

Joining lengths of optical fibre is more complex than joining electrical wires or cables. The ends of the fibres must be carefully cleaved and then spliced together either mechanically or by fusing them together with heat. Special optical fibre connectors are used to make removable connections.

9.1.1 Applications of Optical Fibre Communication

Optical fibre can be used as a medium for telecommunication and networking because it is flexible and can be bundled as cables [20-27]. It is especially advantageous for long distance communications because light propagates through the fibre with little attenuation when compared to electrical cables. This allows long distances to be spanned with few repeaters. Additionally, the per-channel light signals propagating in the fibre have been modulated at rates as high as 111 Gbps by NTT, although 10 or 40 Gbit/s is typical in deployed systems. Each fibre can carry many independent channels each using a different wavelength of light (wavelength-division multiplexing (WDM)). The net data rate (data rate without overhead bytes) per fibre is the per-channel data rate reduced by the FEC overhead, multiplied by the number of channels (usually up to 80 in commercial dense WDM systems as of 2008). The current laboratory fibre optic data rate record, held by Bell Labs in Villarceaux, France, is multiplexing 155 channels, each carrying 100 Gbit/s over a 7000-km fibre. Nippon Telegraph and Telephone (NTT) Corporation have also managed 69.1 Tbit/s over a single 240-km fibre (multiplexing 432 channels, equating to 171 Gbit/s per channel).

For short-distance applications like creating a network within an office building, fibre optic cabling can be used to save space in cable ducts. This is because a single fibre can often carry much more data than many electrical cables, for example, 4 pair Cat-5 Ethernet cabling. Fibre is also immune to electrical interference; there is no cross-talk between signals in different cables and no pickup of environmental noise. Non-armoured fibre cables do not conduct electricity, which makes fibre a good solution for protecting communications equipment located in high-voltage environments such as power generation facilities or metal communication structures prone to lightning strikes. They can also be used in environments where explosive fumes are present without danger of ignition. Wiretapping is more difficult when compared to electrical connections, and there are concentric dual-core fibres that are said to be tap proof.

9.1.2 Fibre Optic Sensors

Fibres have many uses in remote sensing. In some applications, the sensor is itself an optical fibre. In other cases, fibre is used to connect a non-fibre optic sensor to a measurement system [28-37]. Depending on the application, fibre may be used because of its small size, or the fact that no electrical power is needed at the remote location, or because many sensors can be multiplexed along the length of a fibre by using different wavelengths of light for each sensor, or by sensing the time delay as light passes along the fibre through each sensor. Time delay can be determined using a device like an optical time-domain reflectometer.

Optical fibres can be used as sensors to measure strain, temperature, pressure, and other quantities by modifying a fibre so that the quantity to be measured modulates the intensity, phase, polarisation, wavelength, or transit time of light in the fibre. Sensors that vary the intensity of light are the simplest, since only a simple source and detector are required. A particularly useful feature of such fibre optic sensors is that they can, if required, provide distributed sensing over distances of up to one meter.

Extrinsic fibre optic sensors use an optical fibre cable, normally a multimode one, to transmit modulated light from either a non-fibre optical sensor, or an electronic sensor connected to an optical transmitter. A major benefit of extrinsic sensors is their ability to reach places that are otherwise inaccessible. An example is the measurement of temperature inside aircraft jet engines by using a fibre to transmit radiation into a radiation pyrometer located outside the engine. Extrinsic sensors can also be used in the same way to measure the internal temperature of electrical transformers, where the extreme electromagnetic fields present make other measurement techniques impossible. Extrinsic sensors are used to measure vibration, rotation, displacement, velocity, acceleration, torque, and twisting. A solid-state version of the gyroscope using the interference of light has been developed. The fibre optic gyroscope (FOG) has no moving parts and exploits the Sagnac effect to detect mechanical rotation.

A common use for fibre optic sensors are in advanced intrusion detection security systems, where the light is transmitted along the fibre optic sensor cable, which is placed on a fence, pipeline, or communication cabling and the returned signal is monitored and analysed for disturbances. This return signal is digitally processed to identify if there is a disturbance and if an intrusion has occurred, an alarm is triggered by the fibre optic security system.

9.1.3 Other Uses of Optical Fibres

Fibres are widely used in illumination applications. They are used as light guides in medical and other applications where bright light needs to be shown on a target without a clear line of sight path. In some buildings, optical fibres are used to route sunlight from the roof to other parts of the building [38-42]. Optical fibre illumination is also used for decorative applications, including signs, arts, and artificial Christmas trees. Optical fibre is also used in imaging optics. A coherent bundle of fibres is used, sometimes along with lenses, for a long, thin imaging device called an endoscope, which is used to view objects through a small hole. Medical endoscopes are used for minimally invasive exploratory or surgical procedures (endoscopy). Industrial endoscopes (see fiberscope or borescope) are used for inspecting anything hard to reach like jet engine interiors.

In spectroscopy, optical fibre bundles are used to transmit light from a spectrometer to a substance that cannot be placed inside the spectrometer itself in order to analyse its composition. A spectrometer analyses substances by bouncing light off and through them. By using fibres, a spectrometer can be used to study objects that are too large to fit inside or gasses or reactions that occur in pressure vessels.

An optical fibre doped with certain rare earth elements like erbium can be used as the gain medium of a laser or optical amplifier. Rare earth-doped optical fibres can be used to provide signal amplification by splicing a short section of doped fibre into a regular (un-doped) optical fibre line. The doped fibre is optically pumped with a second laser wavelength that is coupled into the line in addition to the signal wave. Both wavelengths of light are transmitted through the doped fibre, which transfers energy from the second pump wavelength to the signal wave. The process that causes the amplification is stimulated emission. Optical fibres doped with a wavelength shifter are used to collect scintillation light in physics experiments.

Optical fibre can be used to supply a low level of power (around one watt) to electronics situated in a difficult electrical environment. Examples of this are electronics in high-powered antenna elements and measurement devices used in high-voltage transmission equipment. A growing trend in iron sights for arms is the use of short pieces of optical fibre for contrast enhancement dots, made in such a way that ambient light falling on the length of the fibre is concentrated at the tip, making the dots slightly brighter than the surroundings. This method is most commonly used in front sights, but many makers offer sights that use fibre optics on front and rear sights. Fibre optic sights can now be found on handguns, rifles, and shotguns; further, they are used both as aftermarket accessories and a growing number of factory guns.

9.1.4 Special Purpose Fibre

Some special purpose optical fibre is constructed with a non-cylindrical core and/or cladding layer usually with an elliptical or rectangular cross section. These include polarisation maintaining fibre and fibre designed to suppress whispering gallery mode propagation.

Photonic crystal fibre is made with a regular pattern of index variation (often in the form of cylindrical holes that run along the length of the fibre). Such fibre uses diffraction effects instead of or in addition to total internal reflection to confine light to the fibre core. The properties of the fibre can be tailored to a wide variety of applications.

9.2 LAST MILE

Although fibre optic systems excel in high bandwidth applications, optical fibre has been slow to achieve its goal of fibre to the premises (FTTP) or to solve the last mile problem. However, as bandwidth demand increases, more and more progress towards this goal can be observed. In Japan, for instance, EPON has largely replaced DSL as a broadband internet source. South Korea's KT also provides a service called Fibre To the Home (FTTH), which provides 100% fibre optic connections to the subscriber's home. The largest FTTH deployments are in Japan, Korea, and most recently in China.

In the US, Verizon Communications provides an FTTH service called FiOS to select high Average Revenue Per User (ARPU) markets within its existing territory. The other major serving Incumbent Local Exchange Carrier (ILEC), AT&T, uses a Fibre To The Node (FTTN) service called U-verse with twisted pair to the home. Their MSO competitors employ FTTN with coax using HFC. All of the major access networks use fibre for the bulk of the distance from the service provider's network to the customer.

The globally dominant access network technology is Ethernet Passive Optical Network (EPON). In Europe, and among telcos in the United States, ATM-based Broadband PON (BPON) and Gigabit PON (GPON) are favoured because of their roots in the Full-Service Access Network (FSAN) and ITU-T standards organisations under their control.

9.2.1 Fibre to the X

Fibre to the X (**FTTX**) is a generic term for any broadband network architecture that uses optical fibre to replace all or part of the usual metal local loop used for last-mile telecommunications. The generic term originated as a generalisation of several configurations of fibre deployment (FTTN, FTTC, FTTB, FTTH, and so on); all starting by FTT but differentiated by the last letter, which is substituted by an X in the generalisation.

9.2.1.1 Definition of terms

The telecommunications industry differentiates between several distinct configurations. The terms in most widespread use today are as follows:

1. **FTTN**—*Fibre to the node*: fibre is terminated in a street cabinet up to several kilometres away from the customer premises with the final connection being copper.
2. **FTTC**—*Fibre to the cabinet or fibre to the curb*: This is very similar to FTTN, but the street cabinet is closer to the user's premises; typically within 300 m.
3. **FTTB**—*Fibre to the building or fibre to the basement*: Fibre reaches the boundary of the building, such as the basement in a multi-dwelling unit, with the final connection to the individual living space being made via alternative means.
4. **FTTH**—*Fibre to the home*: Fibre reaches the boundary of the living space like a box on the outside wall of a home.
5. **FTTP**—*Fibre to the premises*: This term is used in several contexts; as a blanket term for both FTTH and FTTB, or where the fibre network includes both homes and small businesses.

To promote consistency, especially when comparing FTTH penetration rates between countries, the three FTTH Councils of Europe, North America, and Asia Pacific have agreed upon definitions for FTTH and FTTB. The FTTH Councils do not have formal definitions for FTTC and FTTN.

It is worth pointing out that fibre to the telecom enclosure (FTTE) is not considered to be part of the FTTX group of technologies, despite the similarity in name. The FTTE is a form of structured cabling typically used in the enterprise local area network, where fibre is used to link the main computer equipment room to an enclosure close to the desk or workstation. Similarly, in FTTD, a fibre connection is installed from the main computer room to a terminal at the desk.

9.2.1.2 Benefits of fibre in the access network

The speeds of fibre optic and copper cables are both limited by length, but copper is much more sharply limited in this respect. For example, gigabit Ethernet runs over relatively economical category 5e, category 6, or augmented category 6 unshielded twisted pair copper cabling but only to 100 m. However, over the right kind of fibre, gigabit Ethernet can easily reach distances of tens of kilometres.

Even in the commercial world, most computers have copper communication cables. However, these cables are short, typically tens of meters. Most metropolitan network links (e.g., those based on telephone or cable television services) are several kilometres long in the range where fibre significantly outperforms copper. Replacing at least part of these links with fibre shortens the remaining copper segments and allows them to run much faster.

Fibre configurations that bring fibre right into the building can offer the highest speeds since the remaining segments can use standard Ethernet or coaxial cable. Fibre configurations that transition to copper in a street cabinet are generally too far from the users for standard Ethernet configurations over existing copper cabling. They generally use VDSL at (downstream) speeds of several tens of megabits per second.

Fibre is often said to be future proof because the speed of the broadband connection is usually limited by the terminal equipment rather than the fibre itself, permitting at least some speed improvements by equipment upgrades before the fibre itself must be upgraded. Still, the type and length of employed fibres chosen, for example, multimode vs single mode, are critical for applicability for future high-gigabit connections.

9.2.2 Fibre to the Node

Fibre to the node (FTTN) also called fibre to the neighbourhood or fibre to the cabinet (FTTCab), is a telecommunication architecture based on fibre optic cables run to a cabinet serving a neighbourhood. Customers typically connect to this cabinet using traditional coaxial cable or twisted pair wiring. The area served by the cabinet is usually less than 1500 m in radius and can contain several hundred customers. (If the cabinet serves an area of less than 300 m in radius, then the architecture is typically called fibre to the curb.)

Fibre to the node allows delivery of broadband services like high-speed internet. High-speed communication protocols like broadband cable access (typically DOCSIS) or some form of DSL are used between the cabinet and the customers. The data rates vary according to the exact protocol used and according to how close the customer is to the cabinet.

Unlike the competing FTTP technology, fibre to the node often uses the existing coaxial or twisted pair infrastructure to provide last mile service. For this reason, FFTN is less costly to deploy. In the long term, however, its bandwidth potential is limited relative to implementations that bring the fibre still closer to the subscriber.

9.2.3 Fibre to the Last Amplifier

FTTLA stands for of Fibre To The Last Amplifier. When the network cables being able to use several amplifiers, the FTTLA aims at replacing the coaxial cable to the last amplifier (towards the subscriber) by optical fibre. It acts as a new technology aiming at re-using the network cables existing, in particular, on the final part while installing the optical fibre more closely to the subscriber while using the coaxial cable of the networks cables for the last mile or last meters connected with the subscriber. Fibre to the last amplifier (FTTLA) node is an efficient tool to deploy fibre deeper into the CATV network architecture and add most desirable aspects of scalability (performance and reliability) that are necessary when new services (i.e., triple play, video on demand, gaming) are introduced.

The FTTLA is a technology that assists hybrid fibre coaxial CATV networks to provide their customers more bandwidth. Using a replacement of all coaxial active equipment by nodes (optical receiver) with high-power output. The coaxial is maintained from the node to the customer without any active equipment in between. From the optical sender to the node, it uses fibre that is split by 4 or by 8 depending on the distance and on the output power of the optical sender (from 6 to 16 dBm). Further, IM2, IM3, and C/N are modified for a better network and it also has other benefits such as power saving in the network, as the power consumption is lower than a normal HFC network (up to 40%).

9.2.4 Fibre to the Curb

Fibre to the curb (FTTC) is a telecommunications system based on fibre optic cables run to a platform that serves several customers. Each of these customers has a connection to this platform via coaxial cable or twisted pair.

Fibre to the curb allows delivery of broadband services like high-speed internet. High-speed communications protocols such as broadband cable access or some form of DSL are used between the cabinet and the customers. The data rates vary according to the exact protocol used and according to how close the customer is to the cabinet.

The FTTC is subtly distinct from FTTN or FTTP (all are versions of Fibre in the Loop). The chief difference is the placement of the cabinet. The FTTC will be placed near the curb that differs from FTTN, which is placed far from the customer, and FTTP, which is placed right at the serving location.

Unlike the competing FTTP technology, fibre to the curb can use the existing coaxial or twisted pair infrastructure to provide last mile service. For this reason, fibre to the curb costs less to deploy. However, it also has lower bandwidth potential than fibre to the premises.

In the United States of America and Canada, the largest deployment of FTTC was carried out by BellSouth Telecommunications. With the acquisition of BellSouth by AT&T, deployment of FTTC will end. Future deployments will be based on either FTTN or FTTP. Existing FTTC plant may be removed and replaced with FTTP.

9.2.5 Fibre to the Premises

Fibre to the premises is a form of fibre optic communication delivery in which an optical fibre is run from the central office all the way to the premises occupied by the subscriber. Fibre to the premises is often abbreviated with the acronym FTTP. However, this acronym has become ambiguous and may instead refer to a form of fibre to the curb where the fibre terminates at a utility pole without reaching the premises.

9.2.6 FTTH vs FTTB

Fibre to the premises can be categorised based on the fact where the optical fibre ends:

1. **FTTH** (fibre to the home) is a form of fibre optic communication delivery in which the fibre extends from the central office to the subscriber's living or working space. Once at the subscriber's living or working space, the signal may be conveyed throughout the space using any means, including twisted pair, coaxial cable, wireless, power line communication, or optical fibre.
2. **FTTB** (fibre to the building, also called fibre to the basement) is a form of fibre optic communication delivery in which the optical fibre terminates before actually reaching the subscribers living or working space itself, but does extend to the property containing that living or working space. The signal is conveyed the final distance using any non-optical means, including twisted pair, coaxial cable, wireless, or power line communication. By definition, FTTB necessarily applies only to those properties that contain multiple living or working spaces.

An apartment building may provide an example of the distinction between FTTH and FTTB. If a fibre is run to a panel at each subscriber's apartment, this is FTTH. If instead the fibre goes only as far as the apartment building's shared electrical room, then this is FTTB.

9.2.7 Direct Fibre

The simplest optical distribution network can be called direct fibre. In this architecture, each fibre leaving the central office goes to exactly one customer. Such networks can provide excellent bandwidth since each customer gets their own dedicated fibre extending all the way to the central office. However, this approach is about 10% more costly due to the amount of fibre and central office machinery required. The approach is generally favoured by new entrants and competitive operators.

9.3 SHARED FIBRE

More commonly, each fibre leaving the central office is actually shared by many customers. It is not until such a fibre gets relatively close to the customers that it is split into individual customer-specific fibres. There are two competing optical distribution network architectures that achieve this split: active optical networks (AONs) and passive optical networks (PONs).

9.3.1 Active Optical Network (AON)

Active optical networks rely on some sort of electrically powered equipment to distribute the signal, such as a switch, router, or multiplexer. Each signal leaving the central office is directed only to the customer for whom it is intended. Incoming signals from the customers avoid colliding at the intersection because the powered equipment provides buffering.

As of 2007, the most common type of active optical networks is called active Ethernet, a type of Ethernet in the first mile (EFM). Active Ethernet uses optical Ethernet switches to distribute the signal, thus incorporating the customers' premises and the central office into one giant switched Ethernet network. Such networks are identical to the Ethernet computer networks used in businesses and academic institutions, except that their purpose is to connect homes and buildings to a central office rather than to connect computers and printers within a campus. Each switching cabinet can handle up to 1000 customers, although 400–500 is more typical. This neighbourhood equipment performs layer 2/layer 3 switching and routing, offloading full layer 3 routing to the carrier's central office. The IEEE 802.3ah standard enables service providers to deliver up to 100 Mbit/s full duplex over one single-mode optical fibre to the premises depending on the provider. The speeds of 1 Gbit/s are becoming commercially available. The comparison shows how a typical active optical network handles downstream traffic differently than a typical passive optical network. The type of active optical network example is a star network that is capable of multicasting.

9.3.2 Passive Optical Network

A passive optical network (PON) is a point-to-multipoint, fibre-to-the-premises network architecture in which unpowered optical splitters are used to enable a single optical fibre to serve multiple premises, typically 32 and 128. A PON configuration reduces the amount of fibre and central office equipment required when compared with point to point architectures.

Downstream signal coming from the central office is broadcast to each customer premises sharing a fibre. Encryption is used to prevent eavesdropping. Upstream signals are combined using a multiple access protocol invariably time-division multiple access (TDMA).

9.3.3 Electrical Portion

Once on private property, the signal typically travels the final distance to the end user's equipment using an electrical format. A device called an Optical Network Terminal (ONT) also called an Optical Network Unit (ONU) and converts the optical signal into an electrical signal. (ONT is an ITU term, whereas ONU is an IEEE term, but the two terms mean exactly the same thing.) Optical network terminals require electrical power for their operation, and therefore, some providers connect them to back-up batteries in the case of power outages. Optical network units use thin-film filter technology to convert between optical and electrical signals. For fibre to the home and for some forms of fibre to the building, it is common for the building's existing phone systems, local area networks, and cable TV systems to connect directly to the ONT.

If all three systems cannot directly reach the ONT, it is possible to combine signals and transport them over a common medium. Once closer to the end user, equipment such as a router, modem, and/or network interface module can separate the signals and convert them into the appropriate protocol. For example, one solution for apartment buildings uses very high bit-rate digital subscriber line (VDSL) to combine data or video with voice. With this approach, the combined signal travels through the building over the existing telephone wiring until it reaches the end user's living space. Once there, a VDSL modem copies the data and video signals and converts them into Ethernet protocol. These are then sent over the end user's category 5 cables. A network interface module can then separate the video signal and convert it into an RF signal that is sent over the end-user coaxial cable. The voice signal continues to travel over the phone wiring and is sent through DSL filters to remove the video and data signals. An alternative strategy allows data or voice to be transmitted over coaxial cable. In yet another strategy, some office buildings dispense with the telephone wiring altogether, instead using voice over internet protocol phones that can plug directly into the local area network.

9.4 APPLICATION OF OPTICAL FIBRE IN MEDICAL SCIENCE AND MEDICAL ENGINEERING

9.4.1 Endoscope

An endoscope is a device with a light attached that is used to look inside a body cavity or organ. The scope is inserted through a natural opening, such as the mouth during a bronchoscopy, the rectum for a sigmoidoscopy, or the vagina for a cystoscopy.

An endoscopy involves examining the inside of a person's body using an endoscope. An endoscope is a medical device consisting of a long, thin, flexible (or rigid) tube that has a light and a video camera. Images of the inside of the patient's body can be seen on a screen. The whole endoscopy is recorded so that doctors can check it again. Endoscopy is a minimally invasive diagnostic medical procedure. It is used to examine the interior surfaces of an organ or tissue. The endoscope can also be used for enabling biopsies and retrieving foreign objects. An endoscope can be fitted with surgical instruments; it can send pulses or heat and electricity and destroy small tumours or gallstones.

9.4.1.1 Types of endoscopies

Here is a list of some types of endoscopies and their meanings:

1. Amnioscopy: examination of the amniotic cavity and foetus.
2. Arthroscopy: examination of the joints.
3. Bronchoscopy: examination of the air passages and the lungs.
4. Colonoscopy: examination of the colon.
5. Colposcopy: examination of the cervix and the tissues of the vagina and vulva.
6. Cystoscopy: examination of the urinary bladder.
7. Esophagogastroduodenoscopy (EGD): It is also known as panendoscopy examination of the esophagus, stomach, and duodenum.
8. Endoscopic retrograde cholangiopancreatography (ERCP): examination of the liver, gallbladder, bile ducts, and pancreas.
9. Foetoscopy: examination of the foetus.
10. Laparoscopy: a small incision to examine the abdominal cavity.
11. Laryngoscopy: examination of the back of the throat, including the voice box (larynx) and vocal cords.
12. Proctoscopy: examination of the rectum and the end of the colon.
13. Rhinoscopy: examination of the inside of the nose.
14. Thoracoscopy: examination of the lungs or other structures in the chest cavity.

9.4.1.2 Single-mode fibre optic catheter/endoscope (for optical coherence tomography)

Optical coherence tomography (OCT) is a new optical imaging technique that uses low-coherence interferometer to perform high-resolution, cross-sectional imaging in biological systems. An important application of OCT is imaging the vascular system to identify atherosclerotic lesions that are prone to rupture. The OCT has lox greater resolution than state of the art catheter-based ultrasonography, intravascular ultrasound (IWS). Additional research has shown that OCT can perform high-resolution imaging of other organ systems like the gastrointestinal. In order to apply OCT for imaging of internal organ systems, a flexible, small diameter, catheter/endoscope, which is capable of delivering, focusing, scanning, and collecting a single spatial mode optical beam, must be constructed. In this section, we describe the design and performance of a prototype single-mode fibre optic scanning OCT catheter with a diameter of 1 mm. The OCT imaging may be performed a wave length of 1.3 μm using either a super luminescent laser diode source or a Kerr lens mode locked Cr: Forsterite laser, which provides high powers for high-speed imaging. This technology will permit the OCT imaging of many tissues and organ systems, which were not previously accessible. The OCT catheter consists of a drive motor at its proximal end, a single-mode fibre running the length of the catheter, and optical focusing and beam directing elements at the distal end. During image acquisition, the catheter is inserted into an internal tissue channel and the focused beam is scanned perpendicular to the axis of the catheter. An OCT image is acquired as the beam angle of rotation is varied over some range. The speed of imaging depends on the speed of the rotation and the OCT unit acquisition speed. Preliminary images have been acquired using a super luminescent diode-based OCT system. In order to demonstrate imaging with the catheter, one can perform OCT of an in vitro human saphenous vein.

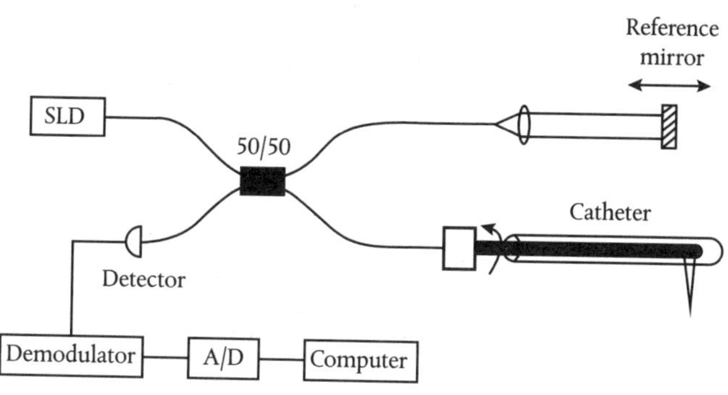

Fig. 9.1 Schematic of the super luminescent diode (SLD)-based OCT system

The intact specimen was taken post-mortem and imaged with the catheter in the centre of the lumen of the vessel. The rectangular raw data image array was displayed in polar coordinates using a linear interpolation algorithm. The image was also processed to correct for the depth-dependent exponential power loss caused by the optical attenuation of the tissue. As demonstrated by Fig. 9.1, the OCT image shows a significant enhancement in resolution and capability to differentiate tissue morphology.

In order to perform high-speed imaging, one can developed an OCT system based on a Kerr lens mode locked, Nd-YAG pumped, and Cr: Forsterite laser, which generates femtosecond pulses at 1.3 μm, as shown in Fig.9.2. The high-average output powers available from mode-locked short-pulsed lasers can provide high-data acquisition rates and high SNR, while the short pulse yields a short coherence length. The Cr:Forsterite laser was set to produce 200 mW of the output power with a bandwidth of 60 nm cantered at 1290 nm. For preliminary studies, the power coupled into the fibre was limited to about 10 mW in order to reduce nonlinear effects. In order to achieve high-speed scanning, the system uses a piezoelectric modulator to stretch the fibre in the reference arm and scan the temporal delay.

Static and dynamic birefringence incurred in the fibre are compensated using Faraday rotators placed in the free-space region of the interferon mm at a rate of 3 m/s, using a triangle wave with a frequency of 500 Hz, resulting in a modulated frequency of 4.5 MHz on the interference.

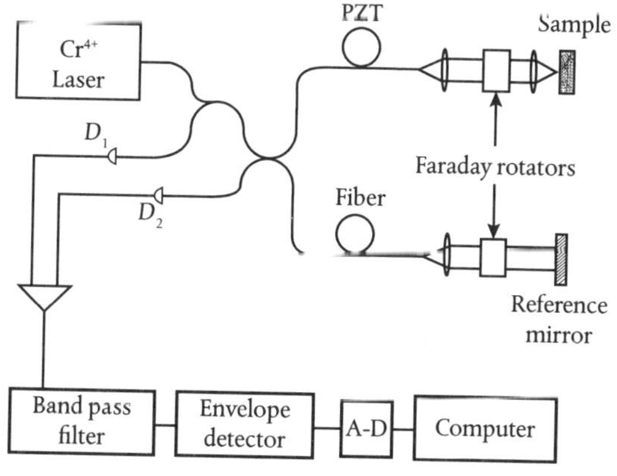

Fig. 9.2 Schematic of the fast scanning, chromium doped forsterite laser OCT.

The system can produce a 500×250 pixel image in 0.5 s. Preliminary studies have been performed with this system and one can report the combination of this high-speed imaging system with the OCT catheter. To conclude, we have described and demonstrated a single-mode fibre scanning OCT catheter/

endoscope. This device is an enabling technology for OCT and will permit minor scale, cross-sectional medical diagnostic imaging in tissues such as the vascular system, the gastrointestinal tract, the urinary tract, and the respiratory tract. The ability to perform optical biopsy to image the microstructure.

Applications of the optical coherence tomography (OCT) as a new method for bioimaging are definitely related to the creation of compact, portable devices suitable for clinical experiments. In this section, we report on developing a compact fast scanning OCT device comprising a movable fibre probe and on using this device in clinics to monitor pathological and postoperative states of human skin in vivo, including inflammation processes, tumours, sutures, and burns. The OCT device is based on a polarisation-preserving optical fibre Michelson interferometer with an integrated unit for fast electro-optic scanning in the depth of the investigated tissue. To modulate the optical path difference in the interferometer arms up to 5 mm, the optical fibre in both arms is fastened to thin piezoceramic plates. When applying a driving voltage to the piezoceramic plates, the linear modulation of the optical paths is achieved being opposite in sign for the sample and reference arms, which doubles the scanning depth. A weak inertia of the piezoelectric scanning system enables us to raise the scanning speed in orders of magnitude as compared to traditional mechanical systems.

9.4.1.3 A photovoltaically modulated micro electromechanical system (MEMS) optical scanner for fibre endoscope

Optical coherent tomography (OCT) diagnosis with a laser beam is practically used for the ophthalmology inspection, in particular for glaucoma diagnosis, and medical instruments of desktop model are commercially available to see the subcutaneous tissue of retina at the microscope resolution. The OCT technology is now being applied to pathologically investigate the internal organs or vascular veins by changing the entire optics into the shape of a fibre optic endoscope for the sake of minimum invasive examination and surgery. Scanning a probe light is necessary to obtain a tomogram images including the cross section of the affected tissue. Mechanically, scanning or rotating the optical fibre is a handy way of scanning the beam but the sensor head cannot be made small enough due to the size of the technical components to be accommodated. In this sense, an MEMS version of optical scanner is attractive to make a miniaturised OCT sensor head. Several different designs of MEMS scanner for the endoscope applications have being reported. Figure 9.3 shows MEMS fibre optic endoscope for in-vivo (vivo refers to within the cell) diagnosis with an OCT system.

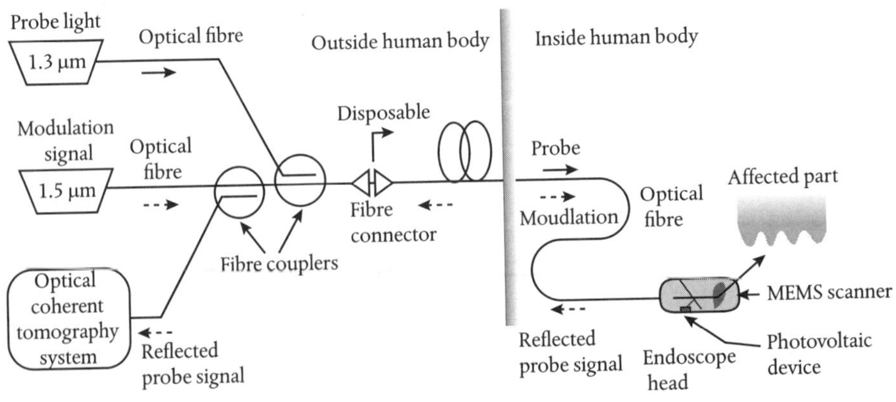

Fig. 9.3 MEMS fibre optic endoscope for in-vivo diagnosis with an OCT system

A drawback is, however, most of such MEMS scanners require relatively high-drive voltage for the electrostatic operation, which may put the patient at risk for electrocution accident in particular for the cardiovascular diagnosis. An attempt to the low voltage operation has been successful to some extent; we reported an optical scanner that could be operated at voltages as low as 5 V or less. However, wired electric control should be avoided because it may cause or suffer from the electromagnetic coupling with other medical instruments.

Where drive voltage for the scanner is generated. Another wavelength at 1.3 μm is used in the identical optical fibre system for medical diagnosis. The new scheme is totally free from the risk of electrocution and electromagnetic interference to/from other medical instruments.

9.4.1.4 Optically modulated endoscope

Figure 9.4 illustrates the schematic view of the endoscope sensor head, in which an MEMS scanner is accommodated. The sensor head in a glass or metal package (currently 5 mm in diameter) is attached at the tail of a single mode optical fibre. One of the infrared light (1.3-μm wavelength) travelling through the fibre is guided through a collimate lens, a beam splitter, and a focus lens to the MEMS scanner, where it is spatially scanned over the affected area.

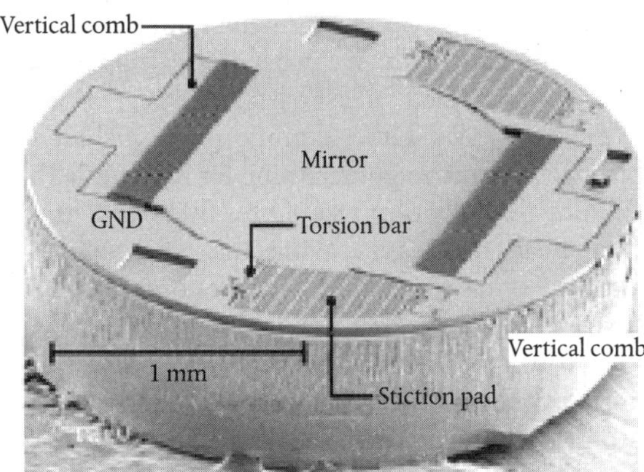

Fig. 9.4 Optically modulated endoscope

The optics is designed to focus the beam spot at a sample on the sensor head window. The reflected signal travels backward through the identical optical system and analysed by optical instruments like the OCT. The optical fibre also carries another light signal at a wavelength of 1.5 microns, which is selectively reflected onto the photovoltaic cells, where drive voltage to the scanner is generated. The actuation signal is intensity-modulated resonant frequency, and therefore, they are already developed and are commercially available. The measurement system located outside was found to be 100 μs and 500 μs for rise and fall, respectively, which was fast enough to drive an MEMS scanner operating at several hundred hertz.

9.4.2 Fibrescope

A fibrescope is a flexible fibre optic bundle with an eyepiece at one end and a lens at the other. It is used for inspection work, often to examine small components in tightly packed equipment, when the inspector cannot easily access the part requiring inspection. The lens is often a wide-angle lens, and the eyepiece is occasionally instead connected to a camera. Some fibrescopes use an additional fibre to carry light from an external source to illuminate the material being inspected for clearer viewing. All fibrescopes introduce a certain amount of image distortion; much of this is similar to the distortion of modern night vision equipment. Quartz fibrescopes can reach lengths of up to about 90 m (300 ft). Fibrescopes are used in medicine, machining, computer repair, espionage, locksmithing, safecracking, and computer forensics, among many other uses.

9.4.3 Borescope

A borescope is an optical device consisting of a rigid or flexible tube with an eyepiece on one end, and an objective lens on the other linked together by a relay optical system in between. The optical system is usually surrounded by optical fibres used for illumination of the remote object. An internal image of the illuminated object is formed by the objective lens and magnified by the eyepiece that presents it to the viewer's eye. Borescopes are used for inspection work where the area to be inspected is inaccessible by other means. Devices for use inside the human body are referred to as endoscopes. Borescopes are commonly used in the visual inspection of aircraft engines, aero-derivative industrial gas turbines, steam turbines, diesel engines, and automotive and truck engines. Gas and steam turbines require particular attention because of safety and maintenance requirements. Borescope inspection of engines can be used to prevent unnecessary maintenance, which can become extremely costly for large turbines. They are also used in manufacturing of machined or cast parts to inspect critical interior surfaces for burrs, surface finish or complete through holes. Other common uses include forensic applications in law enforcement and building inspection, and in gunsmithing for inspecting the interior bore of a firearm.

A flexible borescope includes a bundle of optical fibres that divide the image into pixels. It is also known as a fibrescope and can be used to access cavities that are around a bend, such as a combustion chamber or burner can, in order to view the condition of the compressed air inlets, turbine blades, and seals without disassembling the engine.

Flexible borescopes suffer from pixilation and pixel crosstalk due to the fibre image guide. Image quality varies widely among different models of flexible borescopes depending on the number of fibres and construction used in the fibre image guide. For flexible borescopes, articulation mechanism components, range of articulation, field of view, and angles of view of the objective lens are also important. Fibre content in the flexible relay is also critical to provide the highest possible resolution to the viewer. Minimal quantity is 10 000 pixels, while the best images are obtained with higher numbers of fibres in the 15 000 to 22 000 range for the larger diameter borescopes.

A video borescope or inspection camera is similar to the flexible borescope but uses a miniature video camera at the end of the flexible tube. A display in the handle shows the camera view. Because the complex optical waveguide is replaced with an inexpensive electrical cable, video borescopes can be much less costly and potentially better resolution (depending on the specifications of the camera). Digital models often have an integrated recorder and can save images in common digital formats.

Rigid borescopes are similar to fibrescopes but generally provide a superior image at lower cost when compared to a flexible borescope. Rigid borescopes have the limitation that access to what is to be viewed must be in a straight line. Rigid borescopes are, therefore, better suited to certain tasks such as inspecting automotive cylinders, fuel injectors and hydraulic manifold bodies, and gunsmithing.

Criteria for selecting a borescope are usually image clarity and access. For similar quality instruments, the largest rigid borescope that will fit the hole gives the best image. Optical systems in rigid borescopes can be of 3 basic types: Harold Hopkins rod lenses, achromatic doublets and gradient index rod lenses. For large diameter borescopes, the achromatic doublet relays work quite well, but as the diameter of the borescope tube gets smaller (less than about 4 mm), the designs of Hopkins rod lens and gradient index rod lens provide superior images. For very small rigid borescopes, the gradient index lens relays are better.

9.5 PERFORMANCE LIMITATION OF ELECTRICAL SOURCES

9.5.1 Noise Figure

The noise figure of the microwave photonic (MWP) filter can be defined as the ratio between the total noise power spectral density at the device output N_{out} and the noise power due to only the thermal noise spectral density applied to the input at the reference temperature and affected by the device gain.

The total noise spectral density at the output of the MWP filter is composed of different sources of noise generated along the MWP filter such as (a) optical intensity noise, (b) shot noise, (c) ASE noise, and (d) thermal noise.

9.5.2 Harmonic and Intermodulation Distortion

Harmonic and intermodulation distortion (IMD) features are the other great constraint that should be addressed for a real application of an MWP filter. The main source of signal distortion is normally the E/O conversion stage. If we consider direct IM lasers, it has both static distortion produced by the *P–I* curve and dynamic distortion produced by the laser couple rate equation dynamics.

9.6 OVERVIEW OF OPTICAL FIBRE SENSOR

Over the past 20 years, two major product revolutions have taken place due to the growth of the optoelectronics and fibre optic communications industries. The optoelectronics industry has brought about such products as compact disc players, laser printers, bar code scanners, and laser pointers. The fibre optic communication industry has literally revolutionised the telecommunication industry by providing higher performance and more reliable telecommunication links with ever decreasing bandwidth cost. This revolution is bringing about the benefits of high-volume production to component users and a true information superhighway built of glass.

In parallel with these developments, fibre optic sensor technology has been a major user of technology associated with the optoelectronic and fibre optic communication industry. Many of the components associated with these industries were often developed for fibre optic sensor applications. Fibre optic sensor technology in turn has often been driven by the development and subsequent mass production of components to support these industries. As component prices have fallen and quality improvements have been made, the ability of fibre optic sensors to displace traditional sensors for rotation, acceleration, electric and magnetic field measurement, temperature, pressure, acoustics, vibration, linear and angular position, strain, humidity, viscosity, chemical measurements, and a host of other sensor applications has been enhanced.

In the early days of fibre optic sensor technology, most commercially successful fibre optic sensors were squarely targeted at markets where existing sensor technology was marginal or in many cases, it was nonexistent. The inherent advantages of fibre optic sensors include lightweight, very small size, passive, low power, resistant to electromagnetic interference, high sensitivity, wide bandwidth, and environmental ruggedness; these advantages were heavily used to offset their major disadvantages such as high cost and unfamiliarity to the end user.

The situation is changing. Laser diodes that cost \$3000 in 1979 with lifetimes measured in hours are now being sold for a few dollars in small quantities; they have reliability of tens of thousands of hours and are used widely in compact disc players, laser printers, laser pointers, and barcode readers. Single-mode optical fibre were at a cost of \$20/m in 1979; however, now they cost less than \$0.10/m with vastly improved optical and mechanical properties. Integrated optical devices that were not available in usable form at that time are now commonly used to support production models of fibre optic gyros. Further, they could drop dramatically in price in the future while offering ever more sophisticated optical circuits. As these trends continue, the opportunities for fibre optic sensor designers to produce competitive products will increase and the technology can be expected to assume an ever more prominent position in the sensor marketplace. In the following sections, the basic types of fibre optic sensors that are being developed will be briefly reviewed followed by a discussion of how these sensors work.

9.6.1 Basic Concept and Classification

Fibre optic sensors are often loosely grouped into two basic classes that are referred to as:

1. Extrinsic or hybrid fibre optic sensors
2. Intrinsic or all fibre sensors

9.6.1.1 Extrinsic or hybrid fibre optic sensor

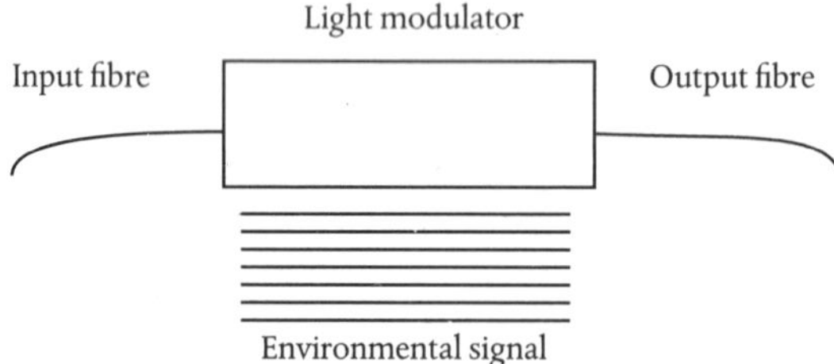

Fig. 9.5 Illustration of the case of an extrinsic or hybrid optic sensor

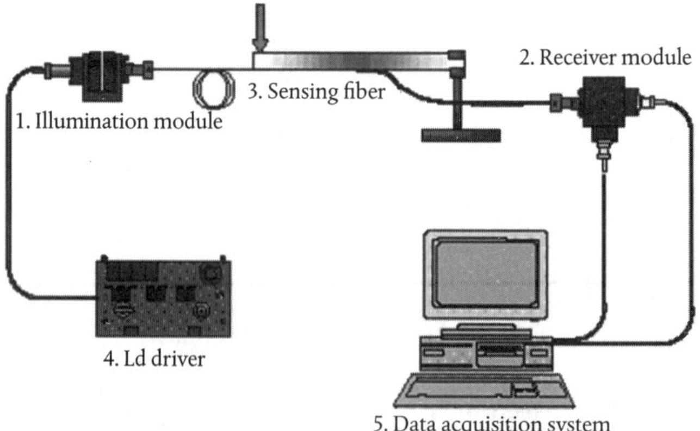

Fig. 9.6 Fibre sensors

Figure 9.5 shows the extrinsic fibre optic sensors consisting of optical fibres that lead up to and out of a black box that modulates the light beam passing through it in response to an environmental effect. In this case, an optical fibre leads up to a black box that impresses information onto the light beam in response to an environmental effect. The information could be impressed in terms of intensity, phase, frequency, polarisation, spectral content, or other methods. An optical fibre then carries the light with the environmentally impressed information back to an optical and/or electronic processor. In some cases, the input optical fibre also acts as the output fibre. Figure 9.6 shows the basic features of fibre sensors.

9.6.1.2 Intrinsic or all-fibre sensors

The intrinsic or all-fibre sensors are shown in Fig. 9.7; they use an optical fibre to carry the light beam and the environmental effect impresses information onto the light beam while it is in the fibre. Each of these classes of fibres in turn has many subclasses; in some cases, sub-subclasses consist of large numbers of fibre sensors.

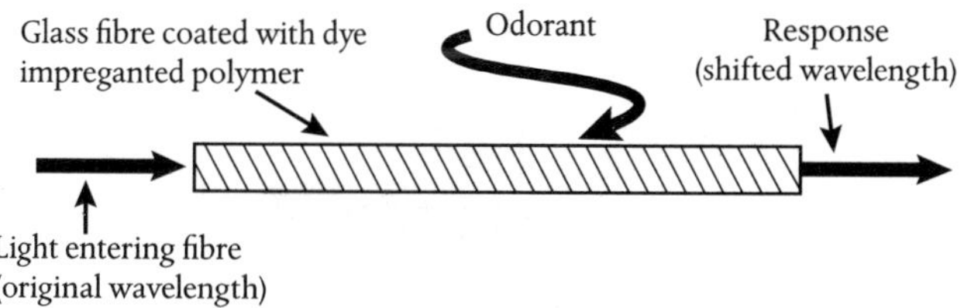

Fig. 9.7 Intrinsic fibre optic sensors rely on the light beam propagating through the optical fibre being modulated by the environmental effect either directly or through environmentally induced optical path length changes in the fibre itself.

There are different types of fibre optic sensors:

1. Intensity-based fibre optic sensor
2. Spectrally based fibre optic sensor
3. Interferometric fibre optic sensors

9.6.2 Intensity-based Fibre Optic Sensor

One of the best examples of intensity-based fibre optic sensor is vibration sensor that consists of two optical fibres that are held in close proximity to each other. Light is injected into one of the optical fibres, and when it exits, the light expands into a cone of light whose angle depends on the difference between the index of refraction of the core and cladding of the optical fibre. The amount of light captured by the second optical fibre depends on its acceptance angle and the distance (d) between the optical fibres. When the distance d is modulated, it, in turn, results in an intensity modulation of the light captured.

A variation on this type of sensor can be done by using a mirror that is flexibly mounted to respond to an external effect like pressure. As the mirror position shifts the effective separation between the optical fibres shift with a resultant intensity modulation. These types of sensors are useful for such applications as door closures where a reflective strip in combination with an optical fibre acting to input and catch the output reflected light can be used.

By arranging two optical fibres in line, a simple translation sensor can be configured as in Fig. 9.8. The output from the two detectors can be proportioned to determine the translational position of the input fibre.

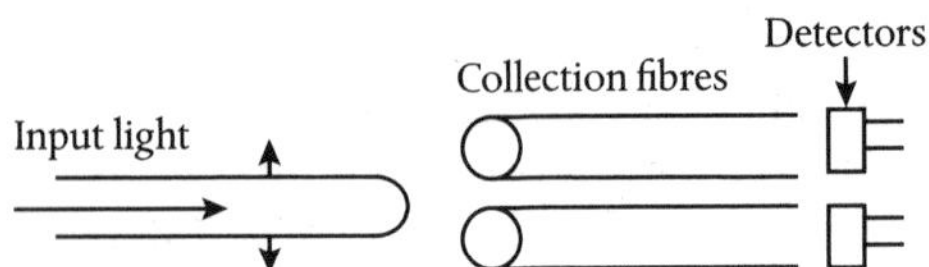

Fig. 9.8 Fibre optic translation sensor based on numerical aperture uses the ratio of the output on the detectors to determine the position of the input fibre

Several companies have developed rotary and linear fibre optic position sensors to support applications like fly by light. These sensors attempt to eliminate electromagnetic interference susceptibility to improve safety, and to reduce shielding in order to reduce weight. Figure 9.9 shows a rotary position sensor that consists of a code plate with variable reflectance patches placed so that each position has a unique code. A series of optical fibres are used to determine the presence or absence of a patch.

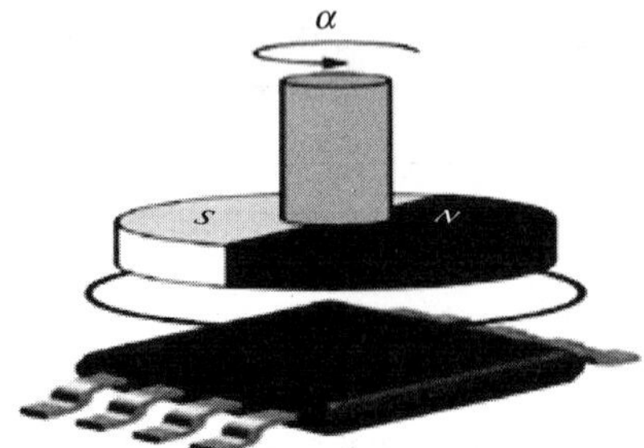

Fig. 9.9 Fibre optic rotary position sensor based on reflectance used to measure the rotational position of the shaft via the amount of light reflected from dark and light patches

An example of a linear position sensor using WDM is illustrated by Fig. 9.10. Here, a broadband light source that might be a light emitting diode is used to couple light into the system. A single optical fibre is used to carry the light beam up to a WDM element that splits the light into separate fibres that are used to interrogate the encoder card and determine linear position. The boxes on the card of Fig.9.10 represent highly reflective patches while the rest of the card has low reflectance. The reflected signals are then recombined and separated out by a second WDM element so that each interrogating fibre signal is read out by a separate detector.

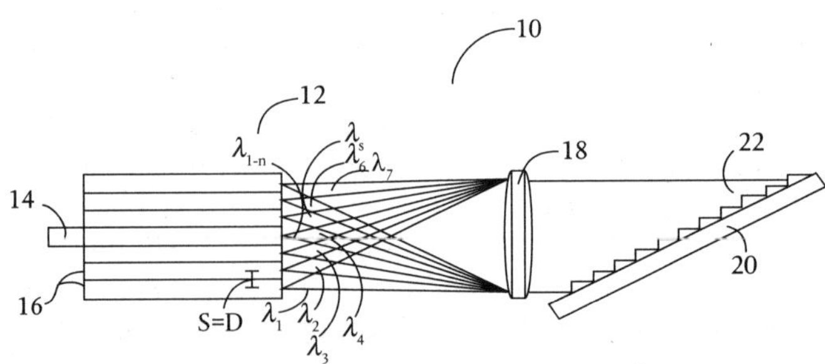

Fig. 9.10 Linear position sensor using WDM decodes position by measuring the presence or absence of reflective patch at each fibre position as the card slides by via independent wavelength separated detectors

A second common method of interrogating a position sensor using a single optical fibre is to use time-division multiplexing methods. In TDM, a light source is pulsed. The light pulse then propagates down the optical fibre and is split into multiple interrogating fibres. Each of these fibres is arranged so that they have delay lines that separate the return signal from the encoder plate by a time that is longer than the pulse duration. When the returned signals are recombined onto the detector, the net result is an encoded signal burst corresponding to the position of the encoded card. These sensors have been used to support tests on military and commercial aircraft that have demonstrated performance comparable to conventional electrical position sensors used for rudder, flap, and throttle position. The principle advantages of the fibre position sensors are immunity to electromagnetic interference and overall weight savings.

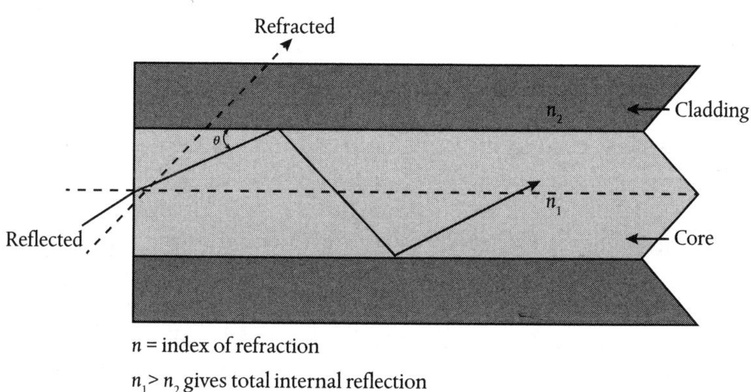

Fig. 9.11 Fibre sensor using critical angle properties of a fibre for pressure/index of refraction measurement via measurements of the light reflected into the fibre

Another class of intensity based fibre optic sensors is based on the principle of total internal reflection. In the case of the sensor in Fig. 9.11, the light propagates down the fibre core and hits the angled end of the fibre. If the medium into which the angled end of the fibre is placed has a low enough index of refraction than virtually all the light is reflected when it hits the mirrored surface and returns via the fibre. If, however, the index of refraction of the medium starts to approach that of the glass, some of the light propagates out of the optical fibre and is lost resulting in an intensity modulation.

This type of sensor can be used for low resolution measurement of pressure or index of refraction changes in a liquid or gel with 1–10% accuracy. Variations on this method have also been used to measure liquid level as shown by the probe configuration. When the liquid level hits the reflecting prism, the light leaks into the liquid greatly attenuating the signal.

9.6.3 Spectrally Based Fibre Optic Sensors

Spectrally based fibre optic sensors depend on a light beam being modulated in wavelength by an environmental effect. Examples of these types of fibre sensors include those based on blackbody radiation, absorption, fluorescence, etalons, and dispersive gratings. One of the simplest of these types of sensors is the blackbody sensor of Fig. 9.12. A blackbody cavity is placed at the end of an optical fibre. When the cavity rises in temperature, it starts to glow and acts as a light source.

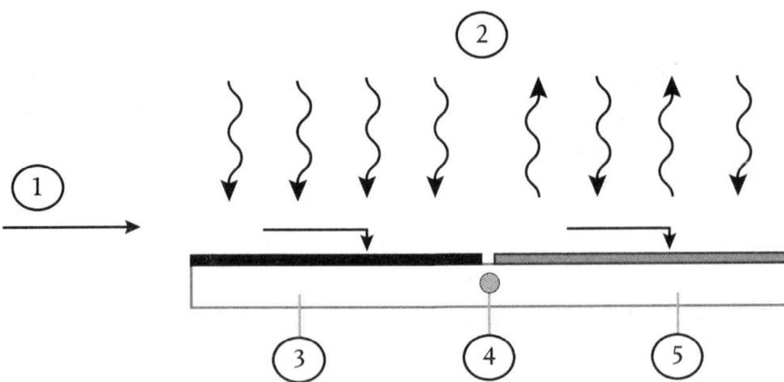

Fig. 9.12 Blackbody fibre optic sensors allow the measurement of temperature at a hot spot and are most effective at temperatures of higher than 300 °C

Detectors in combination with narrow-band filters are then used to determine the profile of the blackbody curve, and in turn, the temperature as in Fig. 9.12. This type of sensor has been successfully commercialised and has been used to measure temperature to within a few degrees Celsius under intense RF fields. The performance and accuracy of this sensor are better at higher temperatures and falls off at temperatures on the order of 200°C because of low signal to noise ratios. Care must be taken to ensure that the hottest spot is the blackbody cavity and not on the optical fibre that leads itself as this can corrupt the integrity of the signal.

Another type of spectrally based temperature sensor is based on the instant absorption. In this case, a Gallium Arsenide (GaAs) sensor probe is used in combination with a broadband light source and input/output optical fibres. The absorption profile of the probe is temperature dependent and may be used to determine temperature. Fluorescent-based fibre sensors are being widely used for medical applications, chemical sensing and can also be used for physical parameter measurements such as temperature, viscosity and humidity. There are a number of configurations for these sensors and illustrates two of the most common ones. In the case of the end-tip sensor, the light propagates down the fibre to a probe of fluorescent material. The resultant fluorescent signal is captured by the same fibre and directed to an output demodulator. The light sources can be pulsed and probes have been made that depend on the time rate of decay of the light pulse.

Fluorescent fibre optic sensor probe configurations can be used to support the measurement of physical parameters as well as the presence or absence of chemical species. These probes may be configured to be single ended or multipoint by using side etches techniques and attaching the fluorescent material to the fibre. In the continuous mode, parameters such as viscosity, water vapour content, and degree of cure in carbon fibre reinforced epoxy, and thermoplastic composite materials can be monitored. An alternative is to use the evanescent properties of the fibre and etch regions of the cladding away and refill them with fluorescent material. By sending a light pulse down the fibre and looking at the resulting fluorescence, a series of sensing regions may be time-division multiplexed. It is also possible to introduce fluorescent dopants into the optical fibre itself. This approach would cause the entire optically activated fibre to fluoresce. By using time-division multiplexing, various regions of the fibre could be used to make a distributed measurement along the fibre length.

In many cases, users of fibre sensors would like to have the fibre optic analogue of conventional electronic sensors. An example is the electrical strain gauge that is used widely by structural engineers. Fibre grating sensors can be configured to have gauge lengths from 1 mm to approximately 1 cm, with sensitivity comparable to conventional strain gauges. This sensor is fabricated by writing a fibre grating onto the core of a Germanium doped optical fibre. This can be done in a number of ways. One method, which is illustrated in Fig. 9.13, uses two short wavelength laser beams that are angled to form an interference pattern through the side of the optical fibre. The interference pattern consists of bright and dark bands that represent local changes in the index of refraction in the core region of the fibre. Exposure time for making these gratings varies from minutes to hours, depending on the dopant concentration in the fibre the wavelengths used, the optical power level, and the imaging optics.

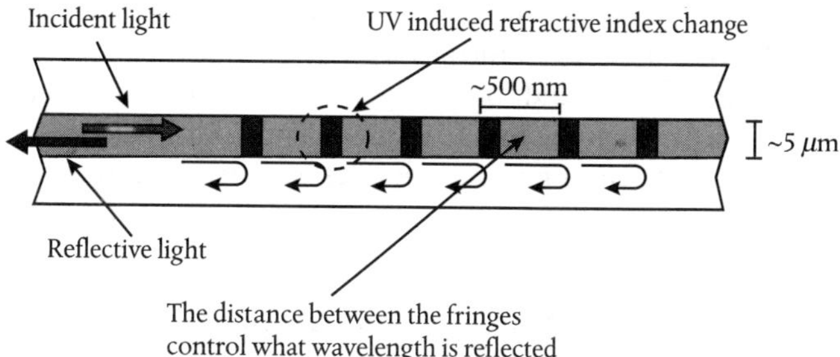

Fig 9.13 Fabrication of a fibre grating sensor can be accomplished by imaging to short wavelength laser beams through the side of the optical fibre to form an interference pattern. The bright and dark fringes that are imaged on the core of the optical fibre induce an index of refraction variation resulting in a grating along the fibre core.

Other methods that have been used include the use of phase masks and interference patterns induced by short high-energy laser pulses. The short duration pulses have the potential to be used to write fibre gratings into the fibre as it is being drawn. Substantial efforts are being made by laboratories around the world to improve the manufacturability of fibre gratings as they have the potential to be used to support optical communication as well as sensing technology. Once the fibre grating has been fabricated the next major issue is how to extract information. When used as a strain sensor the fibre grating is typically attached to, or embedded in, a structure. As the fibre grating is expanded or compressed, the grating period expands or contracts, changing the gratings spectral response.

For a grating operating at 1300 nm, the change in wavelength is about 10nm per micro strain. This type of resolution requires the use of spectral demodulation techniques that are much better than those associated with conventional spectrometers. Several demodulation methods have been suggested using fibre gratings, etalons, and interferometers. The action of the reference fibre grating is to act as a modulator filter. By using similar gratings for the reference and signal gratings and adjusting the reference grating to line up with the active grating, an accurate closed-loop demodulation system may be implemented. Fibre grating demodulation systems require very high resolution spectral measurements. One way to accomplish this is to beat the spectrum of light reflected by the fibre grating against the light transmission characteristics of a reference grating.

An alternative demodulation system would use fibre etalons. One fibre can be mounted on a piezoelectric and the other moved relative to a second fibre end. The spacing of the fibre ends as well as their reflectivity in turn determines the spectral filtering action of the fibre etalon. Intrinsic fibre etalons are formed by in line reflective mirrors that can be embedded into the optical fibre. Extrinsic fibre etalons are formed by two mirrored fibre ends in a capillary tube. A fibre etalon based spectral filter or demodulator is formed by two reflective fibre ends that have a variable spacing.

The fibre etalons can also be used as sensors for measuring strain as the distance between mirrors in the fibre determines their transmission characteristics. The mirrors can be fabricated directly into the fibre by cleaving the fibre, coating the end with titanium dioxide and then replacing. An alternative approach is to cleave the fibre ends and insert them into a capillary tube with an air gap. Both of these approaches are being investigated for applications where multiple in-line fibre sensors are required.

For many applications, a single-point sensor is adequate. In these situations, an etalon can be fabricated independently and attached to the end of the fibre. A series of etalons have been configured to measure pressure, temperature, and refractive index, respectively. Hybrid etalon-based fibre optic sensors often consist of micro-machined cavities that are placed on the end of optical fibres and can be configured so that sensitivity to one environmental effect is optimised.

In the case of pressure, the diaphragm has been designed to deflect. Pressure ranges of 15 to 2000 psi can be accommodated by changing the diaphragm thickness with accuracy of about 0.1% full scale. For temperature, the etalon has been formed by silicon/silicon dioxide interfaces. Temperature ranges of 70 – 500 K can be selected, and for a range of about 100 K, a resolution of about 0.1 K is achievable. For refractive index of liquids, a hole has been formed to allow the flow of the liquid to be measured without the diaphragm deflecting. These devices have been commercialised and are sold with instrument packages.

9.6.4 Interferometric Fibre Optic Sensors

One of the areas of greatest interest has been in the development of high-performance interferometric fibre optic sensors. Substantial efforts have been undertaken on Sagnac interferometers, ring resonators, Mach–Zehnder and Michelson interferometers as well as dual-mode, polarimetric, grating and etalon-based interferometers. In this section, the Sagnac, Mach–Zehnder, and Michelson interferometers are briefly reviewed.

9.7 SAGNAC INTERFEROMETER

The Sagnac interferometer has been principally used to measure rotation and is a replacement for ring laser gyros and mechanical gyros. It may also be employed to measure time varying effects such as acoustics and vibration, and slowly varying phenomenon like strain. By using multiple interferometer configurations, it is possible to employ the Sagnac interferometer as a distributed sensor capable of measuring the amplitude and location of a disturbance.

The single most important application of fibre optic sensors in terms of commercial value is the fibre optic gyro. It was recognised very early that the fibre optic gyro offered the prospect of an all-solid state inertial sensor with no moving parts, unprecedented reliability, and had the prospect of being very low cost.

The potential of the fibre optic gyro is being realised as several manufacturers worldwide are producing them in large quantities to support automobile navigation systems, pointing and tracking of satellite antennas, inertial measurement systems for commuter aircraft and missiles and as the backup guidance system for the Boeing 777. They are also being baseline for such future programs as the Comanche helicopter and are being developed to support long-duration space flights. Other applications where fibre optic gyros are being used include mining operations, tunnelling, attitude control for a radio-controlled helicopter, cleaning robots, antenna pointing and tracking, and guidance for unmanned trucks and carriers.

Two types of fibre optic gyros are being developed. The first type is an open-loop fibre optic gyro with a dynamic range on the order of 1000–5000 (dynamic range is unit less) with scale factor accuracy of about 0.5% (this accuracy number includes nonlinearity and hysteresis effects) and sensitivities that vary from less than 0.01 °/h to 100 °/h and higher. These fibre gyros are generally used for low cost applications where dynamic range and linearity are not the crucial issues. Figure 9.14 shows the open-loop fibre optic gyro rotation sensor. The second type is the closed-loop fibre optic gyro that may have a dynamic range of 106 and scale factor linearity of 10 ppm or better. These types of fibre optic gyros are primarily targeted at medium to high accuracy navigation applications that have high turning rates and require high linearity and large dynamic ranges.

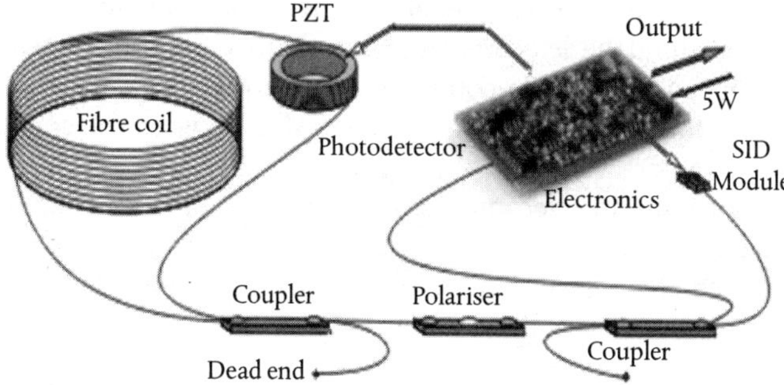

Fig. 9.14 Open-loop fibre optic gyro is the simplest and lowest cost rotation sensor. They are widely used in commercial applications where their dynamic range and linearity limitations are not constraining

When the fibre gyro is rotated in a clockwise direction, the entire coil is displaced slightly increasing the time it takes light to traverse the fibre optic coil. (Remember that the speed of light is invariant with respect to the frame of reference, thus coil rotation increases path length when viewed from outside the fibre). Thus, the clockwise propagating light beam has to go through a slightly longer optical path length than the counter clockwise beam that is moving in a direction opposite to the motion of the fibre coil. The net phase difference between the two beams is proportional to the rotation rate.

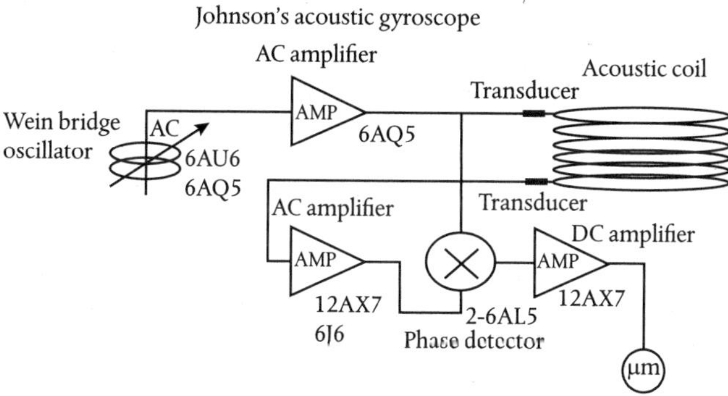

Fig. 9.15 Closed-loop fibre optic gyros use an artificially induced nonreciprocal phase between counter propagating light beams to counterbalance rotationally induced phase shifts. These fibre gyros have wider dynamic range and high linearity needed to support stringent navigation requirements

By including a phase modulator loop offset from the fibre coil, a time difference in the arrival of the two light beams is introduced and an optimised demodulation signal can be realised. Further improvements in dynamic range and linearity can be realised by using a closed-loop configuration where the phase shift induced by rotation is compensated by an equal and opposite artificially imposed phase shift. One way to accomplish this is to introduce a frequency shifter into the loop.

The relative frequency difference of the light beams propagating in the fibre loop can be controlled resulting in a net phase difference that is proportional to the length of the fibre coil and the frequency shift. In Fig. 9.15, this is done by using a modulator in the fibre optic coil to generate a phase shift. When the coil is rotated, a first harmonic signal is induced with a phase that depends on the rotation rate in a manner similar to that described above with respect to open-loop fibre gyros. By using rotationally induced first harmonic signal as an error signal, the frequency shift can be adjusted by using a synchronous demodulator behind the detector to integrate the first harmonic signal into a corresponding voltage. This voltage is applied to a voltage-controlled oscillator whose output frequency is applied to the frequency shifter in the loop so that the phase relationship between the counter propagating light beams is locked to a single value.

It is possible to use the Sagnac interferometer for other sensing and measurement tasks. Examples include slowly varying measurements of strain with 100 μm resolution over distances of about 1 km, spectroscopic measurements of wavelength to about 2 nm, and optical fibre characterisation like thermal expansion to accuracies of about 10 ppm. In each of these applications, frequency shifters are used in the Sagnac loop to obtain controllable frequency offsets between the counter propagating light beams.

Another class of fibre optic sensors based on the Sagnac interferometer can be used to measure rapidly varying environmental signals like sound. The sensitivity of this Sagnac-acoustic sensor depends on the location of the signal. If the signal is in the centre of the loop, the amplification is zero as both counter propagating light beams arrive at the centre of the loop at the same time. As the signal moves away from the centre, the output increases. Several other combinations of interferometers have been tried for a position and amplitude determinations and the first reported success consisted of a combination of the Mach–Zehnder and Sagnac interferometer.

9.8 MACH-ZEHNDER AND MICHELSON INTERFEROMETERS

One of the great advantages of all fibre interferometers such as Mach–Zehnder and Michelson interferometers in particular is that they have extremely flexible geometry and high sensitivity that allow the possibility of a wide variety of high-performance elements and arrays.

The light source module usually consists of a long coherence length isolated laser diode, a beam splitter to produce two light beams, and a means of coupling the beams to the two legs of the transducer. The transducer is configured to sense an environmental effect of isolating one light beam from the environmental effect and using the action of the environmental effect on the transducer is to induce an optical path length difference between the two light beams. Typically, a homodyne demodulator is used to detect the difference in optical path length (various heterodyne schemes have also been used).

One of the basic issues with the Mach–Zehnder interferometer is that the sensitivity will vary as a function of the relative phase of the light beams. One way to solve the signal fading problem is to introduce a piezoelectric fibre stretcher into one of the legs and adjust the relative path length of the two legs for optimum sensitivity. Another approach has the same quadrature solution as the grating-based fibre sensors discussed earlier.

9.9 MULTIPLEXING AND DISTRIBUTED SENSING

Many of the intrinsic and extrinsic sensors may be multiplexed offering the possibility of large numbers of sensors being supported by a single-fibre optic line. The techniques that are most commonly employed are time, frequency, wavelength, coherence, polarisation, and spatial multiplexing.

Time-division multiplexing employs a pulsed light source launching light into an optical fibre and analysing the time delay to discriminate between sensors. This technique is commonly employed to support distributed sensors where measurements of strain, temperature, or other parameters are collected. As the pipe joints are stressed, micro-bending loss increases and the time delay associated with these losses allows the location of faulty joints.

The entire length of the fibre can be made microbends sensitive and Rayleigh scattering loss used to support a distributed sensor that will predominantly measure strain. Other types of scattering from optical pulses propagating down optical fibre have been used to support distributed sensing notably Raman scattering for temperature sensors has been made into a commercial product by York Technology and Hitachi. These units can resolve temperature changes of about $1° C$ with spatial resolution of 1 m for a 1-km sensor using an integration time of about 5 min. Brillouin scattering has been used in the laboratory experiments to support both strain and temperature measurements

9.9.1 Applications

Fibre optic sensors are being developed and used in two major ways. The first is as a direct replacement for existing sensors where the fibre sensor offers significantly improved performance, reliability, safety, or cost advantages to the end user.

The second area is the development and deployment of fibre optic sensors in new market areas. For the case of direct replacement, the inherent value of the fibre sensor to the customer has to be sufficiently high to displace older technology. Because this often involves replacing technology, the customer is familiar with the improvements that must be substantial.

The most obvious example of a fibre optic sensor succeeding in this arena is the fibre optic gyro, which is displacing both mechanical and ring laser gyros for medium accuracy devices. As this technology matures, it can be expected that the fibre gyro will dominate large segments of this market.

Significant development efforts are underway in the United States in the area of fly by light where conventional electronic sensor technology is targeted to be replaced by equivalent fibre optic sensor technology that offers sensors with relative immunity to electromagnetic interference, significant weight savings, and safety improvements.

In manufacturing, fibre sensors are being developed to support process control. Oftentimes, the selling points for these sensors are improvements in environmental ruggedness and safety, especially in areas where electrical discharges could be hazardous.

One other area where fibre optic sensors are being mass produced is the field of medicine, where they are being used to measure blood gas parameters and dosage levels. Because these sensors are completely passive they pose no electrical shock threat to the patient and their inherent safety has led to a relatively rapid introduction.

The automotive industry, construction industry, and other traditional users of sensors remain relatively untouched by fibre sensors mainly because of cost considerations. This can be expected to change as the improvements in optoelectronics and fibre optic communications continue to expand along with the continuing emergence of new fibre optic sensors.

New market areas present opportunities where equivalent sensors do not exist. New sensors, once developed will most likely have a large impact in these areas. A prime example of this is in the area of fibre optic smart structures. Fibre optic sensors are being embedded into or attached to materials during the manufacturing process (1) to enhance process control systems, (2) to augment non-destructive evaluation once parts have been made, (3) to form healthy and damage assessment systems once parts have been assembled into structures, and (4) to enhance control systems.

Fibre optic sensors can be embedded in a panel and multiplexed to minimise the number of leads. The signals from the panel are fed to an optical/electronic processor for decoding. The information is formatted and transmitted to a control system that could be augmenting performance or assessing health. The control system would then act, via a fibre optic link, to modify the structure in response to the environmental effect.

Interesting areas for health and damage assessment systems are on large structures such as buildings, bridges, dams, aircraft and spacecraft. In order to support these types of structures, it will be necessary to have very large numbers of sensors that are rapidly reconfigurable and redundant. It will also be absolutely necessary to demonstrate the value and cost effectiveness of these systems to the end users.

One approach to this problem is to use fibre sensors that have the potential to be manufactured cheaply in very large quantities while offering superior performance characteristics. Two candidates that are under investigation are the fibre gratings and etalons. Both offer the advantages of spectrally based sensors and have the prospect of rapid in-line manufacture. In the case of the fibre grating, the early demonstration of fibre being written into it as it is being pulled has been especially impressive.

These fibre sensors could be folded into the wavelength and time-division multiplexed modular architecture. Here, sensors are multiplexed along fibre strings and an optical switch is used to support the many strings. Potentially, the fibre strings could have tens or hundreds of sensors and the optical switches could support a like number of strings. To avoid overloading the system, the output from the sensors could be slowly scanned to determine status in a continuously updated manner.

A modular architecture for a large smart structure system would consist of strings of fibre sensors accessible via an optical switch and demodulator system that could select key sensors in each string. The information would then be formatted and transmitted after conditioning to a vehicle health management bus.

When an event occurred that required a more detailed assessment, the appropriate strings and the sensors in them could be monitored in a high-performance mode. The information from these sensors would then be formatted and transmitted via a fibre optic link to a subsystem signal processor before the introduction onto a health management bus.

The information from the health management bus could be processed and distributed to the pilot or more likely used to reduce direct workload leaving more time for the necessary control functions.

It is also possible to use fibre optic sensors in combination with fibre optic communication links to monitor stress build-up in critical fault locations and dome build-up of volcanoes. These widely dispersed fibre networks may offer the first real means of gathering information necessary to form prediction models for these natural hazards.

9.10 OPTICAL SPECTRUM ANALYSER

Optical spectrum analysers (OSA) can divide a lightwave signal into its constituent wavelengths. This means that it is possible to see the spectral profile of the signal over a certain wavelength range. The profile is graphically displayed, with wavelength on the horizontal axis and power on the vertical axis. In this way, the many signals combined on a single fibre in a dense wavelength division multiplexing (DWDM) system can be taken apart to perform per-channel analysis of the optical signal and its spectral interaction with the other wavelengths.

One of the most common measurement instruments found in a modern optics laboratory is the optical spectrum analyser (OSA). These instruments measure the optical power spectrum of light emitted from optical sources as well as light generated from optical interactions with matter. The OSAs are based on several technologies. Each technology offers strengths. However, it also has its limitations. Choosing the correct technology for a measurement application requires understanding the concepts of sensitivity (the ability to measure low-power optical signals). Dynamic range (the ability to measure low-power signals in the presence of high-power signals), resolution (the ability to measure two signals separated by a small wavelength difference), amplitude accuracy (the ability to measure the absolute optical power present), and wavelength accuracy.

A spectrum is a collection of sine waves that, when combined properly, produce the time-domain signal under examination. Suppose that we were hoping to see a sine wave. Although the waveform certainly shows us that the signal is not a pure sinusoid, it does not give us a definitive indication of the reason why. Our complex signal in both the time and frequency domains. The frequency domain display plots the amplitude versus the frequency of each sine wave in the spectrum. Let us assume that the spectrum in this case comprises just two sine waves. It contained a second sine wave, the second harmonic in this case. Does this mean we have no need to perform time-domain measurements? Not at all. The time domain is better for many measurements and some can be made only in the time domain. For example, pure time-domain measurements include pulse rise and fall times, overshoot, and ringing.

The frequency domain also has its measurement strengths. The frequency domain is better for determining the harmonic content of a signal. People involved in wireless communications are extremely interested in out of band and spurious emissions. For example, cellular radio systems must be checked for harmonics of the carrier signal that might interfere with other systems operating at the same frequencies as the harmonics. Engineers and technicians are also very concerned about distortion of the message modulated onto a carrier. Third-order inter-modulation (two tones of a complex signal modulating each other) can be particularly troublesome because the distortion components can fall within the band of interest, and therefore, it will not be filtered away.

Spectrum monitoring is another important frequency domain measurement. Government regulatory agencies allocate different frequencies for various radio services, such as broadcast television and radio, mobile phone systems, police and emergency communications, and a host of other applications. It is critical that each of these services operates at the assigned frequency and stays within the allocated channel bandwidth. Transmitters and other intentional radiators can often be required to operate at closely spaced adjacent frequencies. A key performance measure for the power amplifiers and other components used in these systems is the amount of signal energy that spills over into adjacent channels and causes interference.

9.10.1 Fabry-Perot OSA

Two highly reflective mirrors aligned to form an optical resonant cavity is the heart of a Fabry–Perot OSA. Collimated laser light or light exiting an optical fibre is mode matched to the optical cavity. When the mirror spacing is a half-integral number of wavelengths of the incident light, a resonant standing wave occurs and the cavity becomes transmissive. As the mirror spacing is changed slightly from the resonance condition, the cavity is largely reflective. Typically, the optical cavity length is linearly scanned with a high-voltage piezoelectric element, and transmitted optical power is detected by a silicon (Si) photodiode for visible wavelengths or an indium gallium arsenide (InGaAs) photodiode for near-infrared wavelengths. Transmitted optical power versus scan voltage is displayed on an oscilloscope and represents the optical power spectrum of the incident light. When the incident light is of a single wavelength, a resonance occurs when the mirror spacing is a half-integral number of wavelengths. Resonance also occurs at this mirror spacing for wavelengths that have one more or one less standing wave in the cavity. The separation between these resonant wavelengths called the free-spectral range is typically 0.01 to 1 nm depending on the cavity length. Consequently, the Fabry–Perot OSA is useful only for light present over a narrow wavelength range. Sensitivity is generally good because high gain, low bandwidth detector circuits can be used and the mirror cavity has high transmission at resonance. Dynamic range is poor, typically 20–30 dB, caused by the partial transmission of the cavity even when the mirror spacing is non-resonant. Amplitude accuracy is usually poor because cavity transmission is a strong function of how well the incident light is mode matched to the cavity. The absolute wavelength accuracy cannot be meaningfully stated because one does not know how many standing waves are present at resonance. Fabry–Perot OSAs can operate over a wavelength range of about 20% of their design wavelength due to mirror reflectivity limitations. Therefore, several Fabry–Perot cavities are required to cover the entire visible wavelength range or the entire near- infrared range. Their main advantage over other types of OSAs is high resolution and low cost.

9.10.2 Diffraction Grating OSA

Monochromators use a diffraction grating to separate wavelengths spatially. Typically, light is focused on an input slit and is collimated by a lens or mirror before striking a rotatable diffraction grating. Light diffracted in the desired order is focused by a lens or mirror to an output slit. Light passing the slit is detected by a visible or near-infrared detector. As the diffraction grating is rotated, different wavelengths pass through the output slit. Large monochromators can have resolutions approaching 0.01 nm. Sensitivity is often excellent with high-gain detectors. Dynamic range is good due to the sharp filtering properties of a diffraction grating. However, amplitude and wavelength length accuracy are only as good as the user's calibration. The combination of a diffraction grating based monochromatic under computer control with a calibrated amplitude and wavelength display has become a popular OSA in the last decade. The input slit is replaced by an optical fibre launch with the fibre core serving as the input slit. Si and InGaAs photodiode detectors give a large operational wavelength range (up to 350–1700 nm) in one instrument. Resolutions of 0.1 nm are common, and therefore, the only reason to use a standalone monochromatic is for higher resolution.

The sensitivity of grating-based OSAs is excellent. Some instruments have noise floors below $90\,\text{dBm}(1\,\text{pW})$; yet, they can measure strong signals up to $+10\,\text{dBm}(10\,\text{mW})$. Instruments with the best dynamic range use two monochromators in series or take a double pass through one monochromator. Dynamic range is typically specified as the noise floor 1 nm away from a strong peak and can be 60 dB for double-pass monochromators. Such a large dynamic range is necessary to see side modes (other wavelengths) of a distributed feedback (DFB) semiconductor laser, which are commonly 50 dB below the desired oscillation wavelength and separated by about $1\,\text{nm}$. The absolute wavelength accuracy can be below $1\,\text{nm}$ and is limited by small changes that is optical alignment over time and temperature, and by the accuracy of setting and the diffraction grating angle. Amplitude accuracy can be better than 1 dB (about 20%) and is limited by optical alignment changes and accuracy of the factory calibration over the wide operational wavelength range.

9.10.3 Fourier Transform OSA

A scanning Michelson interferometer is a two-arm interferometer where collimated light is split in two directions by a beam splitter, reflected off two mirrors (one of which is moving towards or away from the beam splitter), recombined by the beam splitter and detected. When the input light is monochromatic, the two beams interfere to produce a beat note at a frequency given by the Doppler shift imparted by the moving mirror. This detected beat note is a function of the mirror velocity and the input light wavelength. Light of many frequencies produce a complicated beat note pattern that is hard to analyse in the time domain. In essence, however, the scanning Michelson interferometer performs an autocorrelation of the input light. That is, the optical electric field is mixed with itself after being delayed by a varying amount. The Fourier transform of the autocorrelation function is the power spectral density of the optical electric field, that is, the optical spectrum. For high-wavelength accuracy, another laser of known wavelength is sent through the interferometer simultaneously. The beat note generated by this laser is used as a clock to digitally sample the autocorrelation function of the input light. One of the major advantages of the Fourier transform-based OSA is tremendous wavelength accuracy that can approach one part per million (e.g., 0.001-nm accuracy at 1000-nm wavelength). This accuracy is limited by the stability of the reference laser wavelength and the size of the autocorrelation measured. As with the grating-based OSA, the hardware is under computer control. Amplitude accuracy is comparable to that of the grating based OSA for much the same reasons. Resolution is typically 0.1 nm and is also limited by the size of the autocorrelation measured. A larger autocorrelation requires greater mirror movement and a larger optical platform. Fourier transform-based spectrometers are popular in the 1000 to 10 000 nm wavelength range where detector noise is sizable. The large number of autocorrelation samples taken yields a tremendous averaging effect, where small signals can be seen in the presence of large noise. However, in the visible and near-infrared regime of low-noise Si and InGaAs detectors, the grating-based OSA is more sensitive by about 20 dB. Much of this difference is due to the required high bandwidth (and consequently low gain) detection electronics required with the Fourier transform OSA. Dynamic range is also worse than the grating-based OSA, typically 30 to 40 dB 1-nm away from a strong peak. This limitation is caused by shot noise, as light is always present on the photodetector during the autocorrelation measurement. However, it is only present on the grating OSA photodetector when the grating is positioned at the strong peak. Finally, cost of grating-based OSAs and Fourier transform OSAs are comparable and higher than Fabry–Perot based OSAs. Heterodyne OSA, a heterodyne optical spectrum analyser, works in much the same as an electrical spectrum analyser. The optical electric field waveform of the light to be measured is mixed (multiplied) by the optical electric field of a tunable laser source (the local oscillator). The square-law photodetector will generate photocurrents at the difference frequencies between the tunable laser and the modulated laser carrier and sidebands. Therefore, the photocurrent at 10 GHz represents the modulated laser carrier. While photocurrents at 9 and 11 GHz represent the lower and upper AM sidebands. The spectrum of the photocurrent is displayed on an electrical spectrum analyser and represents the spectrum of the modulated laser. As bandwidths of photodiodes are limited to about 20 GHz, only a 20-GHz slice of the optical spectrum can be measured at one time. The tunable laser can be set to view any 20-GHz spectral range within its typically 100-nm tuning range. The resolution of the measurement is equal to the line width (phase noise) of the tunable laser. A tunable external cavity semiconductor laser has a narrow line width typically below 100 kHz. This corresponds to a wavelength resolution on the order of 1 fm or five orders of magnitude better than a grating-based OSA. The resolution is so good that the line width of the modulated laser can be measured if it is 1 MHz or greater. The sensitivity of a heterodyne OSA is good (about 70 dBm or 100 pW) because a weak signal is mixed by a strong local oscillator (the tunable laser). Dynamic range is limited by the Lorentzian tails of the laser line width and is typically 40 dB. Wavelength accuracy is given by that of the tunable laser and is about 0.05 nm. Amplitude accuracy is not good because the electrical spectrum analyser resolution bandwidth may not be large enough to pass the full peak energy of the photocurrent. However, relative accuracy is good. No complete heterodyne OSA is commercially available now. The required elements (tunable laser, combining coupler, and high-speed photodiode and microwave spectrum analyser) are all available. The combined price of these elements is quite high when compared to the other OSAs discussed here, but the resolution is spectacular.

9.11 FIBRE OPTICS IMAGING

The field of fibre optics depends upon the total internal reflection of light rays travelling through tiny optical fibres. The fibres are so small that once the light is introduced into the fibre with an angle within the confines of the numerical aperture of the fibre, it will continue to reflect almost losslessly off the walls of the fibre and thus, it can travel long distances in the fibre. Bundles of such fibres can accomplish imaging of otherwise inaccessible areas, as shown in Fig.9.16.

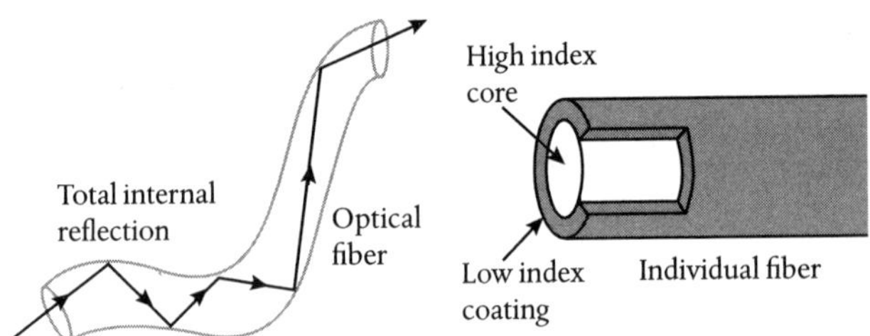

Fig. 9.16 Total internal reflection phenomenon

Fibre optic imaging uses the fact that the light striking the end of an individual fibre will be transmitted to the other end of that fibre. Each fibre acts as a light pipe, transmitting the light from that part of the image along the fibre. If the arrangement of the fibres in the bundle is kept constant, then the transmitted light forms a mosaic image of the light that struck the end of the bundle, as shown in Fig. 9.17.

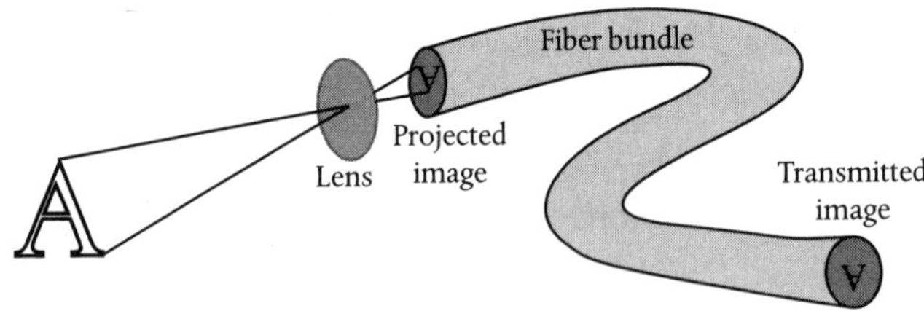

Fig. 9.17 Fibre optic imaging

The use of fibre optics in image transmission is based on the ability of a precisely aligned bundle of optical fibres to transmit an image from one end of the bundle to the other. Each fibre transmits one element of the image so that the complete image is made up of a matrix of dots (pixels) that blend into a recognisable image like a halftone picture on a printed page.

Light is transmitted through individual fibres by means of total internal reflection from the fibre walls. For efficient transmission, each fibre has a highly transparent core (usually glass) coated with a layer (cladding) of lower refractive index (also usually glass). This cladding prevents light from leaking into neighbouring fibres (crosstalk) and protects the core from contamination and wear. Fibre bundles may be either flexible or rigid.

Flexible image bundles are used in a wide variety of industrial and medical fibrescopes or endoscopes to transmit the image from an objective lens at the distal end to an eyepiece in the control handle. Industrial fibrescopes ranging in length from 3 to 8 ft (1 to 2.5 m) are used in aircraft engine inspection, for example, examination of turbine blades. Longer industrial fibrescopes are used for pipe and weld inspection, including those in nuclear power plants.

Medical endoscopes range in size from less than 0.08 in. (2 mm) for viewing inside the arteries (cardioscope) to 0.6 in. (15 mm) for examining the colon (colonoscope). The latter has an operating channel large enough to perform surgery via remotely controlled forceps or electrosurgical snares. Intermediate-sized endoscopes are used to view the bronchi (bronchoscope), the kidneys (nephroscope), the bladder (cystoscope), the throat (laryngoscope), and the stomach (gastroscope).

Fused fibre boules are the starting point for various image transmitting components, including faceplates (windows) for image intensifier tubes and cathode ray tubes, image inverters (twisters), magnifiers, and rigid conduits. Image intensifier tubes are used in military night vision devices, low-light television cameras, and astronomical telescopes. In a cathode ray tube, the fibre optic faceplate is used to transmit the image on the phosphor screen to the exit face for printing directly onto photographic material.

9.11.1 Technical Terms

Coherent Fibre Optics: A fibre optic component consisting of many individual fibre optic elements all precisely aligned so that they can faithfully transmit an image.
Fibre Optic Faceplate: A coherent fibre optic plate that precisely transmits an image from its input surface to its output surface.
Fibre Optic Taper: A coherent fibre optic plate that transmits either a magnified or minified image from its input surface to its output surface.
Image Conduit: A coherent fibre optic rod consisting of thousands of elements that transmits an image from an input surface to the output surface.
Clad Rod: A single-clad fibre that can successfully transmit light, but not images.

SUMMARY

- In 1854, John Tyndall demonstrated to the Royal Society that light could be conducted through a curved stream of water, proving that a light signal could be bent.
- In 1880, Alexander Graham Bell invented his Photophone, which transmitted a voice signal on a beam of light. Bell focused sunlight with a mirror and then talked into a mechanism that vibrated the mirror. At the receiving end, a detector picked up the vibrating beam and decoded it back into a voice the same way a phone did with electrical signals. Many things a cloudy day for instance could interfere with the Photophone causing Bell to stop any further research with this invention.
- In 1880, William Wheeler invented a system of light pipes lined with a highly reflective coating that illuminated homes by using light from an electric arc lamp placed in the basement and directing the light around the home with the pipes.
- In 1888, the medical team of Roth and Reuss of Vienna used bent glass rods to illuminate body cavities.
- In 1895, French engineer Henry Saint Rene designed a system of bent glass rods for guiding light images in an attempt at early television.
- In 1898, American David Smith applied for a patent on a bent glass rod device to be used as a surgical lamp.
- In the 1920s, Englishman John Logie Baird and American Clarence W. Hansell patented the idea of using arrays of transparent rods to transmit images for television and facsimiles, respectively.
- In 1930, German medical student, Heinrich Lamm was the first person to assemble a bundle of optical fibres to carry an image. Lamm's goal was to look inside inaccessible parts of the body. During his experiments, he reported transmitting the image of a light bulb. However, the image was of poor quality,. His effort to file a patent was denied because of Hansell's British patent.
- In 1954, Dutch scientist Abraham Van Heel and British scientist Harold. H. Hopkins separately wrote papers on imaging bundles. Hopkins reported on imaging bundles of unclad fibres while Van Heel reported on simple bundles of clad fibres. He covered a bare fibre with a transparent cladding of a lower refractive index. This protected the fibre reflection surface from outside distortion and greatly reduced interference between fibres. At the time, the greatest obstacle to a viable use of fibre optics was in achieving the lowest signal (light) loss.
- In 1961, Elias Snitzer of American Optical published a theoretical description of single-mode fibres, a fibre with a core so small it could carry light with only one-wave guide mode. Snitzer's

idea was okay for a medical instrument looking inside the human, but the fibre had a light loss of 1 decibel/m. Communications devices needed to operate over much longer distances and required a light loss of no more than 10 or 20 decibels (measurement of light) per kilometre.

- In 1964, a critical (and theoretical) specification was identified by Dr. C.K. Kao for long-range communication devices, with the standard 10 or 20 decibels of light loss per kilometre. Kao also illustrated the need for a purer form of glass to reduce light loss.
- In 1970, one team of researchers began experimenting with fused silica, a material capable of extreme purity with a high melting point and a low refractive index. Corning Glass researchers Robert Maurer, Donald Keck, and Peter Schultz invented fibre optic wire or optical waveguide fibres capable of carrying 65,000 times more information than copper wire, through which information carried by a pattern of light waves could be decoded at a destination even a thousand miles away. The team had solved the problems presented by Dr. Kao.
- In 1975, the United States Government decided to link the computers in the NORAD headquarters at Cheyenne Mountain using fibre optics to reduce interference.
- In 1977, the first optical telephone communication system was installed about 1.5 miles under downtown Chicago, and each optical fibre carried the equivalent of 672 voice channels.
- Today, more than 80% of the world's long distance traffic is carried over optical fibre cables; 25 million km of the cable Maurer, Keck, and Schultz designs has been installed worldwide.

THEORY QUESTIONS AND ANSWERS

Question 1: Write short notes on fibre optic sensors.

Answer:A fibre optic sensor is a sensor that uses optical fibre either as the sensing element (intrinsic sensors), or as a means of relaying signals from a remote sensor to the electronics that process the signals (extrinsic sensors). Fibres have many uses in remote sensing. Depending on the application, fibre may be used because of its small size, or the fact that no electrical power is needed at the remote location, or because many sensors can be multiplexed along the length of a fibre by using different wavelengths of light for each sensor, or by sensing the time delay as light passes along the fibre through each sensor. The time delay can be determined using a device like an optical time-domain reflectometer.

Over the past 20 years, two major product revolutions have taken place due to the growth of the optoelectronics and fibre optic communication industries. The optoelectronics industry has brought products such as compact disc players, laser printers, bar code scanners, and laser pointers. The fibre optic communication industry has literally revolutionised the telecommunication industry by providing higher performance and more reliable telecommunication links with ever decreasing bandwidth cost. This revolution is bringing the benefits of high-volume production to component users and a true information superhighway built of glass.

In parallel with these developments, fibre optic sensor technology has been a major user of technology associated with the optoelectronic and fibre optic communication industry. Many of the components associated with these industries were often developed for fibre optic sensor applications. Fibre optic sensor technology in turn has often been driven by the development and subsequent mass production of components to support these industries. As component prices have fallen and quality improvements have been made, the ability of fibre optic sensors to displace traditional sensors for rotation, acceleration, electric and magnetic field measurement, temperature, pressure, acoustics, vibration, linear and angular position, strain, humidity, viscosity, chemical measurements and a host of other sensor applications, has been enhanced. In the early days of fibre optic sensor technology, most commercially successful fibre optic sensors were squarely targeted at markets where existing sensor technology was marginal or in many cases, it is nonexistent. The inherent advantages of fibre optic sensors that include their ability to be lightweight, of very small size, passive, low power, resistant to electromagnetic interference, high sensitivity, wide bandwidth, and environmental ruggedness were heavily used to offset their major disadvantages of high cost and unfamiliarity to the end user.

The situation is changing. Laser diodes that cost $3000 in 1979 with lifetimes measured in hours are now being sold for a few dollars in small quantities; they have reliability of tens of thousands of hours and are used widely in compact disc players, laser printers, laser pointers, and bar code readers. Single-mode optical fibres were at a cost of $20/m in 1979; however, now they cost less than $0.10/m with vastly improved optical and mechanical properties. Integrated optical devices that were not available in usable form at that time are now commonly used to support production models of fibre optic gyros. Further, they could drop dramatically in price in the future while offering ever more sophisticated optical circuits. As these trends continue, the opportunities for fibre optic sensor designers to produce competitive products will increase and the technology can be expected to assume an ever more prominent position in the sensor marketplace. In the following sections, the basic types of fibre optic sensors that are being developed will be briefly reviewed followed by a discussion of how these sensors are and will be applied.

Question 2: Write a short note on intrinsic sensors or extrinsic sensors.

Answer: Optical fibres can be used as sensors to measure strain, temperature, pressure, and other quantities by modifying a fibre so that the quantity to be measured modulates the intensity, phase, polarisation, and wavelength or transit time of light in the fibre. Sensors that vary the intensity of light are the simplest, since only a simple source and detector are required. A particularly useful feature of intrinsic fibre optic sensors is that they can, if required, provide distributed sensing over very large distances.

Temperature can be measured by using a fibre that has evanescent loss that varies with temperature, or by analysing the Raman scattering of the optical fibre. Electrical voltage can be sensed by nonlinear optical effects in specially doped fibre, which alter the polarisation of light as a function of voltage or electric field. Angle measurement sensors can be based on the Sagnac effect.

Special fibres like long-period fibre grating (LPG) optical fibres can be used for direction recognition. Photonics Research Group of Aston University in UK has some publications on vectorial bend sensor applications.

Optical fibres are used as hydrophones for seismic and sonar applications. Hydrophone systems with more than 100 sensors per fibre cable have been developed. Hydrophone sensor systems are used by the oil industry as well as by the navies of few countries. Both bottom-mounted hydrophone arrays and towed streamer systems are in use. The German company Sennheiser developed a laser microphone for use with optical fibres.

A fibre optic microphone and fibre optic-based headphone are useful in areas with strong electrical or magnetic fields, such as communication amongst the team of people working on a patient inside a magnetic resonance imaging (MRI) machine during MRI-guided surgery. Optical fibre sensors for temperature and pressure have been developed for down hole measurement in oil wells. The fibre optic sensor is well-suited for this environment as it functions at temperatures too high for semiconductor sensors (distributed temperature sensing).

Optical fibres can be made into interferometric sensors like fibre optic gyroscopes, which are used in the Boeing 767 and in some car models (for navigation purposes). They are also used to make hydrogen sensors. Fibre optic sensors have been developed to measure co-located temperature and strain simultaneously with very high accuracy using fibre Bragg gratings. This is particularly useful when acquiring information from small complex structures. Brillouin scattering effects can be used to detect strain and temperature over larger distances (20–30 km).

Extrinsic Sensors

Extrinsic fibre optic sensors use an optical fibre cable, normally a multimode one, to transmit modulated light from either a non-fibre optical sensor or an electronic sensor connected to an optical transmitter. A major benefit of extrinsic sensors is their ability to reach places that are otherwise inaccessible. An example is the measurement of temperature inside aircraft jet by using a fibre to transmit radiation into a radiation pyrometer located outside the engine. Extrinsic sensors can also be used in the same way to measure the internal temperature of electrical transformers, where the extreme electromagnetic fields present make other measurement techniques impossible.

Extrinsic fibre optic sensors provide excellent protection of measurement signals against noise corruption. Unfortunately, many conventional sensors produce electrical output that must be converted into an optical signal for use with fibre. For example, in the case of a platinum resistance thermometer, the temperature changes are translated into resistance changes. The PRT must, therefore, have an electrical power supply. The modulated voltage level at the output of the PRT can then be injected into the optical fibre via the usual type of transmitter. This complicates the measurement process and means that low-voltage power cables must be routed to the transducer. Extrinsic sensors are used to measure vibration, rotation, displacement, velocity, acceleration, torque, and twisting.

Question 3: Discuss the basic concepts and intensity-based fibre optic sensors.

Answer: Fibre optic sensors are often loosely grouped into two basic classes referred to as extrinsic or hybrid fibre optic sensors, and intrinsic or all-fibre sensors. Figure P1 illustrates the case of an extrinsic or hybrid fibre optic sensor.

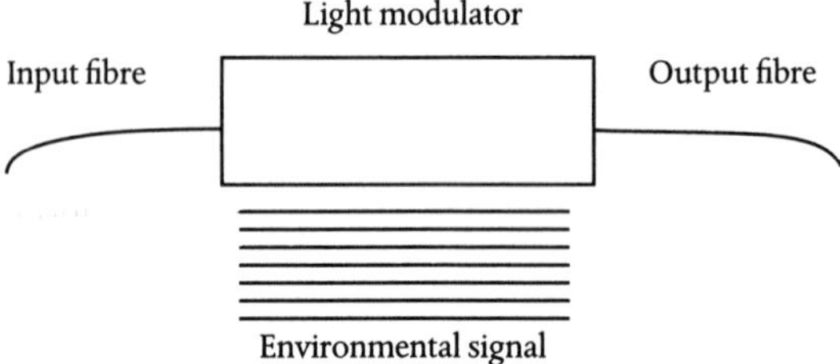

Fig. P1 Extrinsic fibre optic sensors consist of optical fibres that lead up to and out of a "black box" that modulates the light beam passing through it in response to an environmental effect

In this case, an optical fibre leads up to a "black box" that impresses information onto the light beam in response to an environmental effect. The information could be impressed in terms of intensity, phase, frequency, polarisation, spectral content, or other methods. An optical fibre then carries the light with the environmentally impressed information back to an optical and/or electronic processor. In some cases, the input optical fibre also acts as the output fibre. The intrinsic or all-fibre sensors shown in Fig. P2 use optical fibre to carry the light beam and the environmental effect impresses information onto the light beam while it is in the fibre. Each of these classes of fibres in turn has many subclasses; in some cases, sub-subclasses (1) that consist of large number of fibre sensors.

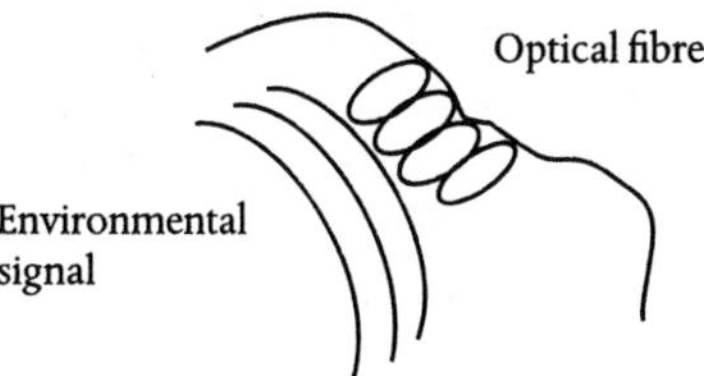

Fig. P2 Intrinsic fibre optic sensors rely on the light beam propagating through the optical fibre being modulated by the environmental effect either directly or through environmentally induced optical path length changes in the fibre itself.

In some prospects, the simplest type of fibre optic sensor is the hybrid type that is based on intensity modulation. Figure P3 shows a simple closure or vibration sensor that consists of two optical fibres that are held in close proximity to each other. Light is injected into one of the optical fibres, and when it exits, the light expands into a cone of light whose angle depends on the difference between the index of refraction of the core and cladding of the optical fibre. The amount of light captured by the second optical fibre depends on its acceptance angle and the distance d between the optical fibres. When the distance d is modulated, it in turn results in an intensity modulation of the light captured.

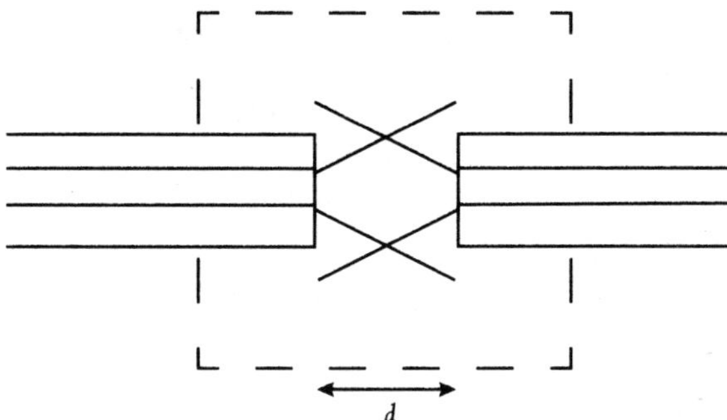

Fig. P3 Closure and vibration fibre optic sensors based on numerical aperture can be used to support door closure indicators and measure levels of vibration in machinery

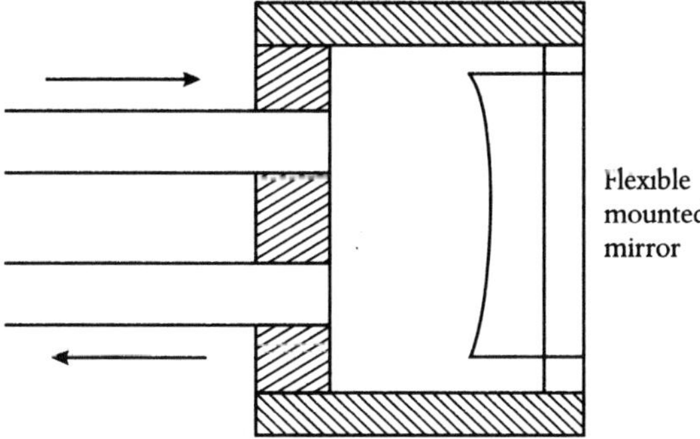

Fig. P4 Numerical aperture fibre sensor based on a flexible mirror can be used to measure small vibrations and displacements.

A variation on this type of sensor is shown in Fig. P4. Here, a mirror is used that is flexibly mounted to respond to an external effect like pressure. As the mirror position shifts the effective separation between the optical fibres shift with a resultant intensity modulation. These types of sensors are useful for such applications as door closures where a reflective strip, in combination with an optical fibre acting to input and catch the output reflected light, can be used.

By arranging two optical fibres in line, a simple translation sensor can be configured as in Fig. P5. The output from the two detectors can be proportioned to determine the translational position of the input fibre.

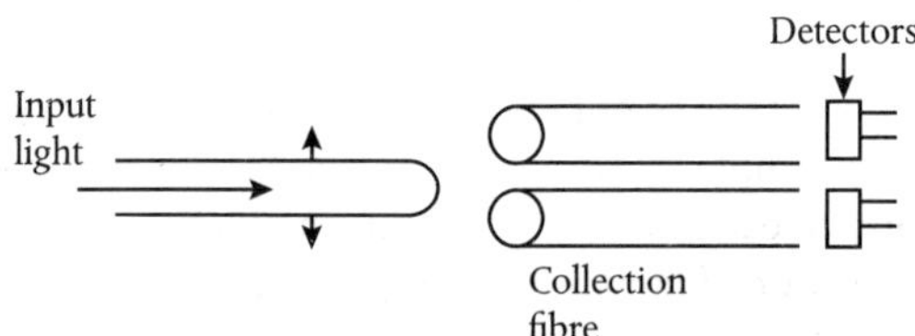

Fig. P5 Fibre optic translation sensor based on numerical aperture uses the ratio of the output on the detectors to determine the position of the input fibre

One of the classes of intensity-based fibre optic sensors is based on the principle of total internal reflection. In the case of the sensor in Fig. P6, light propagates down the fibre core and hits the angled end of the fibre. If the medium into which the angled end of the fibre is placed has a low enough index of refraction, then virtually all the light is reflected when it hits the mirrored surface and returns via the fibre. If, however, the index of refraction of the medium starts to approach that of the glass, some of the light propagates out of the optical fibre and is lost resulting in an intensity modulation.

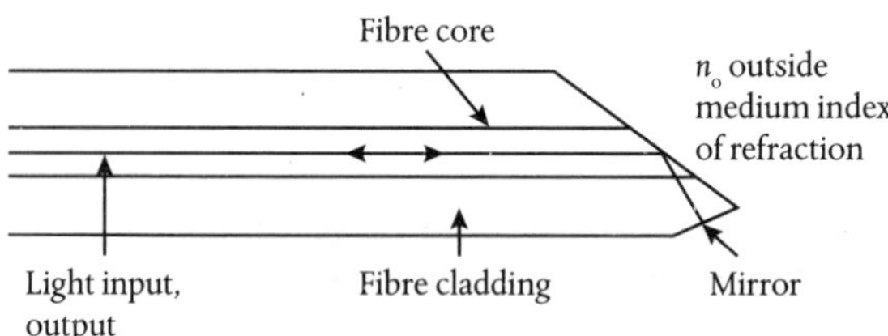

Fig. P6 Fibre sensor using critical angle properties of a fibre for pressure/index of refraction measurement via measurements of the light reflected into the fibre

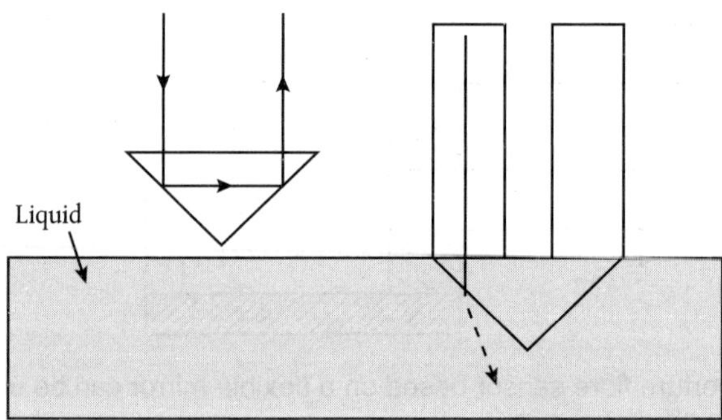

Fig. P7 Liquid-level sensor based on total internal reflection detects the presence or absence of liquid by the presence or absence of a return light signal.

This type of sensor can be used for low-resolution measurement of pressure or index of refraction changes in a liquid or gel with 1 to 10% accuracy. Variations on this method have also been used to measure liquid level as shown by the probe configuration of Fig. P7. When the liquid level hits the reflecting prism, the light leaks into the liquid greatly attenuating the signal.

When single-mode fibre is used there is considerable leakage of the propagating light beam mode beyond the core region into the cladding or medium around it. If a second fibre core is placed nearby, this evanescent tail will tend to cross couple to the adjacent fibre core. The amount of cross coupling depends on a number of parameters including the wavelength of light, the relative index of refraction of the medium in which the fibre cores are placed, the distance between the cores and the interaction length. This type of fibre sensor can be used for the measurement of wavelength, spectral filtering, index of refraction, and environmental effects acting on the medium surrounding the cores (temperature, pressure, and strain). The difficulty with this sensor that is common to many fibre sensors is optimising the design so that only the desired parameters are sensed.

Another way that light may be lost from an optical fibre is when the bend radius of the fibre exceeds the critical angle necessary to confine the light to the core area and there is leakage into the cladding. Microbending of the fibre locally can cause this to result with resultant intensity modulation of light propagating through an optical fibre. A series of microbend-based fibre sensors have been built to sense vibration, pressure, and other environmental effects. Figure P.8 shows a typical layout of this type of device consisting of a light source; a section of optical fibre positioned in a microbend transducer is designed to intensity modulate light in response to an environmental effect and a detector. In some cases, the microbend transducer can be implemented by using special fibre cabling or optical fibre that is simply optimised to be sensitive to microbending loss.

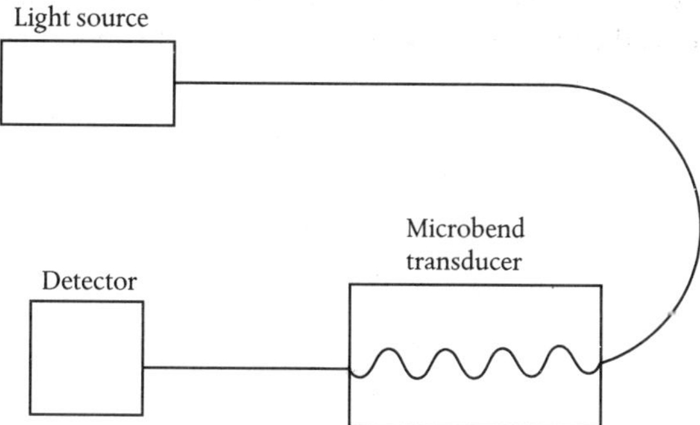

Fig. P8 Microbend fibre sensors are configured so that an environmental effect results in an increase or decrease in loss through the transducer due to light loss resulting from small bends in the fibre

Intensity-based fibre optic sensors have a series of limitations imposed by variable losses in the system that are not related to the environmental effect to be measured. Potential error sources include variable losses due to connectors and splices, microbending loss, macrobending loss, mechanical creep and misalignment of light sources and detectors. To circumvent these problems, many of the successful higher performance intensity-based fibre sensors employ dual wavelengths. One of the wavelengths is used to calibrate out all the errors due to undesired intensity variations by bypassing the sensing region. An alternative approach is to use fibre optic sensors that are inherently resistant to errors induced by intensity variations. In the next section a series of spectrally based fibre sensors that have this characteristic are discussed.

Spectrally Based Fibre Optic Sensors

Spectrally based fibre optic sensors depend on a light beam being modulated in wavelength by an environmental effect. Examples of these types of fibre sensors include those based on blackbody radiation, absorption, fluorescence, etalons, and dispersive gratings.

One of the simplest among these types of sensors is the blackbody sensor as given in Fig. P9. A blackbody cavity is placed at the end of an optical fibre. When the cavity rises in temperature, it starts to glow and acts as a light source.

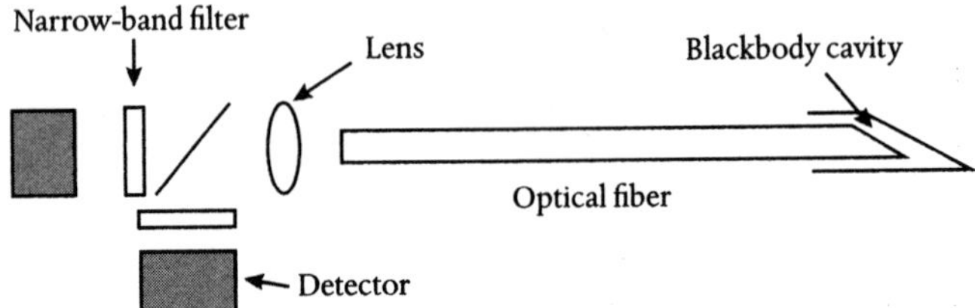

Fig. P9 Blackbody fibre optic sensors allow the measurement of temperature at a hot spot and are most effective at temperatures of higher than 300 °C

Detectors in combination with narrow-band filters are then used to determine the profile of the blackbody curve, and in turn, the temperature as in Fig. P10. This type of sensor has been successfully commercialised and has been used to measure temperature to within a few degrees Celsius under intense RF fields. The performance and accuracy of this sensor is better at higher temperatures and falls off at temperatures on the order of 200 °C because of low signal to noise ratios. Care must be taken to ensure that the hottest spot is the blackbody cavity and not on the optical fibre that leads itself as this can corrupt the integrity of the signal.

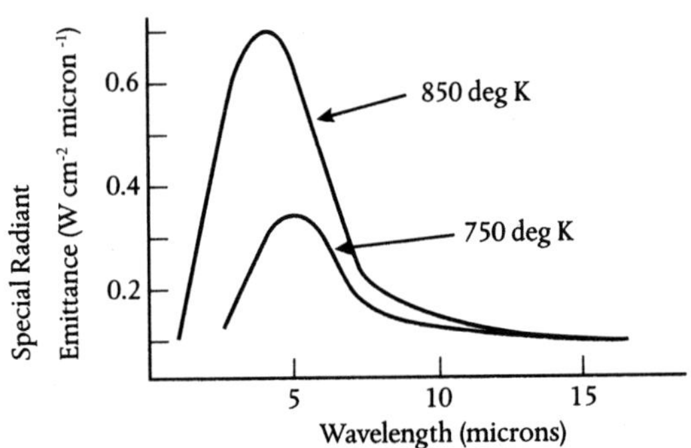

Fig. P10 Blackbody radiation curves provide unique signatures for each temperature

Fluorescent-based fibre sensors are being widely used for medical applications and chemical sensing and they can also be used for physical parameter measurements such as temperature, viscosity, and humidity. There are a number of configurations for these sensors and Fig. P11 illustrates two of the most common ones. In the case of the end-tip sensor, light propagates down the fibre to a probe of fluorescent material. The resultant fluorescent signal is captured by the same fibre and directed back to an output demodulator. The light sources can be pulsed and probes have been made that depend on the time rate of decay of the light pulse.

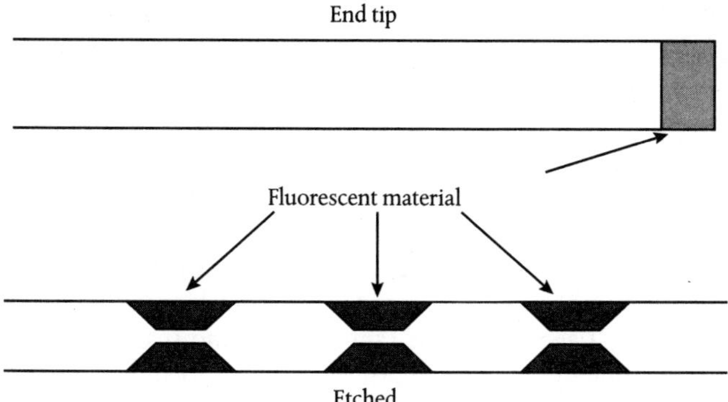

Fig. P11 Fluorescent fibre optic sensor probe configurations can be used to support the measurement of physical parameters as well as the presence or absence of chemical species. These probes may be configured to be single ended or multipoint by using side etches techniques and attaching the fluorescent material to the fibre.

In the continuous mode, parameters such as viscosity, water vapour content, and degree of cure in carbon fibre reinforced epoxy and thermoplastic composite materials can be monitored.

An alternative is to use the evanescent properties of the fibre and etch regions of the cladding away and refill them with fluorescent material. By sending a light pulse down the fibre and looking at the resulting fluorescence, a series of sensing regions may be time-division multiplexed.

It is also possible to introduce fluorescent dopants into the optical fibre itself. This approach would cause the entire optically activated fibre to fluoresce. By using time-division multiplexing, various regions of the fibre could be used to make a distributed measurement along the fibre length. In many cases, users of fibre sensors would like to have the fibre optic analogue of conventional electronic sensors. An example is the electrical strain gauge that is used widely by structural engineers. Fibre grating sensors can be configured to have gauge lengths from 1 mm to approximately 1 cm, with sensitivity comparable to conventional strain gauges.

Question 4: Discuss the interferometric fibre optic sensors and Sagnac interferometer.

Answer: One of the areas of greatest interest has been in the development of high performance interferometric fibre optic sensors. Substantial efforts have been undertaken on Sagnac interferometers, ring resonators, Mach–Zehnder, and Michelson interferometers as well as dual-mode, polarimetric, grating, and etalon-based interferometers. In this section, the Sagnac, Mach–Zehnder, and Michelson interferometers are briefly reviewed.

The Sagnac interferometer has been principally used to measure rotation and is a replacement for ring laser gyros and mechanical gyros. It may also be employed to measure time-varying effects such as acoustics and vibration and slowly varying phenomenon like strain. By using multiple interferometer configurations, it is possible to employ the Sagnac interferometer as a distributed sensor capable of measuring the amplitude and location of a disturbance.

The single most important application of fibre optic sensors in terms of commercial value is the fibre optic gyro. It was recognised very early that the fibre optic gyro offered the prospect of an all solid-state inertial sensor with no moving parts, unprecedented reliability, and very low cost. The potential of the fibre optic gyro is being realised as several manufacturers worldwide are producing them in large quantities to support automobile navigation systems, pointing and tracking of satellite antennas, inertial measurement systems for commuter aircraft and missiles, and as backup guidance system for the Boeing 777. They are also being baselined for such future programs as the Comanche helicopter and developed to support long-duration space flights. Other applications where fibre optic gyros are being used include mining operations, tunnelling, attitude control for a radio-controlled helicopter, cleaning robots, antenna pointing and tracking, and guidance for unmanned trucks and carriers.

Two types of fibre optic gyros are being developed. The first type is an open-loop fibre optic gyro with a dynamic range on the order of 1000–5000 (dynamic range is unitless), with scale factor accuracy of about 0.5% (this accuracy number includes nonlinearity and hysteresis effects) and sensitivities that vary from less than 0.0°/h to 100°/h and higher. These fibre gyros are generally used for low-cost applications where dynamic range and linearity are not the crucial issues. The second type is the closed-loop fibre optic gyro that may have a dynamic range of 106 and scale factor linearity of 10 ppm or better. These types of fibre optic gyros are primarily targeted at medium to high-accuracy navigation applications that have high turning rates and require high linearity and large dynamic ranges.

The basic open-loop fibre optic gyro is illustrated in Fig. P12. A broadband light source like a light-emitting diode is used to couple light into an input/output fibre coupler. The input light beam passes through a polariser that is used to insure the reciprocity of the counter propagating light beams through the fibre coil. The second central coupler splits the two light beams into the fibre optic coil where they pass through a modulator that is used to generate a time-varying output signal indicative of rotation. The modulator is offset from the centre of the coil to impress a relative phase difference between the counter propagating light beams. After passing through the fibre coil, the two light beams recombine and pass through the polariser and are directed onto the output detector.

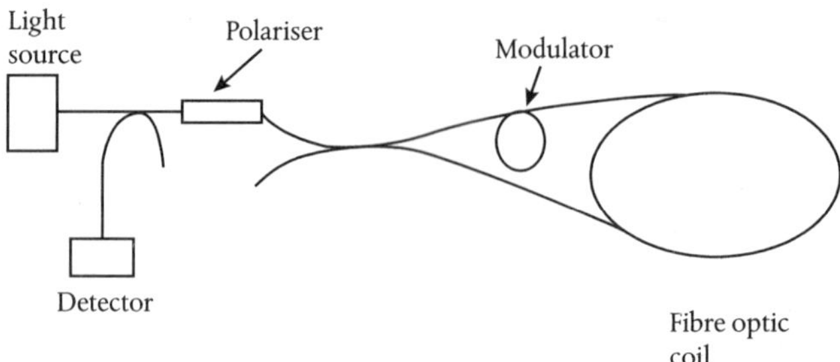

Fig. P12 Open-loop fibre optic gyro is the simplest and lowest cost rotation sensor. They are widely used in commercial applications where their dynamic range and linearity limitations are not constraining

When the fibre gyro is rotated in a clockwise direction, the entire coil is displaced slightly increasing the time it takes light to traverse the fibre optic coil. (Remember that the speed of light is invariant with respect to the frame of reference; thus, coil rotation increases the path length when viewed from outside the fibre.) Thus, the clockwise propagating light beam has to go through a slightly longer optical path length than the counterclockwise beam, which is moving in a direction opposite to the motion of the fibre coil. The net phase difference between the two beams is proportional to the rotation rate.

Question 5: Write a note on Mach–Zehnder and Michelson interferometers.

Answer: One of the greatest advantage of all fibre interferometers, such as Mach–Zehnder and Michelson interferometers in particular, is that they have extremely flexible geometry's and high sensitivity that allow the possibility of a wide variety of high performance elements and arrays, as shown in Fig. P13.

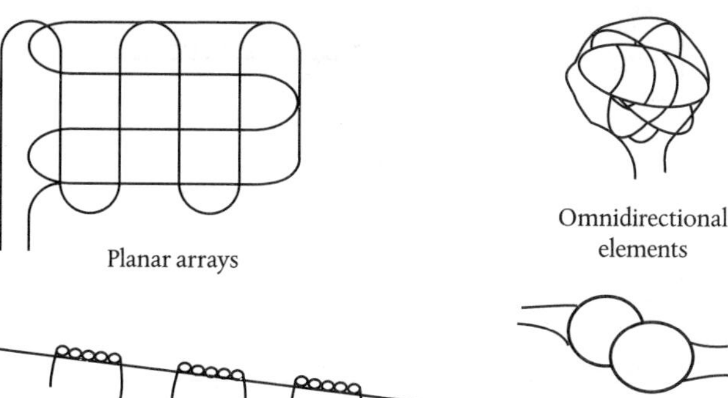

Fig. P13 Flexible geometry's of interferometric fibre optic sensors' transducers are one of the features of fibre sensors that are attractive to designers configuring special purpose sensors

Figure P14 shows the basic elements of a Mach–Zehnder interferometer, which are a light source/coupler module, a transducer, and a homodyne demodulator. The light source module usually consists of a long coherence length isolated laser diode, a beam splitter to produce two light beams, and a means of coupling the beams to the two legs of the transducer. The transducer is configured to sense an environmental effect by isolating one light beam from the environmental effect and using the action of the environmental effect on the transducer is to induce an optical path length difference between the two light beams. Typically, a homodyne demodulator is used to detect the difference in optical path length (various heterodyne schemes have also been used).

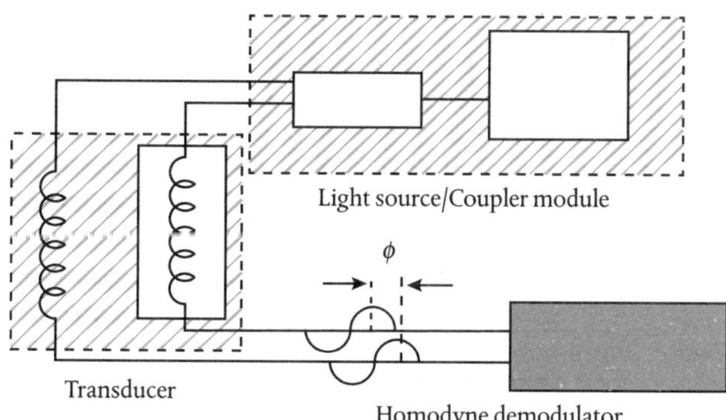

Fig. P14 The basic elements of the fibre optic Mach–Zehnder interferometer are a light source module to split a light beam into two paths, a transducer used to cause an environmentally dependent differential optical path length between the two light beams, and a demodulator that measures the resulting path length difference between the two light beams.

One of the basic issues with the Mach–Zehnder interferometer is that the sensitivity will vary as a function of the relative phase of the light beams in the two legs of the interferometer, as shown in Fig. P15. One way to solve the signal fading problem is to introduce a piezoelectric fibre stretcher into one of the legs and adjust the relative path length of the two legs for optimum sensitivity. Another approach has the same quadrature solution as the grating-based fibre sensors discussed earlier.

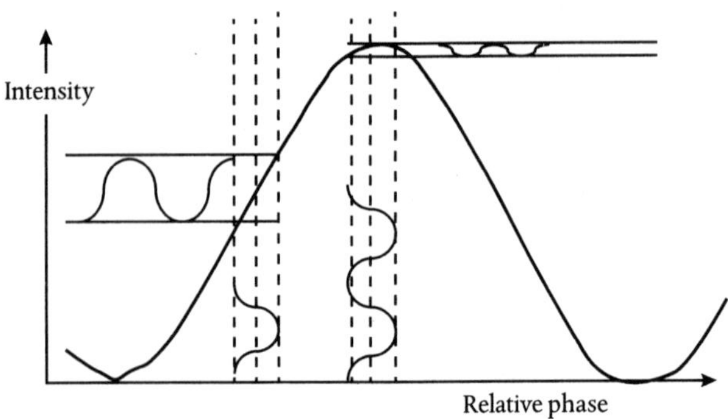

Fig. P15 In the absence of compensating demodulation methods the sensitivity of the Mach–Zehnder varies with the relative phase between the two light beams. It falls to low levels when the light beams are completely in or out of phase.

Question 6: Discuss multiplexing and distributed sensing schemes in optical fibre sensor.

Answer: Many of the intrinsic and extrinsic sensors may be multiplexed offering the possibility of large numbers of sensors being supported by a single-fibre optic line. The techniques that are most commonly employed are time, frequency, wavelength, coherence, polarisation, and spatial multiplexing. Time-division multiplexing employs a pulsed light source launching light into an optical fibre and analysing the time delay to discriminate between sensors. This technique is commonly employed to support distributed sensors where measurements of strain, temperature, or other parameters are collected. Figure P16 illustrates a time-division multiplexed system that uses microbend sensitive areas on pipe joints.

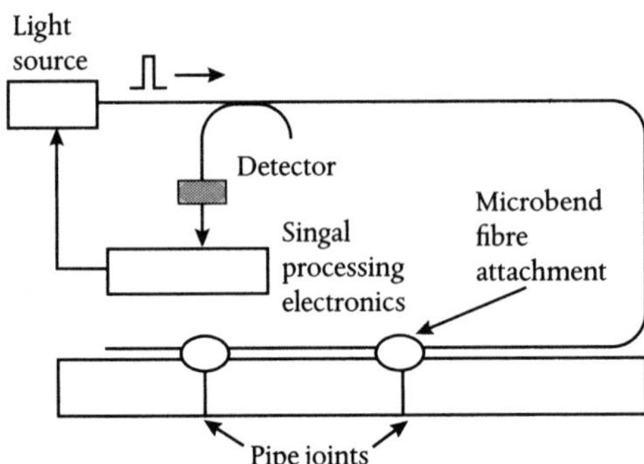

Fig. P16 Time-division multiplexing methods can be used in combination with microbend sensitive optical fibre to locate the position of stress along a pipeline

As the pipe joints are stressed, microbending loss increases and the time delay associated with these losses allows the location of faulty joints. The entire length of the fibre can be made microbend sensitive and Rayleigh scattering loss used to support a distributed sensor that will predominantly measure strain. Other types of scattering from optical pulses propagating down optical fibre have been used to support distributed sensing, notably Raman scattering for temperature sensors has been made into a commercial product by York Technology and Hitachi. These units can resolve temperature changes of about 1 °C with spatial resolution of 1 m for a 1-km sensor using an integration time of about 5 min. Brillioun scattering has been used in laboratory experiments to support both strain and temperature measurements.

A frequency-division multiplexed system is shown in Fig. P17. In this example, a laser diode is frequency chirped by driving it with a saw-tooth current drive. Successive Mach–Zehnder interferometers are offset with incremental lengths $(L-L_1)$, $(L-L_2)$, and $(L-L_3)$, which differ sufficiently that the resultant carrier frequency of each sensor $(dF/dt)(L-L_n)$ is easily separable from the other sensors via electronic filtering of the output of the detector.

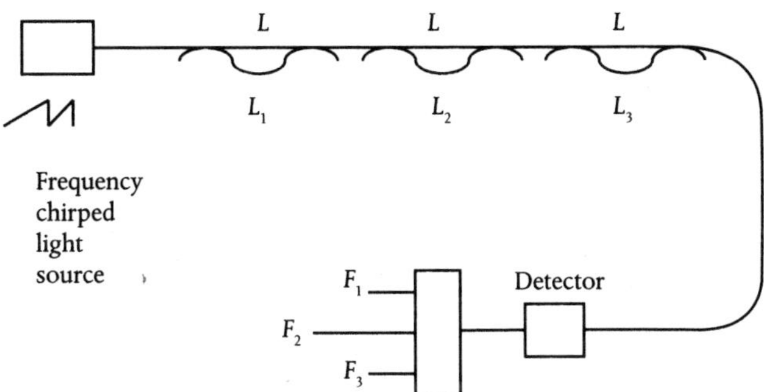

Fig. P17 Frequency-division multiplexing can be used to tag a series of fibre sensors, as in this case the Mach–Zehnder interferometers are shown with a carrier frequency on which the output signal ride.

Wavelength-division multiplexing is one of the best methods of multiplexing as it uses optical power very efficiently. It also has the advantage of being easily integrated into other multiplexing systems allowing the possibility of large numbers of sensors being supported in a single-fibre line. Figure P18 illustrates a system where a broadband light source, like a light emitting diode, is coupled into a series of fibre sensors that reflect signals over wavelength bands that are subsets of the light source spectrum. A dispersive element, such as a grating or prism, is used to separate the signals from the sensors onto separate detectors.

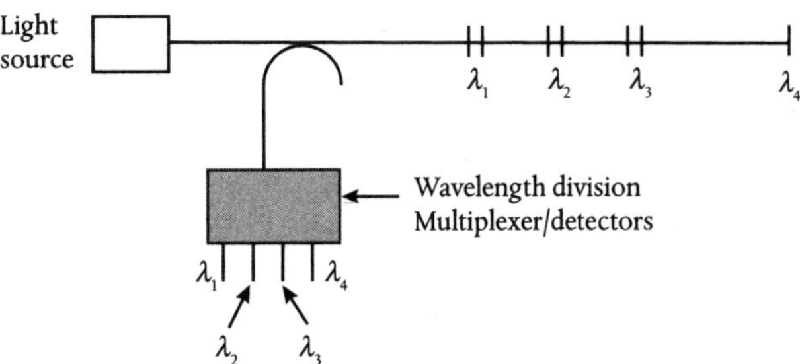

Fig. P18 Wavelength-division multiplexing are often very energy efficient. A series of fibre sensors are multiplexed by being arranged to reflect in a particular spectral band that is split via a dispersive element onto separate detectors

Light sources can have widely varying coherence lengths depending on their spectrum. By using light sources that have coherence lengths that are short when compared to offsets between the reference and signal legs in Mach–Zehnder interferometers and between successive sensors, a coherence multiplexed system similar to Fig. P19 may be set up. The signal is extracted by putting a rebalancing interferometer in front of each detector so that the sensor signals may be processed. Coherence multiplexing is not as commonly used as time-, frequency-, and wavelength-division multiplexing because of optical power budgets and the additional complexities in setting up the optics properly. It is still a potentially powerful technique and may become more widely used as optical component performance and availability continue to improve, especially in the area of integrated optic chips where control of optical path length differences is relatively straightforward.

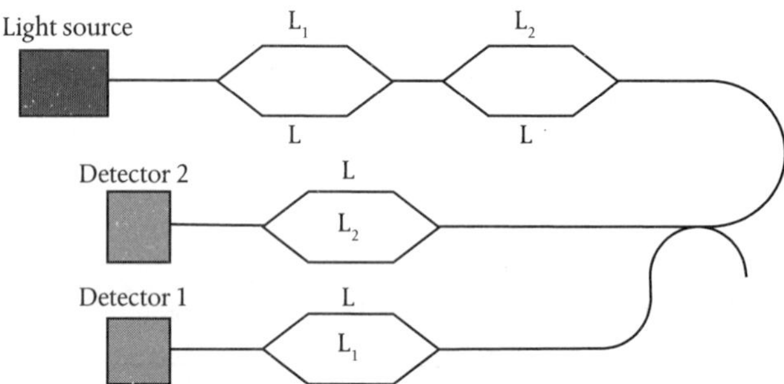

Fig. P19 A low-coherence light source is used to multiplex two Mach–Zehnder interferometers by using offset lengths and counterbalancing interferometers

One of the least commonly used techniques is polarisation multiplexing. In this case, the idea is to launch light with particular polarisation states and extract each state. A possible application is shown in Fig. P20 where light is launched with two orthogonal polarisation modes; preserving fibre and evanescent sensors have been set up along each of the axes. A polarising beam splitter is used to separate the two signals. There is a recent interest in using polarisation preserving fibre in combination with time-domain techniques to form polarisation-based distributed fibre sensors. This has potential to offer multiple sensing parameters along a single-fibre line.

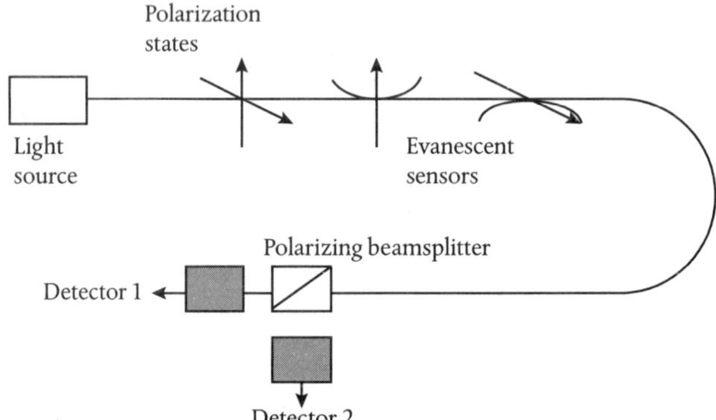

Fig. P20 Polarisation multiplexing is used to support two fibre sensors that access the cross-polarisation states of polarisation preserving optical fibre

Finally, it is possible to use spatial techniques to generate large sensor arrays using relatively few input and output optical fibres. Figure P21 shows a 2×2 array of sensors where two light sources are amplitude modulated at different frequencies. Two sensors are driven at one frequency and two more at the second. The signals from the sensors are put onto two output fibres each carrying a sensor signal from two sensors at different frequencies.

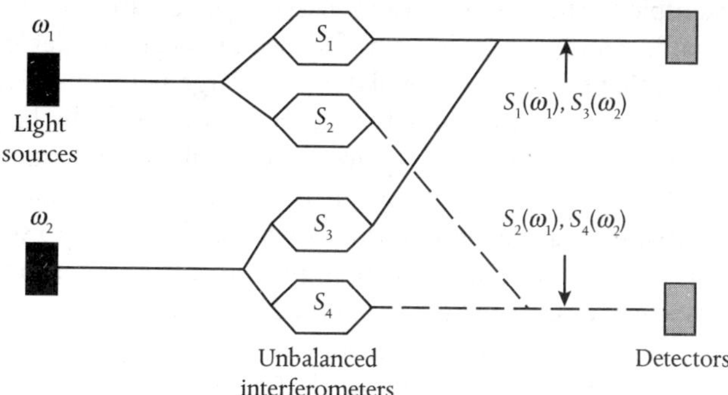

Fig. P21 Spatial multiplexing of four fibre optic sensors may be accomplished by operating two light sources with different carrier frequencies and cross coupling the sensor outputs onto two output fibres

This sort of multiplexing is easily extended to m input fibres and n output fibres to form $m \times n$ arrays of sensors as shown in Fig. P22.

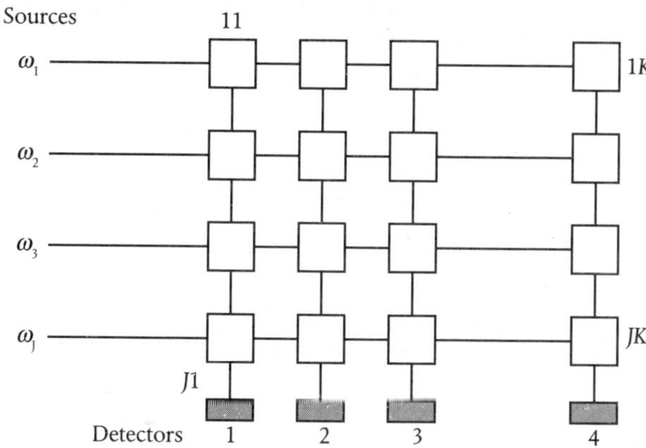

Fig. P22 Extensions of spatial multiplexing the **JK** sensors can be accomplished by operating **J** light sources at **J** different frequencies and cross coupling to **K** output fibres

All of these multiplexing techniques can be used in combination with one another to form extremely large arrays.

Question 7: Discuss the applications of fibre optic sensors.

Answer: Fibre optic sensors are being developed and used in two major ways. The first is used as a direct replacement for existing sensors where the fibre sensor offers significantly improved performance, reliability, safety and/or cost advantages to the end user. The second area is the development and deployment of fibre optic sensors in new market areas. For the case of direct replacement, the inherent value of the fibre sensor, to the customer, has to be sufficiently high to displace the older technology. Since this often involves replacing technology the customer is familiar with, the improvements must be substantial.

The most obvious example of a fibre optic sensor succeeding in this arena is the fibre optic gyro, which is displacing both mechanical and ring laser gyros for medium accuracy devices. As this technology matures, it can be expected that the fibre gyro will dominate large segments of this market. Significant development efforts are underway in the United States in the area of fly-by light where conventional electronic sensor technology are targeted to be replaced by equivalent fibre optic sensor technology that offers sensors with relative immunity to electromagnetic interference, significant weight savings, and safety improvements. In manufacturing, fibre sensors are being developed to support process control. Oftentimes, the selling points for these sensors are improvements in environmental ruggedness and safety, especially in areas where electrical discharges could be hazardous.

Medicine is one other area where fibre optic sensors are being mass produced. Here, they are being used to measure blood gas parameters and dosage levels. Because these sensors are completely passive, they pose no electrical shock threat to the patient and their inherent safety has led to a relatively rapid introduction. The automotive industry, construction industry, and other traditional users of sensors remain relatively untouched by fibre sensors mainly because of cost considerations. This can be expected to change as the improvements in optoelectronics and fibre optic communications continue to expand along with the continuing emergence of new fibre optic sensors.

New market areas present opportunities where equivalent sensors do not exist. New sensors, once developed, will most likely have a large impact in these areas. A prime example of this is in the area of fibre optic smart structures. Fibre optic sensors are being embedded into or attached to materials (1) during the manufacturing process to enhance process control systems, (2) to augment non-destructive evaluation once parts have been made, (3) to form health and damage assessment systems once parts have been assembled into structures, and (4) to enhance control systems. A basic fibre optic smart structure system is shown in Fig. P23.

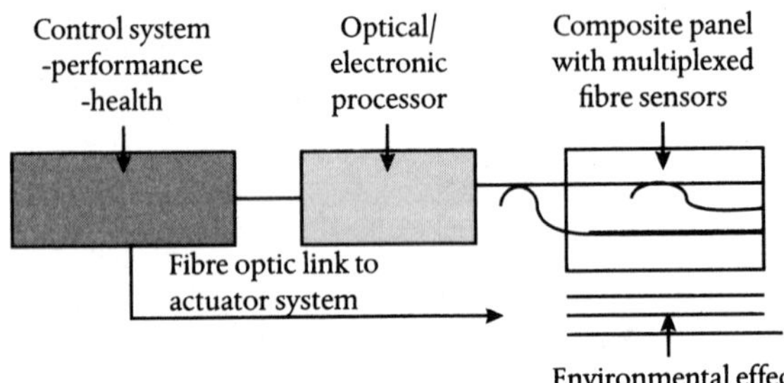

Fig. P23 Fibre optic smart structure systems consist of optical fibre sensors embedded or attached to parts sensing environmental effects that are multiplexed and directed down. The effects are then sent through an optical fibre to an optical/electronic signal processor that in turn feeds the information to a control system that may or may not act on the information via a fibre link to an actuator

Fibre optic sensors can be embedded in a panel and multiplexed to minimise the number of leads. The signals from the panel are fed to an optical/electronic processor for decoding. The information is formatted and transmitted to a control system that could be augmenting performance or assessing health. The control system would then act, via a fibre optic link, to modify the structure in response to the environmental effect. Figure P24 shows how the system might be used in manufacturing. Here, fibre sensors are attached to a part to be processed in an autoclave. Sensors could be used to monitor internal temperature, strain, and degree of cure. These measurements could be used to control the autoclaving process, improving yield and the quality of the parts.

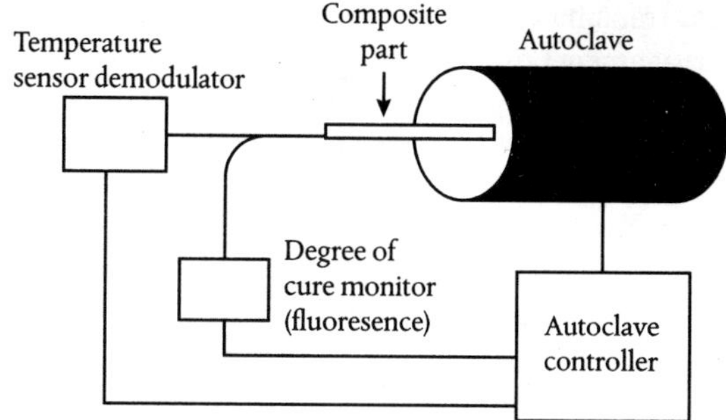

Fig. P24 Smart manufacturing systems offer the prospect of monitoring key parameters of parts as they are being made, which increases yield and lowers overall costs

Interesting areas for health and damage assessment systems are on large structures such as buildings, bridges, dams, aircrafts, and spacecraft. In order to support these types of structures, it will be necessary to have very large numbers of sensors that are rapidly reconfigurable and redundant. It will also be absolutely necessary to demonstrate the value and cost effectiveness of these systems to the end users. One approach to this problem is to use fibre sensors that have the potential to be manufactured cheaply in very large quantities while offering superior performance characteristics. Two candidates that are under investigation are the fibre gratings and etalons described in the prior sections. Both offer the advantages of spectrally based sensors and have the prospect of rapid in-line manufacture. In the case of the fibre grating, the early demonstration of fibre being written into it as it is being pulled has been especially impressive.

These fibre sensors could be folded into the wavelength and time-division multiplexed modular architecture shown in Fig. P25. Here, sensors are multiplexed along fibre strings and an optical switch is used to support the many strings. Potentially, the fibre strings could have tens or hundreds of sensors and the optical switches could support a like number of strings. To avoid overloading the system, the output from the sensors could be slowly scanned to determine status in a continuously updated manner.

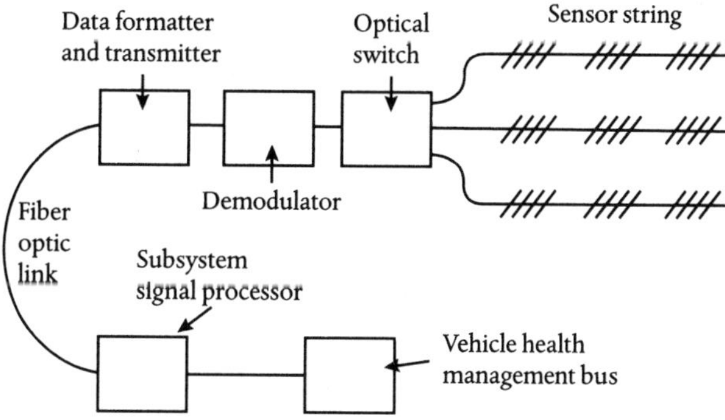

Fig. P25 A modular architecture for a large smart structure system would consist of strings of fibre sensors accessible via an optical switch and demodulator system that could select key sensors in each string. The information would then be formatted and transmitted after conditioning to a vehicle health management bus

When an event occurred that required a more detailed assessment, the appropriate strings and the sensors in them could be monitored in a high-performance mode. The information from these sensors would then be formatted and transmitted via a fibre optic link to a subsystem signal processor before introduction into a health management bus. In the case of avionics, the system architecture might look like Fig. P26. The information from the health management bus could be processed and distributed to the pilot, or more likely, it could reduce its direct workload leaving more time for the necessary control functions.

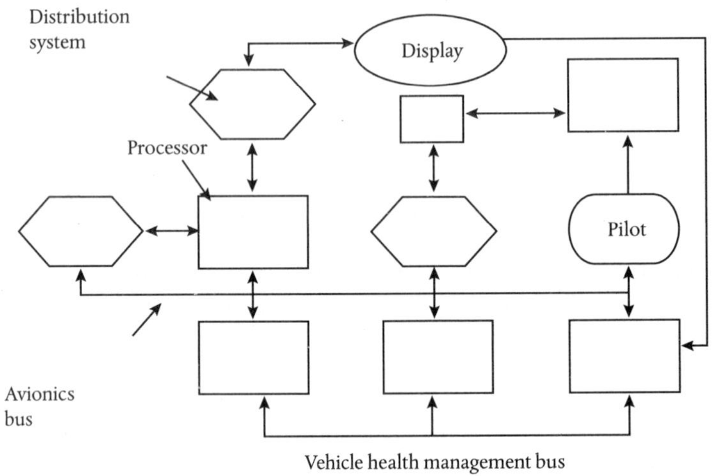

Fig. P26 A typical vehicle health management bus for an avionics system would be the interface point for the fibre optic smart structure modules of Fig. P25

As fibre to the curb and fibre to the home moves closer to reality there is the prospect of merging fibre optic sensor and communication systems into very large systems capable of monitoring the status of buildings, bridges, highways, and factories over widely dispersed areas. Functions such as fire, police, maintenance scheduling, and emergency response to earthquakes, hurricanes, and tornadoes could be readily integrated into very wide area networks of sensors as in Fig. P27.

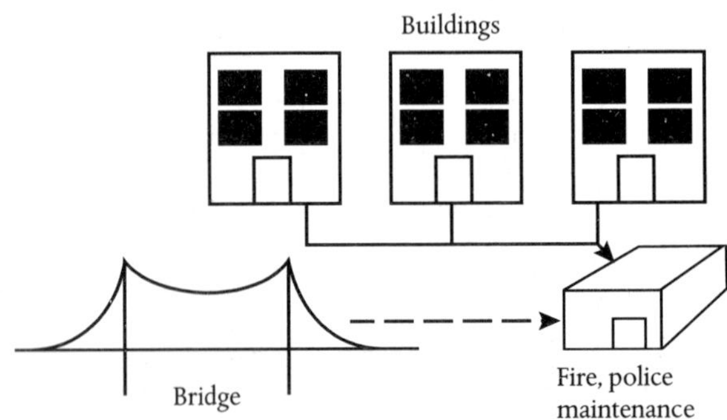

Fig. P27 Fibre optic sensor networks to monitor the status of widely dispersed assets as buildings, bridges, and dams could be used to augment fire, police, and maintenance services.

It is also possible to use fibre optic sensors in combination with fibre optic communication links to monitor stress build-up in critical fault locations and dome build-up of volcanoes. These widely dispersed fibre networks may offer the first real means of gathering information necessary to form prediction models for these natural hazards.

Question 8: Discuss the active and passive optical networks.

Answer: Fibre optics uses light signals to transmit data. As this data moves across a fibre, there needs to be a way to separate it so that it gets to the proper destination. There are two important types of systems that make fibre to the home broadband connections possible. These are active optical networks and passive optical networks. Each offers ways to separate data and route it to the proper place, and each has advantages and disadvantages as compared to the other. An active optical system uses electrically powered switching equipment, such as a router or a switch aggregator, to manage signal distribution and direct signals to specific customers. This switch opens and closes in various ways to direct the incoming and outgoing signals to the proper place. In such a system, a customer may have a dedicated fibre running to his or her house.

A passive optical network, on the other hand, does not include electrically powered switching equipment and instead uses optical splitters to separate and collect optical signals as they move through the network. A passive optical network shares fibre optic strands for portions of the network. Powered equipment is required only at the source and receiving ends of the signal. In some cases, FTTH systems may combine elements of both passive and active architectures to form a hybrid system. Passive optical networks, or PONs, have some distinct advantages. They are efficient, in that each fibre optic strand can serve up to 32 users. The PONs have a low building cost relative to active optical networks along with lower maintenance costs. Because there are few moving or electrical parts, there is simply less that can go wrong in a PON.

Passive optical networks also have some disadvantages. They have less range than an active optical network, meaning subscribers must be geographically closer to the central source of the data. PONs also make it difficult to isolate a failure when they occur. Further, because the bandwidth in a PON is not dedicated to individual subscribers, data transmission speed may slow down during peak usage times in an effect known as latency. Latency quickly degrades services such as audio and video, which need a smooth rate to maintain quality. Active optical networks offer certain advantages as well. Their reliance on Ethernet technology makes interoperability among vendors easy. Subscribers can select hardware that delivers an appropriate data transmission rate and scale up, as their needs increase without restructuring the network. Active optical networks, however, also have their weaknesses. They require at least one switch aggregator for every 48 subscribers. Because it requires power, an active optical network inherently is less reliable than a passive optical network.

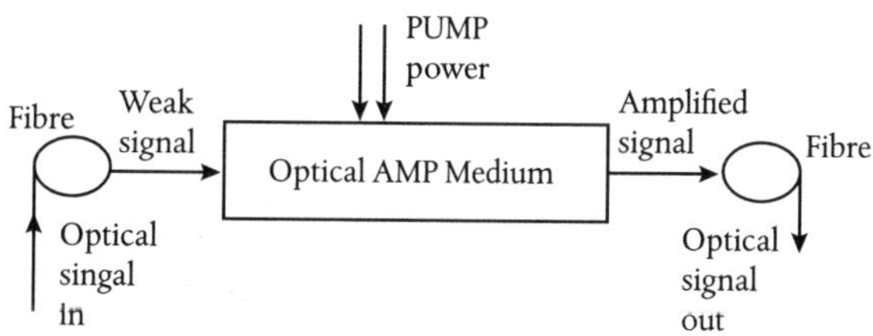

Fig. P28 The general form of an optical amplifier

Question 9: Discuss the optical amplifiers and laser amplifier.

Answer:

Optical Amplifiers

The transmission distance of a fibre optic communication system has traditionally been limited by fibre attenuation and by fibre distortion. By using optoelectronic repeaters, these problems have been eliminated. These repeaters convert the signal into an electrical signal, and then use a transmitter to send the signal again at a higher intensity than it was before. Because of the high complexity with modern wavelength-division multiplexed signals (including the fact that they had to be installed about once every 20 km), the cost of these repeaters is very high. An alternative approach is to use an optical amplifier, which amplifies the optical signal directly without having to convert the signal into the electrical domain. It is made by doping a length of fibre with the rare-earth mineral erbium, and pumping it with light from a laser with a shorter wavelength than the communications signal (typically 980 nm). Amplifiers have largely replaced repeaters in new installations.

An optical amplifier is a device that amplifies an optical signal directly without the need to first convert it to an electrical signal. An optical amplifier may be thought of as a laser without an optical cavity or one in which feedback from the cavity is suppressed. Optical amplifiers are important in optical communication and laser physics.

There are several different physical mechanisms that can be used to amplify a light signal, which correspond to the major types of optical amplifiers. In doped-fibre amplifiers and bulk lasers, stimulated emission in the amplifier's gain medium causes amplification of incoming light. In semiconductor optical amplifiers (SOAs), electron–hole recombination occurs. In Raman amplifiers, Raman scattering of incoming light with photons in the lattice of the gain medium produces photons coherent with the incoming photons. Parametric amplifiers use parametric amplification.

Laser Amplifiers

Almost any laser active gain medium can be pumped to produce gain for light at the wavelength of a laser made with the same material as its gain medium. Such amplifiers are commonly used to produce high-power laser systems. Special types such as regenerative amplifiers and chirped pulse amplifiers are used to amplify ultrashort pulses.

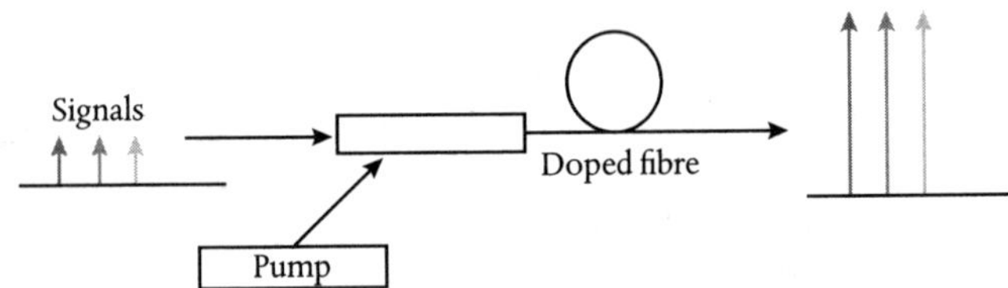

Fig. P29 Schematic diagram of a simple doped fibre amplifier

Doped fibre amplifiers (DFAs) are optical amplifiers that use a doped optical fibre as a gain medium to amplify an optical signal. They are related to fibre lasers. The signal to be amplified and a pump laser are multiplexed into the doped fibre and the signal is amplified through interaction with the doping ions. The most common example is the erbium-doped fibre amplifier (EDFA), where the core of a silica fibre is doped with trivalent Erbium ions and can be efficiently pumped with a laser at a wavelength of 980 nm or 1480 nm, and exhibits gain in the 1550 nm regions.

Amplification is achieved by stimulated emission of photons from dopant ions in the doped fibre. The pump laser excites ions into a higher energy from where they can decay via stimulated emission of a photon at the signal wavelength back to a lower energy level. The excited ions can also decay spontaneously (spontaneous emission) or even through non-radiative processes involving interactions with photons of the glass matrix. These last two decay mechanisms compete with stimulated emission reducing the efficiency of light amplification. The amplification window of an optical amplifier is the range of optical wavelengths for which the amplifier yields a usable gain. The amplification window is determined by the spectroscopic properties of the dopant ions, the glass structure of the optical fibre, and the wavelength and power of the pump laser.

The EDFA is the most deployed fibre amplifier as its amplification window coincides with the third transmission window of silica-based optical fibre. Two bands have developed in the third transmission window the Conventional, or C-band, from approximately 1525–1565 nm, and the Long, or L-band, from approximately 1570–1610 nm. Both of these bands can be amplified by EDFAs, but it is normal to use two different amplifiers, each optimised for one of the bands. The principal difference between C- and L-band amplifiers is that a longer length of doped fibre is used in L-band amplifiers. The longer length of fibre allows a lower inversion level to be used, thereby giving at longer wavelengths (due to the band structure of Erbium in silica) while still providing a useful amount of gain. The EDFAs have two commonly used pumping bands −980 nm and 1480 nm. The 980-nm band has a higher absorption cross section and is generally used where low-noise performance is required. The absorption band is relatively narrow and so wavelength stabilised laser sources are typically needed. The 1480-nm band has a lower, but broader, absorption cross section and is generally used for higher power amplifiers. A combination of 980 nm and 1480 nm pumping is generally utilised in amplifiers.

Basic principle of EDFA

A relatively high-powered beam of light is mixed with the input signal using a wavelength selective coupler. The input signal and the excitation light must of course be at significantly different wavelengths. The mixed light is guided into a section of fibre with erbium ions included in the core. This high-powered light beam excites the erbium ions to their higher energy state. When the photons belonging to the signal at a different wavelength from the pump light meet the excited erbium atoms, the erbium atoms give up some of their energy to the signal and return to their lower energy state. A significant point is that the erbium gives up its energy in the form of additional photons that are exactly in the same phase and direction as the signal being amplified. Therefore, the signal is amplified along its direction of travel only. This is not unusual when an atom "lases"; it always gives up its energy in the same direction and phase as the incoming light. This is just the way lasers work. Thus, all the additional signal power is guided in the same fibre mode as the incoming signal. There is usually an isolator placed at the output to prevent reflections returning from the attached fibre. Such reflections disrupt amplifier operation and in the extreme case can cause the amplifier to become a laser.

Question 10: Discuss the semiconductor optical amplifier.

Answer: Semiconductor optical amplifiers (SOAs) are amplifiers that use a semiconductor to provide the gain medium, as shown in Fig. P.30. These amplifiers have a similar structure to Fabry–Pérot laser diodes but with anti-reflection design elements at the end faces. Recent designs include anti-reflective coatings and tilted waveguide and window regions that can reduce end-face reflection to less than 0.001%. Since this creates a loss of power from the cavity, which is greater than the gain, it prevents the amplifier from acting as a laser.

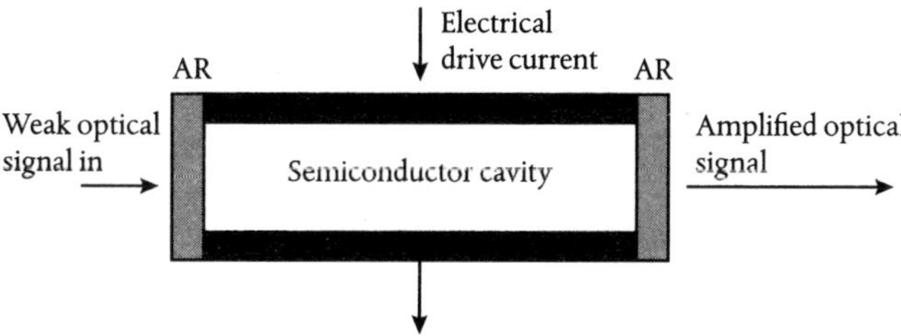

Fig. P30 Basic design of semiconductor optical amplifier

Semiconductor optical amplifiers are typically made from Group III–V compound semiconductors such as GaAs/AlGaAs, InP/InGaAs, InP/InGaAsP, and InP/InAlGaAs, though any direct bandgap semiconductors, such as II VI could conceivably be used. Such amplifiers are often used in telecommunication systems in the form of fibre pigtailed components, operating at signal wavelengths between 0.85 μm and 1.6 μm and generating gains of up to 30 dB.

The semiconductor optical amplifier is of small size and electrically pumped. It can be potentially less expensive than the EDFA and can be integrated with semiconductor lasers, modulators, etc. However, the performance is still not comparable with the EDFA. The SOA has higher noise, lower gain, moderate polarisation dependence, and high nonlinearity with fast transient time. The main advantage of SOA is that all four types of nonlinear operations (cross-gain modulation, cross-phase modulation, wavelength conversion, and four-wave mixing) can be conducted. Furthermore, SOA can be run with a low-power laser. This originates from the short nanosecond or less upper-state lifetime, so that the gain reacts rapidly to changes of pump or signal power and the resultant gain change also brings about phase changes that can distort the signals. This nonlinearity presents the most severe problem for optical communication applications. However, it provides the possibility for gain in different wavelength regions from the EDFA. "Linear optical amplifiers" using gain clamping techniques have been developed.

High-optical nonlinearity makes semiconductor amplifiers attractive for all-optical signals processing like all-optical switching and wavelength conversion. There has been much research on semiconductor optical amplifiers as elements for optical signal processing, wavelength conversion, clock recovery, signal de-multiplexing, and pattern recognition.

Question 11: Discuss the Raman amplifier.

Answer: In a Raman amplifier, the signal is intensified by Raman amplification. Unlike the EDFA and SOA, the amplification effect is achieved by a nonlinear interaction between the signal and a pump laser within an optical fibre, as shown in Fig. P31. There are two types of Raman amplifier: distributed and lumped. A distributed Raman amplifier is one in which the transmission fibre is utilised as the gain medium by multiplexing a pump wavelength with signal wavelength, while a lumped Raman amplifier utilises a dedicated, shorter length of fibre to provide amplification. In the case of a lumped Raman amplifier, highly nonlinear fibre with a small core is utilised to increase the interaction between signals and pump wavelengths, and thereby, it reduces the length of fibre required.

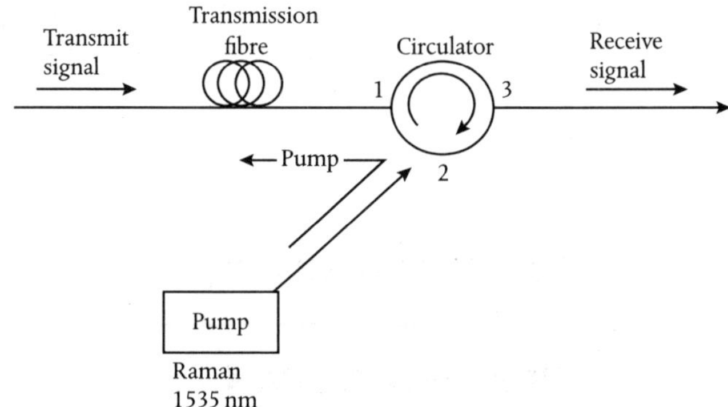

Fig. P31 Typical Raman amplifier configuration

The pump light may be coupled into the transmission fibre in the same direction as the signal (co directional pumping), in the opposite direction (contra directional pumping), or both. Contra directional pumping is more common as the transfer of noise from the pump to the signal is reduced. The pump power required for Raman amplification is higher than that required by the EDFA, with in excess of 500 mW being required to achieve useful levels of gain in a distributed amplifier. Lumped amplifiers, where the pump light can be safely contained to avoid safety implications of high optical powers, may use over 1 W of optical power.

The principle advantage of Raman amplification is its ability to provide distributed amplification within the transmission fibre, thereby increasing the length of spans between amplifier and regeneration sites. The amplification bandwidth of Raman amplifiers is defined by the pump wavelengths utilised, and therefore, the amplification can be provided over wider and different regions than it may be possible with other amplifier types that rely on dopants and device design to define the amplification 'window'.

Question 12: Discuss the function of optical isolators.

Answer: Fibre isolator is used in the networks to block out reflected and unwanted optical light. In fibre optic communication systems, optical lightwave is reflected from any components such as fibre optic connectors and receivers. In addition, scattering at the optical fibre itself and at the components that is made by optical fibre reflects light. As we know, the performance of laser and optical amplifier greatly degrades if back reflected light enters these equipment. The reflected optical light degrades the signal-to-noise ratio and affects the bit-error rate. Fibre optic isolator is such a device to prevent the propagation of back reflected optical light.

An optical isolator, or optical diode, is an optical component that allows the transmission of light in only one direction as shown in Fig. P32. It is typically used to prevent unwanted feedback into an optical oscillator, like a laser cavity. The operation of the device depends on the Faraday effect (which in turn is produced by magneto-optic effect), which is used in the main component, the Faraday rotator.

Fig. P32 The optical circuit symbol for an isolator

MULTIPLE CHOICE QUESTIONS

1. Photodiodes used as fibre optic directors are
   (a) Unbiased to generate a voltage same as a solar cell
   (b) Forward bias
   (c) Reversed bias
   (d) Thermoelectrically cooled
2. The wavelength of light has no role in
   (a) Diffraction
   (b) Interference
   (c) Polarisation
   (d) Reflection

3. What are the advantages of using parallel optical fibre interconnections within board-to-board or frame-to-frame connections in computer systems?
   (a) No bandwidth limit
   (b) Space saving
   (c) Avoids electrical crosstalk
   (d) All the above
4. Which of the following is used as an optical transmitter on the fibre optical communications?
   (a) APD
   (b) LSA diode
   (c) PIN diode
   (d) LED
5. Which of the following is used as an optical receiver in fibre optics communications?
   (a) APD
   (b) Tunnel diode
   (c) Laser diode
   (d) LED
6. In an optical fibre communication system, which among the following is not a typical transmitter function?
   (a) Coding for error protection
   (b) Decoding of input data
   (c) Electrical to optical conversion
   (d) Recoding to match output standard
7. Which among the following is a key process adopted for the laser beam formation as it undergoes the light amplification?
   (a) Spontaneous emission
   (b) Stimulated emission
   (c) Both (a) and (b)
   (d) None of the above
8. Which method determines the dispersion limitation of an optical link?
   (a) Link power budget
   (b) Rise time budget
   (c) Both (a) and (b)
   (d) None of the above
9. Application of fibre-optics communication include:
   (a) Ocean spanning submarine cables.
   (b) long-distance telephone transmission on land.
   (c) connecting telephone company facilities.
   (d) transmitting data to high resolution video terminals.
   (e) All the above
10. Which part of the telephone network carries the highest speed signal?
    (a) Subscriber loop
    (b) Feeder cable
    (c) Trunk cable
    (d) Backbone system

Answers to Multiple Choice Questions

1. (c)
2. (c)
3. (d)
4. (d)
5. (a)
6. (d)
7. (b)
8. (b)
9. (e)
10. (d)

QUESTIONS

1. Discuss absorption losses in optical fibres by comparing and contrasting the intrinsic and extrinsic absorption mechanisms.
2. Compare stimulated Brillouin and stimulated Raman scattering in optical fibres, and indicate the way in which they may be avoided in optical fibre communications.
3. Explain what is meant by the critical bending radius for an optical fibre.
4. Describe the phenomenon of modal noise in optical fibres and suggest how it may be avoided.
5. Discuss dispersion mechanisms with regard to single-mode fibres indicating the dominating effects. Hence, describe how intramodal dispersion may be minimised within the single-mode region?
6. Describe the techniques employed and the fibre structures utilised to provide:
   (a) Modal birefringence;
   (b) The beat lengthin single-mode fibres.
7. Calculate the carrier frequency for optical communication systems operating at $0.88, 1.3,$ and $1.55\,\mu\text{m}$. What is the photon energy (in eV) in each case?
8. Calculate the transmission distance over which the optical power will attenuate by a factor of 10 for three fibres with losses of 0.2, 20, and 2000 dB/km. Assuming that the optical power decreases as $\exp(-\alpha L)$, calculate $\alpha(\text{in cm}^{-1})$ for the three fibres.
9. Assume that a digital communication system can be operated at a bit rate of up to 1% of the carrier frequency. How many audio channels at 64 kb/s can be transmitted over a microwave carrier at 5 GHz and an optical carrier at $1.55\,\mu\text{m}$?
10. A 1-h lecture script is stored on the computer hard disc in the ASCII format. Estimate the total number of bits assuming a delivery rate of 200 words per minute and on average 5 letters per word. How long will it take to transmit the script at a bit rate of 1 Gb/s?
11. A 1.55-µm digital communication system operating at 1 Gb/s receives an average power of $-40\,\text{dBm}$ at the detector. Assuming that 1 and 0 bits are equally likely to occur, calculate the number of photons received within each 1 bit.
12. An analogue voice signal that can vary over the range 0–50 mA is digitised by sampling it at 8 kHz. The first four sample values are 10, 21, 36, and 16 mA. Write the corresponding digital signal (a string of 1 and 0 bits) by using a 4-bit representation for each sample.
13. Sketch the variation of optical power with time for a digital NRZ bit stream 010111101110 by assuming a bit rate of 2.5 Gb/s. What is the duration of the shortest and widest optical pulse?
14. A 1.55-µm fibre-optic communication system is transmitting digital signals over 100 km at 2 Gb/s. The transmitter launches 2 mW of average power into the fibre cable, having a net loss of $0.3\,\text{dB}/\text{km}$. How many photons are incident on the receiver during a single1-bit? Assume that 0 bits carry no power, while 1 bit is in the form of a rectangular pulse occupying the entire bit slot (NRZ format).

15. A 0.8-μm optical receiver needs at least 1000 photons to detect the 1 bits accurately. What is the maximum possible length of the fibre link for a 100-Mb/s optical communication system designed to transmit $-10\,\text{dBm}$ of average power? The fibre loss is $2\,\text{dB}/\text{km}$ at $0.8\,\mu\text{m}$. Assume the NRZ format and a rectangular pulse shape.
16. A 1.3-μm optical transmitter is used to obtain a digital bit stream at a bit rate of 2 Gb/s. Calculate the number of photons contained in a single 1 bit when the average power emitted by the transmitter is 4 mW. Assume that the 0 bits carry no energy.
17. A multimode fibre with a 50-μm core diameter is designed to limit the intermodal dispersion to $10\,\text{ns}/\text{km}$. What is the numerical aperture of this fibre? What is the limiting bit rate for transmission over 10 km at $0.88\,\mu\text{m}$? Use 1.45 for the refractive index of the cladding.
18. Use Maxwell's equations to express the field components E_ρ, E_ϕ, H_ρ and H_ϕ in terms of E_z and H_z.
19. Derive the eigenvalue equation by matching the boundary conditions at the core–cladding interface of a step-index fibre.
20. A single-mode fibre has an index step $n_1 - n_2 = 0.005$. Calculate the core radius if the fibre has a cut-off wavelength of $1\,\mu\text{m}$. Estimate the spot size (FWHM) of the fibre mode and the fraction of the mode power inside the core when this fibre is used at $1.3\,\mu\text{m}$. Use $n_1 = 1.45$.
21. A 1.55-μm unchirped Gaussian pulse of 100-ps width (FWHM) is launched into a single-mode fibre. Calculate its FWHM after 50 km if the fibre has a dispersion of $16\,\text{ps}/(\text{km}-\text{nm})$. Neglect the source spectral width.
22. A single-mode fibre is measured to have $\lambda^2 (d^2 n / d\lambda^2) = 0.02$ at $0.8\,\mu\text{m}$. Calculate the dispersion parameters β_2 and D.
23. Show that a chirped Gaussian pulse is compressed initially inside a single-mode fibre when $\beta_2 C < 0$. Derive expressions for the minimum width and the fibre length at which the minimum occurs.
24. Estimate the limiting bit rate for a 60-km single-mode fibre link at 1.3 and $1.55\,\mu\text{m}$ wavelengths assuming transform limited, 50-ps (FWHM) input pulses. Assume that $\beta_2 = 0$ and $-20\dfrac{\text{ps}^2}{\text{km}}$ and $\beta_3 = 0.1\dfrac{\text{ps}^3}{\text{km}}$ and 0 at 1.3 – and 1.55 – μm wavelengths, respectively. Further, assume that $V\omega \gg 1$.
25. A 0.88-μm communication system transmits data over a 10-km single-mode fibre by using 10-ns (FWHM) pulses. Determine the maximum bit rate if the LED has a spectral FWHM of 30 nm. Use $D = -80\,\text{ps}/(\text{km}-\text{nm})$.
26. A 1.55-μm optical communication system operating at $5\,\text{Gb}/\text{s}$ is using Gaussian pulses of width 100 ps (FWHM) chirped such that $C = -6$. What is the dispersion-limited maximum fibre length? How much will it change if the pulses were unchirped? Neglect laser linewidth and assume that $\beta_2 = -20\,\text{ps}^2/\text{km}$.
27. A 1.3-μm lightwave system uses a 50-km fibre link and requires at least $0.3\,\mu\text{W}$ at the receiver. The fibre loss is 0.5 dB/km. Fibre is spliced every 5 km and has two connectors of 1-dB loss at both ends. Splice loss is only 0.2 dB. Determine the minimum power that must be launched into the fibre.
28. Calculate the threshold power for stimulated Brillouin scattering for a 50-km fibre link operating at $1.3\,\mu\text{m}$ and having a loss of $0.5\,\text{dB}/\text{km}$. How much does the threshold power change if the operating wavelength is changed to $1.55\,\mu\text{m}$, where the fibre loss is only $0.2\,\text{dB}/\text{km}$? Assume that $A_{eff} = 50\,\mu\text{m}$ and $g_b = 5\times10^{-11}$ m/W at both wavelengths.
29. Calculate the power launched into a 40-km-long single-mode fibre for which the SPM-induced nonlinear phase shift becomes 180°. Assume $\lambda = 1.55\,\mu\text{m}, A_{eff} = 40\,\mu\text{m}^2$, $\alpha = 0.2\,\text{dB}/\text{km}$, and $\bar{n}_2 = 2.6\times10^{-20}\,\text{m}^2/\text{W}$.
30. Find the maximum frequency shift occurring because of the SPM-induced chirp imposed on a Gaussian pulse of 20-ps width (FWHM) and 5-mW peak power after it has propagated 100 km. Use the fibre parameters of the preceding problem but assume $\alpha = 0$.

REFERENCES

1. Liu M M K, *Principles and Applications of Optical Communications,* New Delhi: Tata McGraw Hill Edition, 2010.
2. Wipiefewski T, Review of Optical Sources for Polymer Optical Fibres, POF 2005, Hong Kong, September 21–23, 2005.
3. Wu M C, Lin L Y, Lee S S and Pister K S J, Micro machined free space integrated micro-optics, *Sens. Actuator* A, vol. 50, pp. 127–134, 1995.
4. Belluci S, Carbon nanotubes: physics and applications, *Physica Status Solidi (c),* vol. 2, no. 1, pp. 34–47, 19 Jan. 2005.
5. Eiskelenborg M A, *et al.,* Microstructured Polymer Optical Fibre, *Opt. Express,* 9 (7), pp 3319–27 (2001).
6. Fraser D M, Glucose biosensors – the sweet smell of success, *Med. Device Technol.,* vol. 5, no.9, pp.44–47, 1994.
7. Futaba D N and Hata K, Shape engineerable and highly densely packed single-walled carbon nanotubes and their application as super-capacitor electrodes, *Nat. Mater.,* vol. 5, no.12, pp. 987–994. 2006.
8. Higson S P J, Reddy S M and Vadgama P M, Enzyme and other biosensors: evolution of a technology, *Eng. Sci. Educ. J.,* pp.41–48, Feb 1994.
9. Malhotra B D, Singhal R, Chaubey A, Sharma S K and Kumar A, Recent trends in biosensors, *Curr. Appl. Phys.,* vol. 5, pp. 92–97, 2005.
10. Mamyshev P V and Cherinikov S V, Femto second soliton propagation in fibres with slowly decreasing dispersion, *J. Opt Soc America B.,* vol. 8. pp. 1633–1641, 1991.
11. Marcuse D, *Light Transmission Optics,* New York, USA: Van Nostr and Reinhold Company, 1972.
12. Marcuse D, *Principle of Optical Fibre Measurements,* 1st E., USA, 1981.
13. Marquette C A and Blum L J, State of the art and recent advances in immunoanalytical systems, *Biosens. Bioelectron.,* vol. 21, pp. 1424–1433, 2006.
14. Meltz G, *et al.,* Formation of Bragg gratings in optical fibres by a transverse holographic method. *Opt. Lett.,* vol. 14, pp. 823, 1989.
15. Minowa J and Fujii Y, Dielectric multilayer thin-film filter for WDM transmission systems, *IEEE, J. of Lightwave Techno.,* LT-1 (1), pp. 116–121, Mar. 1983.
16. Murphy T E, Design, *Fabrication and Measurement of Integrated Bragg Grating Optical Filters,* Doctor of Philosophy at the Massachusetts Institute of Technology, February, 2001.
17. Naeini S S, Chow Y L, Chaudhuri S K and Goss A, Wide angle phase-corrected Y-junction of dielectric waveguides for low loss application, *J. Lightwave Technol.,* vol. LT-11, pp. 567–675, 1993.
18. Northrup M A and Gonzalez C, *et. al.,* A MEMS-based miniature DNA analysis system, in the *Proceedings of the 8th International Conference on Solid-State Sensors and Actuators, and Euro-sensors IX, TRANSDUCERS'95,* pp.764–767, June25–29, 1995.
19. Olivieria J R and Moura M A, Analytical solution for the Modified NLSE describing optical shock formation, *Phys. Rev. E.,* vol. 57, pp. 4751–4755, 1998.
20. Othonos A and Kalli K, *Fibre Bragg Gratings: Fundamentals and Application Telecommunications and Sensing,* Boston: Artech House, 1999.
21. Paek U C, High speed high strength fibre drawing, *J. Lightwave Technol.,* vol. 4, pp. 1048–1060, Aug. 1986.
22. Papoulis A, *The Fourier Integral and Its Applications,* New York: McGraw-Hill, 1962.
23. Peterson K, "Biomedical Applications of MEMS", *Electron devices meeting,* pp. 239–242, 1996.

24. Pattnaik P K, Selvarajan A and Srinivas T, Guided wave optical MEMS pressure sensor, *Proceedings of ISA/IEEE Conference on Sensors for Industry SIcon/05*, pp. 122–125, 9–10 Feb. 2005, Houston, Texas, USA.
25. Polishuk P, POF Market and Technology Assessment Report IGI Consulting, 2005.
26. Poole S B and Payne D N, Special optical fibres, *SPIE Fibre Optics 87: Fifth International conference on Fibre Optics and Opto Electronics*, pp. 92–103, 1987.
27. Raghuwanshi S K, Ray paths in an Elliptic parabolic refractive index profile fibre, *International Conference on Communication, Computation, Control and NanoTechnology* (ICN 2010), 29–30 Nov. 2010, Bhalki India.
28. Raghuwanshi S K, Kumar V and Pandey R R, Dispersion study of cylindrical dielectric waveguide without computing $\frac{d^2\beta}{dk^2}$ numerically, *13th International Conference of International Academy of Physical Sciences (CONIAPS-XIII)*, June 14–16, 2011, Dehradun, India.
29. Rozzi T E and Intveld G H, Field and network analysis of interacting step discontinuities in planar dielectric waveguides, *IEEE Trans. Microw. Theory Tech.*, vol. MTT-27, no. 4, pp. 303–309, April 1979.
30. Shimizu Y, Shiokawa N, Yamamoto N, Kozuma M and Kuga T, Control of light pulse propagation with only a few cold atoms in a high finesse microcavity, *Phys. Rev. Lett.*, vol. 23, Dec. 2002.
31. Raghuwanshi S K, Basics of MEMS/MOEMS Technology, *National conference on Information and Communication Technology*, NCICT - 2009, Mumbai.
32. Ramo, Simon, John Whinnery R and Theodore van Duzer, *Fields and Waves in Communications Electronics*, 2nd ed., New York: John Wiley and Sons, 1984.
33. Senior J M, *Optical Fibre Communications, Principles and Practice*, 2nd ed., Cambridge: Cambridge Univ. Press, 1992.
34. Singh V, Prajapati Y and Saini J P, Modal analysis and dispersion curves of a new unconventional Bragg waveguide using a very simple method, *Prog. Electromagn. Res.*, PIER 64, 191–204, 2006.
35. Sinnott S B and Rodney A, Carbon nanotubes: synthesis, properties, and applications, *Critical Crit. Rev. Solid State Mater. Sci.*, vol. 26, no. 3, pp. 145–249, July 2001.
36. Somani P R and Umeno M, *Carbon Nanotube: Multifunctional Material*, India: Applied Science Innovations Pvt. Ltd., 2009.
37. Stark H, *Applications of Optical Fourier Transforms*, New York: Academic Press, 1982.
38. Wang X, Li Q, Xie J, Jin Z, Wang J, Li Y, Jiang K and Fan S, Fabrication of ultralong and electrically uniform single-walled carbon nanotubes on clean substrates, *Nano Lett.*, vol. 9, no. 9, pp. 3137–3141, 2009.
39. Special issue on fibre gratings, photosensitivity and Poling, *IEEE J. Lightwave Technol.*, vol. 15, no. 8, August 1997.
40. Andreas O and Kyriacos K, *Fibre Bragg Gratings: Fundamentals and Applications in Telecommunications and Sensing*, Boston, USA: Artech House, 1999.
41. Aston C, Biological warfare canaries, *Cover Story: IEEE Spectrum*, Oct. 2001.
42. Bartic C and Borghs G, Organic thin-film transistors as transducers for (bio) analytical applications, *Anal. Bioanal. Chem.*, vol. 384, no. 2, pp. 354–365, Jan 2006.

Appendix

MISCELLANEOUS QUESTIONS

1. When the mean optical power launched into an 8-km length of fibre is 120 μW, the mean optical power at the fibre output is 3 μW.
   Determine:
   (a) The overall signal attenuation or loss in decibels through the fibre assuming there are no connectors or splices;
   (b) The signal attenuation per kilometre for the fibre.
   (c) The overall signal attenuation for a 10-km optical link using the same fibre with splices at 1-km intervals, each giving an attenuation of 1dB;
   (d) The numerical input/output power ratio in (c).
2. Silica has an estimated fictive temperature of 1400 K with an isothermal compressibility of 7×10^{-11} m^{2}N^{-1}. The refractive index and the photoelastic coefficient for silica are 1.46 and 0.286, respectively. Determine the theoretical attenuation in decibels per kilometre due to the fundamental Rayleigh scattering in silica at optical wavelengths of 0.63, 1.00, and 1.30 μm. Boltzmann's constant is 1.381×10^{-23} JK^{-1}.
3. A long single-mode optical fibre has an attenuation of 0.5 dB km^{-1} when operating at a wavelength of 1.3 μm. The fibre core diameter is 6 μm and the laser source bandwidth is 600 MHz. Compare the threshold optical powers for stimulated Brillouin and Raman scattering within the fibre at the wavelength specified.
4. Two-step index fibres exhibit the following parameters:
   (a) A multimode fibre with a core refractive index of 1.50, a relative refractive index difference of 3%, and an operating wavelength of 0.82 μm.
   (b) An 8 $-\mu$m core diameter single-mode fibre with a core refractive index the same as (a), a relative refractive index difference of 0.3% and an operating wavelength of 1.55 μm.
5. A multimode graded index fibre exhibits total pulse broadening of 0.1 μs over a distance of 15 km.
   Estimate:
   (a) The maximum possible bandwidth on the link assuming no intersymbol interference;
   (b) The pulse dispersion per unit length;
   (c) The bandwidth–length product for the fibre.
6. The material dispersion of a glass is given by $\left|\lambda^2\left(\frac{d^2n_1}{d\lambda^2}\right)\right|$ of 0.025. Determine the material dispersion parameter at a wavelength of 0.85 μm, and estimate the rms pulse broadening per kilometre for a good LED source with an rms spectral width of 20 nm at this wavelength.

7. Estimate the rms pulse broadening per kilometre for the fibre in above-mentioned question when the optical source used in an injection laser with a relative spectral width σ_λ / λ of 0.0012 at a wavelength of 0.85 μm.
8. A 6-km optical link consists of multimode step-index fibre with a core refractive index of 1.5 and a relative refractive index difference of 1%. Estimate:
   (a) The delay difference between the slowest and the fastest modes at the fibre output;
   (b) The rms pulse broadening due to intermodal dispersion on the link;
   (c) The maximum bit rate that may be obtained without substantial errors on the link assuming only intermodal dispersion;
   (d) The bandwidth–length product corresponding to (c).
9. Compare the rms pulse broadening per kilometre due to intermodal dispersion for the multimode step index fibre of Problem 8 with the corresponding rms pulse broadening for an optimum near-parabolic profile graded index fibre with the same core axis refractive index and relative refractive index difference.
10. A multimode step index fibre has a numerical aperture of 0.3 and a core refractive index of 1.45. The material dispersion parameter for the fibre is 250 ps $nm^{-1}km^{-1}$, which makes material dispersion the totally dominating intramodal dispersion mechanism. Estimate (a) the total rms pulse broadening per kilometre when the fibre is used with an LED source of rms spectral width 50 nm and (b) the corresponding bandwidth–length product of the fibre.
11. A typical single-mode fibre has a zero-dispersion wavelength of 1.31 μm with a dispersion slope of 0.09 ps $nm^{-1}km^{-1}$. Compare the total first-order dispersion for the fibre at the wavelengths of 1.28 μm and 1.55 μm. When the material dispersion and profile dispersion at the latter wavelength are 13.5 ps $nm^{-1}km^{-1}$ and 0.4 ps $nm^{-1}km^{-1}$, respectively; determine the waveguide dispersion at this wavelength.
12. The beat length in a single-mode optical fibre is 9 cm when light from an injection laser with a spectral linewidth of 1nm and a peak wavelength of 0.9 μm is launched into it. Determine the modal birefringence and estimate the coherence length in this situation. In addition, calculate the difference between the propagation constants for the two orthogonal modes and check the result.
13. Two polarisation maintaining fibres operating at a wavelength of 1.3 μm have beat lengths of 0.7 mm and 80 m. Determine the modal birefringence in each case and comment on the results.
14. The polarisation mode dispersion in uniformly birefringent single-mode fibre is 13.5 ps km^{-1}. Calculate the maximum bit rate that may be obtained on a 20-km repeaterless link assuming only polarisation mode dispersion to occur.
15. A 3.5-km length of two polarisation mode PM fibre has a polarisation crosstalk of −27 dB at its output end. Determine the mode coupling parameter for the fibre.
16. The mean optical power launched into an optical fibre link is 1.5 mW and the fibre has an attenuation of 0.5 dB km^{-1}. Determine the maximum possible link length without repeaters (assuming lossless connectors) when the minimum mean optical power level required at the detector is 2 μW.
17. The numerical input/output mean optical power ratio in a 1 km length of optical fibre is found to 2.5. Calculate the received mean optical power when a mean optical power of 1mW is launched into a 5-km length of the fibre (assuming no joints or connectors).
18. A 15-km optical link uses fibre with a loss of 1.5 dB km^{-1}. The fibre is jointed every kilometre with connectors that give an attenuation of 0.8 dB each. Determine the minimum mean optical power that must be launched into the fibre in order to maintain mean optical power level of 0.3 μW at the detector.

19. Briefly describe linear scattering losses in optical fibres with regard to:
    (a) Rayleigh scattering;
    (b) Mie scattering.
20. A K_2O-SiO_2 glass core optical fibre has an attenuation resulting from Rayleigh scattering of 0.46 dB km^{-1} at a wavelength of 1 μm. The glass has an estimated fictive temperature of 758 K, isothermal compressibility of $8.4\times10^{-11} m^2N^{-1}$, and a photoelastic coefficient of 0.245. Determine from theoretical considerations the refractive index of the glass.
21. The photoelastic coefficient and the refractive index for silica are 0.286 and 1.46, respectively. Silica has an isothermal compressibility of $7\times10^{-11} m^2N^{-1}$ and an estimated fictive temperature of 1400 K. Determine the theoretical attenuation in decibels per kilometre due to the fundamental Rayleigh scattering in silica at optical wavelengths of 0.85 and 1.55 μm. Boltzmann's constant is $1.381\times10^{-23} JK^{-1}$.
22. The threshold optical power for stimulated Brillouin and Raman scattering in a long 8 –μm core diameter single-mode fibre are found to be 190 mW and 1.70 W, respectively, when using an injection laser source with a bandwidth of 1 GHz. Calculate the operating wavelength of the laser and the attenuation in decibels per kilometre of the fibre at this wavelength.
23. The threshold optical power for stimulated Brillouin scattering at a wavelength of 0.85 μm in a long single-mode fibre using an injection laser source with a bandwidth of 800 MHz is 127 mW. The fibre has an attenuation of 2 dB km^{-1} at this wavelength. Determine the threshold optical power for stimulated Raman scattering within the fibre at a wavelength of 0.9 μm assuming the fibre attenuation is reduced to 1.8 dB km^{-1} at this wavelength.
24. A multimode graded index fibre has a refractive index at the core axis of 1.46 with a cladding refractive index of 1.45. The critical radius of curvature that allows large bending losses to occur is 84 μm when the fibre is transmitting light of a particular wavelength. Determine the wavelength of the transmitted light.
25. A single-mode step index fibre with a core refractive index of 1.49 has a critical bending radius of 10.4 mm when illuminated with light at a wavelength of 1.30 μm. If the cutoff wavelength for the fibreis 1.15 μm, calculate its relative refractive index difference.
26. (a) A multimode step indexfibre gives a total pulse broadening of 95 ns over a 5-km length. Estimate the bandwidth–length product for the fibre when a nonreturn to zero digital code is used. (b) A single-mode step index fibre has a bandwidth–length product of 10GHz km. Estimate the rms pulse broadening over a 40-km digital optical link without repeaters consisting of the fibre and using a return to zero code.
27. An 8-km optical fibre link without repeaters uses multimode graded index fibrethathas a bandwidth–length product of 400 MHz km. Estimate:
    (a) The total pulse broadening on the link;
    (b) The rms pulse broadening on the link.
    It may be assumed that return to zero code is used.
28. The material dispersion in an optical fibre defined by $\left|\left(\frac{d^2n_1}{d\lambda^2}\right)\right|$ is 4.0×10^{-2} μm^{-2}. Estimate the pulse broadening per kilometre due to material dispersion within the fibre when it is illuminated with an LED source with a peak wavelength of 0.9 μm and an rms spectral width of 45 nm.
29. An 11-km optical fibre link consisting of optimum near-parabolic profile graded index fibre exhibits rms intermodal pulse broadening of 346 ps over its length. If the fibre has a relative refractive index difference of 1.5%, estimate the core axis refractive index. Hence, determine the numerical aperture for the fibre.

30. A multimode step index fibre has a relative refractive index difference of 1% and a core refractive index of 1.46. The maximum optical bandwidth that may be obtained with a particular source on a 4.5 km link is 3.1 MHz.
    (a) Determine the rms pulse broadening per kilometre resulting from intramodal dispersion mechanisms.
    (b) Assuming waveguide dispersion may be ignored, estimate the rms spectral width of the source used, if the material dispersion parameter for the fibre at the operating wavelength is 90 $\text{ps nm}^{-1}\text{km}^{-1}$
31. A single-mode step index fibre exhibits material dispersion of 7 $\text{ps nm}^{-1}\text{km}^{-1}$ at an operating wavelength of 1.55 μm. Using the approximation obtained in Problem 31, estimate the fibre core diameter thatwill enable the waveguide dispersion to cancel the material dispersion so that zero intramodal dispersion is obtained at this wavelength. The refractive index of thefibre core is 1.45.
32. A single-mode step index fibre as a zero-dispersion wavelength of 1.29 μm and exhibits total first-order dispersion of 3.5 $\text{ps nm}^{-1}\text{km}^{-1}$ at a wavelength of 1.32 μm. Determine the total first-order dispersion in the fibre at a wavelength of 1.54 μm.
33. The difference between the propagation constants for the two orthogonal modes in a single-mode fibre is 250. It is illuminated with light of peak wavelength 1.55 μm from injection laser source with a spectral linewidth of 0.8 nm. Estimate the coherence length within the fibre.

MATLAB PROGRAMS

PROGRAM 1

In graded index fibre, core refractive index (*n*) vs. radial distance (*r*) is:

$$n(r) = \begin{cases} n_1 \left[1 - 2\Delta\left(\dfrac{r}{a}\right)^{\alpha}\right]^{1/2} & \text{for} \quad 0 \le r \le a \\ n_1 (1-2\Delta)^{1/2} \simeq n_1 (1-\Delta) = n_2 & \text{for} \quad r \ge a \end{cases},$$

where $n_1 = 1.5; n_2 = 1.48$

$$\Delta = \frac{n_1^2 - n_2^2}{2n_1^2} \simeq \frac{n_1 - n_2}{n_1}$$

MATLAB Program:

```
clc;
clear all;
close all;
n1=1.5;
n2=1.48;
D=(n1-n2)/n1;
```

```
a=10;

q=1;

r=0:0.001:20;

n=n1*(sqrt(1-2*D*(abs(r)./a).^q)).*(abs(r)<=a)+n1*(sqrt(1-
2*D)).*(abs(r)>a);

plot(n, '.r');

hold on;

q=2;

r=0:0.001:20;

n=n1*(sqrt(1-2*D*(abs(r)./a).^q)).*(abs(r)<=a)+n1*(sqrt(1-
2*D)).*(abs(r)>a);

plot(n, '.b');

hold on;

q=3;

r=0:0.001:20;

n=n1*(sqrt(1-2*D*(abs(r)./a).^q)).*(abs(r)<=a)+n1*(sqrt(1-
2*D)).*(abs(r)>a);

plot(n, '.g');

hold on;

q=200;

r=0:0.001:20;

n=n1*(sqrt(1-2*D*(abs(r)./a).^q)).*(abs(r)<=a)+n1*(sqrt(1-
2*D)).*(abs(r)>a);

plot(n, '.black-');

hold on;

grid on;

legend('triangular', 'graded index', 'gaussian', 'step-index');
```

RESULT

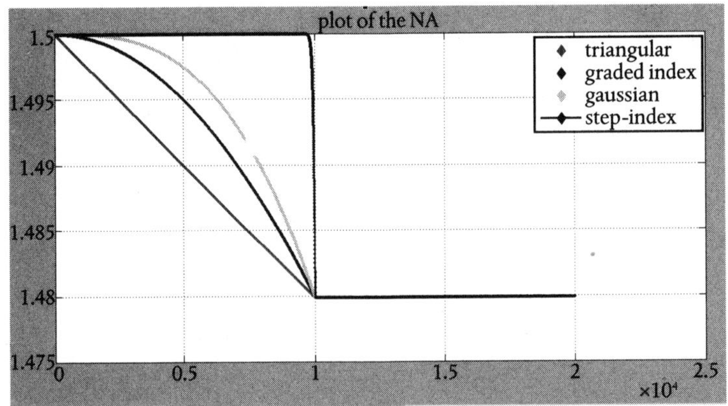

PROGRAM 2

In graded Index fibre, numerical aperture (NA) vs. radial distance (*r*) is:

$$\mathrm{NA}(\mathrm{r}) = \begin{cases} \left[n^2(r) - n_2^2\right]^{1/2} \simeq \mathrm{NA}(0)\sqrt{1-(r/a)^{\alpha}} & \text{for} \quad r \le a \\ 0 & \text{for} \quad r > a \end{cases},$$

where $n_1 = 1.5; n_2 = 1.48$ and axial numerical aperture is:

$$\mathrm{NA}(0) = \left[n^2(0) - n_2^2\right]^{1/2} = \left(n_1^2 - n_2^2\right)^{1/2} \simeq n_1\sqrt{2\Delta}.$$

```
clc;

clear all;

close all;

n1=1.5;

n2=1.48;

D=(n1-n2)/n1;

a=10;

NA0=n1*sqrt(2*D);

q=1;

r=-20:0.001:20;

NA=NA0*(sqrt(1-(abs(r)./a).^q)).*(abs(r)<=a)+0.*(abs(r)>a);

plot(NA, '.r');
```

```
hold on;
title('plot of the NA');
q=2;
r=-20:0.001:20;
NA=NA0*(sqrt(1-(abs(r)./a).^q)).*(abs(r)<=a)+0.*(abs(r)>a);
plot(NA, '.b');
hold on;
q=3;
r=-20:0.001:20;
NA=NA0*(sqrt(1-(abs(r)./a).^q)).*(abs(r)<=a)+0.*(abs(r)>a);
plot(NA, '.g');
hold on;
q=200;
r=-20:0.001:20;
NA=NA0*(sqrt(1-(abs(r)./a).^q)).*(abs(r)<=a)+0.*(abs(r)>a);
plot(NA, '.black');
hold on;
grid on;
legend('triangular', 'graded index', 'gaussian', 'step-index');
```

RESULT

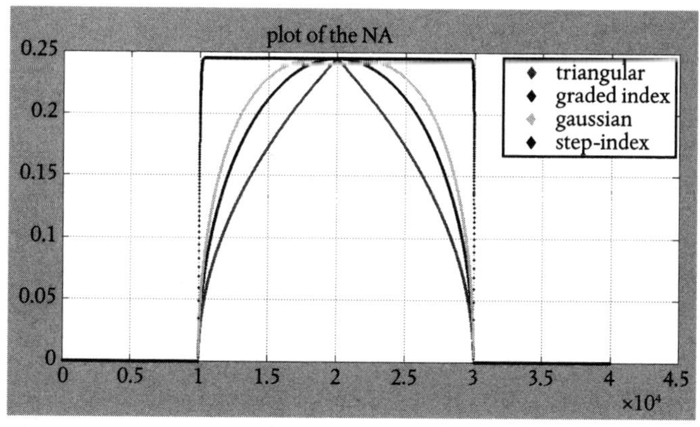

PROGRAM 3

```
clc;
clear all;
close all;
r=20*10^(-6);
a=50*10^(-6);
n1=1.5;
delta=0.04;
angle=60;
n=n1*sqrt(1-2*delta*(r/a)^2);
theta=0:0.01:4*pi;
x=r.*cos(theta);
y=(a*n*sin(deg2rad(angle)))/(n1*sqrt(2*delta))*sin(theta);
plot3(theta,x,y);
grid on;
```

RESULT

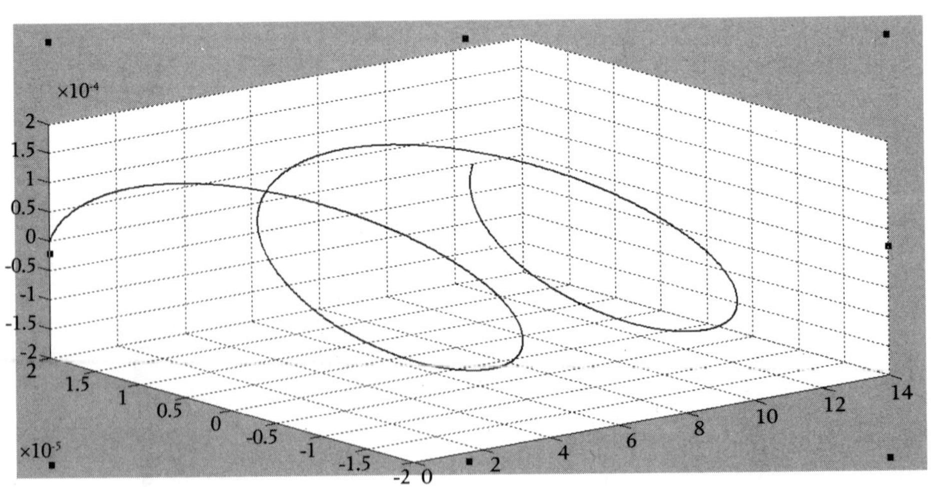

PROGRAM 4

```
clc;

clear all;

close all;

r=20*10^(-6);

a=50*10^(-6);

n1=1.5;

delta=0.04;

angle=0;

n=n1*sqrt(1-2*delta*(r/a)^2);

theta=0:0.01:4*pi;

x=r.*cos(theta);

y=(a*n*sin(deg2rad(angle)))/(n1*sqrt(2*delta))*sin(theta);

plot3(theta,x,y);

grid on;
```

RESULT

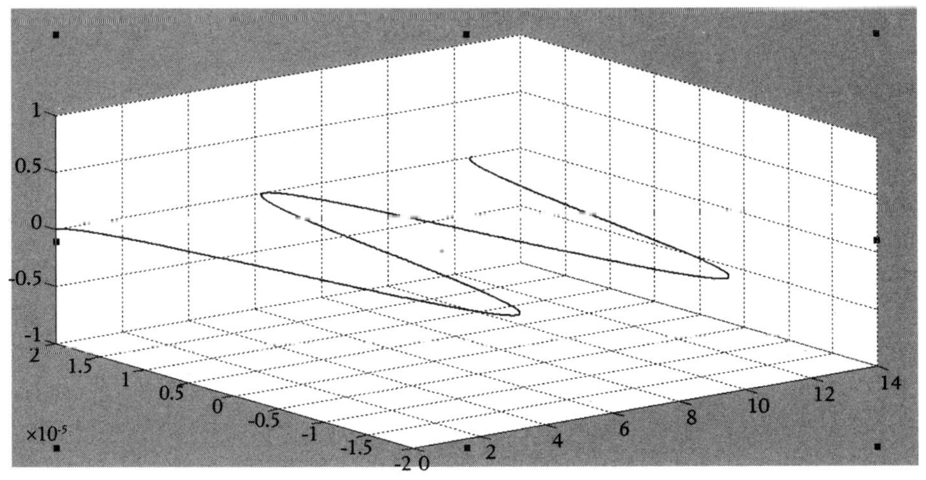

PROGRAM 5

```
clc;

clear all;

close all;

r=20*10^(-6);

a=50*10^(-6);

n1=1.5;

delta=0.04;

angle=60;

n=n1*sqrt(1-2*delta*(r/a)^2);

theta=0:0.01:7*pi;

x=r.*cos(theta);

y=(a*n*sin(deg2rad(angle)))/(n1*sqrt(2*delta))*sin(theta);

plot3(theta,x,y);

grid on;
```

RESULT

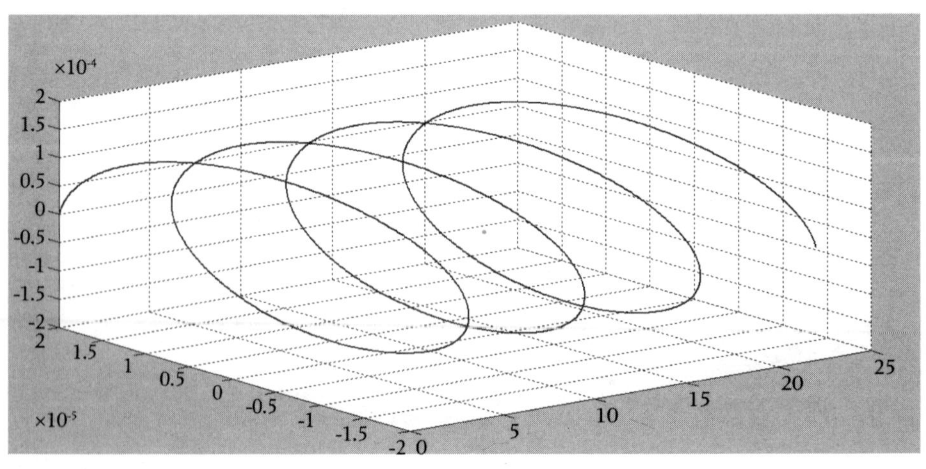

PROGRAM 6

```
clc;

clear all;

close all;

n1=1.5;

n2=1.48;

D=(n1-n2)/n1;

a=10;

q=1;

r=0:0.001:20;

n=n1*(sqrt(1-2*D*(abs(r)./a).^q)).*(abs(r)<=a)+n1*(sqrt(1-
2*D)).*(abs(r)>a);

plot(n, '.r');

hold on;

title('plot of the NA');

q=2;

r=0:0.001:20;

n=n1*(sqrt(1-2*D*(abs(r)./a).^q)).*(abs(r)<=a)+n1*(sqrt(1-
2*D)).*(abs(r)>a);

plot(n, '.b');

hold on;

q=3;

r=0:0.001:20;

n=n1*(sqrt(1-2*D*(abs(r)./a).^q)).*(abs(r)<=a)+n1*(sqrt(1-
2*D)).*(abs(r)>a);

plot(n, '.g');

hold on;

q=200;

r=0:0.001:20;
```

```
n=n1*(sqrt(1-2*D*(abs(r)./a).^q)).*(abs(r)<=a)+n1*(sqrt(1-
2*D)).*(abs(r)>a);

plot(n, '.black-');

hold on;

grid on;

legend('triangular', 'graded index', 'gaussian', 'step-index');
```

RESULT

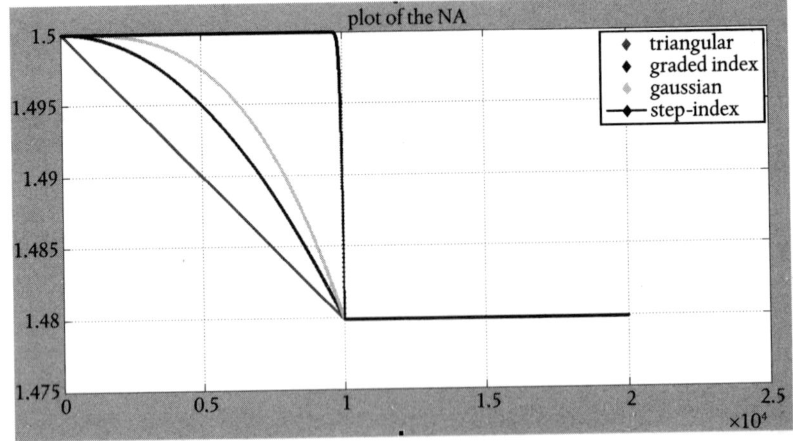

PROGRAM 7

```
clc;

clear all;

close all;

n1=1.5;

n2=1.48;

D=(n1-n2)/n1;

a=10;

NA0=n1*sqrt(2*D);

q=1;

r=-20:0.001:20;

NA=NA0*(sqrt(1-(abs(r)./a).^q)).*(abs(r)<=a)+0.*(abs(r)>a);

plot(NA, '.r');
```

```
hold on;
title('plot of the NA');
q=2;
r=-20:0.001:20;
NA=NA0*(sqrt(1-(abs(r)./a).^q)).*(abs(r)<=a)+0.*(abs(r)>a);
plot(NA, '.b');
hold on;
q=3;
r=-20:0.001:20;
NA=NA0*(sqrt(1-(abs(r)./a).^q)).*(abs(r)<=a)+0.*(abs(r)>a);
plot(NA, '.g');
hold on;
q=200;
r=-20:0.001:20;
NA=NA0*(sqrt(1-(abs(r)./a).^q)).*(abs(r)<=a)+0.*(abs(r)>a);
plot(NA, '.black');
hold on;
grid on;
legend('triangular', 'graded index', 'gaussian', 'step-index');
```

RESULT

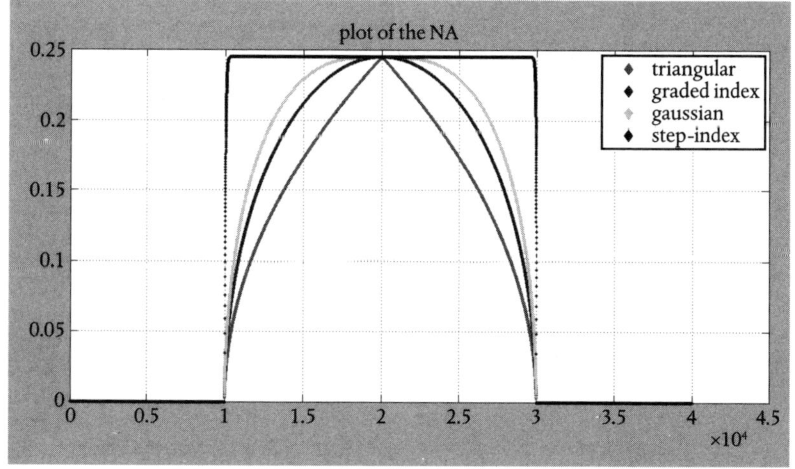

PROGRAM 8

```
clc;
clear;
c=3e5;
n=1;
la=0.8:0.001:2;
B=sqrt(4*pi.^2./la.^2-2*n);
dB=-4*pi.^2./(la.^3.*B);
d2B=(-16*pi.^4./(la.^6.*B.^3))+12*pi.^2./(la.^4.*B);
groupdelay=2./(c.*la.*B);
Dw=((-4*pi.^2./(la.^4.*B))+16*pi.^2./(la.^4.*B.^3))./(2*pi.*c);
plot(la, groupdelay*10e4 );
grid on;
axis([0.8 2 0 0.2]);
title ('group delay');
xlabel('wavelenght/\ (um)--------->');
ylabel('groupdelay in x10us------>');
figure;
plot(la, Dw);
title ('waveguide dispersion');
xlabel('wavelenght/\ (um)--------->');
ylabel('groupdelay in x10us------>');
axis([0.8 2 0 0.2]);
grid on;
```

RESULT

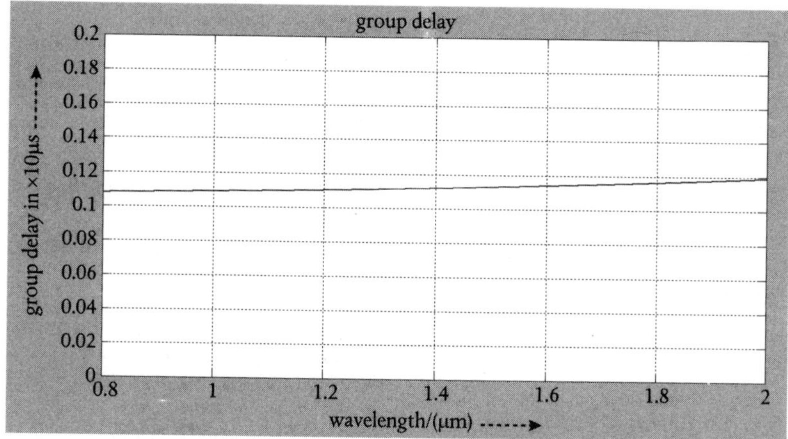
group delay
group delay in ×10μs ------▶
0.2
0.18
0.16
0.14
0.12
0.1
0.08
0.06
0.04
0.02
0
0.8
1
1.2
1.4
1.6
1.8
2
wavelength/(μm) ------▶

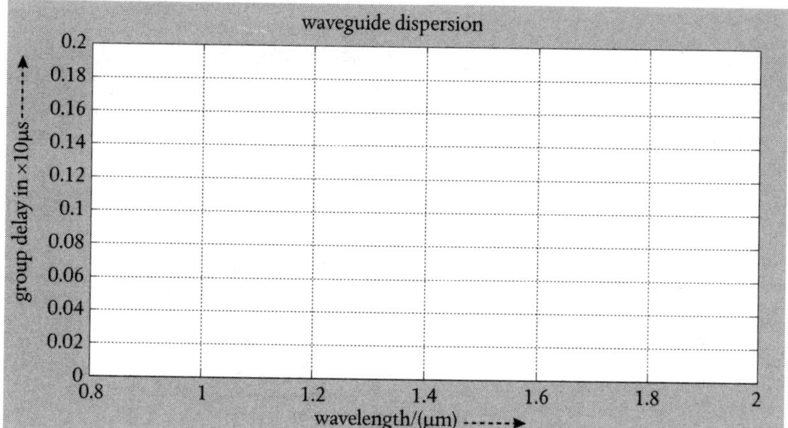
waveguide dispersion
group delay in ×10μs ------▶
0.2
0.18
0.16
0.14
0.12
0.1
0.08
0.06
0.04
0.02
0
0.8
1
1.2
1.4
1.6
1.8
2
wavelength/(μm) ------▶

Index

Figure entries are identified with a suffix *f* against the respective page numbers.